FUNDAMENTALS OF MECHANICAL VIBRATIONS

McGraw-Hill Series in Mechanical Engineering

Consulting Editors

***Jack P. Holman**, Southern Methodist University*
***John R. Lloyd**, Michigan State University*

Anderson: *Modern Compressible Flow: With Historical Perspective*
Arora: *Introduction to Optimum Design*
Bray and Stanley: *Nondestructive Evaluation: A Tool for Design, Manufacturing, and Service*
Culp: *Principles of Energy Conversion*
Dally: *Packaging of Electronic Systems: A Mechanical Engineering Approach*
Dieter: *Engineering Design: A Materials and Processing Approach*
Eckert and Drake: *Analysis of Heat and Mass Transfer*
Edwards and McKee: *Fundamentals of Mechanical Component Design*
Gebhart: *Heat Conduction and Mass Diffusion*
Heywood: *Internal Combustion Engine Fundamentals*
Hinze: *Turbulence*
Howell and Buckius: *Fundamentals of Engineering Thermodynamics*
Hutton: *Applied Mechanical Vibrations*
Juvinall: *Engineering Considerations of Stress, Strain, and Strength*
Kane and Levinson: *Dynamics: Theory and Applications*
Kays and Crawford: *Convective Heat and Mass Transfer*
Kelly: *Fundamentals of Mechanical Vibrations*
Kimbrell: *Kinematics Analysis and Synthesis*
Martin: *Kinematics and Dynamics of Machines*
Modest: *Radiative Heat Transfer*
Norton: *Design of Machinery*
Phelan: *Fundamentals of Mechanical Design*
Raven: *Automatic Control Engineering*
Reddy: *An Introduction to the Finite Element Method*
Rosenberg and Karnopp: *Introduction to Physics*
Schlichting: *Boundary-Layer Theory*
Shames: *Mechanics of Fluids*
Sherman: *Viscous Flow*
Shigley: *Kinematic Analysis of Mechanisms*
Shigley and Mischke: *Mechanical Engineering Design*
Shigley and Uicker: *Theory of Machines and Mechanisms*
Stiffler: *Design with Microprocessors for Mechanical Engineers*
Stoecker and Jones: *Refrigeration and Air Conditioning*
Ullman: *The Mechanical Design Process*
Vanderplaats: *Numerical Optimization: Techniques for Engineering Design, with Applications*
White: *Viscous Fluid Flow*
Zeid: *CAD/CAM Theory and Practice*

FUNDAMENTALS OF MECHANICAL VIBRATIONS

S. Graham Kelly

The University of Akron

McGraw-Hill, Inc.

New York St. Louis San Francisco Auckland Bogotá
Caracas Lisbon London Madrid Mexico Milan Montreal
New Delhi Paris San Juan Singapore Sydney Tokyo Toronto

This book was set in Times Roman by Science Typographers, Inc.
The editors were John J. Corrigan and John M. Morriss;
the production supervisor was Louise Karam.
The cover was designed by Carla Bauer.
Project supervision was done by Science Typographers, Inc.
R. R. Donnelley & Sons Company was printer and binder.

FUNDAMENTALS OF MECHANICAL VIBRATIONS

2 3 4 5 6 7 8 9 0 DOC DOC 9 0 9 8 7 6 5 4 3

P/N 034023-4
PART OF
ISBN 0-07-911533-0

Library of Congress Cataloging-in-Publication Data

Kelly, S. Graham.
Fundamentals of mechanical vibrations / S. Graham Kelly.
p. cm. —(McGraw-Hill series in mechanical engineering)
Includes bibliographical references and index.
ISBN 0-07-911533-0 (set)
1. Vibration. I. Title. II. Series.
QA935.K38 1993
531′.32—dc20 92-35797

ABOUT THE AUTHOR

S. Graham Kelly received an M.S. in 1977 and a Ph.D. in 1979, both in Engineering Mechanics, from Virginia Polytechnic Institute and State University. He studied with Ali Nayfeh and Dean Mook. At Virginia Tech, Kelly understood that all branches of engineering mechanics are derived from the same limited basic laws of nature, with constitutive equations and mathematics tying them all together. With this in mind, he found it hard to specialize in a single area of mechanics, having interest in fluid mechanics, dynamics, and solid mechanics.

Dr. Kelly was fortunate to begin his career at the University of Notre Dame, teaching statics, dynamics, fluid mechanics, and applied mathematics to undergraduates. When he was given the opportunity to develop and teach graduate courses in applied engineering mathematics, Dr. Kelly developed a course with the objectives of preparing graduate students for advanced engineering course work and research in any field of engineering.

When he arrived at The University of Akron in 1982, Dr. Kelly was assigned to teach a course in vibrations to undergraduates. Dr. Kelly soon realized that vibrations is a transition course in a mechanical engineering student's curriculum. He realized that the effective teaching of vibrations requires showing students how to model engineering systems and that students should be provided with many practical applications. It is from these ten years of teaching vibrations that this book is derived.

During his tenure at The University of Akron, Dr. Kelly has also taught undergraduate courses in stress analysis, fluid mechanics, compressible fluid mechanics, and numerical analysis. He has taught graduate courses in vibrations, continuum mechanics, hydrodynamic stability, and applied engineering mathematics. His research areas include hydrodynamic stability and nonlinear vibrations. Dr. Kelly also serves the College of Engineering as Associate Dean for Undergraduate Studies.

Dr. Kelly resides near Akron with his wife, Seala Fletcher-Kelly and his son Samuel Graham Kelly IV. In his spare time, he enjoys tournament bridge and is a Life Master of the American Contract Bridge League.

To Seala

CONTENTS

PREFACE

Engineering is the application of mathematics and science to the solution of practical problems. A mechanical engineering undergraduate starts his/her academic career by taking basic mathematics and science courses and continuing with basic engineering science courses. These include courses such as statics, dynamics, mechanics of solids, fluid mechanics, and thermodynamics. These courses teach students the basic laws of nature, laws specific to the system at hand (i.e., constitutive equations), and how to apply them to simple problems.

Vibrations is one of the first courses in which a student is taught to apply the basic laws of nature to an engineering system to solve a practical problem. Thus vibrations is a very important course in an undergraduate's curriculum. Students do not just learn vibrations for its own sake, but learn how to apply all the engineering knowledge filling their heads (heat transfer is a similar course). A good vibrations course teaches students how to synthesize their knowledge. Students can then use this acquired talent in design courses.

Thus it is important for an undergraduate vibrations textbook to emphasize topics such as mathematical modeling of an engineering system, while keeping a focus on the fundamentals of vibrations. It is this philosophy which has guided me in the development of this book.

This book is intended as a text for a junior or senior undergraduate course in vibrations or for a dual-level undergraduate/graduate course in which the graduate students have a limited previous knowledge of vibrations. I have emphasized topics usually covered in undergraduate courses, and have minimized material that is extraneous to undergraduate courses. The focus of this book is free and forced vibrations of linear one- and multi-degree-of-freedom systems. A chapter on continuous systems is included primarily for reference and for the sake of completeness. A short chapter on nonlinear vibrations is included because of the great current interest in the topic and because nature is inherently nonlinear. These chapters may be used in dual-level courses. During a one-semester dual-level course in which most of the undergraduates are juniors, at The University of Akron, I cover most of the material in Chaps. 1 through 8 and 11. The prerequisites for the course are courses in dynamics, mechanics of solids, and differential equations.

Since vibration analysis involves the application of the basic principles of dynamics, it is essential students are well founded in dynamics. Thus Chap. 1 includes a brief, yet thorough, review of dynamics. A clear concise method for application of the basic principles of dynamics is presented and is consistent with most popular dynamics textbooks (e.g., Beer and Johnston, Meriam and Kraige, and Hibbeler). In keeping with the book's philosophy, Chap. 1 also includes a discussion of mathematical modeling of engineering systems and dimensional analysis.

Chapter 2 introduces the elements of vibrating systems. Chapters 3 through 5 focus on vibrations of one-degree-of-freedom systems, while Chaps. 6 through 8 focus on multi-degree-of-freedom systems. Chapter 9 presents the fundamentals of vibration control. Physical interpretation of equations and results are emphasized throughout the text. Energy methods often provide these explanations.

Numerical methods are an important tool for the vibration analyst. The methods presented here are limited to those that undergraduates might learn before taking a vibrations course and those that are derived from principles of vibration. The former methods include numerical integration and numerical solution of linear ordinary differential equations, while the latter include matrix iteration and the Rayleigh-Ritz method. The finite-element method is a powerful tool for the analyst, but it cannot be done justice in a vibrations text. Its discussion is best left to books and courses devoted solely to the topic.

The accompanying diskette with the VIBES software is designed to aid in understanding the principles of mechanical vibrations. The programs are of three types: (1) illustrative, (2) problems for solution, and (3) numerical methods. The illustrative programs include animation to demonstrate the motion of vibrating systems. The problem programs can be used for laboratory simulations or may require a direct solution of a vibration problem. The numerical methods programs include BASIC versions of many of the programs presented in the text. Some have a menu of built-in excitations while others allow the user to provide their own through a subprogram.

The author acknowledges the support and encouragement of John Corrigan, senior engineering editor at McGraw-Hill. The author also acknowledges Dr. Benjamin Chung, department head of mechanical engineering at The University of Akron, for his support throughout the project, Ken Kuhlmann, Mark Pixley, and Ashish Choski for their help with some of the details, and Mrs. Peggy Duckworth for help with many of the clerical details. Many valuable comments and suggestions were provided by Donald Adams, University of Wyoming; Atila Ertas, Texas Tech University; Andrew Hansen, University of Wyoming; Eugene I. Rivin, Wayne State University; S. C. Sinha, Auburn University; Robert Steidel, University of California–Berkeley; and J. Kim Vandiver, Massachusetts Institute of Technology. Finally, the author expresses his appreciation to his wife, Seala Fletcher-Kelly, and his son, Graham, for their patience and support.

S. Graham Kelly

FUNDAMENTALS OF MECHANICAL VIBRATIONS

CHAPTER 1

INTRODUCTION

1.1 THE STUDY OF VIBRATIONS

Vibrations are the fluctuations of a mechanical system about an equilibrium position. In order for vibrations to occur, the mechanical system must be subject to a restoring force or restoring moment that continually pulls the system toward its equilibrium position. Vibrations are initiated when energy is imparted to the mechanical system by an external source.

The spring, an elastic element, provides the restoring force in the dynamic system of Fig. 1.1*a*. When work is performed on the block to displace it from equilibrium, the spring is stretched and it stores potential energy and a force is developed in the spring. When the block is released, the spring force pulls the block toward equilibrium as the potential energy is converted to kinetic energy. In the absence of nonconservative forces, this process of energy transfer is continual and causes the block to oscillate about its equilibrium position.

Gravity provides a restoring moment to the simple pendulum of Fig. 1.1*b*. When the pendulum is rotated away from the vertical equilibrium position, the moment of the gravity force about the support pulls the pendulum back toward equilibrium.

Vibrations occur in many common engineering systems, and if uncontrolled can lead to catastrophic results. For example, vibrations of a structure induced during an earthquake lead to large stresses and can result in structural failure. Vibrations resulting from the rotating unbalance in a helicopter blade can lead to the pilot's losing control of the helicopter and crashing. Vibrations of machine tools lead to improper machining of a part. Excessive vibrations of industrial compressors or pumps increase the noise level in the machine's

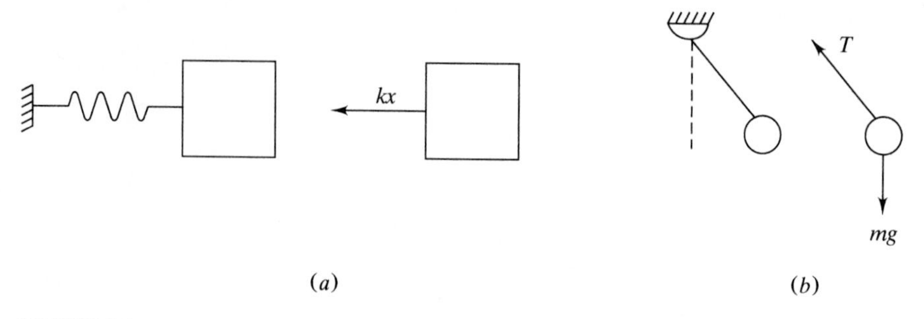

FIGURE 1.1
(*a*) When the block is displaced from equilibrium, the force developed in the spring as a result of stored potential energy pulls the block back toward its equilibrium position; (*b*) when the pendulum is rotated away from the vertical equilibrium position, the moment of the gravity force about the support pulls the pendulum back toward the equilibrium position.

surrounding, can induce vibrations of the surrounding structure, and cause inefficient operation of the machine.

Vibrations can also be introduced beneficially in systems in which they would not normally occur. The suspension system of an automobile protects the passengers and the automobile from rough terrain. Cushioning is used in packaging fragile items to prevent breakage when dropped.

In view of the previous discussion, it is clear that engineers must understand the theory of vibrations and its practical application.

The study of vibrations requires synthesis of basic engineering sciences and mathematics. Vibration theory is developed by applying basic laws of nature and appropriate constitutive equations to dynamic systems. This application requires knowledge of the basic principles of dynamics, mechanics of solids, and occasionally fluid mechanics. Application of basic laws of nature to dynamic systems with restoring forces usually leads to differential equations. The solutions of the differential equations are used to develop theories which can be used to analyze general systems.

1.2 REVIEW OF DYNAMICS

A brief review of rigid-body dynamics is presented to familiarize the reader with notation and methods. The reader is also encouraged to review the basic concepts of mechanics of solids and the solution of second-order ordinary differential equations.

1.2.1 Kinematics

The location of a particle on a rigid body at any instant of time can be referenced to a fixed cartesian reference frame, as shown in Fig. 1.2. The

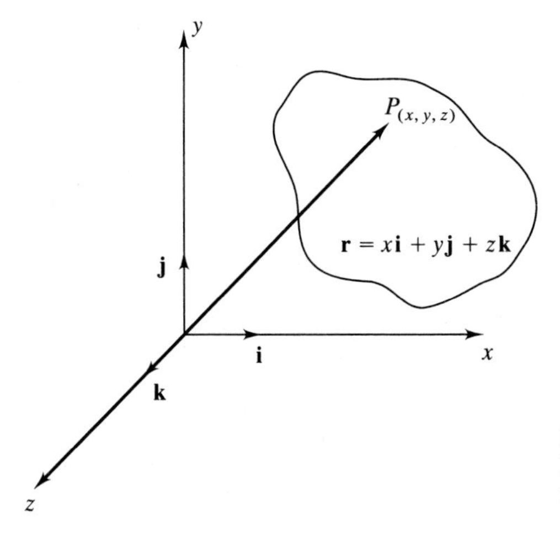

FIGURE 1.2
Position vector of a particle, P, on a rigid body, in a cartesian reference frame.

particle's position vector is given by

$$\mathbf{r} = x(t)\mathbf{i} + y(t)\mathbf{j} + z(t)\mathbf{k} \tag{1.1}$$

from which the particle's velocity and acceleration are determined

$$\mathbf{v} = \frac{d\mathbf{r}}{dt} = \dot{x}(t)\mathbf{i} + \dot{y}(t)\mathbf{j} + \dot{z}(t)\mathbf{k} \tag{1.2}$$

$$\mathbf{a} = \frac{d\mathbf{v}}{dt} = \ddot{x}(t)\mathbf{i} + \ddot{y}(t)\mathbf{j} + \ddot{z}(t)\mathbf{k} \tag{1.3}$$

where a dot above a quantity indicates differentiation of that quantity with respect to time.

In general, a rigid body is rotating and translating. Assume that at an arbitrary instant of time the rigid body is rotating about an axis defined by a unit vector $\mathbf{e}$ with an angular speed ω. The angular velocity vector is

$$\boldsymbol{\omega} = \omega\mathbf{e} \tag{1.4}$$

from which the angular acceleration vector is calculated

$$\boldsymbol{\alpha} = \frac{d\boldsymbol{\omega}}{dt} \tag{1.5}$$

Consider two particles, A and B, fixed to the same rigid body. Let $\mathbf{r}_{B/A}$ be the position vector of B relative to A. The velocity of B relative to A is

$$\mathbf{v}_{B/A} = \mathbf{v}_A + \boldsymbol{\omega} \times \mathbf{r}_{B/A} \tag{1.6}$$

The acceleration of B relative to A is

$$\mathbf{a}_{B/A} = \mathbf{a}_A + \boldsymbol{\alpha} \times \mathbf{r}_{B/A} + \boldsymbol{\omega} \times (\boldsymbol{\omega} \times \mathbf{r}_{B/A}) \tag{1.7}$$

Consider a particle on a rigid body rotating about a fixed axis with an angular displacement θ, measured in a plane normal to the axis of rotation. Every particle on the rigid body travels on a circle centered on the axis of rotation. The velocity of a point of the rigid body, a distance r from the axis of rotation, is

$$\mathbf{v} = r\dot{\theta}\mathbf{i}_t \tag{1.8}$$

where $\mathbf{i}_t$ is a unit vector instantaneously tangent to the circle. The particle's acceleration is given by

$$\mathbf{a} = r\ddot{\theta}\mathbf{i}_t - r\dot{\theta}^2\mathbf{i}_n \tag{1.9}$$

where $\mathbf{i}_n$ is a unit vector instantaneously normal to the circle directed away from the axis of rotation, as shown in Fig. 1.3.

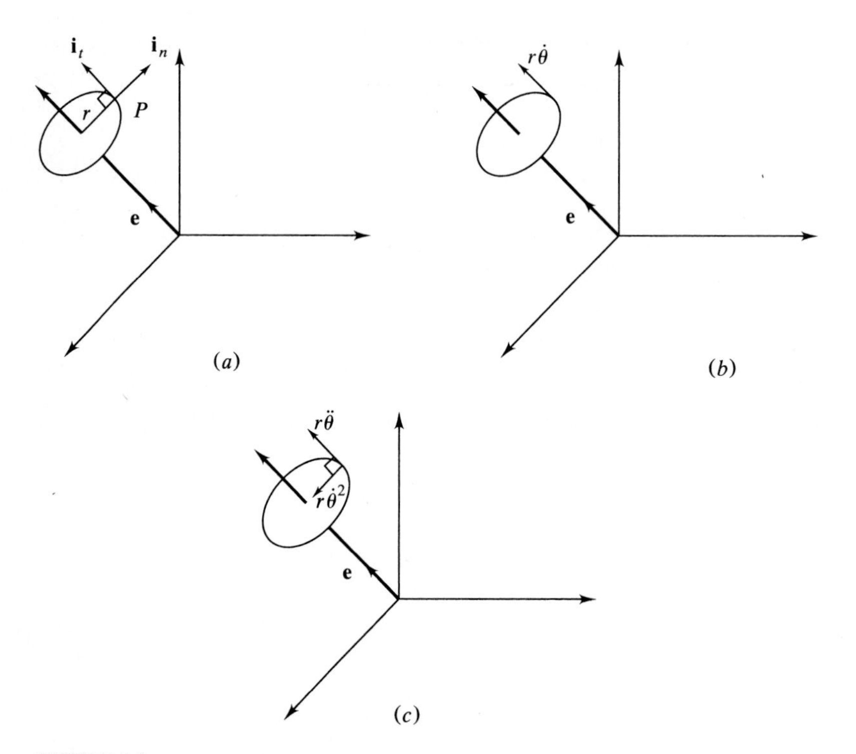

FIGURE 1.3
(a) The particle P is on a rigid body rotating about a constant axis defined by the unit vector, $\mathbf{e}$. P moves in a circle of radius r about the axis. $\mathbf{i}_n$ is instantaneously normal to the circle, directed away from the axis of rotation. $\mathbf{i}_t$ is instantaneously tangent to the circle, in the direction of rotation; (b) $\mathbf{v}_P = r\dot{\theta}\mathbf{i}_t$; (c) $\mathbf{a}_P = r\ddot{\theta}\mathbf{i}_t - r\dot{\theta}^2\mathbf{i}_n$.

1.2.2 Basic Principles of Rigid-Body Kinetics for Planar Motion

A rigid body undergoes planar motion when its mass center moves on a plane and the body rotates about a fixed axis. The principles governing rigid-body kinetics of a body undergoing planar motion are obtained by applying the basic laws of particle kinetics to a system of particles and taking the limit as the number of particles in the system grows large. Applying Newton's second law for a particle and using the limiting process, it can be shown that for a rigid body in plane motion

$$\sum \mathbf{F} = m\bar{\mathbf{a}} \tag{1.10}$$

and

$$\sum \mathbf{M}_G = \bar{I}\boldsymbol{\alpha} \tag{1.11}$$

where $\bar{I}$ is the moment of inertia of the body about an axis through its mass center and parallel to the axis of rotation. In general, a bar above a quantity refers to the quantity being evaluated for the body's mass center, G.

Recall that a system of forces and moments acting on a rigid body can be replaced by a force equal to the resultant of the force system applied at any point on the body and a moment equal to the resultant moment of the system about the point where the resultant force is applied. The resultant force and moment act equivalently to the original system of forces and moments. Thus Eqs. (1.10) and (1.11) imply that the system of external forces and moments acting on a rigid body is equivalent to a force equal to $m\bar{\mathbf{a}}$ applied at the body's mass center and a resultant moment equal to $\bar{I}\boldsymbol{\alpha}$. This latter resultant system is called the system of effective forces. The equivalence of the external forces and the effective forces is illustrated in Fig. 1.4.

The previous discussion suggests the solution procedure for rigid-body kinetics problems that is used throughout this book. Two free-body diagrams are drawn for a rigid body. One free-body diagram shows all external forces and moments acting on the rigid body. The second free-body diagram shows the

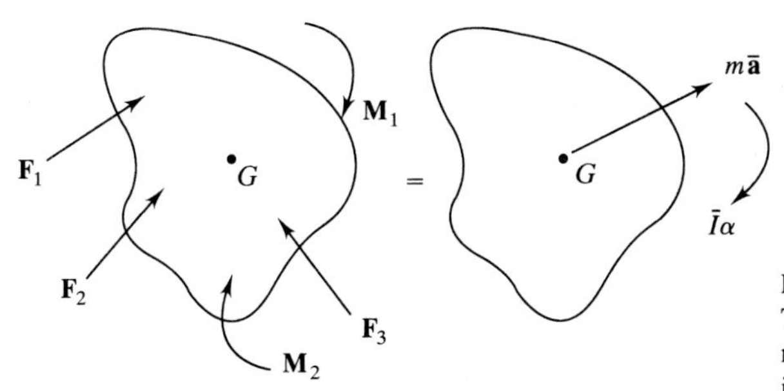

FIGURE 1.4
The system of external forces and moments acting on a rigid body in plane motion is equivalent to a force $m\bar{\mathbf{a}}$ applied at the body's mass center and a moment $\bar{I}\boldsymbol{\alpha}$.

effective forces. If the problem involves a system of rigid bodies, it may be possible to draw a single free-body diagram showing the external forces acting on the system of rigid bodies and one free-body diagram showing the effective forces of all of the rigid bodies. Equations (1.10) and (1.11) are equivalent to

$$\sum \mathbf{F}_{\text{ext}} = \sum \mathbf{F}_{\text{eff}} \tag{1.12}$$

and

$$\sum \mathbf{M}_{O_{\text{ext}}} = \sum \mathbf{M}_{O_{\text{eff}}} \tag{1.13}$$

taken about any point O on the rigid body.

Example 1.1. The slender rod AC of Fig. 1.5 of mass m is pinned at B and held horizontally by a cable at C. Determine the angular acceleration of the bar immediately after the cable is cut.

Immediately after the cable is cut, the angular velocity of the bar is zero. Equation (1.9) is used to determine the acceleration of the mass center in terms of the bar's angular acceleration, $\boldsymbol{\alpha}$.

Summing moments about B, using the free-body diagrams of Fig. 1.5b, gives

$$\sum \overset{\curvearrowleft +}{M}_{B_{\text{ext}}} = \sum \overset{\curvearrowleft +}{M}_{B_{\text{eff}}}$$

$$mg\frac{L}{4} = \left(m\frac{L}{4}\alpha\right)\left(\frac{L}{4}\right) + \frac{1}{12}mL^2\alpha$$

$$\alpha = \frac{12g}{7L}$$

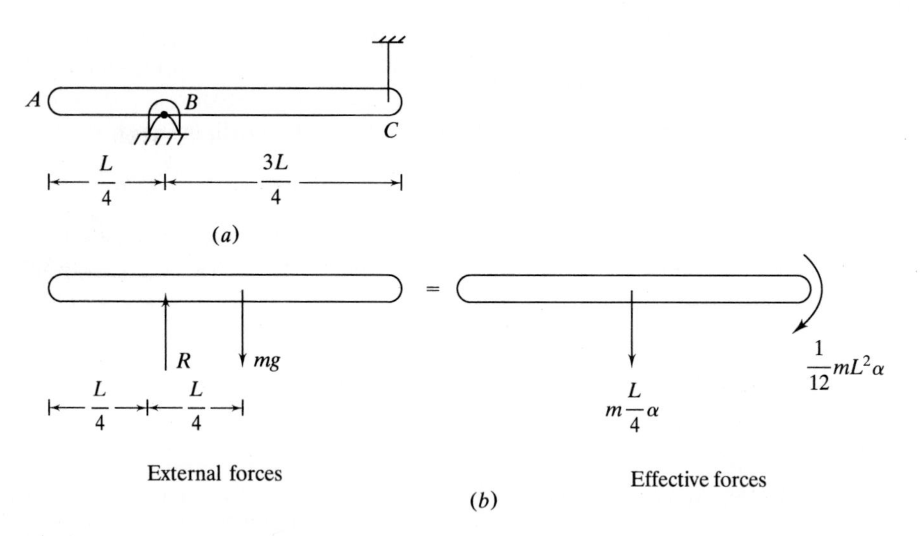

FIGURE 1.5
(a) Slender rod of Example 1.1 is pinned at B and held by cable at C; (b) free-body diagrams immediately after cable is cut.

Example 1.2. Determine the angular acceleration of the pulley of Fig. 1.6.

Consider the system of rigid bodies composed of the pulley and the two blocks. If $\boldsymbol{\alpha}$ is the counterclockwise angular acceleration of the pulley, then, assuming no slip between the pulley and the cables, block A has a downward acceleration of $r_A\boldsymbol{\alpha}$ and block B has an upward acceleration of $r_B\boldsymbol{\alpha}$.

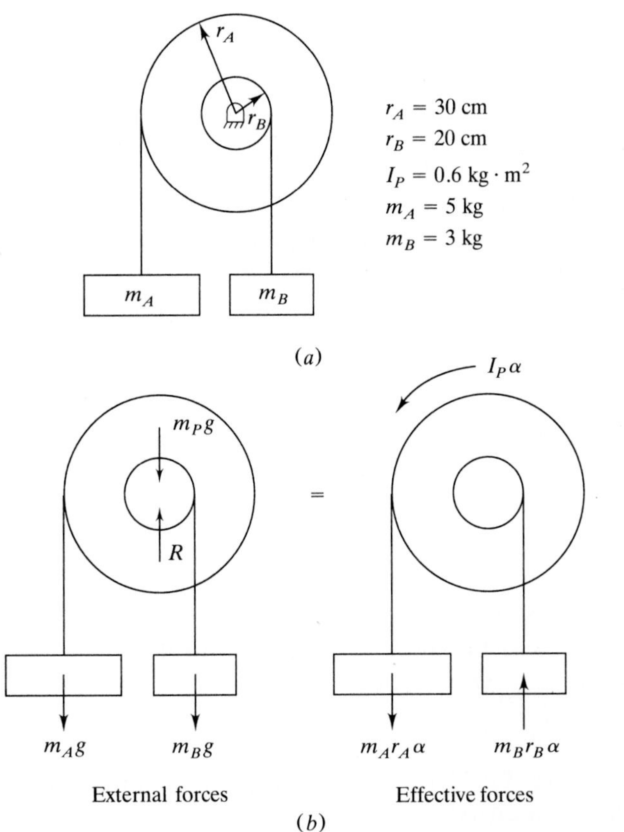

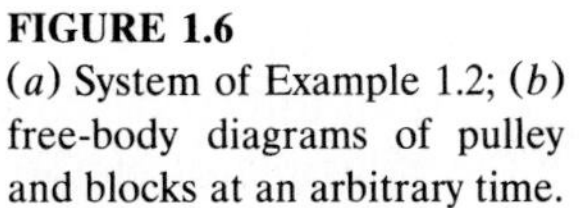

FIGURE 1.6
(*a*) System of Example 1.2; (*b*) free-body diagrams of pulley and blocks at an arbitrary time.

Summing moments about the center of the pulley, neglecting axle friction in the pulley, and using the free-body diagrams of Fig. 1.6*b* yields

$$\sum \overset{\curvearrowleft +}{M}_{O_{\text{ext}}} = \sum \overset{\curvearrowleft +}{M}_{O_{\text{eff}}}$$

$$m_A g r_A - m_B g r_B = I_P\alpha + m_A r_A^2\alpha + m_B r_B^2\alpha$$

Substituting given values leads to $\boldsymbol{\alpha} = 7.55$ rad/s^2.

1.2.3 Principle of Work-Energy

The kinetic energy of a rigid body is

$$T = \tfrac{1}{2}m\bar{v}^2 + \tfrac{1}{2}\bar{I}\omega^2 \tag{1.14}$$

The work done by a force, F, acting on a rigid body as the point of application of the force travels between two points described by position vectors $\mathbf{r}_A$ and $\mathbf{r}_B$ is

$$U_{A \to B} = \int_{\mathbf{r}_A}^{\mathbf{r}_B} \mathbf{F} \cdot d\mathbf{r} \tag{1.15}$$

where $d\mathbf{r}$ is a differential position vector in the direction of motion. The work done by a moment acting on a rigid body in plane motion is

$$U_{A \to B} = \int_{\theta_A}^{\theta_B} M \, d\theta \tag{1.16}$$

If the work of a force is independent of the path taken from A to B, the force is called conservative. Examples of conservative forces are spring forces, gravity forces, and normal forces. A potential energy function, $V(\mathbf{r})$, can be defined for conservative forces. The work done by a conservative force can be expressed as a difference in potential energies

$$U_{A \to B} = V_A - V_B \tag{1.17}$$

Since the system of external forces is equivalent to the system of effective forces, the total work done on a rigid body in planar motion is

$$U_{A \to B} = \int_{\mathbf{r}_A}^{\mathbf{r}_B} m\bar{\mathbf{a}} \cdot d\mathbf{r} + \int_{\theta_A}^{\theta_B} \bar{I}\alpha \, d\theta \tag{1.18}$$

When integrated the right-hand side of Eq. (1.18) is equal to the difference in the kinetic energy of the rigid body between A and B. Thus Eq. (1.18) yields the principle of work-energy,

$$T_B - T_A = U_{A \to B} \tag{1.19}$$

If all forces are conservative, Eq. (1.17) is used in Eq. (1.19) and the result is the principle of conservation of energy

$$T_A + V_A = T_B + V_B \tag{1.20}$$

1.2.4 Principle of Impulse and Momentum

The impulse of the force $\mathbf{F}$ between t_1 and t_2 is defined as

$$\mathbf{I} = \int_{t_1}^{t_2} \mathbf{F} \, dt \tag{1.21}$$

The total angular impulse of a system of forces and moments about a point O is

$$\mathbf{J}_O = \int_{t_1}^{t_2} \sum \mathbf{M}_O \, dt \tag{1.22}$$

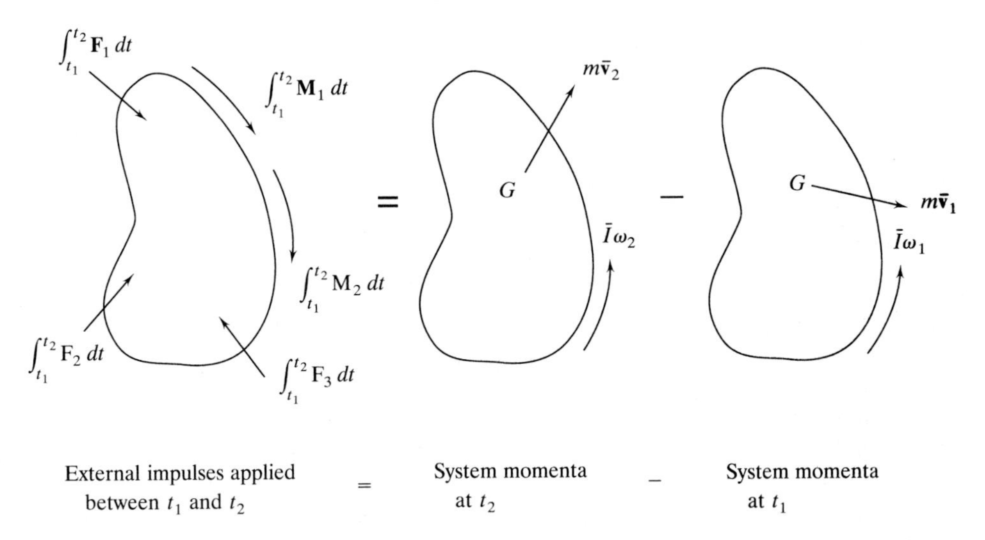

FIGURE 1.7
Illustration of the principle of impulse and momentum.

The system momenta at a given time are defined by the system's linear momentum

$$\mathbf{L} = m\bar{\mathbf{v}} \tag{1.23}$$

and its angular momentum about its mass center

$$\mathbf{H}_G = \bar{I}\boldsymbol{\omega} \tag{1.24}$$

Integrating Eqs. (1.10) and (1.11) between arbitrary times t_1 and t_2 leads to

$$\mathbf{L}_1 + \mathbf{I}_{1\to 2} = \mathbf{L}_2 \tag{1.25}$$

and

$$\mathbf{H}_{G_1} + \mathbf{J}_{G_{1\to 2}} = \mathbf{H}_{G_2} \tag{1.26}$$

Using an equivalent force system argument similar to that used to obtain Eqs. (1.12) and (1.13), it is deduced from Eqs. (1.25) and (1.26) that the system of applied impulses is equivalent to the difference between the system momenta at t_1 and the system momenta at t_2. This form of the principle of impulse and momentum, convenient for problem solution, is illustrated in Fig. 1.7.

Example 1.3. The slender rod of mass m of Fig. 1.8 is swinging through a vertical position with an angular velocity ω_1 when it is struck at A by a particle of mass $m/4$ moving with a velocity v_p. Upon impact the particle sticks to the bar. Determine (a) the angular velocity of the bar and particle immediately after impact, (b) the maximum angle through which the bar and particle will swing after impact, and (c) the angular acceleration of the bar and particle when they reach the maximum angle.

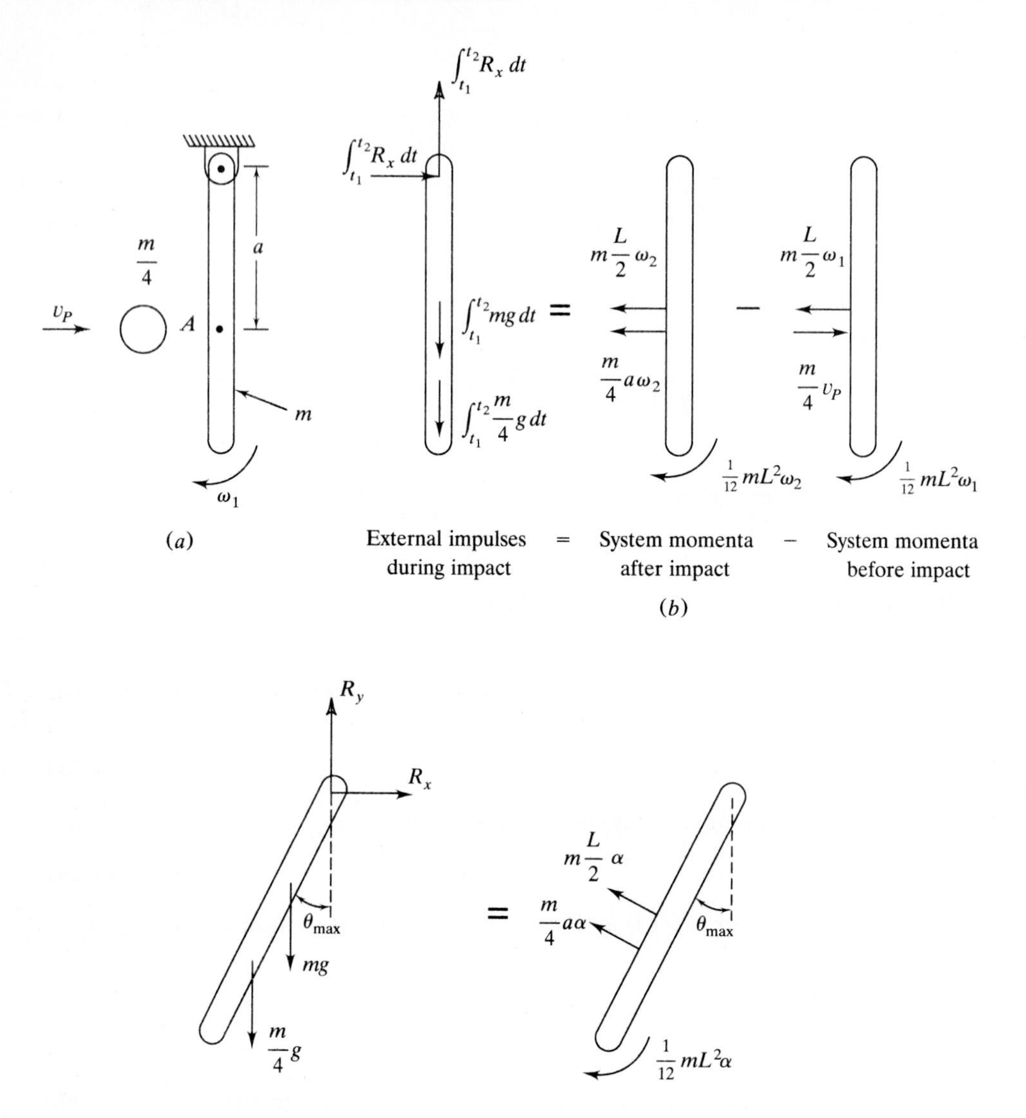

FIGURE 1.8
(*a*) Slender rod of Example 1.3 is swinging through vertical with angular velocity ω_1 when struck at A by particle moving with horizontal velocity v_P; (*b*) impulse and momentum diagrams the time immediately before impact and the time immediately after impact; (*c*) free-body diagrams of bar as it swings through maximum angle.

(a) Let t_1 occur immediately before impact and t_2 occur immediately after impact. Consider the bar and the particle as a system. During the time of impact, the only external impulses are due to gravity and the reactions at the pin support. The principle of impulse and momentum is used in the following form:

$$\begin{pmatrix}\text{External angular}\\ \text{impulses about } O\\ \text{between } t_1 \text{ and } t_2\end{pmatrix} = \begin{pmatrix}\text{Angular momentum}\\ \text{about } O\\ \text{at } t_2\end{pmatrix} - \begin{pmatrix}\text{Angular momentum}\\ \text{about } O\\ \text{at } t_1\end{pmatrix}$$

Using the momentum diagrams of Fig. 1.8b, this becomes

$$0 = \left(m\frac{L}{2}\omega_2\right)\left(\frac{L}{2}\right) + \left(\frac{m}{4}a\omega_2\right)(a) + \frac{1}{12}mL^2\omega_2$$
$$-\left[\left(m\frac{L}{2}\omega_1\right)\left(\frac{L}{2}\right) - \left(\frac{m}{4}v_p\right)(a) + \frac{1}{12}mL^2\omega_1\right]$$

which is solved yielding

$$\omega_2 = \frac{4L^2\omega_1 - 3v_p a}{4L^2 + 3a^2}$$

(b) Let t_3 be the time when the bar and particle assembly attains its maximum angle. Gravity forces are the only external forces that do work; hence conservation of energy applies between t_2 and t_3. Thus from Eq. (1.20)

$$T_2 + V_2 = T_3 + V_3$$

The potential energy of a gravity force is the magnitude of the force times the distance its point of application is above a horizontal datum plane. Choosing the datum as the horizontal plane through the support, using Eq. (1.14) for the kinetic energy of a rigid body, and noting $T_3 = 0$ yields

$$\frac{1}{2}m\left(\frac{L}{2}\omega_2\right)^2 + \frac{1}{2}\frac{1}{12}mL^2\omega_2^2 + \frac{1}{2}\frac{m}{4}(a\omega_2)^2 - mg\frac{L}{2} - \frac{mg}{4}a$$
$$= -mg\frac{L}{2}\cos\theta_{\max} - \frac{m}{4}ga\cos\theta_{\max}$$

which is solved to yield

$$\theta_{\max} = \cos^{-1}\left[1 - \frac{(4L^2 + 3a^2)\omega_2^2}{g(12L + 6a)}\right]$$

(c) Since the bar attains its maximum angle at t_3, $\omega_3 = 0$. Summing moments about O using the free-body diagrams of Fig. 1.8c gives

$$\sum \overset{\curvearrowleft +}{M}_{O_{\text{ext}}} = \sum \overset{\curvearrowleft +}{M}_{O_{\text{eff}}}$$

$$-(mg)\left(\frac{L}{2}\sin\theta_{\max}\right) - \left(\frac{mg}{4}\right)(a\sin\theta_{\max})$$
$$= \left(m\frac{L}{2}\alpha\right)\left(\frac{L}{2}\right) + \left(\frac{m}{4}a\alpha\right)(a) + \frac{1}{12}mL^2\alpha$$

which is solved to yield

$$\alpha = -\frac{(6L + 3a)g \sin \theta_{max}}{4L^2 + 3a^2}$$

1.3 CLASSIFICATION OF VIBRATION

Vibrations are classified in a number of ways according to a number of possible factors.

The number of *degrees of freedom* of a system is the number of kinematically independent variables necessary to describe entirely the motion of the system. If the values of these variables are known at any instant of time, kinematics can be used to determine the displacement, velocity, and acceleration of any particle in the system.

Any set of kinematically independent coordinates is called a set of *generalized coordinates*. Basic laws of nature and applicable constitutive equations are applied to derive a system of differential equations in which the generalized coordinates are the dependent variables and time is the independent variable. The motion of an n-degree-of-freedom system is governed by n ordinary differential equations. A system with a finite number of degrees of freedom is called a *discrete* system.

Example 1.4. Each of the systems of Fig. 1.9 in equilibrium in the position shown. For each system, determine the number of degrees of freedom and recommend a set of generalized coordinates.

(a) The system only has one degree of freedom. If θ is the clockwise angular displacement of the bar from the horizontal, then a particle initially a distance l from the support has a horizontal position $l \cos \theta$ and a vertical displacement $l \sin \theta$. The vibrations are motion about a fixed axis and Eqs. (1.8) and (1.9) are used to calculate the velocity and acceleration of any point in terms of θ, $\dot{\theta}$, and $\ddot{\theta}$.

(b) The system has two degrees of freedom. Let x be the displacement of the mass center of the bar, measured from equilibrium, and θ the clockwise angular displacement of the bar, measured from the horizontal. Then the displacement from equilibrium of a particle a distance d to the right of the mass center is given by $x + d \sin \theta$. The relative velocity and acceleration relations, Eqs. (1.6) and (1.7), are used to calculate the velocity and acceleration of the particle in terms of x and θ and their derivatives.

(c) The system has only one degree of freedom. If θ is the counterclockwise angular displacement of the pulley, measured from equilibrium and assuming no slip between the pulley and the cable, the displacement of the block of mass m_1 is $r\theta$ downward and the displacement of the block of mass m_2 is $2r\theta$ upward.

(d) This system has three degrees of freedom. Since an elastic element connects each of the blocks to the pulley, no kinematic relationship exists between their displacements and the angular displacement of the pulley. A set of three generalized coordinates is: θ, the clockwise angular displacement of the pulley; x_1, the downward displacement of the block of mass m_1; and x_2, the upward displacement of the block of mass m_2.

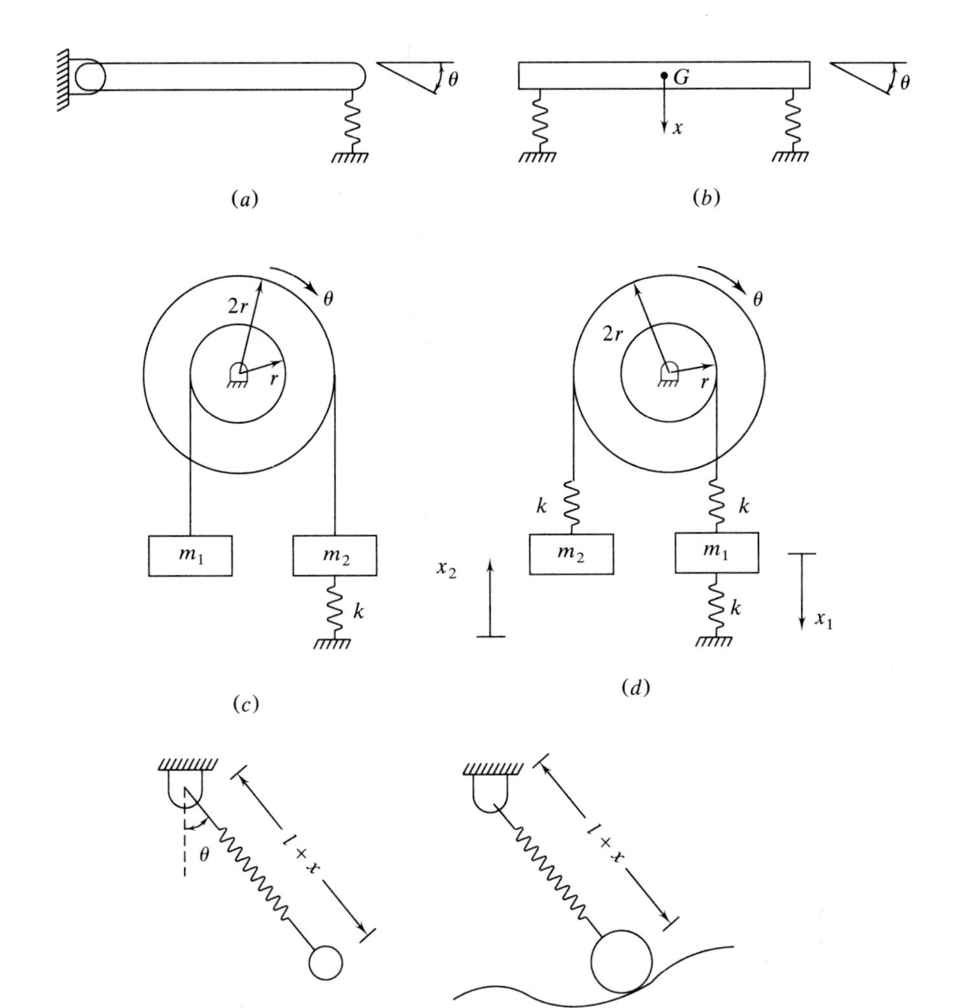

FIGURE 1.9
Systems for Example 1.4. One possible choice of a set of generalized coordinates is illustrated for each system.

(*e*) The swinging spring has two degrees of freedom. Let x be the change in length of the spring from when the system is in equilibrium and θ the counterclockwise angular displacement of the mass.

(*f*) This system only has one degree of freedom. Let x be the change in length of the spring from when the system is in equilibrium. Then, since the mass is constrained to move in a track of known shape, its location is determined using geometry.

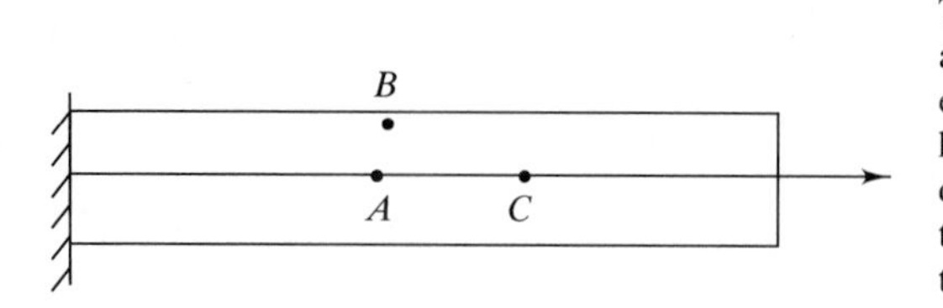

FIGURE 1.10
The displacements of particles A and B are related through the assumptions of elementary beam theory. However, no kinematic relationship exists between the displacements of particles A and C. Thus the cantilever beam is a continuous system.

Particles A and C are along the neutral axis of the cantilever beam of Fig. 1.10 While B is in the cross section obtained by passing a perpendicular plane through the neutral axis at A. A basic assumption of beam theory is that plane cross sections remain plane; thus the displacements of A and B are kinematically related. However, knowledge of the loading is required to determine the relation between the displacement of A and the displacement of C. Thus the displacements of no two points along the neutral axis of the beam are kinematically related. The beam has an infinite number of degrees of freedom. Deformable bodies such as beams, plates, and columns are called *continuous* or *distributed parameter* systems. The displacement of a particle on the beam is a function of the particle's location on the beam as well as time. Thus the motion of a continuous system is governed by a partial differential equation. Continuous systems are often approximated by discrete systems for easier analysis.

A system is *linear* if its motion is described by a linear system of differential equations. A system is *nonlinear* if its motion is described by a nonlinear system of differential equations. All systems are inherently nonlinear. However, simplifying assumptions are often made such that nonlinear terms are negligible compared to linear terms or can be replaced by linear approximations. Linear systems are much easier to analyze than nonlinear systems.

If an external energy source is applied to initiate vibrations and then removed, the resulting vibrations are *free vibrations*. In the absence of nonconservative forces, free vibrations sustain themselves and are periodic. Free vibrations decay when a nonconservative force is present. Free vibrations of one-degree-of-freedom systems are considered in Chap. 3, while free vibrations of multi-degree of freedom systems are considered in Chap. 7 and free vibrations of continuous system are described in Chap. 9.

If vibrations occur during the presence of an external energy source, the vibrations are called *forced vibrations*. The behavior of a system undergoing forced vibrations is dependent on the type of excitation. If the excitation is periodic, the forced vibrations of a linear system are also periodic. One-degree-of-freedom systems subject to periodic excitations are analyzed in Chap. 4. Forced nonperiodic vibrations are called transient vibrations and are analyzed in Chap. 5. Forced vibrations of multi-degree-of-freedom systems are considered in Chap. 8 and forced vibrations of continuous systems are considered in Chap. 9.

Under certain conditions nonlinear systems subject to periodic excitations have a nonperiodic response. Such systems are said to be *chaotic* and are the focus of the ongoing research in vibrations.

If the value of the excitation force is known at any instant of time, the excitation is said to be *deterministic*. If the excitation force is unknown for all time, but averages and deviations are known, the excitation is said to be *random*. In this case the response is also random, and cannot be determined exactly at any instant of time. Only statistical values of the response can be determined. Random vibration analysis is used to analyze the earthquake excitation of structures and the behavior of automobile suspension systems due to random road terrain, and is important in vibration measurement.

1.4 MATHEMATICAL MODELING OF PHYSICAL SYSTEMS

Solution of an engineering problem often requires mathematical modeling of a physical system. The modeling procedure is the same for different engineering disciplines, although the details of the modeling vary between disciplines. The steps in this procedure are presented and the details specialized for application to mechanical vibration problems. The procedure is illustrated through application to a dynamics problem.

1.4.1 Problem Identification

The system to be modeled is abstracted from its surroundings. The information to be obtained from the modeling is specified. Known parameters are listed.

1.4.2 Assumptions

Assumptions are made to simplify the modeling. If all effects are included in the modeling of a physical system, the resulting equations are usually so complex that a mathematical solution is impossible. When assumptions are used, an approximate physical system is modeled. Approximations should only be made if the solution to the approximate problems is easier than the solution to the original problem and the results are accurate enough for whatever use they are intended.

Certain implicit assumptions are used in the modeling of most physical systems. These assumptions are taken for granted and rarely mentioned explicitly. Implicit assumptions used throughout this book include:

1. Physical properties are continuous functions of spatial variables. This is the continuum assumption and implies that a system can be treated as a continuous piece of matter.

2. The earth is an inertial reference frame. This assumption allows application of Newton's laws.
3. Relativistic effects are ignored.
4. Gravity is the only external force field. The acceleration due to gravity is 9.81 m/s^2 on the surface of the earth.
5. The systems considered are not subject to nuclear reactions, chemical reactions, external heat transfer, or any other source of thermal energy.
6. All materials are linear, elastic, homogeneous, and isotropic.
7. The usual assumptions from mechanics of materials apply (i.e., plane sections for beams in bending remain plane, circular shafts under torsional loads do not warp).

Explicit assumptions are those specific to a particular problem. Explicit assumptions are used to eliminate negligible effects from the analysis and/or to simplify the problems.

All physical systems are inherently nonlinear. Accurate mathematical modeling of any physical system leads to nonlinear differential equations which often have no analytical solution. Assumptions are often made to linearize a system. Consider the simple pendulum of Fig. 1.11. As the pendulum moves in air, it encounters friction in the form of aerodynamic drag. The drag force leads to a nonlinear term in the governing differential equation. However, the effect of drag on certain aspects of the motion of the pendulum is negligible. Even when aerodynamic drag is neglected, the exact differential equation governing the motion of the pendulum is nonlinear. As shown in Chap. 3, if the maximum displacement of the pendulum is small, an approximate linear differential equation can be substituted for the exact nonlinear differential equation without a great loss in accuracy.

When analyzing the results of mathematical modeling, one has to keep in mind that the mathematical model is only an approximation of the true physical system and the actual system behavior may be somewhat different. If the pendulum of Fig. 1.11 is given an initial displacement and then released, oscillations about the vertical equilibrium position ensue. An analysis that does not include aerodynamic drag or some other dissipative mechanism will predict perpetual motion. Such perpetual motion is obviously impossible. Even though the absence of aerodynamic drag in the model leads to an incorrect prediction

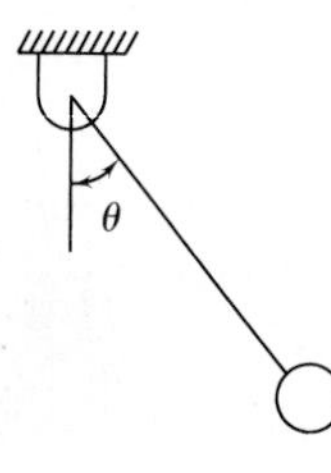

FIGURE 1.11
Aerodynamic drag leads to a nonlinear term in the differential equation describing the motion of the pendulum. If drag is neglected and small θ is assumed, then a linear differential equation is obtained representing an approximation to the motion of the pendulum.

of the steady state, the model is still useful in providing approximations to information about the system such as the time it takes the pendulum to execute one cycle.

One has to be careful when linearizing inherently nonlinear systems. The qualitative behavior of nonlinear systems is significantly different from the behavior of linear systems. It is well documented that even small geometric and material nonlinearities can lead to unwanted and unexpected modes of vibration.

All assumptions made in modeling a physical system should be checked for their validity before the results of the modeling are used.

1.4.3 Basic Laws of Nature

A basic law of nature is a physical law that applies to all physical systems regardless of the material from which the system is constructed. These laws are observable, but cannot be derived from any more fundamental law. They are empirical. There exist only a few basic laws of nature: conservation of mass, conservation of momentum, conservation of energy, and the second and third laws of thermodynamics.

Conservation of momentum, both linear and angular, is usually the only physical law that is of significance in application to vibrating systems. Its application is as described in Sec. 1.2. Application of conservation of mass to vibrations problems is trivial. Applications of the second and third laws of thermodynamics do not yield any useful information. In the absence of thermal energy, the principle of conservation of energy reduces to the mechanical work-energy principle which is derived from Newton's laws and explained in Sec. 1.2.3.

1.4.4 Constitutive Equations

Constitutive equations provide information about the materials of which a system is made. Different materials behave differently under different conditions. Steel and rubber behave differently because their constitutive equations have different forms. While the constitutive equations for steel and aluminum are of the same form, the constants involved in the equations are different. Constitutive equations are used in Chap. 2 to develop laws for mechanical components that are used in modeling vibrating systems.

1.4.5 Geometric Constraints

Application of geometric constraints is often necessary to complete the mathematical modeling of an engineering system. Geometric constraints can be in the form of kinematic relationships between displacement, velocity, and acceleration. When application of basic laws of nature and constitutive equations lead to differential equations, the use of geometric constraints is often necessary to formulate the requisite boundary and/or initial conditions.

1.4.6 Mathematical Solution

The mathematical modeling of a physical system results in the formulation of a mathematical problem. The modeling is not complete until the appropriate mathematics is applied and a solution obtained.

The type of mathematics required is different for different types of problems. Modeling of many static, dynamics, and mechanics of solids problems leads only to algebraic equations. Mathematical modeling of vibrations problems leads to differential equations. Vibrations of a one-degree-of-freedom discrete system are governed by one ordinary differential equation. Vibrations of multi-degree-of-freedom system are governed by a system of ordinary differential equations. Vibrations of a continuous system are governed by a partial differential equation.

Exact analytical solutions, when they exist, are preferable to numerical or approximate solutions. Exact solutions are available for many linear problems, but for only a few nonlinear problems.

1.4.7 Physical Interpretation of Results

After the mathematical solution is complete, the results are formulated. Physical interpretation of the results is an important final step in the modeling procedure. In certain situations this may involve drawing general conclusions from the mathematical solution, it may involve development of design curves, or it may require only simple arithmetic to arrive at a conclusion for the specific problem.

Example 1.5. A motor-driven winch, shown in Fig. 1.12, is used in an industrial plant to hoist heavy machinery from ground level. A motor drives a drum of radius

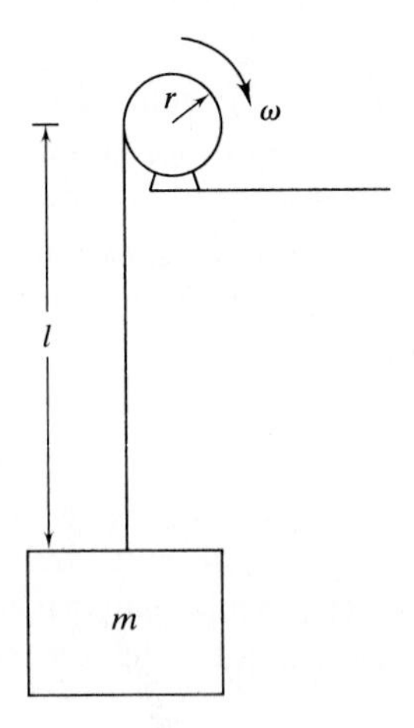

FIGURE 1.12
The winch of Example 1.5 is used to hoist a package connected to an elastic cable.

r which rotates at a constant speed ω. A cable of cross-sectional area A and elastic modulus E is attached to a package. As the drum rotates the cable is taken up and the package lifted. If the motor stops suddenly, what is the maximum stress developed in the cable as a function of height? In particular, suppose the winch is used to hoist a 2000-kg package 45 m. A motor drives the winch at a constant speed of 2 rad/s. The drum radius is 50 cm. What is the minimum radius of a steel ($E = 210 \times 10^9$ N/m^2, tensile strength $= 4.3 \times 10^8$ N/m^2) cable such that, if the motor stops suddenly, the tensile stress is not exceeded in the cable if the package has been lifted less than 30 m?

1. Problem formulation. The problem is to determine the stress in the cable as a function of height if the winch stops suddenly. It is also desirable to find the minimum cable radius for the particular application.

When the drum is rotating at a constant speed, the stress in the cable is simply the stress due to the weight of the package. When the winch suddenly stops, the package has a kinetic energy which is converted to potential energy, causing additional stress in the cable. A study will be made to relate the maximum stress in the cable to the drum radius, the rotational speed of the drum, the height to which the package must be hoisted, the mass of the package, the cross-sectional area of the cable, and the elastic modulus of the cable.

2. Assumptions. In addition to the implicit assumptions, the following explicit assumptions are made in modeling the motion of the package and determining the stress in the cable after the winch stops.

1. All sources of friction are neglected. This includes aerodynamic drag, internal friction in the cable, and friction due to part of the cable being wrapped around the drum.
2. The mass of the cable is small in comparison to the mass of the package. This assumption implies that the kinetic energy of the cable is negligible.
3. When the drum is rotating there is no slipping between the drum and the cable and when the drum stops its velocity is instantaneously zero. This assumption allows the velocity of the package immediately after the winch stops to be specified as $r\omega$.

3. Application of basic laws. The appropriate basic law is the mechanical work-energy equation, a spatially integrated form of Newton's law. In view of the assumptions, this simplifies to the principle of conservation of energy, Eq. (1.20).

Consider the system consisting of the package and the cable. Let state A refer to the system immediately after the drum stops. State B refers to the system when the maximum stress occurs in the cable. Let Δ be the displacement of the package between state A and state B. In view of the assumptions the kinetic energy of the system at A is

$$T_A = \tfrac{1}{2}mr^2\omega^2$$

The system has two forms of potential energy: potential energy due to gravity and stored strain energy. The datum for potential energy calculations is taken as the position of the package immediately after the drum stops. Hence the potential energy at A due to gravity is zero, and the potential energy due to gravity at B is

$-mg\Delta$. Let S represent strain energy. Then Eq. (1.20) gives

$$\tfrac{1}{2}mr^2\omega^2 + S_A = T_B - mg\Delta + S_B$$

4. Constitutive equations. It is implicitly assumed the cable is made of an elastic material and has a linear stress-strain relation. The cable is modeled as an axially loaded structural member.

When the winch stops, the cable has a stress due to the weight of the package,

$$\sigma_A = \frac{mg}{A}$$

When the winch stops the vertical length of the cable is l, which is the sum of its unstretched length, z, plus its stretching due to the weight of the package. Using basic principles of strength of materials,

$$l = z\left(1 + \frac{mg}{AE}\right)$$

The strain energy per unit volume due to a normal stress σ is

$$e = \frac{\sigma^2}{2E}$$

The total strain energy in the cable is

$$S = ezA$$

The cable's normal stress is its change in length from its unstretched length times E over z. Thus

$$\sigma_B = \frac{mg}{A} + \frac{\Delta E}{z}$$

5. Geometric constraints. The equation for σ_B shows that maximum stress occurs when Δ is a maximum. Kinematics then implies $T_B = 0$.

6. Mathematics. Substituting for σ_A and σ_B into the expression for total strain energy and in turn substituting into the work-energy equation leads to

$$\frac{1}{2}mr^2\omega^2 + \frac{m^2g^2z}{2AE} = -mg\Delta + \frac{zA}{2E}\left(\frac{mg}{A} + \frac{\Delta E}{z}\right)^2$$

which is solved to yield

$$\Delta = r\omega\sqrt{\frac{mz}{AE}}$$

The maximum stress is calculated as

$$\sigma_{\max} = \sigma_B = \frac{mg}{A} + r\omega\sqrt{\frac{mE}{Al}\left(1 + \frac{mg}{AE}\right.}$$

Numerical values can now be used to determine the
for the specific application. With all other values known,

from the preceding equation requires an iterative solution. However, it is noted that $mg/A\sigma_{max} < 1$. Since $E \gg \sigma_{max}$, $mg/AE \ll mg/A\sigma_{max} < 1$. Thus a good approximation for the minimum area is obtained by solving

$$\sigma_{max} = \frac{mg}{A} + r\omega\sqrt{\frac{mE}{Al}}$$

which can be rearranged to

$$\sigma_{max}\sqrt{A}^2 = mg + r\omega\sqrt{\frac{mE}{l}}\sqrt{A}$$

Substitution of given values, with $l = 15$ m, leads to the following quadratic equation:

$$4.3 \times 10^8\sqrt{A}^2 - 5.3 \times 10^6\sqrt{A} - 1.96 \times 10^4 = 0$$

whose relevant solution is

$$A = 2.34 \times 10^{-4} \text{ m}^2$$

which gives a minimum cable diameter of 17.3 mm. The magnitude of the neglected term is 4×10^{-4}, which is indeed much less than 1.

7. Interpretation of results. The minimum cable diameter for the specific problem is 17.3 mm. A general discussion of the results requires use of dimensional analysis and is delayed until Sec. 1.5.

1.5 DIMENSIONAL ANALYSIS

An engineer using the results of the modeling of the system of Example 1.5 to design a winch for applications needs to know the effect of how the maximum stress changes with certain parameters. Indeed, the engineer may want a set of design curves, showing how the stress varies with the parameters. The maximum stress in the cable is determined as a function of seven parameters (r, ω, E, A, m, l, and g). Six parameters are truly independent. The seventh, g, is fixed. Suppose the engineer calculates minimum stress for 10 values of each independent parameter. If the engineer only changes the value of one parameter between successive calculations, then he will make 1,000,000 calculations. If the engineer investigates this relationship experimentally, 1,000,000 experiments might be performed. These calculations and experiments are time-consuming and unnecessary. The volume of data generated obscures the results and makes it impossible to draw accurate conclusions.

An alternative is to reformulate the relationship between the maximum stress and the independent parameters nondimensionally. Nondimensional formulation of the relationship between a dependent variable and a set of independent parameters leads to a better understanding of the relationship. General conclusions are easier to formulate.

In general, suppose that a dependent variable, u, is a function of n independent parameters, $v_1, v_2, \ldots, v_n$,

$$u = f(v_1, v_2, \ldots, v_n) \tag{1.27}$$

The dimensions of these $n + 1$ parameters (n independent + 1 dependent) involve m basic dimensions. Possible basic dimensions are mass, length, time, temperature, and electric current. The Buckingham Π theorem states there are at most $k = n + 1 - m$ independent dimensionless groups, or Π groups, involved in a nondimensional reformulation of Eq. (1.27). The Π's are related by

$$\Pi_1 = h(\Pi_2, \Pi_3, \ldots, \Pi_k)$$

The dimensionless group Π_1 involves the dependent variable u, but no other Π contains u.

Several methods for determining a set of Π groups for a particular problem are explained in detail in fluid mechanics textbooks. An explicit functional relationship between the dimensional parameters does not have to be known in order to develop the Π groups. Any set of k independent Π groups can be used. However, certain choices of Π groups make ensuing analysis simpler.

Example 1.6. Determine a set of Π groups that can be used in a nondimensional formulation of the relationship between maximum stress and the independent parameters for the system of Example 1.5.

The appropriate functional relationship of the form of Eq. (1.27) is

$$\sigma_{\max} = f(r, \omega, E, A, l, m, g)$$

The velocity of the package immediately after the winch stops is $v = r\omega$. Neither r nor ω have any other effect in the problem. Thus the two parameters r and ω are replaced by the single parameter v resulting in the simpler relationship.

$$\sigma_{\max} = f(v, A, E, l, m, g)$$

Without loss of generality the following functional relationship between the parameters is assumed:

$$\sigma_{\max} = C v^a A^b E^c l^d m^e g^f$$

where C is an arbitrary constant, and a, b, c, d, e, and f are exponents determined such that the equation is dimensionally homogeneous. Replacing parameters with their basic dimensions in the preceding equation leads to

$$\left[\frac{M}{LT^2}\right] = C\left[\frac{L}{T}\right]^a [L^2]^b \left[\frac{M}{LT^2}\right]^c [L]^c [M]^e \left[\frac{L}{T^2}\right]^f$$

Note that force is not a basic dimension, but a derived dimension. There are seven parameters which involve three basic dimensions. Thus the Buckingham Π theorem predicts a nondimensional functional relationship involving $7 - 3 = 4$ Π's. Invoking the principle of dimensional homogeneity leads to the following equations:

$$M: \quad 1 = c + e$$

$$L: \quad -1 = a + 2b - c + d + f$$

$$T: \quad -2 = -a - 2c - 2f$$

which are solved for b, e, and f in terms of a, c, and d,

$$b = -1 - \frac{a}{4} - \frac{d}{2} + c$$

$$e = 1 - c$$

$$f = 1 - \frac{a}{2} - c$$

Equation (1.27) becomes

$$\sigma_{\max} = Cv^a A^{-1-a/4-d/2+c} E^c l^d m^{1-c} g^{1-a/2-c}$$

Collecting like exponents leads to

$$\frac{\sigma_{\max} A}{mg} = C\left(\frac{v}{A^{1/4}g^{1/2}}\right)^a \left(\frac{EA}{mg}\right)^c \left(\frac{l}{A^{1/2}}\right)^d$$

Thus

$$\Pi_1 = \frac{\sigma_{\max} A}{mg}$$

$$\Pi_2 = \frac{v}{A^{1/4}g^{1/2}}$$

$$\Pi_3 = \frac{EA}{mg}$$

$$\Pi_4 = \frac{l}{A^{1/2}}$$

The functional relationship between the maximum stress and the independent parameters, determined from the mathematical modeling, is rewritten in terms of the Π groups as

$$\Pi_1 = 1 + \Pi_2 \sqrt{\frac{\Pi_3}{\Pi_4}\left(1 + \frac{1}{\Pi_3}\right)}$$

As noted in Example 1.5, $1/\Pi_3$ is usually very small, and this term is neglected. In this case an alternate form of the preceding equation is

$$\Pi_1 = 1 + \Pi_5$$

where

$$\Pi_5 = \frac{v}{g}\sqrt{\frac{AE}{ml}}$$

The functional relationship is plotted in Fig. 1.13. Since A is involved in both Π_1 and Π_5, Fig. 1.13 can be used iteratively to solve the particular situation of Example 1.5. A value of A is guessed, and the corresponding value of Π_5 calculated. Figure 1.13 is used to determine the corresponding value of Π_1, which is used to check A. If A is not close enough to the previous guess, the new value of A is used to calculate Π_5 and the procedure repeated until convergence is achieved. This method is useful if the exact relation between Π_1 and Π_5 is not known.

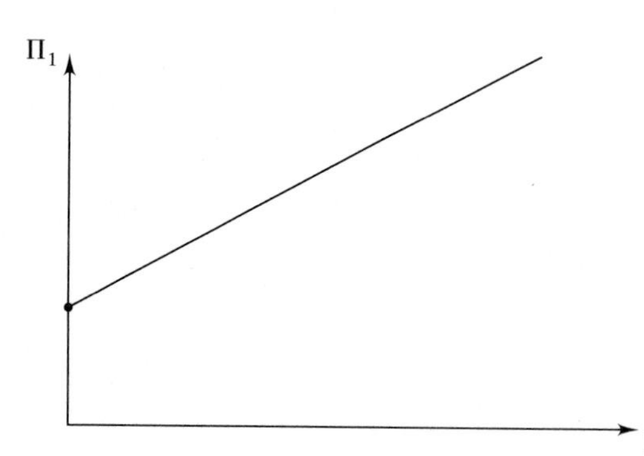

FIGURE 1.13
Plot of relation between nondimensional parameters is used to iteratively determine required cross-sectional area of the cable of Example 1.5.

The Π groups often have identifiable physical meaning. For example, Π_1 is the ratio of the maximum total stress in the cable to the stress simply due to the weight of the package. Π_3 is the inverse of the normal strain in the cable in the static situation. Π_5 is proportional to the ratio of the time it takes the package to accelerate to the velocity v, due to gravity only, to the period of oscillation.

Nondimensional parameters governing a problem can be obtained by directly nondimensionalizing its governing differential equation, when it is known. The corresponding nondimensional boundary conditions are also nondimensionalized.

Example 1.7. The methods of Chaps. 2 and 3 can be applied to derive the following differential equation governing motion of the package:

$$m\ddot{x} + \frac{AE}{l}x = 0$$

subject to the following initial conditions:

$$x(0) = 0$$

$$\dot{x}(0) = v$$

Nondimensionalize the differential equations and initial conditions.

Nondimensional variables of the form

$$x^* = \frac{x}{l} \qquad t^* = \sqrt{\frac{AE}{ml}}\,t$$

are introduced. Substitution of the nondimensional variables into the differential equation and boundary conditions lead to

$$\ddot{x}^* + x^* = 0$$
$$x^*(0) = 0$$
$$\dot{x}^*(0) = r\omega\sqrt{\frac{ml}{AE}}$$

where a dot now represents differentiation with respect to t^*. It is common practice to drop the * from nondimensional variables with an understanding that all variables now used are nondimensional.

PROBLEMS

1.1. A particle starts at the origin of a cartesian coordinate system and moves with a velocity vector

$$\mathbf{v} = 3\cos 2t\,\mathbf{i} + 3\sin 2t\,\mathbf{j} + 0.4t\,\mathbf{k}\ \text{m/s}$$

where t is in seconds.
(a) Determine the magnitude of the particle's acceleration at $t = \pi$ s.
(b) Determine the particle's position vector at $t = \pi$ s.

1.2. The one-dimensional displacement of a particle is

$$x(t) = 0.5e^{-0.2t}\sin 15t\ \text{m}$$

What is the maximum acceleration of this particle?

1.3. The one-dimensional displacement of a particle is

$$x(t) = 0.5e^{-1.2t}\sin(15t + 0.24)\ \text{m}$$

What is the maximum acceleration of this particle?

1.4. At the instant shown in Fig. P1.4 the slender rod has a clockwise angular velocity of 5 rad/s and a counterclockwise angular acceleration of 14 rad/s^2. At this instant what is the acceleration of the particle at P?

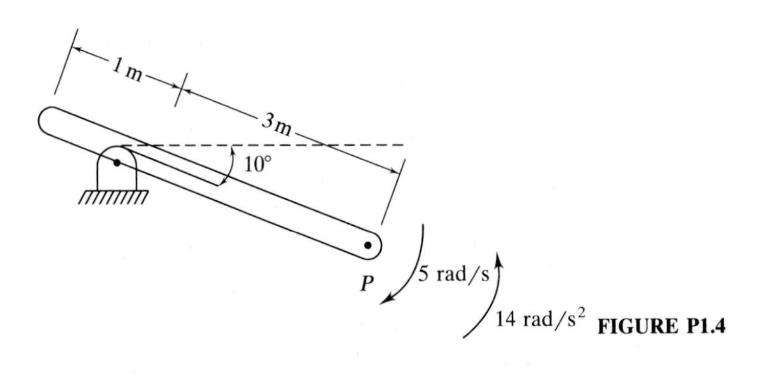

FIGURE P1.4

1.5. At $t = 0$ a particle of mass 1.2 kg is moving with a speed of 10 m/s that is increasing at a constant rate of 0.5 m/s^2. The local radius of curvature of the path of the particle at this instant is 50 m. After the particle travels 100 m, the radius of curvature of the particle's path is 25 m.
(a) What is the speed of the particle after it travels 100 m?
(b) What is the magnitude of the acceleration of the particle after it travels 100 m?
(c) How long does it take the particle to travel 100 m?

1.6. The machine of Fig. P1.6 has a vertical displacement $x(t)$. The machine has a component that rotates with a constant angular speed ω relative to the machine. The center of mass of the rotating component is a distance e from the axis of rotation. If the center of mass is as shown at $t = 0$, determine its vertical component of acceleration.

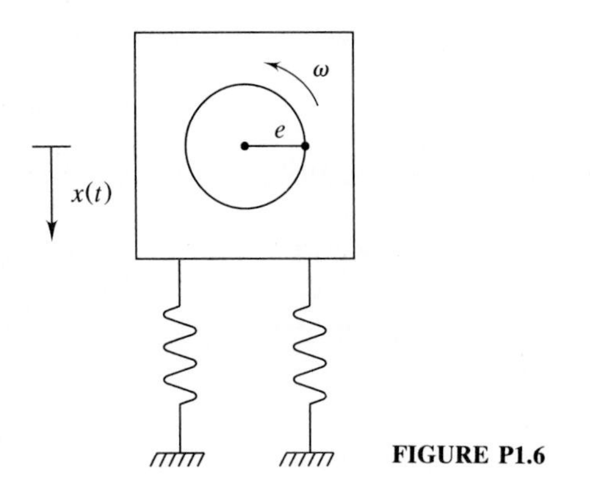

FIGURE P1.6

1.7. The rotor of Fig. P1.7 consists of a disk mounted on a shaft. Unfortunately, the disk is unbalanced and its center of mass is a distance e from the center of the

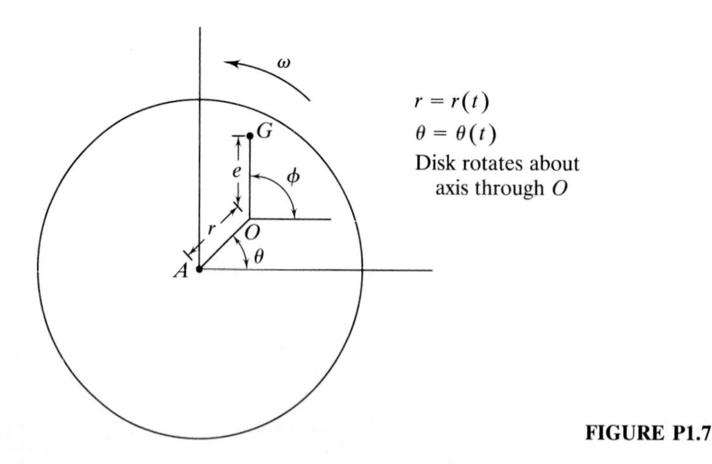

FIGURE P1.7

shaft. As the rotor rotates, this causes a phenomenon, called whirl, where the shaft bows out. Let r be the instantaneous distance from the center of the shaft to the original axis of the shaft and let θ be the angle made by a given radius with the horizontal plane. Determine the acceleration of the mass center of the disk.

1.8. An automobile is traveling with a horizontal velocity of 40 m/s when it encounters a pothole whose depth is approximated by

$$y(x) = 0.02(x^2 - 6x)\text{ m}$$

where x is the distance in meters from the leading edge of the pothole. When the driver encounters the pothole, he begins a constant deceleration of 10 m/s^2. What is the maximum vertical velocity and acceleration attained by the automobile as it traverses the pothole?

1.9. A 2-ton truck is traveling down an icy 10° hill at 50 mph when the driver sees a car stalled at the bottom of the hill, 250 ft away. Due to icy conditions, a braking force of only 1200 lb is generated, when the driver applies his brakes. Does the truck stop before hitting the car?

1.10. A 60-lb block is connected by an inextensible cable through the pulley to the fixed surface. A 40-lb weight is attached to the pulley which is free to move vertically. A force of magnitude $P = (70 + 30e^{-t})$ lb tows the block. The system is released from rest at $t = 0$.

(a) What is the acceleration of the 60-lb block as a function of time?

(b) How far will the block travel up the incline before it attains a velocity of 20 ft/s?

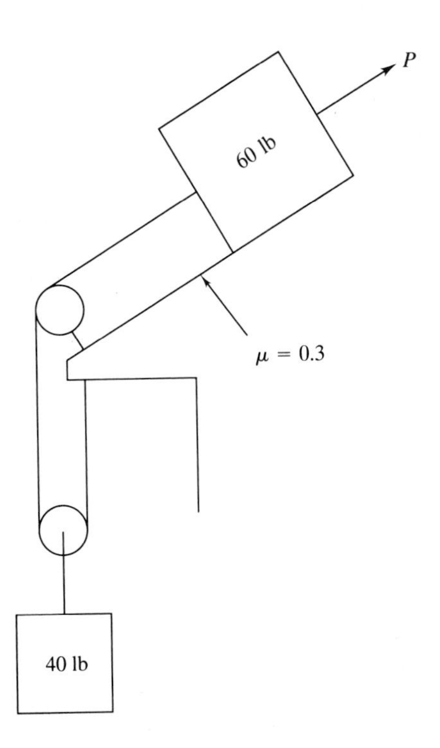

FIGURE P1.10

1.11. Repeat Prob. 1.10 for $P = 80t$ lb.

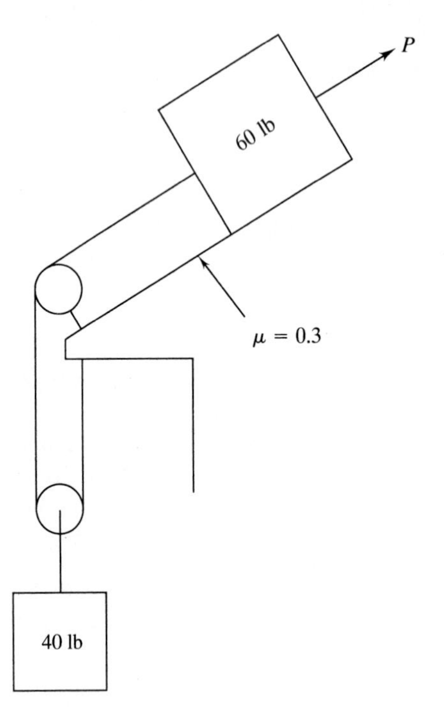

FIGURE P1.11

1.12. Figure P1.12 shows a schematic of a one-cylinder reciprocating engine. If the piston has a velocity v and an acceleration a, determine the angular acceleration

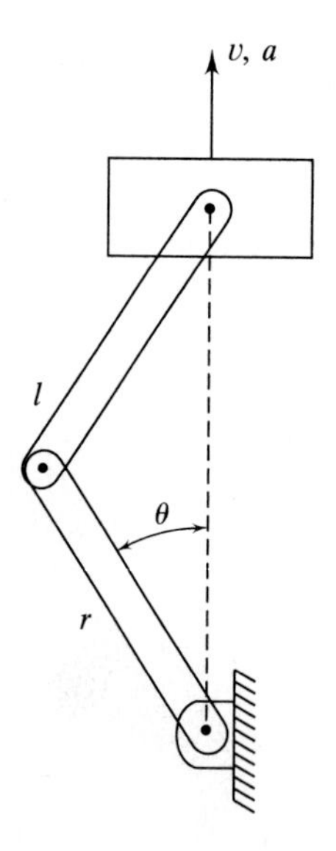

FIGURE P1.12

of the crank in terms of v, a, the crank radius r, the connecting rod length l, and the crank angle θ.

1.13. Determine the reactions at A for the two-link mechanism of Fig. P1.13. The roller at C rolls on a frictionless surface.

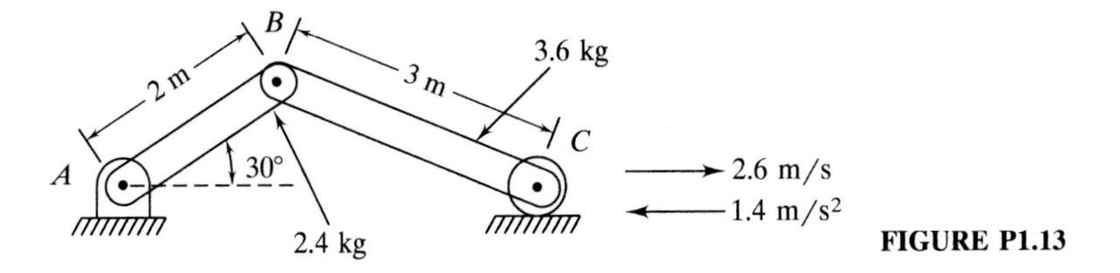

FIGURE P1.13

1.14. Determine the angular acceleration of each of the disks in Fig. P1.14.

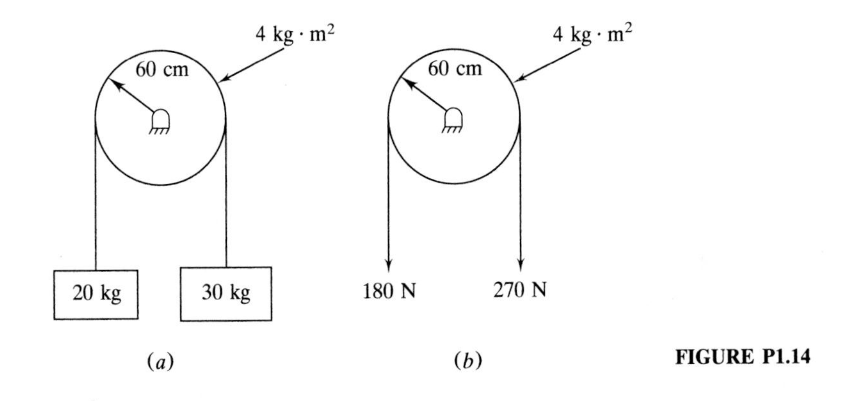

FIGURE P1.14

1.15. The slender rod of Fig. P1.15 has a mass of 50 kg. Determine the applied moment M and reactions at the pin support at the instant shown.

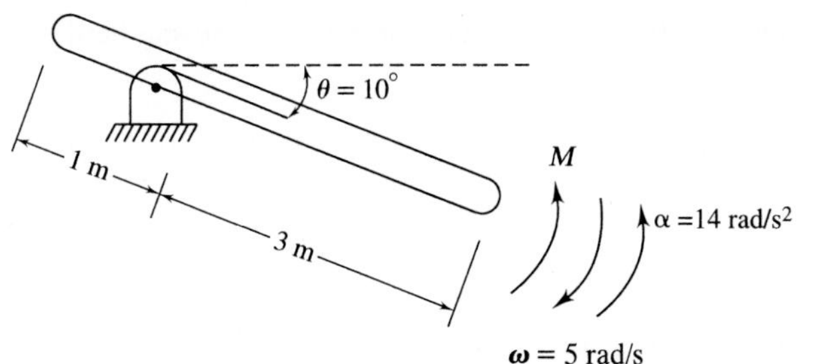

FIGURE P1.15

1.16. The disk of Fig. P1.16 rolls without slip. Determine the acceleration of the mass center of the disk if $P = 18$ N.

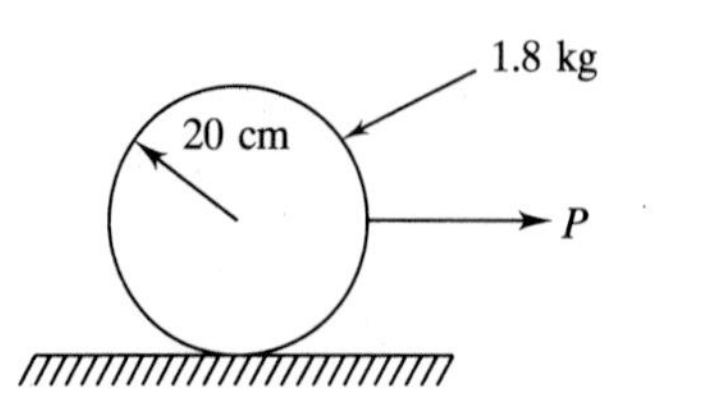

FIGURE P1.16

1.17. The coefficient of friction between the surface and the disk of Fig. P1.16 is 0.12. What is the largest force that can be applied such that the disk rolls without slip?

1.18. The coefficient of friction between the surface and the disk of Fig. P1.16 is 0.12. Determine the angular acceleration of the disk if $P = 15$ N.

1.19. What is the maximum angular velocity attained by the disk of Fig. P1.19 if the 3-kg block is displaced 10 mm and released?

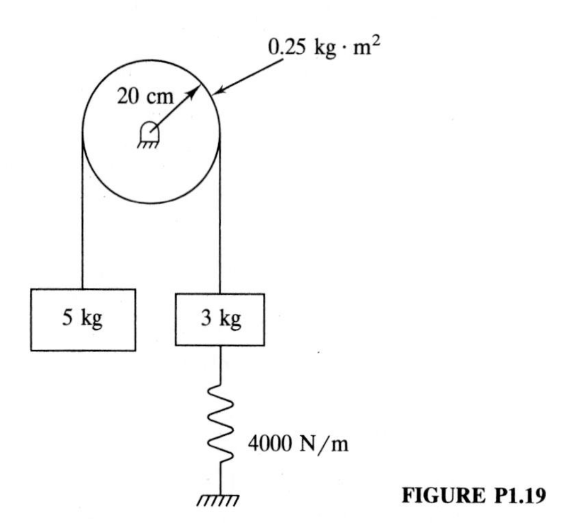

FIGURE P1.19

1.20. The center of the thin disk of Fig. P1.20 is displaced 15 mm and released. What is the maximum velocity attained by the center of the disk assuming no slip between the disk and the surface.

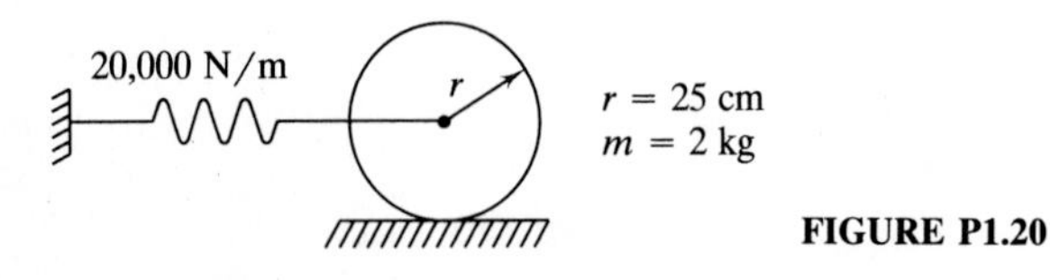

FIGURE P1.20

1.21. The block of Fig. P1.21 is given a displacement δ and then released.
(*a*) What is the minimum value of δ such that motion ensues?
(*b*) What is the minimum value of δ such that the block returns to its original equilibrium position before stopping?

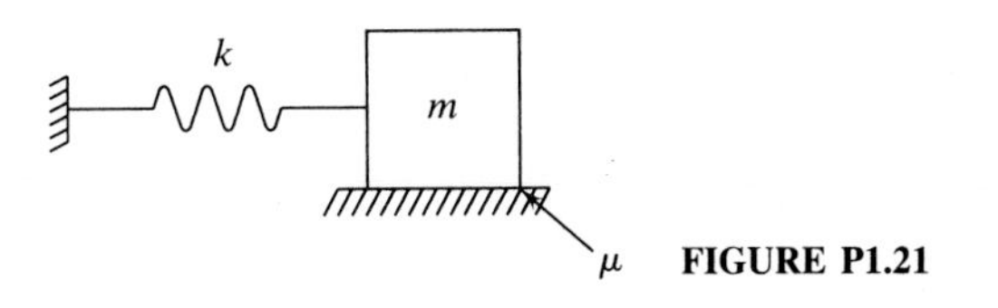

FIGURE P1.21

1.22. The disk of Fig. P1.22 rolls without slip when released after its center is displaced a distance δ. What is the maximum velocity of the ensuing motion?

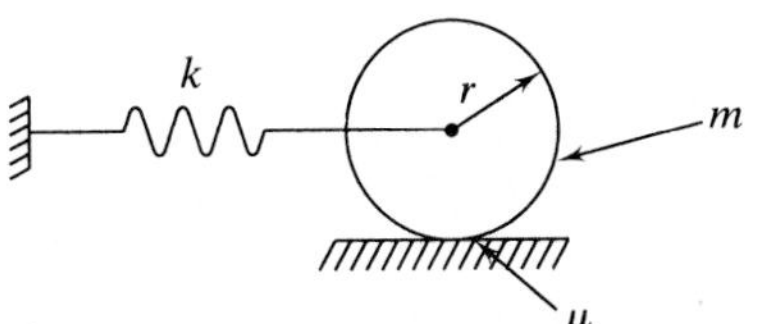

FIGURE P1.22

1.23. The five-blade ceiling fan of Fig. P1.23 operates at 60 rpm. What is its total kinetic energy?

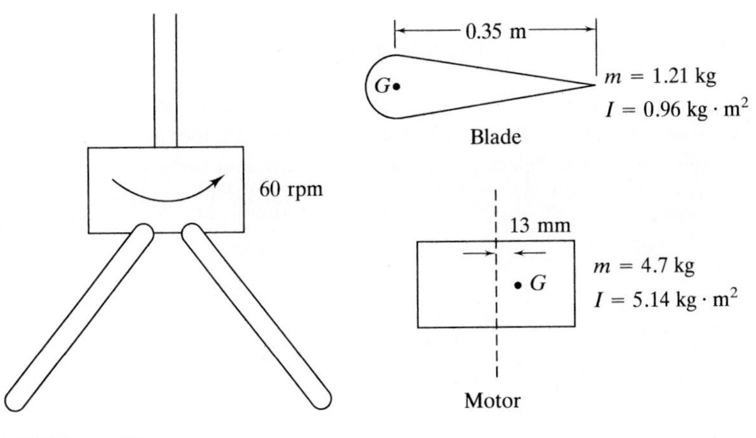

FIGURE P1.23

1.24. The U-tube manometer of Fig. P1.24 rotates about axis A–A at a speed of 40 rad/s. During the configuration shown, the column of liquid moves in the manometer with a velocity of 20 m/s relative to the manometer. Calculate the total kinetic energy of the column of liquid at this instant.

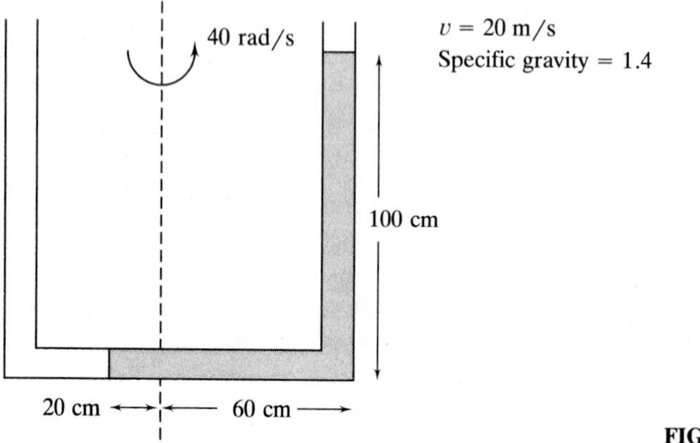

FIGURE P1.24

1.25. The 2.5-kg block of Fig. P1.25 is released from rest from the position shown.
(*a*) What is the maximum deflection of the spring if the block slides on a frictionless surface?
(*b*) What is the acceleration of the block when the spring reaches its maximum deflection?

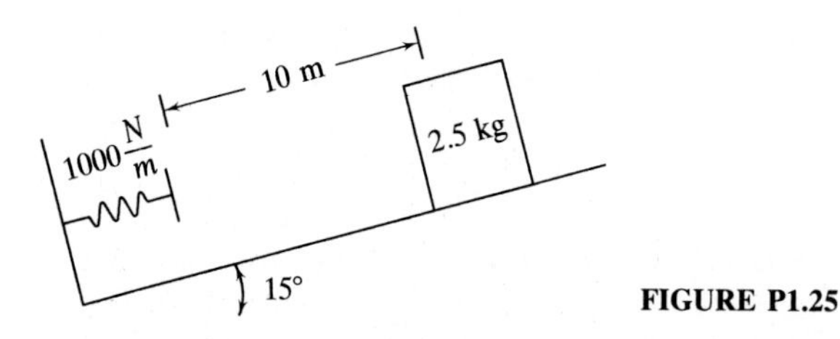

FIGURE P1.25

1.26. The slender rods of Fig. P1.26 are released from rest in the position shown. Assume the roller at C is massless and rolls on a frictionless surface. What is the angular velocity of BC when the angle between AB and the horizontal is 30°?

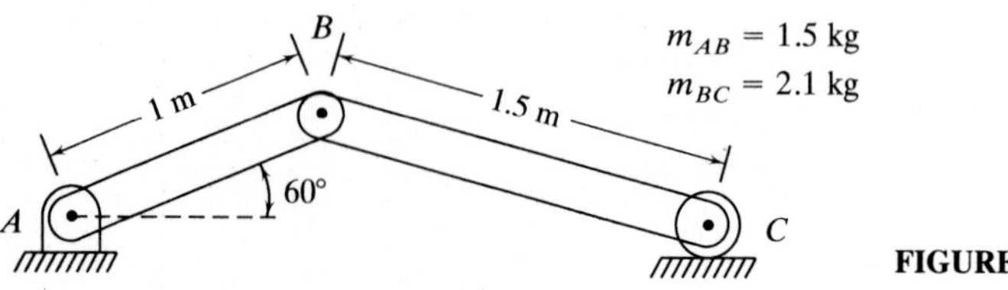

FIGURE P1.26

1.27. The slender rod of Fig. P1.27 is moved to the horizontal position and released. When the bar is horizontal the spring is compressed 22 mm. What is the maximum angle through which the bar will swing?

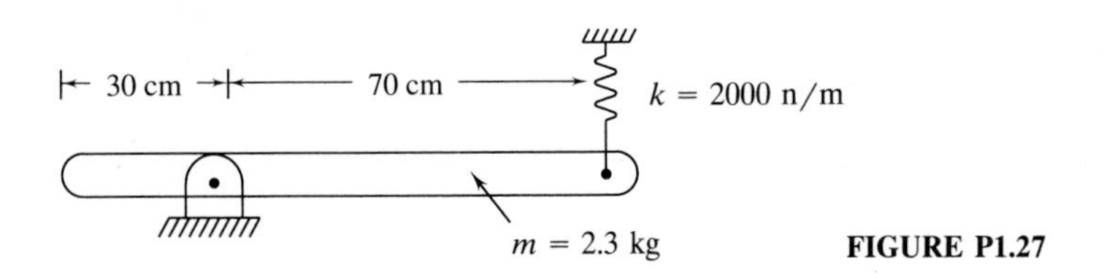

FIGURE P1.27

1.28. The slender rod of Fig. P1.28 is released from the horizontal position where the spring attached at A is stretched 10 mm and the spring attached at B is unstretched.

(a) What is the maximum value of y?

(b) What is the maximum angular velocity attained by the bar?

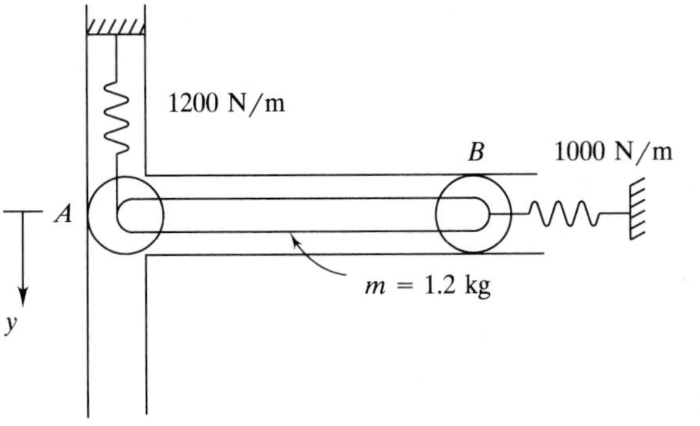

FIGURE P1.28

1.29. A 20-ton railroad car is coupled to a 15-ton car by moving the 20-ton car at 5 mph toward the stationary 15-ton car. What is the speed of the two-car coupling?

1.30. The motion of a baseball bat in a ballplayer's hands is approximated as a rigid-body rotation about an axis through the player's hands. The bat has a centroidal moment

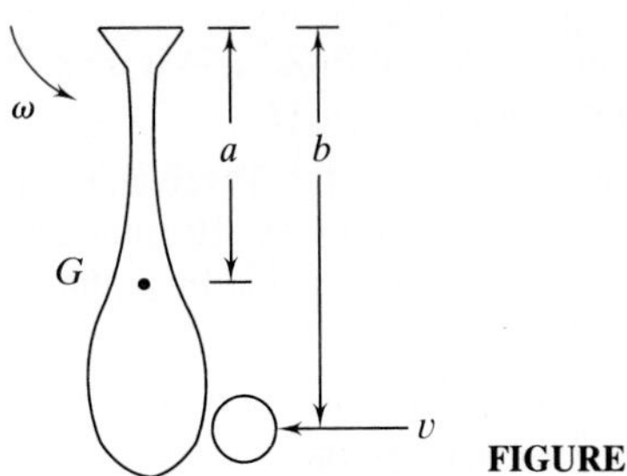

FIGURE P1.30

of inertia I. The player's "bat speed" is ω and the velocity of the pitched ball is v. Determine the distance from the player's hands along the bat where the batter should strike the ball to minimize the impulse felt by the player's hands.

1.31. The 4-kg block of Fig. P1.31 has a velocity of 10 m/s when it impacts the 2-kg block. After impact, the blocks are attached. The surface is frictionless. What is the maximum deflection of the spring?

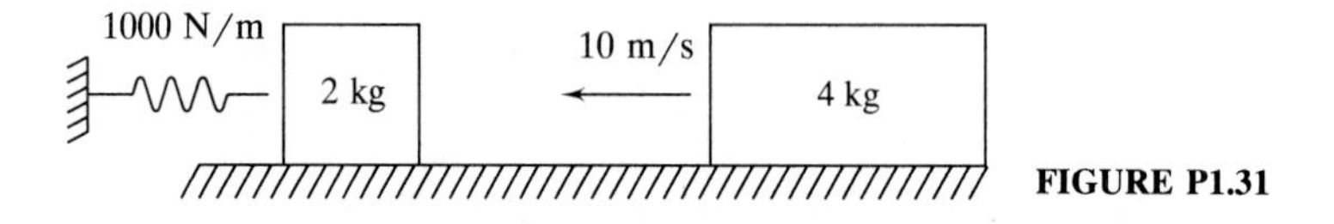

FIGURE P1.31

1.32. Repeat Prob. 1.31 if the coefficient of restitution between the blocks is 0.7.

1.33. Repeat Prob. 1.31 if the coefficient of restitution between the blocks is 0.7 and the coefficient of friction between the surface and block is 0.1.

1.34. A 2-g bullet is fired at the slender rod of Fig. P1.34. The bullet becomes embedded in the rod and the assembly swings to an angle of 17.4°. What is the velocity of the bullet before impact?

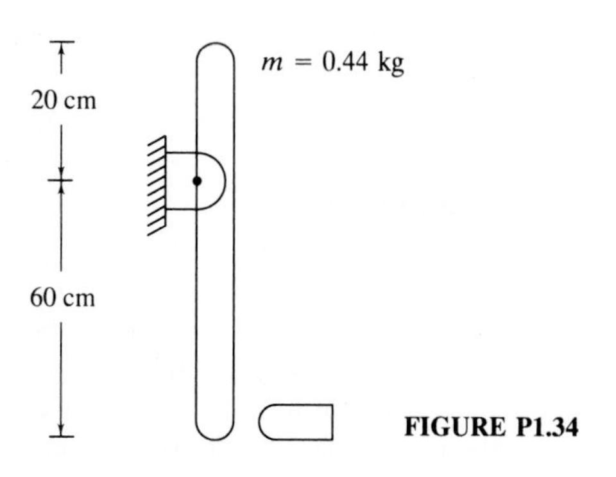

FIGURE P1.34

1.35. The coefficient of restitution between the bullet and the bar of Prob. 1.34 is 0.68. If the bar swings through 14°, what is the firing velocity of the bullet?

1.36. A playground ride has a centroidal moment of inertia of 17 slug-ft^2. Three children of weight 40 lb, 50 lb, and 55 lb are on the ride which is rotating at 110 rpm. The children are 20 in. from the center of rotation. A father stops the ride by grabbing it with his hands. What angular impulse is felt by the father?

1.37–1.43. How many degrees of freedom are required to specify the motion of each system? Identify a set of generalized coordinates for each system.

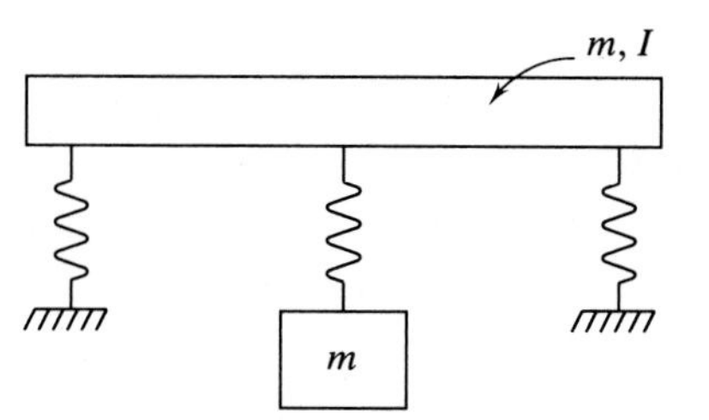

FIGURE P1.37

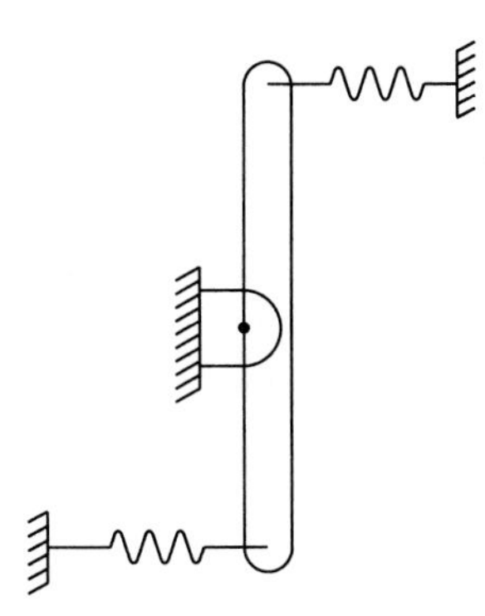

FIGURE P1.38

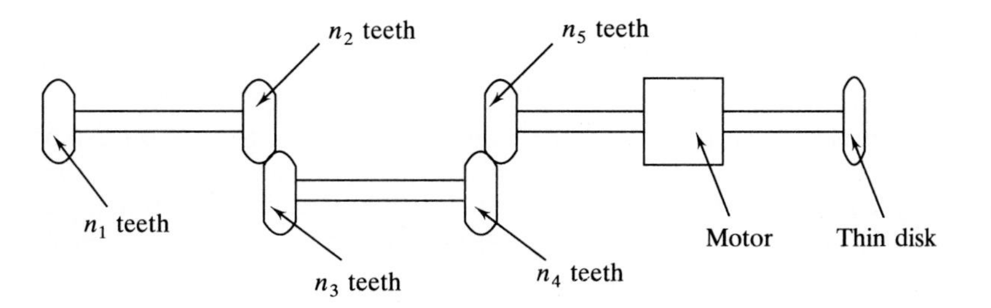

FIGURE P1.39

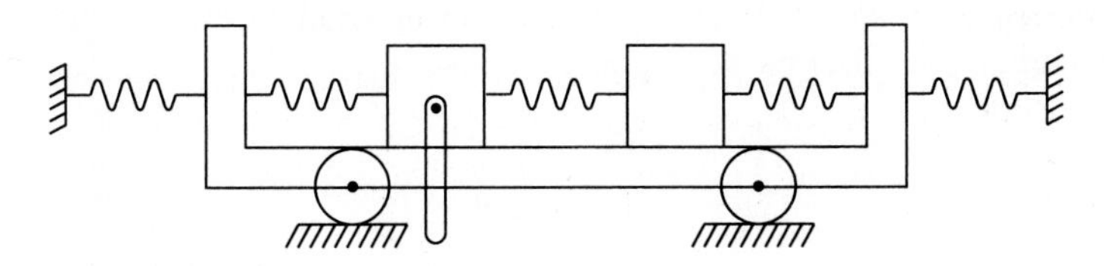

FIGURE P1.40

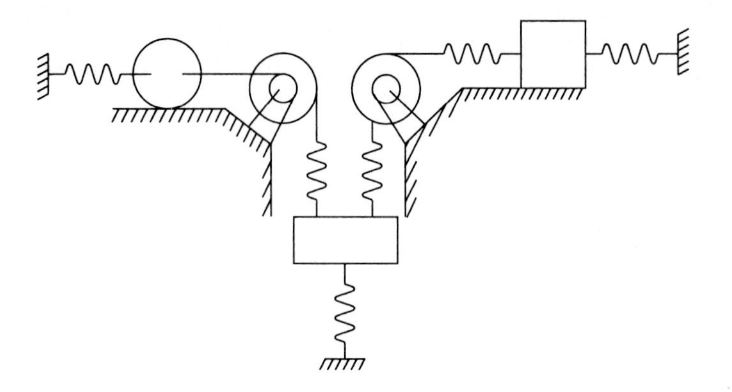

FIGURE P1.41

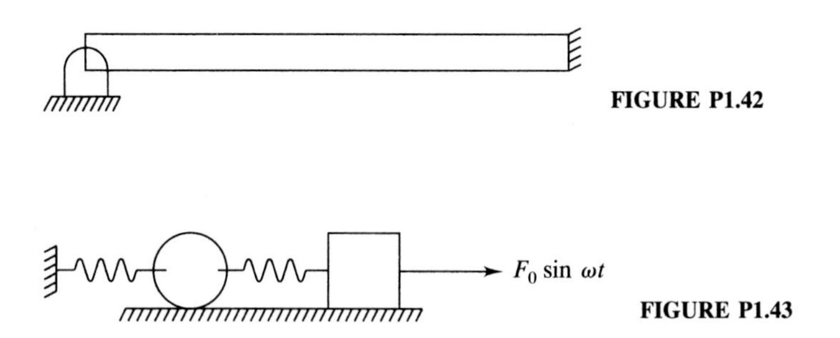

FIGURE P1.42

FIGURE P1.43

1.44. The deflection of the midspan of a simply supported beam due to a concentrated load P applied a distance a from the beam's left support is a function of P, a, the beam's length L, its moment of inertia I, and its elastic modulus E. Determine a set of Π groups which can be used to express the relationship in nondimensional form.

1.45. The natural frequency of a cantilever beam is dependent upon the length of the beam, its mass density, its cross-sectional area, its cross-sectional moment of inertia, and its elastic modulus. Determine a set of Π groups which can be used to express the relationship in nondimensional form.

1.46. The maximum displacement of the system of Fig. P1.46 is a function of the spring stiffness, the mass of the block, the magnitude of the applied harmonic load, and its frequency. Determine a set of Π groups which can be used to express the relationship in nondimensional form.

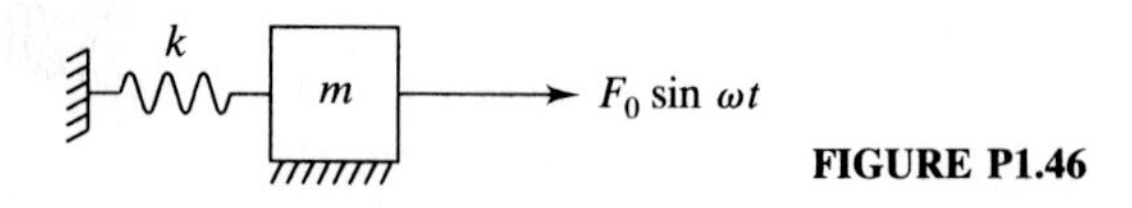

FIGURE P1.46

1.47. Determine a set of Π groups which can be used to express the relationship between the maximum angle to which the slender rod of Fig. P1.47 will swing after being hit by the particle. The coefficient of restitution between the rod and particle is e.

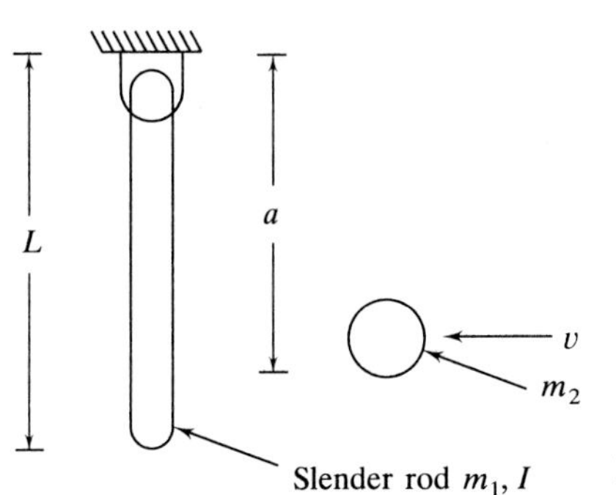

FIGURE P1.47

1.48. Use the nondimensional variables

$$x^* = \frac{x}{X}$$

$$t^* = \sqrt{\frac{k}{m}}\,t$$

to nondimensionalize the differential equation

$$m\ddot{x} + c\dot{x} + kx = F_0 \sin \omega t$$

where F_0 has dimension of force and c has dimensions of force × length/time.

1.49. Use the nondimensional variables introduced in Problem 1.48 to nondimensionalize the differential equation

$$m\ddot{x} + kx = \mu mg + F_0 \sin \omega t$$

where μ is dimensionless and F_0 has dimension of force.

REFERENCES

1. Beer, F. P., and E. R. Johnston: *Vector Mechanics for Engineers, Statics and Dynamics*, 4th ed., McGraw-Hill, New York, 1984.
2. Gerhart, P. M., and R. J. Gross: *Fundamentals of Fluid Mechanics*, Addison-Wesley, Reading, Mass., 1985.
3. Hibbeler, R. C.: *Engineering Mechanics, Dynamics*, 3rd ed., Macmillan, New York, 1983.
4. Meriam, and L. G. Kraige: *Engineering Mechanics, Dynamics*, 2nd ed., Wiley, New York, 1986.
5. White, F. M.: *Fluid Mechanics*, 2nd ed., McGraw-Hill, New York, 1986.

CHAPTER 2

ELEMENTS OF VIBRATING SYSTEMS

2.1 INTRODUCTION

A schematic of a simple mass-spring-dashpot system is shown in Fig. 2.1. Free vibrations occur when the mass is displaced from equilibrium and released from rest. Forced vibrations occur when a time-dependent force is applied to the mass.

The specific system shown in Fig. 2.1 is rarely encountered in application, but yet its motion is the subject of Chaps. 3 through 5. The behavior of many one-degree-of-freedom linear systems is identical to the behavior of the system of Fig. 2.1. Hence the system is used as a model to analyze the vibrations of one-degree-of-freedom systems. Indeed, under appropriate assumptions and after appropriate analysis, each of the systems of Fig. 2.2 can be modeled as a one-degree-of freedom mass-spring system, perhaps with a viscous damper. The appropriate model may be a mass hanging from a spring as for the system of Fig. 2.2*c* or may be a disk attached to a torsional spring as for the system of Fig. 2.2*d*.

The differential equations governing the time history of motion of the systems of Figs. 2.1 and 2.2 all have the same form. The coefficients in these equations are specific to the system and depend upon system parameters. The vibrations of the system of Fig. 2.1 are studied and the results extended to analyze vibrations of all one-degree-of-freedom systems. The general parame-

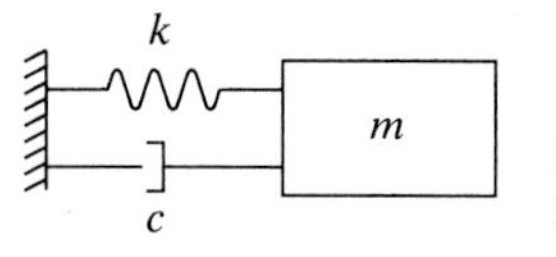

FIGURE 2.1
Schematic of mass-spring-dashpot system.

ters of the system of Fig. 2.1 are replaced in the results by the parameters specific to the system at hand.

It is shown in this chapter how the systems of Fig. 2.2 and others can be modeled by the simple system of Fig. 2.1. Each of the basic components of the system, the spring, the mass, and the dashpot are examined. The methods developed in this chapter can also be used in modeling multi-degree-of-freedom systems.

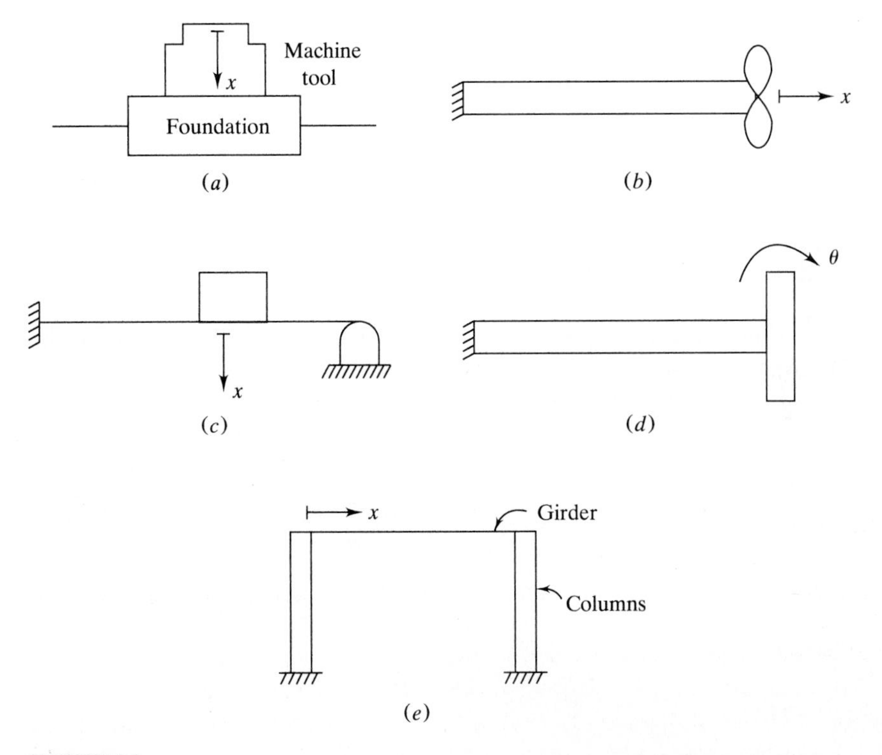

FIGURE 2.2
(*a*) Motion of machine tool on elastic foundation is modeled by system of Fig. 2.1; (*b*) longitudinal motion of propellor attached to shaft can be modeled using the system of Fig. 2.1; (*c*) transverse motion of machine placed on beam is modeled by system of Fig. 2.1 when block is suspended; (*d*) angular oscillations of flywheel attached to shaft are modeled by a disk attached to torsional spring which is analogous to system of Fig. 2.1; (*e*) lateral motion of one-story frame building is modeled by system of Fig. 2.1 when girder is very rigid compared to columns.

2.2 SPRINGS

2.2.1 Introduction

A spring is a flexible mechanical link between two particles in a mechanical system. Since the mass of a spring is usually small, the force at each end of the spring is approximately the same. The coil spring of Fig. 2.3 serves as a model for the following discussion.

The length of a spring when it is not subject to external forces is called its unstretched length. Let x be the change in length of a spring from its unstretched length. Since the spring is made of a flexible material, the following relationship between the spring force, F, and x exists:

$$F = f(x) \tag{2.1}$$

Equation (2.1) is developed using the constitutive equation for the spring's material. Since $f(x)$ must be infinitely differentiable at $x = 0$, it can be expanded in a Taylor series expansion about $x = 0$:

$$F = k_0 + k_1 x + k_2 x^2 + k_3 x^3 + \cdots \tag{2.2}$$

By definition, when $x = 0$, $F = 0$, hence $k_0 = 0$. When x is positive the length of the spring is greater than its unstretched length and the spring is in tension. When x is negative the length of the spring is less than its unstretched length and the spring is in compression. Many materials have the same properties in tension and compression. That is, if a tensile force F is required to lengthen the spring by δ, then a compressive force of the same magnitude F is required to shorten the spring by δ. For these materials the force-displacement relationship, Eq. (2.2), cannot contain even powers of x. Then Eq. (2.2) becomes

$$F = k_1 x + k_3 x^3 + k_5 x^5 + \cdots \tag{2.3}$$

All springs with the same properties in tension and compression really obey a nonlinear force-displacement law of the form of Eq. (2.3). However, for many springs the coefficients multiplying the higher powers of x are much smaller

FIGURE 2.3
Photograph of helical coil spring.

than the coefficient multiplying x. Higher-order terms are often ignored resulting in

$$F = kx \tag{2.4}$$

A spring that obeys the force-displacement relationship of Eq. (2.4) is called a linear spring.

2.2.2 Potential Energy of Linear Springs

Consider the system of Fig. 2.1 when $c = 0$ and the spring is linear and obeys the force-displacement law of Eq. (2.4). As the block moves a force is developed between it and the spring. As the block moves between two positions defined by displacements x_1 and x_2, measured from the spring's unstretched length, the work done by the spring force is

$$U_{1\to 2} = \int_{x_1}^{x_2} kx\,dx = \tfrac{1}{2}kx_2^2 - \tfrac{1}{2}kx_1^2 \tag{2.5}$$

Now consider a path where the block moves between x_1 and $x_3 > x_2$ and then back to x_2. The total work done by the spring force is

$$\begin{aligned} U_{1\to 2} &= U_{1\to 3} + U_{3\to 2} \\ &= \int_{x_1}^{x_3} kx\,dx + \int_{x_3}^{x_2} kx\,dx \\ &= \tfrac{1}{2}kx_3^2 - \tfrac{1}{2}kx_1^2 + \tfrac{1}{2}kx_2^2 - \tfrac{1}{2}kx_3^2 \\ &= \tfrac{1}{2}kx_2^2 - \tfrac{1}{2}kx_1^2 \end{aligned} \tag{2.6}$$

Since the results of Eqs. (2.5) and (2.6) are the same, the work done by the force between the mass and the spring is independent of path, and thus the force is conservative.

A potential energy function can thus be defined for a linear spring as

$$V = \tfrac{1}{2}kx^2 \tag{2.7}$$

where x is the change in length of the spring from its unstretched length.

2.2.3 Helical Coil Springs

The helical coil spring is used in applications such as industrial machines and vehicle suspension systems. The following analysis applies to helical springs that are manufactured from rods of circular cross section, whose coil radius, r, is much larger than the radius of the rod, $D/2$, and that the normal to the plane of one coil nearly coincides with the axis of the spring (the spring is tightly wound).

Consider the helical spring of Fig. 2.3 when subject to an axial load, F, at each end. Imagine cutting the rod with a knife at an arbitrary location in a coil, slicing the spring in two sections. In order for each section to be in static

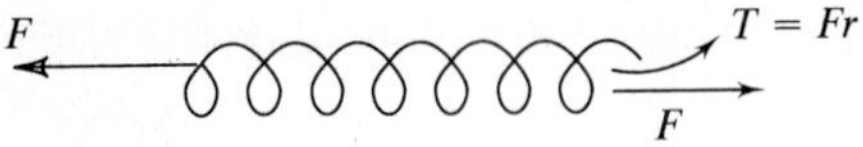

FIGURE 2.4
Free-body diagram of cut coil spring exposes resultant shear force and resultant torque.

equilibrium, the cut exposes an internal shear force, equal to F, and an internal resisting torque, equal to Fr., as shown in Fig. 2.4. These are the resultants of shear stress distributions in the cross section of the spring, as shown in Fig. 2.5. The shear stress causing the resisting torque is circumferential and varies linearly from the center of the rod to a maximum of

$$(\tau_{\max})_T = \frac{TD}{2J} = \frac{16Fr}{\pi D^3}$$

The shear stress causing the internal shear force is vertical and varies nonlinearly with y. This shear stress is zero at C_1 and C_2, as shown in Fig. 2.5, and has a maximum of

$$(\tau_{\max})_V = \frac{8F}{3\pi D^2}$$

along the shaft's neutral axis. The ratio of the maximum shear stresses is

$$\frac{(\tau_{\max})_T}{(\tau_{\max})_V} = \frac{6r}{D}$$

For $r/D \gg 1$ the maximum shear stress due to the shear force is much smaller than the shear stress due to the resisting torque. Thus the effect of the internal shear force is ignored.

The resisting torque in section C–C would cause a horizontal and vertical displacement of point B relative to point A. However, the horizontal displacement is balanced by an equal and opposite horizontal displacement from the

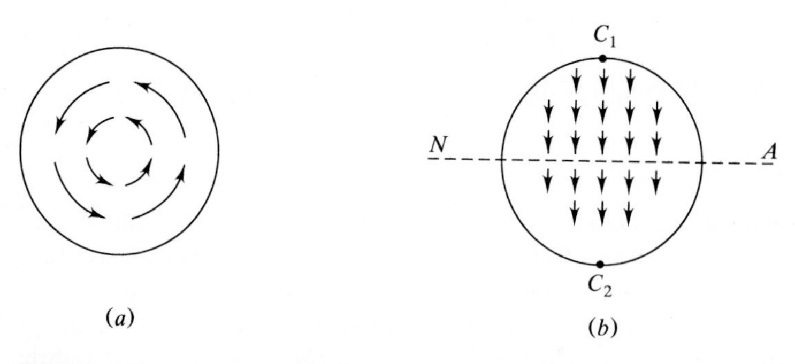

FIGURE 2.5
(a) The radial shear stress due to resisting torque varies linearly with distance from the center of shaft; (b) vertical shear stress due to shear force is zero at C_1 and C_2 and a maximum at neutral axis.

opposite side of the spring. Then the vertical displacement of point B relative to point A is the coil radius times the angular rotation of B relative to A due to the internal resisting torque. This becomes

$$x_{B/A} = r\frac{TL}{JG} \tag{2.8}$$

where $T = (r)(F)$ is the resisting torque, L is the total length of coil, $J = \pi D^4/32$ is the polar moment of inertia of the coil, and G is the shear modulus of the material. Noting that

$$L = 2\pi rN \tag{2.9}$$

where N is the number of coils in the spring, Eq. (2.8) becomes

$$x_{B/A} = \frac{64Fr^3N}{GD^4} \tag{2.10}$$

However, $x_{B/A}$ is the total change in length of the spring due to application of the force F. Equation (2.10) can be rewritten in the form of Eq. (2.4) with

$$k = \frac{GD^4}{64r^3N} \tag{2.11}$$

The helical coil spring can be approximated as a linear spring under the assumption that the change in length due to the internal shear force is small compared with the change in length due to the internal resisting torque.

Example 2.1. A tightly wound spring is made from a 20-mm-diameter bar of 0.2% C-hardened steel ($G = 80 \times 10^9$ N/m^2). The coil diameter is 20 cm. The spring has 30 coils. What is the largest force that can be applied such that the elastic strength in shear of 220×10^6 N/m^2 is not exceeded? What is the change in length of the spring when this force is applied?

Assuming the shear stress due to the shear force is negligible, the maximum shear stress in the spring when a force F is applied is

$$\tau = \frac{FrD}{2J} = F\frac{(0.1\text{ m})(0.02\text{ m})}{2\dfrac{\pi}{32}(0.02\text{ m})^4} = 6.37 \times 10^4 F$$

Thus the maximum allowable force is

$$F_{\text{max}} = \frac{\tau_{\text{max}}}{6.37 \times 10^4} = 3460\text{ N}$$

The stiffness of this spring is calculated using Eq. (2.11)

$$k = \frac{(80 \times 10^9\text{ N/m}^2)(0.02\text{ m})^4}{(64)(30)(0.1\text{ m})^3} = 6670\ \frac{\text{N}}{\text{m}}$$

The total change in length of the spring due to application of the preceding force is

$$\Delta = \frac{F}{k} = 0.519 \text{ m}$$

2.3 STRUCTURAL ELEMENTS AS SPRINGS

When a force is applied to the block of mass m, of Fig. 2.6, the block has a displacement, x_0. A uniform normal strain

$$\epsilon = \frac{F}{AE} = \frac{x_0}{L} \tag{2.12}$$

results in the thin uniform rod to which the block is attached, where E is the elastic modulus of the rod, A is its cross-sectional area, and L is its unstretched length. The work done by the force is converted into strain energy given by

$$S = \tfrac{1}{2}EAL\epsilon^2 \tag{2.13}$$

If the force is maintained and then suddenly removed, the block will oscillate about its equilibrium position. Strain energy is converted to kinetic energy and vice versa, a process which occurs continually.

The mass of the rod is assumed small compared to the mass of the block; thus its inertia is neglected. Hence, when the block is displaced a distance x from equilibrium, the force acting on the block from the rod is the force required to change the length of the rod by x. From strength of materials

$$F = \frac{AE}{L}x \tag{2.14}$$

Thus the force applied to the block from the rod is equal to the force applied to the block when attached to a linear spring of stiffness

$$k = \frac{AE}{L} \tag{2.15}$$

Indeed, if the mass of the rod is small compared to the mass of the block, the motion of the system of Fig. 2.6 can be modeled using the system of Fig. 2.1 with $c = 0$ and k given by Eq. (2.15).

The motion of a block attached to a structural element can be modeled using the mass-spring system of Fig. 2.1 provided the mass of the structural element is small compared to the mass of the block and a linear relationship

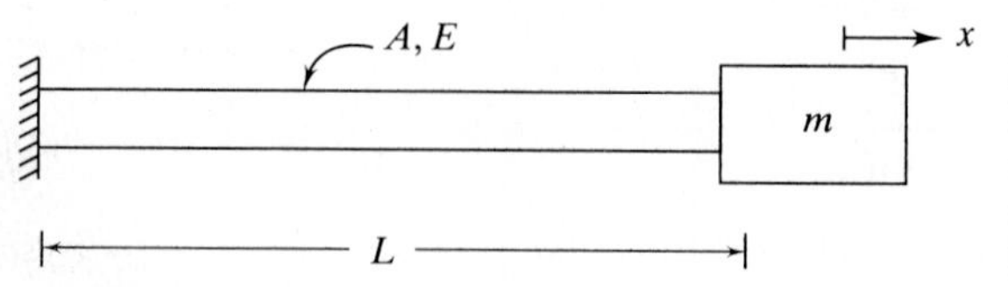

FIGURE 2.6
Longitudinal vibrations of mass attached to end of uniform thin rod can be modeled by system of Fig. 2.1 with $c = 0$ and $k = AE/L$.

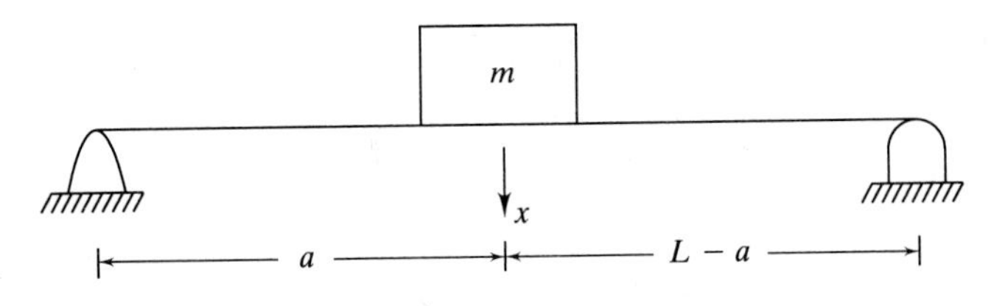

FIGURE 2.7
The transverse vibrations of a block attached to a simply supported beam are modeled by the mass-spring system of Fig. 2.8, provided the mass of the beam is small compared to the mass of the block.

exists between a static force applied to the point where the block is attached and the displacement of that point.

Another example of a structural element that can be modeled as a spring is a beam. Consider the simply supported beam of Fig. 2.7. The displacement due to a static load F at the point where the mass is attached is

$$y = \frac{Fa^2(L-a)^2}{3EIL} \tag{2.16}$$

If the mass of the beam is much smaller than m, the system can be modeled by the system of Fig. 2.8 where the spring has a stiffness

$$k = \frac{3EIL}{a^2(L-a)^2} \tag{2.17}$$

In general, the transverse vibrations of a block attached to a beam at $x = a$ can be modeled by the hanging mass-spring system of Fig. 2.8 with

$$k = \frac{1}{y(a)} \tag{2.18}$$

where $y(x)$ is the deflected shape of the beam due to a concentrated unit force applied at $x = a$.

A torsional spring is a flexible element in which one end rotates relative to the other end when a moment is applied. A linear torsional spring has a linear

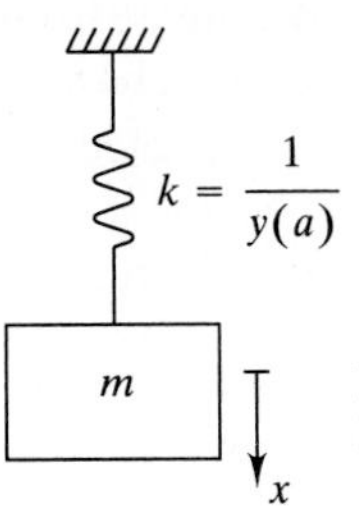

FIGURE 2.8
Hanging mass-spring system used to model transverse vibrations of a block attached to beam of negligible mass at $x = a$.

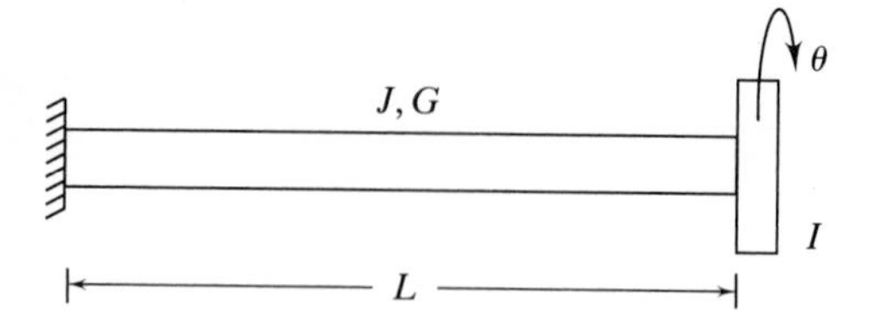

FIGURE 2.9
When a moment is applied to the disk, torsional oscillations occur. The shaft behaves as a torsional spring of stiffness JG/L.

relation between applied moment and angular displacement

$$M = k_t \theta \tag{2.19}$$

where k_t is the torsional spring constant. When a disk at the end of the torsional spring is rotated away from an equilibrium position and then released, the disk rotates about the equilibrium position. These oscillations are called torsional oscillations.

Torsional oscillations occur in the system of Fig. 2.9. The mass moment of inertia of the shaft is assumed small in comparison to the mass moment of inertia of the disk. When the disk has rotated an angle θ from the system equilibrium position, a moment

$$M = \frac{JG}{L}\theta \tag{2.20}$$

develops between the disk and the shaft where G is the shear modulus of the shaft and J is the polar moment of inertia of the cross section of the shaft. Thus the shaft acts as a torsional spring of stiffness

$$k_t = \frac{JG}{L} \tag{2.21}$$

2.4 EQUIVALENT SPRINGS

Often, in applications, springs are placed in combinations. It is possible, for analysis purposes, to replace the combination of springs with a single spring of an equivalent stiffness such that the behavior of the system with the equivalent spring is identical to the behavior of the actual system.

The springs connected to the block of Fig. 2.10 are in parallel. The displacement of each spring is the same, but the force developed in each spring is different and depends upon the spring's stiffness.

It is desired to replace the parallel combination by a single spring of equivalent stiffness, k_{eq}, as shown in Fig. 2.11. The behavior of the system of Fig. 2.11 is identical to the behavior of the system of Fig. 2.10. That is, the displacement of the block of each system is the same, say x, when subject to the same resultant force, F. The resultant force acting on the block attached to the parallel combination of springs is the sum of the individual spring forces

$$F = k_1 x + k_2 x + k_3 x + \cdots + k_n x = \left(\sum_{i=1}^{n} k_i \right) x \tag{2.22}$$

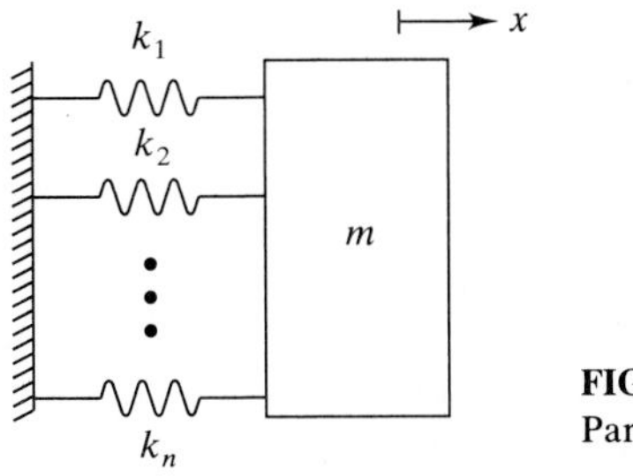

FIGURE 2.10
Parallel combination of springs.

The force acting on the block attached to the spring of an equivalent stiffness is

$$F = k_{eq}x \tag{2.23}$$

Equating the force from Eqs. (2.22) and (2.23) gives

$$k_{eq} = \sum_{i=1}^{n} k_i \tag{2.24}$$

The springs in Fig. 2.12 are in series. The same force is developed in each spring and is equal to the force acting on the block. However, the change in length of each spring is different and is dependent upon the spring stiffness. The displacement of the mass from equilibrium is the sum of the changes in lengths of the springs

$$x = x_1 + x_2 + x_3 + \cdots + x_n = \sum_{i=1}^{n} x_i \tag{2.25}$$

Since the force is the same in each spring, $x_i = F/k_i$ and

$$x = \sum_{i=1}^{n} \frac{F}{k_i} \tag{2.26}$$

The displacement of the block attached to a single spring of an equivalent stiffness must equal the displacement of a block attached to the series combination of springs when each block is subject to the same resultant force. Equations

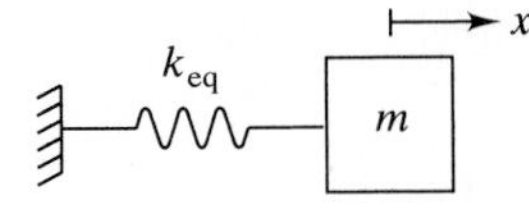

FIGURE 2.11
Combination of springs is replaced by a single spring such that the behavior of the system with an equivalent spring is identical to the behavior of the original system.

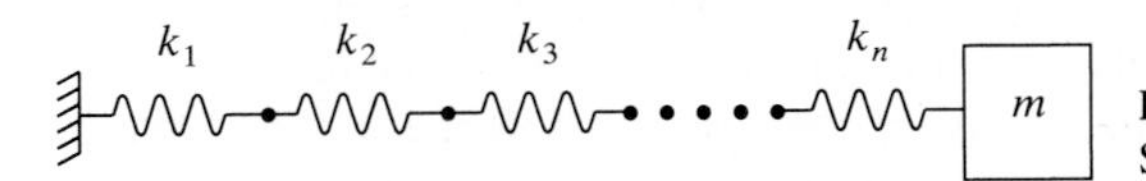

FIGURE 2.12
Series combination of springs.

(2.23) and (2.26) are used to give

$$k_{\mathrm{eq}} = \frac{1}{\sum_{i=1}^{n} \frac{1}{k_i}} \tag{2.27}$$

Electrical circuit components can also be placed in series and parallel and the effect of the combinations replaced by a single component with an equivalent value. The equivalent resistance of resistors in series is the sum of the resistances. The equivalent resistance of resistors in parallel is calculated using an equation similar to Eq. (2.27). The equivalent capacitance of capacitors in series is calculated using an equation similar to Eq. (2.27), while the equivalent capacitance of capacitors in parallel is the sum of the capacitances.

Example 2.2. Model each of the systems of Fig. 2.13 by a mass attached to a single spring of an equivalent stiffness. The system of Fig. 2.13*c* is to be modeled by a disk attached to a torsional spring of an equivalent stiffness.

(*a*) The steps involved in modeling the system of Fig. 2.13*a* by the system of Fig. 2.11 are shown in Fig. 2.14. Equation (2.24) is used to replace the two parallel springs by an equivalent spring of stiffness $3k$. The three springs on the left of the mass are then in series and Eq. (2.27) is used to obtain an equivalent stiffness.

If the mass in Fig. 2.14*b* is given a displacement x to the right, then the spring on the left of the mass will increase in length by x, while the spring on the right of the mass will decrease in length by x. Thus each spring will exert a force to the left on the mass. The spring forces add; the springs behave as if they are in parallel. Hence Eq. (2.24) is used to replace these springs by the equivalent spring shown in Fig. 2.14*c*.

(*b*) The deflection of the simply supported beam due to a unit load at $x = 2m$ is calculated using Table D.2

$$y(x = 2m) = y\left(\frac{2L}{3}\right) = \frac{4L^3}{243EI}$$

from which the equivalent stiffness is obtained

$$k_1 = \frac{243EI}{4L^3} = \frac{243(210 \times 10^9\ \mathrm{N/m^2})(5 \times 10^{-4}\ \mathrm{m^4})}{4(3\ \mathrm{m})^3} = 2.36 \times 10^8\ \frac{\mathrm{N}}{\mathrm{m}}$$

The system is replaced by a block attached to two springs in series as shown in Fig. 2.15*a*. Equation (2.27) is used to calculate the equivalent stiffness

$$k_{\mathrm{eq}} = \frac{1}{\dfrac{1}{2.36 \times 10^8\ \mathrm{N/m}} + \dfrac{1}{1 \times 10^8\ \mathrm{N/m}}} = 7.02 \times 10^7\ \frac{\mathrm{N}}{\mathrm{m}}$$

The behavior of the system of Fig. 2.13*b* is identical to the behavior of the system of Fig. 2.15*b*.

(*c*) The aluminum core of shaft AB is rigidly bonded to the steel shell. Thus the angular rotation at B is the same for both materials. The total resisting torque transmitted to section BC is the sum of the torque developed in the aluminum core and the torque developed in the steel shell. Thus the aluminum core and steel shell of shaft AB behave as two torsional springs in parallel. The resisting torque

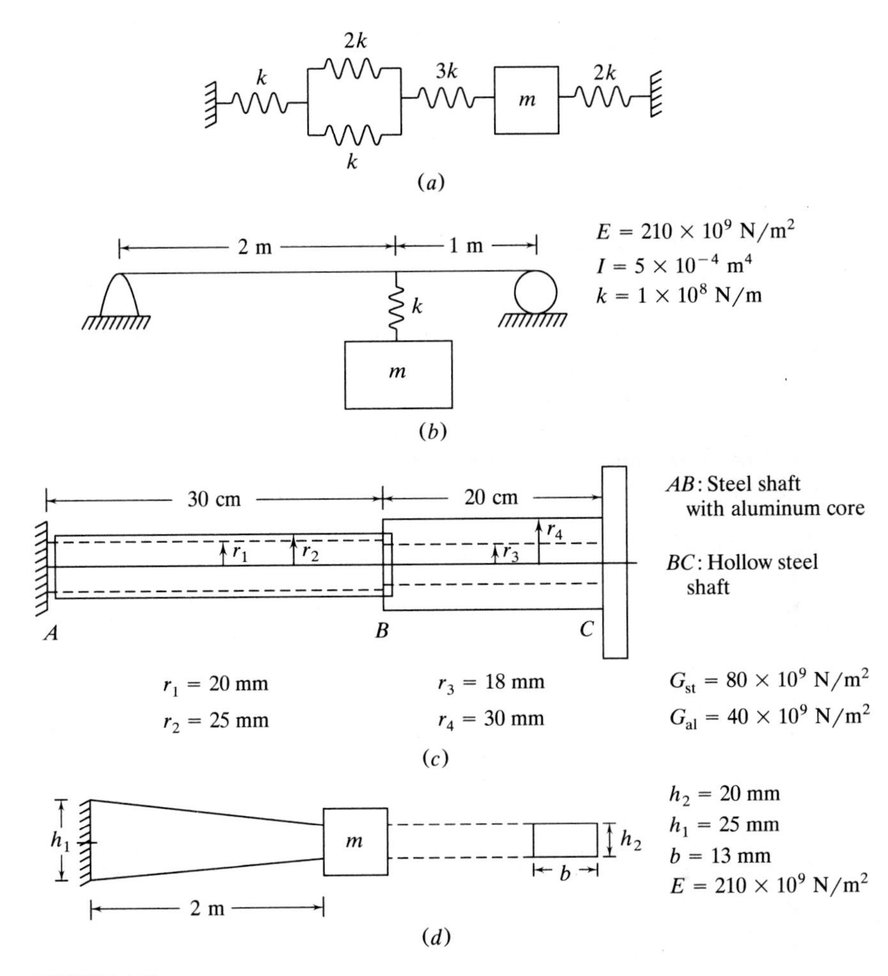

FIGURE 2.13
Systems for Example 2.2.

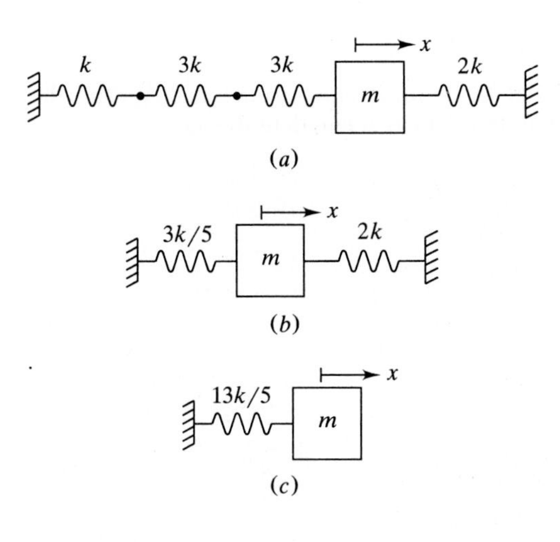

FIGURE 2.14
Steps in replacing the combination of springs of Fig. 2.13*a* by a single spring of an equivalent stiffness.

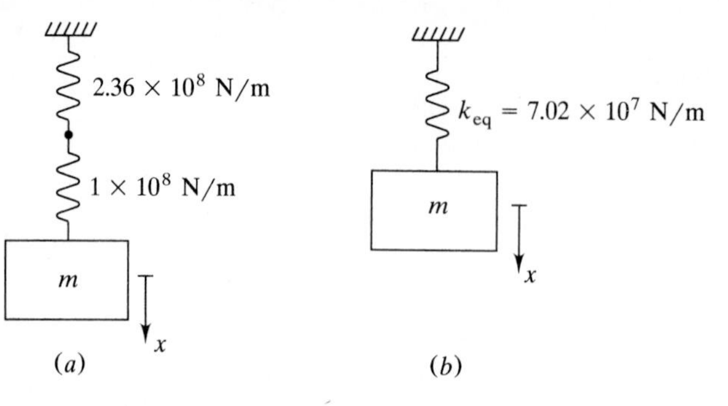

FIGURE 2.15
Steps in reducing the system of Fig. 2.13*b* to an equivalent system. Consisting of a block of mass m attached to a spring of equivalent stiffness.

in shaft AB is the same as the resisting torque in shaft BC. The angular displacement at C is the angular displacement of B plus the angular displacement of C relative to B. Thus shafts AB and BC behave as two torsional springs in series. In view of the preceding discussion and using Eqs. (2.24) and (2.27), the equivalent stiffness of shaft AC is

$$k_{t_{eq}} = \frac{1}{\dfrac{1}{k_{t_{AB_{al}}} + k_{t_{AB_{st}}}} + \dfrac{1}{k_{t_{BC}}}}$$

where

$$k_{t_{AB_{al}}} = \frac{\dfrac{\pi}{32}(0.04 \text{ m})^4\left(40 \times 10^9 \dfrac{\text{N}}{\text{m}^2}\right)}{0.3 \text{ m}} = 3.35 \times 10^4 \frac{\text{N} \cdot \text{m}}{\text{rad}}$$

$$k_{t_{AB_{st}}} = \frac{\dfrac{\pi}{32}\left[(0.05 \text{ m})^4 - (0.04 \text{ m})^4\right]\left(80 \times 10^9 \dfrac{\text{N}}{\text{m}^2}\right)}{0.3 \text{ m}} = 9.66 \times 10^4 \frac{\text{N} \cdot \text{m}}{\text{rad}}$$

$$k_{t_{BC}} = \frac{\dfrac{\pi}{32}\left[(0.06 \text{ m})^4 - (0.036 \text{ m})^4\right]\left(80 \times 10^9 \dfrac{\text{N}}{\text{m}^2}\right)}{0.2 \text{ m}} = 4.43 \times 10^5 \frac{\text{N} \cdot \text{m}}{\text{rad}}$$

Substitution of these values into the equation for k_{eq} gives

$$k_{t_{eq}} = 1.01 \times 10^5 \frac{\text{N} \cdot \text{m}}{\text{rad}}$$

(*d*) Under the assumption that the rate of taper of the bar is small the following equation is used to calculate the change in length of the bar due to a unit load applied at its end:

$$\Delta = \int_0^L \frac{dz}{AE}$$

where the cross-sectional area A is given by

$$A(z) = \left(\frac{h_2 - h_1}{L} z + h_1\right) b$$

Substituting and integrating yields

$$\Delta = \frac{L}{Eb(h_2 - h_1)} \ln\left(\frac{h_2}{h_1}\right) = 3.26 \times 10^{-8} \text{ m}$$

The longitudinal vibrations of the block are identical to the vibrations of a block attached to a linear spring of stiffness

$$k_{\text{eq}} = \frac{1}{\Delta} = 3.06 \times 10^7 \frac{\text{N}}{\text{m}}$$

2.5 STATIC-EQUILIBRIUM POSITION

The static-equilibrium position of a mechanical system is a position in which the system will remain in equilibrium in the absence of accelerations. Vibrations are oscillations about a static-equilibrium position caused by either the presence of an initial kinetic or potential energy or by application of external work.

Each of the systems of Fig. 2.16 has a unique static-equilibrium position. The system of Fig. 2.17, a slender rod pinned at one end, has two static-equilibrium positions. However, the equilibrium position shown in Fig. 2.17*b* is unstable. Any small perturbation will cause the rod to rotate far away from this equilibrium position. Unrestrained systems such as those shown in Fig. 2.18 have an infinite number of static-equilibrium positions. However, the unrestrained system of Fig. 2.19 has no static-equilibrium position unless the moments due to the gravity forces about the center of the pulley are equal. In the latter case the system has an infinite number of static-equilibrium positions.

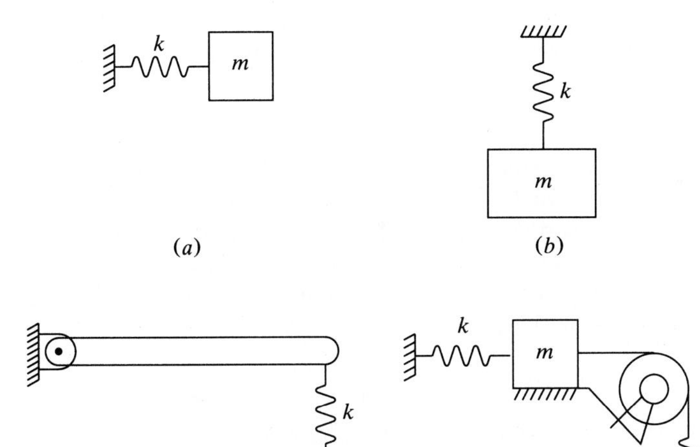

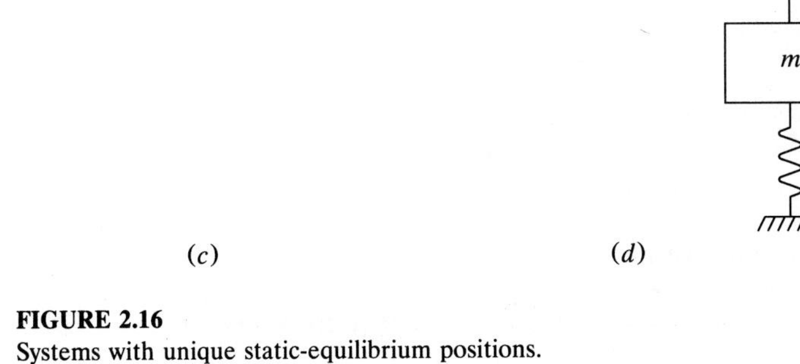

FIGURE 2.16
Systems with unique static-equilibrium positions.

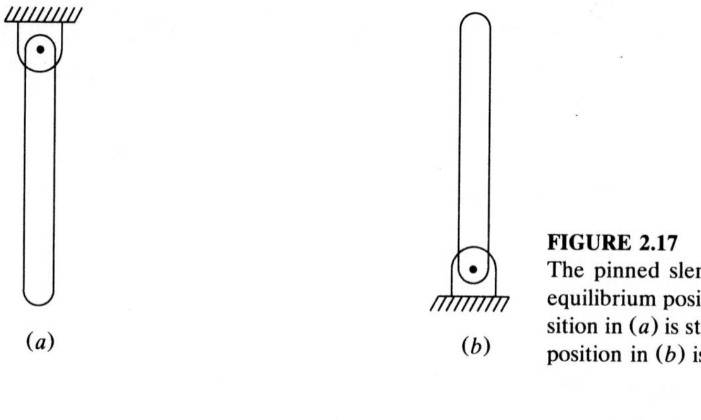

FIGURE 2.17
The pinned slender rod has two possible equilibrium positions. The equilibrium position in (*a*) is stable while the equilibrium position in (*b*) is unstable.

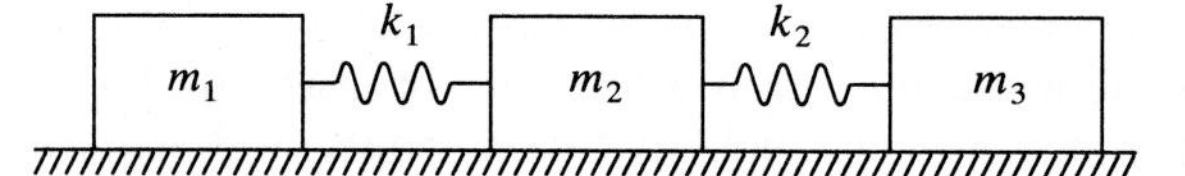

FIGURE 2.18
This unrestrained system has an infinite number of static-equilibrium positions.

The spring in the system of Fig. 2.16*a* retains its unstretched length when the system is in static equilibrium. When the system of Fig. 2.16*b* is in equilibrium, a force is present in the spring to balance the block's gravity force. The spring is thus stretched from its unstretched length. The spring's change in length is called its static deflection and is calculated by applying the equation of static equilibrium to a free-body diagram of the block, resulting in

$$\Delta_{st} = \frac{mg}{k} \tag{2.28}$$

Static deflections occur for springs in systems where the gravity force is not balanced by a constraint force (e.g., a normal force). If a spring were not present, the gravity force would do work and equilibrium could not be maintained. This potential for the gravity force to do work is captured as potential energy in the spring. The springs in the systems of Fig. 2.16*c* and *d* also have nonzero static deflections.

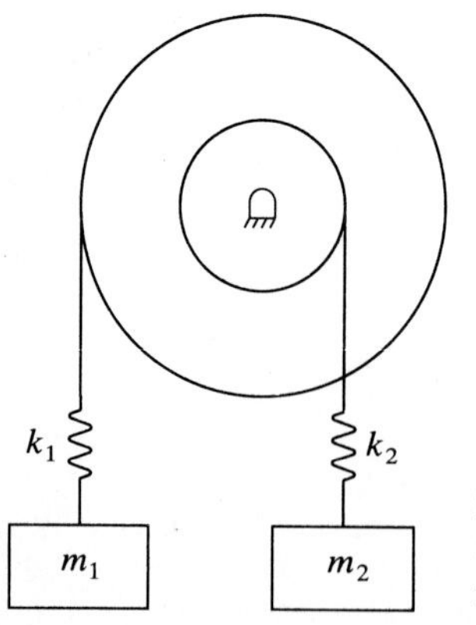

FIGURE 2.19
Static equilibrium is attained only when the moment of gravity forces about the center of the pulley sum to zero.

Static deflections are easy to measure. This measurement and Eq. (2.28) or a static analysis of the spring's system are often used to calculate a spring's stiffness.

Example 2.3. A 100-kg mass is hung on the end of a 50-cm variable-area bar. The length of the bar after the mass is hung is measured as 51 cm. What is the equivalent stiffness of the bar?

The mass hung from the bar can be modeled by the system of Fig. 2.16*b*. The static deflection due to the 100-kg mass is measured as 1 cm. Equation (2.28) is used to calculate the equivalent stiffness as

$$k = \frac{mg}{\Delta} = \frac{(100\text{ kg})\,(9.81\text{ m/s}^2)}{0.01\text{ m}} = 9.81 \times 10^4\ \frac{\text{N}}{\text{m}}$$

Example 2.4. A uniform rigid bar of mass 150 kg is pinned at B and connected to identical springs of stiffness 1×10^5 N/m at A and C, as shown in Fig. 2.20*a*. When the system is in equilibrium, the angle between the bar and the horizontal is 1°. What are the static deflections of the springs?

The free-body diagram of the static-equilibrium position is shown in Fig. 2.20*b*. Summing moments about B gives

$$\sum \overset{\curvearrowleft}{M_B} = 0 = mg\frac{L}{4}\cos 1^\circ + k\Delta_1\frac{L}{4}\cos 1^\circ - k\Delta_2\frac{3L}{4}\cos 1^\circ$$

Assuming the springs are identical, with ends of each attached to a fixed support at the same vertical position, the following geometric relation exists between the

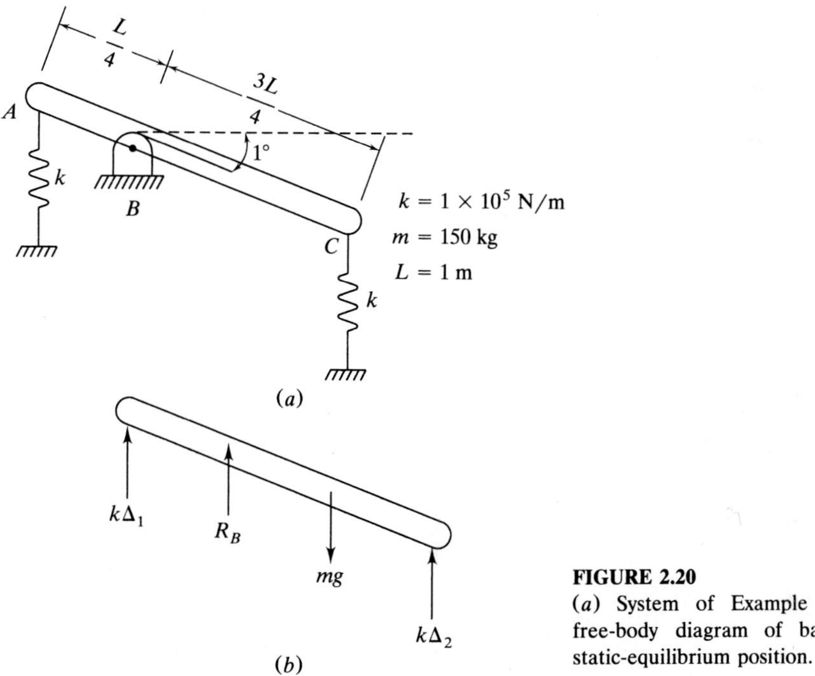

FIGURE 2.20
(*a*) System of Example 2.4; (*b*) free-body diagram of bar in its static-equilibrium position.

static deflections:

$$\Delta_2 = \Delta_1 + L \sin 1^\circ$$

Substitution of the geometric relationship into the moment equation leads to

$$\Delta_1 = \frac{mg}{2k} - \frac{3}{2} L \sin 1^\circ = -18.8 \text{ mm}$$

and $\Delta_2 = -1.35$ mm.

Springs used in multi-degree-of-freedom systems are also subject to static deflections, as evidenced by the following example.

Example 2.5. Determine the static deflection in each of the springs in the two-degree-of-freedom system of Fig. 2.21*a*.

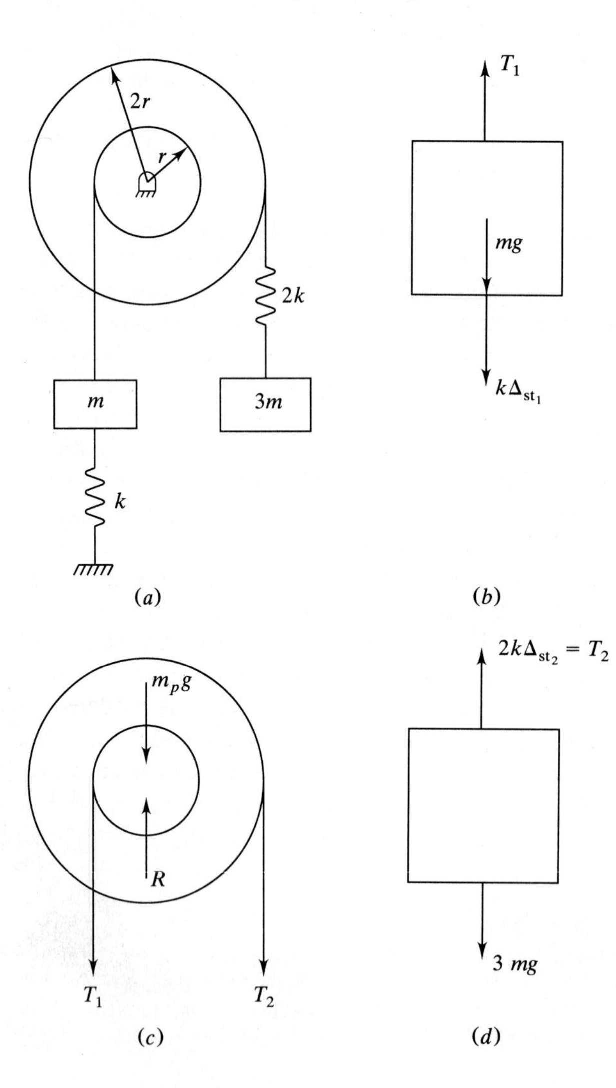

FIGURE 2.21
(*a*) Two-degree-of-freedom system of Example 2.5; (*b*)–(*d*) free-body diagrams of system components shown when system is in equilibrium.

Summing forces in the vertical direction using the free-body diagrams of Fig. 2.21*b* and *d* gives

$$T_1 = mg + k\Delta_{st_1}$$

and

$$T_2 = 2k\Delta_{st_2} = 3mg$$

Summing moments about the center of the pulley using the free-body diagram of Fig. 2.21*c* gives

$$T_1 r - T_2(2r) = 0$$

Substituting from the preceding equation and rearranging leads to

$$\Delta_{st_1} = \frac{5mg}{k}$$

and

$$\Delta_{st_2} = \frac{3mg}{2k}$$

2.6 INERTIA EFFECTS OF SPRINGS

When a force is applied to displace the block of Fig. 2.11 from its equilibrium position, the work done by the force is converted into strain energy stored in the spring. If the mass is held in this position and then released, the strain energy is converted to kinetic energy of both the block and the spring. If the mass of the spring is much smaller than the mass of the block, its kinetic energy is negligible. In this case the inertia of the spring has negligible effect on the motion of the block, and the system is modeled using one degree of freedom. The generalized coordinate is usually chosen as the displacement of the block.

If the mass of the spring is comparable to the mass of the block, the one-degree-of-freedom assumption is not valid. The particles along the axis of the spring are kinematically independent from each other and from the block. The spring should be modeled as a continuous system. The analysis of the vibrations of the block requires solution of a partial differential equation.

If the mass of the spring is much smaller than the mass of the block, but not negligible, a reasonable one-degree-of-freedom approximation can be made by approximating the spring's inertia effects. The actual system is modeled by the ideal system of Fig. 2.22. The spring in Fig. 2.22 is massless. The mass of the block in Fig. 2.22 is greater than the mass of the actual block to account for inertia effects of the spring. The value of m_{eq} is calculated such that the kinetic energy of the system of Fig. 2.22 is the same as the kinetic energy of the system

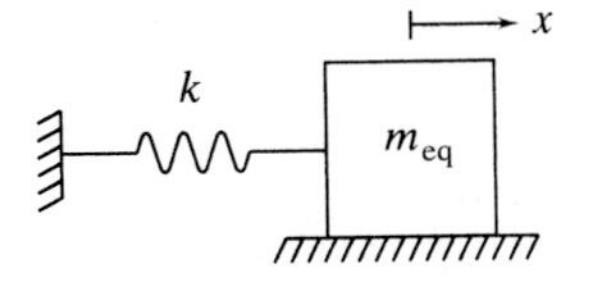

FIGURE 2.22
The mass of the block of Fig. 2.11 is increased to account for the inertia effects of the spring. The equivalent mass is calculated by setting the kinetic energy of this system equal to the kinetic energy of the system of Fig. 2.11, including the kinetic energy of the spring.

of Fig. 2.11, including the kinetic energy of the spring, when the velocities of both blocks are equal. Unfortunately, calculation of the exact kinetic energy of the spring requires a continuous system analysis. Thus an approximation to the spring's kinetic energy is used.

Let $x(t)$ be the generalized coordinate describing the motion of both the block of Fig. 2.11 and the block of Fig. 2.22. The kinetic energy of the system of Fig. 2.11 is

$$T = T_s + \tfrac{1}{2}m\dot{x}^2 \tag{2.29}$$

where T_s is the kinetic energy of the spring. The kinetic energy of the system of Fig. 2.22 is

$$T = \tfrac{1}{2}m_{\text{eq}}\dot{x}^2 \tag{2.30}$$

The spring in Fig. 2.11 is uniform, has an unstretched length l, and a total mass m_s. Define the coordinate z along the axis of the spring, measured from its fixed end, as defined in Fig. 2.23. The coordinate z measures the distance of a particle from the fixed end in the spring's unstretched state. The displacement of a particle on the spring, $u(z)$, is assumed explicitly independent of time and a linear function of z such that $u(0) = 0$ and $u(l) = x$,

$$u(z) = \frac{x}{l}z \tag{2.31}$$

Equation (2.31) represents the displacement function of a uniform spring when it is statically stretched. Consider a differential element of length dz, located a distance z from the spring's fixed end. The kinetic energy of the differential element is

$$dT_s = \frac{1}{2}\dot{u}^2(z)\,dm = \frac{1}{2}\dot{u}^2(z)\frac{m_s}{l}\,dz \tag{2.32}$$

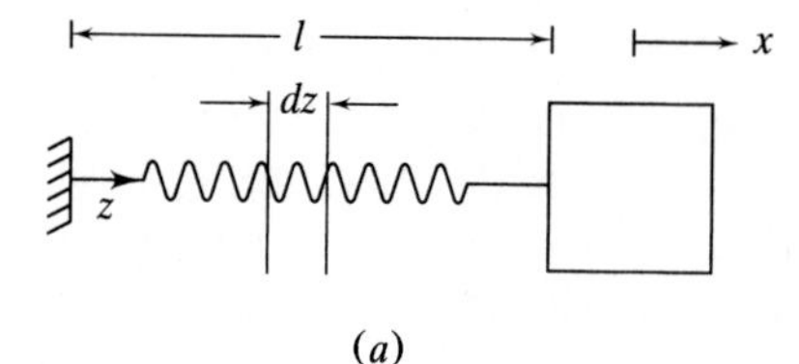

(a)

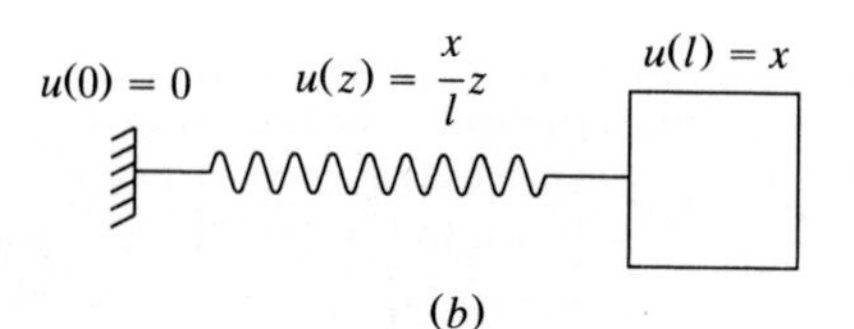

(b)

FIGURE 2.23
(*a*) The coordinate z is measured along axis of spring from its fixed end; (*b*) the displacement in the spring is assumed as a linear function of z.

The total kinetic energy of the spring is

$$T_s = \int dT_s = \int_0^l \frac{1}{2}\frac{m_s}{l}\left(\frac{\dot{x}z}{l}\right)^2 dz$$

$$= \frac{1}{2}\frac{m_s}{l^3}\dot{x}^2\frac{z^3}{3}\bigg|_0^l$$

$$= \frac{1}{2}\left(\frac{m_s}{3}\right)\dot{x}^2 \tag{2.33}$$

Equating T from Eqs. (2.29) and (2.30) and using T_s from Eq. (2.33) gives

$$m_{\text{eq}} = m + \frac{m_s}{3} \tag{2.34}$$

Equation (2.34) can be interpreted as follows: The inertia effects of a linear spring with one end fixed and the other end connected to a moving body can be approximated by placing a particle whose mass is one-third of the mass of the spring at the point where the spring is connected to the body.

The preceding statement is true for all springs where use of a linear displacement function of the form of Eq. (2.31) is justified. This is valid for helical coil springs, bars that are modeled as springs for longitudinal vibrations, and shafts acting as torsional springs.

Example 2.6. The springs in the system of Fig. 2.24*a* are all identical, with stiffness k and mass m_s. Calculate the kinetic energy of the system in terms of $\theta(t)$, including the inertia effects of the springs.

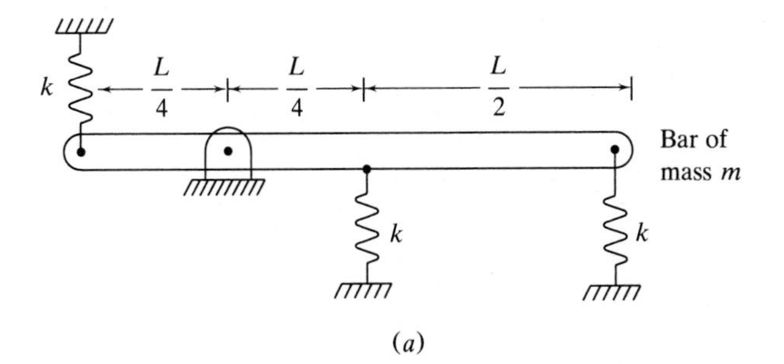

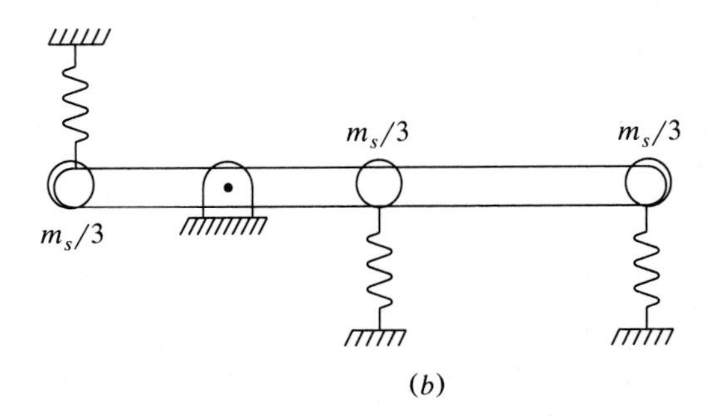

FIGURE 2.24
(*a*) System of Example 2.6; (*b*) inertia effects of springs are considered by placing particles of mass $m_s/3$ at locations on bar where springs are attached.

Each spring is replaced by a massless spring and a particle of mass $m_s/3$ at the point on the bar where the spring is attached, as shown in Fig. 2.24*b*. The total kinetic energy of the system of Fig. 2.24*b* is the kinetic energy of the bar plus the kinetic energy of each of the particles

$$T = \frac{1}{2}m\bar{v}^2 + \frac{1}{2}\bar{I}\dot{\theta}^2 + T_1 + T_2 + T_3$$

$$= \frac{1}{2}m\left(\frac{L}{4}\dot{\theta}\right)^2 + \frac{1}{2}\frac{1}{12}mL^2\dot{\theta}^2 + \frac{1}{2}\frac{m_s}{3}\left(\frac{L}{4}\dot{\theta}\right)^2 + \frac{1}{2}\frac{m_s}{3}\left(\frac{L}{4}\dot{\theta}\right)^2 + \frac{1}{2}\frac{m_s}{3}\left(\frac{3L}{4}\dot{\theta}\right)^2$$

$$= \frac{1}{2}\frac{7m + 11m_s}{48}L^2\dot{\theta}^2$$

Example 2.7. The simply supported beam of Fig. 2.25 is uniform and has a total mass of 100 kg. A machine of mass 350 kg is attached at B, as shown. What is the mass of a particle that should be placed at B to approximate the beam's inertia effects?

The displacement function for the beam does not satisfy a linear displacement relation of the form of Eq. (2.31). Since the exact expression for the dynamic beam deflection is not known, an approximate displacement function must be used in the calculation of the beam's kinetic energy. Let z be a coordinate along the beam's neutral axis. Assume that the time-dependent displacement of any particle can be expressed as

$$w(z,t) = w_B(t)y(z)$$

where $w_B(t)$ is the deflection of B. An appropriate approximation for $y(z)$ is the static deflection of the beam due to a concentrated load, P, applied at B, such that B has a unit deflection.

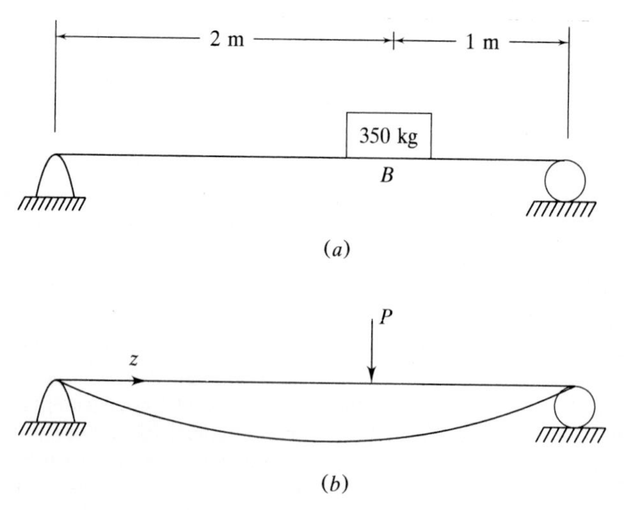

FIGURE 2.25
(*a*) System of Example 2.7; (*b*) static deflection of beam due to concentrated load at B.

Using the methods of App. D, the static deflection due to a concentrated load at B is

$$y(z) = \begin{cases} \dfrac{P}{18EI} z\left(\dfrac{8L^2}{9} - z^2\right) & 0 \le z \le \dfrac{2L}{3} \\ \dfrac{P}{18EI}\left(2z^3 - 6z^2L + \dfrac{44}{9}zL^2 - \dfrac{8}{9}L^3\right) & \dfrac{2L}{3} \le z \le L \end{cases}$$

The load required to cause a unit deflection at B is

$$P = \frac{243EI}{4L^3}$$

Consider a differential element of length dz, located a distance z from the left support. The kinetic energy of the element is

$$dT = \tfrac{1}{2}\dot{w}^2(z,t)\rho A\,dz$$

where ρ is the mass density of the beam and A is its cross-sectional area. The beam's total kinetic energy is calculated by integrating dT over the entire beam. Substituting the previous results for $w(x,t)$ in this integral leads to

$$T = \frac{1}{2}\rho A\left(\frac{27}{8L^3}\right)^2 \dot{w}_B^2\left[\int_0^{2L/3} z^2\left(\frac{8L^2}{9} - z^2\right)^2 dz\right.$$
$$\left. + \int_{2L/3}^{L}\left(2z^3 - 6z^2L + \frac{44}{9}zL^2 - \frac{8}{9}L^3\right)^2 dz\right]$$

which after considerable algebra gives

$$T = \frac{1}{2}0.598\rho AL\dot{w}_B^2$$

Noting that the total mass of the beam is ρAL, a particle of mass 59.8 kg should be added at B to approximate the inertia effects of the beam. The system of Fig. 2.25a is modeled as a one-degree-of-freedom system with a particle of 409.8 kg located at B.

Example 2.8. A block of mass m is attached to two identical springs in series. Each spring has a mass m_s and a stiffness k. Determine the mass of a particle that should be attached to the block to approximate the inertia effects of the springs.

Consider the system when the mass is displaced a distance x, from equilibrium. Since the springs are in series, the same force is developed in each spring. Also, since the springs are identical, each spring undergoes the same change in length, which must be $x/2$.

Since one end of the left spring is fixed, its inertia effects can be approximated in an equivalent system by placing a particle of mass $m_s/3$ at the right end of the spring, where it is connected to the right spring, as shown in Fig. 2.26. The kinetic

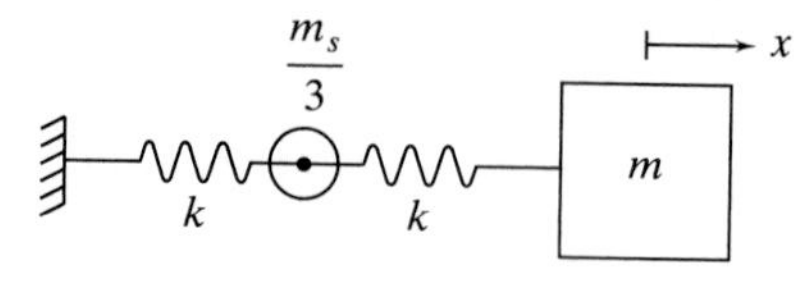

FIGURE 2.26
The inertia effects of left spring are approximated by the inertia of a particle of mass $m_s/3$ at its right end.

energy of this particle is

$$T_l = \frac{1}{2}\frac{m_s}{3}\left(\frac{\dot{x}}{2}\right)^2 = \frac{1}{2}\frac{m_s}{12}\dot{x}^2$$

The displacement in the right spring is assumed to be linear. Let z be a coordinate measured from the left end of the right spring, along the right spring in its unstretched configuration. Noting that the left end of the spring has a displacement $x/2$ and the right end of the spring has a displacement x,

$$u(z) = \frac{x}{2}\left(1 + \frac{z}{L}\right)$$

The kinetic energy of the right spring is calculated by integrating the kinetic energy of a differential spring element over the length of the spring

$$T_r = \int_0^L \frac{1}{2}\frac{m_s}{L}\left(\frac{\dot{x}}{2}\right)^2\left(1 + \frac{z}{L}\right)^2 dz$$

$$= \frac{1}{2}\frac{7}{12}m_s\dot{x}^2$$

The total kinetic energy of the two springs is

$$T = T_l + T_r = \tfrac{1}{2}\tfrac{2}{3}m_s\dot{x}^2$$

Hence a particle of mass $2m_s/3$ should be attached to the block to approximate the inertia effects of the springs.

2.7 VISCOUS DAMPERS

Viscous damping occurs in a mechanical system when a component of the system is in contact with a viscous fluid. Damping occurs as a result of the viscous friction between the component and the fluid. When a rigid body is in contact with a fluid, the damping force is usually proportional to the velocity of the body

$$F = cv \tag{2.35}$$

where c is called the damping coefficient. The damping coefficient has dimensions of mass per time.

Viscous damping may occur naturally, as when a buoyant body oscillates on the surface of a lake or in the oscillations of a column of liquid in a manometer, as shown in Fig. 2.27.

Viscous damping often has desirable effects in mechanical systems. Its presence causes the amplitude of free vibrations to decay. It can also cause the

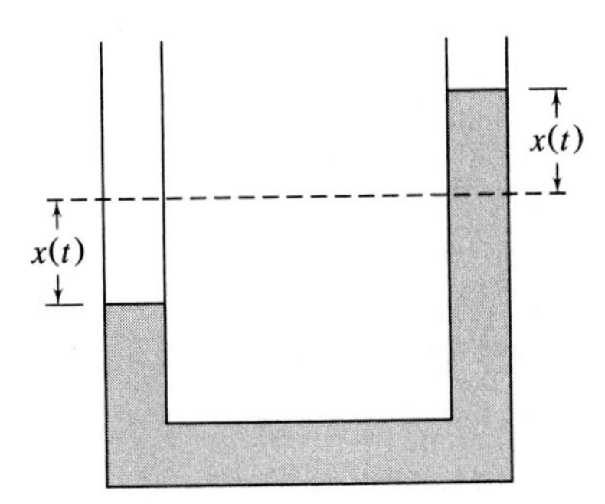

FIGURE 2.27
Viscous damping occurs when the column of liquid oscillates in a U-tube manometer.

reduction in amplitude of vibrations due to harmonic excitations. The presence of viscous damping in a system only gives rise to linear terms in the governing differential equations.

For the reasons given previously, viscous damping is often introduced in systems where it may not otherwise be present. A mechanical device called a dashpot is often added to a system to provide viscous damping. Several possible dashpot configurations are shown in Fig. 2.28.

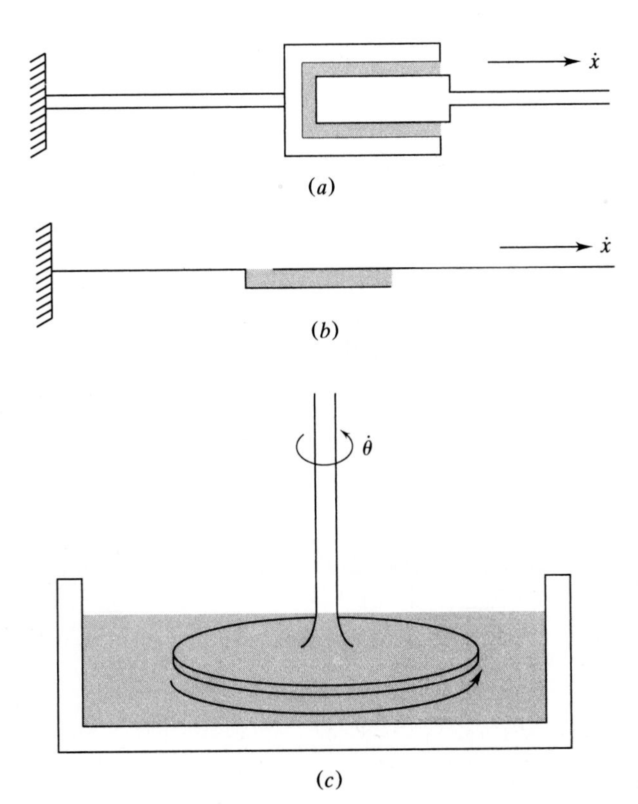

FIGURE 2.28
Devices that act as viscous dampers: (*a*) piston slides in cylinder with viscous liquid; (*b*) plate slides over reservoir of viscous liquid; (*c*) disk rotates in plate of viscous liquid producing a moment about axis of shaft and acts as a torsional viscous damper.

The end of the piston rod of Fig. 2.28a is rigidly connected to a point on a vibrating body. As the body moves with a velocity v, the piston slides in the cylinder and is lubricated by a viscous fluid. The viscous fluid exerts a friction force on the piston, which is turn is felt by the mass.

The upper plate of the dashpot of Fig. 2.28b is connected to the vibrating body. As the body moves the plate slides over a reservoir of a viscous fluid, which resists the motion of the plate. The resisting friction force is felt by the vibrating body.

Figure 2.28c shows a viscous torsional damper. The shaft is rigidly connected to a point on a body undergoing torsional oscillations. The shaft and disk oscillate with the same angular velocity as the point to which the shaft is attached. As the disk rotates in a plate of viscous liquid, a resisting moment is developed about the axis of the shaft, which in turn is felt by the vibrating body.

Although awkward to use the dashpot of Fig. 2.28b is the simplest to analyze and is used to illustrate the principle upon which a dashpot operates. It is shown in further detail in Fig. 2.29. The upper plate moves with a velocity v over a stationary plate filled with a viscous liquid of dynamic viscosity μ and mass density ρ. The depth of liquid in the plate is h. Let y be a coordinate measured positive upward from the lower plate, into the liquid. Let $u(y,t)$ be the one-dimensional velocity distribution developed in the liquid. There is no pressure gradient in the liquid. Basic principles of fluid mechanics are used to derive the governing equation for the velocity distribution as

$$\rho \frac{\partial u}{\partial t} = \mu \frac{\partial^2 u}{\partial^2 y} \tag{2.36}$$

The no-slip condition is used to develop the boundary conditions

$$u(0,t) = 0 \tag{2.37a}$$

and

$$u(h,t) = v \tag{2.37b}$$

The problem is truly unsteady as the velocity of the upper plate changes with time. Nondimensional variables are introduced as

$$u^* = \frac{u}{v_m} \qquad y^* = \frac{y}{h} \qquad t^* = \frac{v}{h}t \tag{2.38}$$

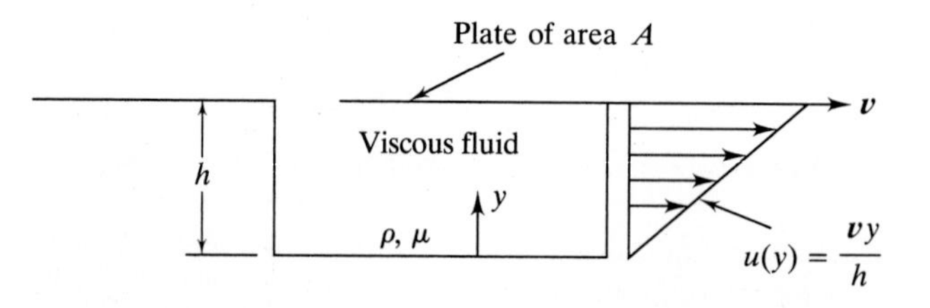

FIGURE 2.29
Geometry of plate-reservoir dashpot.

where v_m is the maximum velocity. Nondimensionalization of Eq. (2.36) using Eqs. (2.38) leads to

$$\mathrm{Re}\,\frac{\partial u^*}{\partial t^*} = \frac{\partial^2 u^*}{\partial^2 y^*}$$

where the Reynolds number Re is defined as

$$\mathrm{Re} = \frac{\rho v_m h}{\mu} \tag{2.39}$$

It is assumed that h is small enough such that the unsteady fluid acceleration term can be ignored. Then Eq. (2.36) becomes

$$\frac{d^2 u}{d^2 y} = 0 \tag{2.40}$$

The solution of Eq. (2.40) subject to Eq. (2.37) is

$$u(y) = \frac{v}{h} y \tag{2.41}$$

The uniform shear stress exerted on the plate from the fluid is determined using Newton's viscosity law

$$\tau = \mu \left(\frac{du}{dy} \right)_{y=h} = \mu \frac{v}{h} \tag{2.42}$$

If the area of the plate in contact with the fluid is A, then the friction force exerted on the plate from the fluid is

$$F = \frac{\mu A}{h} v \tag{2.43}$$

Comparison of Eq. (2.35) with (2.43) shows that the damping coefficient for this dashpot is

$$c = \frac{\mu A}{h} \tag{2.44}$$

Equation (2.44) shows that large damping is achieved with a very viscous fluid, a small h, and a large A. This is often impractical, and thus the device of Fig. 2.28*b* is rarely used as a dashpot.

Calculations show that the unsteady acceleration term of Eq. (2.36) may not be negligible for this problem. However, the no-slip boundary conditions, Eq. (2.37), still apply. Thus, even if the instantaneous velocity profile is nonlinear, if h is small enough, Eq. (2.42) still provides a reasonable approximation to the shear stress.

When a free-body diagram of a system connected to a dashpot is drawn, the force provided by the dashpot should be shown as the damping coefficient times the velocity of the point to which the dashpot is attached. The force should be drawn opposite the direction of a positive velocity.

The force provided by a viscous damper is nonconservative. Since the damping force always opposes the direction of motion, it does negative work. Thus, as motion continues, energy is used to do work against the viscous damper. Since the damping force is proportional to the velocity, the amount of energy dissipated over one cycle of motion is dependent on the velocity and energy is continually dissipated. Motion does not cease due to viscous damping.

2.8 EQUIVALENT SYSTEMS

All one-degree-of-freedom conservative linear systems can be modeled using one of the equivalent systems of Fig. 2.30. The elastic elements are massless, with their inertia effects being included in the equivalent mass, if necessary. The system of Fig. 2.30*a* is the appropriate model if the generalized coordinate is a linear displacement and the elastic elements are unstretched when the system is in equilibrium. The system of Fig. 2.30*b* is the appropriate model if the generalized coordinate is a linear displacement and an elastic element is stretched when the system is in equilibrium. Fig. 2.30*c* is the appropriate model if the generalized coordinate is an angular displacement. The potential energy of the system of Fig. 2.30*a* is

$$V = \frac{1}{2}k_{\text{eq}}x^2 \tag{2.45}$$

where x is the displacement of the mass from equilibrium. The kinetic energy of

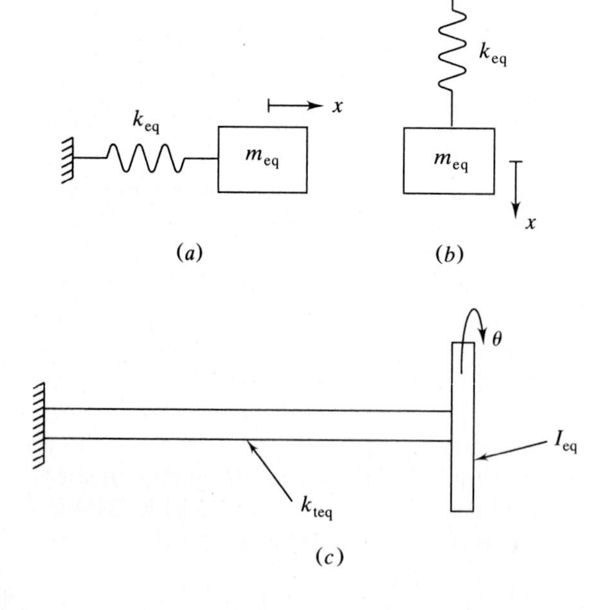

FIGURE 2.30
All linear one-degree-of freedom conservative systems can be modeled using one of the systems shown.

the system is

$$T = \tfrac{1}{2} m_{\text{eq}} \dot{x}^2 \tag{2.46}$$

Consider a one-degree-of-freedom conservative system. Assume that the only source of potential energy is due to energy stored in linear springs. Let x be a generalized coordinate that represents the displacement of a particle in the system. Suppose n springs are attached to the one-degree-of-freedom system. The system's total potential energy is

$$V = \tfrac{1}{2} \sum_{i=1}^{n} k_i x_i^2 \tag{2.47}$$

where x_i is the displacement of the particle to which the spring of stiffness k_i is attached. Since the system only has one degree of freedom, the displacement of any particle can be determined from x, using only geometry. Thus for a linear system

$$x_i = \alpha_i x \tag{2.48}$$

Substituting Eq. (2.48) in Eq. (2.47) yields

$$V = \tfrac{1}{2} \left(\sum_{i=1}^{n} k_i \alpha_i^2 \right) x^2 \tag{2.49}$$

The potential energies in Eqs. (2.45) and (2.49) are equal if

$$k_{\text{eq}} = \sum_{i=1}^{n} k_i \alpha_i^2 \tag{2.50}$$

A similar argument is used to show that the kinetic energy of any one-degree-of-freedom linear system can be written in the form of Eq. (2.46) with an appropriate choice of m_{eq}.

The potential energy for any linear one-degree-of-freedom system has the form of Eq. (2.45). If the potential energy is due to a source other than an elastic element, the equivalent stiffness for use in a model of Fig. 2.30 is obtained by writing the potential energy of the system in the form of Eq. (2.45).

When gravity provides a source of potential energy, the system is truly nonlinear. If the motion is small, a linearization assumption introduced in Chapter 3 is used. Development of an equation of the form of Eq. (2.45) requires use of and truncation of a Taylor series expansion for trigonometric terms.

If any spring in the system has a nonzero static deflection when the system is in equilibrium, then the more appropriate model is the system of Fig. 2.30*b*. If x is the displacement of the mass from the system's equilibrium position, the potential energy of the system of Fig. 2.30*b* is

$$V = \tfrac{1}{2}k_{eq}(x + \Delta_{st})^2 - m_{eq}gx$$

$$= \tfrac{1}{2}k_{eq}\left(x + \frac{m_{eq}g}{k_{eq}}\right)^2 - m_{eq}gx$$

$$= \tfrac{1}{2}k_{eq}x^2 + \tfrac{1}{2}k_{eq}\Delta_{st}^2 \tag{2.51}$$

Note that the potential energy due to gravity cancels with the change in potential energy stored in the spring between the static equilibrium position of the system and its arbitrary position described by x. The second term on the right-hand side of Eq. (2.51) represents the potential energy in the spring when the system is in equilibrium. Thus static deflection has no effect on the equivalent stiffness of a linear system. The change in potential energy due to the static deflection can always be assumed to cancel with the potential energy due to the gravitational force, which causes the static deflection.

Example 2.9. Consider the one-degree-of-freedom system of Fig. 2.31. The springs each have a mass m_s. The disk rolls without slip, no slip occurs at the pulley, and the pulley is frictionless. Let x be the displacement of the mass center of the disk. Determine an equivalent stiffness and equivalent mass such that the system of Fig. 2.30*b* can be used to model the oscillations of the center of the disk about its equilibrium position.

Let θ be the angular rotation of the disk from equilibrium. Assuming no slip, the displacement of the mass center of the disk is

$$x = r\theta$$

and the displacement of the block is

$$x_B = 2r\theta$$

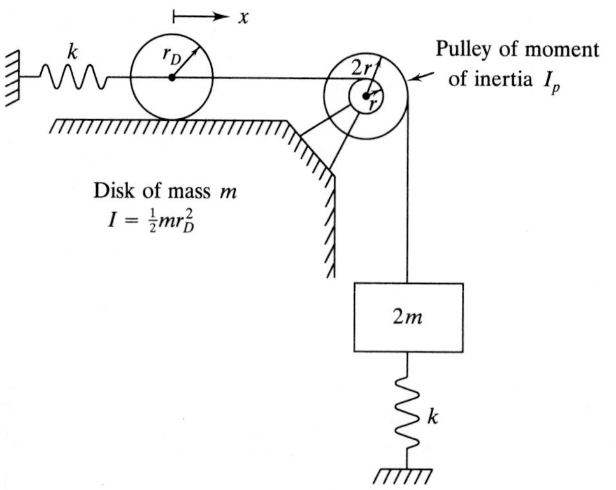

FIGURE 2.31
System of Example 2.9. The disk rolls without slip and the pulley is frictionless. The system can be modeled by the system of Fig. 2.30*b* with appropriate choices of k_{eq} and m_{eq}.

Eliminating θ between these equations shows that $x_B = 2x$. Noting that the effect of static deflection cancels the effect of gravity, the change in potential energy from the equilibrium position is

$$V = \tfrac{1}{2}kx^2 + \tfrac{1}{2}k(2x)^2 = \tfrac{1}{2}5kx^2$$

which leads to an equivalent stiffness of

$$k_{\text{eq}} = 5k$$

Since the disk rolls without slip, its angular velocity is given by

$$\omega_D = \frac{\dot{x}}{r_D}$$

The inertia effects of the springs are included by placing particles of mass $m_s/3$ at the center of the disk and the center of the block. The total kinetic energy is the sum of the kinetic energies of the disk, the pulley, the block, and the particles representing the inertia effects of the springs

$$T = \frac{1}{2}m\dot{x}^2 + \frac{1}{2}\left(\frac{1}{2}mr_D^2\right)\left(\frac{\dot{x}}{r_D}\right) + \frac{1}{2}2m(2\dot{x})^2$$

$$+ \frac{1}{2}I_p\left(\frac{\dot{x}}{r}\right)^2 + \frac{1}{2}\frac{m_s}{3}\dot{x}^2 + \frac{1}{2}\frac{m_s}{3}(2\dot{x})^2$$

$$= \frac{1}{2}\left(\frac{19}{2}m + \frac{I_p}{r^2} + \frac{5}{3}m_s\right)\dot{x}^2$$

Thus the system's equivalent mass is

$$m_{\text{eq}} = \frac{19}{2}m + \frac{I_p}{r^2} + \frac{5}{3}m_s$$

Example 2.10. The slender rod of Fig. 2.32 will be subject only to small displacements from the vertical. Under this condition determine an equivalent stiffness and equivalent mass such that the motion of this system can be modeled by the system of Fig. 2.30*c*.

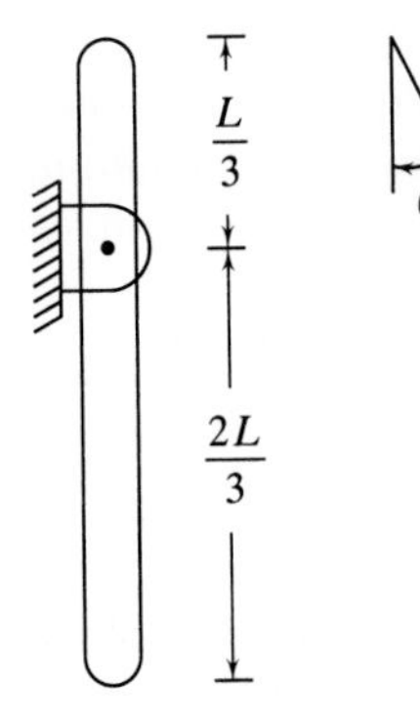

FIGURE 2.32
For small θ, the nonlinear system can be approximated by the linear system of Fig. 2.30*c*.

Gravity is the only source of potential energy as the bar rotates about its pin support. Choosing the datum as the pin support, the potential energy is

$$V = -\frac{mgL}{6}\cos\theta$$

If the rod is subject to only small displacements,

$$\cos\theta = 1 - \frac{\theta^2}{2} + \frac{\theta^4}{24} - \cdots \approx 1 - \frac{\theta^2}{2}$$

Then the potential energy is approximated as

$$V = -\frac{mgL}{6} + \frac{1}{2}\frac{mgL}{6}\theta^2$$

The term independent of θ is the potential energy when the system is in equilibrium, analogous to the last term of Eq. (2.51), and is of no consequence. Comparing the potential energy to Eq. (2.51), but for a system where an angular coordinate is the generalized coordinate

$$k_{\text{teq}} = \frac{mgL}{6}$$

The kinetic energy of the system is

$$T = \frac{1}{2}m\left(\frac{L}{6}\dot{\theta}\right)^2 + \frac{1}{2}\frac{1}{12}mL^2\dot{\theta}^2 = \frac{1}{2}\frac{mL^2}{9}\dot{\theta}^2$$

Comparing the kinetic energy to Eq. (2.46), but for a torsional system

$$I_{\text{eq}} = \tfrac{1}{9}mL^2$$

The equivalent model of a one-degree-of-freedom system with viscous damping is shown in Fig. 2.33. The work done by the viscous damping force as the block moves between x_1 and x_2 is

$$U_{1\to 2} = -\int_{x_1}^{x_2} c_{\text{eq}}\dot{x}\,dx \tag{2.52}$$

If two systems have equivalent kinetic and potential energies for the same displacement and the work done by the nonconservative forces is the same for both systems, the two systems have the same behavior. An equivalent viscous damping coefficient for a system with viscous damping is obtained by calculating the work done by the viscous damping forces and writing it in the form of Eq. (2.52).

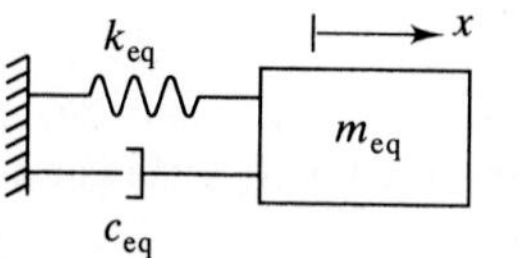

FIGURE 2.33
The equivalent mass-spring-dashpot system is used as a model for a one-degree-of-freedom linear system with viscous damping.

Example 2.11. The system of Example 2.9 has been modified by adding two viscous dampers as shown in Fig. 2.34. Again, using x as the displacement of the mass center of the disk, derive the equivalent damping coefficient such that the system of Fig. 2.34 is modeled by the system of Fig. 2.33.

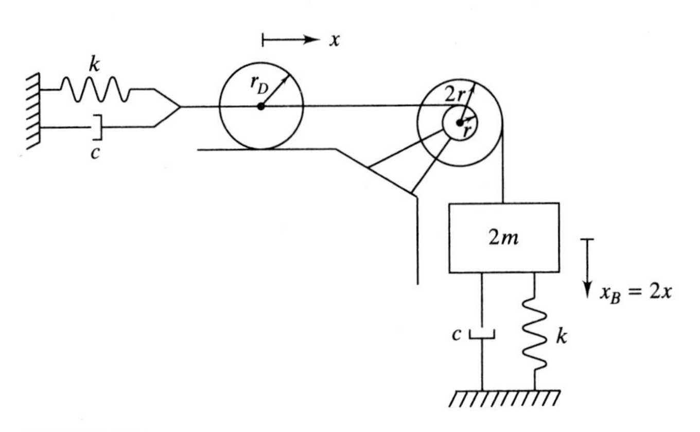

FIGURE 2.34
System of Example 2.11.

The work done by the dashpots as the center of the disk moves between x_1 and x_2 is

$$U_{1\to 2} = -c\int_{x_1}^{x_2} \dot{x}\,dx - c\int_{x_{B_1}}^{x_{B_2}} \dot{x}_B\,dx_B$$

As shown in Example 2.9, $x_B = 2x$. Thus

$$U_{1\to 2} = -c\int_{x_1}^{x_2} \dot{x}\,dx - c\int_{x_1}^{x_2} (2\dot{x})2\,dx$$

$$= -5c\int_{x_1}^{x_2} \dot{x}\,dx$$

Comparing the preceding equation to Eq. (2.52),

$$c_{\text{eq}} = 5c$$

2.9 FLOATING AND IMMERSED BODIES

Vibrations of a body immersed in a liquid or floating on the interface of a liquid and a gas can be modeled using the methods of this chapter with special considerations.

2.9.1 Buoyancy

When a solid body is submerged in a liquid or floating on the interface of a liquid and air, a force acts vertically upward on the body due to the variation of hydrostatic pressure. This force is called the buoyant force. Archimedes' principle states that the buoyant force acting on a floating or submerged body is equal to the weight of the liquid displaced by the body.

Example 2.12. A sphere of mass 2.5 kg and radius 10 cm is hanging from a spring of stiffness 1000 $\mathrm{N/m^2}$ in a fluid of mass density 1200 $\mathrm{kg/m^3}$. What is the static deflection of the spring?

The spring force must balance with the gravity force and the buoyancy force as shown on the free-body diagram in Fig. 2.35

$$k\Delta_{st} + F_B - mg = 0$$

Archimedes' principle is used to calculate the buoyant force as

$$F_B = \tfrac{4}{3}\rho g \pi r^3 = \tfrac{4}{3}(1200\ \mathrm{kg/m^3})\pi(9.81\ \mathrm{m/s^2})(0.1\ \mathrm{m})^3 = 49.31\ \mathrm{N}$$

The static deflection is calculated as -24.8 mm.

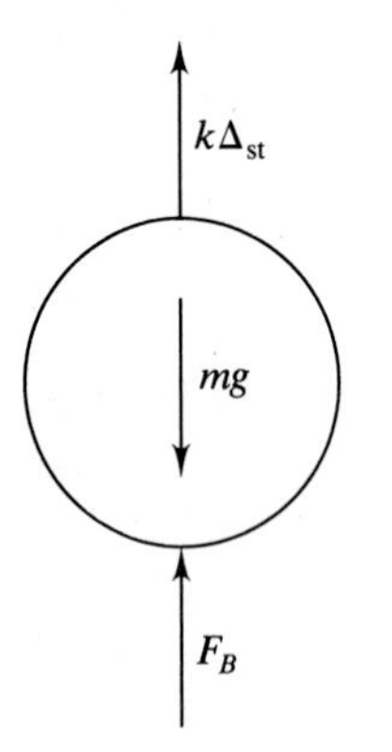

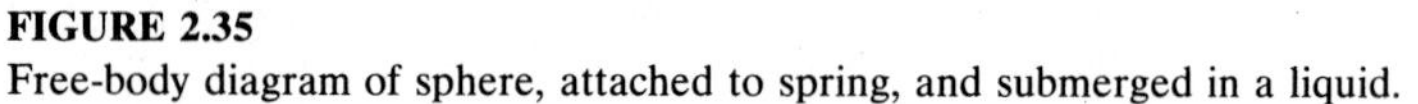

FIGURE 2.35
Free-body diagram of sphere, attached to spring, and submerged in a liquid.

Consider a body floating stably on a liquid-air interface. The buoyant force balances with the gravity force. If the body is pushed further into the liquid, the buoyant force increases. If the body is then released, it seeks to return to its equilibrium configuration. The buoyant force does work, which is converted into kinetic energy and oscillations about the equilibrium position ensue.

The circular cylinder of Fig. 2.36 has a cross-sectional area A and floats stably on the surface of a fluid of density ρ. When the cylinder is in equilibrium, it is subject to a buoyant force mg and its center of gravity is a distance Δ from the surface. Let x be the vertical displacement of the center of gravity of the cylinder from this position. The additional volume displaced by the cylinder is xA. According to Archimedes' principle, the buoyant force is

$$F_B = mg + \rho g A x$$

Calculations show that the work done by the buoyant force as the cylinder's center of gravity moves between positions x_1 and x_2 is

$$U_{1\to 2} = \tfrac{1}{2}\rho g A x_2^2 - \tfrac{1}{2}\rho g A x_1^2$$

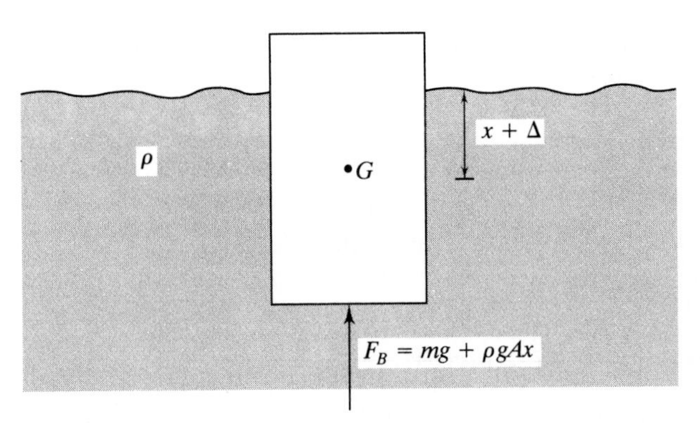

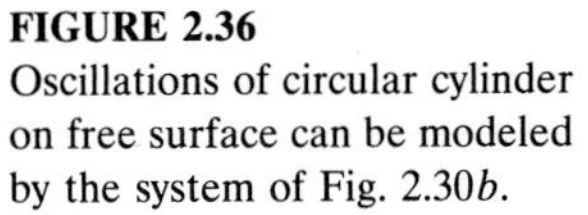
FIGURE 2.36
Oscillations of circular cylinder on free surface can be modeled by the system of Fig. 2.30*b*.

and is independent of path. Hence the buoyant force is conservative. Its effect on the cylinder is the same as that of a linear spring of stiffness $\rho g A$. The oscillations of the cylinder on the liquid-gas interface can be modeled by an equivalent mass-spring system.

2.9.2 Added Mass

Consider a mass-spring system immersed in an inviscid fluid, as shown in Fig. 2.37. The spring is stretched from its equilibrium configuration and the mass released. The ensuing motion of the mass causes motion in the surrounding fluid. The strain energy initially stored in the spring is converted to kinetic energy for both the mass and the fluid. Since the fluid is inviscid, energy is conserved

$$T_m + T_f + V = C \tag{2.53}$$

The inertia effects of the fluid can be included in an analysis by using a method similar to that used in Sec. 2.6 to account for the inertia effects of springs. An imagined particle is attached to the mass such that the kinetic energy of the particle is equal to the total kinetic energy of the fluid. If x is the displacement

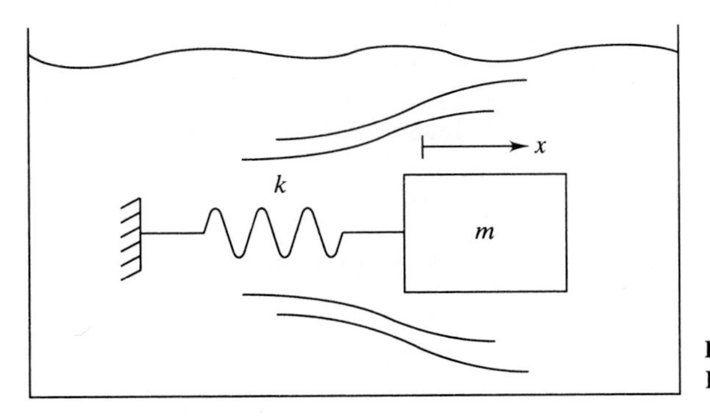

FIGURE 2.37
Immersed mass-spring system.

of the mass, the total kinetic energy of the system is given by Eq. (2.46) where

$$m_{\text{eq}} = m + m_a \tag{2.54}$$

The mass of the particle is called the added mass.

The kinetic energy of the fluid is difficult to quantify. The motion of the body theoretically entrains fluid infinitely far away in all directions. The total kinetic energy of the fluid is calculated from

$$T_f = \frac{1}{2}\iiint \rho v^2\, dV \tag{2.55}$$

where v is the velocity of the fluid set in motion due to the motion of the body. The integration is carried out from the body surface to infinity in all directions. If the integration of Eq. (2.55) is carried out, the added mass is calculated from

$$m_a = \frac{T_f}{\frac{1}{2}\dot{x}^2} \tag{2.56}$$

Potential flow theory can be used to develop the velocity distribution in a fluid for a body moving through the fluid at a constant velocity. This velocity distribution is used in Eqs. (2.55) and (2.56) to calculate the added mass. Table 2.1 is adapted from Wendel (1956) and Patton (1965) and presents the added mass for common body shapes.

Rotational motion of a body in a fluid also imparts motion to the fluid resulting in rotational kinetic energy of the fluid. The inertia effects of the fluid

TABLE 2.1
Added Mass for Common Two- and Three-Dimensional Bodies. Note that ρ is the Mass Density of the Fluid

Body	Added Mass
Sphere of diameter D	$\frac{1}{12}\pi\rho D^3$
Thin Circular disk of diameter D	$\frac{1}{3}\rho D^3$
Thin square plate of side h	$0.1195\pi\rho h^3$
Circular cylinder of length L, diameter D	$\frac{1}{4}\pi\rho D^2 L$
Thin flat plate of length L, width w	$\frac{1}{4}\pi\rho w^2 L$
Square cylinder of side h, length L	$0.3775\rho\pi h^2 L$
Cube of side h	$2.33\rho h^3$

TABLE 2.2
Added moments of inertia for common bodies (Note that ρ is the mass density of the fluid)

Body	Added moment of inertia
Sphere	0
Circular cylinder	0
Any body rotating about axis of symmetry	0
Thin plate of length L, rotating about axis in the plane of the surface area of plate, perpendicular to direction for which L is defined	$0.0078125\pi\rho L^4$
Disk of diameter D rotating about a diameter	$\frac{1}{90}\rho D^5$

are taken into account by adding a disk of an appropriate moment of inertia to the rotating body. If ω is the angular velocity of the body, the added mass moment of inertia is calculated from

$$I_a = \frac{T_f}{\frac{1}{2}\omega^2} \tag{2.57}$$

Note that the added mass moment of inertia is zero if the body is rotating about an axis of symmetry. Both the added mass and added moment of inertia terms are negligible for bodies moving in gases. Table 2.2 presents added moments of inertia for a few common bodies. It is adapted from Wendel (1956).

2.9.3 Other Fluid Effects

There are other fluid effects acting on bodies which greatly influence the vibration characteristics of submerged bodies. Some of these are briefly considered in the following discussion. Many of the results considered are empirical.

If a body is placed in an accelerating flow, a fluid inertia force, separate from the added mass effect, acts on the body. The motion of the fluid causes a pressure variation over the surface of the body. This in turn results in a buoyant force acting in the direction of the fluid acceleration and given by

$$F_B = m_D a_f \tag{2.58}$$

where m_D is the mass of the fluid displaced by the body and a_f is the fluid acceleration.

Bluff bodies in steady flows are subject to unsteady forces due to vortex shedding. The flow over the body separates, on both the upper and lower surfaces, producing a trailing wake. Vortices are shed alternately from these surfaces, producing an unsteady pressure variation in the wake. This leads to a drag force, which over a wide range of flow parameters, can be approximated as harmonic in time.

A body placed in a real fluid is subject to a force, opposing the motion, called the drag force. The total drag is due to the pressure variation over the body (form drag) and friction (friction drag). The mathematical form of the drag force depends upon the flow parameters.

For slow-creeping flows (low Reynolds number), the drag force is proportional to the flow velocity and is given by

$$D = C_D^* \mu v l \tag{2.59}$$

where C_D^* is a calculated coefficient, v is the fluid (or body) velocity, and l is a reference length.

For a sphere of diameter d in a slow-creeping flow the drag force is given by Stokes' law

$$D = 3\pi \mu v d \tag{2.60}$$

and hence $C_D^* = 3\pi$. For a thin disk of diameter d placed normal to the flow

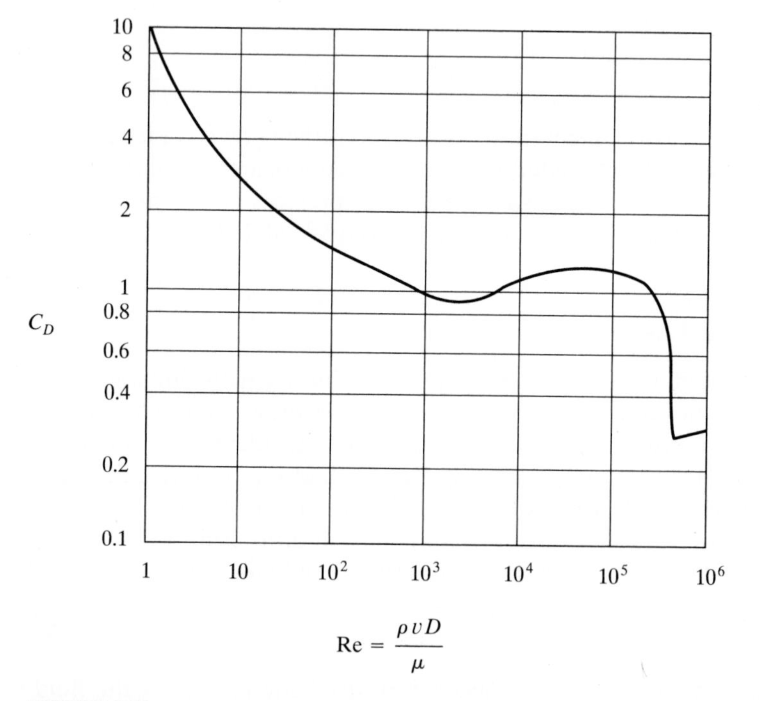

FIGURE 2.38
Drag coefficient of cylinder as a function of Reynolds number. Drag force is $\frac{1}{2}\rho C_D v^2 A$, where A is the projected area in the plane normal to velocity. (*From Shames.*)

$C_D^* = 8$. For a circular cylinder of length L and diameter d

$$C_D^* = \frac{4\pi}{\ln\left(\frac{2L}{D}\right) + \frac{1}{2}} \tag{2.61}$$

but the reference length for use in Eq. (2.59) is the length L. Note that the drag coefficient decreases as the cylinder becomes more slender. A comprehensive table of drag coefficients at low Reynolds numbers is found in Gerhart and Gross (1985).

For high-speed flows (high Reynolds numbers), the drag force is proportional to the square of the velocity and is given by

$$D = \tfrac{1}{2} C_D \rho v^2 A \tag{2.62}$$

where C_D is an empirical drag coefficient, and A is the area of the body projected onto a plane normal to the direction of the flow. The drag coefficient could be a function of Reynolds number, as for the circular cylinder (see Fig. 2.38). Drag coefficients for other bodies are given in Table 2.3. Inclusion of the

TABLE 2.3
Drag Coefficients at High Reynolds Numbers (Re > 10^4) Adapted from White, *Fluid Mechanics*

Body	C_D
Thin plate normal to flow	2.0
Square cylinder normal to flow	2.1
Cube, side normal to flow	1.07
Thin circular disk, normal to flow	1.17
Circular cylinder, flat edge normal to flow (L = length, d = diameter)	
$L/d = 0.5$	1.15
$L/d = 4$	0.87
Circular cylinder, axis normal to flow	See Fig. 2.38
Sphere	Varies with Re—similar to circular cylinder, see White or Gerhart and Gross for details

Source: Adapted from White, *Fluid Mechanics*.

drag force at high Reynolds numbers in a vibrations analysis leads to nonlinear differential equations.

PROBLEMS

All springs are assumed to be linear helical coil springs with negligible mass, unless otherwise indicated.

2.1. A spring with a cubic nonlinearity has a force-displacement relation

$$F = k_1 x + k_3 x^3$$

(*a*) Calculate the work done by the spring force as the spring is stretched from its equilibrium length by a length δ.

(*b*) Is the spring force in a spring with a cubic nonlinearity conservative?

2.2. A 5-kg block is attached to the end of a spring with a cubic nonlinearity (Prob. 2.1) with $k_1 = 3 \times 10^5$ N/m and $k_3 = 1 \times 10^8$ N/m^3. The other end of the spring is fixed. The block, which slides on a frictionless surface, is displaced 22 mm and released from rest. Calculate the maximum velocity attained by the block during its subsequent motion.

2.3. If the coefficient of friction between the block and the surface of the system of Prob. 2.2 is 0.15, calculate the maximum compression of the spring if it is initially stretched 22 mm.

2.4. The block of Prob. 2.2 is hit by a 20-g particle traveling at 100 m/s. After impact, the particle becomes embedded in the block. Determine the maximum compression of the spring assuming the block slides on a frictionless surface.

2.5. A 10-kg mass is hung from a spring whose force-displacement relationship is

$$F = 2 \times 10^4 x - 4 \times 10^7 x^3 \text{ N}$$

where x is in meters. What is the static deflection of the spring?

2.6. The springs in Fig. P2.6 are identical and have the force-displacement relation of the spring of Prob. 2.5. Determine the static deflection of each spring.

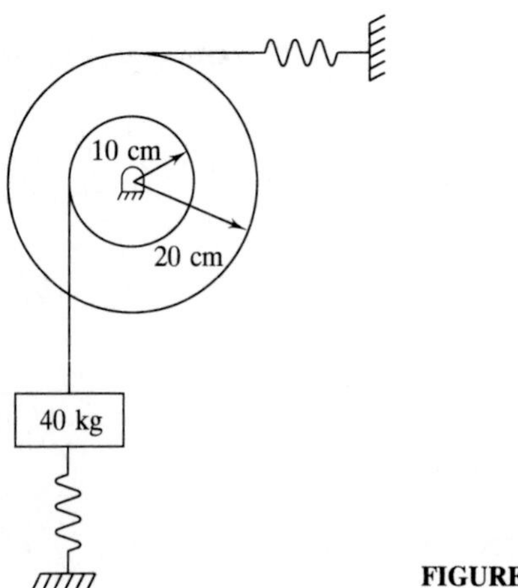

FIGURE P2.6

2.7. A coil spring is made of a steel with a shear modulus of 80×10^9 N/m^2 and an elastic shear strength of 200×10^6 N/m^2. The spring is to have a coil radius of 20 cm. A 10-kg block is to be suspended from the spring. The block is subject to a displacement not to exceed 20 mm. Design a spring (specify the diameter of each coil and the number of active turns) such that the spring's elastic strength is not exceeded.

2.8. The ends of a 20-kg bar are connected to collars which slide along tracks as shown in Fig. P2.8. The coefficient of friction between the collars and the tracks is 0.15. The collar sliding on the vertical track is to be attached to a spring such that the system is in equilibrium when $\theta = 30°$ and the spring is compressed 10 cm when $\theta = 0$. Design a steel spring ($G = 80 \times 10^9$ N/m^2) to meet these specifications. The coil radius of the spring should be 5 cm. Specify the radius of the bar from which the spring is made and the number of active turns. If the maximum displacement from equilibrium of the collar at B is 10 cm, what is the maximum shear stress developed in the spring?

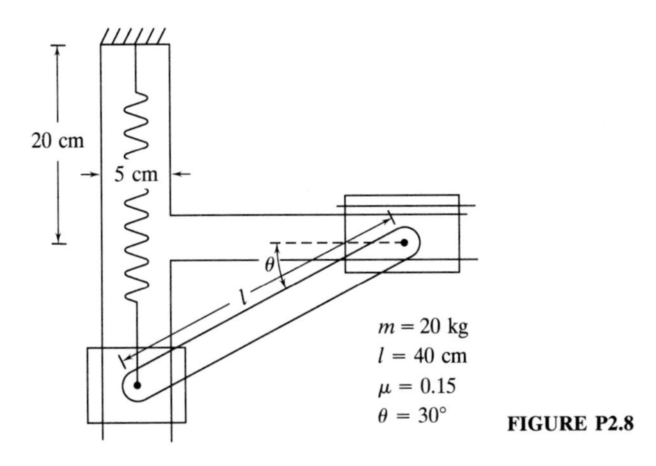

FIGURE P2.8

2.9. Vibration isolation of a milling machine is achieved by mounting the machine on identical elastic elements at four locations. Using the methods of Chap. 11, it has been determined that the maximum allowable equivalent stiffness of the isolation system is 3.6×10^5 N/m. However, the only elastic elements immediately available each have a stiffness of 4×10^5 N/m. These elements can be connected in series at each mounting location. What is the minimum number of elastic elements that can be used?

2.10. To achieve effective isolation, a 100-kg loudspeaker system is to be mounted on elastic elements at three locations. Using the vibration isolation theory of Chap. 11, it is determined that the minimum static deflection of the elastic elements is 2.5 mm. What is the maximum allowable stiffness of each elastic element?

2.11–2.16. Determine the deflection of each spring from its free length when the system shown is in equilibrium.

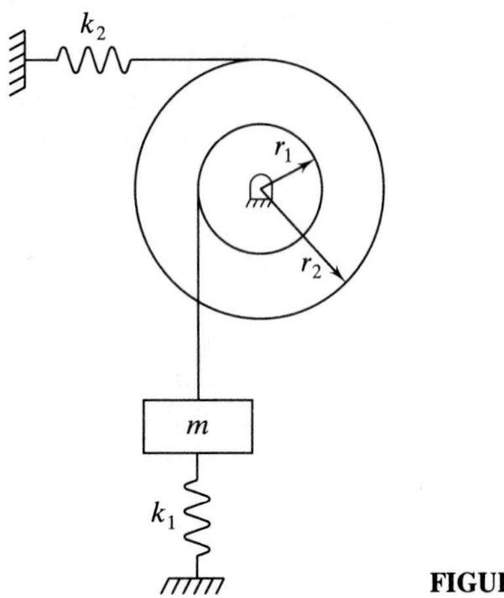

FIGURE P2.11

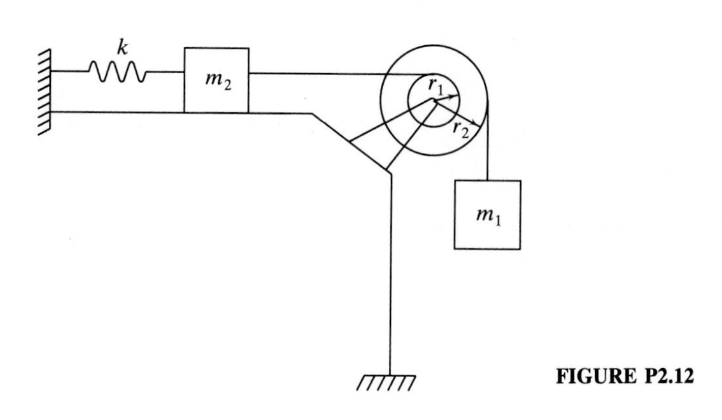

FIGURE P2.12

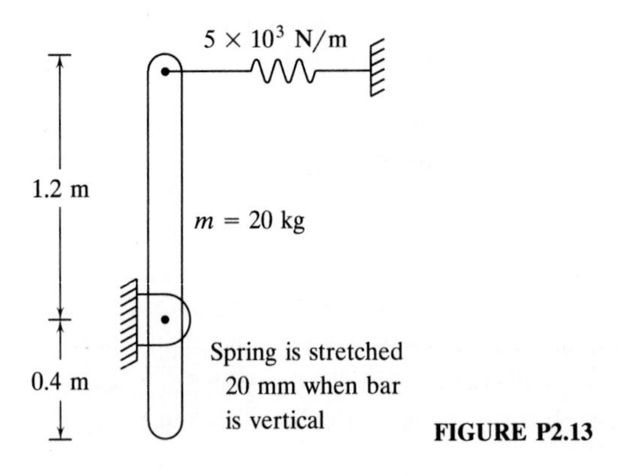

FIGURE P2.13

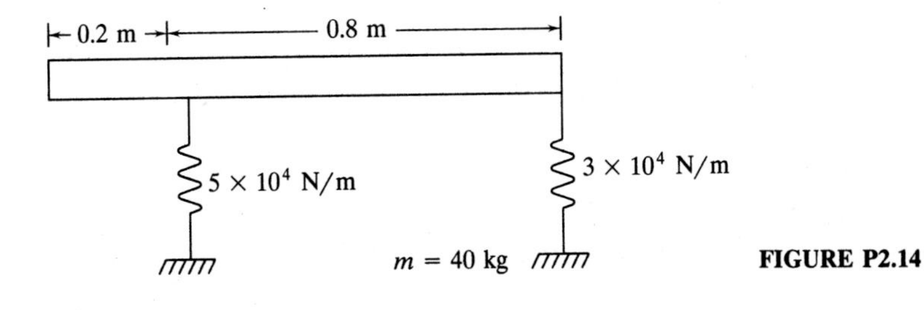

FIGURE P2.14

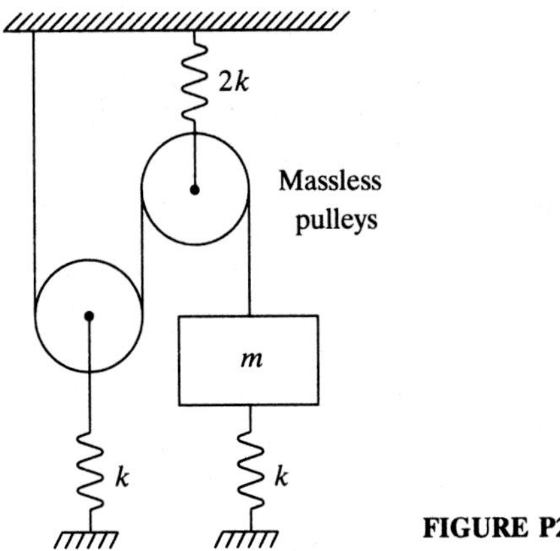

FIGURE P2.15

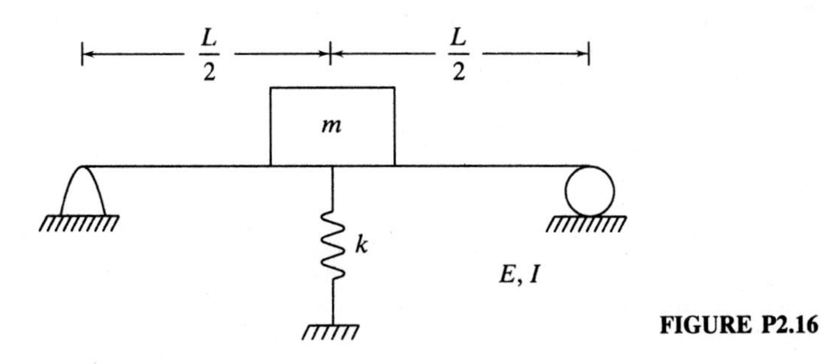

FIGURE P2.16

2.17–2.19. Calculate the equivalent torsional spring constant when the system shown is modeled using a one-degree-of-freedom system with a disk attached to a torsional spring and θ is the generalized coordinate.

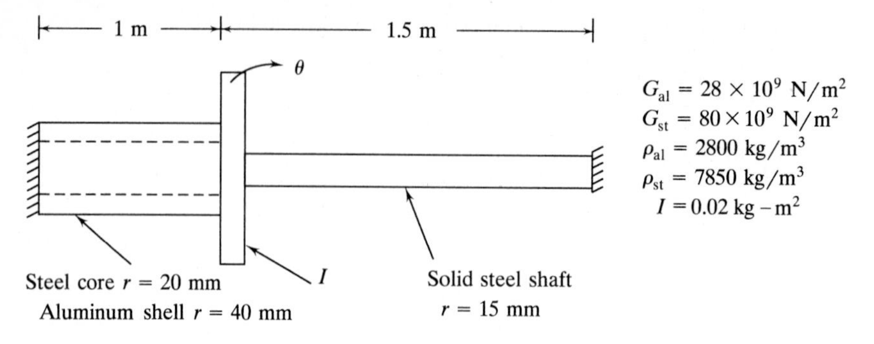

FIGURE P2.17

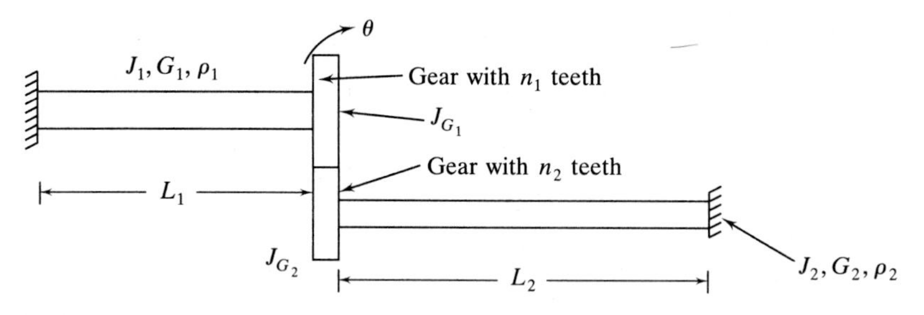

FIGURE P2.18

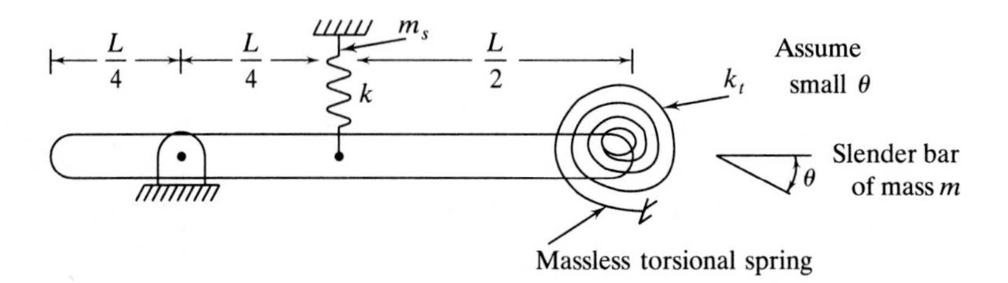

FIGURE P2.19

2.20–2.26. Calculate the equivalent stiffness of a linear spring when a one-degree-of-freedom model of a mass attached to a linear spring is used to model the system shown and x is used as the generalized coordinate.

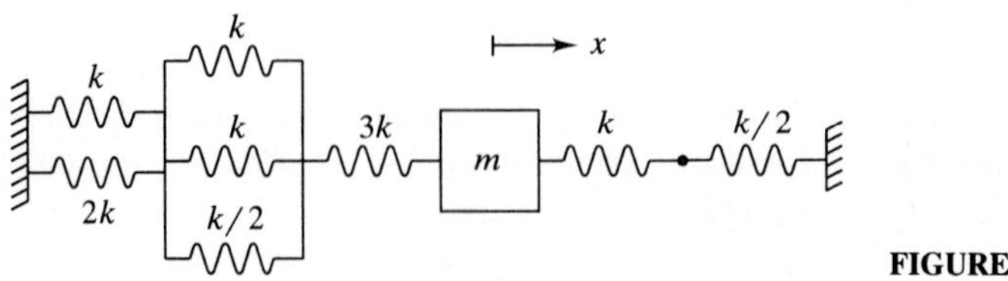

FIGURE P2.20

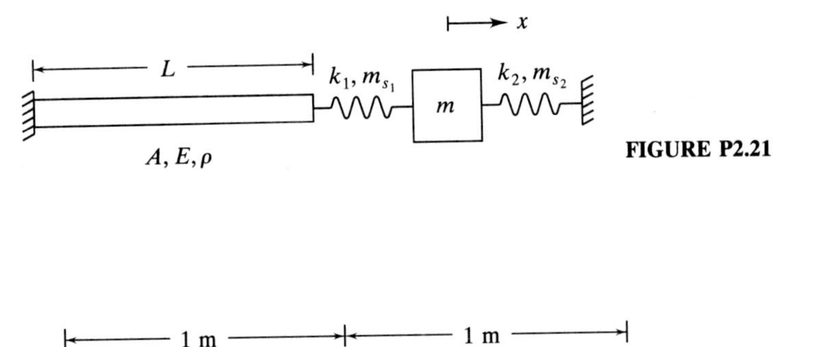

FIGURE P2.21

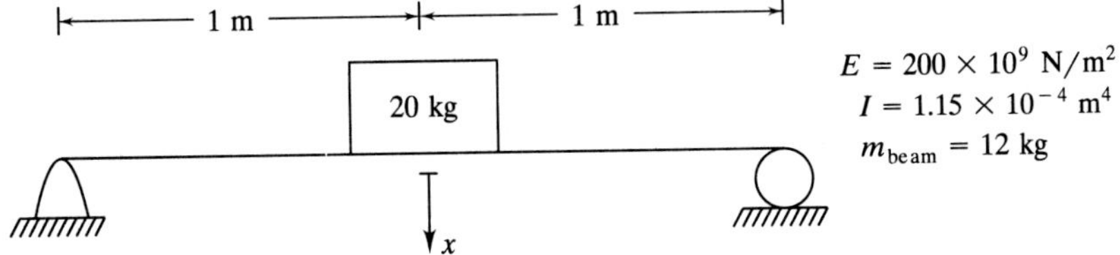

FIGURE P2.22

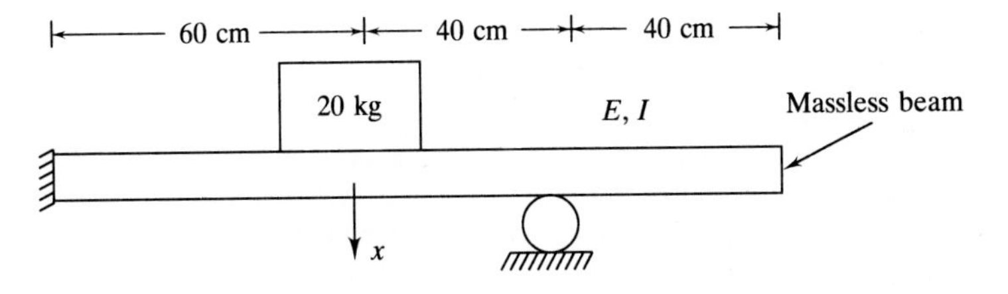

FIGURE P2.23

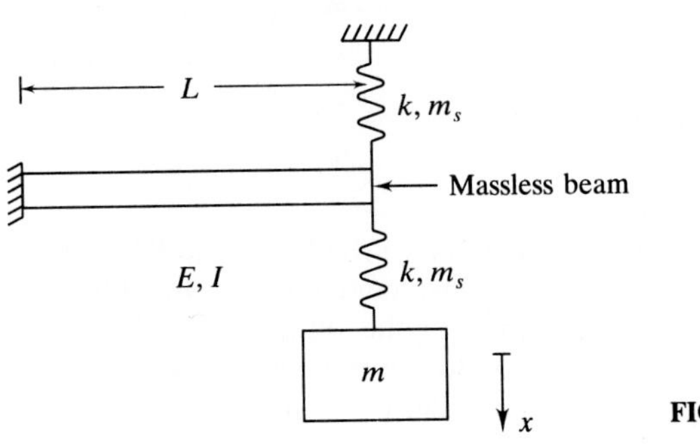

FIGURE P2.24

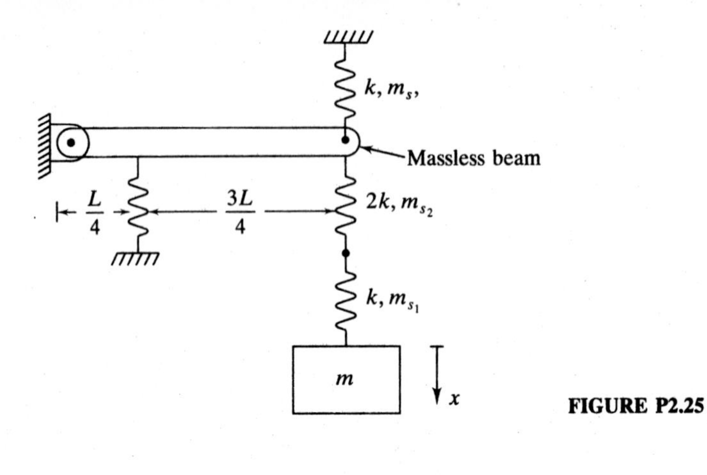

FIGURE P2.25

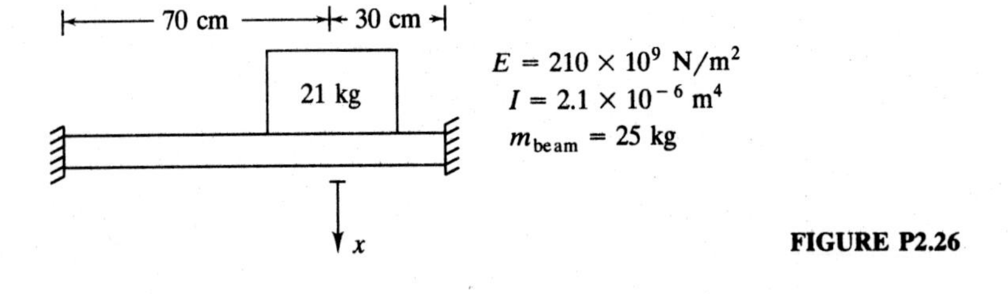

FIGURE P2.26

2.27. The disk attached to the end of the circular beam of Fig. P2.27 has three degrees of freedom. The longitudinal displacement, transverse deflection, and angular rotation are kinematically independent. In fact, the degrees of freedom are also kinetically independent. For example, application of a torque does not induce longitudinal or transverse displacement of the disk. Calculate the longitudinal stiffness, torsional stiffness, and transverse stiffness for this beam.

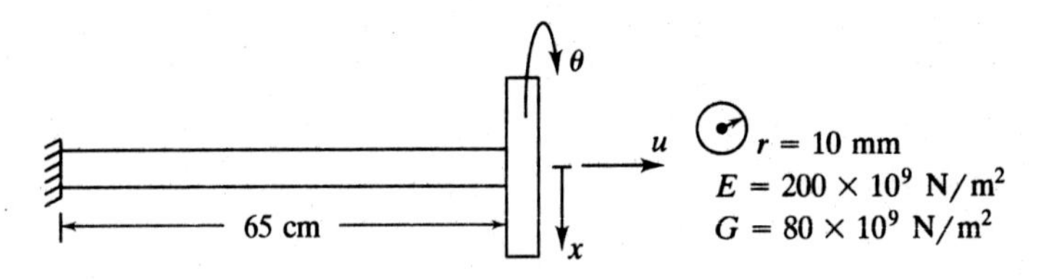

FIGURE P2.27

2.28. The propeller shaft of a ship is the tapered circular cylinder of Fig. P2.28. When installed in the ship, one end is constrained against longitudinal motion relative to the ship while a 500-kg propeller is attached to the other end. Calculate the equivalent spring stiffness and equivalent mass to use in a one-degree-of-freedom model of the longitudinal vibrations of the propeller on its shaft.

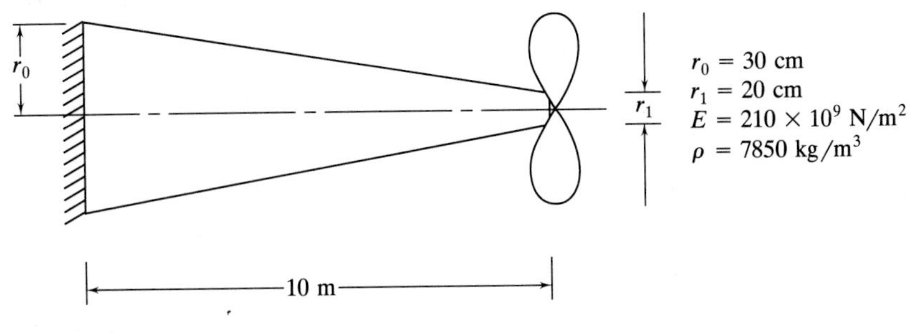

FIGURE P2.28

2.29. One end of a spring of stiffness k_1 and total mass m_{s_1} is attached to a wall while its other end is connected to a spring of stiffness k_2 and mass m_{s_1}. The end of the second spring is attached to a mass which moves with a displacement $x(t)$. Calculate the equivalent mass of these two springs in series.

2.30. Use the equivalent-systems approach to show that the equivalent stiffness of n springs in series is given by Eq. (2.27). That is, equate the potential energy of n springs in series to the potential energy of an equivalent spring.

2.31–2.33. Calculate the equivalent moment of inertia of the systems of Figs. P2.17–P2.19 when the system is modeled with one degree of freedom using θ as the generalized coordinate. Include the inertia effects of all elastic elements.

2.34–2.39. Calculate the equivalent mass of the systems of Figs. P2.21–P2.26 when the system is modeled with one degree of freedom using x as the generalized coordinate. Include the inertia effects of elastic elements, unless otherwise specified.

2.40–2.41. Determine the torsional viscous damping coefficient for the torsional damper shown. Assume a linear velocity profile between the fixed surface and the surface of the rotating object.

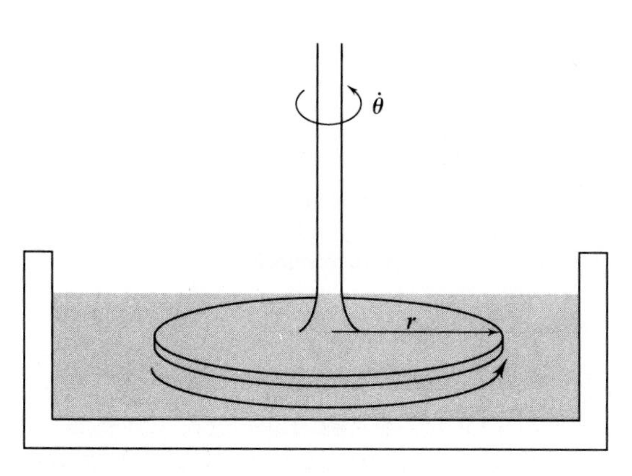

Disk of radius r
Oil of density ρ, viscosity μ
Depth of oil $= h$

FIGURE P2.40

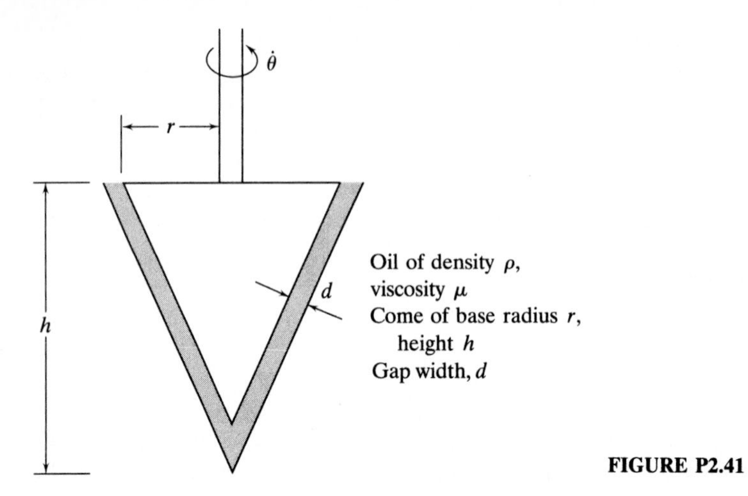

FIGURE P2.41

2.42. The column of liquid of length l, mass density ρ, and dynamic viscosity μ moves in the manometer of Fig. P2.42 relative to any motion of the manometer. The manometer has a circular cross section of radius R. The velocity profile of the fluid relative to the manometer is assumed to be parabolic over the cross section of the manometer. That is,

$$v(r) = ar^2 + br + c$$

where r is the distance from the center of the tube to a point in the liquid and a, b, and c are constants. The damping coefficient is related to the damping force by

$$F = c\bar{v}$$

where $\bar{v}$ is the average velocity over the cross section. Determine the damping coefficient.

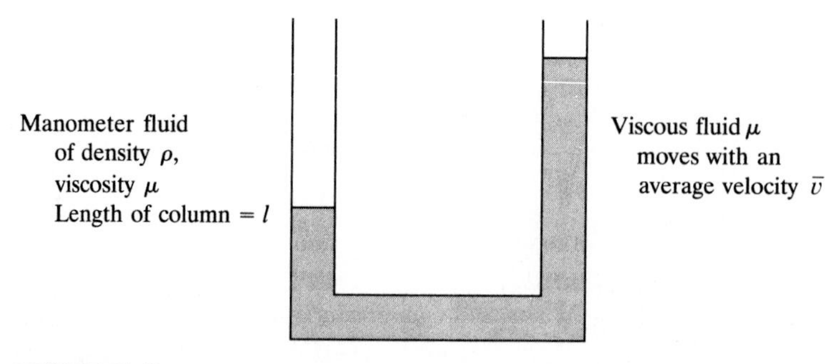

FIGURE P2.42

2.43. Shock absorbers and many other forms of viscous dampers use a piston moving in a cylinder of a viscous liquid as shown in Fig. P2.43. For this configuration the force developed in the piston is the sum of the viscous forces acting on the side of the piston and the force due to the pressure difference between the top and bottom surface of the piston.

(*a*) Assume the piston moves with a constant velocity v_P. Draw a free-body diagram of the piston and mathematically relate the damping force, the viscous force, and the pressure force.

(*b*) Assume steady flow between the side of the piston and the side of the cylinder. Show that the equation governing the velocity profile between the piston and cylinder is

$$\frac{dp}{dx} = \mu \frac{\partial^2 v}{\partial r^2}$$

(*c*) Assume the vertical pressure gradient is constant. Use the preceding results to determine the velocity profile in terms of the damping force and the shear stress on the side of the piston.

(*d*) Use the results of part (*c*) and Eq. (2.42) to determine the wall shear stress in terms of the damping force.

(*e*) Note that the flow rate between the piston and cylinder is equal to the rate at which liquid is displaced by the piston. Use this information to determine the damping force in terms of the velocity and hence the damping coefficient.

(*f*) Use the results of part (*e*) to design a shock absorber for a motorcycle that uses SAE 10/40 oil and requires a damping coefficient of 1000 N · m/s.

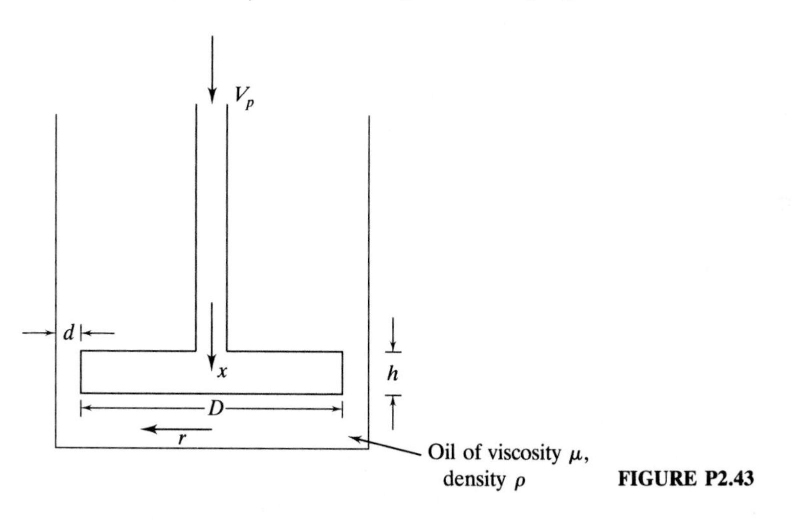

FIGURE P2.43

2.44. Determine the equivalent viscous damping coefficient for two viscous dampers placed in series.

2.45–2.53. Replace the system shown by a one-degree-of-freedom system with an equivalent mass (or moment of inertia), stiffness (or torsional stiffness), and damping coefficient (or torsional damping coefficient). Use the generalized coordinate shown.

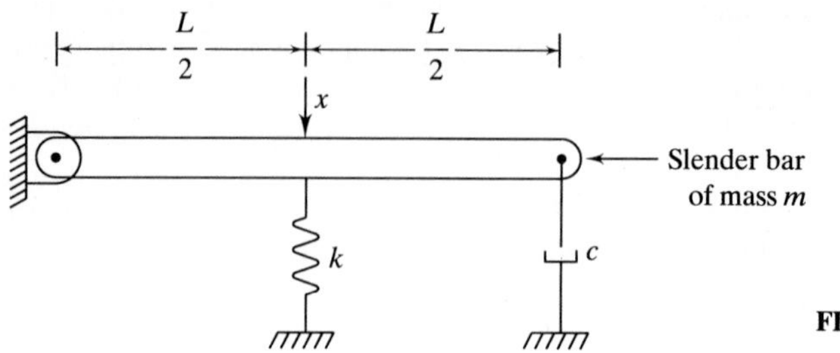

FIGURE P2.45

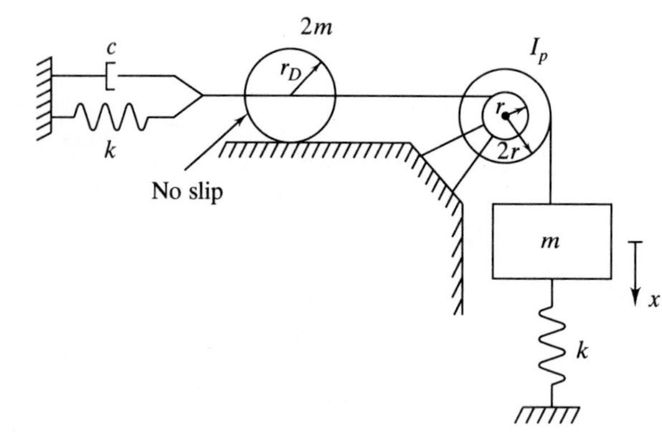

FIGURE P2.46

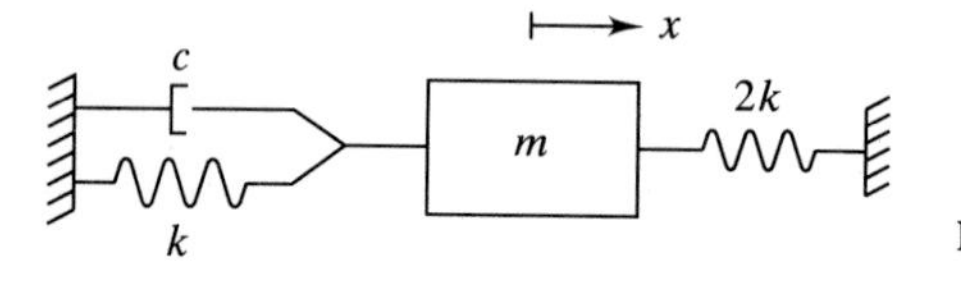

FIGURE P2.47

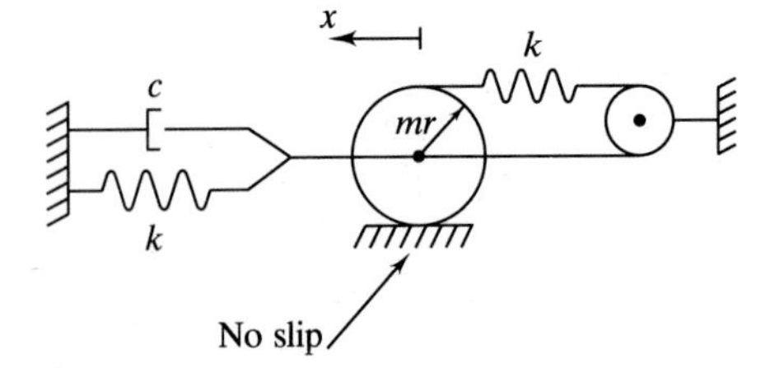

FIGURE P2.48

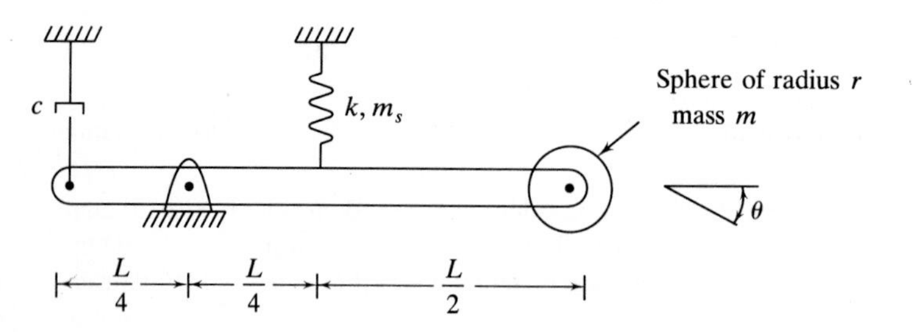

FIGURE P2.49

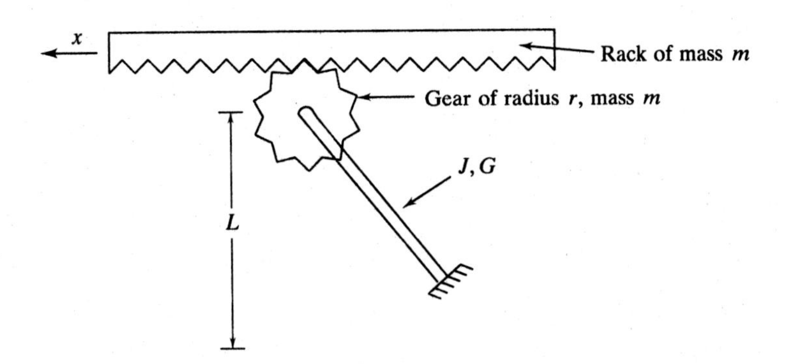

FIGURE P2.50

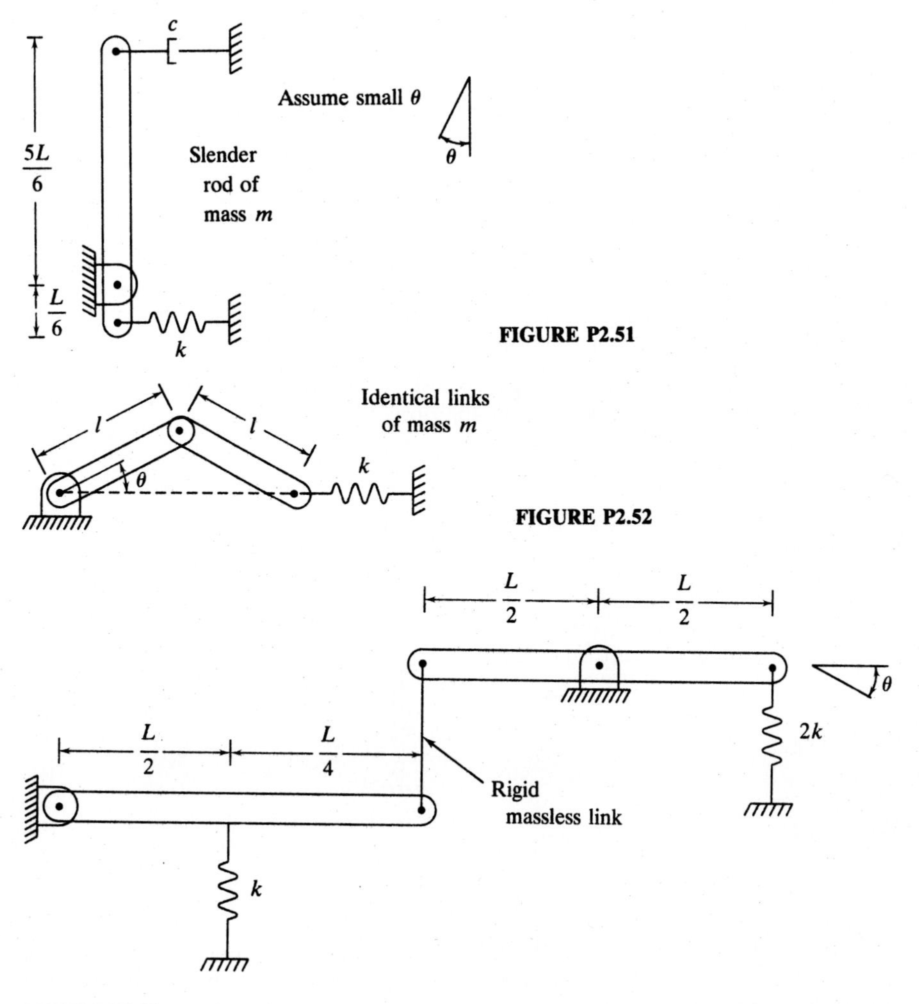

FIGURE P2.51

FIGURE P2.52

FIGURE P2.53

2.54. The spring in Fig. P2.54 is unstretched in the position shown. What is the deflection of the spring when the system is in its equilibrium position?

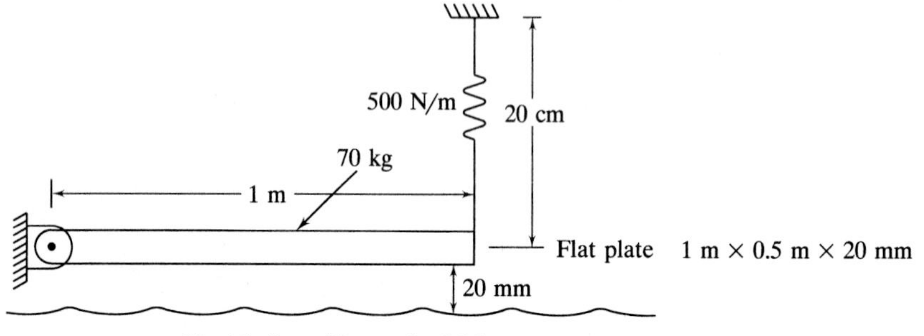

FIGURE P2.54

2.55. A 1-mm-diameter ball of mass density 500 kg/m^3 is dropped in a pool of SAE 30 oil at 20°C. What is the terminal velocity of the ball?

2.56. A 20 mm × 20 mm × 80 mm block is attached to a spring of stiffness 5×10^4 N/m. The assembly is immersed in a liquid of specific gravity 1.05. What is the added mass required to account for the inertia of the fluid?

2.57. Model the system of Fig. P2.57 by a disk of an appropriate moment of inertia attached to a torsional spring of an appropriate stiffness.

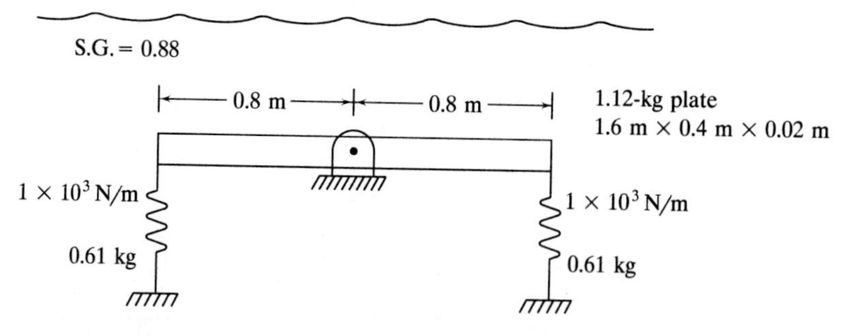

FIGURE P2.57

2.58. A wedge is floating stably on the interface between a liquid of mass density ρ, as shown in Fig. P2.58. When disturbed from equilibrium, let x be the displacement of the wedge's mass center from equilibrium.

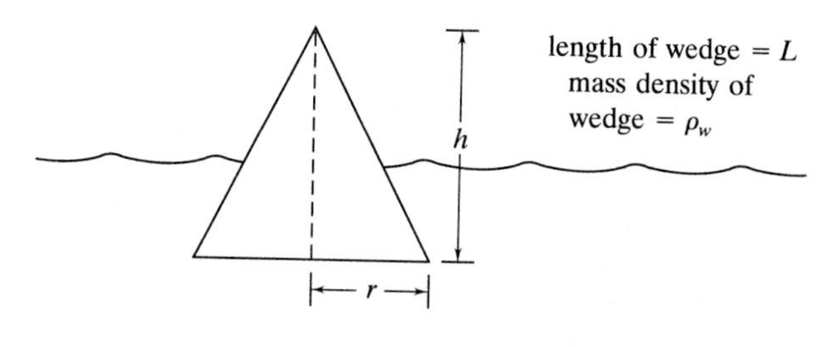

FIGURE P2.58

(*a*) What is the buoyant force acting on the wedge?
(*b*) What is the work done by the buoyant force as its mass center moves between x_1 and x_2?
(*c*) Can the oscillations of the wedge on the surface be modeled by a mass attached to a linear spring?

2.59. A bar of length L and cross-sectional area A is made of a material whose stress-strain diagram is shown in Fig. P2.59. If the internal force developed in the bar is such that $\sigma < \sigma_p$, then the bar's stiffness for a one-degree-of-freedom model is given by Eq. (2.15). Consider the case where $\sigma > \sigma_p$. Let $P = \sigma_p A + \delta P$ be the applied load which results in a deflection $\Delta = \sigma_p L/E + \delta\Delta$.
(*a*) The work done ($W = P\Delta/2$) by the applied force is equal to the strain energy developed in the bar. The strain energy per unit volume is the area under the stress-strain curve. Use this information to relate δP to $\delta\Delta$.
(*b*) What is an approximation to the linear stiffness for small $\Delta\delta$?

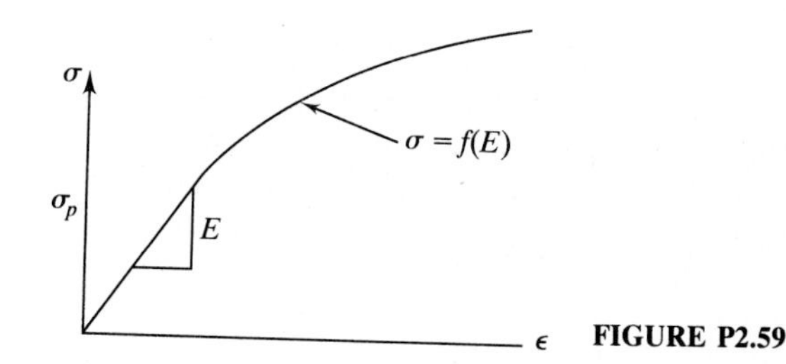

FIGURE P2.59

2.60. Consider a solid circular shaft of length L and radius c made of an elastoplastic material whose shear stress—shear strain diagram is given in Fig. 2.60. If the applied torque is such that the shear stress at the outer radius of the shaft is less than τ_p, a linear relationship exists between the torque and the angular displacement resulting in Eq. (2.21). When the applied torque is large enough to cause plastic behavior, a plastic shell develops around an elastic core of radius $r < c$. Let $T = \pi\tau_p c^3/2 + \delta T$ be the applied torque which results in an angular displacement of $\theta = \tau_p L/(cG) + \delta\theta$.
(*a*) The shear strain at the outer radius of the shaft is related to the angular displacement by $\theta = \gamma_c L/c$. The shear strain distribution is linear over a given cross section. Show that this implies

$$\theta = \frac{\tau_p L}{rG}$$

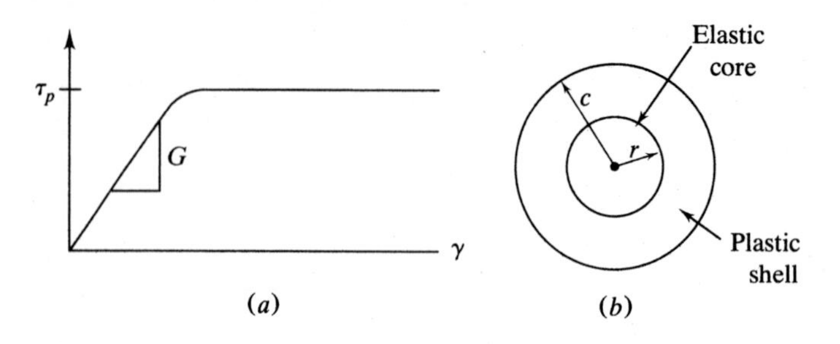

FIGURE P2.60

(*b*) The torque is the resultant moment of the shear stress distribution over the cross section of the shaft,

$$T = \int_0^c 2\pi\tau\rho^2 \, d\rho$$

Use this to relate the torque to the radius of the elastic core.

(*c*) Determine the relationship between δT and $\delta\theta$.

(*d*) Determine a linear approximation to the stiffness for small $\delta\theta$.

2.61. A gas spring consists of a piston of area A moving in a cylinder of gas. As the cylinder moves, the gas expands and contracts changing the pressure exerted on the piston. If the process occurs adiabatically (without heat transfer), then

$$p = C\rho^\gamma$$

where p is the gas pressure, ρ is the gas density, γ is the constant ratio of specific heats, and C is a constant dependent on the initial state. Consider a spring where the initial pressure is p_0 at a temperature T_0. At this pressure the height of the gas column in the cylinder is h. Let $F = p_0 A + \delta F$ be the pressure force on the piston when the piston has displaced a distance x into the gas from its initial height.

(*a*) Determine the relation between δF and x.

(*b*) Linearize the relationship of part (*a*) to approximate the air spring by a linear spring. What is the equivalent stiffness of the spring?

(*c*) What is the required piston area for an air spring ($\gamma = 1.4$) to have a stiffness of 300 N/m for a pressure of 150 kPa (absolute) with $h = 30$ cm?

2.62. The static-deflection relation is used to calculate the equivalent mass of a beam because the dynamic deflection is not known and its use is consistent with the determination of the stiffness. A better approximation may be obtained by using a different expression for the deflection in approximating the beam's kinetic energy. Consider a simply supported beam. The function

$$y(z) = A \sin \frac{\pi z}{L}$$

satisfies all boundary conditions. Use this function to resolve Example 2.7.

2.63. Develop a trigonometric function similar to that used in Prob. 2.61 that satisfies the boundary conditions for a fixed-fixed beam. Use this shape function to approximate the equivalent mass of the beam to use in a one-degree-of-freedom model of a mass attached to the midspan of the beam.

REFERENCES

1. Blevins, R. D.: *Flow Induced Vibrations*, 2nd ed., Van Nostrand Reinhold, New York, 1990.
2. Clough, R. W., and J. Penzien: *Dynamics of Structures*, McGraw-Hill, New York, 1975.
3. Fertis, D. G.: *Dynamics and Vibrations of Structures*, Wiley-Interscience, New York, 1973.
4. Gerhart, P. M., and R. G. Gross: *Fundamentals of Fluid Mechanics*, Addison-Wesley, Reading, Mass., 1985.
5. Higdon, A., E. Ohlsen, W. B. Stiles, J. A. Weese, and W. F. Riley: *Mechanics of Materials*, 4th ed., Wiley, New York, 1975.
6. Shames, I. H.: *Mechanics of Fluids*, 3rd ed., McGraw Hill, New York, 1992.
7. Shigley, J. E.: *Mechanical Engineering Design*, McGraw-Hill, New York, 1963.
8. Patton, K. T.: "Tables of Hydrodynamic Mass Factors for Translating Motion" ASME Paper 65-WA/UNT-2 1965.
9. Wahl: *Mechanical Springs*, 2nd ed., McGraw-Hill, New York, 1963.
10. Wendel, K.: "Hydrodynamic Masses and Hydrodynamic Moments of Inertia," U.S. Navy David Taylor Model Basin Transactions 260, 1956.
11. White, F. M.: *Fluid Mechanics*, 2nd ed., McGraw-Hill, New York, 1986.

CHAPTER 3

FREE VIBRATIONS OF ONE-DEGREE-OF-FREEDOM SYSTEMS

3.1 INTRODUCTION

3.1.1 Cause of Free Vibrations

Free vibrations occur in a system in the absence of any external excitation as a result of a kinetic energy or potential energy initially present in the system. These vibrations are oscillations about one of the system's static-equilibrium positions.

Consider the mass-spring system of Fig. 3.1. When the block is displaced a distance x_0 from equilibrium, a potential energy, $kx_0^2/2$, develops in the spring. A force kx_0 acts on the block from the spring. When the system is released from this position, the spring force draws the block toward the system's equilibrium position with the potential energy being converted to kinetic energy. When the system reaches its equilibrium position, the kinetic energy reaches a maximum and motion continues until the spring is compressed a distance x_0 when the velocity of the block is again zero. This process of transfer of potential energy to kinetic energy is continual in the absence of nonconservative forces. Thus vibrations will theoretically continue forever. In a physical system, such perpetual motion is impossible. Dry friction, internal friction in the spring, aerodynamic drag, and other nonconservative mechanisms eventually dissipate the potential energy.

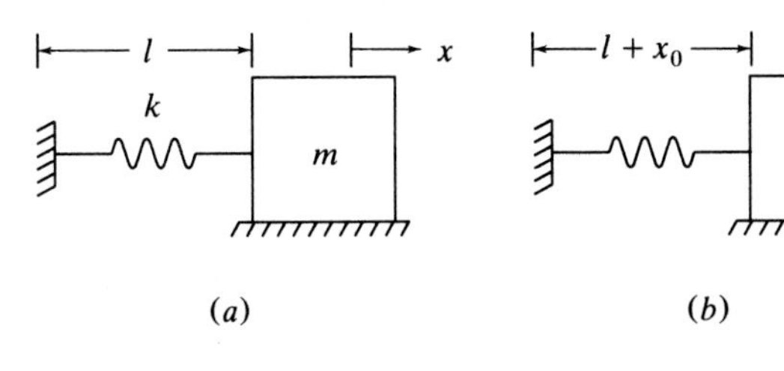

FIGURE 3.1
(a) When the mass-spring system is at rest in equilibrium the spring has an unstretched length l; (b) when the mass is displaced a distance x_0, a force kx_0 and potential energy $\frac{1}{2}kx_0^2$ develops in the spring.

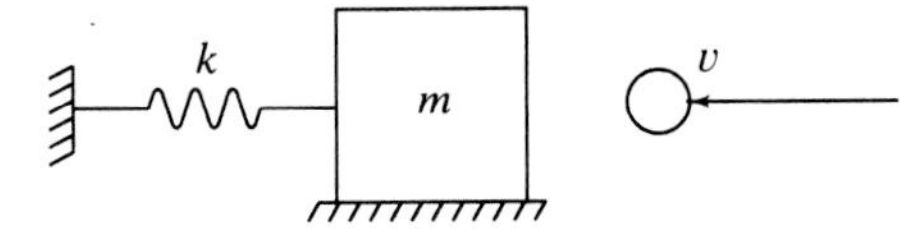

FIGURE 3.2
The impact between the particle and the block initiates free vibrations by imparting an initial kinetic energy to the block.

Free vibrations may also be initiated as the result of an initial kinetic energy. The mass-spring system of Fig. 3.2 is at rest in equilibrium when the block is struck by a moving particle. The impact imparts an initial kinetic energy to the block and starts the system in motion. Free vibrations begin with a continual transfer of kinetic energy to potential energy and vice versa.

3.1.2 Examples of Systems Undergoing Free Vibrations

The old-fashioned mechanism of a cuckoo clock requires a correctly tuned pendulum to keep accurate time. See Fig. 3.3. The pendulum is a slender rod with a concentrated mass located along the rod. The location of the mass is

FIGURE 3.3
The distance between the mass and the axis of rotation affects the tuning of the clock. The mass must be correctly placed in order for the clock to be accurately tuned.

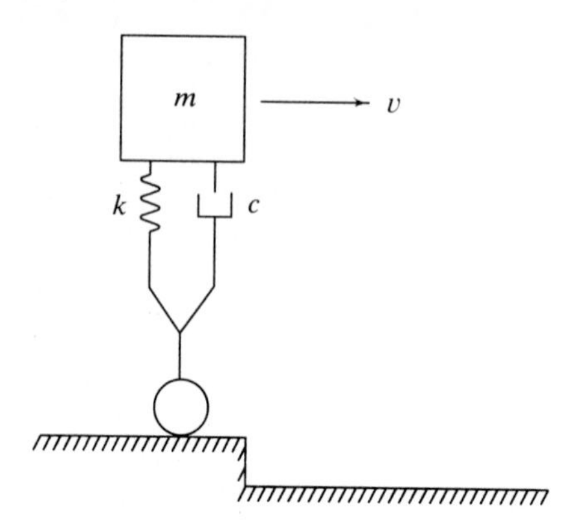

FIGURE 3.4
Simplified one-degree-of-freedom model of vehicle suspension system. When the vehicle encounters a bump, potential energy develops initiating free vibrations.

adjusted to tune the clock. The pendulum is subject to an impulse at the end of each cycle to restore energy dissipated. The period is independent of the impulse's magnitude. Since the pendulum is not subject to any external excitation, as it executes each cycle between application of the impulse, it is undergoing free vibrations.

A model of a simplified vehicle suspension system is shown in Fig. 3.4. The suspension system is modeled as a dashpot and a spring in parallel that connect the body to its axle. The wheels are assumed to be rigid. The elastic element allows a relative displacement between the wheels and the body of the vehicle such that the vehicle's occupants do not fully feel all changes in road contour. The vehicle in Fig. 3.4 is traveling along a smooth road and encounters a bump where the wheels are given a sudden vertical displacement, causing an instantaneous stretching of the spring. After horizontal motion is resumed, the suspension system undergoes free vibrations.

When the rope swing of Fig. 3.5 is pulled back from its vertical equilibrium position, a potential energy due to gravity develops. When released, the swing executes free oscillations about its vertical equilibrium position. As the swing rotates, the rope is in contact with the tree limbs. This resulting friction may be modeled by belt friction, a form of Coulomb damping. If the swing is not given pushes, the initial potential energy is continually dissipated. Eventually, the energy is insufficient to overcome friction and motion ceases.

FIGURE 3.5
Free oscillations occur when the rope swing is pulled back and released. Friction between the ropes and the tree causes dissipation of the initial potential energy. Free oscillations eventually cease if no additional pushes are given.

3.2 DERIVATION OF DIFFERENTIAL EQUATIONS

The free vibrations of a one-degree-of-freedom system are usually described by a second-order homogeneous ordinary differential equation. Time is the independent variable. The dependent variable is the chosen generalized coordinate. The generalized coordinate is selected to represent the displacement of some particle in the system. The dependent variable is usually chosen measured from the particle's position when the system is in static equilibrium. If the system is subject to viscous damping, and no other nonconservative mechanisms, the differential equation is homogeneous. If Coulomb damping is present the differential equation is nonhomogeneous.

For a linear system the term proportional to the second time derivative of the generalized coordinate represents the inertia of the system. If viscous damping is present the differential equation includes a term proportional to the generalized coordinate's first derivative. A term proportional to the zeroth derivative represents restoring forces.

Nonlinear differential equations can occur as a result of geometric or material nonlinearities or may be due to nonlinear conservative forces. A variety of nonlinear terms can appear in the differential equations. The focus of this text is on linear systems. Nonlinear systems are examined quantitatively only when the nonlinearities are small and the governing differential equation can be satisfactorily approximated by a linear differential equation. Qualitative aspects of nonlinear systems are briefly discussed in Chap. 10.

Two methods are presented for derivation of governing differential equations for one-degree-of-freedom systems. In the first method, Newton's laws are applied to appropriate free-body diagrams, as described in Chap. 1. The alternative is to use the principle of work-energy and the equivalent-systems analysis of Sec. 2.8. In either case the procedure for mathematical modeling of a physical system outlined in Sec. 1.4 is followed.

3.2.1 Application of Newton's Laws

Newton's laws, as formulated in Chap. 1, are applied to free-body diagrams of vibrating systems to derive the governing differential equation. The following steps are used in application to a one-degree-of-freedom system.

1. A generalized coordinate is chosen. This variable should represent the displacement of a particle in the system. If rotational motion is involved, the generalized coordinate could represent an angular displacement.
2. Free-body diagrams are drawn showing the system at an arbitrary instant of time. In line with the methods of Chap. 1, two free-body diagrams are drawn. One free-body diagram shows all external forces acting on the system. The second free-body diagram shows all effective forces acting on the system.

Recall that the effective forces are a force equal to $m\bar{a}$, applied at the mass center and a couple equal to $\bar{I}\alpha$.

The forces drawn on each free-body diagram represent the force at an arbitrary instant of time. The direction of each force and moment are drawn consistent with the positive direction of the generalized coordinate. Geometry, kinematics, constitutive equations, and other laws valid for specific systems can be used to specify the external and effective forces.

3. The appropriate form of Newton's law is applied to the free-body diagrams.
4. Applicable assumptions are used along with algebraic manipulation. The result is the governing differential equation.

Example 3.1. Derive the differential equation governing the motion of the block of Fig. 3.6*a* assuming the spring is massless.

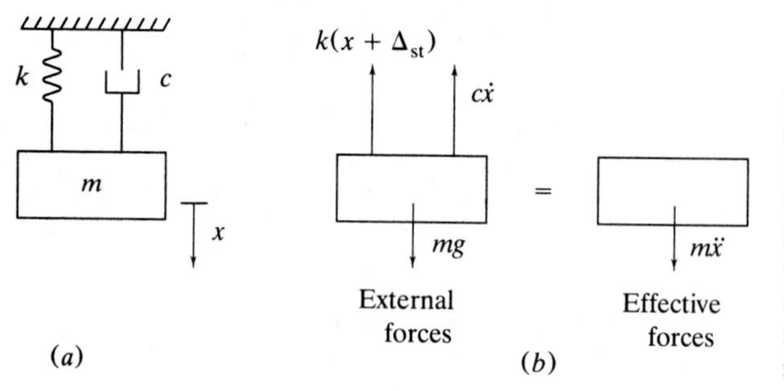

FIGURE 3.6
(*a*) Mass-spring-dashpot system of Example 3.1; (*b*) free-body diagrams at an arbitrary instant of time. Directions of external and effective forces are consistent with positive direction of generalized coordinate x.

Let $x(t)$ represent the displacement of the block, measured positive downward, from its static-equilibrium position. Free-body diagrams showing the external and effective forces acting on the block at an arbitrary instant of time are given in Fig. 3.6*b*. The spring is assumed to be linear. Thus the force developed in the spring is given by Eq. (2.4) where x, in that equation, represents the change in length of the spring from its unstretched length. Since x is measured from the static-equilibrium position of the system, the spring force developed for the system of Fig. 3.6 is

$$F_s = k(x + \Delta_{st})$$

where Δ_{st} is the static deflection of the spring.

Note that since x is measured positive downward, when x is positive the spring is stretched further from its equilibrium position and pulls on the mass, as illustrated on the free-body diagram of external forces. The effective force is also drawn downward to be consistent with the choice of positive x.

The appropriate form of Newton's law for this problem is

$$\sum F_{\text{ext}} = \sum F_{\text{eff}}$$

which when applied to the free-body diagrams of Fig. 3.6*b* gives

$$mg - k(x + \Delta_{st}) - c\dot{x} = m\ddot{x}$$

Analysis of the static-equilibrium position reveals

$$\Delta_{st} = \frac{mg}{k}$$

When this result is substituted into the previous equation, the static-deflection term cancels with the gravity term leaving

$$m\ddot{x} + c\dot{x} + kx = 0$$

The time history of motion of the system in Fig. 3.6 is obtained by solving the preceding second-order linear homogeneous ordinary differential equation subject to appropriate initial conditions.

Example 3.2. Derive the differential equation governing the angular oscillations of the compound pendulum of Fig. 3.7*a*.

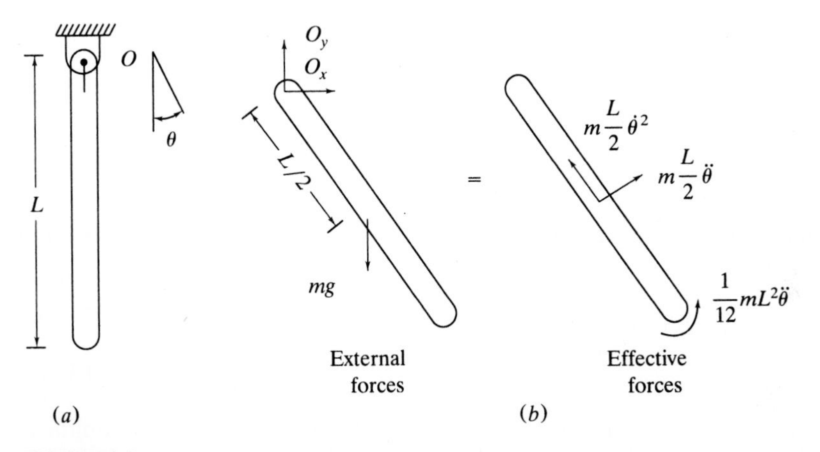

FIGURE 3.7
(*a*) Compound pendulum of Example 3.2 is a slender rod pinned at one end. The generalized coordinate, θ, is the counterclockwise angular displacement from equilibrium; (*b*) free-body diagrams at an arbitrary instant of time.

Let $\theta(t)$ be the counterclockwise angular displacement of the rod measured from its vertical equilibrium position. Summing moments about O using the free-body diagrams of Fig. 3.7*b*,

$$\left(\sum \overset{\curvearrowleft +}{M_O}\right)_{ext} = \left(\sum \overset{\curvearrowleft +}{M_O}\right)_{eff}$$

yields

$$-mg\frac{L}{2}\sin\theta = m\frac{L^2}{12}\ddot{\theta} + m\frac{L}{2}\ddot{\theta}\frac{L}{2}$$

which becomes

$$m\frac{L^2}{3}\ddot{\theta} + mg\frac{L}{2}\sin\theta = 0$$

The differential equation obtained in Example 3.2 is a second-order nonlinear ordinary differential equation. While an exact solution exists in terms of elliptic integrals for this equation, exact solutions for most nonlinear equations have not been found.

Approximations for solutions of problems governed by nonlinear differential equations are obtained by one of two approaches. An approximate solution of the exact equation can be obtained by a numerical method, or, if conditions are right, the differential equation can be approximated by a linear equation whose exact solution is easily obtained. The latter approach is used here.

Consider the Taylor series expansion for $\sin\theta$ about $\theta = 0$:

$$\sin\theta = \theta - \frac{\theta^3}{6} + \frac{\theta^5}{120} - \cdots \tag{3.1}$$

For small θ,

$$\sin\theta \approx \theta \tag{3.2a}$$

Similar truncations of the Taylor series expansions for $\cos\theta$ and $\tan\theta$ for small θ lead to

$$\cos\theta \approx 1 \tag{3.2b}$$

$$\tan\theta \approx \theta \tag{3.2c}$$

The small-angle approximations of Eq. (3.2) are used to linearize nonlinear differential equations. When the small-angle assumption is made for Example 3.2, the resulting linearized differential equation is

$$\ddot{\theta} + \frac{3g}{2L}\theta = 0$$

Example 3.3. A flywheel of mass moment of inertia I is attached to the end of a solid circular shaft of radius r, length L, shear modulus G, and mass m, as shown in Fig. 3.8a. A moment is applied to the disk, rotating it from its static-equilibrium configuration. The disk is released and torsional oscillations about the equilibrium position ensue. Derive the differential equation governing the torsional oscillations. Include the inertia effects of the shaft.

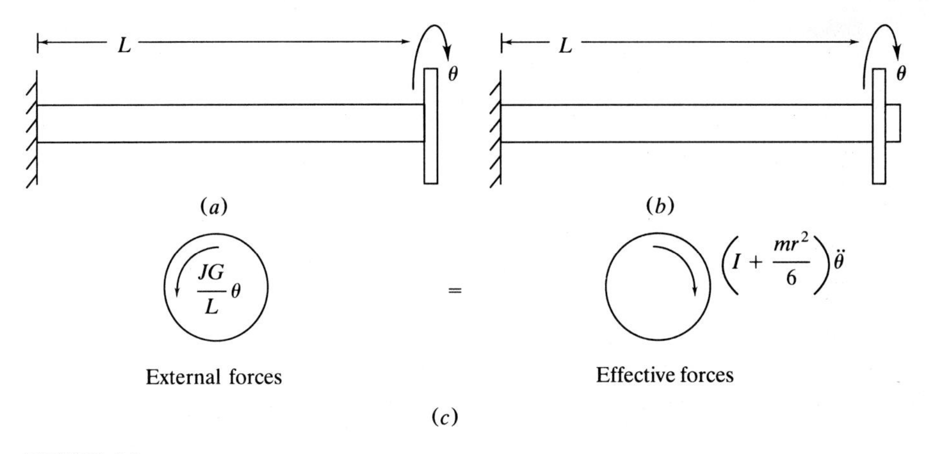

FIGURE 3.8
(a) The system of Example 3.3 is a flywheel of mass moment of inertia I attached to the end of a shaft of length L, radius r, and mass m; (b) inertia effects of the shaft are approximated by attaching a disk of mass moment of inertia $mr^2/6$ to flywheel; (c) free-body diagrams at an arbitrary instant of time.

Let $\theta(t)$ be the angular displacement of the disk from its equilibrium position. The shaft acts as a torsional spring. As the disk oscillates, a moment

$$M = \frac{JG}{L}\theta$$

is developed between the shaft and the disk. The moment acting on the shaft is in the direction of the rotation. From Newton's third law the moment from the shaft on the disk resists the rotation.

Using the results of Sec. 2.6, the inertia effects of the shaft are approximated by adding an extra disk at the end of the shaft. The mass moment of inertia of the added disk is one-third of the mass moment of inertia of the shaft.

$$I_{D_1} = \tfrac{1}{6}mr^2$$

The equivalent system is shown in Fig. 3.8b.

Summing moments about the center of the disk

$$\left(\sum \overset{\curvearrowleft +}{M_C}\right)_{\text{ext}} = \left(\sum \overset{\curvearrowleft +}{M_C}\right)_{\text{eff}}$$

using the free-body diagrams of Fig. 3.8c leads to

$$-\frac{JG}{L}\theta = \left(I + \frac{mr^2}{6}\right)\ddot{\theta}$$

or

$$\ddot{\theta} + \frac{JG}{\left(I + \dfrac{mr^2}{6}\right)L}\theta = 0$$

Example 3.4. A slender rod of length L and mass m is pinned at O, as shown in Fig. 3.9. A spring of stiffness k is connected to the rod at point P while a dashpot of damping coefficient c is connected at point Q. Assuming small displacements, derive a linear differential equation governing the free vibrations of this system.

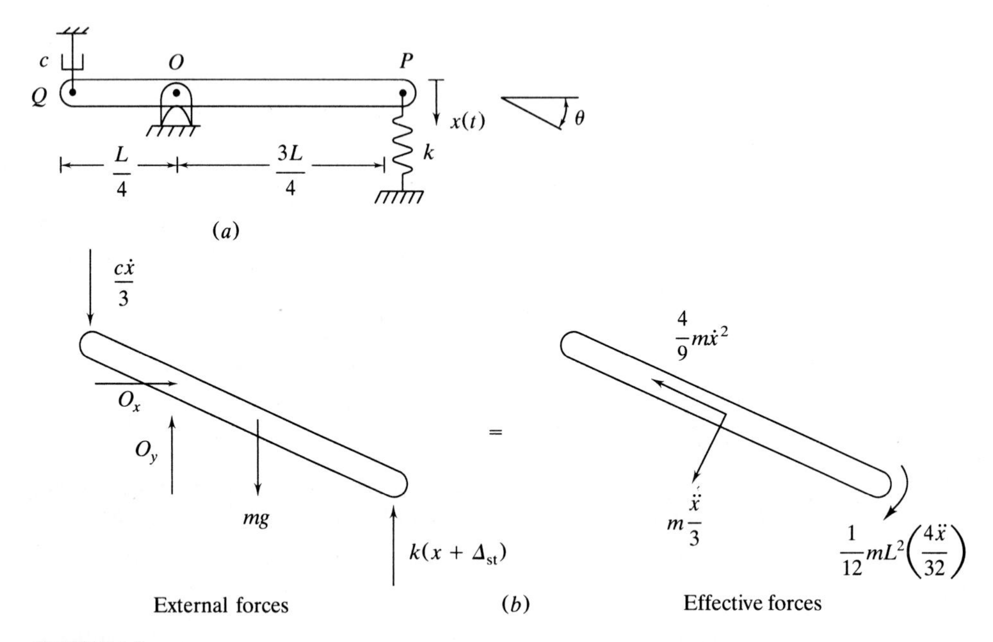

FIGURE 3.9
(a) System of Example 3.4; (b) free-body diagrams at an arbitrary instant of time.

Use x, the displacement of particle P, measured from the system's equilibrium position, as the generalized coordinate.

An analysis of the static-equilibrium position of the system reveals

$$-3k\frac{L}{4}\Delta_{\text{st}} + mg\frac{L}{4} = 0$$

Consistent with the assumption that x is small, the lines of action of the damping force and the spring force are assumed to be vertical. From the geometry of Fig. 3.9

$$x = \frac{3L}{4}\sin\theta$$

Using the small-angle approximation, and thus Eq. (3.2),

$$x \approx \frac{3L}{4}\theta$$

The appropriate equation for summation of moments about O is

$$\left(\sum \overset{\curvearrowleft}{\underset{}{+}}M_O\right)_{\text{ext}} = \left(\sum \overset{\curvearrowleft}{\underset{}{+}}M_O\right)_{\text{eff}}$$

which when applied to the free-body diagrams of Fig. 3.9b gives

$$mg\frac{L}{4}\cos\left(\frac{4x}{3L}\right) - k(x + \Delta_{\text{st}})\frac{3L}{4}\cos\left(\frac{4x}{3L}\right) - c\frac{\dot{x}}{3}\frac{L}{4}\cos\left(\frac{4x}{3L}\right)$$

$$= \frac{mL^2}{12}\frac{4\ddot{x}}{3L} + m\frac{\ddot{x}}{3}\frac{L}{4}$$

The small-angle approximation and Eq. (3.2) are used to approximate the cosine terms by one. The static-equilibrium condition is used to cancel the static-deflection terms with the gravity terms. The resulting differential equation becomes

$$\frac{7m}{36}\ddot{x} + \frac{c}{12}\dot{x} + \frac{3k}{4}x = 0$$

The preceding equation is a linear approximation to the exact differential equation.

Example 3.5. Using x, the displacement of the block measured positive downward from its equilibrium position as the generalized coordinate, derive the differential equation governing the free vibrations of the system of Fig. 3.10a.

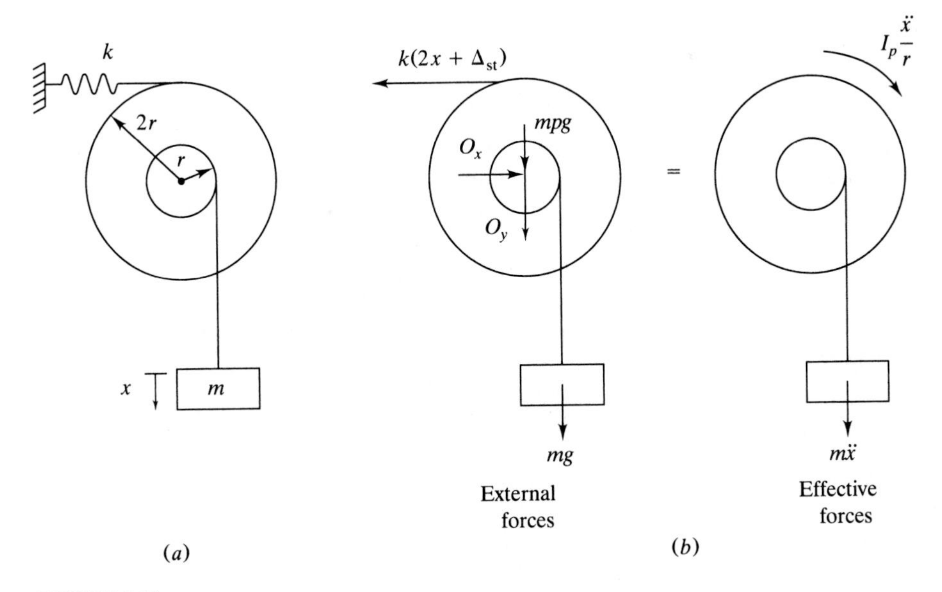

FIGURE 3.10
(a) System of Example 3.5; (b) free-body diagrams at an arbitrary instant of time.

The static-equilibrium position of the system is analyzed and yields the following relation between gravity and static deflection of the spring:

$$2k\Delta_{st} - mg = 0$$

Free-body diagrams showing the external and effective forces acting on the mass-pulley system at an arbitrary instant of time are shown in Fig. 3.10*b*. No slip is assumed between the pulley and the cables, and friction is neglected. Kinematics is used to express the relation between x and the change in length of the spring.

The appropriate form of Newton's law is

$$\left(\sum \overset{\curvearrowright +}{M_O}\right)_{\text{ext}} = \left(\sum \overset{\curvearrowright +}{M_O}\right)_{\text{eff}}$$

which applied to the free-body diagrams gives

$$-k(2x + \Delta_{st})2r + mgr = \frac{I_p}{r}\ddot{x} + m\ddot{x}r$$

The static-equilibrium condition is used to eliminate gravity and static deflection from the equation resulting in

$$\left(\frac{I_p}{r} + mr\right)\ddot{x} + 4krx = 0$$

Example 3.6. A sphere of radius r and mass m is attached to a spring of stiffness k. The assembly is placed in a highly viscous fluid of dynamic viscosity μ and mass density ρ. The sphere is displaced from its equilibrium configuration and released from rest. Oscillations about the equilibrium configuration ensue. Derive the differential equation governing the oscillations.

When the system is in equilibrium, a balance exists between the gravity force, the buoyant force, and the force in the spring due to its static deflection

$$mg - F_B - k\Delta_{st} = 0$$

As the sphere oscillates, fluid surrounding the sphere is set in motion, resulting in kinetic energy. Using the results of Sec. 2.9 and Table 2.1, the kinetic energy of the fluid can be taken into account by adding a particle of mass

$$m_a = \tfrac{3}{4}\pi\rho r^3$$

to the sphere.

If the fluid is very viscous and the initial displacement of the sphere is small, the Reynolds number is small. Assume that the Reynolds number is low enough such that Eq. (2.59) applies where the coefficient of drag is

$$D = 6\pi\mu r v$$

where v is the velocity of the sphere.

Let $x(t)$ be the displacement of the center of the sphere from its equilibrium position. The free-body diagrams of Fig. 3.11*b* show the forces acting on the sphere at an arbitrary instant of time. The appropriate form of Newton's law is

$$\sum F_{\text{ext}} = \sum F_{\text{eff}}$$

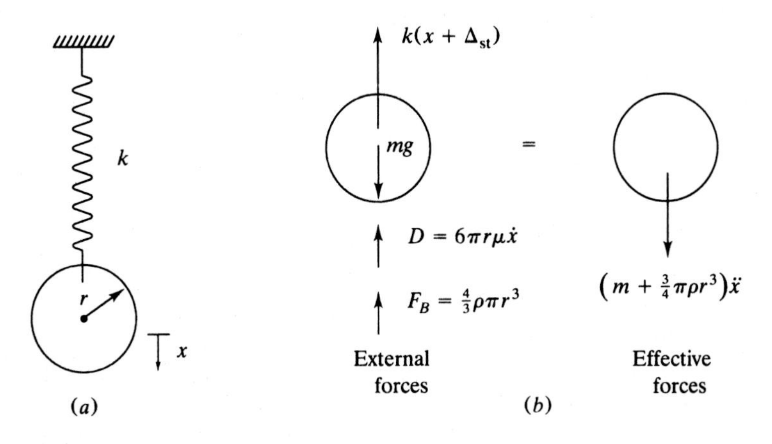

FIGURE 3.11
(a) Solid sphere of radius r is suspended from spring of stiffness k in a fluid of mass density ρ and viscosity μ; (b) free-body diagrams at an arbitrary instant of time. Drag force assumes low Reynolds number. Effective force includes added mass to account for inertia effects of entrained fluid.

Its application to the free-body diagrams yields

$$-k(x + \Delta_{st}) - 6\pi\mu r\dot{x} + mg - F_B = (m + m_a)\ddot{x}$$

Use of the static-equilibrium condition eliminates gravity, buoyancy, and static-deflection terms resulting in

$$(m + \tfrac{3}{4}\rho\pi r^3)\ddot{x} + 6\pi\mu r\dot{x} + kx = 0$$

In the previous examples and for all linear systems the static deflection of springs cancel with the gravity or buoyancy forces causing the static deflection when the governing differential equation is simplified. This is true whenever the generalized coordinate is chosen to be measured from the system's static equilibrium position. Thus static deflections of springs and the forces that cause them have no effect on the vibrations of a linear system. Since this is now recognized, it can be stated a priori when deriving a differential equation that static spring forces cancel with other forces and will not be included in the analysis.

In many cases the force system composed of the static spring force and the forces that cause the static deflection is equivalent to the zero force. Consider the system of Ex. 3.4. The static equilibrium condition is obtained by summing moments about the support. The moment of the gravity force balances with the moment of the static spring force, but the resultant of these forces is not the zero-force. These forces can be ignored when deriving the governing differential equation, as it requires summing moments about the support. However, if the values of the reactions at the support are subsequently required, the static spring force and the gravity force must be included when summing forces.

3.2.2 Energy Methods: Equivalent-Systems Approach

The use of energy methods is an alternative to the direct application of Newton's laws to derive the differential equations governing the vibrations of a one-degree-of-freedom system. The principle of work-energy is actually a spatially integrated form of Newton's law. If all forces in the system are conservative, the principle of conservation of energy applies.

It is shown in Sec. 2.8 that all conservative one-degree-of-freedom systems, where energy stored in springs is the only source of potential energy, can be modeled by the mass-spring system of Fig. 3.12*a*. Actually all linear one-degree-of-freedom systems can be modeled by the system of Fig. 3.12*a*. The equivalent mass and stiffness terms are calculated using the system's kinetic and potential energies, respectively. If viscous damping is present the appropriate model is the mass-spring-dashpot system of Fig. 3.12*b*. The equivalent damping coefficient is calculated from the work done by the viscous damping forces.

The system of Fig. 3.12*b* is considered in Example 3.1. The following differential equation governs the displacement of the equivalent mass from the system's equilibrium position:

$$m_{\text{eq}}\ddot{x} + c_{\text{eq}}\dot{x} + k_{\text{eq}}x = 0 \tag{3.3}$$

Systems in which angular oscillations or torsional oscillations occur can be modeled by the system of Fig. 3.13. This differential equation governing the angular displacement of the disk from its equilibrium position is derived in

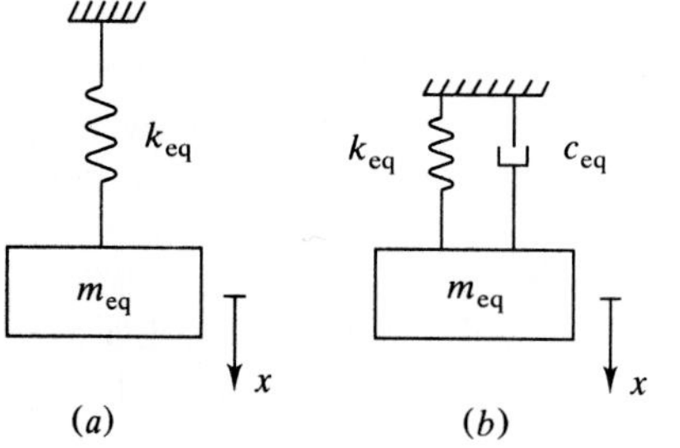

FIGURE 3.12
(*a*) Equivalent mass-spring system is used to model conservative linear one-degree-of-freedom systems; (*b*) equivalent mass-spring-dashpot system is used to model linear one-degree-of-freedom systems with viscous damping.

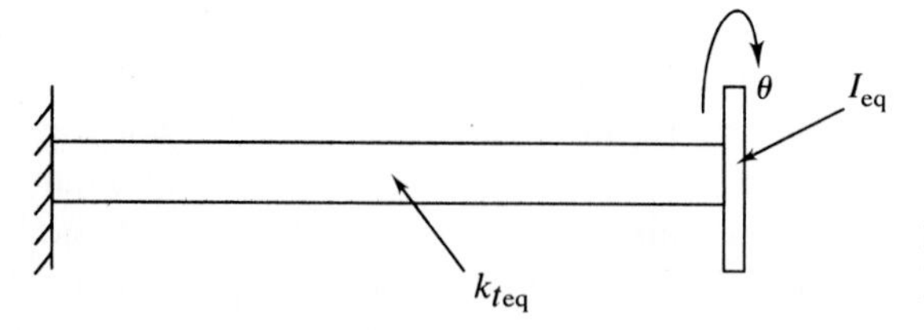

FIGURE 3.13
Equivalent shaft-disk system is used to model linear torsional systems.

Example 3.3 as

$$I_{\text{eq}}\ddot{\theta} + k_{t_{\text{eq}}}\theta = 0 \tag{3.4}$$

Thus the governing differential equations for one-degree-of-freedom systems can be obtained by replacing the system with its equivalent model, Fig. 3.12 or 3.13. The equivalent values of the inertia, damping, and stiffness coefficients are found using work and energy. The governing differential equations can be written using Eq. (3.3) or (3.4).

The following is noted about this procedure:

1. While the equivalent-systems method is only valid for linear systems, it can be used for linear approximations of nonlinear systems.
2. The appropriate model for a mass-spring-dashpot system may be either the hanging system of Fig. 3.12 or a system where the mass is sliding on a frictionless surface. Example 3.1 shows how the static deflection cancels with the gravity force for the hanging system. This result is also obtained in Sec. 2.8. Thus the differential equations governing the motion of the hanging mass and the sliding mass are the same.

   The differential equations for the systems of Examples 3.1 through 3.6 can be derived using energy methods.
3. The equivalent-systems approach, as outlined, is not applicable for systems subject to forced excitations. The method of virtual work must be used to determine the appropriate nonhomogeneous terms due to forced excitations when the equivalent-systems approach is used.

Example 3.7. Write the differential equation governing free vibrations of the one-degree-of-freedom system of Fig. 2.34. Use x, the displacement of the mass center of the disk measured from its equilibrium position, as the dependent variable.

The equivalent-systems approach is used in Examples 2.9 and 2.10 to replace the system of Fig. 2.34 by the equivalent system of Fig. 3.12. The governing differential equation is Eq. (3.3) where the equivalent mass, stiffness, and damping coefficients are given in Examples 2.9 and 2.11.

$$\left(\frac{19}{2}m + \frac{5}{3}m_s + \frac{I_p}{r^2}\right)\ddot{x} + 5c\dot{x} + 5kx = 0$$

Example 3.8. An air compressor of mass m is mounted on four identical elastic isolators. When the compressor is placed on the isolators, each has a static deflection δ. Use the equivalent-systems approach to model the free vibrations of the air compressor on the elastic foundation. Assumed damping is negligible.

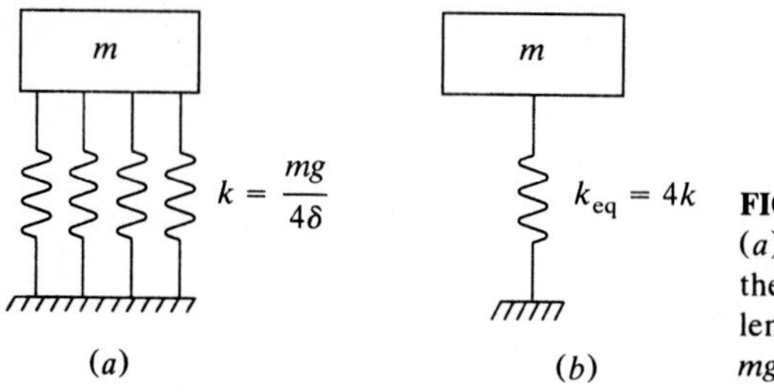

FIGURE 3.14
(*a*) The behavior of each isolator attached to the air compressor of Example 3.8 is equivalent to that of a linear spring of stiffness $mg/4\delta$; (*b*) equivalent systems model.

Assume that the relationship between the force in the isolator and the deflection of the isolator is linear. Then each isolator can be modeled as a linear spring, as shown in Fig. 3.14*a*. A static analysis shows that the equivalent stiffness of each isolator is

$$k = \frac{mg}{4\delta}$$

The four isolators behave as four springs in parallel. The equivalent model of the compressor on the isolators is shown in Fig. 3.14*b*. The governing differential equation is

$$m\ddot{x} + \frac{mg}{\delta}x = 0$$

Example 3.9. Use the principle of conservation of energy to derive the differential equation governing free vibrations of the system of Fig. 3.15. Use θ, the counterclockwise angular rotation of the slender rod from the vertical equilibrium position, as the dependent variable.

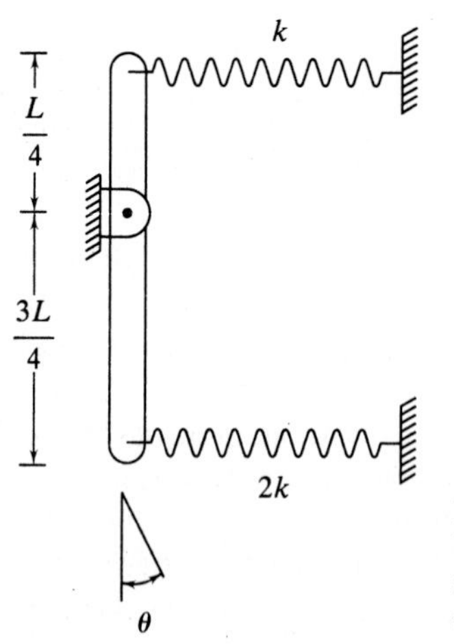

FIGURE 3.15
Using a small-angle assumption, the system of Example 3.9 is modeled by mass-spring system. Conservation of energy can be used to derive governing differential equation.

Since all external forces acting on the rod are conservative, conservation of energy applies

$$T + V = C$$

where C is a constant determined from the initial conditions. At an arbitrary instant of time, the kinetic energy of the bar is

$$T = \frac{1}{2}m\bar{v}^2 + \frac{1}{2}\bar{I}\omega^2 = \frac{1}{2}m\left(\frac{L}{4}\dot{\theta}\right)^2 + \frac{1}{2}\frac{1}{12}mL^2\dot{\theta}^2 = \frac{1}{2}\left(\frac{7}{48}mL^2\right)\dot{\theta}^2$$

The total potential energy is the sum of the potential energy in each of the springs and the potential energy due to gravity. The springs are assumed to be unstretched when the bar is vertical, the spring forces are assumed to remain vertical, and the datum for potential energy calculations is taken as the position of the mass center of the rod when the system is in equilibrium

$$V = \frac{1}{2}k\left(\frac{L}{4}\sin\theta\right)^2 + \frac{1}{2}2k\left(\frac{3L}{4}\sin\theta\right)^2 + mg\frac{L}{4}(1-\cos\theta) = \frac{1}{2}\frac{19}{16}kL^2\sin^2\theta + \frac{mgL}{4}(1-\cos\theta)$$

Conservation of energy thus implies

$$\frac{1}{2}\left(\frac{7}{48}mL^2\right)\dot{\theta}^2 + \frac{1}{2}\left(\frac{19}{16}kL^2\right)\sin^2\theta + mg\frac{L}{4}(1-\cos\theta) = C$$

Differentiating with respect to time, using the chain rule yields

$$\frac{7}{48}mL^2\dot{\theta}\ddot{\theta} + \frac{19}{16}kL^2\sin\theta\cos\theta\dot{\theta} + mg\frac{L}{4}\sin\theta\dot{\theta} = 0$$

Using the small-angle assumption, Eq. (3.2), in the preceding equation gives

$$\frac{7}{48}mL^2\ddot{\theta} + \left(\frac{19}{16}kL^2 + mg\frac{L}{4}\right)\theta = 0$$

3.3 INITIAL CONDITIONS

The differential equation governing the vibrations of a one-degree-of-freedom system is second order. Its homogeneous solution is a linear combination of two linearly independent solutions. The two constants in the linear combination are arbitrary and are called constants of integration.

The constants of integration are uniquely determined by application of conditions satisfied by the system at an initial time. Such conditions are called initial conditions. Unique determination of the constants requires two initial conditions. The form of the solution is independent of the initial conditions. The initial conditions uniquely specify the solution. The mathematical formulation of a problem is not complete until the initial conditions are specified.

The initial conditions necessary to determine the constants of integration are the initial values of the generalized coordinate and its first time derivative. These usually represent the initial displacement and velocity of some particle in the system. If an angular displacement is chosen as the generalized coordinate, the initial values of the angular displacement and angular velocity are required.

The initial conditions for some problems can be specified by observation. Determination of the initial conditions for other problems require a static or dynamic analysis of the initial situation.

Example 3.10. Reconsider the compound pendulum of Example 3.2 and Fig. 3.7. Determine the conditions that the dependent variable $\theta(t)$ satisfies at $t = 0$ if

(a) A clockwise moment, M, is applied statically to the rod and removed, releasing the rod from rest.

(b) The rod is at rest in equilibrium when a particle of mass $m/2$ strikes the rod squarely at its lower end. The coefficient of restitution between the rod and the particle is e.

(a) The initial configuration of the rod is shown in Fig. 3.16a. A free-body diagram of the initial state is shown in Fig. 3.16b. The initial angular displacement is negative because θ is defined in Example 3.2 as positive counterclockwise. Static

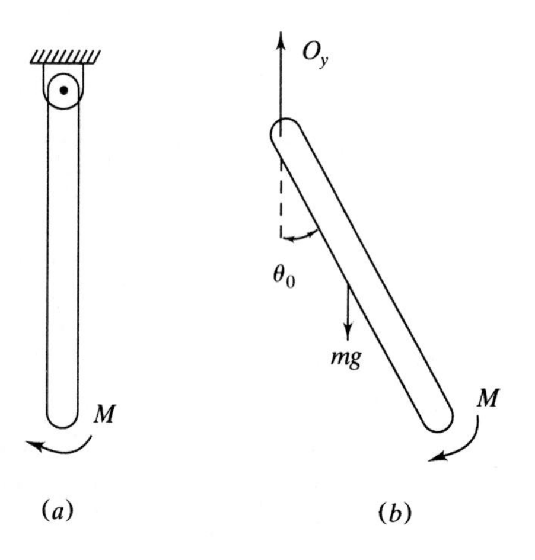

FIGURE 3.16
(a) Moment M is statically applied to compound pendulum and then removed, initiating oscillations about system's equilibrium position. (b) free-body diagram of pendulum before moment is removed. Since θ is measured positive counterclockwise, θ_0 is assumed counterclockwise.

equilibrium requires

$$\sum \overset{\curvearrowleft}{\underset{+}{}} M_O = 0$$

which, when applied to the free-body diagram of Fig. 3.16b, gives

$$mg\frac{L}{2}\sin\theta(0) + M = 0$$

which leads to

$$\theta(0) = -\sin^{-1}\left(\frac{2M}{mgL}\right)$$

If M is small, the small-angle approximation may be applied. Using Eq. (3.2) in the preceding equation yields

$$\theta(0) = -\frac{2M}{mgL}$$

Since the rod is released from rest,

$$\dot{\theta}(0) = 0$$

(*b*) The initial situation is shown in Fig. 3.17*a*. Define $t = 0$ immediately after the particle impacts the rod. Consider the system composed of the rod and the particle. Impulse and momentum diagrams for the system are shown in Fig. 3.17*b*. During impact, angular momentum about O is conserved

$$\left(H_{O_{sys}}\right)_1 = \left(H_{O_{sys}}\right)_2$$

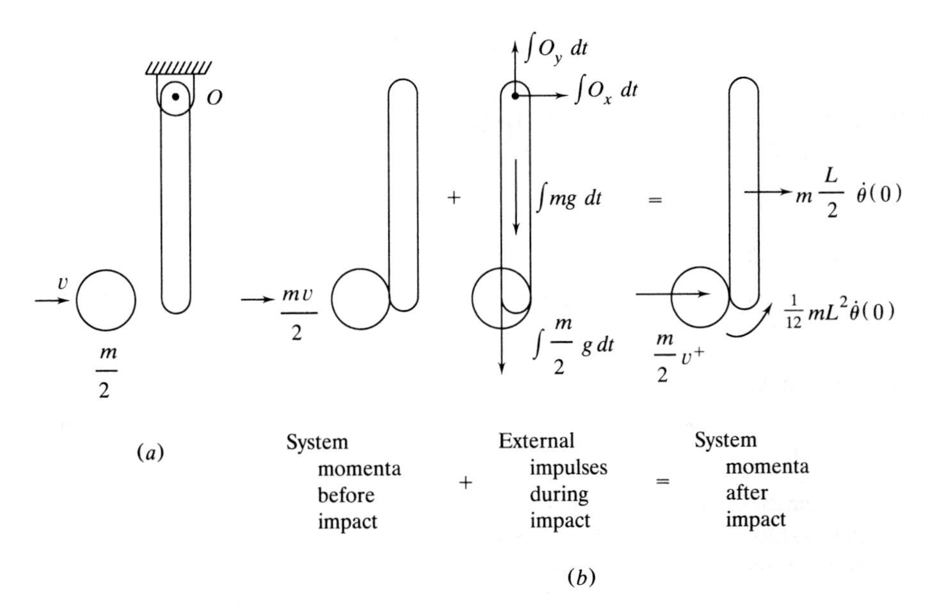

FIGURE 3.17
(*a*) Free oscillations of pendulum are initiated by impact with particle; (*b*) impulse-momentum diagrams used to determine initial velocity.

When applied to the impulse momentum diagrams, this gives

$$\frac{m}{2}vL = \frac{m}{2}v^{+}L + \frac{1}{12}mL^2\dot{\theta}(0) + m\frac{L}{2}\dot{\theta}(0)\frac{L}{2}$$

The definition of the coefficient of restitution gives

$$e = \frac{L\dot{\theta}(0) - v^{+}}{v}$$

Simultaneous solution of the previous two equations gives

$$\dot{\theta}(0) = \frac{3v}{5L}(1 + e)$$

Since the rod is in its equilibrium position before impact,

$$\theta(0) = 0$$

Example 3.11. Figure 3.18a shows a simplified model of a toy vehicle's suspension system. The body is connected to an assumed rigid wheel by an elastic element. The vehicle is traveling at a constant horizontal speed with no vertical motion. The wheel suddenly drops a distance h when it encounters a dip. Horizontal motion is maintained after the dip. The sudden change initiates vertical vibrations of the body. Let $x(t)$ be the displacement of the body, measured from the position it would maintain if the system were in static equilibrium at the lower level. Formulate the mathematical problem whose solution describes the vibrations of the system after the vehicle encounters the dip.

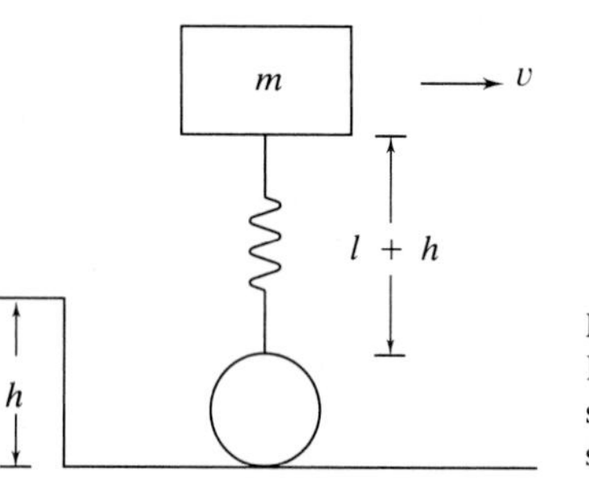

FIGURE 3.18
Immediately after vehicle encounters dip, spring is stretched a distance h from its unstretched length.

Application of Newton's law to appropriate free-body diagrams leads to

$$m\ddot{x} + kx = 0$$

Define $t = 0$ immediately after the dip is encountered. At this time the wheel is in contact with the lower surface, while the mass is still at the upper level. Thus the mass is a distance h above the equilibrium position it would attain on the lower surface. Since x is measured positive downward,

$$x(0) = -h$$

The vertical velocity of the mass is zero when the dip is encountered; thus

$$\dot{x}(0) = 0$$

3.4 FREE VIBRATIONS OF UNDAMPED ONE-DEGREE-OF-FREEDOM SYSTEMS

The general form of the differential equation for undamped free vibrations of a one-degree-of-freedom system is

$$\tilde{m}\ddot{x} + \tilde{k}x = 0 \tag{3.5}$$

where $\tilde{m}$ and $\tilde{k}$ are coefficients specific to the system determined during the derivation of the differential equation. Equation (3.5) is subject to initial conditions of the form

$$x(0) = x_0 \tag{3.6a}$$

and

$$\dot{x}(0) = \dot{x}_0 \tag{3.6b}$$

The solution of Eq. (3.5) subject to Eq. (3.6) is

$$x(t) = x_0 \cos \omega_n t + \frac{\dot{x}_0}{\omega_n} \sin \omega_n t \tag{3.7}$$

where

$$\omega_n = \sqrt{\frac{\tilde{k}}{\tilde{m}}} \tag{3.8}$$

An alternate and more instructive form of Eq. (3.7) is

$$x(t) = A \sin(\omega_n t + \phi) \tag{3.9}$$

Expanding Eq. (3.9) using the trigonometric identity for the sine of the sum of angles

$$\sin(a + b) = \sin a \cos b + \cos a \sin b \tag{3.10}$$

gives

$$x(t) = A \cos \phi \sin \omega_n t + A \sin \phi \cos \omega_n t \tag{3.11}$$

Equating coefficients of like trigonometric terms of Eqs. (3.7) and (3.11) leads to

$$A = \sqrt{x_0^2 + \left(\frac{\dot{x}_0}{\omega_n}\right)^2} \tag{3.12}$$

and

$$\phi = \tan^{-1}\left(\frac{\omega_n x_0}{\dot{x}_0}\right) \tag{3.13}$$

The free-vibration response of a one-degree-of-freedom system, described mathematically by Eq. (3.9), is plotted in Fig. 3.19. The initial conditions determine the energy initially present in the system. Potential energy is continually converted to kinetic energy and vice versa. Since energy is conserved, the system eventually returns to its initial state with its original kinetic and potential energies, completing the first cycle of motion. The subsequent motion duplicates the previous motion. The system takes the same amount of time to execute its second cycle as it does its first. Since no energy is dissipated from the system, the system executes cycles of motion indefinitely.

A motion which exactly repeats after some time is said to be periodic. The period is the amount of time it takes the system to execute one cycle. The frequency is the number of cycles the system executes in a period of time and is the reciprocal of the period.

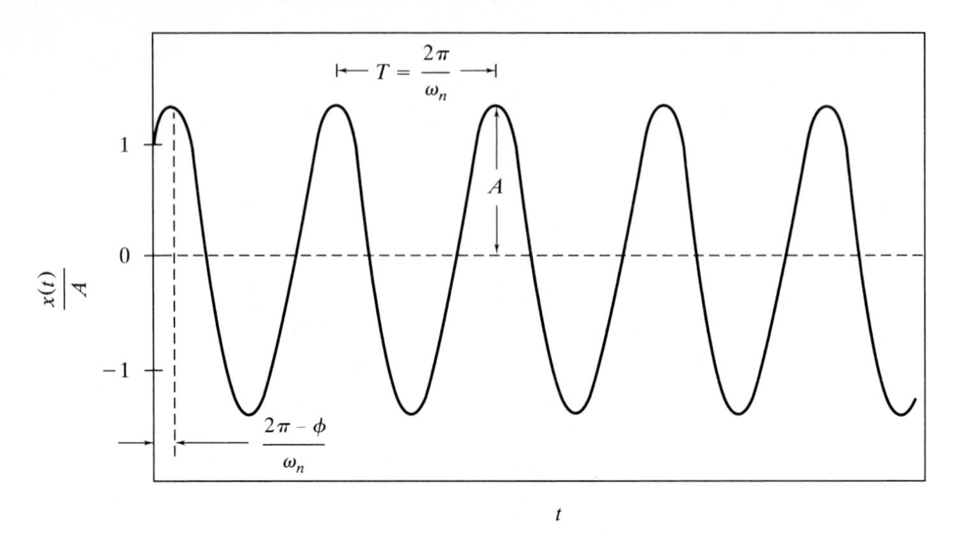

FIGURE 3.19
Free-vibration response of an undamped one-degree-of-freedom system is periodic of period $2\pi/\omega_n$. The phase angle depends upon initial conditions.

Figure 3.19 shows that the free undamped vibrations of a one-degree-of-freedom system are periodic of period

$$T = \frac{2\pi}{\omega_n} \tag{3.14}$$

and frequency

$$f = \frac{\omega_n}{2\pi} \tag{3.15}$$

The frequency of Eq. (3.15) is usually calculated in units of cycles per second or hertz (Hz). The frequency is converted into radians per second by noting

$$1 \text{ cycle/s} = 2\pi \text{ rad/s}$$

Thus ω_n is the frequency in radians per second. The parameter ω_n is called the natural frequency of free vibration. It is a property of the system and is determined directly from the governing differential equation and Eq. (3.8). The natural frequency is a function of system parameters and independent of initial conditions.

Equation (3.5) is divided by $\tilde{m}$ yielding

$$\ddot{x} + \omega_n^2 x = 0 \tag{3.16}$$

Equation (3.16) is the standard form of the differential equation for free vibrations of an undamped one-degree-of-freedom system. The derived differential equation for any system can be put in this standard form. The natural frequency is determined directly from the equation.

The amplitude A, defined by Eq. (3.12), is the maximum displacement from equilibrium. The amplitude is a function of the system parameters and the initial conditions. The amplitude is a measure of the energy imparted to the system through the initial conditions. For a linear system

$$A = \sqrt{\frac{2E}{\tilde{k}}} \tag{3.17}$$

where E is the sum of kinetic and potential energies.

The phase angle, ϕ, calculated from Eq. (3.13), is an indication of the lead or lag between the response and a pure sinusoidal response. The response is purely sinusoidal with $\phi = 0$ if $x_0 = 0$. The response leads a pure sinusoidal response by $\pi/2$ rad if $\dot{x}_0 = 0$. The system takes a time of

$$t = \begin{cases} \dfrac{2\pi - \phi}{\omega_n} & \phi > 0 \\ -\dfrac{\phi}{\omega_n} & \phi \leq 0 \end{cases}$$

to reach its equilibrium position from its initial position.

Example 3.12. An engine of mass 500 kg is mounted on an elastic foundation of equivalent stiffness 7×10^5 N/m. Determine the natural frequency of the system.

The system is modeled as a hanging mass-spring system. Equation (3.5) governs the displacement of the engine from its static-equilibrium position. The natural frequency is determined using Eq. (3.8)

$$\omega_n = \sqrt{\frac{7 \times 10^5 \text{ N/m}}{500 \text{ kg}}} = 37.42 \frac{\text{rad}}{\text{s}} = 5.96 \text{ Hz}$$

Example 3.13. A wheel is mounted on a steel shaft ($G = 83 \times 10^9$ N/m^2) of length 1.5 m and radius 0.80 cm. The wheel is rotated 5° and released. The period of oscillation is observed as 2.3 s. Determine the mass moment of inertia of the wheel.

The oscillations of the wheel about its equilibrium position are modeled as the torsional oscillations of a disk on a massless shaft, as in Example 3.3. The differential equation derived in Example 3.3 is put into standard form, the form of Eq. (3.16), by dividing by I, yielding

$$\ddot{\theta} + \frac{JG}{IL}\theta = 0$$

Comparison of the preceding equation with Eq. (3.16) gives

$$\omega_n = \sqrt{\frac{JG}{IL}}$$

The observed natural frequency is

$$\omega_n = \frac{2\pi}{T} = \frac{2\pi \text{ rad/cycle}}{2.3 \text{ s/cycle}} = 2.73 \frac{\text{rad}}{\text{s}}$$

Thus the moment of inertia of the wheel is calculated from

$$I = \frac{JG}{L\omega_n^2} = \frac{\frac{\pi}{2}(0.008 \text{ m})^4\left(83 \times 10^9 \frac{\text{N}}{\text{m}^2}\right)}{(1.5 \text{ m})\left(2.73 \frac{\text{rad}}{\text{s}}\right)^2} = 47.8 \text{ kg} \cdot \text{m}^2$$

Example 3.14. A mass of 45 kg is dropped onto the end of a cantilever beam with a velocity of 0.5 m/s, as shown in Fig. 3.20. The mass sticks to the beam and vibrates with the beam. The I beam is made of steel ($E = 210 \times 10^9$ N/m^2), is 3.5 m long, has a depth of 20 cm, and a cross-sectional moment of inertia of 3×10^{-5} m^4. Determine the maximum stress developed in the beam as it vibrates. Neglect the inertia of the beam.

Let $x(t)$ be the displacement of the mass, measured positive downward from the static-equilibrium position of the mass after it is attached to the beam. The

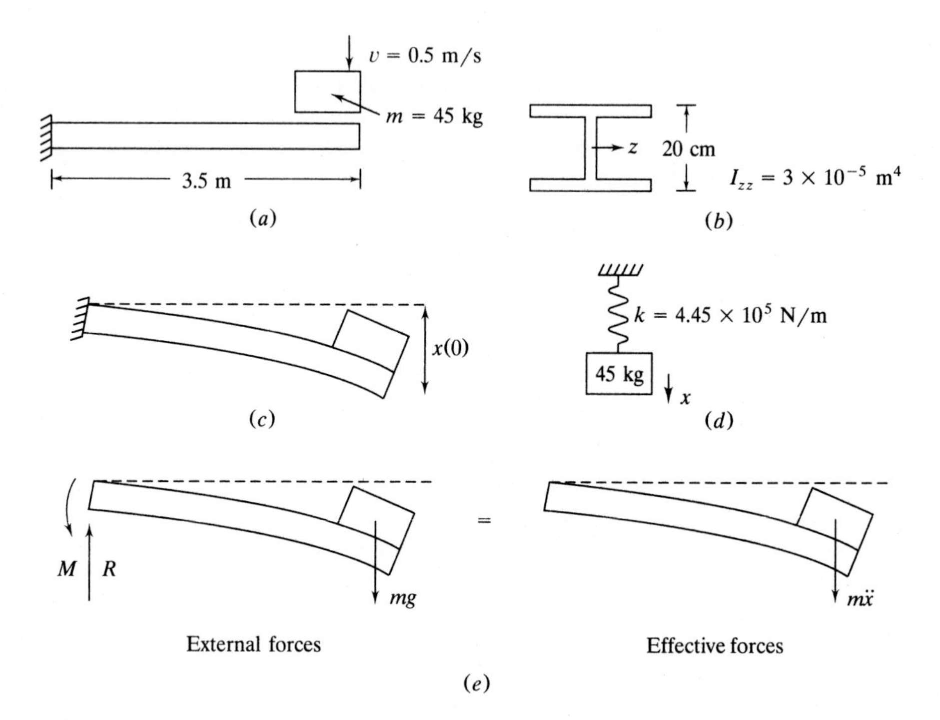

FIGURE 3.20
(*a*) System of Example 3.14; (*b*) *I* cross section of beam; (*c*) $x(t)$ is measured from static-equilibrium position when mass is attached to beam. Thus $x(0) = -\Delta_{st}$; (*d*) equivalent mass-spring model; (*e*) free-body diagrams at an arbitrary instant of time.

system is modeled as a mass of 45 kg hanging from a spring of stiffness

$$k_{eq} = \frac{3EI}{L^3} = \frac{3(210 \times 10^9 \text{ N/m}^2)(3 \times 10^{-5} \text{ m}^4)}{(3.5 \text{ m})^3} = 4.4 \times 10^5 \frac{\text{N}}{\text{m}}$$

The natural frequency of free vibrations is

$$\omega_n = \sqrt{\frac{k_{eq}}{m}} = \sqrt{\frac{4.4 \times 10^5 \text{ N/m}}{45 \text{ kg}}} = 99.0 \frac{\text{rad}}{\text{s}}$$

The beam is undeflected at $t = 0$, when the mass strikes. If the system were in static equilibrium, the end of the beam would have a static deflection. Thus

$$x(0) = -\Delta_{st} = -\frac{mg}{k_{eq}} = -\frac{g}{\omega_n^2} = -\frac{9.81 \text{ m/s}^2}{(99.0 \text{ rad/s})^2} = -1.00 \times 10^{-3} \text{ m}$$

The initial velocity is the velocity at which the mass strikes the beam

$$\dot{x}(0) = 0.5 \text{ m/s}$$

The time history of the mass's displacement is calculated using Eq. (3.9)

$$x(t) = A \sin(99.0t + \phi)$$

where the amplitude A and the phase angle ϕ are calculated from Eqs. (3.12) and (3.13) as

$$A = \sqrt{(-1.00 \times 10^{-3} \text{ m})^2 + \left(\frac{0.5 \text{ m/s}}{99.0 \text{ rad/s}}\right)^2} = 5.15 \times 10^{-3} \text{ m}$$

$$\phi = \tan^{-1}\left(\frac{(99.0 \text{ rad/s})(-1 \times 10^{-3} \text{ m})}{0.5 \text{ m/s}}\right) = -0.195 \text{ rad} = -11.2°$$

The maximum normal stress due to bending occurs in the outer fibers of the beam at its fixed end. The reaction moment at the support is obtained by summing moments on the free-body diagrams of Fig. 3.20*e*.

$$\left(\sum \overset{\curvearrowleft +}{M_A}\right)_{ext} = \left(\sum \overset{\curvearrowleft +}{M_A}\right)_{eff}$$

$$-M + mgL = m\ddot{x}L$$

The maximum acceleration is calculated by

$$\ddot{x}_{max} = \omega_n^2 A = (99.0 \text{ rad/s})^2 (5.15 \times 10^{-3} \text{ m}) = 50.5 \text{ m/s}^2$$

The maximum absolute value of the reaction moment is

$$|M| = mgL + m\ddot{x}_{max}L = 45 \text{ kg}(9.81 \text{ m/s}^2 + 50.5 \text{ m/s}^2)(3.5 \text{ m})$$
$$= 9.5 \times 10^3 \text{ N} \cdot \text{m}$$

The maximum normal stress is calculated using the elastic flexure formula

$$\sigma_{max} = \frac{Mc}{I} = \frac{(9.5 \times 10^3 \text{ N} \cdot \text{m})(0.1 \text{ m})}{3 \times 10^{-5} \text{ m}^4} = 3.2 \times 10^7 \frac{\text{N}}{\text{m}^2}$$

Example 3.15. The pendulum of a cuckoo clock consists of a slender rod on which an aesthetically designed mass slides. If the clock gains time, should the mass be moved closer to or farther away from the support to correct the tuning?

The pendulum is modeled as a particle of mass m on a rigid, massless rod. The particle is assumed to be a distance l from its axis of rotation. Summing moments about the point of support on the free-body diagrams of Fig. 3.21 and

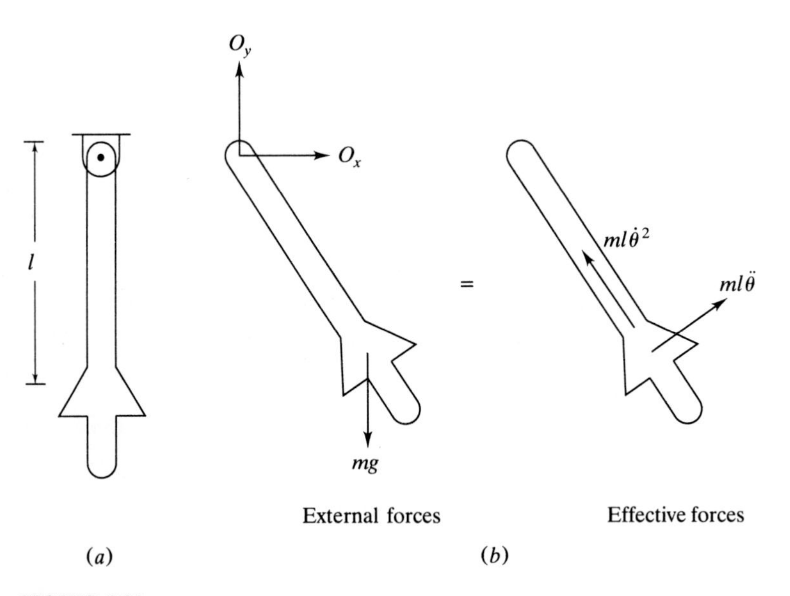

FIGURE 3.21
(*a*) Pendulum of cuckoo clock; (*b*) free-body diagrams at an arbitrary instant of time.

using the small-angle assumption yields the differential equation of motion

$$\ddot{\theta} + \frac{g}{l}\theta = 0$$

The differential equation is compared with the standard form, Eq. (3.16), yielding

$$\omega_n = \sqrt{\frac{g}{l}}$$

The period of oscillation is

$$T = 2\pi\sqrt{\frac{l}{g}}$$

Since the clock is running fast, the period of the pendulum needs to be lengthened. Thus l should be increased and the mass moved farther away from the axis of rotation.

The validity of the small-angle approximation can now be checked. The differential equation of motion of the pendulum of Example 3.15 is truly nonlinear. The equation is linearized for small angular oscillations using Eq. (3.2).

The differential equation governing the oscillations of the pendulum,

$$\ddot{\theta} + \omega_n^2 \sin\theta = 0 \tag{3.18}$$

is one of the few nonlinear equations for which an exact solution is known. The solution of Eq. (3.18) subject to

$$\theta(0) = \theta_0$$

and

$$\dot{\theta}(0) = 0$$

can be developed in terms of elliptic integrals, which are well-known tabulated functions.

TABLE 3.1
Ratio of period of simple pendulum calculated from exact nonlinear solution to period calculated from linearized equation as a function of initial angle, θ_0. Nonlinear period is $4K$ where K is the complete elliptic integral of the first kind with a parameter of $\sin(\theta_0/2)$

θ_0 (°)	$\frac{T}{2\pi}\sqrt{g/l}$	θ_0 (°)	$\frac{T}{2\pi}\sqrt{g/l}$
2	1.00007	48	1.04571
4	1.00032	50	1.04978
6	1.00070	52	1.05405
8	1.00120	54	1.05851
10	1.00191	56	1.06328
12	1.00274	58	1.06806
14	1.00376	60	1.07321
16	1.00490	62	1.07850
18	1.00618	64	1.08404
20	1.00764	66	1.08982
22	1.00930	68	1.09588
24	1.01108	70	1.10211
26	1.01305	72	1.10867
28	1.01515	74	1.11548
30	1.01738	76	1.12255
32	1.01987	78	1.12987
34	1.02248	80	1.13751
36	1.02528	82	1.14540
38	1.02821	84	1.15368
40	1.03132	86	1.16221
42	1.03463	88	1.17112
44	1.03814	90	1.18035
46	1.04183		

The period of motion of a nonlinear system is dependent upon the initial conditions, while the period of a linear system is independent of initial conditions. One method of assessing the validity of the small-angle approximation for a given amplitude is to compare the period calculated using the exact solution to the period calculated using the linearized differential equations for different initial displacements. This comparison is given in Table 3.1, which shows that the small-angle approximation leads to accurate prediction of the period for amplitudes as large as 40°. For an initial angular displacement of 40°, the error in the period from using the small-angle approximation is only 3.1%.

The success of the use of the small-angle approximation in the pendulum example should give confidence to its use in other problems, where an exact solution is not available. The alternative to its use, for most problems, is to find a numerical solution to the exact nonlinear equation.

3.5 FREE VIBRATIONS OF ONE-DEGREE-OF-FREEDOM SYSTEMS WITH VISCOUS DAMPING

The general form of the differential equation for the displacement of a particle in a one-degree-of-freedom linear system where viscous damping is present is

$$\tilde{m}\ddot{x} + \tilde{c}\dot{x} + \tilde{k}x = 0 \tag{3.19}$$

where the coefficients are determined during the derivation of the differential equation. Dividing Eq. (3.19) by $\tilde{m}$ gives

$$\ddot{x} + \frac{\tilde{c}}{\tilde{m}}\dot{x} + \frac{\tilde{k}}{\tilde{m}}x = 0 \tag{3.20}$$

The general solution of Eq. (3.20) is obtained by assuming

$$x(t) = Be^{\alpha t} \tag{3.21}$$

Substitution of Eq. (3.21) into Eq. (3.20) leads to the following quadratic equation for α:

$$\alpha^2 + \frac{\tilde{c}}{\tilde{m}}\alpha + \frac{\tilde{k}}{\tilde{m}} = 0 \tag{3.22}$$

The quadratic formula is used to obtain the roots of Eq. (3.22) as

$$\alpha_{1,2} = -\frac{\tilde{c}}{2\tilde{m}} \pm \sqrt{\left(\frac{\tilde{c}}{2\tilde{m}}\right)^2 - \frac{\tilde{k}}{\tilde{m}}} \tag{3.23}$$

The mathematical form of the solution of Eq. (3.20) and the physical behavior of the system depends on the sign of the discriminant of Eq. (3.23). If the discriminant is positive, Eq. (3.22) has two real roots. If the discriminant is negative, Eq. (3.22) has two complex conjugate roots. If the discriminant is zero, Eq. (3.22) has two equal real roots.

The physical nature of the vibrations is dependent on the sign of the discriminant. The case when the discriminant is zero is a special case and occurs only for a certain combination of parameters. When this occurs the system is said to be critically damped. For fixed values of $\tilde{k}$ and $\tilde{m}$, the value of $\tilde{c}$ which causes critical damping is called the critical damping coefficient, $\tilde{c}_c$. From Eq. (3.23)

$$\tilde{c}_c = 2\sqrt{\tilde{k}\tilde{m}} \tag{3.24}$$

The nondimensional damping ratio, ζ, is defined as the ratio of the actual value of $\tilde{c}$, to the critical damping coefficient,

$$\zeta = \frac{\tilde{c}}{\tilde{c}_c} = \frac{\tilde{c}}{2\sqrt{\tilde{k}\tilde{m}}} \tag{3.25}$$

The damping ratio is a property of the system parameters.

Using Eqs. (3.25) and (3.8), Eq. (3.23) is rewritten in terms of ζ and ω_n as

$$\alpha_{1,2} = -\zeta\omega_n \pm \omega_n\sqrt{\zeta^2 - 1} \tag{3.26}$$

For $\zeta \neq 1$, the general solution of Eq. (3.20) is

$$x(t) = e^{-\zeta\omega_n t}\left(C_1 e^{\omega_n\sqrt{\zeta^2-1}\,t} + C_2 e^{-\omega_n\sqrt{\zeta^2-1}\,t}\right) \tag{3.27}$$

where C_1 and C_2 are arbitrary constants of integration. From Eq. (3.27) it is evident that the nature of the motion depends upon the value of ζ. Using Eqs. (3.8) and (3.25), Eq. (3.20) becomes

$$\ddot{x} + 2\zeta\omega_n\dot{x} + \omega_n^2 x = 0 \tag{3.28}$$

Equation (3.28) is the standard form of the differential equation governing the free vibrations of a one-degree-of-freedom system with viscous damping.

Three cases must be examined to explore completely the behavior of a one-degree-of-freedom system with viscous damping.

Case 1: $\zeta < 1$ (underdamped free vibrations). For $\zeta < 1$ the roots of Eq. (3.22) exist as a complex conjugate pair

$$\alpha_{1,2} = \omega_n\left(-\zeta \pm i\sqrt{1-\zeta^2}\right) \tag{3.29}$$

Euler's identity is used to replace the complex exponentials that occur in Eq. (3.27) by a linear combination of trigonometric terms of real argument,

$$x(t) = e^{-\zeta\omega_n t}\left(c_1 \cos \omega_n\sqrt{1-\zeta^2}\,t + C_2 \sin \omega_n\sqrt{1-\zeta^2}\,t\right) \tag{3.30}$$

The constants of integration are solved by applying the initial conditions, Eq. (3.6), resulting in

$$x(t) = e^{-\zeta\omega_n t}\left(x_0 \cos \omega_n\sqrt{1-\zeta^2}\,t + \frac{\dot{x}_0 + \zeta\omega_n x_0}{\omega_n\sqrt{1-\zeta^2}} \sin \omega_n\sqrt{1-\zeta^2}\,t\right) \tag{3.31}$$

An alternate form of the solution is developed using the trigonometric identity, Eq. (3.10),

$$x(t) = Ae^{-\zeta\omega_n t}\sin(\omega_d t + \phi_d) \tag{3.32}$$

where

$$A = \sqrt{x_0^2 + \left(\frac{\dot{x}_0 + \zeta\omega_n x_0}{\omega_d}\right)^2} \tag{3.33}$$

$$\phi_d = \tan^{-1}\left(\frac{x_0\omega_d}{\dot{x}_0 + \zeta\omega_n x_0}\right) \tag{3.34}$$

and

$$\omega_d = \omega_n\sqrt{1 - \zeta^2} \tag{3.35}$$

Equation (3.32) is plotted in Fig. 3.22. Once free oscillations of a viscously damped system commence, the nonconservative viscous damping force continually dissipates energy. Since no work is being done on the system, this leads to a continual decrease in the sum of the potential and kinetic energies. For underdamped free vibrations, the system still oscillates about an equilibrium position. However, each time it reaches equilibrium, the system's total energy level is less than the previous time. The maximum deflection on each cycle of motion is continually decreasing. Equation (3.32) and Fig. 3.22 show that the amplitude decreases exponentially with time.

The free vibrations of an underdamped system are oscillatory, but not periodic. The vibrations would be periodic if it were not for the decay in amplitude. Even though the amplitude decreases between cycles, the system

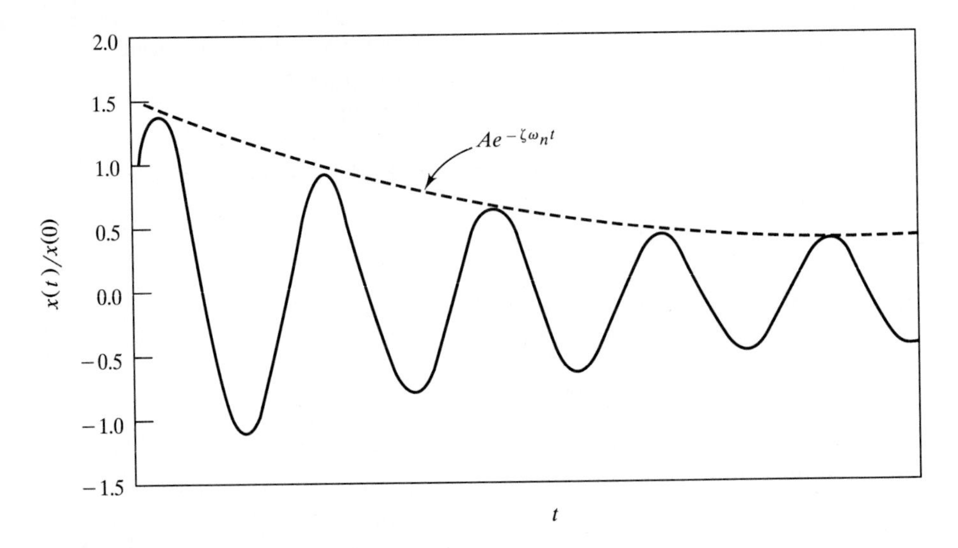

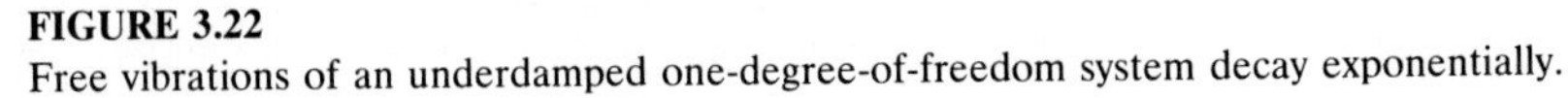

FIGURE 3.22
Free vibrations of an underdamped one-degree-of-freedom system decay exponentially.

takes the same amount of time to execute each cycle. This time is called the period of free underdamped vibrations or the damped period and is given by

$$T_d = \frac{2\pi}{\omega_d} \tag{3.36}$$

Thus ω_d is called the damped natural frequency. Note that $\omega_d < \omega_n$.

Consider a mass-spring-dashpot system with a nonzero initial displacement and a zero initial velocity. Then $\phi = 0$ and $A = x_0/\sqrt{1-\zeta^2}$. Using Eq. (3.32), the sum of the kinetic and potential energies at any instant of time for an underdamped system is

$$\begin{aligned} E &= \frac{1}{2}kx^2 + \frac{1}{2}m\dot{x}^2 \\ &= \frac{1}{2}\frac{kx_0^2 e^{-2\zeta\omega_n t}}{\sqrt{1-\zeta^2}}\left[(1+\zeta^2)\sin^2\omega_d t - 2\frac{\zeta}{\sqrt{1-\zeta^2}}\sin\omega_d t\cos\omega_d t \right. \\ &\qquad \left. + \sqrt{1-\zeta^2}\cos^2\omega_d t\right] \end{aligned} \tag{3.37}$$

The total energy at the end of the first period, $t = 2\pi/\omega_d$, is

$$E\left(t = \frac{2\pi}{\omega_d}\right) = \frac{1}{2}kx_0^2 e^{-4\pi\zeta/\sqrt{1-\zeta^2}} \tag{3.38}$$

The energy dissipated over one cycle is given by

$$\begin{aligned} \Delta E_n &= E\left(\frac{2n\pi}{\omega_d}\right) - E\left(\frac{2(n+1)\pi}{\omega_d}\right) \\ &= \frac{1}{2}x_0^2 e^{-4n\pi\zeta/\sqrt{1-\zeta^2}}\left(1 - e^{-4\pi\zeta/\sqrt{1-\zeta^2}}\right) \end{aligned} \tag{3.39}$$

The ratio of the energy dissipated between successive cycles is

$$\frac{\Delta E_n}{\Delta E_{n+1}} = e^{4\pi\zeta/\sqrt{1-\zeta^2}} \tag{3.40}$$

Equations (3.39) and (3.40) show that the energy dissipated over one cycle of motion is a fraction of the total energy at the beginning of the cycle. The ratio of energy dissipated over a cycle is constant and depends only upon the damping ratio. The larger the damping ratio, the larger the fraction of energy dissipated over a single cycle. Since the energy dissipated over a given cycle is a fixed fraction of the energy at the beginning of the cycle, the total remaining energy is never completely dissipated. This indicates that free vibrations for an undamped system continue indefinitely, with exponential decaying amplitude.

If the system is underdamped, the damping is insufficient to dissipate the initial energy. However, Eq. (3.38) shows that if $\zeta = 1$, all energy is dissipated within the first cycle. This implies that oscillatory motion is not possible for $\zeta \geq 1$.

The logarithmic decrement, δ, is defined for underdamped free vibrations as the natural logarithm of the ratio of the amplitudes of vibration on successive cycles,

$$\begin{aligned}\delta &= \ln\left(\frac{x(t)}{x(t+T_d)}\right) \\ &= \ln\left(\frac{Ae^{-\zeta\omega_n t}\sin(\omega_d t + \phi_d)}{Ae^{-\zeta\omega_n(t+T_d)}\sin[\omega_d(t+T_d)+\phi_d]}\right) \\ &= \zeta\omega_n T_d \\ &= \frac{2\pi\zeta}{\sqrt{1-\zeta^2}} \end{aligned} \tag{3.41}$$

For small ζ,

$$\delta = 2\pi\zeta \tag{3.42}$$

The logarithmic decrement is easily measured by experiment and then used to determine the damping ratio

$$\zeta = \frac{\delta}{\sqrt{4\pi^2 + \delta^2}} \tag{3.43}$$

It can be shown that the following are equivalent to the logarithmic decrement equations:

$$\delta = \frac{1}{n}\ln\left(\frac{x(t)}{x(t+nT_d)}\right) \tag{3.44}$$

for any integer n and

$$\delta = \ln\left(\frac{\dot{x}(t)}{\dot{x}(t+T_d)}\right) \tag{3.45}$$

$$\delta = \ln\left(\frac{\ddot{x}(t)}{\ddot{x}(t+T_d)}\right) \tag{3.46}$$

Equation (3.44) implies that the logarithmic decrement can be determined from amplitudes measured on nonsuccessive cycles, while Eqs. (3.45) and (3.46) imply that velocity and acceleration data can also be used to determine the logarithmic decrement.

Example 3.16. The slender rod of Example 3.4 has a mass of 31 kg and a length of 2.6 m. A 50-N force is statically applied to the bar at P and then removed. The ensuing oscillations of P are monitored and an oscilloscope provides the acceleration data shown in Fig. 3.23b where the time scale is calibrated but the acceleration scale is not calibrated. Use the data to find the spring constant, k, and the damping coefficient, c. Also calibrate the acceleration scale.

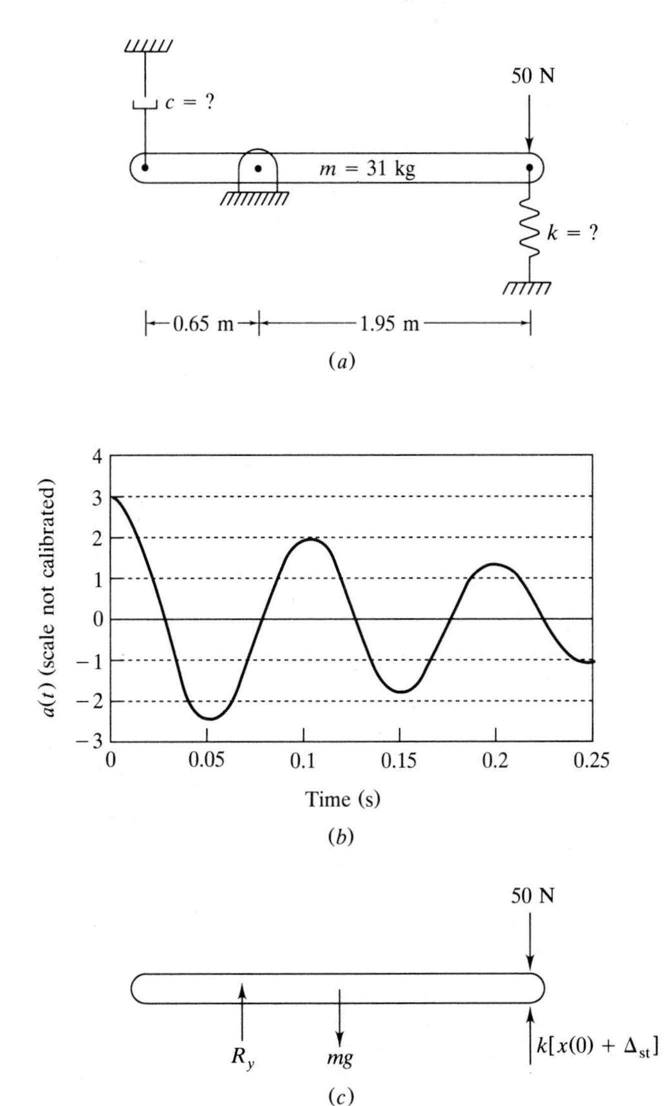

FIGURE 3.23
(*a*) Initial position of bar of Example 3.16; (*b*) oscilloscope data for Example 3.16; (*c*) free-body diagram of initial position.

The differential equation of Example 3.4 is divided by the coefficient of its highest derivative

$$\ddot{x} + \frac{3c}{7m}\dot{x} + \frac{27k}{7m}x = 0$$

The natural frequency and damping ratio are determined by comparing the preceding equation with the standard form of the differential equation for damped free vibrations, Eq. (3.28),

$$\omega_n = \sqrt{\frac{27k}{7m}}$$

$$\zeta = \frac{3c}{14m\omega_n}$$

The period of damped free vibrations is determined from the oscilloscope data as 0.1 s. The value of the logarithmic decrement is determined from the oscilloscope data and Eq. (3.46)

$$\delta = \ln\left(\frac{\ddot{x}(0)}{\ddot{x}(0.1\text{ s})}\right) = \ln\frac{3}{2} = 0.405$$

$$\zeta = \frac{0.405}{\sqrt{4\pi^2 + (0.405)^2}} = 0.0643$$

The damping ratio is calculated using Eq. (3.43).

Equation (3.36) is used to calculate the damped natural frequency as 62.83 rad/s. Equation (3.35) is used to calculate the natural frequency as 62.96 rad/s.

The spring stiffness and damping coefficient are determined to be

$$k = \frac{7m\omega_n^2}{27} = \frac{7(31\text{ kg})(62.96\text{ rad/s})^2}{27} = 3.19 \times 10^4\ \frac{\text{N}}{\text{m}}$$

$$c = \frac{14m\omega_n\zeta}{3} = \frac{14(31\text{ kg})(62.96\text{ rad/s})(0.0643)}{3} = 585.6\ \frac{\text{N}\cdot\text{s}}{\text{m}}$$

A static analysis of the equilibrium position provides the initial displacement of P as

$$x(0) = \frac{F}{k} = \frac{50\text{ N}}{3.19 \times 10^4\text{ N/m}} = 1.6\text{ mm}$$

The initial acceleration is calculated using the governing differential equation

$$\ddot{x}(0) = -\frac{3c}{7m}\dot{x}(0) - \frac{27k}{7m}x(0)$$

$$= -\frac{27(3.19 \times 10^4\text{ N/m})}{7(31\text{ kg})}(0.0016\text{ m})$$

$$= -6.35\ \frac{\text{m}}{\text{s}^2}$$

The acceleration scale is then calibrated.

$$1\text{ unit} = \frac{6.35\text{ m/s}^2}{3} = 2.12\ \frac{\text{m}}{\text{s}^2}$$

Case 2: $\zeta = 1$ (critically damped vibrations). For $\zeta = 1$, Eq. (3.22) has two equal roots

$$\alpha_1 = \alpha_2 = -\omega_n \tag{3.47}$$

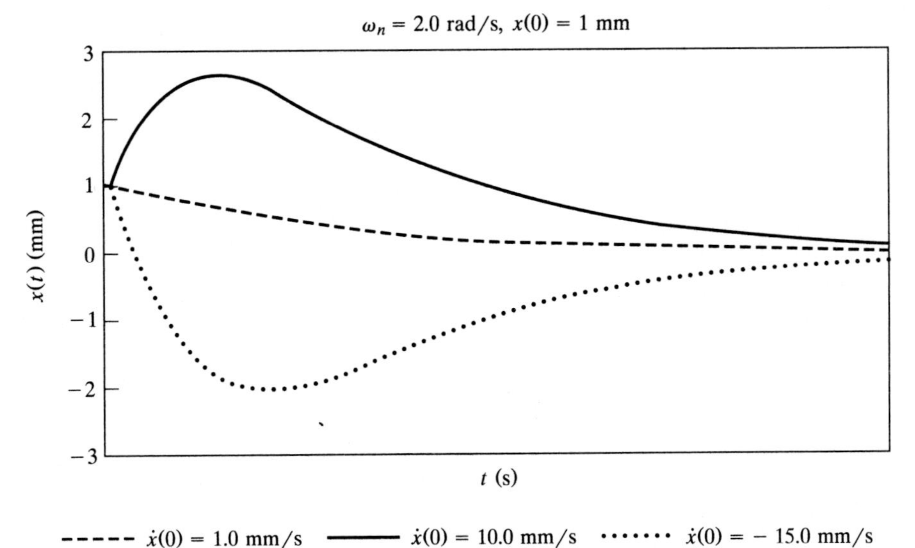

FIGURE 3.24
The free-vibration response of a critically damped one-degree-of-freedom system is aperiodic and rapidly decays. The system may pass through its equilibrium position if initial conditions are of opposite sign.

The general solution of Eq. (3.20) is

$$x(t) = e^{-\omega_n t}(C_1 + C_2 t)$$

Application of the initial conditions, Eq. (3.6), leads to

$$x(t) = e^{-\omega_n t}\left[x_0 + (\dot{x}_0 + \omega_n x_0)t\right] \tag{3.48}$$

The response of a one-degree-of-freedom system subject to critical viscous damping is plotted in Fig. 3.24 for different initial conditions. If the initial conditions are of opposite sign or if $\dot{x}_0 = 0$, the motion decays immediately. If both initial conditions have the same sign or if $x_0 = 0$, the absolute value of x initially increases and reaches a maximum value of

$$x_{\max} = e^{-\dot{x}_0/(\dot{x}_0 + \omega_n x_0)}\left(x_0 + \frac{\dot{x}_0}{\omega_n}\right) \tag{3.49}$$

at

$$t = \frac{\dot{x}_0}{\omega_n(\dot{x}_0 + \omega_n x_0)} \tag{3.50}$$

If the signs of the initial conditions are opposite and

$$\frac{x_0}{\dot{x}_0 + \omega_n x_0} < 0$$

then the response overshoots the equilibrium position before eventually decaying.

Equation (3.38) shows that a damping force that leads to critical damping is sufficient to dissipate all of a system's initial energy before one cycle of motion is complete. A critically damped system can thus pass through equilibrium at most once before the motion decays. The total energy decays exponentially, but never reaches zero. Thus critically damped motion is predicted to continue indefinitely.

Case 3: $\zeta > 1$ (overdamped free vibrations). For $\zeta > 1$, Eq. (3.22) has two real, distinct roots

$$\alpha_{1,2} = \omega_n\left(-\zeta \pm \sqrt{\zeta^2 - 1}\right) \tag{3.51}$$

The general solution of Eq. (3.22) is given by Eq. (3.29). Application of Eq. (3.8) leads to

$$x(t) = \frac{e^{-\zeta\omega_n t}}{2\sqrt{\zeta^2 - 1}}\left\{\left[\frac{\dot{x}_0}{\omega_n} + x_0\left(\zeta + \sqrt{\zeta^2 - 1}\right)\right]e^{\omega_n\sqrt{\zeta^2-1}\,t} + \left[-\frac{\dot{x}_0}{\omega_n} + x_0\left(-\zeta + \sqrt{\zeta^2 - 1}\right)\right]e^{-\omega_n\sqrt{\zeta^2-1}\,t}\right\} \tag{3.52}$$

Equation (3.52) is plotted in Fig. 3.25. The response of an overdamped one-degree-of-freedom system is not periodic. It attains its maximum either at

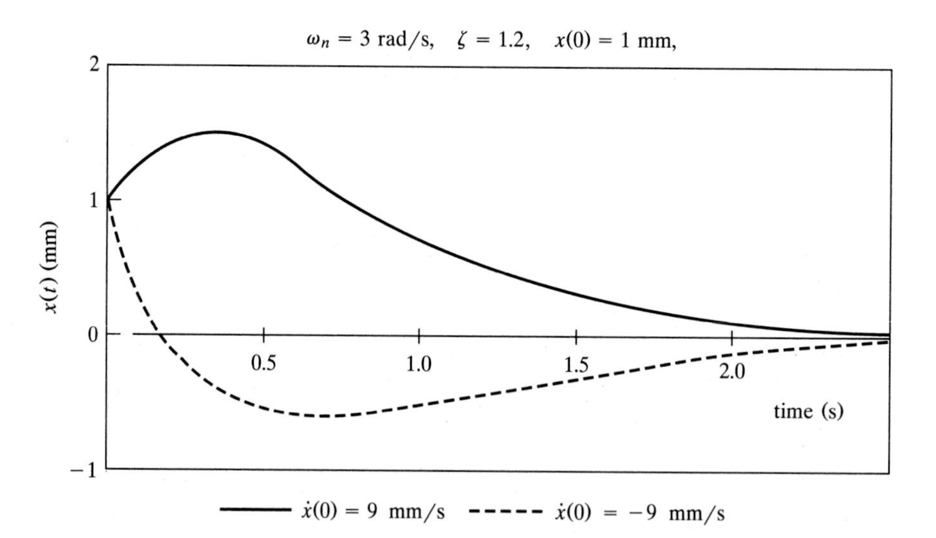

FIGURE 3.25
The free-vibration response of an overdamped one-degree-of-freedom system is aperiodic. The response quickly decays after reaching a maximum.

$t = 0$ or at

$$t = -\frac{1}{2\omega_n\sqrt{\zeta^2 - 1}} \ln\left[\frac{\zeta - \sqrt{\zeta^2 - 1}}{\zeta + \sqrt{\zeta^2 - 1}} \frac{\dfrac{\dot{x}_0}{\omega_n} + x_0\left(\zeta + \sqrt{\zeta^2 - 1}\right)}{\dfrac{\dot{x}_0}{\omega_n} + x_0\left(\zeta - \sqrt{\zeta^2 - 1}\right)}\right] \tag{3.53}$$

Viscous damping leads to the decay of free oscillations. When added to a linear one-degree-of-freedom system, it adds a force linearly proportional to the velocity, and the differential equation remains linear. Viscous damping also leads to positive effects when added to systems undergoing forced excitation. For these reasons viscous damping is often artificially added to many mechanical systems.

Many systems are designed with viscous damping such that the free vibrations are underdamped because of its favorable effect in reducing amplitudes of forced vibration. When a system is subject to free vibrations only, the system may be designed with critical damping or overdamping to alleviate the oscillatory motion. A critically damped system returns to its equilibrium quicker than an overdamped system. Thus recoil mechanisms of guns are designed with critical damping to allow rapid firing.

Automobile suspension systems are often subject to both free and forced vibrations. Free vibrations occur when the vehicle is subject to a sudden change in road contour. In this case the shock absorbers should have a damping ratio near one. However, if the vehicle is traveling on a bumpy road, the vehicle is subject to a possible random excitation. In this case the system should be underdamped. For these reasons vehicle shock absorbers are often self-adaptive.

Example 3.17. The restroom door of Fig. 3.26 is equipped with a torsional spring and a torsional viscous damper so that it automatically returns to its closed position after being opened. The door has a mass of 60 kg and a centroidal moment of inertia about an axis parallel to the axis of the door's rotation of 7.2 $\text{kg} \cdot \text{m}^2$. The torsional spring has a stiffness of 25 $\text{N} \cdot \text{m/rad}$.

(*a*) What is the damping coefficient such that the system is critically damped?
(*b*) A man with an armload of packages, but in a hurry, kicks the door to cause it to open. What angular velocity must his kick impart to cause the door to open 70°?
(*c*) How long after his kick will the door return to within 5° of completely closing?
(*d*) Repeat parts *a*–*c* if the door is designed with a damping ratio, $\zeta = 1.3$.

The differential equation is derived from the free-body diagrams of Fig. 3.26*b*,

$$(\bar{I} + md^2)\ddot{\theta} + c_t\dot{\theta} + k_t\theta = 0$$

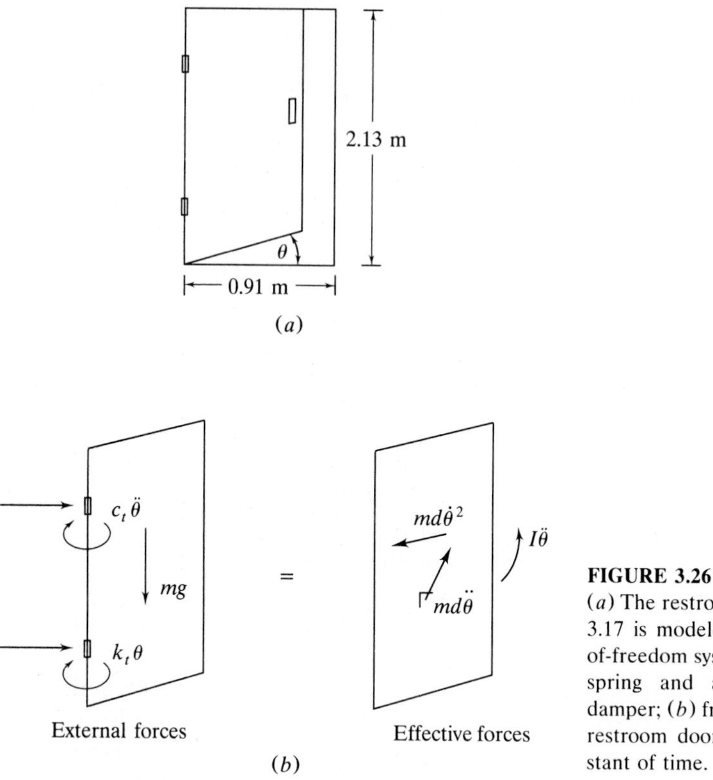

FIGURE 3.26
(*a*) The restroom door of Example 3.17 is modeled as a one-degree-of-freedom system with a torsional spring and a torsional viscous damper; (*b*) free-body diagrams of restroom door at an arbitrary instant of time.

The differential equation is put in the standard form of Eq. (3.22) by dividing by $\bar{I} + md^2$. Then it is evident that

$$\omega_n = \sqrt{\frac{k_t}{\bar{I} + md^2}} = \sqrt{\frac{25 \text{ N} \cdot \text{m/rad}}{7.2 \text{ kg} \cdot \text{m}^2 + (60 \text{ kg})(0.45 \text{ m})^2}} = 1.14 \frac{\text{rad}}{\text{s}}$$

and

$$\zeta = \frac{c_t}{2\omega_n(\bar{I} + md^2)}$$

(*a*) For critical damping the damping ratio is 1. Thus

$$c_t = 2\omega_n(\bar{I} + md^2) = 44.1 \text{ N} \cdot \text{m} \cdot \text{s}$$

(*b*) If the kick is given when the door is closed, the maximum displacement occurs, using Eq. (3.50), at

$$t = \frac{1}{\omega_n} = 0.88 \text{ s}$$

and is given by

$$\theta_{\max} = \frac{\dot{\theta}_0}{e\omega_n}$$

Requiring $\theta_{max} = 70°$ yields

$$\dot{\theta}_0 = 70°\left(\frac{2\pi \text{ rad}}{360°}\right)\left(1.14\frac{\text{rad}}{\text{s}}\right)e = 3.78\frac{\text{rad}}{\text{s}}$$

(*c*) Applying Eq. (3.48) with $\theta = 5°$ gives

$$5°\left(\frac{2\pi \text{ rad}}{360°}\right) = e^{-(1.14 \text{ rad/s})t}\left(3.78\frac{\text{rad}}{s}\right)t$$

which is solved by trial and error to yield $t = 4.6$ s.

(*d*) Setting $\zeta = 1.3$ yields

$$c_t = 1.3(2\zeta\omega_n) = 57.5 \text{ N}\cdot\text{m}\cdot\text{s}$$

From Eq. (3.53) the maximum displacement occurs at

$$t = -\frac{1}{2(1.14 \text{ rad/s})\sqrt{(1.3)^2 - 1}}\ln\left(\frac{1.3 - \sqrt{(1.3)^2 - 1}}{1.3 + \sqrt{(0.13)^2 - 1}}\right) = 0.80 \text{ s}$$

Substituting the preceding result in Eq. (3.52) and setting $\theta = 70°$ yields

$$70°\left(\frac{2\pi \text{ rad}}{360°}\right) = \left(\frac{\dot{\theta}_0}{1.14 \text{ rad/s}}\right)\frac{1}{2\sqrt{(1.3)^2 - 1}}e^{-1.3(1.14 \text{ rad/s})(0.8 \text{ s})}$$
$$\times\left(e^{1.14 \text{ rad/s}\sqrt{(1.3)^2-1}(0.8 \text{ s})} - e^{-1.14 \text{ rad/s}\sqrt{(1.3)^2-1}(0.8 \text{ s})}\right)$$

which gives

$$\dot{\theta}_0 = 4.56 \text{ rad/s}$$

Applying Eq. (3.52) with $\theta = 5°$ yields

$$5°\left(\frac{2\pi \text{ rad}}{360°}\right) = \left(\frac{e^{-1.14(1.3)t}}{2\sqrt{(1.3)^2 - 1}}\right)\left(\frac{4.56 \text{ rad/s}}{1.14 \text{ rad/s}}\right)$$
$$\times\left(e^{1.14\sqrt{(1.3)^2-1}t} - e^{-1.14\sqrt{(1.3)^2-1}t}\right)$$

This equation could be solved by trial and error. However, a good approximation is obtained by neglecting the smaller exponential to give $t = 6.2$ s. The neglected term at this time is 0.00081 rad which is only 0.9% of the total angular displacement.

Note that a harder kick is required to open the door when the system is overdamped than when the system is critically damped even though the time required to open the door is approximately the same. This reflects the increase in the viscous resistance moment.

3.6 FREE VIBRATIONS WITH OTHER FORMS OF DAMPING

Much of the discussion in this text focuses on systems subject to viscous damping, yet viscous damping occurs naturally in very few systems. The presence of viscous damping in mechanical systems leads to beneficial effects. In addition, it introduces a linear term in the governing differential equation making the system easy to analyze. Hence, for these reasons, dashpots are often added to mechanical systems where viscous damping is not naturally present.

Other forms of damping such as Coulomb damping, hysteretic damping, and aerodynamic drag occur naturally. However, their presence leads to nonlinear terms in the governing differential equations, making systems with their presence difficult to analyze. An analytical solution is available to describe the vibrations of a system subject to Coulomb damping. However, linear approximations must be made to obtain analytical solutions for systems subject to other forms of damping.

3.6.1 Coulomb Damping

Coulomb damping is the damping that occurs due to dry friction when two surfaces slide against one another. Coulomb damping can be the result of a mass sliding on a dry surface, axle friction in a journal bearing, belt friction, or rolling resistance. The case of a mass sliding on a dry surface is analyzed here, but the qualitative results apply to all forms of Coulomb damping.

As the mass of Fig. 3.27 slides on a dry surface, a friction force that resists the motion develops between the masss and the surface. Coulomb's law states that the friction force is proportional to the normal force developed between the mass and the surface. The constant of proportionality, μ_k, is called the

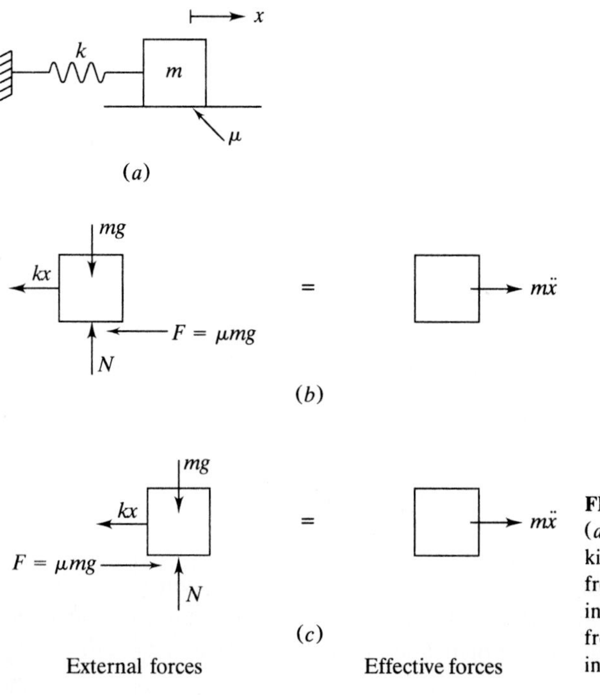

FIGURE 3.27
(*a*) Mass slides on a surface with a kinetic coefficient of friction μ; (*b*) free-body diagrams at an arbitrary instant of time when $\dot{x} > 0$; (*c*) free-body diagrams at an arbitrary instant of time with $\dot{x} < 0$.

kinetic coefficient of friction. Since the friction force always resists the motion, its sign depends upon the sign of the velocity.

Application of Newton's law to the free-body diagrams of Fig. 3.27*b* yields the following differential equations:

$$m\ddot{x} + kx = -\mu mg \qquad \dot{x} > 0 \tag{3.54a}$$

$$m\ddot{x} + kx = \mu mg \qquad \dot{x} < 0 \tag{3.54b}$$

Equation (3.54) is generalized using a single equation

$$m\ddot{x} + kx = -\mu mg\frac{|\dot{x}|}{\dot{x}} \tag{3.55}$$

The right-hand side of Eq. (3.55) is a nonlinear function of the generalized coordinate. Thus the free vibrations of a one-degree-of-freedom system with Coulomb damping are governed by a nonlinear differential equation. However, an analytical solution exists and is obtained by solving Eq. (3.54).

Without loss of generality, assume that free vibrations of the system of Fig. 3.27 are initiated by displacing the mass a distance δ to the right, from equilibrium, and releasing it from rest. The spring force draws the mass toward equilibrium; thus the velocity is initially negative. Equation (3.54*b*) applies over the first half-cycle of motion.

The solution of Eq. (3.54*b*) subject to

$$x(0) = \delta$$

and

$$\dot{x}(0) = 0$$

is

$$x(t) = \left(\delta - \frac{\mu mg}{k}\right)\cos\omega_n t + \frac{\mu mg}{k} \tag{3.56}$$

Equation (3.56) describes the motion until the velocity changes sign at $t = \pi/\omega_n$ when

$$x\left(\frac{\pi}{\omega_n}\right) = -\delta + \frac{2\mu mg}{k} \tag{3.57}$$

Equation (3.54*a*) governs the motion until the velocity next changes sign. The solution of Eq. (3.54*a*) using Eq. (3.57) and

$$\dot{x}\left(\frac{\pi}{\omega_n}\right) = 0 \tag{3.58}$$

as initial conditions is

$$x(t) = \left(\delta - \frac{3\mu mg}{k}\right)\cos\omega_n t - \frac{\mu mg}{k} \qquad \frac{\pi}{\omega_n} \le t \le \frac{2\pi}{\omega_n} \tag{3.59}$$

The velocity again changes sign at $t = 2\pi/\omega_n$ when

$$x\left(\frac{2\pi}{\omega_n}\right) = \delta - \frac{4\mu mg}{k} \tag{3.60}$$

The motion during the first complete cycle is described by Eqs. (3.56) and (3.59). The amplitude change between the beginning and the end of the cycle is

$$x(0) - x\left(\frac{2\pi}{\omega_n}\right) = \frac{4\mu mg}{k} \tag{3.61}$$

The analysis of the subsequent and each successive cycle continues in the same fashion. Equation (3.54*b*) governs during the first half of the cycle, while Eq. (3.54*a*) governs during the second half of the cycle. The initial conditions used to solve for the displacement during a half-cycle are that the velocity is zero and the displacement is the displacement calculated at the end of the previous half-cycle.

The period of each cycle is

$$T = \frac{2\pi}{\omega_n} \tag{3.62}$$

Thus Coulomb damping has no effect on the natural frequency.

Mathematical induction is used to develop the following expressions for the displacement of the mass during each half-cycle:

$$x(t) = \left[\delta - (4n - 3)\frac{\mu mg}{k}\right]\cos\omega_n t + \frac{\mu mg}{k}$$

$$2(n-1)\frac{\pi}{\omega_n} \le t \le 2\left(n - \frac{1}{2}\right)\frac{\pi}{\omega_n} \tag{3.63}$$

$$x(t) = \left[\delta - (4n - 1)\frac{\mu mg}{k}\right]\cos\omega_n t - \frac{\mu mg}{k}$$

$$2\left(n - \frac{1}{2}\right)\frac{\pi}{\omega_n} \le t \le 2n\frac{\pi}{\omega_n} \tag{3.64}$$

$$x\left(2n\frac{\pi}{\omega_n}\right) = \delta - 4n\frac{\mu mg}{k} \tag{3.65}$$

Equation (3.65) shows that the displacement at the end of each cycle is $4\mu mg/k$ less than the displacement at the end of the previous cycle. Thus the amplitude of free vibration decays linearly as shown, when Eqs. (3.63) and (3.64) are plotted in Fig. 3.28.

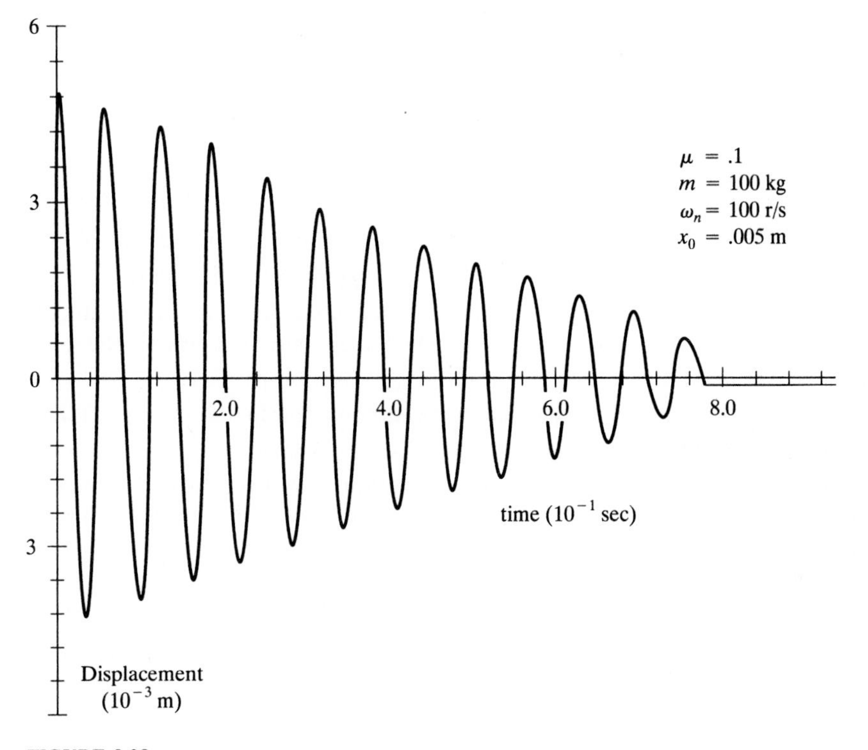

FIGURE 3.28
Coulomb damping does not alter the natural period of free vibration of a one-degree-of-freedom system, but causes a linear decay in amplitude. The motion of the system ceases when the amplitude is too small for the spring force to overcome the friction force.

The motion continues with this constant decrease in amplitude as long as the restoring force is sufficient to overcome the resisting friction force. However, since the friction causes a decrease in amplitude, the restoring force eventually becomes less than the friction force. This occurs when

$$k\left|x\left(2n\frac{\pi}{\omega_n}\right)\right| \leq \mu mg \tag{3.66}$$

Motion ceases during the nth cycle where n is the smallest integer such that

$$n > \frac{k\delta}{4\mu mg} - \frac{1}{4} \tag{3.67}$$

When motion ceases a constant displacement from equilibrium of $\mu mg/k$ is maintained.

The effect of Coulomb damping differs from the effect of viscous damping in these respects:

1. Viscous damping causes a linear term proportional to the velocity in the governing differential equation, while Coulomb damping gives rise to a nonlinear term.
2. The natural frequency of an undamped system is unchanged when Coulomb damping is added, but is decreased when viscous damping is added.
3. Motion is not periodic if the viscous damping coefficient is large enough, whereas the motion is always periodic when Coulomb damping is the only source of damping.
4. The amplitude decreases linearly due to Coulomb damping and exponentially due to viscous damping
5. Coulomb damping leads to a cessation of motion with a resulting permanent displacement from equilibrium, while motion of a system with only viscous damping continues indefinitely with a decaying amplitude.

Since the motion of all physical systems ceases in the absence of continuing external excitation, Coulomb damping is always present. Coulomb damping appears in many forms, such as axle friction in journal bearings and friction due to belts in contact with pulleys or flywheels. The response of systems to these and other forms of Coulomb damping can be obtained in the same manner as the response for dry sliding friction.

The general form of the differential equation governing the free vibrations of a linear system where Coulomb damping is the only source of damping is

$$\ddot{x} + \omega_n^2 x = \begin{cases} \dfrac{F_f}{\tilde{m}} & \dot{x} < 0 \\ -\dfrac{F_f}{\tilde{m}} & \dot{x} > 0 \end{cases} \tag{3.68}$$

where F_f is the magnitude of the Coulomb damping force. The decrease in amplitude per cycle of motion is

$$\Delta A = \frac{4F_f}{\tilde{m}\omega_n^2} \tag{3.69}$$

Example 3.18. An experiment is run to determine the kinetic coefficient of friction between a block and a surface. The block is attached to a spring and displaced 150 mm from equilibrium. It is observed that the period of motion is 0.5 s and that the amplitude decreases by 10 mm on successive cycles. Determine the coefficient of friction and how many cycles of motion the block executes before motion ceases.

The natural frequency is calculated as

$$\omega_n = \frac{2\pi}{T} = \frac{2\pi}{0.5\ s} = 12.57\ \frac{\text{rad}}{\text{s}}$$

The decrease in amplitude is expressed as

$$\Delta A = \frac{4\mu mg}{k} = \frac{4\mu g}{\omega_n^2}$$

which is rearranged to yield

$$\mu = \frac{\Delta A}{4g}\omega_n^2 = \frac{(0.01\ \text{m})(12.57\ \text{rad/s})^2}{4(9.81\ \text{m/s}^2)} = 0.04$$

From Eq. (3.67) the motion ceases during the 15th cycle. The mass has a permanent displacement of 2.5 mm from its original equilibrium position.

Example 3.19. A father builds a swing for his children. The swing consists of a board attached to two ropes, as shown in Fig. 3.29. The swing is mounted on a tree branch, with the board 3.5 m below the branch. The diameter of the branch is 8.2 cm and the kinetic coefficient of friction between the ropes and the branch is 0.1. After the swing is installed and his child is seated, the father pulls the swing back 10°, and releases. What is the decrease in angle of each swing and how many swings will the child receive before Dad needs to give another push?

Due to the friction between the tree branch and the ropes, the tension on opposite sides of a rope will be different. These tensions can be related using the principles of belt friction. When the swing is swinging clockwise, $T_2 > T_1$, and

$$T_2 = T_1 e^{\mu\beta}$$

where β is the angle of contact between the tree branch and the rope. As the child swings the angle of contact may vary. However, this complication is too much to handle with a simplified analysis. A good approximation is to assume β is constant and $\beta = \pi$ rads. When the swing is swinging counterclockwise $T_1 > T_2$ and

$$T_1 = T_2 e^{\mu\beta}$$

Let θ be the clockwise angular displacement of the swing from equilibrium. Summing forces in the direction of the tensions gives

$$\sum F_{\text{ext}} = \sum F_{\text{eff}}$$

$$2T_1 + 2T_2 - mg\cos\theta = ml\dot{\theta}^2$$

The swing is only pulled back 10°. Thus the usual small-angle approximation is valid. Equation (3.2*b*) is used to approximate the cosine term and the nonlinear inertia term is ignored in comparison to the tensions and gravity. The belt friction

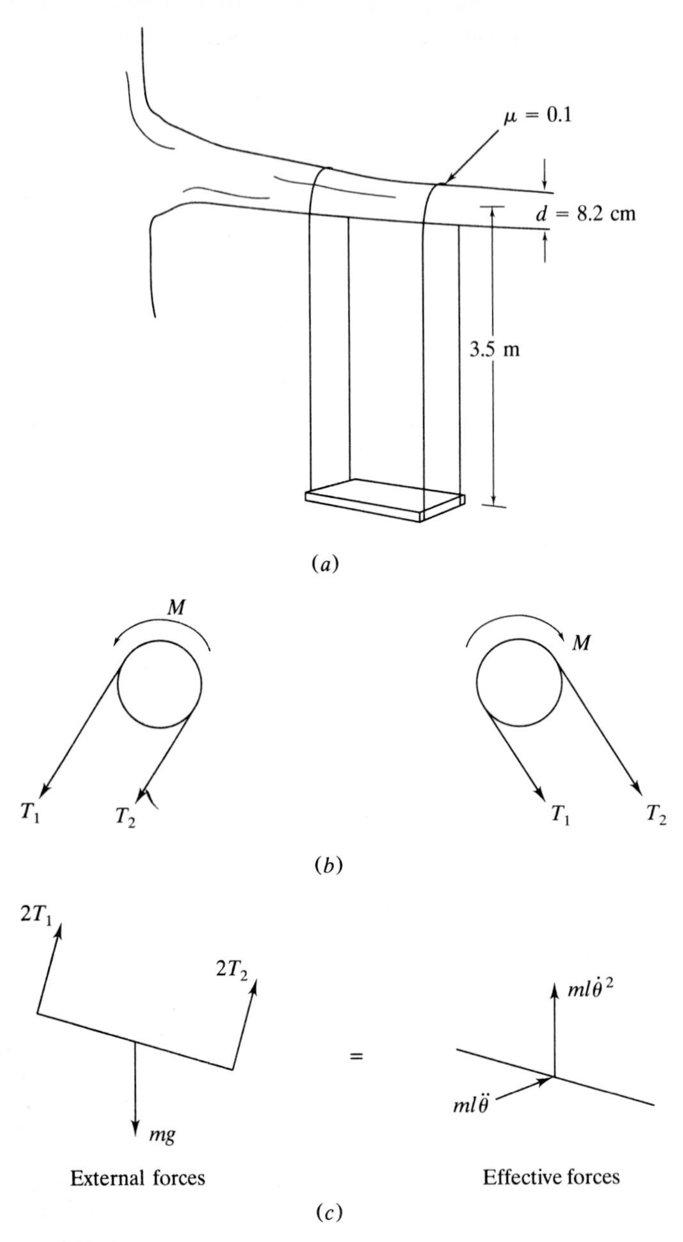

FIGURE 3.29
(a) Tree swing of Example 3.19; (b) due to friction the tension developed in opposite sides of a rope are unequal; (c) free-body diagrams at an arbitrary instant of time.

relations and the normal force equation are solved simultaneously to yield

$$\dot{\theta}>0, \qquad T_1 = \frac{mg}{2(1+e^{\mu\pi})}$$

$$T_2 = \frac{mge^{\mu\pi}}{2(1+e^{\mu\pi})}$$

$$\dot{\theta}<0, \qquad T_1 = \frac{mge^{\mu\pi}}{2(1+e^{\mu\pi})}$$

$$T_2 = \frac{mg}{2(1+e^{\mu\pi})}$$

Summing moments about the center of the tree branch, using the free-body diagrams of Fig. 3.29c, and the small-angle assumption yields

$$\left(\sum M_O\right)_{\text{ext}} = \left(\sum M_O\right)_{\text{eff}}$$

$$(2T_1 - 2T_2)\frac{d}{2} - mgl\theta = ml^2\ddot{\theta}$$

Substituting for the tensions into the preceding equation and rearranging leads to

$$\ddot{\theta} + \frac{g}{l}\theta = \begin{cases} \dfrac{gd}{2l^2}\dfrac{1-e^{\mu\pi}}{1+e^{\mu\pi}} & \dot{\theta}>0 \\ -\dfrac{gd}{2l^2}\dfrac{1-e^{\mu\pi}}{1+e^{\mu\pi}} & \dot{\theta}<0 \end{cases}$$

The frequency of the swinging is

$$\omega_n = \sqrt{\frac{g}{l}} = 1.67\ \frac{\text{rad}}{\text{s}}$$

which is the same as it would be in the absence of friction.

The governing differential equation is of the form of Eq. (3.68). Thus, from Eq. (3.69), the decrease in amplitude per swing is

$$\frac{2d}{l}\frac{e^{\mu\pi}-1}{e^{\mu\pi}+1} = 2\left(\frac{0.082\text{ m}}{3.5\text{ m}}\right)\frac{e^{0.1\pi}-1}{e^{0.1\pi}+1} = 0.0073\text{ rad} = 0.42°$$

Motion ceases when, at the end of a cycle, the moment of the gravity force about the center of the branch is insufficient to overcome the frictional moment. This occurs when

$$mgl\theta < |T_2 - T_1|d$$

or

$$\theta < \frac{d}{2l}\frac{e^{\mu\pi}-1}{e^{\mu\pi}+1} = 0.10°$$

Thus if Dad does not give the swing another push after 24 swings, the swing will come to rest with an angle of repose of 0.1°.

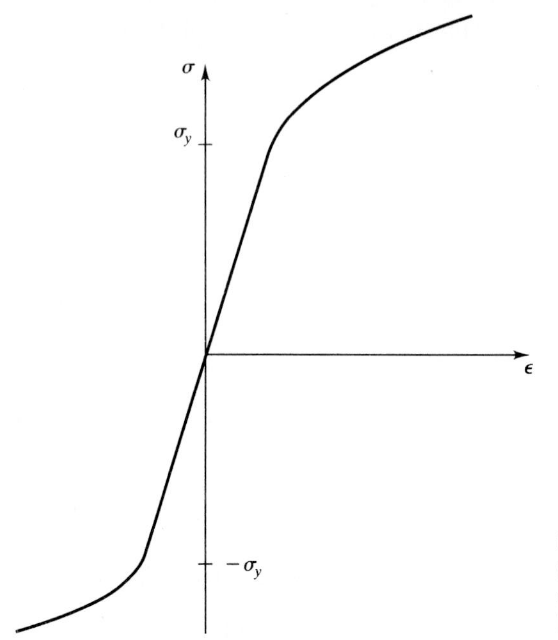

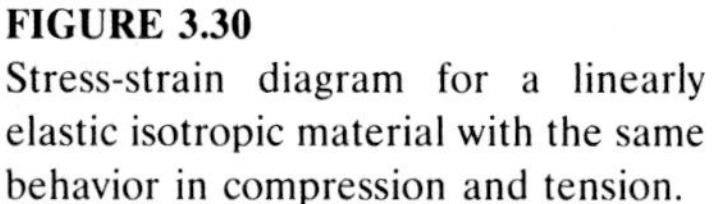
FIGURE 3.30
Stress-strain diagram for a linearly elastic isotropic material with the same behavior in compression and tension.

3.6.2 Hysteretic Damping

The stress-strain diagram for a typical linearly elastic material is shown in Fig. 3.30. Ideally, if the material is stressed below its yield point and then unloaded the stress-strain curve for the unloading follows the same curve for the loading. However, in a real engineering material, internal planes slide relative to one another and molecular bonds are broken causing conversion of strain energy into thermal energy and causing the process to be irreversible. A more realistic stress–strain curve for the loading-unloading process is shown in Fig. 3.31.

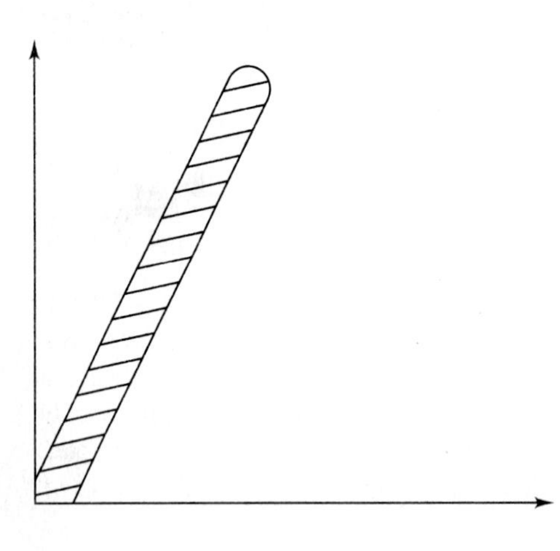

FIGURE 3.31
Imperfections such as dislocations cause the loading process in most real materials to be irreversible resulting in a hysteresis loop. The energy dissipated is the area within the hysteresis loop.

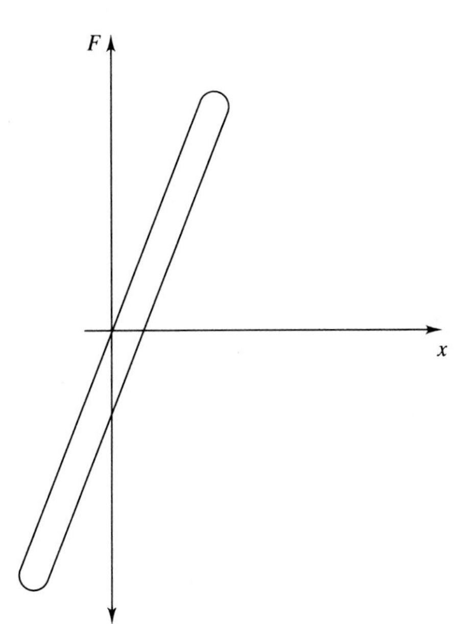

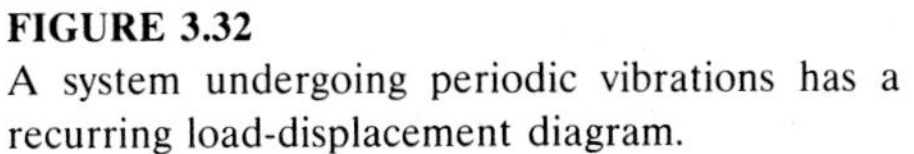

FIGURE 3.32
A system undergoing periodic vibrations has a recurring load-displacement diagram.

The curve in Fig. 3.31 is a hysteresis loop. The area enclosed by the curve is the dissipated strain energy per unit volume. The area enclosed by the hysteresis loop from a force-displacement curve is the total strain energy dissipated during a loading-unloading cycle. In general, the area under a hysteresis curve is independent of the rate of the loading-unloading cycle.

In a vibrating mechanical system an elastic member undergoes a cyclic load-displacement relationship as shown in Fig. 3.32. The loading is repeated over each cycle. The existence of the hysteresis loop leads to energy dissipation from the system during each cycle which causes natural damping, called hysteretic damping. It has been shown experimentally that the energy dissipated per cycle of motion is independent of the frequency and proportional to the square of the amplitude. An empirical relationship is

$$\Delta E = \pi k h X^2 \tag{3.70}$$

where X is the amplitude of motion during the cycle and h is a constant, called the hysteretic damping coefficient.

The hysteretic damping coefficient cannot be simply specified for a given material. It is dependent upon other considerations such as how the material is prepared and the geometry of the structure under consideration. Data are available for the specific damping capacity; the percentage of strain energy dissipated over one cycle of motion, for a solid circular cylinder in torsional, for certain materials at a given stress level. The data cannot be extended to meet every situation. Thus empirical results are necessary for every particular case to determine the hysteretic damping coefficient.

Mathematical modeling of hysteretic damping is developed from a work-energy analysis. Consider a simple mass-spring system with hysteretic damping. Let X_1 be the amplitude at a time when the velocity is zero and all energy is potential energy stored in the spring. Hysteretic damping dissipates some of that energy over the next cycle of motion. Let X_2 be the displacement of the mass at the next time when the velocity is zero, after the system executes one half-cycle of motion. Let X_3 be the displacement at the subsequent time when the velocity is zero, one full cycle later. Application of the work-energy principle over the first half-cycle of motion gives

$$T_1 + V_1 + U_{1\to 2} = T_2 + V_2$$

$$T_1 + V_1 = T_2 + V_2 + \frac{\Delta E}{2}$$

The energy dissipated by hysteretic damping is approximated by Eq. (3.70) using X as the amplitude at the beginning of the half-cycle.

$$\tfrac{1}{2}kX_1^2 = \tfrac{1}{2}kX_2^2 + \tfrac{1}{2}\pi khX_1^2$$

This yields

$$X_2 = \sqrt{1 - \pi h}\, X_1$$

A work-energy analysis over the second half-cycle leads to

$$X_3 = \sqrt{1 - \pi h}\, X_2 = (1 - \pi h) X_1 \tag{3.71}$$

Thus the rate of decrease of amplitude on successive cycles is constant, as it is for viscous damping. By analogy a logarithmic decrement is defined for hysteretic damping as

$$\delta = \ln \frac{X_1}{X_3} = -\ln(1 - \pi h) \tag{3.72}$$

which for small h is approximated as

$$\delta = \pi h \tag{3.73}$$

By analogy with viscous damping an equivalent damping ratio for hysteretic damping is defined as

$$\zeta = \frac{\delta}{2\pi} = \frac{h}{2} \tag{3.74}$$

and an equivalent viscous damping coefficient is defined as

$$c_{eq} = 2\zeta\sqrt{\tilde{m}\tilde{k}} = \frac{h\tilde{k}}{\omega_n} \tag{3.75}$$

The response of a system subject to hysteretic damping is the same as the response of the system when subject to viscous damping with an equivalent viscous damping coefficient given by Eq. (3.75). This is true only for small hysteretic damping as subsequent plastic behavior leads to a highly nonlinear system. The analogy between viscous damping and hysteretic damping is also only true for linearly elastic material and for materials where the energy dissipated per unit cycle is proportional to the square of the amplitude. In addition, the hysteretic damping coefficient is a function of geometry as well as the material.

The response of a system subject to hysteretic or viscous damping continues indefinitely with exponentially decaying amplitude. However, hysteretic damping is significantly different from viscous damping in that the energy dissipated per cycle for hysteretic damping is independent of frequency, whereas the energy dissipated per cycle increases with frequency for viscous damping. Thus while the mathematical treatments of viscous damping and hysteretic damping are the same, they have significant physical difference.

Example 3.20. The force-displacement curve for a structure is shown in Fig. 3.33. The structure is modeled as a one-degree-of-freedom system with an equivalent mass 500 kg located at the position where the measurements are made. Describe the response of this structure when a shock imparts a velocity of 20 m/s to this point on the structure.

The area under the hysteresis curve is approximated by counting the squares inside the hysteresis loop. Each square represents 20 N · m of dissipated energy. There are approximately 38.5 squares inside the hysteresis loop resulting in 770 N · m dissipated over one cycle of motion with an amplitude of 20 mm.

The equivalent stiffness is the slope of the force deflection curve and is determined as 5,000,000 N/m. Application of Eq. (3.70) leads to

$$h = \frac{\Delta E}{\pi k X^2} = \frac{770 \text{ N} \cdot \text{m}}{\pi(5 \times 10^6 \text{ N/m})(0.02 \text{ m})^2} = 0.123$$

The logarithmic decrement, damping ratio, and natural frequency are calculated using Eqs. (3.73) and (3.74)

$$\delta = \pi h = 0.386$$

$$\zeta = \frac{h}{2} = 0.0615$$

$$\omega_n = \sqrt{\frac{\tilde{k}}{\tilde{m}}} = \sqrt{\frac{5 \times 10^6 \text{ N/m}}{500 \text{ kg}}} = 100 \ \frac{\text{rad}}{\text{s}}$$

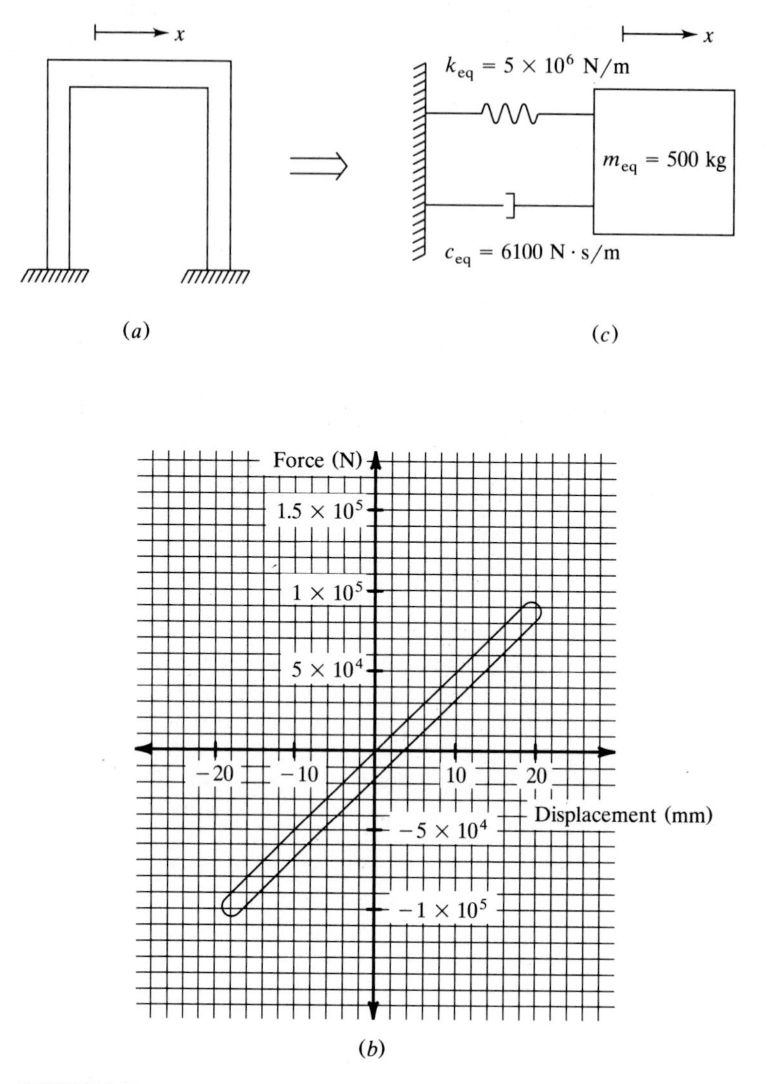

FIGURE 3.33
(a) One-story frame structure modeled as a one-degree-of-freedom system; (b) force-displacement curve for the structure of Example 3.20; (c) hysteretic damping leads to an equivalent viscous damping coefficient of 6100 N · s/m for this example.

The response of this structure with hysteretic damping is approximately the same as the response of a simple mass-spring-dashpot system with a damping ratio of 0.0615 and a natural frequency of 100 rad/s. Then from Eq. (3.32) with $\dot{x}_0 = 20$ m/s, and $x_0 = 0$, the response is

$$x(t) = 0.20e^{-6.15t}\sin(99.81t) \text{ m}$$

3.6.3. Other Forms of Damping

A mechanical or structural system may be subject to other forms of damping such as aerodynamic drag, radiation damping, or anelastic damping. However, these give rise to nonlinear terms in the governing differential equations. Exact solutions do not exist for these forms of damping. The periodic motion of systems subject to these forms of damping can be approximated by developing an equivalent viscous damping coefficient. The equivalent viscous damping coefficient is obtained by equating the energy dissipated over one cycle of motion, assuming harmonic motion at a specific amplitude and frequency, for the particular form of damping with the energy dissipated over one cycle of motion due to the force in a dashpot of the equivalent viscous damping coefficient.

For a harmonic motion of the form

$$x(t) = X \sin \omega t$$

the energy dissipated over one cycle of motion due to a damping force F_D is

$$\begin{aligned} \Delta E &= \int_0^{2\pi/\omega} F_D \dot{x}\, dt \\ &= \int_0^{2\pi/\omega} F_D X\omega \cos \omega t\, dt \end{aligned} \tag{3.76}$$

For viscous damping Eq. (3.76) yields

$$\begin{aligned} \Delta E &= \int_0^{2\pi/\omega} c\dot{x}^2\, dt \\ &= \int_0^{2\pi/\omega} c\omega^2 X^2 \cos^2 \omega t\, dt \\ &= c\omega\pi X^2 \end{aligned} \tag{3.77}$$

Thus, by analogy, the equivalent viscous damping coefficient for another form of damping is

$$c_{eq} = \frac{\Delta E}{\pi\omega X^2} \tag{3.78}$$

Aerodynamic drag is present in all real problems. However, its effect is often ignored. The determination of the correct form of the drag force is a problem in fluid mechanics. At high Reynolds numbers the drag is very nearly proportional to the square of the velocity and can be written as

$$F_D = C_D \dot{x}|\dot{x}| \tag{3.79}$$

where C_D is a coefficient that is a function of body geometry and air properties.

For moderate Reynolds numbers appropriate forms of the drag force have been proposed as

$$F_D = C_D|\dot{x}|^{\alpha}\dot{x} \tag{3.80}$$

where $0 < \alpha \leq 1$. In either case the resulting differential equation is nonlinear.

Some materials (e.g., rubber) are viscoelastic and obey a constitutive equation in which stress is related to strain and strain rate. It is shown in Chap. 4 that for an undamped system the forced response is in phase with a harmonic excitation, whereas a phase lag occurs for a damped system. This phase lag also occurs for many viscoelastic materials. Indeed, many viscoelastic materials have constitutive equations that are derived by modeling the material as a spring in parallel with a dashpot. This is called a Kelvin model. The phase lag results in energy dissipation and the resulting damping is called anelastic damping.

Damping occurs when energy is dissipated form a vibrating body by any means. Another example is radiation damping that occurs for a body vibrating on the free surface between two fluids. The vibrating body causes pressure waves to be radiated outward causing energy transfer from the body to the surrounding fluids.

Most physical systems are subject to a combination of forms of damping. Indeed, a simple mass-spring-dashpot system is subject to viscous damping from the dashpot, Coulomb damping from the dry sliding friction, hysteretic damping from the spring, and aerodynamic drag. The presence of Coulomb damping leads to cessation of free vibrations after a finite time. The aerodynamic drag is usually neglected in an analysis as its effect is negligible and it leads to a nonlinear differential equation. The hysteretic damping acts in parallel with the viscous damping. The equivalent damping coefficient is the sum of the viscous damping coefficient for the dashpot and the equivalent viscous damping coefficient for the hysteretic damping. For small amplitudes the effect of viscous damping is much greater than the effect of hysteretic damping. For large amplitudes the hysteretic damping can be dominant. However, if the amplitude is not large the hysteretic damping model is not applicable.

Example 3.21. A block of mass 1 kg is attached to a spring of stiffness 3×10^5 N/m. The block is displaced 20 mm from equilibrium and released from rest. The block is in a fluid where the drag force is given by Eq. (3.79) with $C_D = 0.86$ N · s^2/m. Approximate the number of cycles before the amplitude is reduced to 15 mm.

The energy lost per cycle of motion due to aerodynamic drag is calculated using Eq. (3.76)

$$\begin{aligned}\Delta E &= \int_0^{2\pi/\omega} C_D X^3\omega^3 \cos^2 \omega t|\cos \omega t|\, dt \\ &= 4\int_0^{\pi/2\omega} C_D X^3\omega^3 \cos^3 \omega t\, dt \\ &= \tfrac{8}{3}C_D\omega^2 X^3\end{aligned}$$

TABLE 3.2
Viscous Approximation Used to Predict Decay in Amplitude for Ex. 3.21

Cycle	Amplitude at beginning of cycle $X_n = X_{n-1}e^{-2.32X_{n-1}}$
1	20.0
2	19.09
3	18.26
4	17.50
5	16.81
6	16.16
7	15.56
8	15.00

From Eq. (3.78) the equivalent viscous damping coefficient is calculated as

$$C_{\text{eq}} = 0.730\omega X$$

If the equivalent viscous damping is small, the frequency is approximately equal to the natural frequency of free undamped vibrations

$$\omega = \sqrt{\frac{k}{m}} = 547.7\ \frac{\text{rad}}{\text{s}}$$

The damping ratio on a given cycle is

$$\zeta = \frac{c_{\text{eq}}}{2\sqrt{km}} = \frac{0.73(547.7\ \text{rad/s})X}{2\sqrt{(1\ \text{kg})(3 \times 10^5\ \text{N}/m)}} = 0.37X$$

Using Eq. (3.40) the logarithmic decrement is

$$\delta = 2\pi\zeta = 2.32X$$

Since the equivalent viscous damping coefficient, and hence the damping ratio and the logarithmic decrement depend upon the amplitude, the decrease in amplitude is not constant on each cycle. Using an amplitude of 20 mm for the first cycle, the amplitude at the beginning of the second cycle is obtained using the logarithmic decrement, which in turn is used to predict the amplitude at the beginning of the third cycle. Table 3.2 is developed in this fashion. The amplitude of vibration is reduced to 15 mm in seven cycles.

3.7 DYNAMIC STABILITY

The equilibrium position of a system is unstable if, when slightly displaced from equilibrium, the system moves farther away from the equilibrium position. If the system is stable, the system will either return to its equilibrium position or oscillate with a bounded amplitude about the equilibrium position.

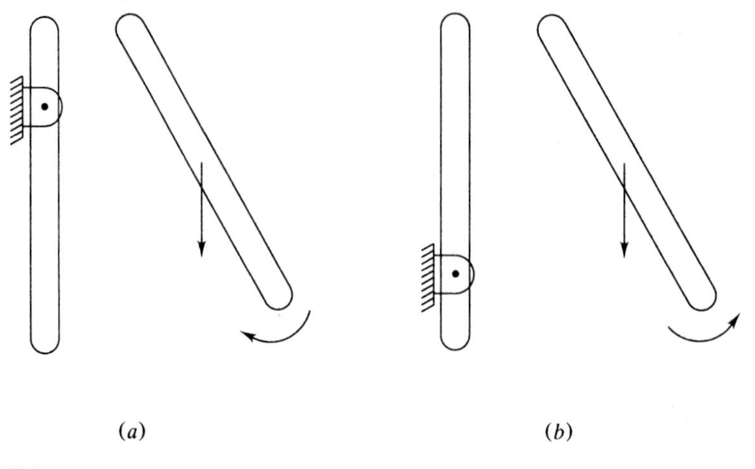

FIGURE 3.34
(*a*) Compound pendulum pinned above center of gravity is stable. Moment due to gravity is restoring; (*b*) compound pendulum pinned below center of gravity is unstable. Moment due to gravity pulls pendulum away from equilibrium.

Reconsider Eq. (3.23), which defines the exponents in the free-vibration response of a linear one-degree-of-freedom system with viscous damping:

$$\alpha_{1,2} = -\frac{\tilde{c}}{2\tilde{m}} \pm \sqrt{\left(\frac{\tilde{c}}{2\tilde{m}}\right)^2 - \frac{\tilde{k}}{\tilde{m}}} \tag{3.23}$$

If $\tilde{k} < 0$ then both roots are real with

$$\alpha_1 > 0 \qquad \alpha_2 < 0$$

The general solution of Eq. (3.19) which describes the system response is

$$x(t) = C_1 e^{\alpha_1 t} + C_2 e^{\alpha_2 t} \tag{3.81}$$

where C_1 and C_2 are arbitrary constants, determined by application of the initial conditions. Equation (3.81) shows, that for this special case, the motion is not oscillatory for any value of $\tilde{c}$. The system moves farther from its equilibrium position at t increases. Thus this system is unstable.

Occasionally, a system may have multiple equilibrium positions. Such is the case for the compound pendulum of Fig. 3.34. If the pendulum is pinned at any point other than its center of gravity, it has two vertical static-equilibrium positions.

The equilibrium position where the center of gravity is below the pin is stable. If the bar is perturbed slightly from this position, the moment of the gravity force about the pin support pushes the pendulum back toward equilibrium.

The equilibrium position where the center of gravity is above the pendulum is unstable. When perturbed, the moment of the gravity force pulls the pendulum farther away from equilibrium.

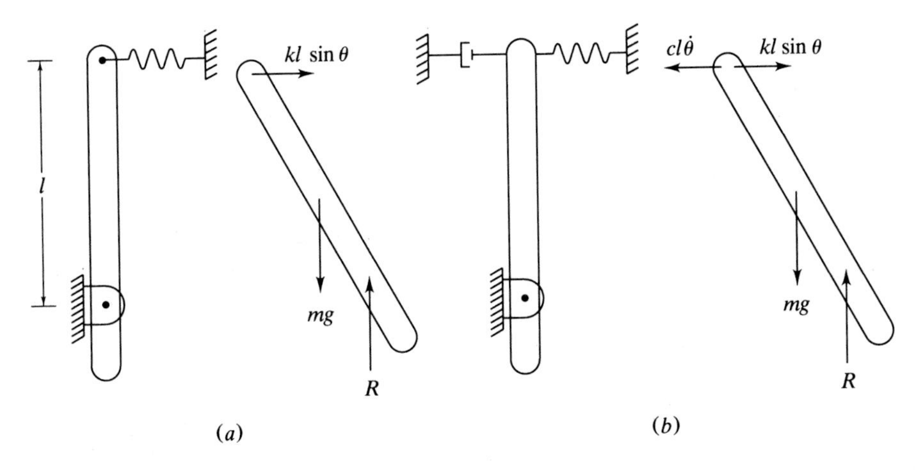

FIGURE 3.35
(*a*) If moment of spring force is greater than moment of gravity force, spring stabilizes compound pendulum; (*b*) the addition of viscous damping will not stabilize an unstable equilibrium position.

If the bar is pinned below its center of gravity, the addition of springs tends to stabilize the equilibrium position. If the system of Fig. 3.35 is perturbed from equilibrium, the moment due to gravity tends to pull the system farther from equilibrium while the moment due to the force developed in the spring tends to return the system to its equilibrium position. If the moment due to gravity is greater, the system is unstable. Otherwise the system is stable.

The addition of a viscous damper does not affect the stability of the equilibrium position. If the equilibrium position is unstable, the damper slows the rate of movement away from the equilibrium position. If the equilibrium position is stable, but without damping, the pendulum, when perturbed, oscillates about the equilibrium position with constant amplitude. The addition of the viscous damper causes the amplitude to decay.

Example 3.22. What is the minimum spring stiffness, k, such that the system of Fig. 3.36*a* is stable?

Summing moments on the free-body diagrams of Fig. 3.36*b* and using the small-angle approximation, the approximate linear differential equation governing θ, the counterclockwise angular displacement from equilibrium is

$$\frac{mL^2}{9}\ddot{\theta} + \left(\frac{kL^2}{9} - \frac{mgL}{6}\right)\theta = 0$$

The system is stable if

$$\frac{kL^2}{9} - \frac{mgL}{6} > 0$$

or

$$k < \frac{3mg}{2L}$$

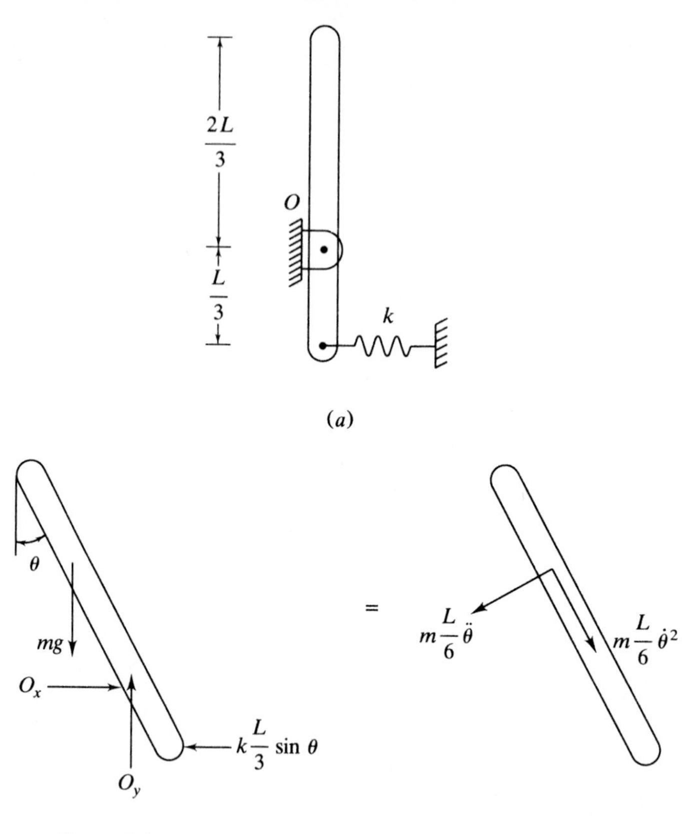

FIGURE 3.36
(a) System of Example 3.22; (b) free-body diagrams at an arbitrary instant of time.

Substituting in given values

$$k < \frac{3(10\text{ kg})(9.81\text{ m/s}^2)}{2(1.2\text{ m})} = 122.6\ \frac{\text{N}}{\text{m}}$$

PROBLEMS

3.1–3.17. Using the generalized coordinate shown as the dependent variable, derive the differential equation governing vibrations of the one-degree-of-freedom system by applying the appropriate form(s) of Newton's laws to the appropriate free-body diagrams. Linearize nonlinear differential equations by assuming small displacements. Determine the undamped natural frequency for each system.

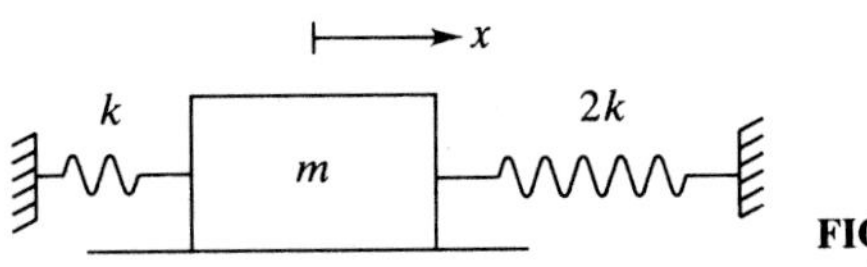

FIGURE P3.1

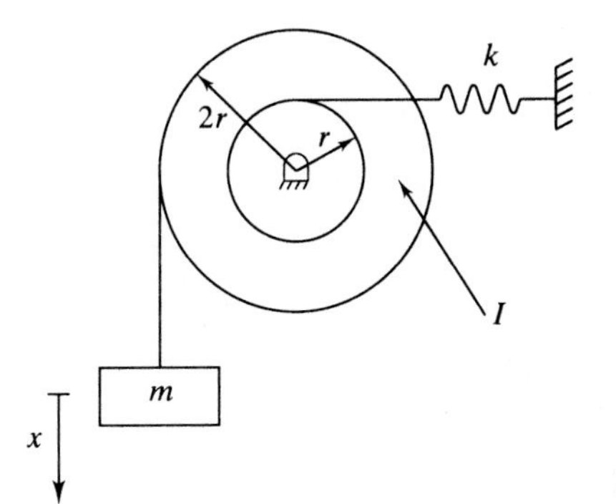

FIGURE P3.2

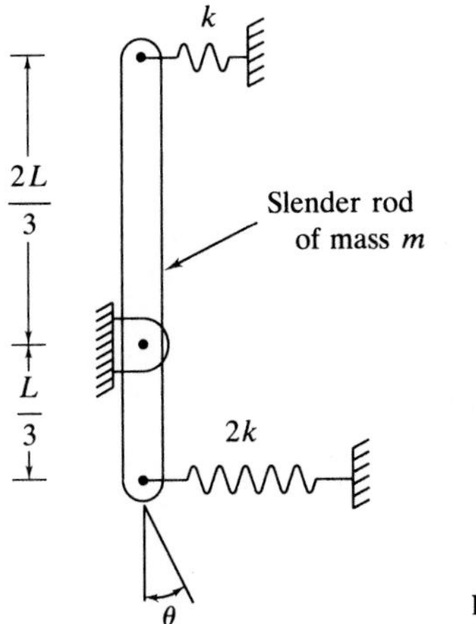

FIGURE P3.3

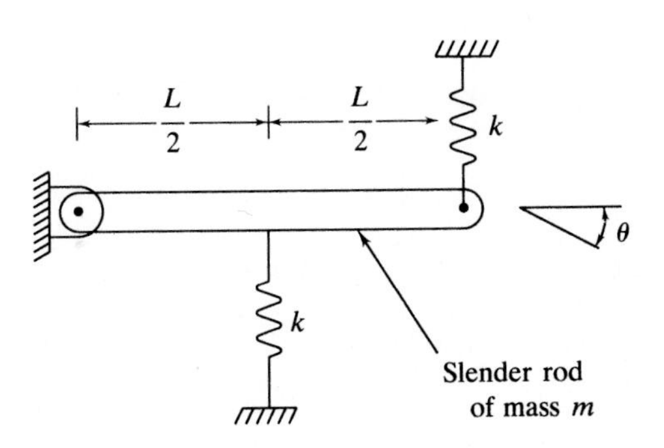

FIGURE P3.4

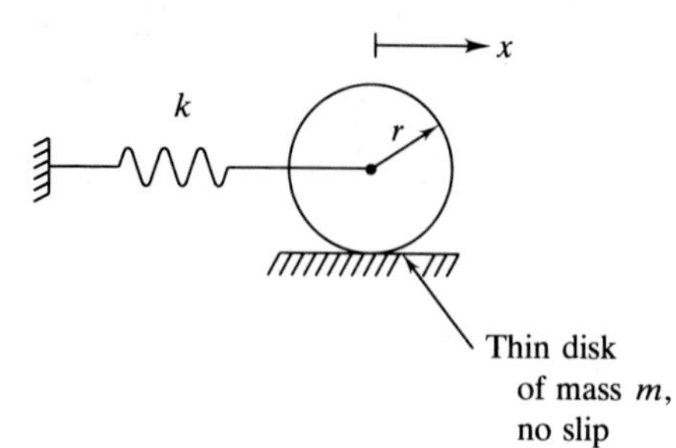

FIGURE P3.5

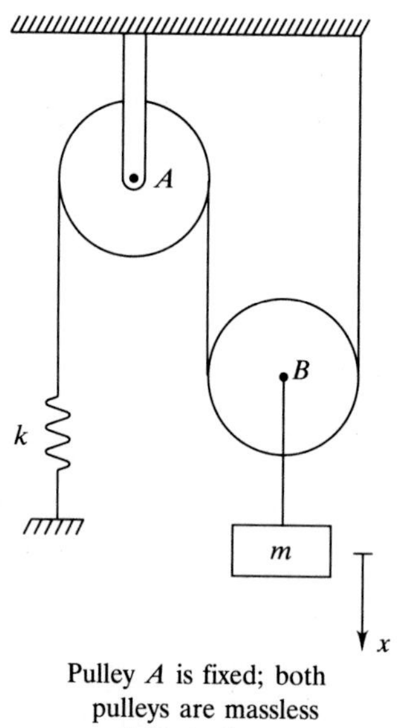

FIGURE P3.6

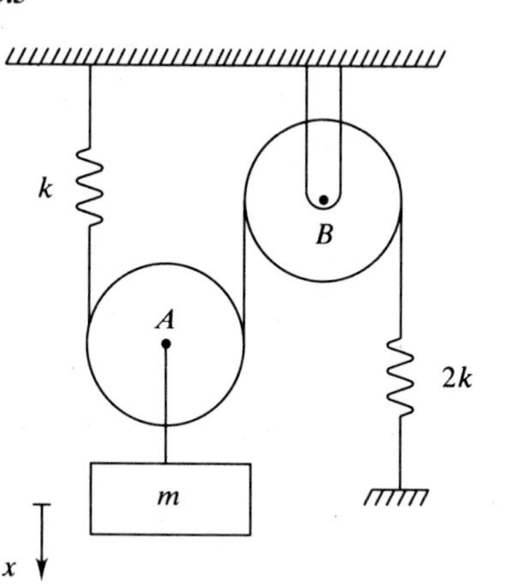

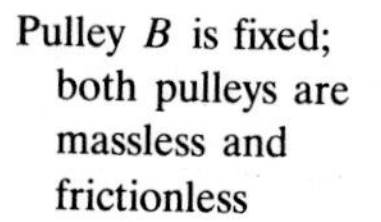

FIGURE P3.7

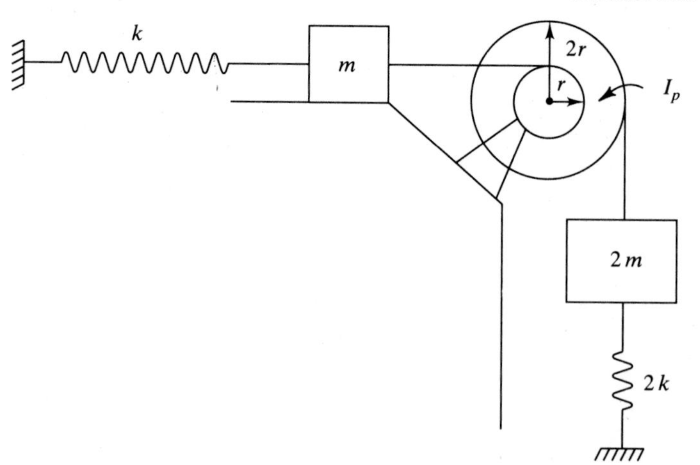

FIGURE P3.8

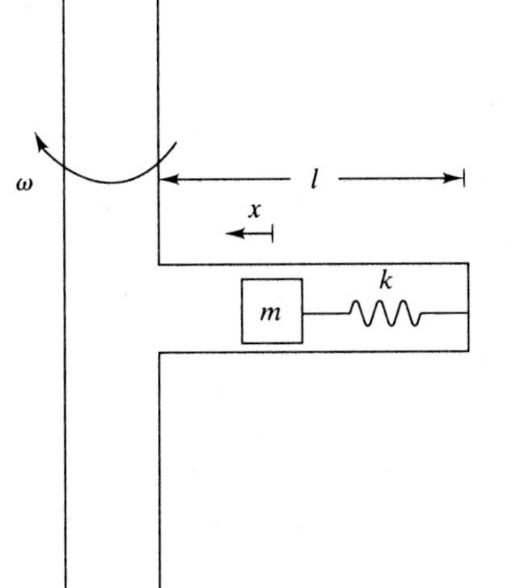

FIGURE P3.9

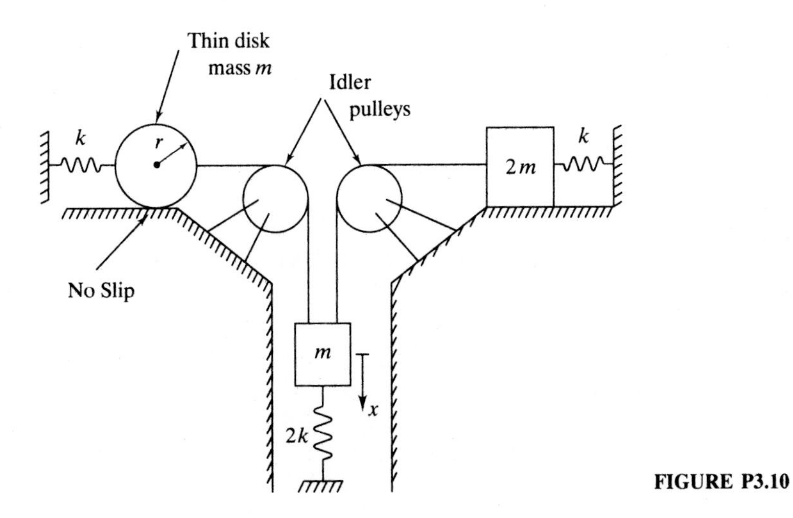

FIGURE P3.10

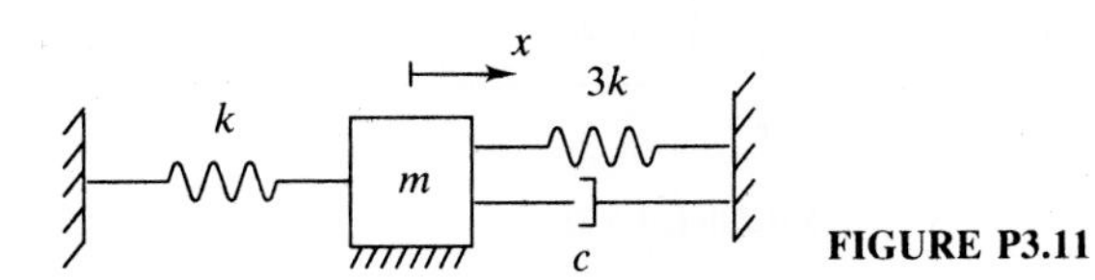

FIGURE P3.11

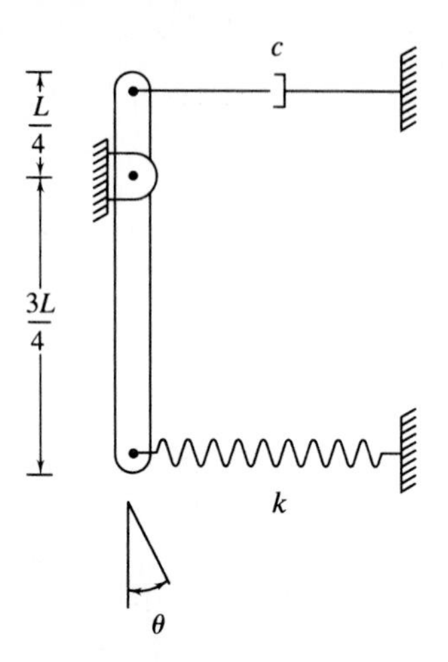

FIGURE P3.12

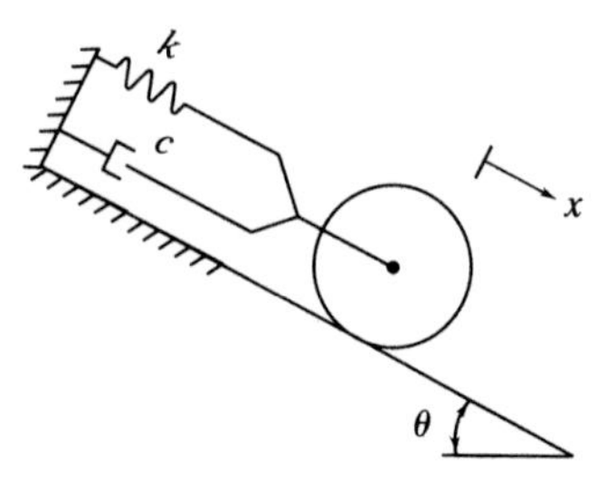

Thin disk of mass m
radius r rolls
without slip

FIGURE P3.13

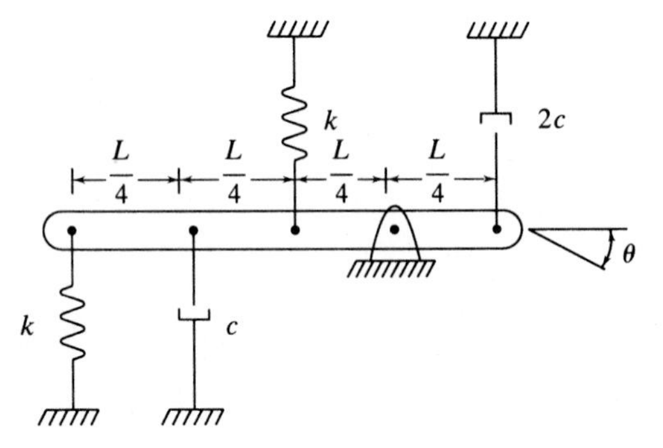

FIGURE P3.14

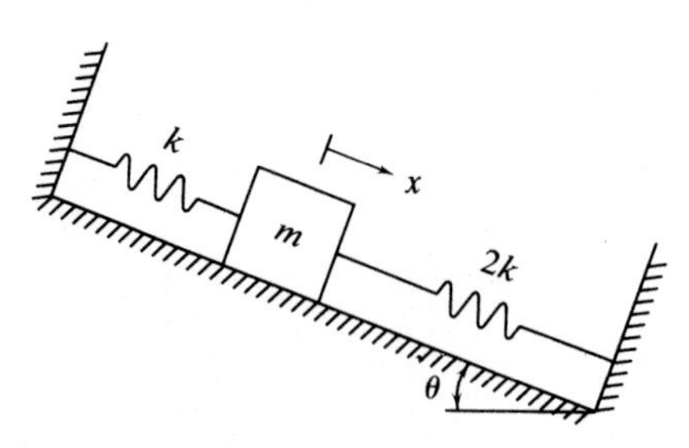

FIGURE P3.15

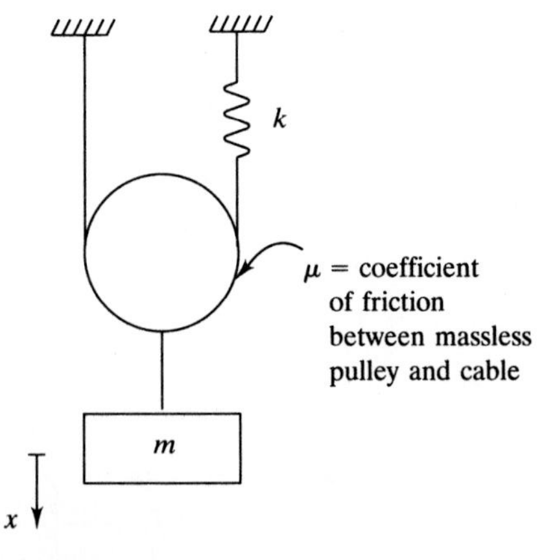

FIGURE P3.16

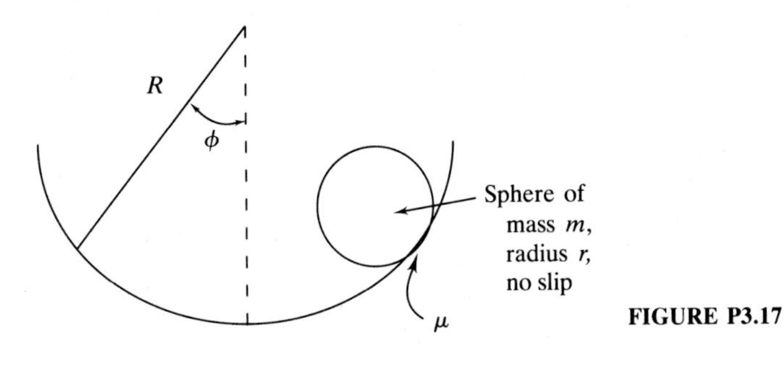

FIGURE P3.17

3.18–3.31. Using the generalized coordinate shown as the dependent variable, derive the differential equation governing vibrations of the one-degree-of-freedom system using an energy approach. The equivalent-systems method or direct application of the principle of work-energy may be used. Linearize nonlinear differential equations by assuming small displacements. Determine the undamped natural frequency for each system.

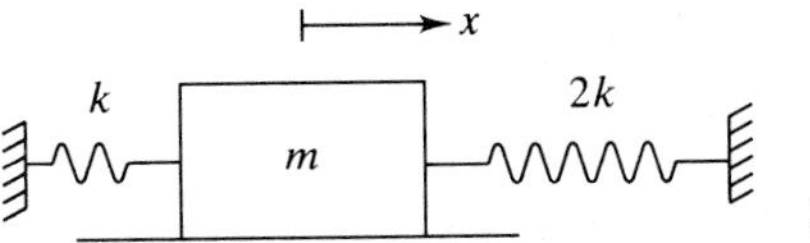

FIGURE P3.18

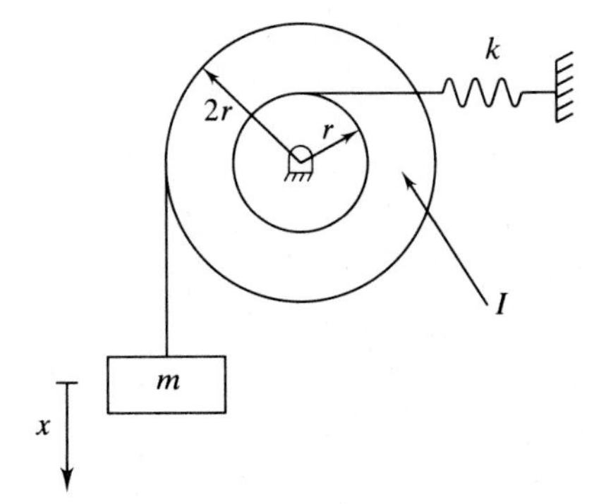

FIGURE P3.19

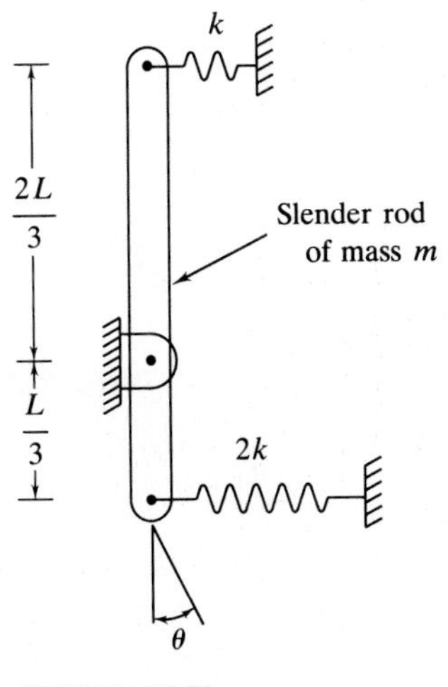

FIGURE P3.20

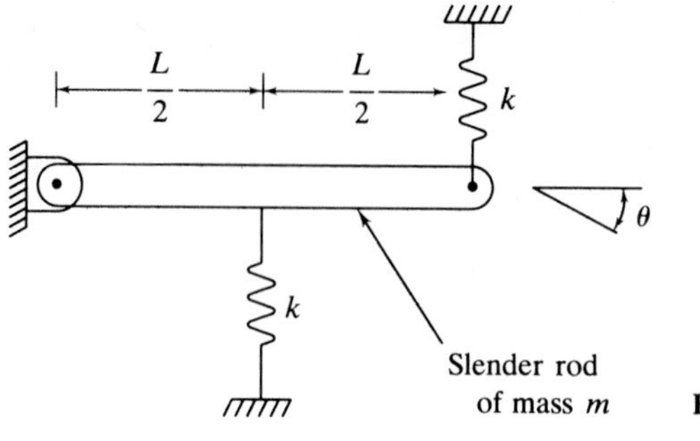

FIGURE P3.21

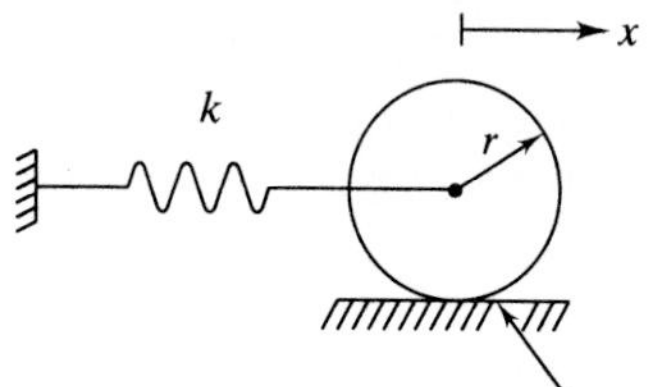

Thin disk of mass m, no slip

FIGURE P3.22

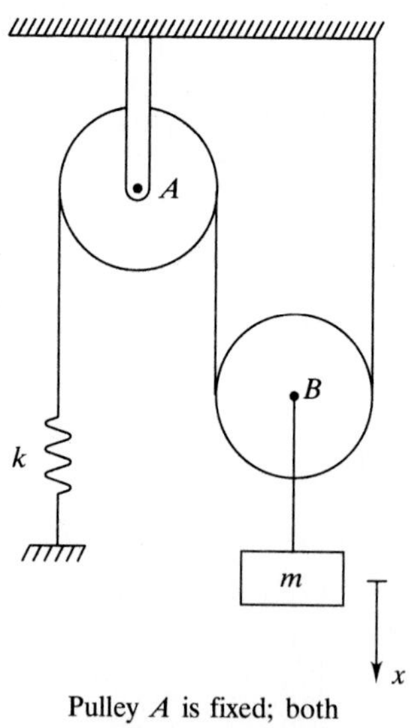

Pulley A is fixed; both pulleys are massless and frictionless

FIGURE P3.23

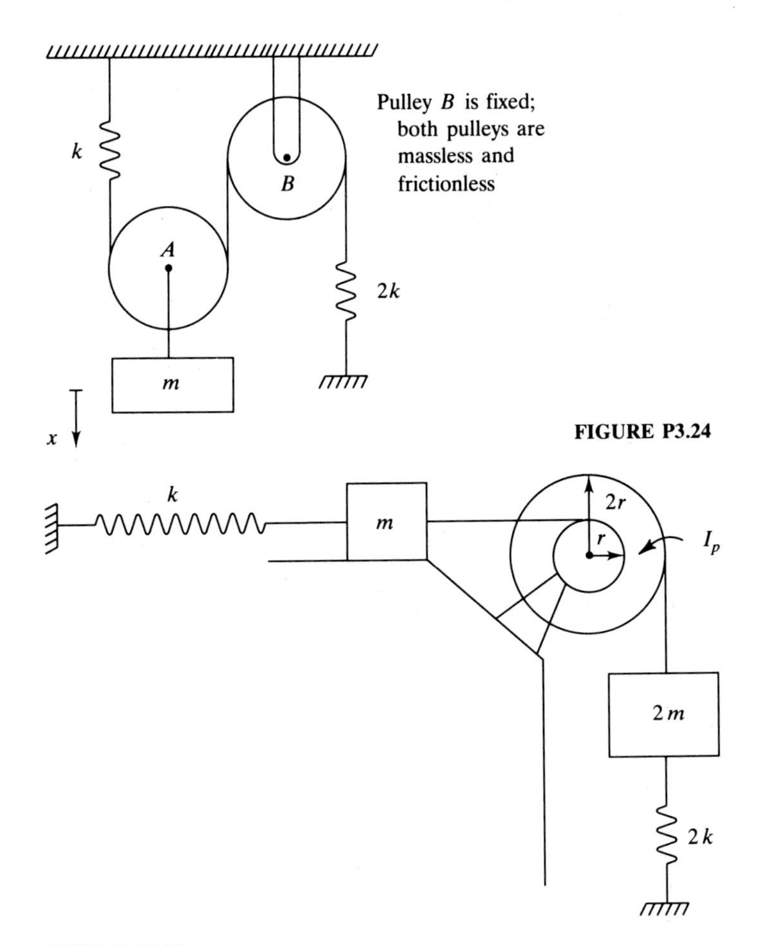

FIGURE P3.24

FIGURE P3.25

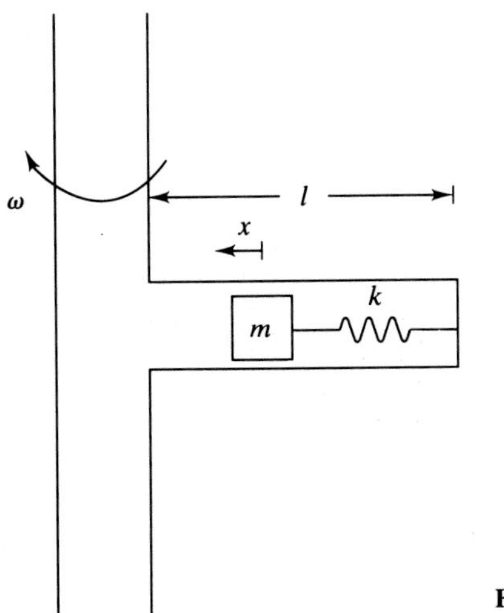

FIGURE P3.26

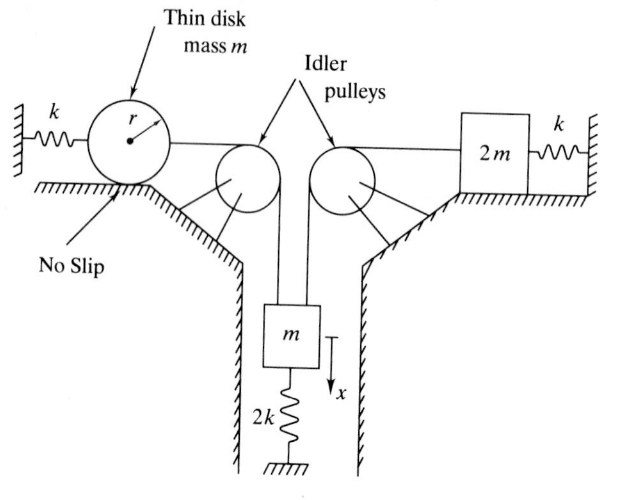

FIGURE P3.27

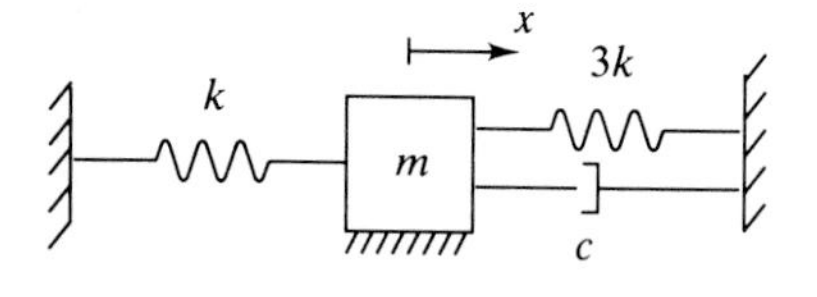

FIGURE P3.28

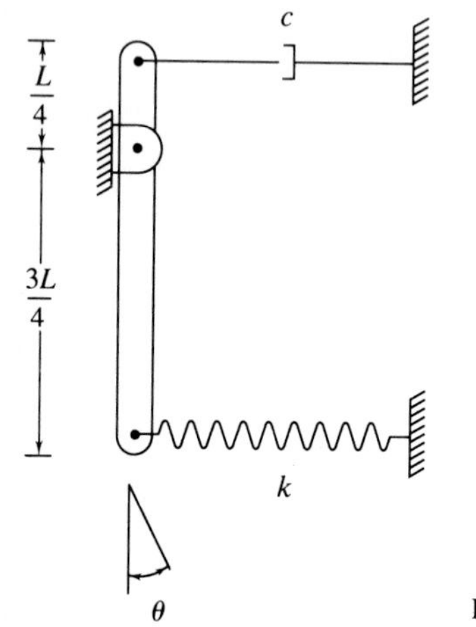

FIGURE P3.29

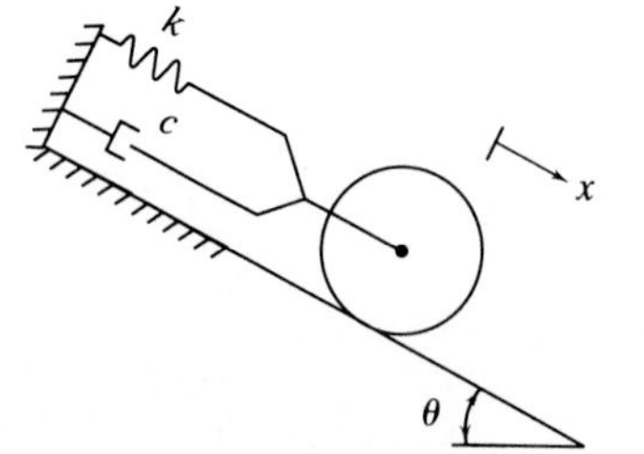

Thin disk of mass m
radius r rolls
without slip

FIGURE P3.30

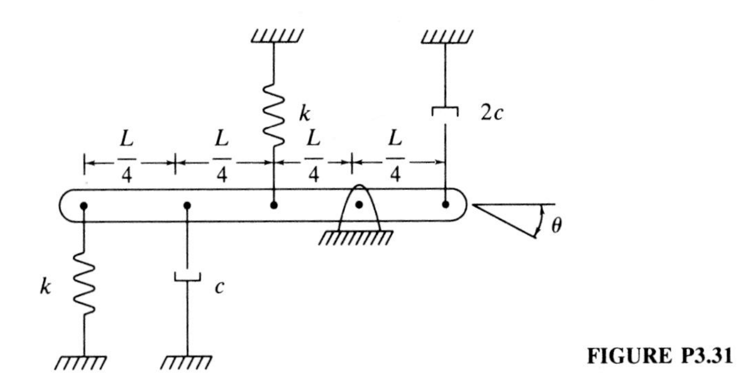

FIGURE P3.31

3.32. When an engine of mass 400 kg is placed on an elastic foundation, the foundation deflects 5 mm. What is the natural frequency of the system in rpm?

3.33. A cylindrical container of mass 25 kg floats stably in an unknown fluid as shown in Fig. P3.33 When disturbed, the period of free oscillations is 0.2 s. What is the specific gravity of the fluid?

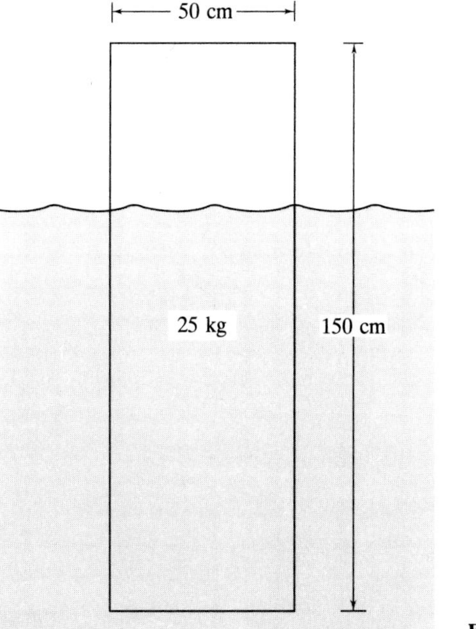

FIGURE P3.33

3.34. A ceiling fan is an assembly of five blades driven by a motor. The assembly is attached to the end of a thin shaft whose other end is fixed to the ceiling. What is the natural frequency of torsional oscillations of the fan shown in Fig. P3.34?

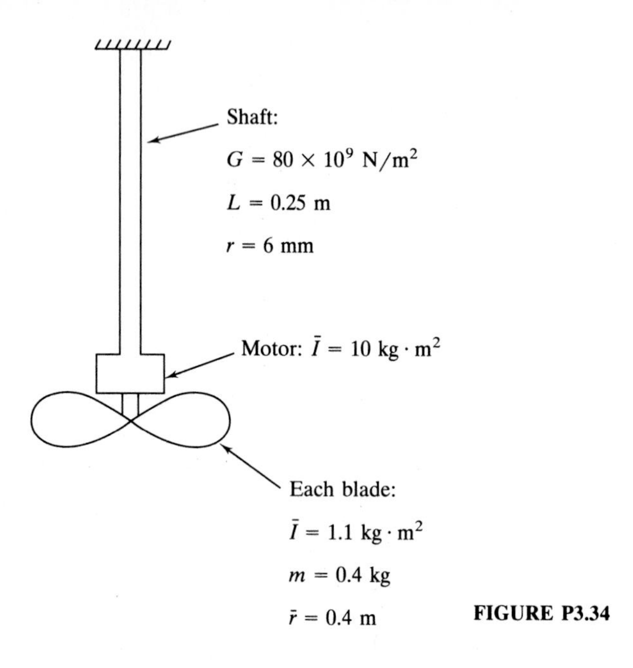

FIGURE P3.34

3.35–3.40. Write the differential equations governing the free vibrations of the mass attached to the elastic member shown. Ignore the inertia effects of the elastic elements. What is the natural frequency of the system?

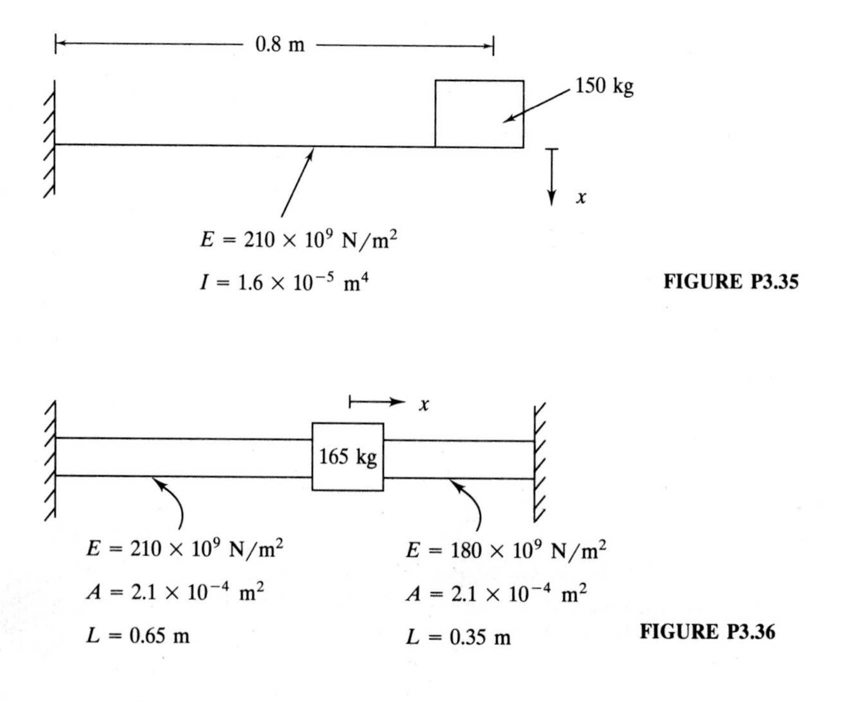

FIGURE P3.35

FIGURE P3.36

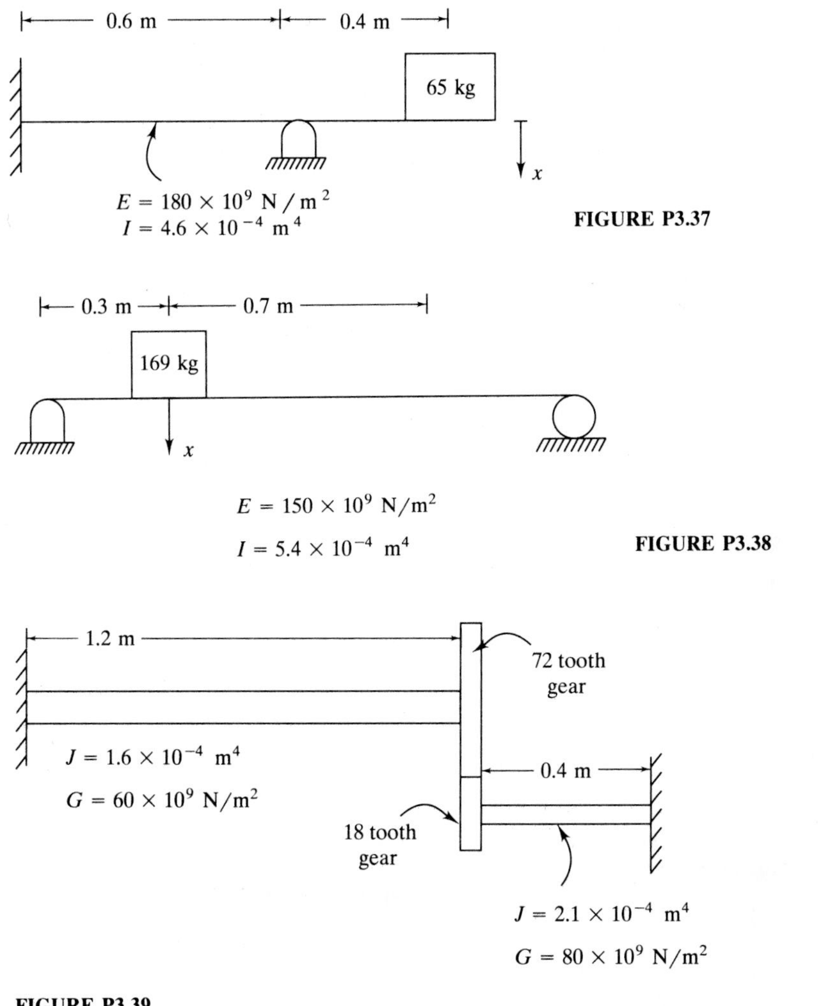

FIGURE P3.37

FIGURE P3.38

FIGURE P3.39

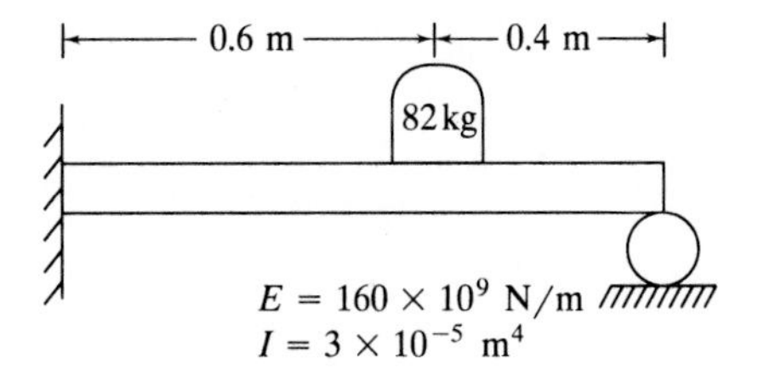

3.41. The mass of the pendulum bob of a cuckoo clock is 0.49 kg. How far from the pin should the bob be placed such that the pendulum's period of oscillation is 1.0 s?

3.42. When a 400-kg mass is placed on the end of a cantilever beam, it deflects 5.0 mm. What is the period of free oscillations of the mass?

3.43. When the 5.1-kg connecting rod of Fig. P3.43 is placed in the position shown, the spring deflects 0.5 mm. When the end of the rod is displaced and released, the resulting period of oscillation is observed as 0.15 s. Determine the location of the center of gravity of the connecting rod and the centroidal mass moment of inertia of the rod.

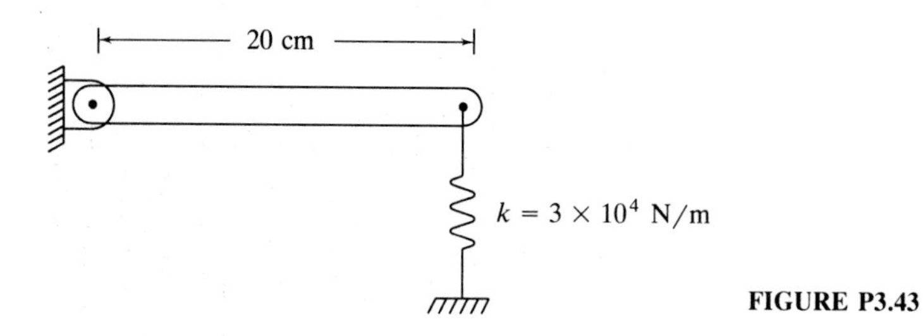

FIGURE P3.43

3.44. One end of the mercury-filled U-tube manometer of Fig. P3.44 is open to the atmosphere while the other end is capped and under a pressure of 20 psig. The cap is suddenly removed. Determine $x(t)$, the displacement of the mercury-air interface from the column's equilibrium position.

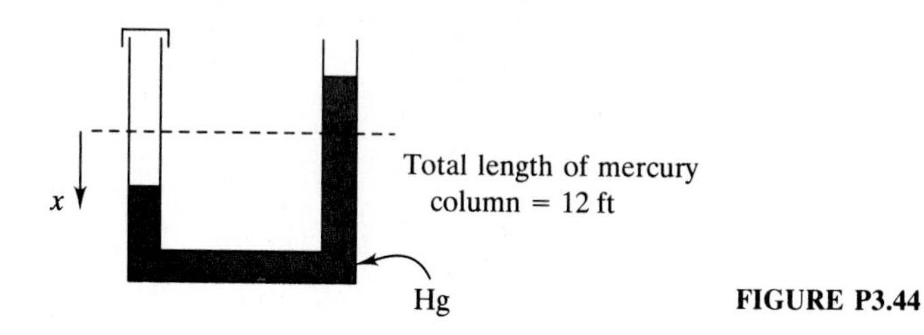

FIGURE P3.44

3.45. A moment M is applied to the pulley of Fig. P3.2. Determine $x(t)$ after the moment is suddenly removed.

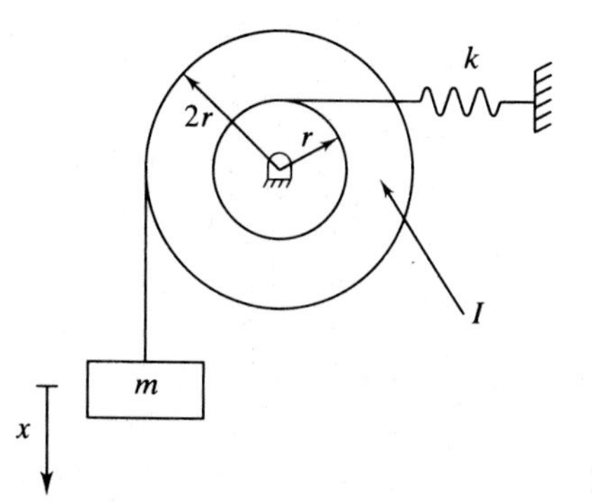

FIGURE P3.45

3.46. The disk of Fig. P.46 is displaced a distance δ and the system released. Determine $x(t)$.

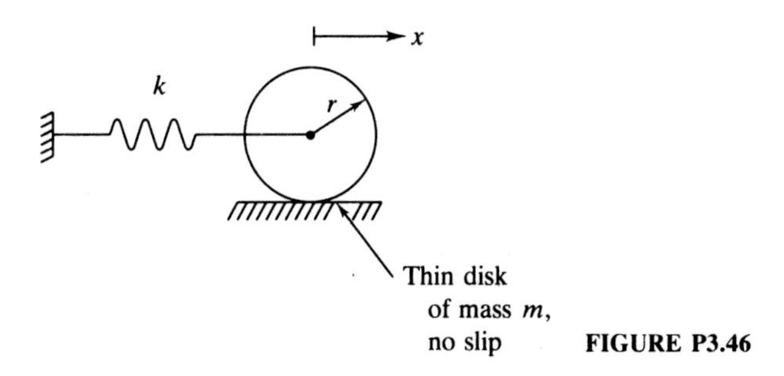

FIGURE P3.46

3.47. A force F is statically applied to the mass of Fig. P3.47. Determine $x(t)$ when the force is suddenly removed.

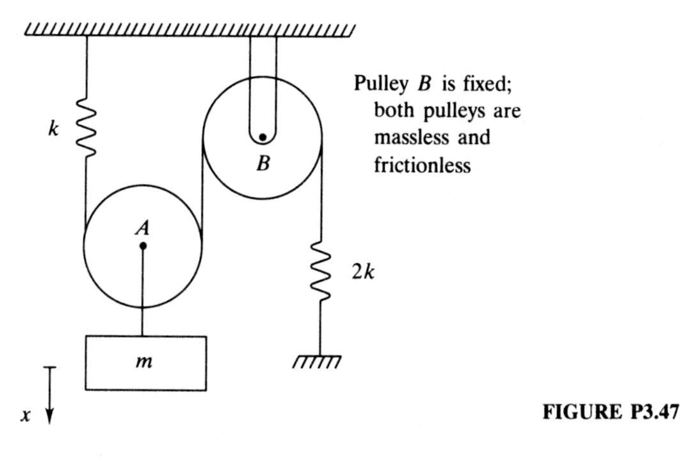

FIGURE P3.47

3.48. An angular impulse of magnitude J is applied to the end furthest from the pin of the slender rod of Fig. P3.48. Determine $\theta(t)$.

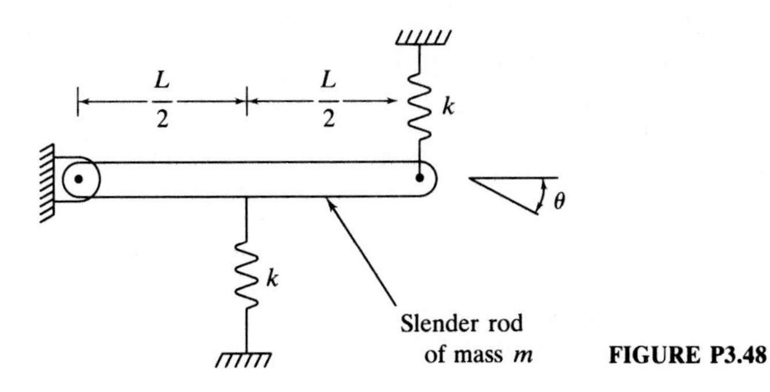

FIGURE P3.48

3.49. The coefficient of friction between the disk and surface of Fig. P3.49 is μ. What is the largest initial velocity of the mass center that can be imparted such that the disk rolls without slip over its entire motion?

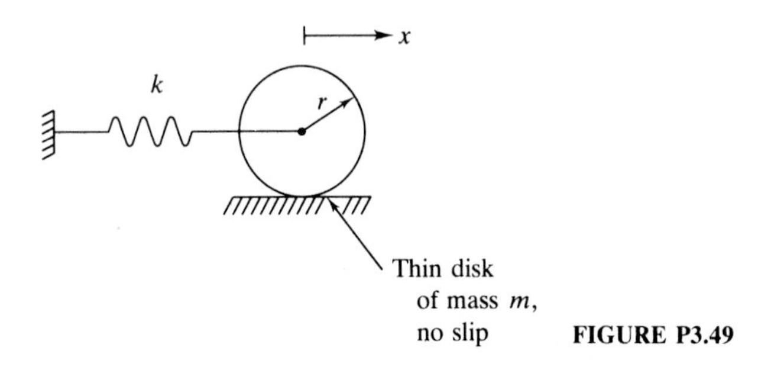

FIGURE P3.49

3.50. A 200-kg package is being hoisted 50 m by a 120-mm-diameter steel ($E = 210 \times 10^9$ N/m^2) cable at a constant velocity v. What is the largest value of v such that the cable's elastic strength of 560×10^6 N/m^2 is not exceeded if the hoisting mechanism suddenly fails and the package is not within 10 m of its destination.

3.51. A 3-kg block is hanging in equilibrium from a coil spring of diameter 5 cm made from a steel rod of diameter 5 mm. The shear modulus of the spring's material is 80×10^9 N/m^2 and the spring has 25 active coils. A weight W is dropped from 4 m onto the block. Upon impact, the weight attaches itself to the block. What is the largest weight that can be dropped such that the spring's elastic shear strength 100×10^6 N/m^2 is not exceeded?

3.52. The system shown in Fig. P3.52 is in static equilibrium when the slender rod is horizontal. A 1000-N · m moment is applied to the end of the rod and suddenly removed. Assuming small displacements,

(*a*) Determine $\theta(t)$.

(*b*) What is the maximum angular acceleration of the rod?

(*c*) Comment on the validity of the small-angle assumption.

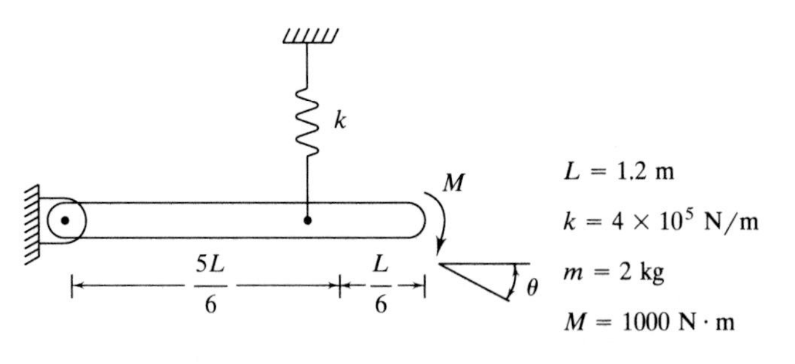

FIGURE P3.52

3.53. When a 40-kg machine is placed on an elastic foundation, its free vibrations appear to decay exponentially with a frequency of 91.7 rad/s. When a 60-kg machine is placed on the same foundation, the frequency of the exponentially decaying oscillations is 75.5 rad/s. Determine the equivalent stiffness and equivalent viscous damping coefficient for the foundation.

3.54. The amplitude of vibration of the system of Fig. P3.54 decays to half of its initial value in 11 cycles with a period of 0.3 s. Determine the spring stiffness and the viscous damping coefficient.

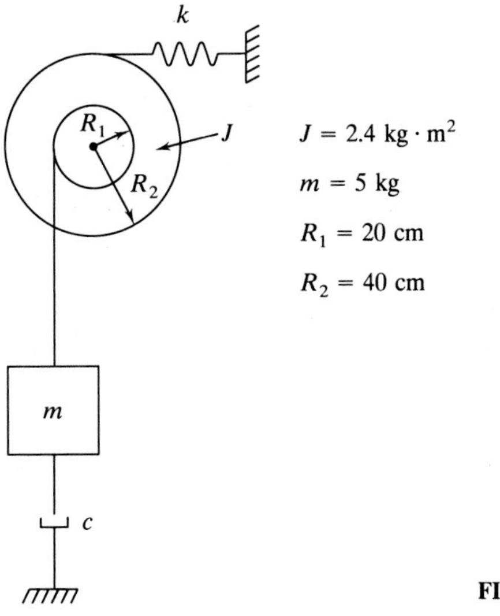

$J = 2.4 \text{ kg} \cdot \text{m}^2$

$m = 5 \text{ kg}$

$R_1 = 20 \text{ cm}$

$R_2 = 40 \text{ cm}$

FIGURE P3.54

3.55. If the stiffness of the spring in Fig. P3.54 is 10,000 N/m, what is the damping coefficient c that gives critical damping?

3.56. The 40-cm radius, 1-kg-thin disk of Fig. P3.56 is connected to a spring of stiffness 4×10^3 N/m in parallel with a viscous damper. The disk rolls without slip.

(*a*) What is the critical damping coefficient of the system?

(*b*) If $c = c_c/2$ plot the response of the system when the center of the disk is displaced 5 mm from equilibrium and released.

(*c*) Repeat part (*b*) if $c = 3c_c/2$.

(*d*) If the coefficient of friction between the disk is 0.15, is the no-slip assumption valid for parts (*b*) and (*c*)?

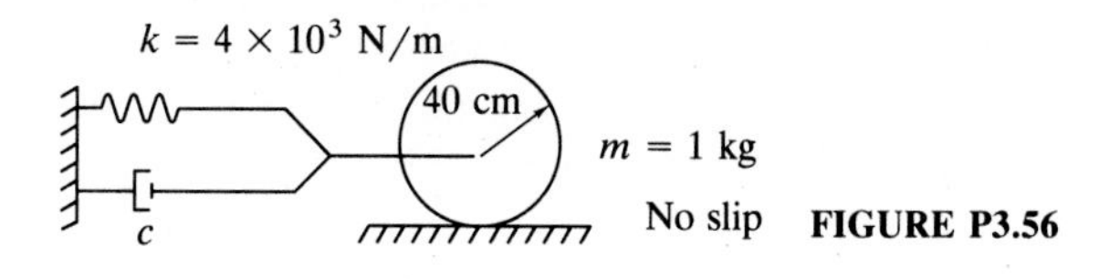

FIGURE P3.56

3.57. The block in the system of Fig. P3.57 is displaced and released from rest. For what values of c is the resulting motion nonoscillatory?

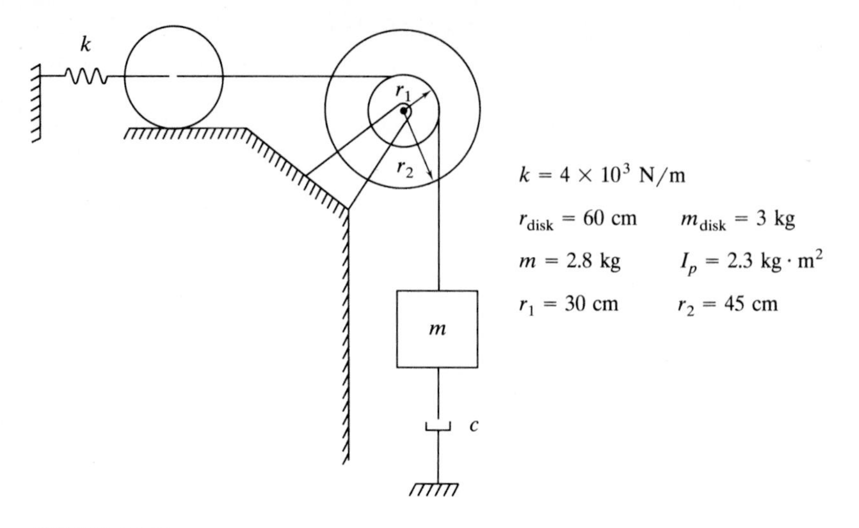

FIGURE P3.57

3.58. The damping ratio of the system of Fig. P3.58 is 1/3. How long will it take for an amplitude of free oscillation to be reduced to 2% of its initial value?

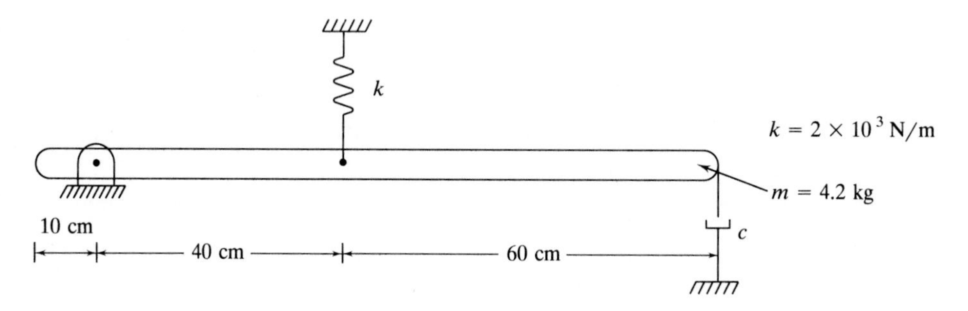

FIGURE P3.58

3.59. A recoil mechanism of a gun is designed as a spring and viscous damper in parallel such that critical damping is attained. A 52-kg gun has a maximum recoil of 50 cm during firing. Specify the stiffness and the damping coefficient of a recoil mechanism that returns to within 5 mm of firing position 0.5 s after firing.

3.60. The response of an underdamped system is given by Eq. (3.32). Determine the maximum displacement in terms of x_0, $\dot{x}_0$, ω_n, and ζ for

(a) $x_0 \neq 0$, $\dot{x}_0 = 0$.
(b) $x_0 = 0$, $\dot{x}_0 \neq 0$.
(c) $x_0 \neq 0$, $\dot{x}_0 \neq 0$.

3.61. Derive Eq. (3.49).

3.62. Derive Eq. (3.53).

3.63. Without calculating the system response, sketch the free-vibration response of a system with $\zeta = 1.0$, $\omega = 10$ rad/s, and
(*a*) $x_0 = 0$, $\dot{x}_0 = 0.5$ m/s.
(*b*) $x_0 = 0.05$ m, $\dot{x}_0 = 0$.
(*c*) $x_0 = 0.1$ m, $\dot{x}_0 = 1.1$ m/s.
(*d*) $x_0 = -0.1$ m, $\dot{x}_0 = 0.5$ m/s.
(*e*) $x_0 = -0.1$ m, $\dot{x}_0 = 2.0$ m/s.

3.64. A 20-kg block is attached to a spring of stiffness 4×10^5 N/m in parallel with a viscous damper. The block is displaced 10 mm from equilibrium and released. How long will it take for the block to become permanently within 1 mm of equilibrium if
(*a*) $c = 2200$ N · s/m?
(*b*) $c = 7500$ N · s/m?

3.65. A block of mass m is attached to a spring of stiffness k and slides on a horizontal surface with a coefficient of friction μ. At some time t, the velocity is zero and the block is displaced a distance δ from equilibrium. Use the principle of work-energy to calculate the spring deflection at the next instant when the velocity is zero. Can this result be generalized to determine the decrease in amplitude between successive cycles?

3.66. Reconsider Example 3.19 using a work-energy analysis. That is, assume the amplitude of the swing is θ at the end of an arbitrary cycle. Use the principle of work-energy to determine the amplitude at the end of the next half-cycle.

3.67. The center of the thin disk of Fig. 3.67 is displaced a distance δ and the disk released. The coefficient of friction between the disk and the surface is μ. The initial displacement is sufficient to cause the disk to roll and slip.
(*a*) Derive the differential equation governing the motion when the disk rolls and slips.
(*b*) When the displacement of the mass center from equilibrium becomes small enough, the disk rolls without slip. At what displacement does this occur?
(*c*) Derive the differential equation governing the motion when the disk rolls without slip.
(*d*) What is the change in amplitude per cycle of motion?

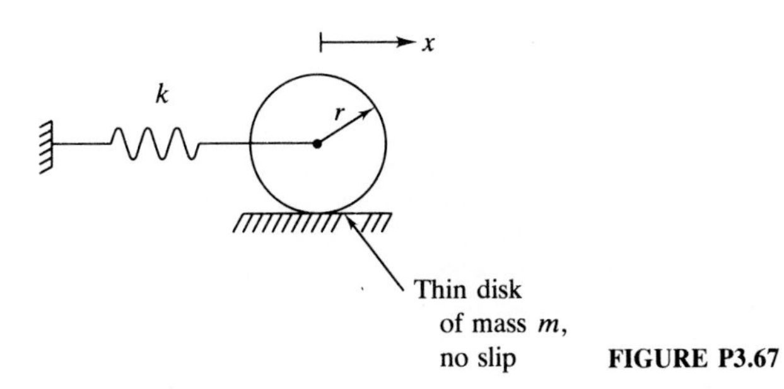

FIGURE P3.67

3.68. A 10-kg block is attached to a spring of stiffness 3×10^4 N/m. The block slides on a horizontal surface with a coefficient of friction of 0.2. The block is displaced 30 mm and released. How long will it take before the block returns to rest?

3.69. The block of Prob. 3.68 is displaced 30 mm and released. What is the range of values of the coefficient of friction such that the block comes to rest during the 14th cycle?

3.70. A 2.2-kg block is attached to a spring of stiffness 1000 N/m and slides on a surface that makes an angle of 7° with the horizontal. When displaced from equilibrium and released, the decrease in amplitude per cycle of motion is observed to be 2 mm. Determine the coefficient of friction.

3.71. The coefficient of friction between the block and the horizontal surface of Fig. P3.71 is μ. If the vertically moving mass is displaced a distance δ from equilibrium and released, how many cycles will the system execute before coming to rest? What is the permanent displacement of the horizontally moving mass?

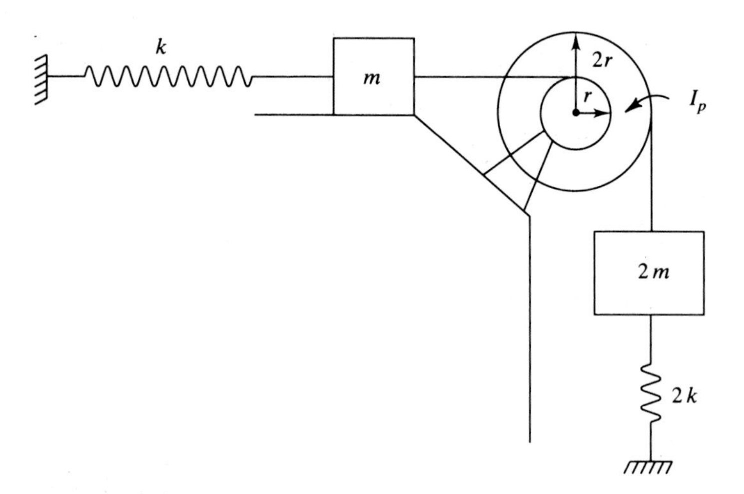

FIGURE P3.71

3.72. An impulse is applied to give the vertically moving block of Fig. P3.72 a velocity v. If the coefficient of friction between the block and the surface is μ, how long will it take for the system to return to rest?

3.73. A block of mass m is attached to a spring of stiffness k and viscous damper of damping coefficient c and slides on a horizontal surface with a coefficient of friction μ. Let $x(t)$ represent the displacement of the block from equilibrium.
(a) Derive the differential equation governing $x(t)$.
(b) Solve the equation and sketch the response over two periods of motion.

3.74. A connecting rod is fitted around a cylinder with a connecting rod between the cylinder and bearing. The coefficient of friction between the cylinder and bearing is 0.08. If the rod is rotated 12° counterclockwise and then released, how many cycles of motion will it execute before it comes to rest?

3.75. If slipping occurs between the belt and pulley of Fig. P3.16, what is the decrease in amplitude per cycle of motion?

3.76. The radius of the glass tube of Fig. P3.76 is 2 cm. What is the damping ratio for the column of liquid as it moves through the manometer? Resolve Prob. 3.44 including the effect of viscous damping. Refer to Prob. 2.42.

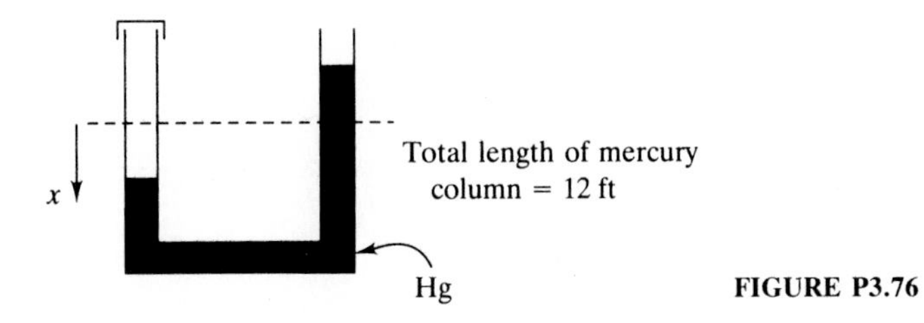

FIGURE P3.76

3.77. A one-degree-of-freedom structure has a mass of 65 kg and a stiffness of 238 N/m. After 10 cycles of motion the amplitude of free vibrations is decreased by 75%. Calculate the hysteretic damping coefficient and the total energy lost during the first 10 cycles if the initial amplitude is 20 mm.

3.78. The end of a steel cantilever beam ($E = 210 \times 10^9\ \text{N/m}^2$) of $I = 1.5 \times 10^{-4}\ \text{m}^4$ is given an initial amplitude of 4.5 mm. After 20 cycles of motion the amplitude is observed as 3.7 mm. Determine the hysteretic damping coefficient and the equivalent viscous damping ratio for the beam.

3.79. Is the system of Fig. P3.79 stable?

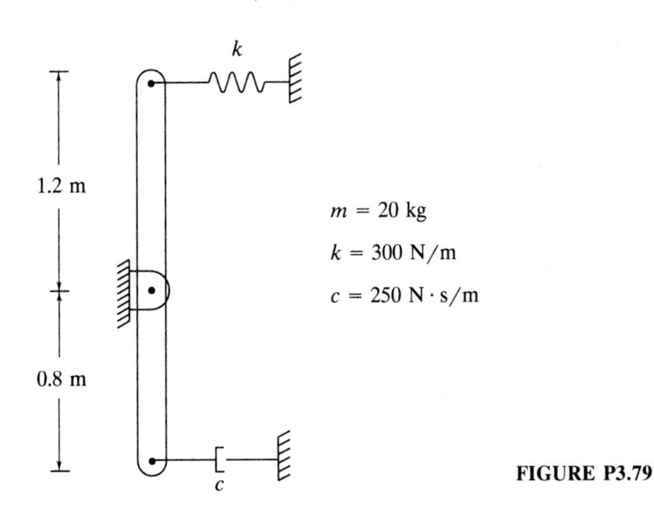

FIGURE P3.79

3.80. Under what condition is the system of Fig. P3.80 stable?

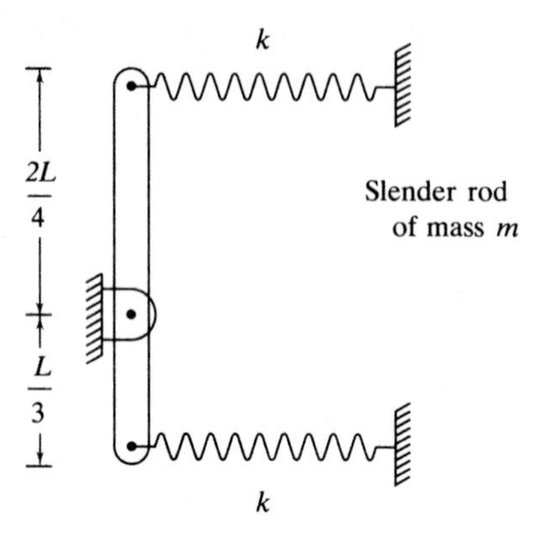

FIGURE P3.80

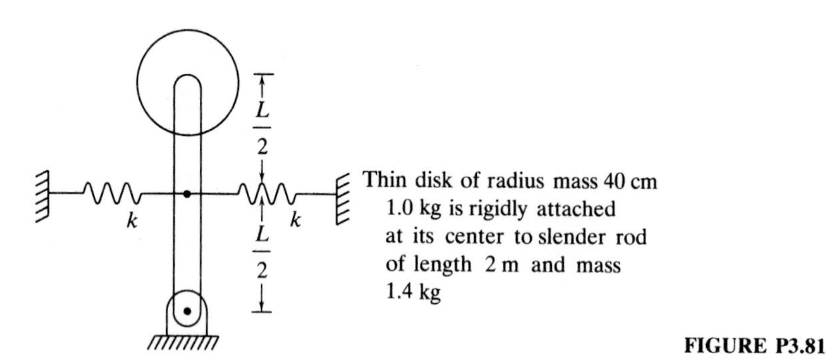

FIGURE P3.81

3.81. For what values of k will the system of Fig. P3.81 be stable?

FREE.BAS is a program included on the accompanying diskette that simulates the free vibration response of a mass-spring-dashpot system. Problems 3.82–3.89 require the use of FREE.BAS.

3.82. Solve Prob. 3.63 using FREE.BAS.

3.83. Solve Probs. 3.56(*b*) and 3.56(*c*) using FREE.BAS.

3.84. Solve Prob. 3.58 using FREE.BAS.

3.85. A 25-kg block is suspended from a spring of stiffness 10,000 N/m in parallel with a dashpot of damping coefficient 100 N · m/s. The block is displaced 5 mm from equilibrium and released. Use FREE.BAS to determine the period of the ensuing motion and the logarithmic decrement.

3.86. The block of Prob. 3.85 is displaced 5 mm from equilibrium and given an initial velocity of 1.2 m/s. Use FREE.BAS to determine the maximum displacement of the block from the equilibrium and the phase angle of the motion.

3.87. Repeat parts (*b*) and (*c*) of Ex. 3.17 using FREE.BAS if $c = 1.5c_c$.

3.88. The mass of the bar of Fig. P3.14 is 15 kg, the springs each have stiffness 40,000 N/m and $c = 250$ N · s/m. The bar is rotated 5° clockwise from equilibrium and released. Use FREE.BAS to determine the time-dependent displacement of the mass center of the bar.

3.89. A 25-kg block is suspended from a spring of unknown stiffness in parallel with a dashpot of unknown damping coefficient. A free-vibration test is run to determine the unknown values. The block is displaced 5 mm from equilibrium and released.
(*a*) Run FREE.BAS, selecting "(4) random choice of system constants" from the main menu. When prompted for a seed number, input the last four digits of your social security number. The resulting plot simulates the system response as reported by an oscilloscope. Use this information to determine k and c.
(*b*) Repeat part (*a*) using twice the last four digits of your social security number as the seed number.

3.90. A 52-kg gun has a recoil of 50 cm during firing. The mechanism is to be designed with a stiffness of 40,000 N/m. Use FREE.BAS to develop a table of the time required to return to within 60 mm of equilibrium versus the damping ratio, ζ, for $1.0 \le \zeta \le 1.5$ for ζ in increments of 0.1.

COULOMB.BAS is a program included on the accompanying diskette that simulates the free and forced response of a one-degree-of-freedom system subject to Coulomb damping. Use COULOMB.BAS to solve Probs. 3.91–3.93.

3.91. A 20-kg block is attached to a spring of stiffness 10,000 N/m. The block slides on a surface with $\mu = 0.15$. The block is displaced 25 mm from equilibrium and released. How long will it take before the ensuing motion ceases?

3.92. Use COULOMB.BAS to solve Ex. 3.19 if $\mu = 0.05$.

3.93. A 40-kg block is attached to spring of unknown stiffness and slides on a surface for which the friction coefficient is unknown. Run COULOMB.BAS. Select a "random choice of system constants" from the main menu. Devise an experiment to be simulated using COULOMB.BAS to determine k and μ.

REFERENCES

1. Bert, C. W.: "Material Damping: An Introductory Review of Mathematical Models, Measurements, and Experimental Techniques," *Journal of Sound and Vibration*, vol. 29, pp. 129–153, 1973.
2. Boyce, C. E., and R. C. DiPrima: *An Introduction to Ordinary Differential Equations*, Wiley, New York, 1970.
3. Crandall, S.H.: The Role of Damping in Vibration Theory," *Journal of Sound and Vibration*, vol. 11, pp. 3–18, 1970.
4. DenHartog, J. P.: *Mechanical Vibrations*, 4th ed., McGraw-Hill, 1956.
5. Dimarogonas, A. D., and S. Haddad: *Vibrations for Engineers*, Prentice-Hall, Englewood Cliffs, N.J., 1992.
6. James, M. L., G. M. Smith, J. C. Wolford, and P. W. Whaley: *Vibrations of Mechanical and Structural Systems with Microcomputer Applications*, Harper and Row, New York, 1989.
7. Kreider, D. L., R. G. Kuller, D. R. Ostberg, and F. W. Perkins: *An Introduction to Linear Analysis*, 2nd ed., Addison-Wesley, Reading, Mass., 1966.
8. Meirovitch, L.: *Elements of Vibration Analysis*, McGraw-Hill, New York, 1975.

9. Rabenstein, A. L.: *Elementary Differential Equations with Linear Algebra*, 3rd ed., Academic Press, New York, 1982.
10. Rao, S. S.: *Mechanical Vibrations*, 2nd ed., Addison-Wesley, Reading, Mass., 1990.
11. Shabana, A. A.: *Theory of Vibration: An Introduction*, vol. 1, Springer-Verlag, New York, 1991.
12. Shabana, A. A.: *Theory of Vibration: Discrete and Continuous Systems*, vol. 2, Springer-Verlag, New York, 1991.
13. Steidel, R. F.: *An Introduction to Mechanical Vibrations*, 3rd ed., Wiley, New York, 1988.
14. Thomson, W. T.: *Theory of Vibrations with Applications*, 3rd ed., Prentice-Hall, Englewood Cliffs, N.J., 1988.
15. Weaver, W., S. P. Timoshenko, and D. H. Young: *Vibration Problems in Engineering*, 5th ed., Wiley-Interscience, New York, 1990.
16. Wylie, C. R., and R. Barrett: *Advanced Engineering Mathematics*, McGraw-Hill, New York, 1982.

CHAPTER 4

HARMONIC EXCITATION OF ONE-DEGREE-OF-FREEDOM SYSTEMS

4.1 INTRODUCTION

Vibrations of a mechanical system occur as a result of an external energy supplied to the system. Free vibrations occur due to an energy source which is removed while the vibrations occur. Forced vibrations occur when work is being done on a system while the vibrations occur. The external work can be in many forms. The ground motion during an earthquake provides an external excitation to an elastic structure. A machine with rotating components can be excited by a harmonic torque provided by a motor.

Vibrations due to periodic (or harmonic) excitations are considered in this chapter. The time history of a harmonic excitation can look like the single-frequency sinusoid of Fig. 4.1*a* or the more complicated periodic form of Fig. 4.1*b*. Vibrations of systems subject to both excitations can be analyzed using the method of this chapter. Vibrations resulting from more general excitations are considered in Chap. 5. Harmonic excitations are considered separately and first because the solution for the forced response is simple, easy to obtain, and can be written in a general nondimensional form. Periodic excitations are usually applied over a long period, while nonperiodic excitations are often of short duration.

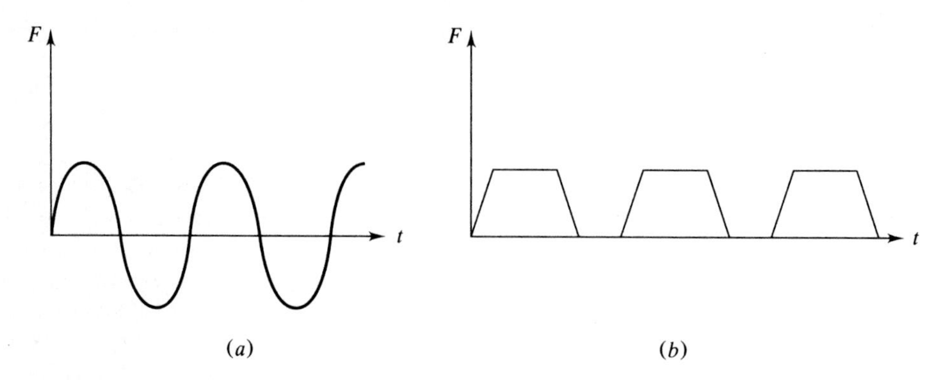

FIGURE 4.1
(*a*) Single-frequency sinusoidal excitation; (*b*) periodic excitation.

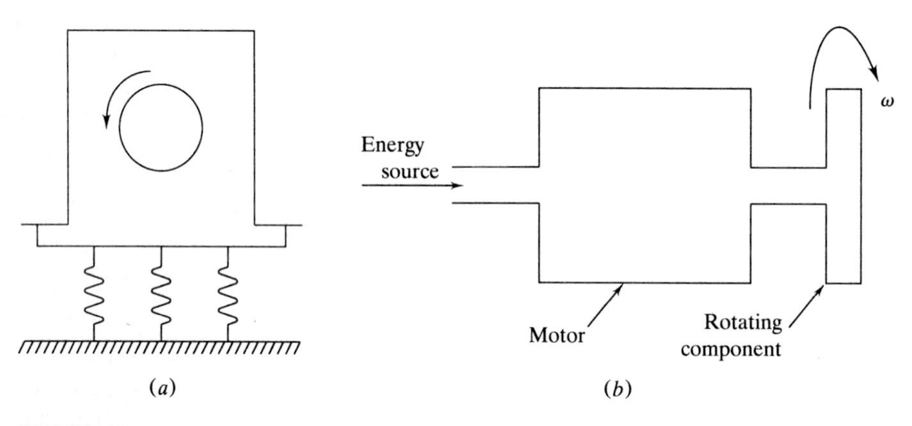

FIGURE 4.2
(*a*) Machine with unbalanced rotating component set on an elastic foundation undergoes harmonic vibrations; (*b*) schematic of rotating component.

Examples of mechanical systems subject to harmonic excitations are shown in Figs. 4.2 through 4.4. If the speed of rotation of the machine of Fig. 4.2 is constant, then, as shown in Sec. 4.6, an inertia force developed due to a rotating unbalance provides a harmonic excitation. If the machine is rigidly mounted to its support, the force transmitted to the support equals the unbalance force. The machine is placed on an elastic support to reduce the transmitted force. This is an application of vibration isolation which is considered in Sec. 4.9.

Vortex shedding due to a steady wind from the light post of Fig. 4.3 leads to harmonic excitation of the light post. Using appropriate assumptions, the vibrations of the light can be modeled as a one-degree-of-freedom system subject to a harmonic excitation.

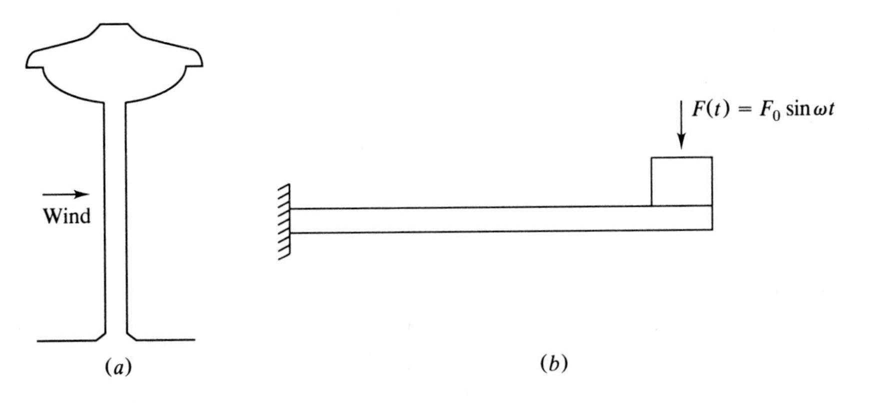

FIGURE 4.3
(a) Steady wind leads to vortex shedding from street light which can lead to harmonic excitation; (b) one-degree-of-freedom model of wind-induced vibrations.

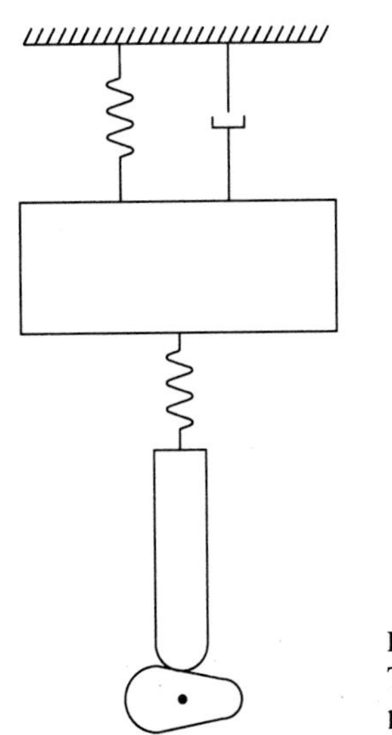

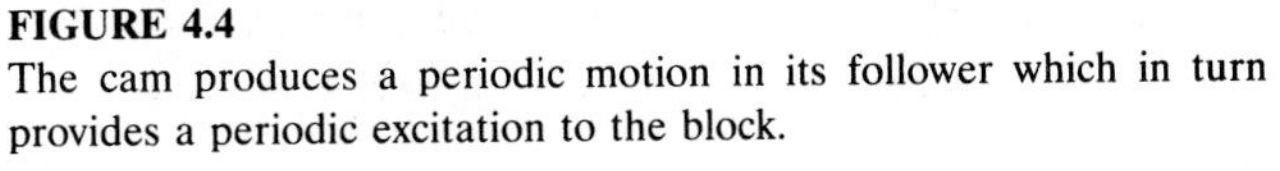

FIGURE 4.4
The cam produces a periodic motion in its follower which in turn provides a periodic excitation to the block.

A cam is used in many machines to provide a desired periodic displacement to its follower. The motion of the follower, in turn, leads to periodic excitation of other parts of the mechanism as shown in Fig. 4.4. The motion of the follower is periodic, but not a single-frequency sinusoid. Vibrations of systems due to general periodic excitations are considered in Sec. 4.13.

The focus of this chapter is the vibration response of a one-degree-of-freedom system under a harmonic excitation. The theory is presented first for a single-frequency excitation. Qualitative behavior of the response due to various forms of single-frequency excitations is studied using nondimensional functions. Specific applications such as the vibrations of unbalanced rotating machinery, design of vibration-measuring instruments, and vibration isolation are considered. The responses of one-degree-of-freedom systems with Coulomb damping and hysteretic damping are obtained. Finally, multifrequency excitations and general periodic excitations are considered.

4.2 DIFFERENTIAL EQUATIONS GOVERNING FORCED VIBRATIONS

Application of Newton's laws to appropriate free-body diagrams is usually the most convenient way to derive the differential equation governing forced vibrations of a one-degree-of-freedom system. Newton's laws are applied to systems with an external excitation in the same manner they are applied to systems undergoing free vibrations. Free-body diagrams showing the external forces acting on the system and the system's effective forces are drawn at an arbitrary instant of time, as illustrated in Chap. 3. The appropriate form of Newton's law is applied to the free-body diagrams, resulting in the governing differential equation.

The mass-spring-dashpot system of Fig. 4.5*a* is a model for one-degree-of-freedom linear systems. Application of Newton's law to the free-body diagrams of Fig. 4.5*b* leads to the following differential equation:

$$m\ddot{x} + c\dot{x} + kx = F(t) \tag{4.1}$$

An alternate method to derive the governing differential equation is to use the equivalent-systems method of Chap. 2 in conjunction with the method of virtual work. It is shown in Chap. 2 that any linear one-degree-of-freedom system can be modeled by the equivalent system of Fig. 4.6. The values of $\tilde{m}$, $\tilde{c}$,

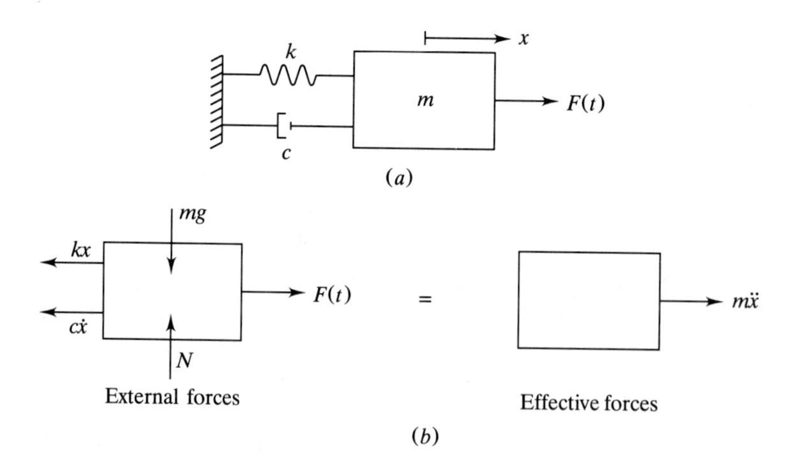

FIGURE 4.5
(*a*) Mass-spring-dashpot system slides on frictionless surface; (*b*) free-body diagrams at an arbitrary instant of time.

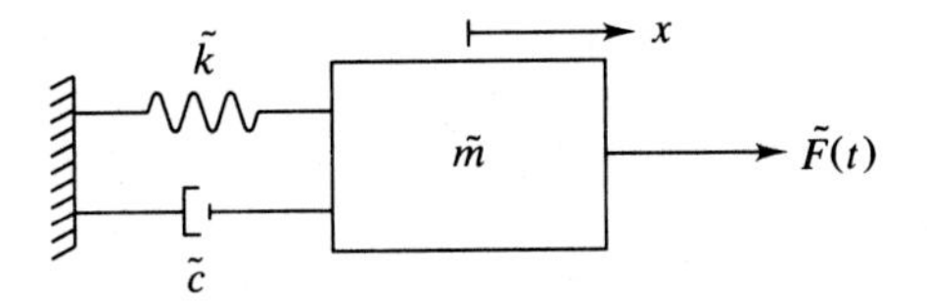

FIGURE 4.6
Equivalent model for a linear one-degree-of-freedom system.

and $\tilde{k}$ are determined using energy methods as described in Chap. 2. The appropriate form of F is obtained by application of the method of virtual work.

Let x be the chosen generalized coordinate. Let δx be a variation of x, which is a small virtual displacement. If the system is given a virtual displacement, then the virtual work done by all external forces can be expressed as

$$\delta W = \tilde{F}\delta x \tag{4.2}$$

The force $\tilde{F}(t)$ appearing in Eq. (4.2) is called the generalized force and is the appropriate force to use when modeling the one-degree-of-freedom system by the system of Fig. 4.6.

By analogy with the system of Fig. 4.5, the differential equation governing the vibrations of a system modeled by the system in Fig. 4.6 is

$$\tilde{m}\ddot{x} + \tilde{c}\dot{x} + \tilde{k}x = \tilde{F}(t) \tag{4.3}$$

Example 4.1. Assuming small displacements and using θ as the generalized coordinate, derive the differential equation governing forced vibrations of the

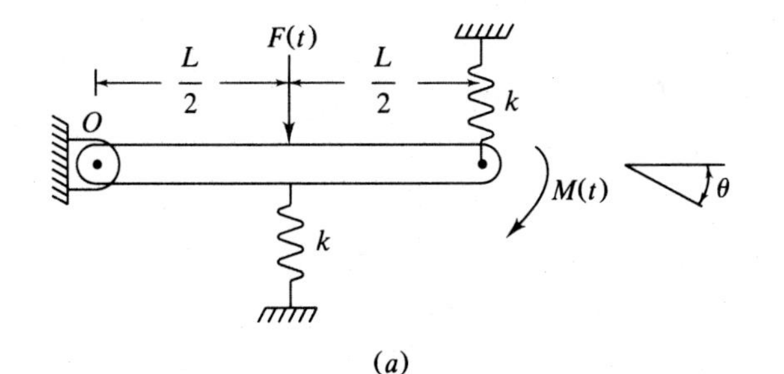

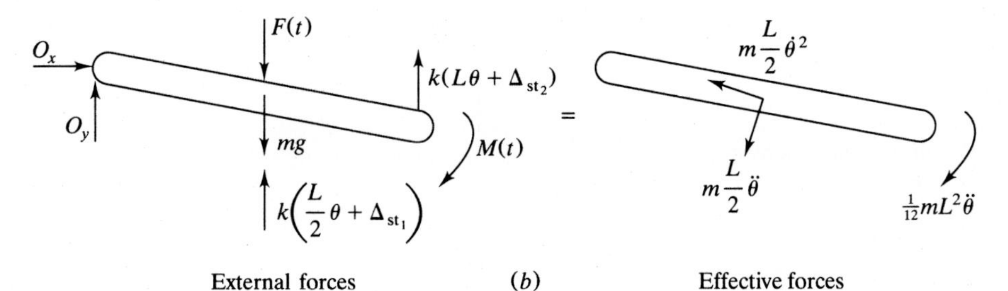

FIGURE 4.7
(a) System of Example 4.1; (b) free-body diagrams at an arbitrary instant of time.

system of Fig. 4.7 using

(*a*) Newton's laws applied to appropriate free-body diagrams.
(*b*) The equivalent-systems approach.

(*a*) Summing moments about the point of support on the free-body diagrams of Fig. 4.7*b*

$$\left(\sum M_O\right)_{\text{ext}} = \left(\sum M_O\right)_{\text{eff}}$$

and using the small-angle assumption leads to

$$F(t)\frac{L}{2} + mg\frac{L}{2} - k\left(\frac{L}{2}\theta + \Delta_{\text{st}_1}\right)\frac{L}{2} - k(L\theta + \Delta_{\text{st}_2})L + M(t)$$

$$= m\frac{L}{2}\ddot{\theta}\frac{L}{2} + \frac{1}{12}mL^2\ddot{\theta}$$

Noting from static equilibrium that the static-deflection terms cancel with gravity and rearranging leads to

$$m\frac{L^2}{3}\ddot{\theta} + \frac{5}{4}kL^2\theta = M(t) + \frac{L}{2}F(t)$$

(*b*) Since θ represents an angular displacement, the appropriate model analogous to the system of Fig. 4.6 is a disk of moment of inertia $\tilde{I}$ attached to a torsional spring of torsional stiffness $\tilde{k}$, with a moment $\tilde{M}$ applied to the disk. Equating the kinetic energy of such a system with the kinetic energy of the system of Fig. 4.6,

$$\frac{1}{2}\tilde{I}\dot{\theta}^2 = \frac{1}{2}m\left(\frac{L}{2}\dot{\theta}\right)^2 + \frac{1}{2}\frac{1}{12}mL^2\dot{\theta}^2$$

which leads to

$$\tilde{I} = \tfrac{1}{3}mL^2$$

Equating the potential energies of the two systems,

$$\tfrac{1}{2}\tilde{k}\theta^2 = \tfrac{1}{2}k\left(\frac{L}{2}\theta\right)^2 + \tfrac{1}{2}k(L\theta)^2$$

which leads to

$$\tilde{k} = \tfrac{5}{4}kL^2$$

If the bar is given a virtual rotation $\delta\theta$, the work done by the external force and moment is

$$\delta W = F\delta\left(\frac{L}{2}\theta\right) + M\delta\theta$$

$$= \left(F\frac{L}{2} + M\right)\delta\theta$$

Thus

$$\tilde{M} = F\frac{L}{2} + M$$

The differential equation governing vibrations of the equivalent system is

$$\tilde{I}\ddot{\theta} + \tilde{k}\theta = \tilde{M}$$

or

$$\frac{1}{3}mL^2\ddot{\theta} + \frac{5}{4}kL^2\theta = F\frac{L}{2} + M$$

which is the same equation derived using Newton's laws.

The forced motion of any one-degree-of-freedom system can be studied by studying the solution of Eq. (4.3). Equation (4.3) is divided by $\tilde{m}$ yielding

$$\ddot{x} + 2\zeta\omega_n\dot{x} + \omega_n^2 x = \frac{\tilde{F}(t)}{\tilde{m}} \tag{4.4}$$

where ζ is the damping ratio and ω_n is the natural frequency of free vibration. Unless otherwise specified, it is assumed that free vibrations of the system are underdamped, $\zeta < 1$.

Equation (4.4) is the general equation governing forced vibrations of a one-degree-of-freedom system with viscous damping. The homogeneous solution of this second-order linear nonhomogeneous differential equation is independent of $\tilde{F}(t)$ and is

$$x_h(t) = \exp(-\zeta\omega_n t)\left[C_1\cos\left(\omega_n\sqrt{1-\zeta^2}\,t\right) + C_2\sin\left(\omega_n\sqrt{1-\zeta^2}\,t\right)\right] \tag{4.5}$$

The general solution is the homogeneous solution plus the particular solution. The particular solution depends upon the specific form of $\tilde{F}(t)$. As t grows large, the homogeneous solution of Eq. (4.5) goes to zero. Thus only the particular solution has a contribution to the long-term behavior of the system which is called the steady state.

4.3 FORCED RESPONSE OF AN UNDAMPED SYSTEM DUE TO A SINGLE-FREQUENCY EXCITATION

The generalized force corresponding to a single-frequency harmonic excitation takes the form

$$F(t) = F_0\sin(\omega t + \psi) \tag{4.6}$$

where F_0 is the magnitude of the excitation, ω is its frequency, and ψ is the phase difference between the excitation and a purely sinusoidal excitation. If $\psi = \pi/2$, then the excitation given by Eq. (4.6) is a pure cosine excitation.

The differential equation for undamped forced vibrations subject to an excitation of the form of Eq. (4.6) is

$$\ddot{x} + \omega_n^2 x = \frac{F_0}{\tilde{m}} \sin(\omega t + \psi) \tag{4.7}$$

If $\omega \neq \omega_n$ the method of undetermined coefficients is used to obtain the particular solution of Eq. (4.7)

$$x_p(t) = \frac{F_0}{\tilde{m}(\omega_n^2 - \omega^2)} \sin(\omega t + \psi) \tag{4.8}$$

The homogeneous solution, Eq. (4.5), is added to the particular solution and the initial conditions of Eq. (3.6) applied, yielding

$$x(t) = \left[x_0 - \frac{F_0 \sin \psi}{\tilde{m}(\omega_n^2 - \omega^2)}\right] \cos(\omega_n t) + \frac{1}{\omega_n}\left[\dot{x}_0 - \frac{F_0 \omega \cos \psi}{\tilde{m}(\omega_n^2 - \omega^2)}\right] \sin(\omega_n t)$$
$$+ \frac{F_0}{\tilde{m}(\omega_n^2 - \omega^2)} \sin(\omega t + \psi) \tag{4.9}$$

The response, plotted in Fig. 4.8, is the sum of two trigonometric terms of different frequencies.

The case when $\omega = \omega_n$ is special. The nonhomogeneous term in Eq. (4.7) and the homogeneous solution are not linearly independent. Thus the method of undetermined coefficients is used to determine the particular solution, and the appropriate trial solution is

$$x_p(t) = At \sin(\omega_n t + \psi) + Bt \cos(\omega_n t + \psi) \tag{4.10}$$

Substitution of Eq. (4.10) in Eq. (4.7) leads to

$$x_p(t) = -\frac{F_0}{2\tilde{m}\omega_n} t \cos(\omega_n t + \psi) \tag{4.11}$$

Application of initial conditions to the sum of the homogeneous and particular solution yields

$$x(t) = x_0 \cos(\omega_n t) + \left(\frac{\dot{x}_0}{\omega_n} + \frac{F_0 \cos \psi}{2\tilde{m}\omega_n^2}\right) \sin(\omega_n t) - \frac{F_0}{2\tilde{m}\omega_n} t \cos(\omega_n t + \psi) \tag{4.12}$$

The response of a system in which the excitation frequency equals the natural frequency grows without bounds, as shown in Fig. 4.9. This condition,

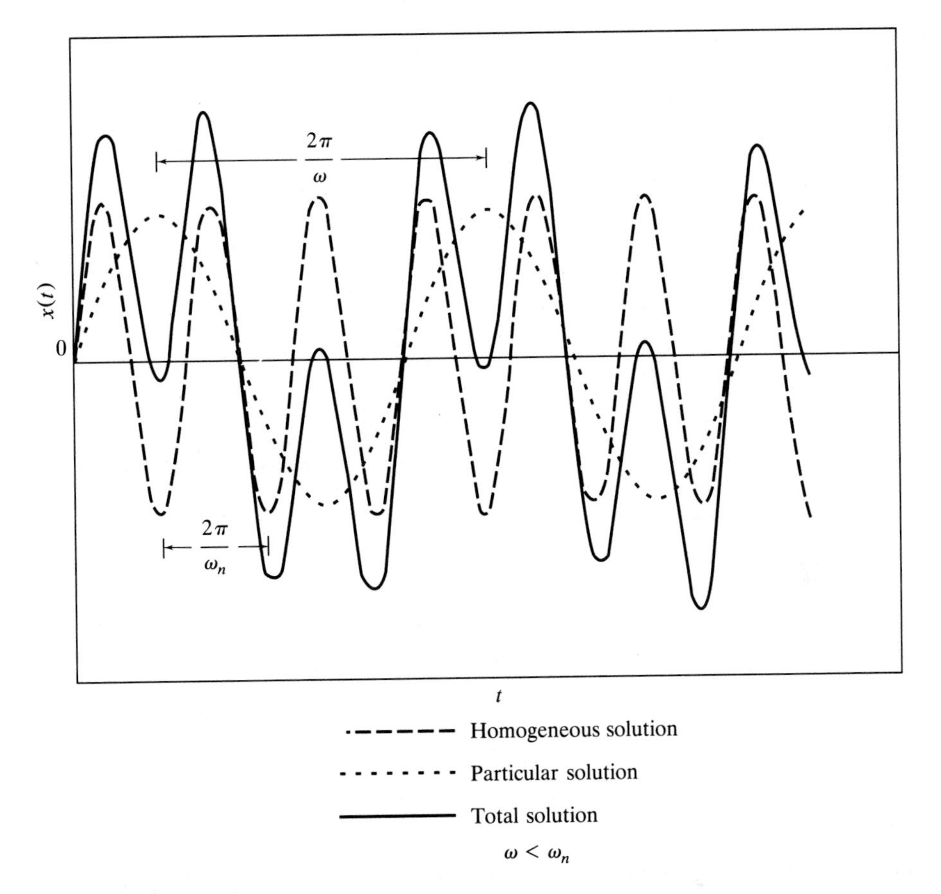

FIGURE 4.8
Response of an undamped system to a single-frequency harmonic excitation.

where the amplitude increases without bound, is called resonance. In a real physical system the amplitude is bounded. In many cases as the amplitude grows, assumptions used in modeling the physical system become invalid. In a system with an elastic element such as a coil spring, the proportional limit of the spring's material is eventually reached. After this time the motion of the system is governed by a nonlinear differential equation reflecting the nonlinear force-displacement relation in the spring. For a system such as a pendulum in which a small displacement assumption is used to linearize the governing differential equation, the assumption becomes invalid when resonance occurs. The original nonlinear differential equation must be used.

Resonance is a dangerous condition in a mechanical or structural system and will produce unwanted large displacements or lead to failure. Resonant vibrations were initially suspected as the cause of the famous Tacoma Narrows

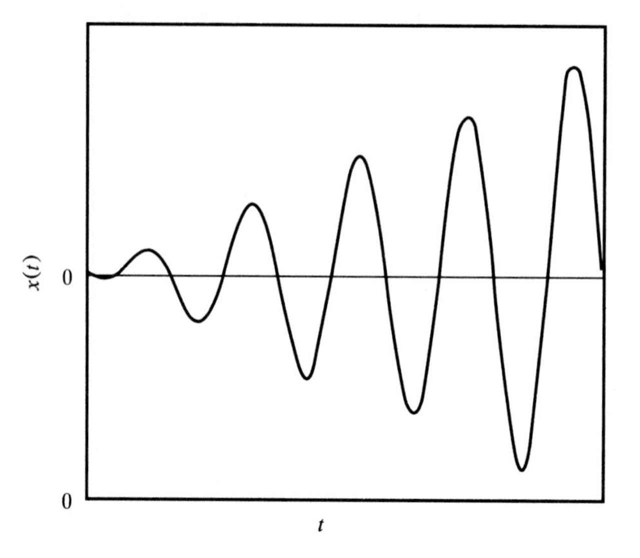

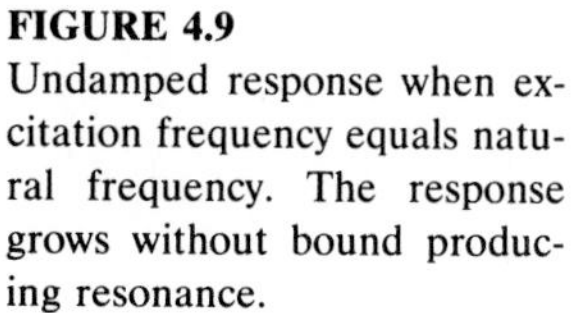
FIGURE 4.9
Undamped response when excitation frequency equals natural frequency. The response grows without bound producing resonance.

Bridge disaster. It was suspected that the frequency at which vortices were shed from the bridge coincided with a torsional natural frequency, leading to oscillations that grew without bound. Recent investigations have shown that nonlinear phenomena resulting from improper installation may have also played a role in the disaster. Resonance can also lead to unwanted large-amplitude vibrations in machinery.

When a conservative system is started in motion, no additional energy input is required to continue periodic motion at its natural frequency. Thus, when the frequency of excitation is the same as the natural frequency, the work done by the external force is not needed to maintain motion. Thus the total energy increases due to the work input and leads to a continual increase in amplitude. When the frequency of excitation is different from the natural frequency, the work done by the external force is necessary to maintain motion at the excitation frequency.

When the excitation frequency is close, but not exactly equal, to the natural frequency, an interesting phenomenon called beating occurs. Beating is a continuous buildup and decrease of amplitude as shown in Fig. 4.10. When ω is very close to ω_n and $x_0 = \dot{x}_0 = 0$ and $\psi = 0$, Eq. (4.9) is approximated by

$$x(t) = \frac{2F_0}{\tilde{m}(\omega_n^2 - \omega^2)} \sin\left(\frac{\omega - \omega_n}{2}\right)t \cos\left(\frac{\omega + \omega_n}{2}\right)t \tag{4.13}$$

From Eq. (4.13), since $|\omega - \omega_n|$ is small, the solution can be viewed as a cosine wave with a slowly varying amplitude. The period of the amplitude is called the period of the beating and equals $2\pi/\,|\omega - \omega_n|$. The period of vibration is $4\pi/(\omega + \omega_n)$.

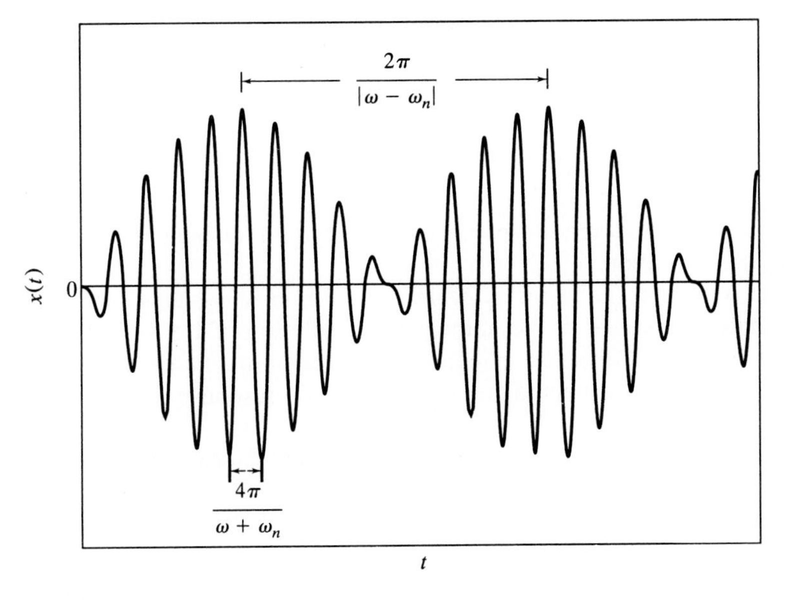

FIGURE 4.10
Beating phenomenon occurs when $\omega \approx \omega_n$.

4.4 FORCED RESPONSE OF A VISCOUSLY DAMPED SYSTEM SUBJECT TO A SINGLE-FREQUENCY HARMONIC EXCITATION

The standard form of the differential equation governing the motion of a viscously damped one-degree-of-freedom system with the single-frequency harmonic excitation of Eq. (4.6) is

$$\ddot{x} + 2\zeta\omega_n\dot{x} + \omega_n^2 x = \frac{F_0}{\tilde{m}}\sin(\omega t + \psi) \tag{4.14}$$

The particular solution of Eq. (4.14) is

$$x_p(t) = \frac{F_0}{\tilde{m}\left[\left(\omega_n^2 - \omega^2\right)^2 + \left(2\zeta\omega\omega_n\right)^2\right]}\left[-2\zeta\omega\omega_n\cos(\omega t + \psi) + \left(\omega_n^2 - \omega^2\right)\sin(\omega t + \psi)\right] \tag{4.15}$$

Use of the trigonometric identity for the sine of the difference of angles and algebraic manipulation leads to the following alternate form of Eq. (4.15):

$$x_p(t) = X\sin(\omega t + \psi - \phi) \tag{4.16}$$

where
$$X = \frac{F_0}{\tilde{m}\left[\left(\omega_n^2 - \omega^2\right)^2 + \left(2\zeta\omega\omega_n\right)^2\right]^{1/2}} \tag{4.17}$$

and
$$\phi = \tan^{-1}\left(\frac{2\zeta\omega\omega_n}{\omega_n^2 - \omega^2}\right) \tag{4.18}$$

X is the amplitude of the forced response and ϕ is the phase angle between the response and the excitation.

Only the long-term behavior is of interest for most systems subject to harmonic excitation. As $t \to \infty$, the homogeneous solution of Eq. (4.5) goes to zero and only the forced response remains. Thus, for harmonic excitation, the free-vibration response is neglected and only the forced response or the steady-state response considered.

The amplitude and phase angle provide important information about the forced response. Formulation of Eqs. (4.17) and (4.18) in nondimensional form allows better qualitative interpretation of the response. It is noted from these equations that

$$X = f(F_0, \tilde{m}, \omega, \omega_n, \zeta) \tag{4.19}$$

and
$$\phi = g(\omega, \omega_n, \zeta) \tag{4.20}$$

The parameters use three basic dimensions: mass, length, and time. Thus the Buckingham Π theorem implies that Eq. (4.17) can be rearranged to yield a relationship between $6 - 3 = 3$ dimensionless parameters. Indeed, multiplying Eq. (4.17) by $\tilde{m}\omega_n^2/F_0$ gives

$$\frac{\tilde{m}\omega_n^2 X}{F_0} = \frac{1}{\left[\left(1 - r^2\right)^2 + \left(2\zeta r\right)^2\right]^{1/2}} \tag{4.21}$$

where
$$r = \frac{\omega}{\omega_n}$$

is the frequency ratio. The ratio

$$M = \frac{\tilde{m}\omega_n^2 X}{F_0} \tag{4.22}$$

is dimensionless and is often called the amplitude ratio or magnification factor. While it can be construed as the ratio of the amplitude of forced response to the amplitude of excitation, a more direct physical interpretation is that of the maximum force developed in the spring of a mass-spring-dashpot system to the maximum value of the exciting force. Thus the nondimensional form of Eq. (4.17) is

$$M(r, \zeta) = \frac{1}{\sqrt{\left(1 - r^2\right)^2 + \left(2\zeta r\right)^2}} \tag{4.23}$$

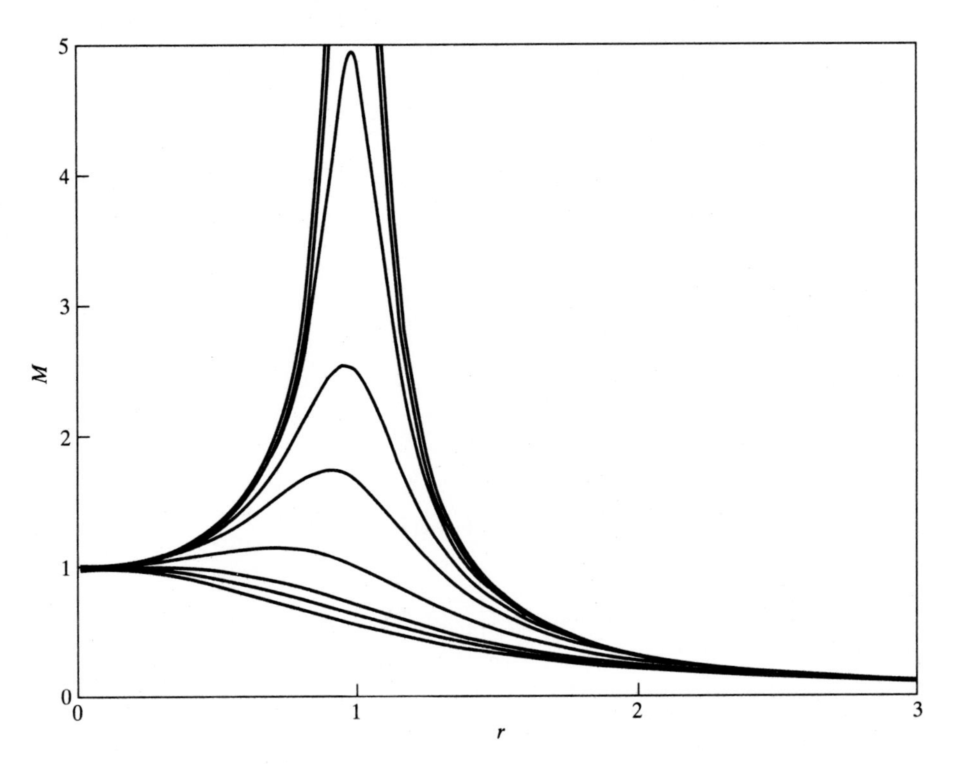

FIGURE 4.11
Magnification factor vs. frequency ratio for different damping ratios.

The magnification factor as a function of frequency ratio for different values of the damping ratio is shown in Fig. 4.11. These curves are called frequency response curves. The following are noted about Eq. (4.23) and Fig. 4.11:

1. $M = 1$ when $r = 0$. In this case the excitation force is a constant and the maximum force developed in the spring of a mass-spring-dashpot system is equal to the value of the exciting force.
2. $M \to 0$ as $r \to \infty$. The amplitude of the forced response is very small for high-frequency excitations.
3. For a given value of r, M decreases with increasing ζ.
4. The magnification factor grows without bound only for $\zeta = 0$. For $0 < \zeta \leq 1/\sqrt{2}$, the magnification factor has a maximum for some value of ζ.
5. For $0 < \zeta \leq 1/\sqrt{2}$, the maximum value of the magnification factor occurs for a frequency ratio of

$$r_m = \sqrt{1 - 2\zeta^2} \tag{4.24}$$

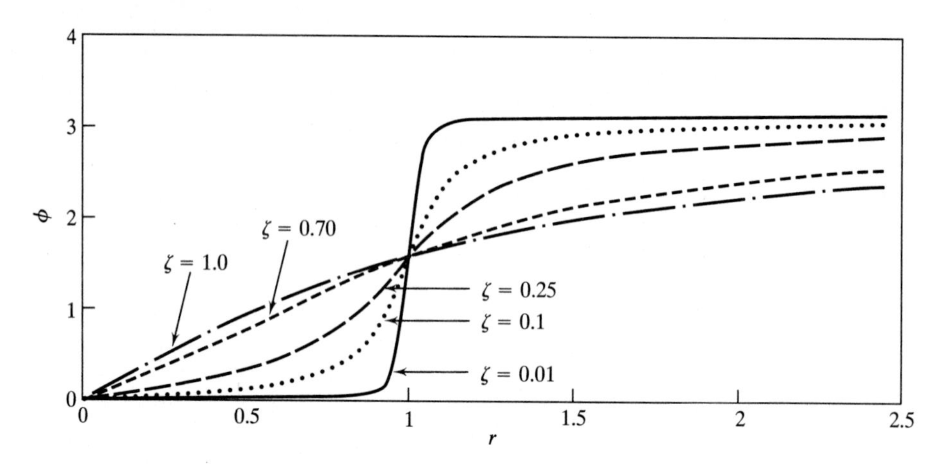

FIGURE 4.12
Phase angle vs. frequency ratio for different damping ratios.

Equation (4.24) is obtained using Eq. (4.23) by determining the value of r such that $dM/dr = 0$.

6. The corresponding maximum value of M is

$$M_{\max} = \frac{1}{2\zeta(1 - \zeta^2)^{1/2}} \tag{4.25}$$

7. For $\zeta = 1\backslash\sqrt{2}$, $dM/dr = 0$ for $r = 0$. For $\zeta > 1/\sqrt{2}$, M monotonically decreases with increasing r.

The nondimensional form of Eq. (4.18) is

$$\phi = \tan^{-1}\left(\frac{2\zeta r}{1 - r^2}\right) \tag{4.26}$$

The phase angle from Eq. (4.26) is plotted as a function of frequency ratio for different values of the damping ratio in Fig. 4.12. The following are noted from Eq. (4.26) and Fig. 4.12:

1. The forced response and the excitation force are in phase for $\zeta = 0$. For $\zeta > 0$, the response and excitation are in phase only for $r = 0$.
2. If $\zeta > 0$ and $0 < r < 1$, then $0 < \phi < \pi/2$. The response lags the excitation as shown in Fig. 4.13.
3. If $\zeta > 0$ and $r = 1$, then $\phi = \pi/2$. If $\psi = 0$, then the excitation is a pure sine wave while the steady-state response is a pure cosine wave. The excitation is in phase with the velocity. The direction of the excitation is always the same as the direction of motion.
4. If $\zeta > 0$ and $r > 1$, then $\pi/2 < \phi < \pi$. The response leads the excitation as shown in Fig. 4.14.

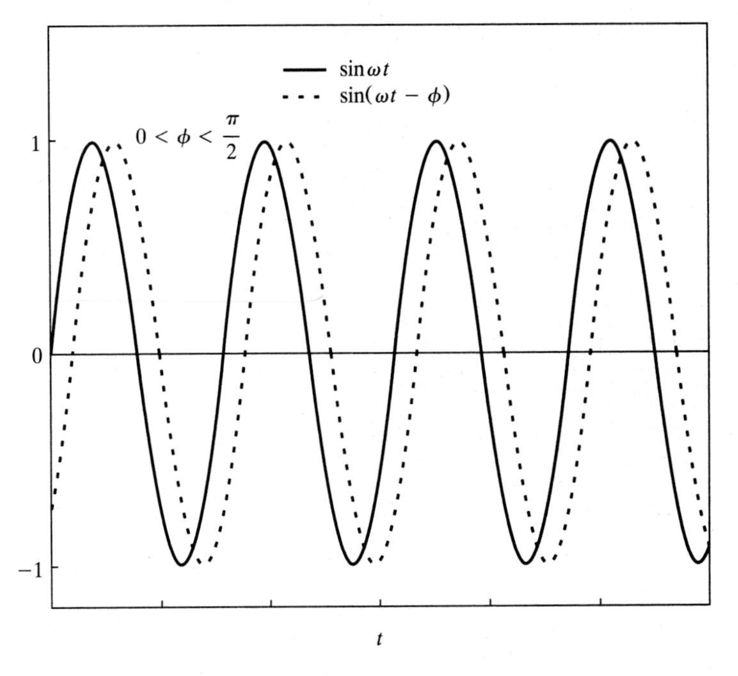

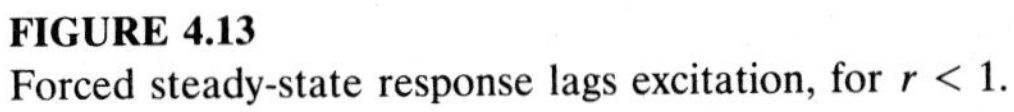

FIGURE 4.13
Forced steady-state response lags excitation, for $r < 1$.

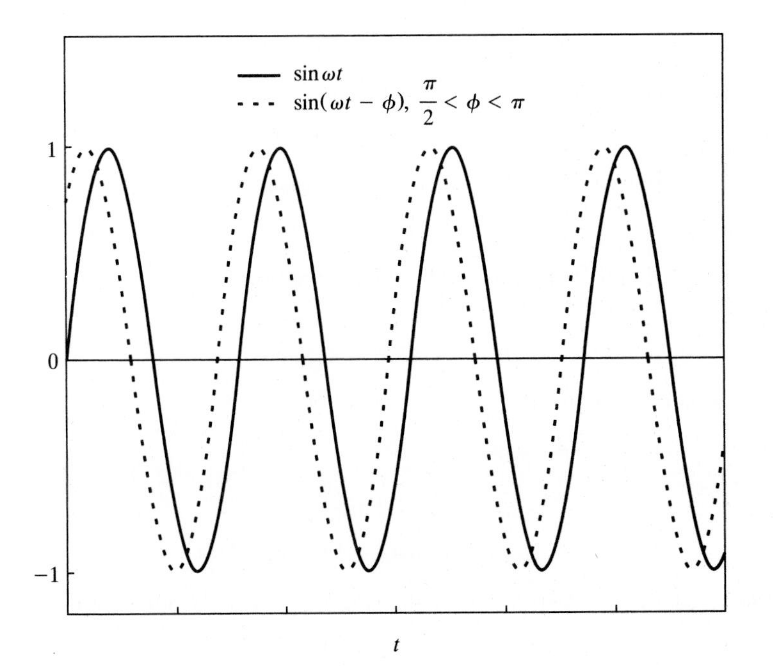

FIGURE 4.14
Response leads excitation when $r > 1$.

5. If $\zeta > 0$ and $r \gg 1$, then $\phi \approx \pi$. The sign of the steady-state response is opposite that of the excitation.

Example 4.2. A moment, $M_0 \sin \omega t$, is applied to the end of the bar of Fig. 4.15. Determine the maximum value of M_0 such that the steady-state amplitude of angular oscillation does not exceed 10° if $\omega = 500$ rpm, $k = 7000$ N/m, $c = 650$ N · s/m, $L = 1.2$ m, and the mass of the bar is 15 kg.

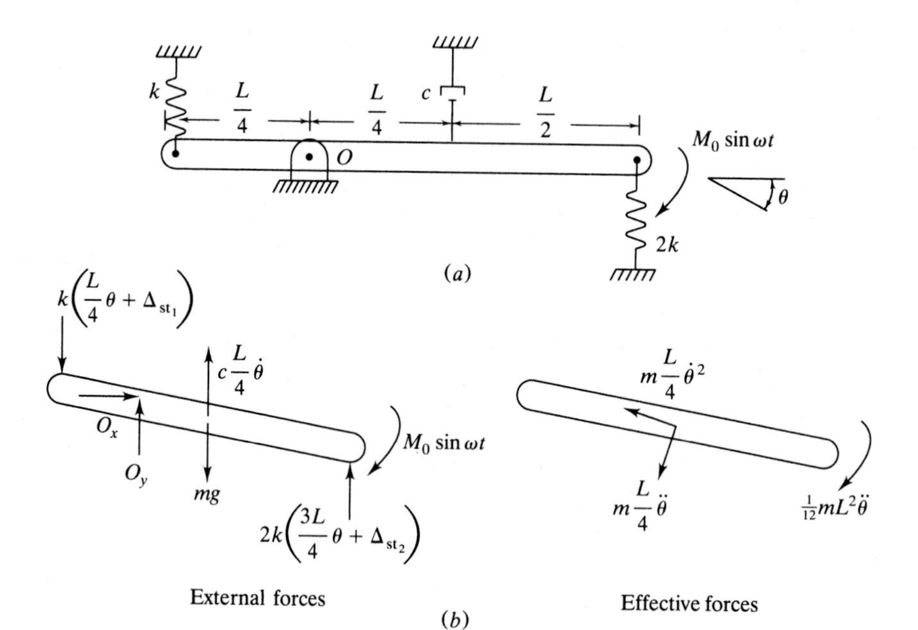

FIGURE 4.15
(*a*) System of Example 4.2; (*b*) free-body diagrams at an arbitrary instant of time.

The differential equation obtained by summing moments about 0 using the free-body diagrams of Fig. 4.15*b* is

$$\tfrac{7}{48}mL^2\ddot{\theta} + \tfrac{1}{16}cL^2\dot{\theta} + \tfrac{19}{16}kL^2\theta = M_0 \sin \omega t$$

Using the notation of Eq. (4.3),

$$\tilde{m} = \tfrac{7}{48}mL^2$$

$$= \tfrac{7}{48}(15 \text{ kg})(1.2 \text{ m})^2$$

$$= 3.15 \text{ kg} \cdot \text{m}^2$$

The differential equation is rewritten in the form of Eq. (4.4) by dividing by $\tilde{m}$:

$$\ddot{\theta} + \frac{3}{7}\frac{c}{m}\dot{\theta} + \frac{57}{7}\frac{k}{m}\theta = \frac{M_0}{\tilde{m}} \sin \omega t$$

The natural frequency and damping ratio are obtained by comparing the preceding equation to Eq. (4.4)

$$\omega_n = \sqrt{\frac{57}{7}\frac{k}{m}}$$

$$= \sqrt{\frac{(57)(7000 \text{ N/m})}{(7)(15 \text{ kg})}}$$

$$= 61.6 \frac{\text{rad}}{\text{s}}$$

$$\zeta = \frac{3}{14}\frac{c}{m\omega_n}$$

$$= \frac{(3)(650 \text{ N}\cdot\text{s/m})}{(14)(15 \text{ kg})(61.6 \text{ rad/s})}$$

$$= 0.15$$

The frequency ratio is

$$r = \frac{(500 \text{ rev/min})(2\pi \text{ rad/rev})(1 \text{ min/60 s})}{61.6 \text{ rad/s}}$$

$$= 0.85$$

The magnification factor is calculated from Eq. (4.21)

$$M = \frac{1}{\sqrt{\left[1-(0.85)^2\right]^2 + [2(0.15)(0.85)]^2}}$$

$$= 2.65$$

The maximum allowable magnitude of the applied moment is calculated using the definition of the magnification factor, Eq. (4.22),

$$\frac{\tilde{m}\omega_n^2\Theta}{M_0} = 2.65$$

where Θ is the amplitude of the steady-state angular oscillations. Requiring $\Theta < 10°$ leads to

$$M_0 < \frac{(3.15 \text{ kg}\cdot\text{m}^2)(61.6 \text{ rad/s})^2(10°)(2\pi \text{ rad/360°})}{2.65}$$

$$= 784.3 \text{ N}\cdot\text{m}$$

Example 4.3. A machine of mass 25.0 kg is placed on an elastic foundation. A sinusoidal force of magnitude 25 N is applied to the machine. A frequency sweep reveals that the maximum steady-state amplitude of 1.3 mm occurs when the period of response is 0.22 s. Determine the equivalent stiffness and damping ratio of the foundation.

The system is modeled as a mass attached to a spring and dashpot in parallel with an applied sinusoidal force of magnitude 25 N. For a linear system the frequency of response is the same as the frequency of excitation. Thus the maximum response occurs for a frequency of $(2\pi)/\omega = 0.22$ s or $\omega = 28.6$ rad/s. From Eq. (4.24) the maximum response occurs when

$$\frac{\omega}{\omega_n} = \sqrt{1 - 2\zeta^2}$$

or

$$\omega_n = \frac{\omega}{\sqrt{1 - 2\zeta^2}}$$

$$= \frac{28.6 \text{ rad/s}}{\sqrt{1 - 2\zeta^2}}$$

Equation (4.25) implies

$$\frac{(25.0 \text{ kg})(0.0013 \text{ m})\omega_n^2}{25 \text{ N}} = \frac{1}{2\zeta\sqrt{1 - \zeta^2}}$$

which upon substitution for ω_n becomes

$$\frac{1.066}{1 - 2\zeta^2} = \frac{1}{2\zeta\sqrt{1 - \zeta^2}}$$

Squaring the preceding equation and rearranging leads to

$$\zeta^4 - \zeta^2 + 0.117 = 0$$

The quadratic formula is used to solve for ζ^2, yielding

$$\zeta = 0.368, 0.930$$

The larger value is discarded because a frequency sweep would yield a maximum only for $\zeta < 1/\sqrt{2}$. Thus $\zeta = 0.368$. The natural frequency is calculated as

$$\omega_n = \frac{28.6 \text{ rad/s}}{\sqrt{1 - 2(0.368)^2}}$$

$$= 33.5 \frac{\text{rad}}{\text{s}}$$

The stiffness of the elastic foundation is

$$k = m\omega_n^2$$

$$= (25 \text{ kg})(33.5 \text{ rad/s})^2$$

$$= 2.81 \times 10^4 \text{ N/m}$$

4.5 FORCED RESPONSE DUE TO EXCITATION WHOSE AMPLITUDE IS PROPORTIONAL TO THE SQUARE OF THE EXCITATION FREQUENCY

Many one-degree-of-freedom systems are subject to a single-frequency harmonic excitation whose amplitude is proportional to the square of its frequency.

Applications such as the operation of unbalanced rotating machinery and vortex shedding from circular cylinders are considered in Sec. 4.6, while the general theory for the steady-state response due to such excitations is considered in this section.

The general form of an excitation whose amplitude is proportional to the square of its frequency is

$$\tilde{F}(t) = A\omega^2 \sin(\omega t + \psi) \tag{4.27}$$

When $\tilde{F}(t)$ represents a force, A has dimensions of $[F][T^2]$ or $[M][L]$. When $\tilde{F}(t)$ represents a moment, A has dimensions of $[F][L][T^2]$ or $[M][L^2]$. The steady-state response due to this type of excitation is developed by applying equations developed in Sec. 4.4 with

$$F_0 = A\omega^2 \tag{4.28}$$

Substitution of Eq. (4.28) into Eq. (4.21) yields

$$\left(\frac{\tilde{m}X}{A}\right)\left(\frac{\omega_n}{\omega}\right)^2 = \frac{1}{\sqrt{\left[1 - \left(\frac{\omega}{\omega_n}\right)^2\right]^2 + \left(2\zeta\frac{\omega}{\omega_n}\right)^2}}$$

or

$$\tilde{m}\frac{X}{A} = \Lambda \tag{4.29}$$

where

$$\Lambda = \frac{r^2}{\sqrt{(1 - r^2)^2 + (2\zeta r)^2}} \tag{4.30}$$

Λ is, like M, a nondimensional function of the frequency ratio and the damping ratio. Equation (4.29) provides its relationship to the parameters in a problem where the magnitude of the excitation force is given by Eq. (4.28). Λ may be related differently to parameters in other problems, but it always has the same functional dependence on r and ζ, given by Eq. (4.30). Λ is related to M by

$$\Lambda = r^2 M \tag{4.31}$$

The steady-state response is given by Eq. (4.16) where X is determined from Eqs. (4.29) and (4.30) and ϕ is still determined using Eq. (4.26).

Λ is plotted as a function of r for various values of ζ in Fig. 4.16. The following are noted from Eq. (4.30) and Fig. 4.16:

1. $\Lambda = 0$ if and only if $r = 0$ for all values of ζ.
2. $\Lambda \approx 1$ for large r for all values of ζ.
3. Λ grows without bound near $r = 1$ for $\zeta = 0$.

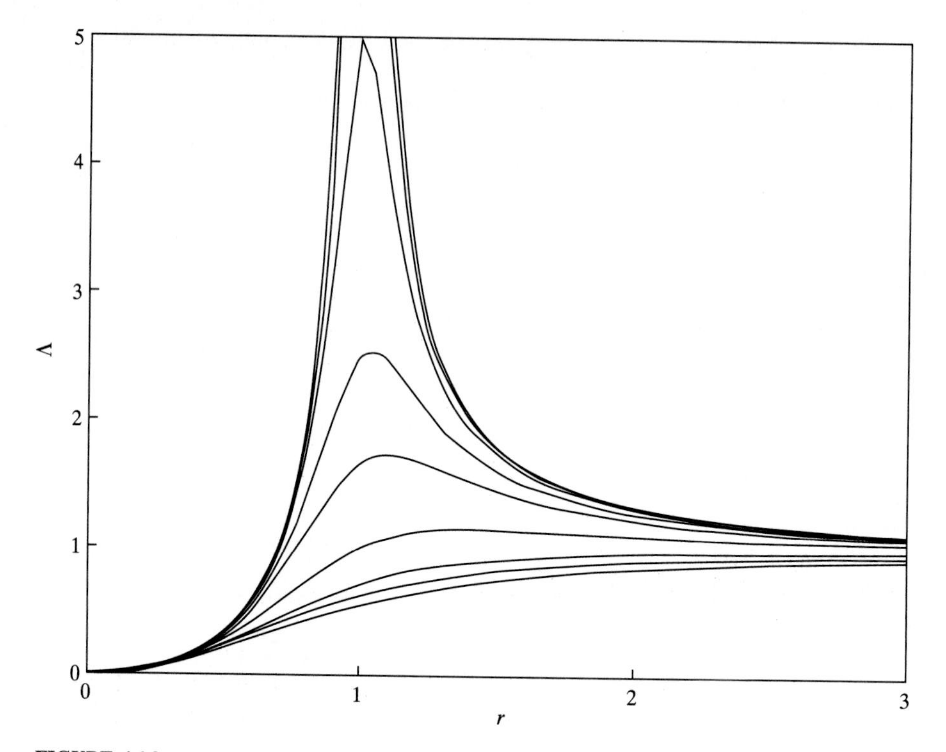

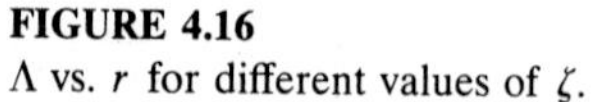

FIGURE 4.16
Λ vs. r for different values of ζ.

4. For $0 < \zeta < 1/\sqrt{2}$, Λ has a maximum for a frequency ratio of

$$r_m = \frac{1}{\sqrt{1 - 2\zeta^2}} \tag{4.32}$$

   Equation (4.32) is derived by finding the value of r such that $d\Lambda/dr = 0$.
5. For a given $0 < \zeta < 1/\sqrt{2}$, the maximum value of Λ corresponds to the frequency ratio of Eq. (4.32) and is given by

$$\Lambda_{\max} = \frac{1}{2\zeta\sqrt{1 - \zeta^2}} \tag{4.33}$$

6. For $\zeta > 1/\sqrt{2}$, Λ does not reach a maximum. Λ grows slowly from zero for $r = 0$ and approaches one for large r.

Example 4.4. A one-degree-of-freedom system is subject to a harmonic excitation whose magnitude is proportional to the square of its frequency. The frequency of excitation is varied and the steady-state amplitude noted. A maximum amplitude of 8.5 mm occurs at a frequency of 200 Hz. When the frequency is much higher than 200 Hz, the steady-state amplitude is 1.5 mm. Determine the damping ratio for the system.

From Fig. 4.16, $\Lambda \to 1$ as $r \to \infty$. Thus, from Eq. (4.29) and the given information,

$$\frac{\tilde{m}}{A} = \frac{1}{1.5\ \text{mm}}$$

Substituting the preceding equation into Eq. (4.33) yields

$$\frac{8.5\ \text{mm}}{1.5\ \text{mm}} = \frac{1}{2\zeta\sqrt{1-\zeta^2}}$$

Inverting, squaring, and rearranging leads to

$$\zeta^4 - \zeta^2 + 0.00778 = 0$$

The appropriate root of the preceding equation is $\zeta = 0.089$.

4.6 RESPONSE DUE TO EXCITATIONS WHOSE AMPLITUDE IS PROPORTIONAL TO THE SQUARE OF THE FREQUENCY: APPLICATIONS

4.6.1 Rotating Unbalance

The machine of Fig. 4.17 has a component which rotates at a constant speed, ω. Its center of mass is located a distance e, called the eccentricity, from the center of rotation. The mass of the rotating component is m_0, while the total mass of the machine, including the rotating component, is m. The machine is constrained to move vertically. However, the rotating component has a component of acceleration equal to $m_0 e\omega^2$, directed between the center of mass of the rotating component and the center of rotation. Since the location of the center of mass of the rotating component moves as the component rotates, the actual direction of this component of acceleration also changes.

Summation of forces applied to the free-body diagrams of Fig. 4.18 yields

$$\sum \overset{\downarrow +}{F}_{\text{ext}} = \sum \overset{\downarrow +}{F}_{\text{eff}}$$

$$-k(x + \Delta_{\text{st}}) - c\dot{x} + mg = m\ddot{x} + m_0 e\omega^2 \sin\theta \tag{4.34}$$

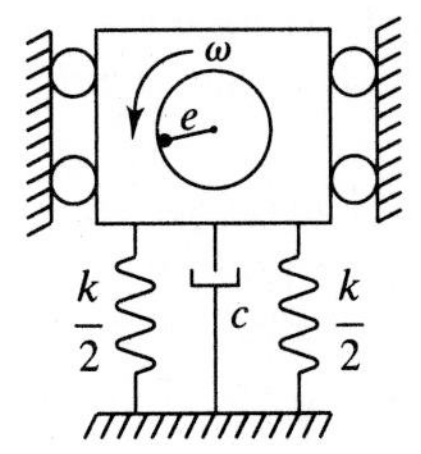

FIGURE 4.17
Rotating unbalance produces harmonic excitation of machine on elastic foundation.

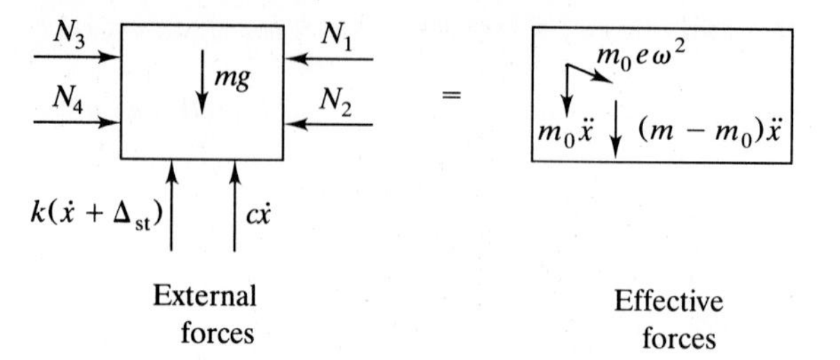

FIGURE 4.18
Free-body diagrams of machine with rotating unbalance at an arbitrary instant of time.

For constant ω,

$$\theta = \omega t + \theta_0 \tag{4.35}$$

where θ_0 is an angle between the initial position of the center of mass of the rotating component and the horizontal. Canceling gravity with static deflection, using Eq. (4.35) in Eq. (4.34), and rearranging yields

$$m\ddot{x} + c\dot{x} + kx = -m_0 e \omega^2 \sin(\omega t + \theta_0) \tag{4.36}$$

The negative sign is eliminated by defining

$$\psi = \theta_0 + \pi$$

Then Eq. (4.36) becomes

$$m\ddot{x} + c\dot{x} + kx = m_0 e \omega^2 \sin(\omega t + \psi) \tag{4.37}$$

It is apparent from Eq. (4.37) that the unbalanced rotating component leads to a harmonic excitation whose amplitude is proportional to the square of its frequency. The constant of proportionality is

$$A = m_0 e \tag{4.38}$$

Using Eq. (4.29),

$$\Lambda = \frac{mX}{m_0 e} \tag{4.39}$$

Example 4.5. A 150-kg electric motor has a rotating unbalance of 0.5 kg, 0.2 m from the center of rotation. The motor is to be mounted at the end of a steel ($E = 210 \times 10^9$ N/m^2) cantilever beam of length 1 m. The operating range of the motor is from 500 to 1200 rpm. For what values of I, the beam's cross-sectional moment of inertia, will the steady-state amplitude of vibration be less than 1 mm? Assume the damping ratio is 0.1.

The maximum allowable value of Λ is

$$\Lambda_{\text{allow}} = \frac{(150 \text{ kg})(0.001 \text{ m})}{(0.5 \text{ kg})(0.2 \text{ m})} = 1.5$$

Since $\Lambda_{\text{allow}} > 1$ and $\zeta < 1/\sqrt{2}$, Fig. 4.16 shows that two values of r correspond to $\Lambda = \Lambda_{\text{allow}}$. These are determined by substituting into Eq. (4.30)

$$1.5 = \frac{r^2}{\sqrt{(1 - r^2)^2 + (0.2r)^2}}$$

Rearrangement leads to the following equation:

$$0.556r^4 - 1.96r^2 + 1 = 0$$

whose positive roots are

$$r = 0.787, 1.71$$

Thus $\Lambda < 1.5$ if $r < 0.787$ or $r > 1.706$. Requiring $r < 0.787$ over the entire operating range yields

$$\frac{(1200 \text{ rev/min})(2\pi \text{ rad/rev})(1 \text{ min}/60 \text{ s})}{\omega_n} < 0.787$$

or $\omega_n > 159.7$ rad/s. The one-degree-of-freedom approximation for the natural frequency of the motor attached to the end of a cantilever beam of negligible mass is

$$\omega_n = \sqrt{\frac{3EI}{mL^3}}$$

Thus

$$I > \frac{(159.7 \text{ rad/s})^2 L^3 m}{3E}$$

$$= \frac{(159.7 \text{ rad/s})^2 (1 \text{ m})^3 (150 \text{ kg})}{3(210 \times 10^9 \text{ N/m}^2)}$$

$$= 6.07 \times 10^{-6} \text{ m}^4$$

Requiring $r > 1.706$ over the entire operating range

$$\frac{(500 \text{ rev/min})(2\pi \text{ rad/rev})(1 \text{ min}/60 \text{ s})}{\omega_n} > 1.706$$

or $\omega_n < 30.7$ rad/s. This requirement leads to $I < 2.24 \times 10^{-7}$ m^4.

Thus the amplitude of vibration will be limited to 1 mm if $I > 6.08 \times 10^{-6}$ m^4 or $I < 2.24 \times 10^{-7}$ m^4. However, other considerations limit the design of the beam. The smaller the moment of inertia, the larger the bending stress in the outer fibers of the beam at the support.

4.6.2 Vortex Shedding from Circular Cylinders

When a circular cylinder is placed in a steady uniform stream at sufficient velocity, flow separation occurs on the cylinder's surface, as illustrated in Fig. 4.19. The separation leads to vortex shedding from the cylinder and the formation of a wake behind the cylinder. Vortices are shed alternately from the upper and lower surfaces of the cylinder at a constant frequency. The alternate

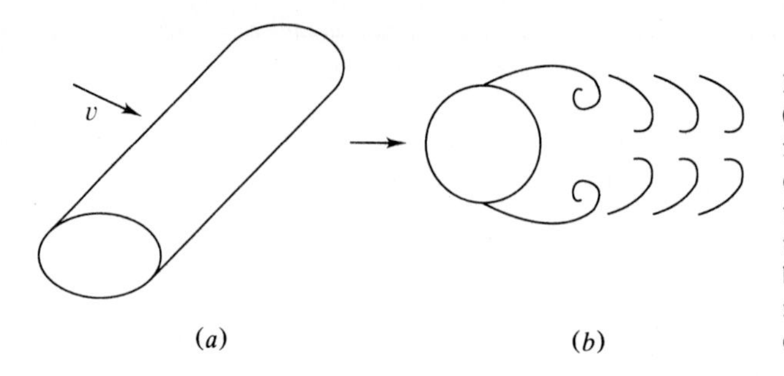

FIGURE 4.19
(*a*) Circular cylinder in steady flow; (*b*) cross section of cylinder, showing vortices shed alternately from each surface of the cylinder resulting in a wake behind the cylinder and a harmonic force acting on the cylinder.

shedding of vortices causes oscillating streamlines in the wake which, in turn, leads to an oscillating pressure distribution. The oscillating pressure distribution, in turn, gives rise to an oscillating force acting normal to the cylinder. The force is given by

$$F(t) = F_0 \sin(\omega t) \tag{4.40}$$

where F_0 is the magnitude of the force and ω is the frequency of vortex shedding. These parameters are dependent upon the fluid properties and the geometry of the cylinder. That is,

$$F_0 = F_0(v, \rho, \mu, D, L) \tag{4.41}$$

and

$$\omega = \omega(v, \rho, \mu, D, L) \tag{4.42}$$

where v = the magnitude of fluid velocity, $[L]/[T]$
ρ = the fluid density, $[M]/[L]^3$
μ = the dynamic viscosity of fluid, $[M]/([L][T])$
D = the diameter of cylinder, $[L]$
L = the length of cylinder, $[L]$

The dependent parameters F_0 and ω are both functions of five independent parameters. Thus the Buckingham Π theorem implies that Eqs. (4.41) and (4.42) can be rewritten as relationships between three dimensionless parameters. Indeed, nondimensional forms of Eqs. (4.41) and (4.42) are

$$C_D = f\left(\text{Re}, \frac{D}{L}\right) \tag{4.43}$$

$$S = f\left(\text{Re}, \frac{D}{L}\right) \tag{4.44}$$

where $C_D = F_0/(\frac{1}{2}\rho v^2 \hat{A})$, the drag coefficient, the ratio of the drag force to the inertia force

where $\hat{A} = DL$, projected area of cylinder normal to direction of flow
Re $= \rho v D/\mu$, the Reynolds number, the ratio of inertia forces to viscous forces

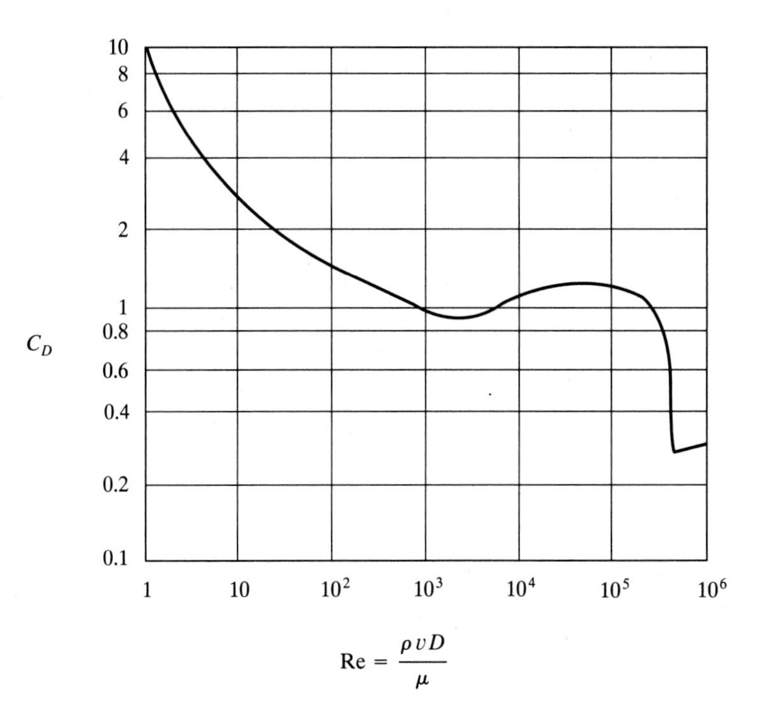

FIGURE 4.20
Drag coefficient vs. Reynolds number for a circular cylinder $C_D \approx 1$ for $1 \times 10^3 < \mathrm{Re} < 2 \times 10^5$. (*From Shames.*)

$S = \omega D/(2\pi v)$, the Strouhal number, a ratio of the inertia forces due to the local acceleration of the flow to inertia forces due to convective acceleration.

D/L = diameter to length ratio

For long cylinders ($D/L \ll 1$), a two-dimensional approximation is used. Then the effect of D/L on the drag coefficient and Strouhal number is negligible. Empirical data are used to determine the forms of Eqs. (4.43) and (4.44), assuming that both the drag coefficient and Strouhal number are independent of D/L. Empirical curves are shown in Figs. 4.20 and 4.21.

The density and dynamic viscosity of air at 20°C are 1.204 $\mathrm{kg/m^3}$ and 1.82×10^{-5} $\mathrm{N \cdot s/m}$, respectively. Thus for air at 20°C the Reynolds number for flow over a 10-cm-diameter circular cylinder at 20 m/s is

$$\mathrm{Re} = \frac{(1.204\ \mathrm{kg/m^3})(20\ \mathrm{m/s})(0.1\ \mathrm{m})}{1.82 \times 10^{-5}\ \mathrm{N \cdot s/m}} = 1.3 \times 10^5$$

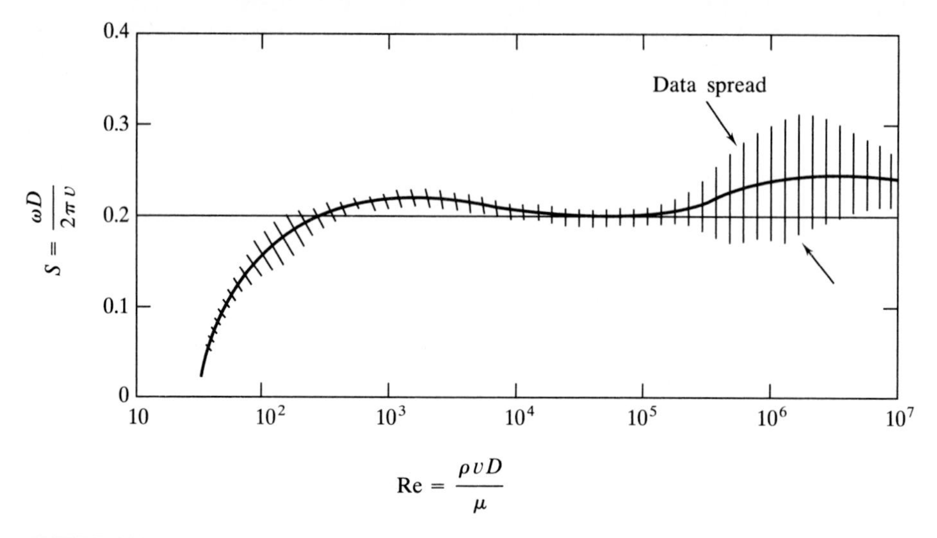

FIGURE 4.21
Strouhal number as a function of Reynolds number. The Strouhal number is approximately 0.2 for $1 \times 10^3 < \text{Re} < 2 \times 10^5$. [*From White, based on experimental data in Roshko* (1954) *and Jones* (1968).]

The Reynolds number for many situations involving wind-induced oscillations is between 1×10^3 and 2×10^5. Over this Reynolds number regime, both the drag coefficient and the Strouhal number are approximately constant. From Figs. 4.20 and 4.21,

$$C_D \approx 1 \qquad 1 \times 10^3 < \text{Re} < 2 \times 10^5 \tag{4.45}$$

$$S \approx 0.2 \qquad 1 \times 10^3 < \text{Re} < 2 \times 10^5 \tag{4.46}$$

From Eq. (4.46) and the definition of the Strouhal number,

$$v = \frac{\omega D}{0.4\pi} \tag{4.47}$$

Then using Eqs. (4.45) and (4.47) and the definition of C_D

$$F_0 = 0.317\rho D^3 L \omega^2 \tag{4.48}$$

Hence the harmonic excitation to a circular cylinder provided by vortex shedding when the Reynolds number is between 1×10^3 and 2×10^5 has a magnitude that is proportional to the square of its frequency. Using the notation of Sec. 4.5,

$$A = 0.317\rho D^3 L \tag{4.49}$$

and

$$\Lambda = \frac{3.16\tilde{m}X}{\rho D^3 L} \tag{4.50}$$

The theory is presented previously for vortex shedding from circular cylinders. If the frequency at which the vortices are shed is near the natural frequency of the structure, then large-amplitude vibrations exist. The effects of vortex shedding must be taken into account when designing structures such as street lamp posts, transmission towers, chimneys, and tall buildings. Vortex shedding also occurs from noncircular structures such as buildings and bridges. Vortex shedding at a frequency near a torsional natural frequency is thought to be partially responsible for the famous Tacoma Narrows Bridge disaster. Observed amplitudes of a torsional mode were as large as 45°.

Example 4.6. A street lamp consists of a 60-kg light fixture attached at the end of a 3-m-tall solid steel ($E = 210 \times 10^9 \text{ N/m}^2$) cylinder with a diameter of 20 cm. Use a one-degree-of-freedom model consisting of a cantilever beam with a concentrated mass at its end to analyze the response of the light fixture to wind excitation. Assume the beam has an equivalent viscous damping ratio of 0.2.

(*a*) At what wind speed will the maximum steady-state amplitude of vibration due to vortex shedding occur?
(*b*) What is the corresponding maximum amplitude?
(*c*) Redesign the light by changing its diameter such that the maximum amplitude of vibration does not exceed 0.50 mm for any wind speed.

Before proceeding with the analysis, there are several questions associated with the modeling that must be addressed. Vortices are shed along the entire length of the cylinder. The two-dimensional assumption implies that the force per unit length is constant along the entire length of the light post. Thus the force given by Eq. (4.40) is really the resultant of this force per unit length distribution. Its point of application should be the midpoint of the light post. However, the problem is not really two dimensional due to, among other things, the boundary layer of the earth. The presence of a boundary layer causes a varying wind velocity over the length of the light post, which, in turn, causes a nonuniform force per unit length distribution, as shown in Fig. 4.22*a*. Thus the actual point of application of the resultant force will be somewhat higher than the midpoint of the light post. In addition, the mass is assumed to be lumped at the end of the beam, while the point of application of the applied force is elsewhere. The resultant force can be

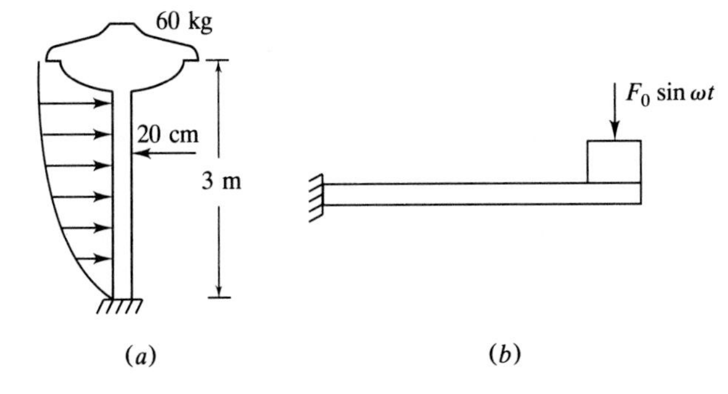

FIGURE 4.22
(*a*) Street light post in steady wind is subject to harmonic excitation whose amplitude is proportional to the square of the frequency due to vortex shedding; (*b*) the problem is modeled as a mass concentrated at the end of a cantilever beam.

replaced by a force of the same magnitude located at the end of the beam and a moment. However, the moment causes rotational effects which are not adequately taken into account in a one-degree-of-freedom model. At least a two-degree-of-freedom model should be used. In order to attain an approximate result, these effects are neglected. A one-degree-of-freedom model is used where the excitation is provided by a concentrated harmonic load located at the light of fixture, as shown in Fig. 4.22*b*.

Assume air at 20°C. The Reynolds number for a velocity of 20 m/s is

$$\mathrm{Re} = \frac{(1.204\ \mathrm{kg/m^3})(20\ \mathrm{m/s})(0.20\ \mathrm{m})}{(1.82 \times 10^{-5}\ \mathrm{N\cdot s/m})} = 2.6 \times 10^5$$

This Reynolds number is higher than the 2×10^5 upper limit on the range of strict applicability of the theory presented previously. However, from Fig. 4.21, the Strouhal number is only slightly higher than 0.2. Using 0.2 as an approximation for the Strouhal number is in line with other approximations made in the modeling.

(*a*) Using a one-degree-of-freedom model, the natural frequency of the cantilever beam is

$$\omega_n = \sqrt{\frac{3EI}{mL^3}}$$

$$= \sqrt{\frac{3(210 \times 10^9\ \mathrm{N/m^2})(\pi/64)(0.2\ \mathrm{m})^4}{(60\ \mathrm{kg})(3\ \mathrm{m})^3}}$$

$$= 174.8\ \frac{\mathrm{rad}}{\mathrm{s}}$$

Since the magnitude of the excitation force is proportional to the square of its frequency, the theory of Sec. 4.5 applies. Thus, from Eq. (4.32), the maximum steady-state amplitude occurs for a frequency ratio of

$$r_{\max} = \frac{1}{\sqrt{1 - 2\zeta^2}} = 1.043$$

Thus the frequency at which the maximum amplitude occurs is

$$\omega = 1.043(174.8\ \mathrm{rad/s}) = 182.2\ \mathrm{rad/s}$$

The wind velocity that gives rise to this frequency is calculated using the definition of the Strouhal number

$$v = \frac{\omega D}{2\pi S}$$

$$= \frac{(182.2\ \mathrm{rad/s})(0.2\ \mathrm{m})}{2\pi(0.2)}$$

$$= 29.0\ \frac{\mathrm{m}}{\mathrm{s}}$$

(*b*) The value of Λ corresponding to this frequency ratio is calculated from Eq. (4.33)

$$\Lambda_{\max} = \frac{1}{2\zeta\sqrt{1-\zeta^2}} = 2.55$$

The corresponding maximum amplitude is calculated using Eq. (4.50)

$$X = \frac{\rho D^3 L \Lambda}{3.16m}$$

$$= \frac{(1.204\ \text{kg/m}^3)(0.2\ \text{m})^3(3\ \text{m})(2.55)}{3.16(60\ \text{kg})}$$

$$= 3.9 \times 10^{-4}\ \text{m}$$

(*c*) The maximum value of Λ is a function of ζ only and does not change with ω_n. The steady-state amplitude can be limited to 0.1 mm for all wind speeds by requiring that $\Lambda = 2.55$ for $X = 0.1$ mm. This leads to

$$D = \left(\frac{3.16mX}{\rho L \Lambda}\right)^{1/3}$$

$$= 12.7\ \text{cm}$$

4.7 RESPONSE DUE TO HARMONIC EXCITATION OF SUPPORT

Consider the mass-spring-dashpot system of Fig. 4.23. The spring and dashpot are in parallel with one end of each connected to the mass and the other end of each connected to a moveable support. Let $y(t)$ denote the known displacement

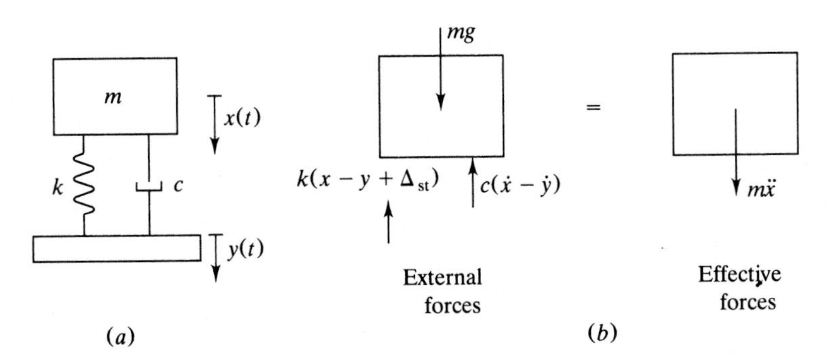

FIGURE 4.23
(*a*) Block is connected through spring and viscous damper in parallel to moveable support; (*b*) free-body diagrams at an arbitrary instant of time after support is given a displacement $y(t)$.

of the support and let $x(t)$ denote the absolute displacement of the mass. Application of Newton's law to the free-body diagrams of Fig. 4.23b yields

$$-k(x - y + \Delta_{st}) - c(\dot{x} - \dot{y}) + mg = m\ddot{x} \tag{4.51}$$

Canceling the static-deflection term with gravity and rearranging yields

$$m\ddot{x} + c\dot{x} + kx = c\dot{y} + ky \tag{4.52}$$

Define

$$z(t) = x(t) - y(t) \tag{4.53}$$

as the displacement of the mass relative to the displacement of its support. Equation (4.52) is rewritten using z as the dependent variable

$$m\ddot{z} + c\dot{z} + kz = -m\ddot{y} \tag{4.54}$$

Dividing Eqs. (4.52) and (4.54) by m yields

$$\ddot{x} + 2\zeta\omega_n\dot{x} + \omega_n^2 x = 2\zeta\omega_n\dot{y} + \omega_n^2 y \tag{4.55}$$

and

$$\ddot{z} + 2\zeta\omega_n\dot{z} + \omega_n^2 z = -\ddot{y} \tag{4.56}$$

If the base displacement is given by a single-frequency harmonic of the form

$$y(t) = Y \sin \omega t \tag{4.57}$$

then Eqs. (4.55) and (4.56) become

$$\ddot{x} + 2\zeta\omega_n\dot{x} + \omega_n^2 x = 2\zeta\omega_n\omega Y \cos \omega t + \omega_n^2 Y \sin \omega t \tag{4.58}$$

and

$$\ddot{z} + 2\zeta\omega_n\dot{z} + \omega_n^2 z = \omega^2 Y \sin \omega t \tag{4.59}$$

Equation (4.59) shows that a mass-spring-dashpot system subject to harmonic base motion is yet another example in which the magnitude of a harmonic excitation is proportional to the square of its frequency. Using the theory of Sec. 4.5,

$$z(t) = Z \sin(\omega t - \phi) \tag{4.60}$$

where

$$Z = \Lambda Y \tag{4.61}$$

where Λ is defined by Eq. (4.30) and ϕ defined by Eq. (4.26).

When Eqs. (4.60) and (4.61) are substituted into Eq. (4.53), the absolute displacement becomes

$$x(t) = Y[\Lambda \sin(\omega t - \phi) + \sin \omega t] \tag{4.62}$$

Using the trigonometric relationship for the sine of the difference of two angles, it is possible to express Eq. (4.62) in the form

$$x(t) = X \sin(\omega t - \lambda) \tag{4.63}$$

where
$$\frac{X}{Y} = T(r, \zeta) \tag{4.64}$$

and
$$\lambda = \tan^{-1}\left[\frac{2\zeta r^3}{1 + (4\zeta^2 - 1)r^2}\right] \tag{4.65}$$

where $T(r, \zeta)$ is yet another nondimensional function of the frequency ratio and the damping ratio defined by

$$T(r, \zeta) = \sqrt{\frac{1 + (2\zeta r)^2}{(1 - r^2)^2 + (2\zeta r)^2}} \tag{4.66}$$

X/Y is the amplitude of the absolute displacement of the mass to the amplitude of displacement of the base. Multiplying the numerator and denominator of this ratio by ω^2 shows that $T(r, \zeta)$ also represents the ratio of the amplitude of absolute acceleration of the mass to the amplitude of acceleration of the base. Equation (4.66) is plotted in Fig. 4.24. The following are noted about $T(r, \zeta)$:

1. The amplitude (acceleration) ratio is near unity for small r.
2. For all ζ, $0 < \zeta < 1$, the amplitude ratio grows until it reaches a maximum for a frequency ratio of

$$r_{\max} = \frac{1}{2\zeta}\left(\sqrt{1 + 8\zeta^2} - 1\right)^{1/2} \tag{4.67}$$

3. The maximum amplitude corresponding to the frequency ratio of Eq. (4.67) is

$$T_{\max} = 4\zeta^2\left[\frac{\sqrt{1 + 8\zeta^2}}{2 + 16\zeta^2 + (16\zeta^4 - 8\zeta^2 - 2)\sqrt{1 + 8\zeta^2}}\right]^{1/2} \tag{4.68}$$

4. The absolute amplitude ratio has a value of one for $r = \sqrt{2}$, independent of the value of ζ.
5. For $r < \sqrt{2}$, the amplitude ratio is larger for smaller values of ζ. However, for $r > \sqrt{2}$, the amplitude ratio is smaller for smaller values of ζ.
6. For all values of ζ, the amplitude ratio is less than one when and only when $r > \sqrt{2}$.

The phase angle between the absolute displacement and the displacement of the base is given by Eq. (4.65).

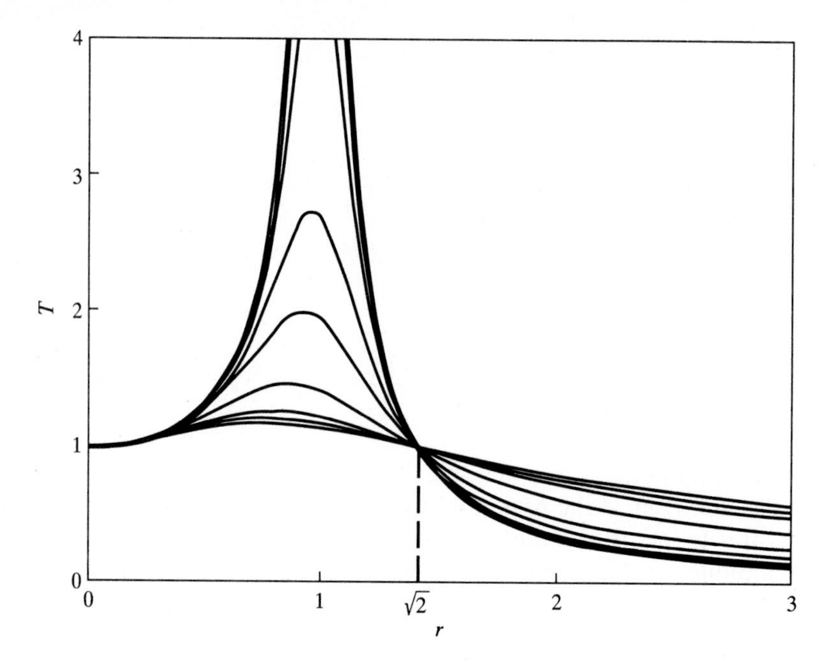

FIGURE 4.24
T vs. r for different values of ζ.

Example 4.7. A simple one-degree-of-freedom model of a vehicle suspension system is shown in Fig. 4.25. The vehicle is traveling over a rough road with a horizontal speed v. The road contour is approximated by a sinusoid as shown in Fig. 4.25*b*. The mass of the vehicle is 1500 kg. The equivalent spring constant of its suspension system is 3×10^7 N/m, and it has a damping ratio of 0.35.

(*a*) If $v = 80$ m/s, what is the maximum displacement and maximum acceleration of the body of the vehicle?
(*b*) For what values of v, in m/s, will the maximum displacement be limited to 10 mm?

Let ξ be a coordinate, measured in meters, that measures horizontal distance from a reference location. The equation for the road contour is

$$y(\xi) = 0.030 \sin(\pi \xi) \text{ m}$$

If the vehicle is at the reference location at $t = 0$, the instantaneous horizontal location of the vehicle is

$$\xi = vt$$

Thus

$$y(t) = 0.30 \sin(\pi v t)$$

The motion of the wheel over the road provides a vertical input which excites the body of the vehicle. If the suspension system were not present, the amplitude of

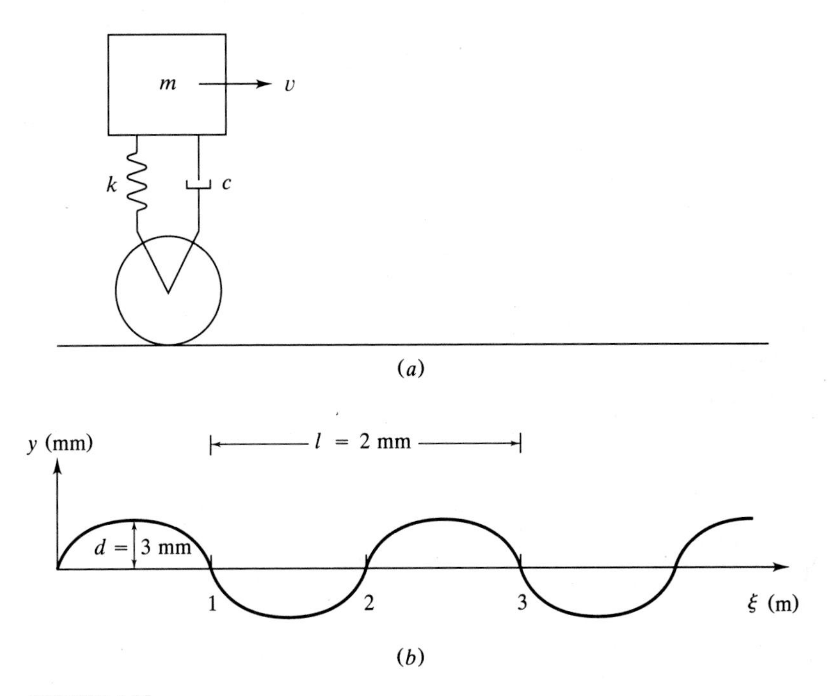

FIGURE 4.25
(*a*) Simplified one-degree-of-freedom model of vehicle suspension system; (*b*) sinusoidal road contour of Example 4.7.

the base displacement would be 30 mm with an acceleration amplitude of 1895 m/s^2 for a vehicle speed of 80 m/s. If properly designed the suspension system will isolate the vehicle from these large accelerations.

The system's natural frequency is

$$\omega_n = \sqrt{\frac{3 \times 10^7 \text{ N/m}}{1500 \text{ kg}}} = 141.4 \frac{\text{rad}}{\text{s}}$$

(*a*) For a vehicle speed of 80 m/s, the frequency at which the wheel traverses the bumps is

$$\omega = \pi(80 \text{ m/s}) = 251.3 \text{ rad/s}$$

leading to a frequency ratio of 1.78. From Eqs. (4.64) and (4.66) the amplitude of absolute displacement is

$$X = 0.030\left\{\frac{1 + [2(0.35)(1.78)]^2}{\left[1 - (1.78)^2\right]^2 + [2(0.35)(1.78)]^2}\right\}^{1/2}$$

$$= 0.019 \text{ m}$$

The acceleration amplitude is

$$A = \omega^2 X = 1200 \text{ m/s}^2$$

(*b*) The maximum allowable amplitude ratio is

$$\frac{X}{Y} = \frac{0.010 \text{ m}}{0.030 \text{ m}} = 0.333$$

From Fig. 4.24 the amplitude ratio is less than 0.333 only for frequency ratios greater than a certain value. The minimum allowable frequency ratio is determined using Eq. (4.66) by setting

$$0.333 = \sqrt{\frac{1 + (0.7r)^2}{(1 - r^2)^2 + (0.7r)^2}}$$

Squaring and rearranging leads to

$$r^4 - 5.93r^2 - 8 = 0$$

for which the appropriate solution is $r = 2.66$. Thus the minimum velocity such that the vehicle's absolute amplitude is less than 10 mm is

$$v_{\min} = \frac{r_{\min}\omega_n}{\pi} = 119.7 \frac{\text{m}}{\text{s}}$$

Mechanisms can be used to produce harmonic base excitations. One simple example is the eccentric circular cam of Fig. 4.26. When rotating at a constant speed, the cam produces a displacement of $e \sin \omega t$ to its follower, which, in turn, produces a harmonic base excitation in the arrangement shown. The Scotch yoke of Fig. 4.27 is another mechanism that produces simple harmonic motion. When the crank is rotating at a constant speed the base is given a displacement of $l \sin \omega t$.

Example 4.8. A Scotch yoke mechanism provides a harmonic base excitation for the mass-spring-dashpot system of Fig. 4.27. The crank arm is 80 mm long. The speed of rotation of the crank arm is varied and the resulting steady-state amplitude is recorded at each speed. The maximum recorded amplitude of the

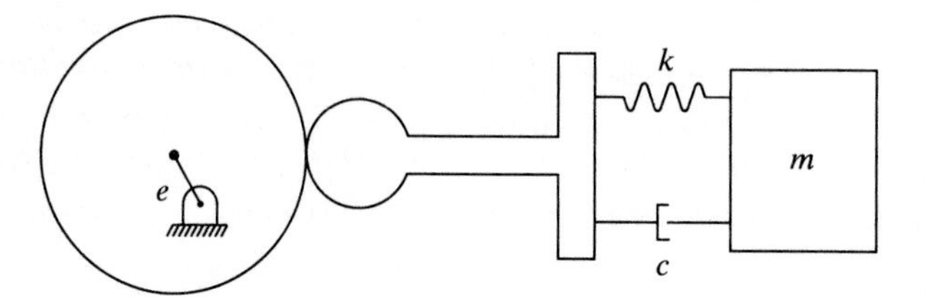

FIGURE 4.26
Eccentric circular cam produces harmonic motion of follower which provides support excitation to mass-spring-dash pot system.

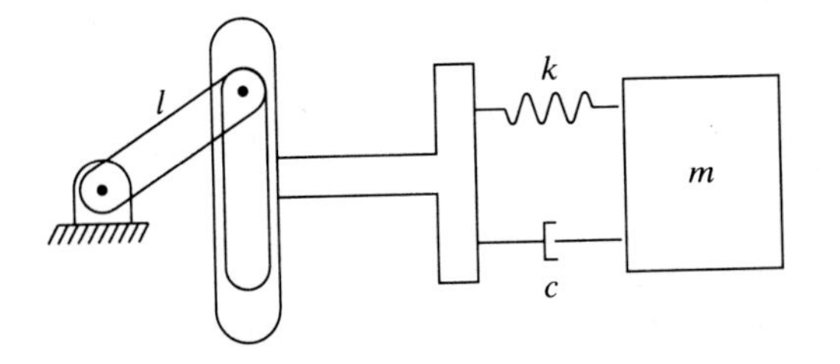

FIGURE 4.27
Scotch yoke mechanism produces simple harmonic motion and provides support excitation to mass-spring-dashpot system.

14.73-kg block is 13 cm at 1000 rpm. Determine the spring stiffness and damping coefficient.

The amplitude of the base displacement is 0.08 m. The maximum displacement of the mass is 0.13 m. Thus

$$T_{\max} = \frac{X_{\max}}{Y} = 1.625$$

The value of ζ which corresponds to this $T_{\max}$ is determined by solving Eq. (4.68). However, algebraic manipulation of Eq. (4.68) yields a fifth-order polynomial equation for ζ^2. A numerical method must be used to find ζ. An easier trial-and-error approach is outlined in the following discussion, and then used to find the value of ζ for this example.

Equation (4.67) is rearranged as

$$\zeta = \sqrt{\frac{1 - r_{\max}^2}{2r_{\max}^4}}$$

A value of $r_{\max} < 1$ is guessed and its corresponding value of ζ calculated from the preceding equation. Equation (4.68) or (4.66) is then used to calculate the value of $T_{\max}$ corresponding to the guessed value of $r_{\max}$. However, small changes in the accuracy of an intermediate calculation using Eq. (4.68) lead to large changes in the result. Thus Eq. (4.66) is usually used. The calculated value of $T_{\max}$ is compared against the desired value of 1.625. If $T_{\max} > 1.625$ another guess for $r_{\max}$, smaller than the previous one, should be made. Other iteration schemes are possible, but the method presented is the most direct using the equations as presented. The trial-and-error scheme is illustrated in the following table:

$r_{\max}$ (guess)	ζ	$T_{\max}$ [from Eq. (4.64)]
0.98	0.147	3.180
0.90	0.381	1.702
0.89	0.407	1.640
0.88	0.437	1.573

Then

$$\omega_n = \frac{\omega}{r_{\max}} = \left(1000\ \frac{\text{rev}}{\text{min}}\right)\left(2\pi\ \frac{\text{rad}}{\text{rev}}\right)\left(\frac{1\ \text{min}}{60\ \text{s}}\right)\frac{1}{0.89}$$

$$= 117.7\ \frac{\text{rad}}{\text{s}}$$

and

$$k = m\omega_n^2 = 2.04 \times 10^5\ \text{N/m}$$

4.8 VIBRATION-MEASURING INSTRUMENTS

Vibration measurement itself is an extensive topic to which entire books have been devoted. Some components of an extensive vibration measurement system require knowledge of other topics such as electronics and microprocessors to be fully understood. The present discussion is limited to the components of a vibration measurement system which work on principles developed using the basic concepts of vibrations of one-degree-of-freedom systems.

A simple frequency-measuring instrument is the cantilever beam shown in Fig. 4.28. The beam is attached to the vibrating body. The length of the beam is adjusted until resonance is achieved. The resonant frequency is the frequency of excitation and is calculated using the equation for the lowest natural frequency of a cantilever beam

$$\omega_n = 3.52\sqrt{\frac{EI}{mL^3}} \tag{4.69}$$

These instruments, called Fullarton tachometers, are usually calibrated such that when the length corresponding to the resonance is found the frequency is read directly off the instrument.

Another form of tachometer is the Frahm tachometer. Multiple-cantilever strips with masses are attached to the vibrating body. Each strip is calibrated to a different natural frequency which is clearly marked. The frequency of vibration of the body to which the tachometer is attached is closest to the natural frequency of the strip with the largest vibration amplitude.

Knowledge of the time history of vibration is often required. For example, the damping ratio of a system can be determined using the ratio of amplitudes on two successive cycles of free vibration. An impact hammer, shown in Fig. 4.29, can be used to initiate vibrations. A vibration-measuring instrument is attached to the body which, in some manner, senses the vibrations. The vibration-measuring instrument is connected to some type of output device, possibly an oscilloscope, where a record is made of the history of the vibrations.

FIGURE 4.28
A Fullarton tachometer is a cantilever beam of adjustable length used to measure frequencies.

FIGURE 4.29
An impact hammer is used to initiate free vibrations.

(A typical vibration-measuring system is shown in Fig. 4.30.) This record can be used to determine properties of the vibrating system such as the damping ratio. Vibrations of multi-degree-of-freedom systems are more complex with multiple frequencies involved in their response. The time history may be obtained and fed into a signal analyzer to determine the frequencies involved in the response and their relative magnitudes.

FIGURE 4.30
Components of experimental setup to measure vibrations: (*a*) a shaker is used to generate harmonic excitation of a bar with a spring support. The experiment is set up to measure the effectiveness of the vibration absorber attached at the end of the bar. An accelerometer is attached to the surface of the bar; (*b*) An audio generator provides a harmonic signal, which passes through an amplifier to the shaker. (*c*) The signal from the accelerometer passes through a charge amplifier. The signal is displayed on a recording oscilloscope. (*d*) A signal analyzer can be used to determine the frequency components of a multifrequency signal.

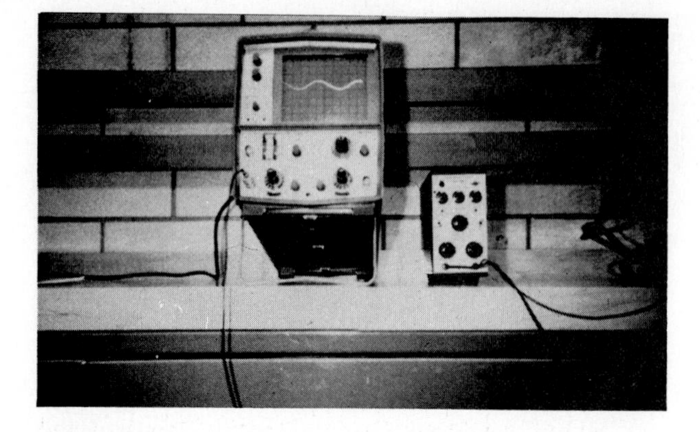

FIGURE 4.30 (*Continued*)

Time histories of vibrations are sensed using seismic instruments. A schematic of a seismic instrument is shown in Fig. 4.31. The housing of the seismic instrument is mounted to the body whose vibrations are to be measured. A seismic mass in the housing is connected to the housing through a spring and viscous damper. In practice, the housing may be filled with a viscous fluid to provide the damping. A transducer is attached to the housing. As the body vibrates, the seismic mass moves relative to the housing. Since the transducer is attached to the housing, it monitors the motion of the seismic mass relative to the housing.

The motion of the seismic mass is analyzed by considering the vibrating body to provide a harmonic excitation to the housing. It is assumed that the motion of the seismic mass has no effect on the motion of the body. That is, the addition of the seismic instrument to the system does not add another degree of freedom to the system. This assumption is true if the seismic mass is much smaller than the equivalent mass of the vibrating body.

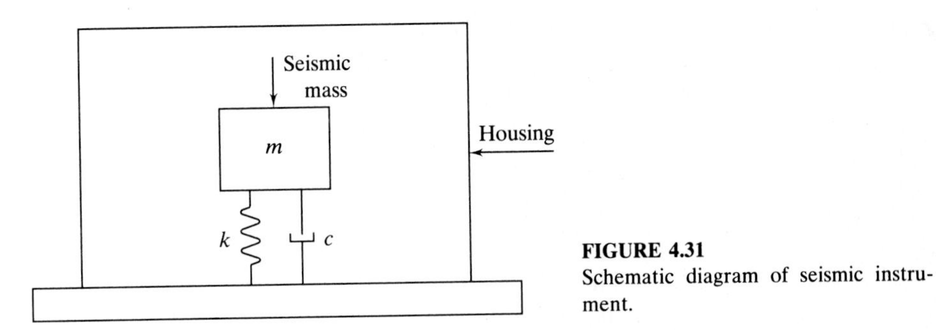

FIGURE 4.31
Schematic diagram of seismic instrument.

In view of the preceding discussion, the analysis of the vibrations of the seismic mass is performed using the support motion theory. Assume the vibrations of the body, the vibrations to be measured, consist of a single-frequency harmonic of the form

$$y(t) = Y \sin \omega t \tag{4.70}$$

The displacement of the seismic mass relative to the displacement of the vibrating body is

$$z(t) = Z \sin(\omega t - \phi) \tag{4.71}$$

where Z is calculated from Eqs. (4.29) and (4.61) and ϕ is calculated from Eq. (4.26). The frequency ratio is $r = \omega/\omega_n$ where ω_n is the natural frequency of the seismic mass.

Figure 4.16 shows that $\Lambda = Z/Y$ is approximately 1 for large r ($r > 3$). That is, the amplitude of the relative displacement which is monitored by the transducer is approximately the same as the amplitude of the displacement of the vibrating body. Comparison of Eqs. (4.70) and (4.71) shows that the measurement reproduces the displacement of the vibrating body, with only a difference in phase. The difference in phase is not important for a single-frequency motion. The difference in the transient motion of the base and the transient motion of the seismic mass also causes an apparent time difference between the displacement measured and the displacement being measured.

A seismic instrument that requires a large frequency ratio for accurate measurement is called a seismometer. A large frequency ratio requires a small natural frequency of the seismic mass. This in turn requires a large seismic mass and a small value of k, or a very flexible spring. Thus seismometers are often large and are not practical for many applications. The percentage error in using a seismometer to measure the amplitude of vibrations is

$$E = \left(\frac{Y_{\text{actual}} - Y_{\text{measured}}}{Y_{\text{actual}}} \right) 100 \tag{4.72}$$

Since the seismometer actually measures Z, then using Eq. (4.60), Eq. (4.72) becomes

$$E = 100|1 - \Lambda| \tag{4.73}$$

It is noted that the error of Eq. (4.73) is not the same as the uncertainty in the measurement. Uncertainty in measurement is introduced because of, among other things, inaccuracy in making the measurement. The error of Eq. (4.73) is introduced because Eq. (4.71) is not an exact reproduction of Eq. (4.70).

The acceleration of the particle to which the seismic instrument is attached is obtained by differentiating Eq. (4.70) twice with respect to time

$$\ddot{y} = -\omega^2 Y \sin \omega t \tag{4.74}$$

Equation (4.61) is used to relate Y to Z and Λ. Then Eq. (4.31) is used to relate Λ to M. Thus Eq. (4.74) becomes

$$\ddot{y} = -\frac{\omega_n^2 Z}{M} \sin \omega t \tag{4.75}$$

Using Eq. (4.71), it is apparent

$$\ddot{y} = \frac{\omega_n^2}{M} z\left(t + \frac{\phi}{\omega} - \frac{\pi}{\omega}\right) \tag{4.76}$$

The negative sign appearing in Eq. (4.75) is taken into account in Eq. (4.76) by subtracting π from the phase angle. For small r, $M \approx 1$ and

$$\ddot{y}(t) \approx \omega_n^2 z\left(t + \frac{\phi}{\omega} - \frac{\pi}{\omega}\right) \tag{4.77}$$

Thus for small r the acceleration of the particle to which the seismic instrument is attached is approximately proportional to the relative displacement between the particle and the seismic mass on a shifted time scale. This shift is irrelevant for single-frequency vibrations. Thus for small r measurement of the relative displacement also provides an approximation to the acceleration. A vibration-measuring instrument that works on this principle is called an accelerometer.

The transducer in an accelerometer records the relative displacement. This is electronically multiplied by ω_n^2 to yield the acceleration of the vibrating body. The acceleration is twice integrated yielding the displacement.

The natural frequency of an accelerometer must be high to measure accurately vibrations over a wide range of frequencies. Thus the seismic mass must be small and the spring stiffness must be large. Thus accelerometers are small instruments. Many accelerometers, as the one in Fig. 4.32, use piezoelectric transducers, which are inexpensive and lightweight.

The error in using an accelerometer is

$$E = 100\frac{M-1}{M} \tag{4.78}$$

Measurement of a response composed of multifrequencies presents a special problem. Assume the vibration to be measured has the representation

$$y(t) = \sum_{i=1}^{n} Y_i \sin(\omega_i t + \psi_i) \tag{4.79}$$

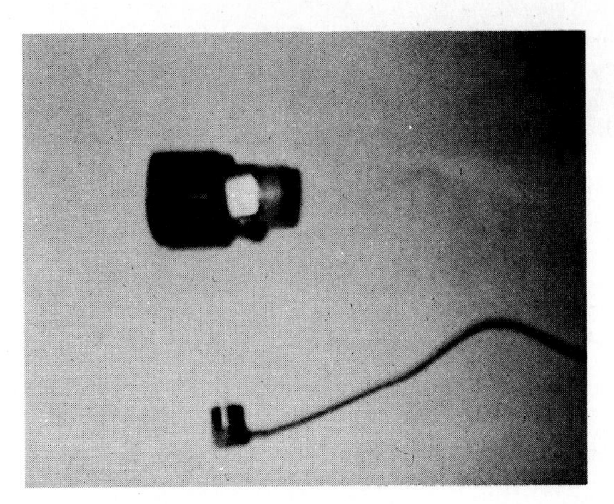

FIGURE 4.32
A piezoelectric accelerometer.

If an accelerometer is used to measure the vibrations, its output device records

$$z(t) \approx -\frac{1}{\omega_n^2}\sum_{i=1}^{n}\omega_i^2 Y_i \sin(\omega_i t + \psi_i - \phi_i) \tag{4.80}$$

where

$$\phi_i = \tan^{-1}\left(\frac{2\zeta r_i}{1 - r_i^2}\right)$$

Since the phase shift for each frequency component is different, a true representation of the acceleration is not recorded. This phenomenon is called phase distortion. The phase shift for each frequency term is zero if the damping ratio is zero, but it is impractical to design an accelerometer without damping. Note that for $r < 1$ and $\zeta = 0.70$, Fig. 4.12 shows that the phase angle is approximately linearly proportional to r. That is,

$$\phi_i = \alpha\frac{\omega_i}{\omega_n}$$

where α is the constant of proportionality. Thus, if $\zeta = 0.70$, Eq. (4.80) becomes

$$z(t) = -\frac{1}{\omega_n^2}\sum_{i=1}^{n}\omega_i^2 Y_i \sin\left[\omega_i\left(t - \frac{\alpha}{\omega_n}\right) - \psi_i\right]$$

or

$$z(t) = -\frac{1}{\omega_n^2}\ddot{y}\left(t - \frac{\alpha}{\omega_n}\right) \tag{4.81}$$

Thus, when an accelerometer with $\zeta = 0.7$ is used, its output device duplicates the actual acceleration, but on a shifted time scale.

Most accelerometers are designed with $\zeta = 0.70$ to eliminate phase distortion. It is left as an exercise to show that phase distortion is not a problem for a seismometer.

Example 4.9. A seismometer has a natural frequency of 15 Hz and a damping ratio of 0.5. What is the lowest frequency response that the seismometer can measure with an error of less than 2%?

Applying Eq. (4.73) with $E = 2$ yields

$$1.02 = \frac{r^2}{\sqrt{(1 - r^2)^2 + (2\zeta r)^2}}$$

Squaring and rearranging leads to the following quadratic equation for r^2:

$$0.03883r^4 - r^2 + 1 = 0$$

whose solutions for r are

$$r = 4.97, 1.022$$

The larger value is the appropriate solution. The lowest frequency that should be measured is

$$\omega = r\omega_n = 4.97(15 \text{ Hz}) = 74.6 \text{ Hz}$$

4.9 VIBRATION ISOLATION

The machine of Fig. 4.33 is bolted directly to the floor in an industrial plant. When the machine is subject to a harmonic excitation, perhaps from a rotating unbalance, the harmonic force is directly transmitted to the bolts and the foundation. If the amplitude is large enough, this repeated loading leads to fatigue damage in the bolts and the structure. The transmitted force could possibly be reduced by mounting the machine on an elastic foundation, as shown in Fig. 4.34. The elastic foundation is called an isolator and this method of reducing the transmitted force is called vibration isolation. The theory of vibration isolation is introduced in this section as it is a direct application of the theory for harmonic excitation. The application of vibration isolation is covered in more detail in Chap. 11.

The elastic foundation is modeled as a spring and viscous damper in parallel. The force transmitted to the floor from the elastic foundation is the sum of the elastic force and the damping force

$$F(t) = kX(t) + c\dot{x}(t) \tag{4.82}$$

Using the response of a one-degree-of-freedom system due to a single-frequency excitation, Eq. (4.16), in Eq. (4.82) leads to

$$F(t) = kX\sin(\omega t + \psi - \phi) + c\omega X\cos(\omega t + \psi - \phi) \tag{4.83}$$

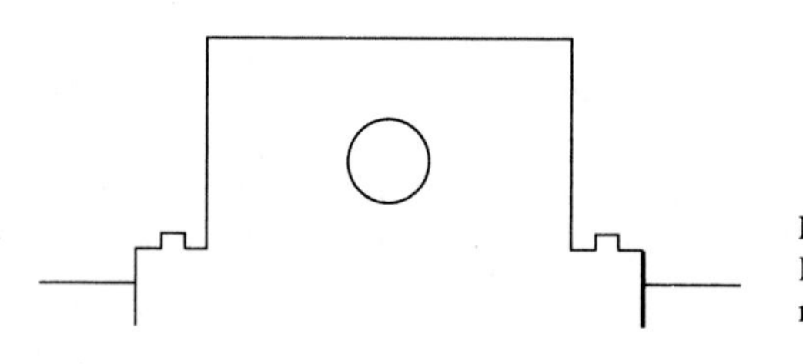

FIGURE 4.33
Machine with rotating unbalance bolted directly to floor.

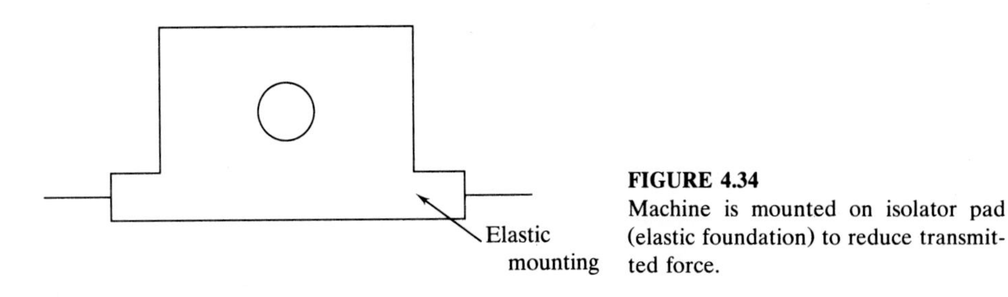

FIGURE 4.34
Machine is mounted on isolator pad (elastic foundation) to reduce transmitted force.

Use of trigonometric identities allows Eq. (4.83) to be written in the form

$$F(t) = F_T \sin(\omega t + \psi - \lambda) \tag{4.84}$$

where

$$F_T = F_0 T(r, \zeta) \tag{4.85}$$

and λ is defined in Eq. (4.65). The function $T(r, \zeta)$ is defined in Eq. (4.66) and repeated for convenience

$$T(r, \zeta) = \sqrt{\frac{1 + (2\zeta r)^2}{(1 - r^2)^2 + (2\zeta r)^2}} \tag{4.66}$$

The ratio F_T/F_0 is called the transmissibility ratio. It is the ratio of the magnitude of the repeating component of the force transmitted to the support to the magnitude of the excitation force. Vibration isolation is achieved if the transmissibility ratio is less than one. From the discussion in Sec. 4.7, the following are revealed about vibration isolation:

1. Vibration isolation is possible only for $r > \sqrt{2}$.
2. For $r > \sqrt{2}$, a larger percentage isolation is achieved for larger r.
3. Increased damping actually hinders isolation. For $r > \sqrt{2}$, a larger value of ζ leads to a larger value of T.
4. Even though damping hinders isolation, it is still necessary. Damping limits the amplitude and transmitted forces as resonance is passed and causes the transient (free) vibrations to decay.

The following examples illustrate the basic application of the theory. More detailed examples as well as a discussion of the design of isolators appear in Chap. 11.

Example 4.10 Undamped isolator design. An air conditioner weighs 250 lb and is driven by a motor at 500 rpm. What is the required static deflection of an undamped isolator to achieve 80% isolation.

Eighty percent isolation requires a transmissibility ratio of 0.2. For an undamped isolator

$$0.2 = \sqrt{\frac{1}{(1 - r^2)^2}}$$

Since $r > \sqrt{2}$ to achieve isolation, and a positive result is required after the square root is taken, the appropriate form of the preceding equation is

$$0.2 = \frac{1}{r^2 - 1}$$

which yields $r = 2.45$. The maximum natural frequency for the air conditioner–isolator system to achieve 80% isolation is

$$\begin{aligned}\omega_n &= \frac{\omega}{r} \\ &= \frac{(500\text{ rev/min})(2\pi\text{ rad/rev})(1\text{ min}/60\text{ s})}{2.45} \\ &= 21.4\,\frac{\text{rad}}{\text{s}}\end{aligned}$$

The required static deflection is calculated from

$$\begin{aligned}\Delta_{st} &= \frac{g}{\omega_n^2} \\ &= \frac{32.2\text{ ft/s}^2}{(21.4\text{ rad/s})^2} \\ &= 0.07\text{ ft}\end{aligned}$$

Example 4.11 Damped isolator design. An industrial sewing machine has a mass of 430 kg and operates at 1500 rpm (157 rad/s). It appears to have a rotating unbalance of magnitude $m_0 e = 0.8\text{ kg}\cdot\text{m}$. Structural engineers suggest that the maximum repeated force transmitted to the floor is 10,000 N. The only isolator available has a stiffness of 7×10^6 N/m and a damping ratio of 0.1. If the isolator is placed between the machine and the floor, will the transmitted force be reduced to an acceptable level? If not, what can be done?

The maximum allowable transmissibility ratio is

$$\begin{aligned}T_{max} &= \frac{F_{T_{max}}}{m_0 e \omega^2} \\ &= \frac{10{,}000\text{ N}}{(0.8\text{ kg}\cdot\text{m})(157\text{ rad/s})^2} \\ &= 0.507\end{aligned}$$

The natural frequency with the isolator in place is

$$\begin{aligned}\omega_n &= \sqrt{\frac{7 \times 10^6\text{ N/m}}{430\text{ kg}}} \\ &= 127.6\,\frac{\text{rad}}{\text{s}}\end{aligned}$$

which leads to a frequency ratio of $1.24 < \sqrt{2}$. Use of this isolator actually amplifies the force transmitted to the floor.

Adequate isolation is achieved only by increasing the frequency ratio, thus decreasing the natural frequency. The maximum allowable natural frequency is obtained by solving for r from

$$0.507 = \sqrt{\frac{1 + (0.2r)^2}{(1 - r^2)^2 + (0.2r)^2}}$$

The preceding equation is squared and rearranged to yield the following quadratic equation for r^2:

$$r^4 - 2.12r^2 - 2.89 = 0$$

The appropriate solution is $r = 1.75$. Thus the maximum natural frequency is

$$\omega_n = \frac{157 \text{ rad/s}}{1.75} = 89.7 \frac{\text{rad}}{\text{s}}$$

If more than one of the described isolator were available, the natural frequency of the system can be decreased by placing isolators in series. The equivalent stiffness for n isolators in series is k/n. Further calculations show that at least two isolator pads in series are necessary to reduce the natural frequency below 89.7 rad/s.

If only one isolator pad is available, the natural frequency is decreased by adding mass to the machine. A mass of at least 530 kg must be rigidly attached to the machine and the assembly placed on the existing isolator.

4.10 FORCED RESPONSE WITH COULOMB DAMPING

The differential equations derived using the free-body diagram of Fig. 4.35, governing the response of a one-degree-of-freedom system with Coulomb damping due to a harmonic excitation are

$$m\ddot{x} + kx = F_0 \sin(\omega t + \psi) - F_f \qquad \dot{x} > 0 \tag{4.86a}$$

$$m\ddot{x} + kx = F_0 \sin(\omega t + \psi) + F_f \qquad \dot{x} < 0 \tag{4.86b}$$

where $F_f = \mu mg$ is the magnitude of the friction force.

If the initial displacement and velocity are both zero, motion commences only when the excitation force is as large as the friction force. Motion will continue until the resultant of the spring force and the excitation force is less than the friction force,

$$|kx - F_0 \sin \omega t| < F_f \Rightarrow \dot{x} = 0 \tag{4.87}$$

The resultant eventually grows large enough such that the inequality in Eq. (4.87) is no longer satisfied, when motion again commences. This process is known as stick-slip and can occur several times during one cycle of motion.

Equation (4.86) is nonlinear. Thus the principles guiding the solution of linear differential equations are not applicable. Specifically, the general solution cannot be written as a homogeneous solution independent of the excitation plus a particular solution. Thus, even though free vibrations of a system with Coulomb damping decay linearly and eventually cease, it is not possible to predict the particular solution as a steady-state solution. Indeed, from the

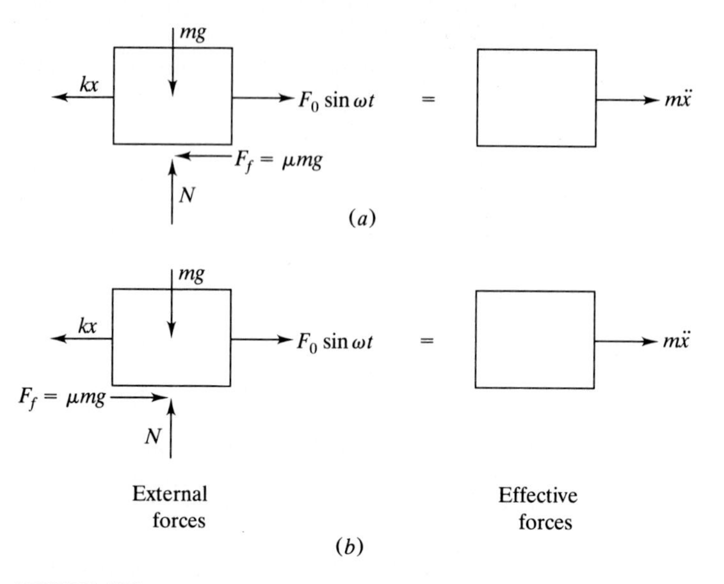

FIGURE 4.35
Free-body diagrams, at an arbitrary instant of time, for block subject to harmonic excitation and Coulomb damping: (*a*) $\dot{x} > 0$; (*b*) $\dot{x} < 0$.

preceding discussion, the stick-slip process should occur for large time and cannot be predicted by a particular solution.

The analytical solution to Eq. (4.86) can be attained using a procedure similar to that of Sec. 3.6 used to obtain the free-vibration response of a system subject to Coulomb damping. The solutions of Eqs. (4.86*a*) and (4.86*b*) are readily available over the time that the equation governs. The constants of integration are determined by noting that the velocity is zero and the displacement is continuous at the time when the equation first begins to govern. Equation (4.87) must be checked over each half-cycle to determine if and when the mass sticks.

The analytical solution is very involved and difficult to use to predict long-term behavior. In many applications only the maximum displacement is of interest. It is a function of five parameters

$$X = f(m, \omega, \omega_n, F_0, F_f) \tag{4.88}$$

The nondimensional form of Eq. (4.88) is

$$\frac{m\omega_n^2 X}{F_0} = f(r, \iota) \tag{4.89}$$

where

$$\iota = \frac{F_f}{F_0} \tag{4.90}$$

For small ι, the friction force is much less than the excitation force, and it is expected that the transient solution will decrease as t increases and a harmonic steady state exists for large time. For this case the effects of Coulomb damping are approximated using an equivalent viscous model. That is, a differential equation of the form of Eq. (4.4) is used to model the steady-state behavior of the system where ζ is replaced by the equivalent viscous damping coefficient, ζ_{eq}. Using the definition of Sec. 3.6, the equivalent viscous damping coefficient for Coulomb damping is

$$C_{eq} = \frac{4F_f}{\pi \omega X} \tag{4.91}$$

which leads to an equivalent damping ratio of

$$\zeta_{eq} = \frac{2\iota}{\pi r M} \tag{4.92}$$

where M is the magnification factor defined by Eq. (4.22). Using ζ_{eq} in place of ζ in Eq. (4.23) leads to

$$M = \frac{1}{\sqrt{(1-r^2)^2 + \left(\dfrac{4\iota}{\pi M}\right)^2}}$$

which is solved for M yielding

$$M = \sqrt{\frac{1 - \left(\dfrac{4\iota}{\pi}\right)^2}{(1-r^2)^2}} \tag{4.93}$$

Equation (4.93) is plotted in Figure 4.36 as a function of r for several values of ι. The following are noted from Eq. (4.93) and Fig. 4.36:

1. The small ι theory predicts that M exists only for $\iota < \pi/4$. The equivalent viscous damping theory cannot be used to predict the maximum displacement for $\iota > \pi/4$.
2. Resonance still occurs for Coulomb damping with small ι. That is, an infinite amplitude is predicted for $r = 1$. This occurs because, for small ι, the excitation provides more energy per cycle of motion than is dissipated by the friction. Since free vibrations sustain themselves, the extra energy leads to an amplitude buildup.
3. For all values of r, M is smaller for larger ι.

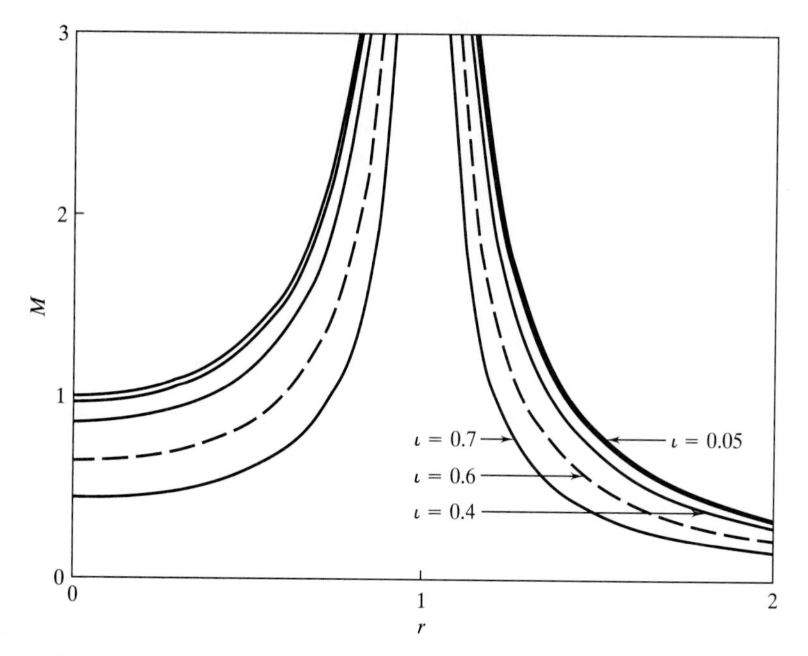

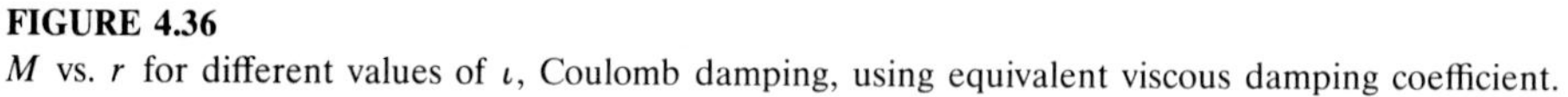
FIGURE 4.36
M vs. r for different values of ι, Coulomb damping, using equivalent viscous damping coefficient.

When Eq. (4.92) is substituted into Eq. (4.26) and the resulting equation manipulated, the following result for the phase angle occurs:

$$\phi = \tan^{-1}\left[\frac{\dfrac{4\iota}{\pi}}{\sqrt{1-\left(\dfrac{4\iota}{\pi}\right)^2}}\right] \qquad r < 1 \tag{4.94a}$$

$$\phi = \tan^{-1}\left[\frac{\dfrac{4\iota}{\pi}}{\sqrt{1-\left(\dfrac{4\iota}{\pi}\right)^2}}\right] \qquad r > 1 \tag{4.94b}$$

The phase angle is constant with r, except that it is positive for $r < 1$ and negative for $r > 1$.

The preceding theory is sufficient for small ι. For larger ι, the equation is truly nonlinear and the results more complex. However, it is expected that larger ι leads to smaller-amplitude vibrations, and less serious problems. In the absence of initial energy, vibrations will not be initiated for $\iota > 1$.

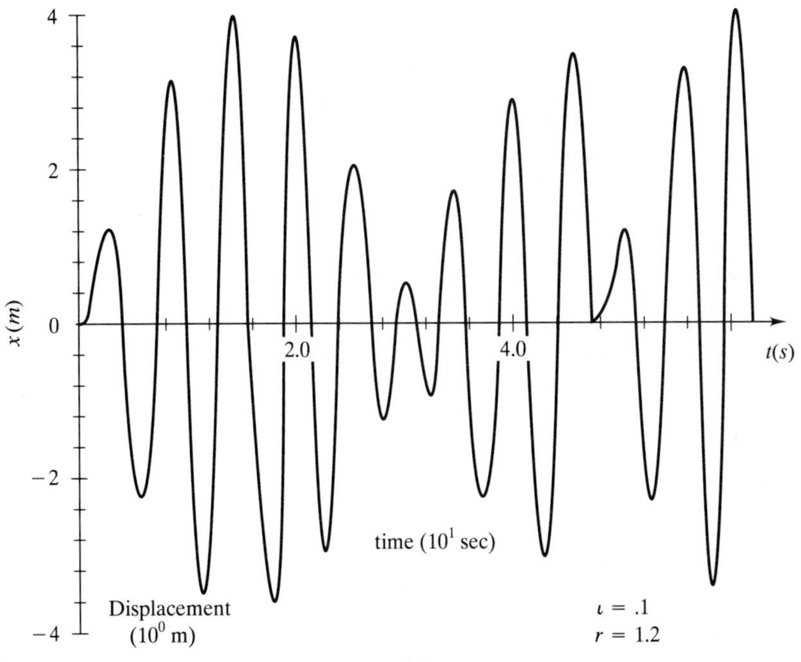

(a)

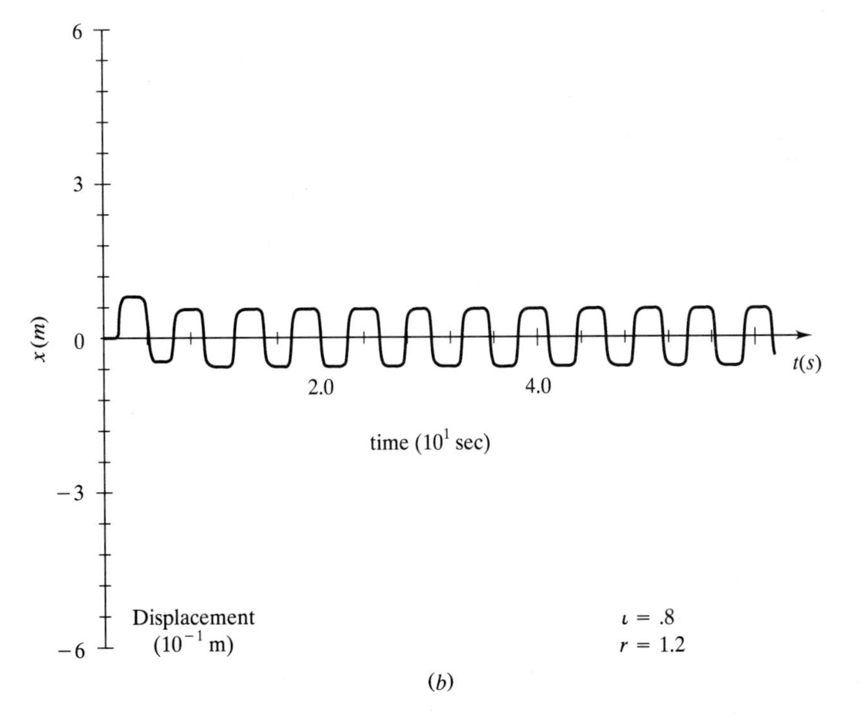

(b)

FIGURE 4.37
Forced response of system with Coulomb damping from COULOMB.BAS: (a) $r = 1.2$, $\iota = 0.1$ (note that $M = 4$); (b) $r = 1.2$, $\iota = 0.8$. The equivalant viscous damping theory does not work for this case. Note that motion stops twice during each cycle and that $M = 0.1$.

The exact solution for the response of a one-degree-of-freedom system with Coulomb damping subject to a harmonic excitation is programmed in COULOMB.BAS, provided on the accompanying diskette. COULOMB.BAS is used to generate the plots of Fig. 4.37. The approximate theory appears to be adequate for small ι; for larger ι the response is more nonlinear, and motion stops when the combination of the excitation force and the spring force are insufficient to overcome friction.

Example 4.12. A Scotch yoke mechanism operating at π rad/s is used to provide base excitation to a block as shown in Fig. 4.38. The block has a mass of 1.5 kg and the coefficient of friction between the block and the surface is 0.08. For what values of k will the steady-state amplitude of vibration be less than 20 cm?

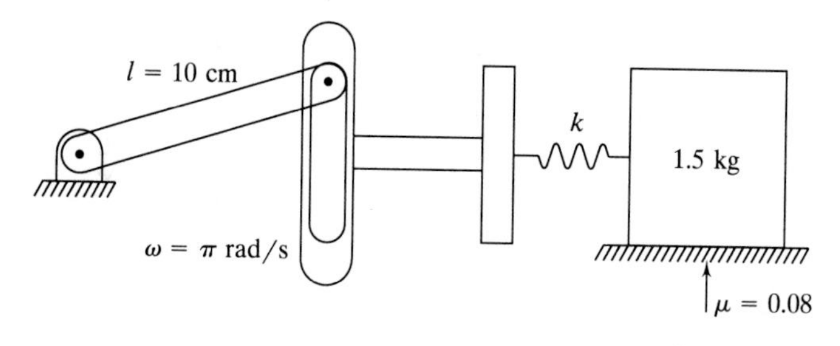

FIGURE 4.38
Scotch yoke mechanism provides base displacement to system with Coulomb damping.

The differential equation for the absolute displacement of the block is

$$m\ddot{x} + kx = kl \sin \omega t \mp \mu mg$$

The magnitude of the exciting force is kl; thus the magnification factor becomes

$$M = \frac{X}{l} = \frac{20 \text{ cm}}{10 \text{ cm}} = 2$$

Knowledge of either the force ratio or the frequency ratio requires knowledge of k. Equation (4.93) is rewritten in dimensional form with k as unknown.

$$2 = \sqrt{\frac{1 - \left(\dfrac{4\mu mg}{\pi kl}\right)^2}{\left(1 - \dfrac{m\omega^2}{k}\right)^2}}$$

Squaring, rearranging, and substituting given values leads to the following quadratic equation:

$$3k^2 - 118.4k + 1100 = 0$$

where k is in Newtons per meter. Its solutions are

$$k = 15.02 \text{ N/m}, 24.44 \text{ N/m}$$

However, if $k < 15.02$ N/m, then $\iota > 0.783$ and the theory is not applicable. If $k > 24.44$ N/m, then $\iota < 0.481$ and the approximate theory is adequate.

4.11 RESPONSE OF SYSTEM WITH HYSTERETIC DAMPING DUE TO HARMONIC EXCITATION

Recall from Sec. 3.6 that the energy dissipated per cycle of motion for a system with hysteretic damping is independent of frequency but proportional to the square of the amplitude. This leads to the direct analogy between viscous damping and hysteretic damping and the development of an equivalent viscous damping coefficient

$$c_{eq} = \frac{hk}{\omega} \tag{4.95}$$

The true forced response of a mass-spring system with hysteretic damping is nonlinear. The equivalent viscous damping coefficient of Eq. (4.95) is valid only when the response consists of a single-frequency harmonic. During the initial part of the response, the transient solution and the particular solution have harmonic terms with different frequencies. Using the viscous damping analogy, it is suspected that the transient solution decays leaving only the steady-state solution after a long time. The differential equation governing the steady-state response of a mass-spring system with hysteretic damping due to a single-frequency harmonic excitation is assumed to be

$$m\ddot{x} + \frac{kh}{\omega}\dot{x} + kx = F_0 \sin(\omega t + \psi) \tag{4.96}$$

It is noted that the generalization of Eq. (4.96) to a more general excitation is not permissible because the damping approximation is only valid for a single-frequency harmonic response. The equation is also nonlinear so that the method of superposition is not applicable to determine particular solutions for multifrequency excitations.

The steady-state solution of Eq. (4.96) is obtained by comparison with Eq. (4.4). The equivalent damping ratio is

$$\zeta_{eq} = \frac{h}{2r} \tag{4.97}$$

The steady-state response is given by Eq. (4.16) where X is related to the magnification factor through Eq. (4.22). Using Eq. (4.97) in Eqs. (4.23) and (4.26) leads to

$$M = \frac{1}{\sqrt{(1 - r^2) + h^2}} \tag{4.98}$$

and

$$\phi = \tan^{-1}\left(\frac{h}{1 - r^2}\right) \tag{4.99}$$

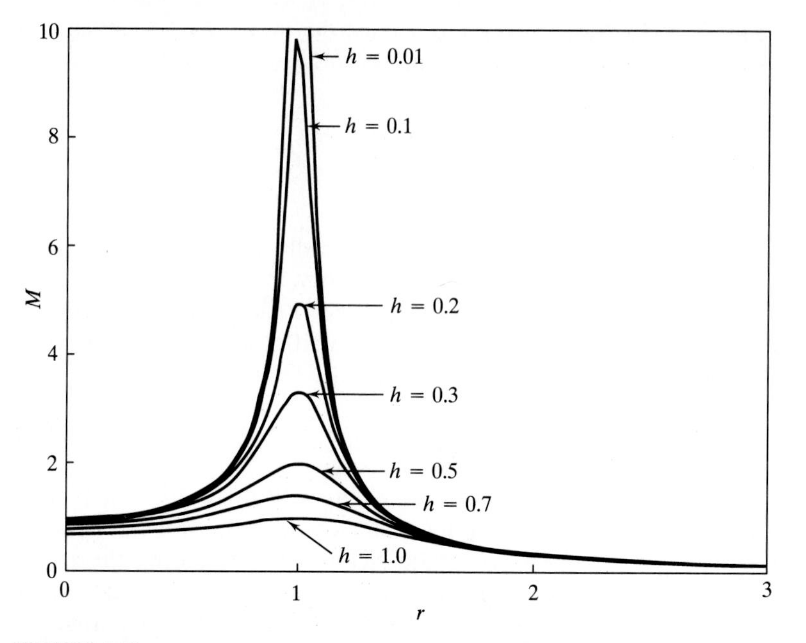

FIGURE 4.39
Magnification factor for hysteretic damping for different values of h.

Equations (4.98) and (4.99) are plotted in Figs. 4.39 and 4.40. The following are noted from these equations and figures:

1. For a given h, the magnification ratio attains a maximum value of $1/h$ for $r = 1$.
2. The phase angle is nonzero for $r = 0$. The response is never in phase with the excitation.

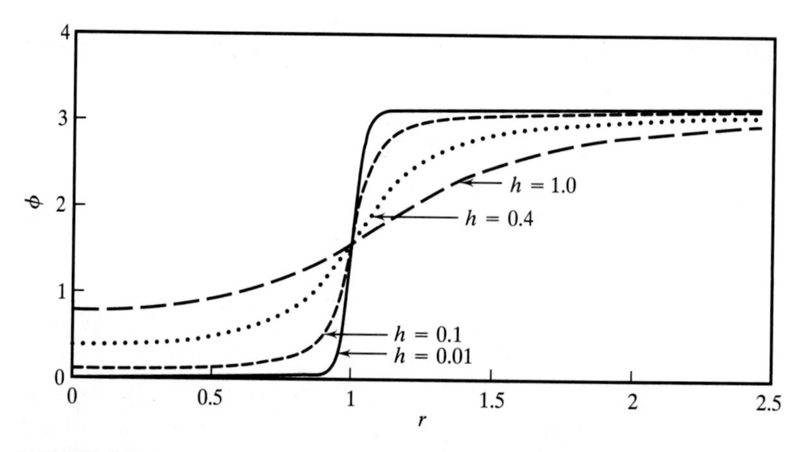

FIGURE 4.40
ϕ vs. r for different values of h.

Most damping is not viscous, but hysteretic. The differences are slight, but noticeable. Viscous damping is often assumed, even when hysteretic damping is present. The viscous damping assumption is easier to use because the damping ratio is independent of frequency. For hysteretic damping the damping ratio is higher for lower frequencies.

Example 4.13. An isolator pad is to be used to isolate vibrations from a milling machine with a rotating unbalance of 1.12 kg · m. The machine has a mass of 510 kg and operates at 1800 rpm. Tests on the isolator have shown that it has a stiffness of 5×10^6 N/m and a hysteretic damping coefficient of 0.4. What is the magnitude of the force transmitted to the floor using this isolator?

The natural frequency of the milling machine on the isolator is

$$\omega_n = \sqrt{\frac{5 \times 10^6 \text{ N/m}}{510 \text{ kg}}} = 99.0 \frac{\text{rad}}{\text{s}}$$

The frequency ratio is

$$r = \frac{(1800 \text{ rev/min})(2\pi \text{ rad/rev})(1 \text{ min}/60 \text{ s})}{99.0 \text{ rad/s}} = 1.90$$

Using Eq. (4.97) for the damping ratio in Eq. (4.66) yields

$$T = \sqrt{\frac{1 + h^2}{(1 - r^2)^2 + h^2}}$$

$$= \sqrt{\frac{1 + (0.4)^2}{\left[1 - (1.90)^2\right]^2 + (0.4)^2}}$$

$$= 0.408$$

The transmitted force becomes

$$F_T = TF_0$$

$$= 0.408(1.12 \text{ kg} \cdot \text{m})\left[\left(1800 \frac{\text{rev}}{\text{min}}\right)\left(2\pi \frac{\text{rad}}{\text{rev}}\right)\left(\frac{1 \text{ min}}{60 \text{ s}}\right)\right]^2$$

$$= 16{,}240 \text{ N}$$

4.12 MULTIFREQUENCY EXCITATIONS

A multifrequency excitation has the form

$$F(t) = \sum_{i=1}^{n} F_i \sin(\omega_i t + \psi_i) \tag{4.100}$$

Without loss of generality, it is assumed that $F_i > 0$ for each i. The steady-state response due to a multifrequency excitation is obtained using the response for a single-frequency excitation and the principle of linear superposition. The total response is the sum of the responses due to each of the individual frequency terms. Thus the solution of Eq. (4.4) with the excitation of Eq. (4.100) is

$$x(t) = \sum_{i=1}^{n} X_i \sin(\omega_i t + \psi_i - \phi_i) \tag{4.101}$$

where

$$X_i = \frac{M_i F_i}{\tilde{m}\omega_n^2} \tag{4.102}$$

$$\phi_i = \tan^{-1}\left(\frac{2\zeta r_i}{1 - r_i^2}\right) \tag{4.103}$$

where

$$r_i = \frac{\omega_i}{\omega_n} \tag{4.104}$$

and

$$M_i = \frac{1}{\sqrt{(1 - r_i^2)^2 + (2\zeta r_i)^2}} \tag{4.105}$$

The maximum displacement from equilibrium is difficult to obtain. The maxima of the trigonometric terms in Eq. (4.101) do not occur simultaneously. An upper bound on the maximum is

$$X_{\max} \leq \sum_{i=1}^{n} X_i \tag{4.106}$$

Example 4.14. A slider-crank mechanism is used to provide a base motion for the block shown in Fig. 4.41. Plot the maximum absolute displacement of the block as a function of frequency ratio for a damping ratio of 0.05. The crank rotates with a constant speed, ω.

The instantaneous position of the block relative to point O is

$$X(t) = \hat{r}\cos\omega t + l\cos\alpha$$

Application of the law of sines gives

$$\sin\alpha = \frac{\hat{r}}{l}\sin\omega t$$

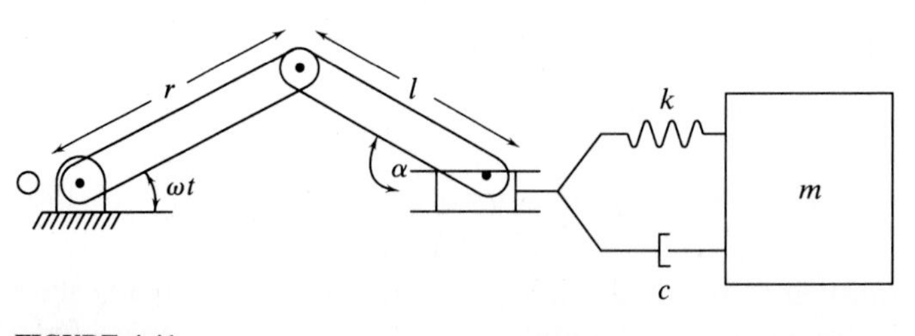

FIGURE 4.41
Slider-crank mechanism produces multifrequency excitation to system of Example 4.14.

Thus

$$y(t) = \hat{r}\cos\omega t + \sqrt{1 - \left(\frac{r}{l}\sin\omega t\right)^2}$$

Assuming r/l is small, the binomial expansion is used to expand the square root

$$y(t) = 1 - \frac{l}{4}\left(\frac{\hat{r}}{l}\right)^2 + r\cos\omega t + \frac{l}{4}\left(\frac{\hat{r}}{l}\right)^2 \cos 2\omega t + \cdots$$

where the expansion has been terminated after the term proportional to $\sin^2 \omega t$ and the double-angle formula is used to replace $\sin^2 \omega t$. The principle of linear superposition and the theory of Sec. 4.7 are used to solve for the absolute displacement of the mass

$$x(t) = l\left[1 - \frac{1}{4}\left(\frac{\hat{r}}{l}\right)^2\right] + \hat{r}T_1 \sin\left(\omega t - \lambda_1 - \frac{\pi}{2}\right) + \frac{l}{4}\left(\frac{\hat{r}}{l}\right)^2 T_2 \cos\left(2\omega t - \lambda_2 - \frac{\pi}{2}\right)$$

where $$T_i = \sqrt{\frac{1 + (2\zeta r_i)^2}{(1 - r_i^2)^2 + (2\zeta r_i)^2}}$$

and $$\lambda_i = \tan^{-1}\left(\frac{2\zeta r_i^3}{1 + (4\zeta^2 - 1)r_i^2}\right)$$

with $$r_1 = \frac{\omega}{\omega_n}$$

and $$r_2 = \frac{2\omega}{\omega_n}$$

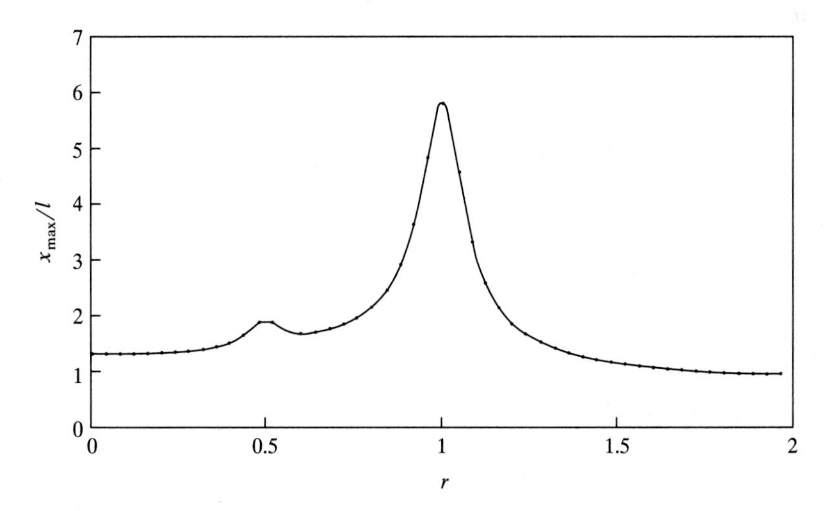

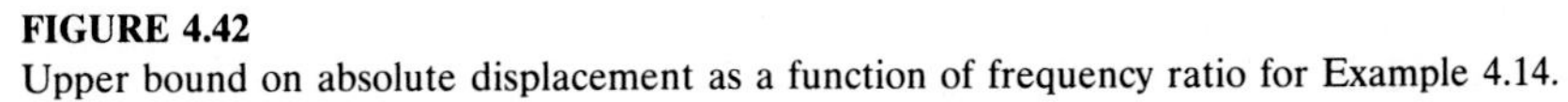

FIGURE 4.42
Upper bound on absolute displacement as a function of frequency ratio for Example 4.14.

The response is the sum of the responses due to each frequency term plus the response due to the constant term. The maximum displacement is difficult to attain. Instead an upper bound is calculated

$$x_{\max} < l\left[1 - \left(\frac{\hat{r}}{l}\right)^2\right] + \hat{r}T_1 + \frac{l}{4}\left(\frac{\hat{r}}{l}\right)^2 T_2$$

$x_{\max}/l$ vs. ω/ω_n is plotted in Fig. 4.42 for $r/l = 1/2$ for $\zeta = 0.05$. The graph has two peaks. The first peak near $\omega/\omega_n = 1/2$ is smaller than the second peak near $\omega/\omega_n = 1$. If additional terms from the binomial expansion were used, higher harmonics would appear in the solution. Small peaks on the frequency response curve will appear near values of $\omega/\omega_n = 1/i$ where i is an even integer. The magnitude of the peaks grows smaller with increasing i.

4.13 FOURIER SERIES REPRESENTATION OF PERIODIC FUNCTIONS

If $F(t)$ is a piecewise continuous function of period T, then the Fourier series representation for $F(t)$ is

$$F(t) = \frac{a_0}{2} + \sum_{l=1}^{\infty}(a_l \cos \omega_l t + b_l \sin \omega_l t) \tag{4.107}$$

where

$$a_l = \frac{2}{T}\int_0^T F(t)\cos \omega_l t\, dt \qquad l = 0, 1, 2, \ldots \tag{4.108}$$

$$b_l = \frac{2}{T}\int_0^T F(t)\sin \omega_l t\, dt \qquad l = 1, 2, \ldots \tag{4.109}$$

and

$$\omega_l = \frac{2l\pi}{T} \tag{4.110}$$

The Fourier series representation has the following properties:

1. The Fourier series converges to $F(t)$ for all t where $F(t)$ is continuous for $0 \le t \le T$.
2. If $F(t)$ has a finite jump discontinuity at t, then the Fourier series converges to $\frac{1}{2}[F(t^+) + F(t^-)]$.
3. The Fourier series converges to the periodic extension of $F(t)$ for $t > T$.
4. If $F(t)$ is an odd function, that is, $F(-t) = -F(t)$ for all t, $0 \le t \le T$, then $a_l = 0, 1, 2, 3, \ldots$.
5. If $F(t)$ is an even function, that is, $F(-t) = F(t)$ for all t, $0 \le t \le T$, then $b_l = 0$, $l = 1, 2, 3, \ldots$.

Example 4.15. Decide whether each of the functions in Fig. 4.43 is an even function, an odd function, or neither and draw the Fourier series representation of each function on the interval [0, 6].

The function in Fig. 4.43*a* is an even function, the function in Fig. 4.43*b* is neither even nor odd, the function in Fig. 4.43*c* is an odd function. The functions to which the Fourier series representation converges are given in Fig. 4.44.

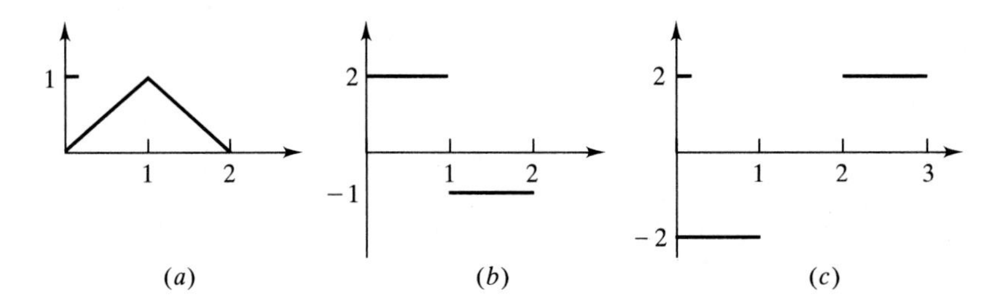

FIGURE 4.43
One period of periodic excitations. Function in (*a*) is even; function in (*c*) is odd; function in (*b*) is neither even nor odd.

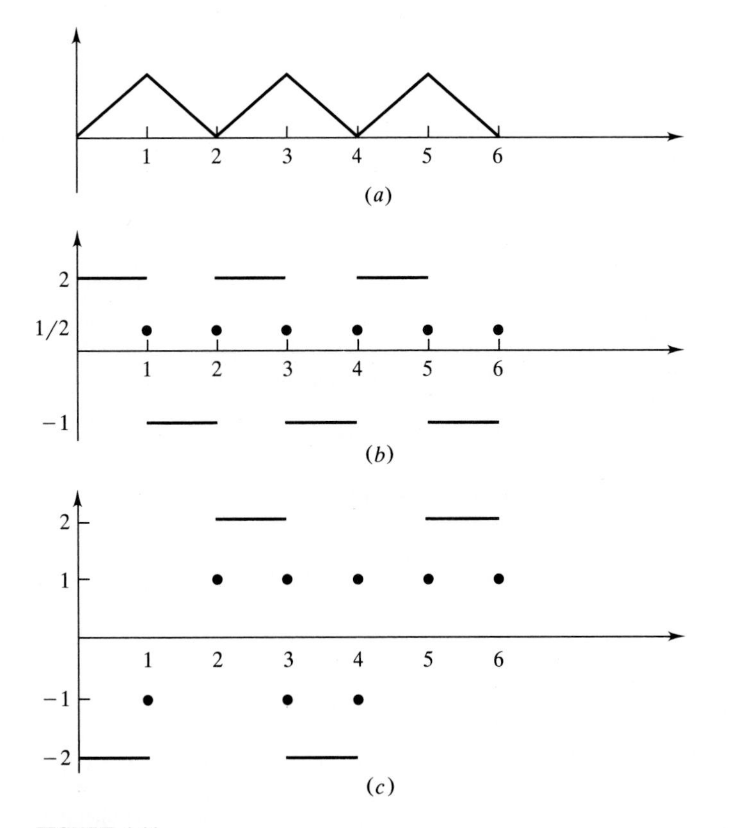

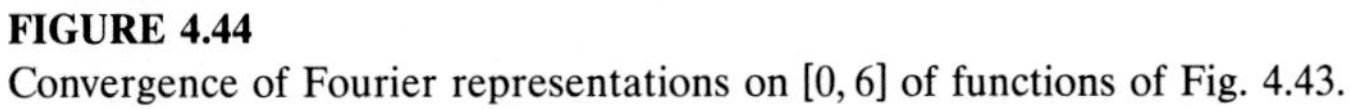

FIGURE 4.44
Convergence of Fourier representations on [0, 6] of functions of Fig. 4.43.

Use of the trigonometric identity for the sine of the sum of two angles and algebraic manipulation leads to an alternate form for the Fourier series representation

$$F(t) = \frac{a_0}{2} + \sum_{l=1}^{\infty} c_l \sin(\omega_l t + \kappa_l) \tag{4.111}$$

where

$$c_l = \sqrt{a_l^2 + b_l^2} \tag{4.112}$$

and

$$\kappa_l = \tan^{-1} \frac{a_l}{b_l} \tag{4.113}$$

If $F(t)$ is a periodic excitation for a one-degree-of-freedom system with viscous damping, the differential equation governing the response of the system is

$$\ddot{x} + 2\zeta\omega_n \dot{x} + \omega_n^2 x = \frac{1}{\tilde{m}}\left[\frac{a_0}{2} + \sum_{l=1}^{\infty} c_l \sin(\omega_l t + \kappa_l)\right] \tag{4.114}$$

The principle of linear superposition is used to determine the response as

$$x(t) = \frac{1}{\tilde{m}\omega_n^2}\left[\frac{a_0}{2} + \sum_{l=1}^{\infty} c_l M_l \sin(\omega_l t + \kappa_l - \phi_l)\right] \tag{4.115}$$

where M_l and ϕ_l are defined in Eqs. (4.105) and (4.103), respectively.

Example 4.16. A punch press of mass 500 kg sits on an elastic foundation of stiffness $k = 1.25 \times 10^6$ N/m and damping ratio $\zeta = 0.1$. The press operates at a speed of 120 rpm. The punching operation occurs over 40% of each cycle and provides a force of 5000 N to the machine. The excitation force is approximated as the periodic function of Fig. 4.45. Estimate the maximum displacement of the elastic foundation.

From the given information, the period of one revolution is 0.5 s and the natural frequency of the system is 50 rad/s.

The excitation force is periodic, but it is neither an even function nor an odd function. Its mathematical representation is

$$F(t) = \begin{cases} 5000 \text{ N} & 0 < t < 0.2 \text{ s} \\ 0 & 0.2 \text{ s} < t < 0.5 \text{ s} \end{cases}$$

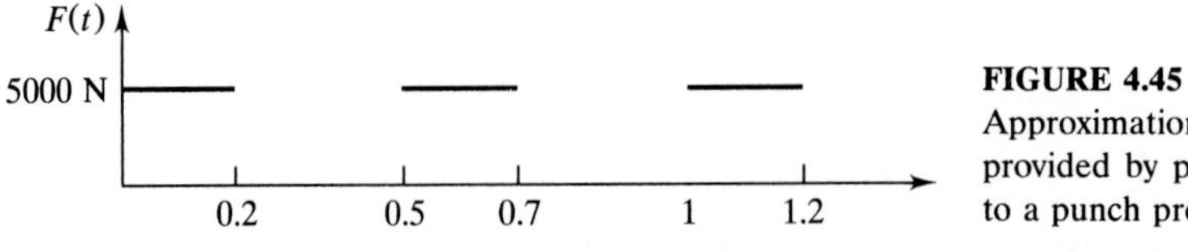

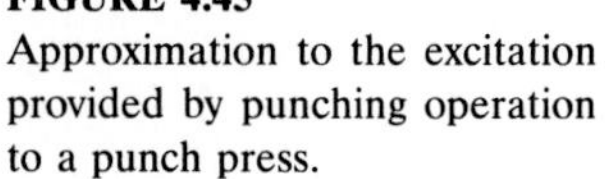

FIGURE 4.45
Approximation to the excitation provided by punching operation to a punch press.

The Fourier coefficients for the Fourier series representation for $F(t)$ are

$$a_0 = \frac{2}{0.5\text{ s}}\left(\int_0^{0.2\text{ s}} 5000\text{ N}\,dt + \int_{0.2\text{ s}}^{0.5\text{ s}} (0)\,dt\right)$$

$$= 4000\text{ N}$$

$$a_l = \frac{2}{0.5\text{ s}}\left(\int_0^{0.2\text{ s}} 5000\text{ N}\cos 4\pi lt\,dt\right)$$

$$= \frac{5000}{\pi l}\sin 4\pi lt\Big|_0^{0.2}\text{ N}$$

$$= \frac{5000}{\pi l}\sin 0.8\pi l\text{ N}$$

and

$$b_l = \frac{2}{0.5\text{ s}}\left(\int_0^{0.2\text{ s}} 5000\text{ N}\sin 4\pi lt\,dt\right)$$

$$= -\frac{5000}{\pi l}\cos 4\pi lt\Big|_0^{0.2\text{ s}}\text{ N}$$

$$= \frac{5000}{\pi l}(1 - \cos 0.8\pi l)\text{ N}$$

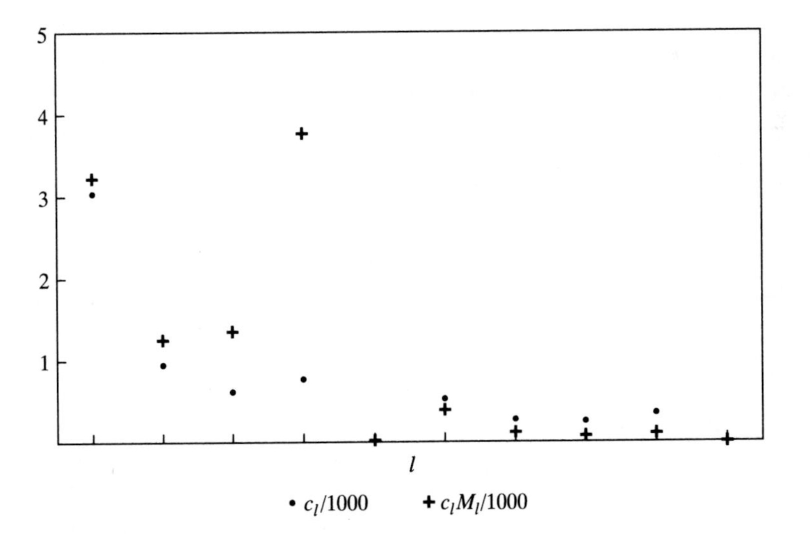

FIGURE 4.46
Fourier coefficients (c_l) and amplitude factor (c_lM_l) as functions of mode number, for Example 4.16.

The Fourier series representation of the excitation force is

$$F(t) = \frac{a_0}{2} + \sum_{l=1}^{\infty} c_l \sin(4\pi l t + \psi_l)$$

where
$$c_l = \frac{5000}{\pi l}\sqrt{2(1 - \cos 0.8\pi l)} \text{ N}$$

and
$$\kappa_l = \tan^{-1}\left(\frac{\sin 0.8\pi l}{1 - \cos 0.8\pi l}\right)$$

An upper bound on the displacement is

$$x_{\max} < \frac{1}{\tilde{m}\omega_n^2}\left(\frac{a_0}{2} + \sum_{l=1}^{\infty} c_l M_l\right)$$

The values of c_l and $c_l M_l$ are plotted vs. l in Fig. 4.46. The magnitude of the Fourier coefficient is largest for $l = 1$, but the product of the Fourier coefficient and the magnification factor is largest for $l = 4$. When the preceding equations are evaluated, an upper bound of 9.9 mm for the displacement of the machine is obtained.

PROBLEMS

4.1. A 40-kg mass is hanging from a spring of stiffness 4×10^4 N/m. A harmonic force of magnitude 100 N and frequency 120 rad/s is applied. Determine the amplitude of the forced response.

4.2. Determine the amplitude of the forced oscillation of the 30-kg block of Fig. P4.2 when a harmonic excitation of amplitude 200 N at a frequency of 10 rad/s is applied to the block.

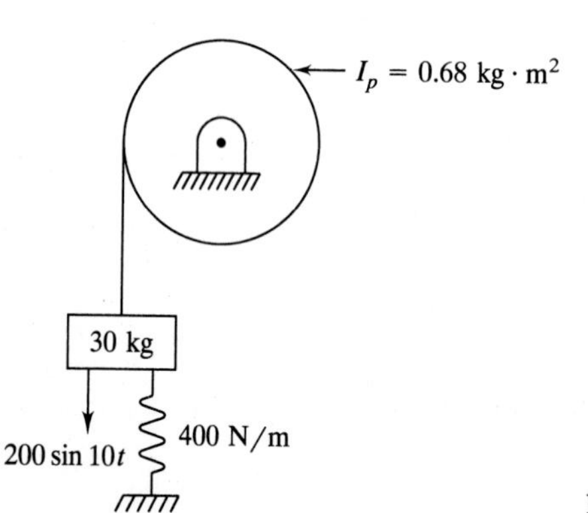

FIGURE P4.2

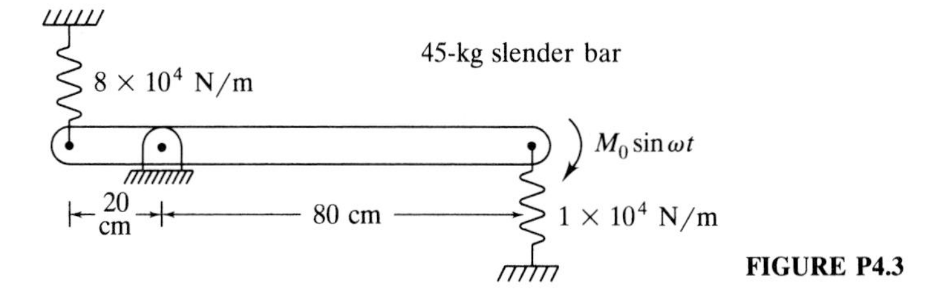

FIGURE P4.3

4.3. For what values of M_0 will the amplitude of the angular displacement of the bar of Fig. P4.3 due to the harmonic excitation be less than 3°?

4.4. A 2-kg gear of radius 20 cm is attached to the end of a 1-m long steel ($G = 80 \times 10^9$ N/m^2) shaft. A moment $M = 100 \sin 150t$ N · m is applied to the gear. For what shaft radii will the amplitude of torsional oscillations be less than 4°?

4.5. A 1-kg block is to be suspended from a coil spring and subject to a harmonic excitation. The magnitude of the excitation will be 360 N, but its frequency ranges from 30 to 120 rad/s. The coil spring is to be made from a 1-cm-diameter steel ($G = 80 \times 10^9$) bar. Design a spring by specifying the coil radius and the number of active turns such that the coil radius is less than 20 cm and the shear stress developed in the spring is less than 50×10^7 N/m^2 at all frequencies.

4.6. A 3-kg mass is placed at the end of a 25-cm steel ($E = 210 \times 10^9$ N/m^2) beam with $I = 1 \times 10^{-8}$ m^4. When excited by a harmonic excitation of magnitude 100 N, a vibration amplitude of 0.2 mm is observed. What is the frequency of the excitation?

4.7. A mass-spring system of natural frequency 22 Hz is subject to a harmonic excitation at a frequency of 24 Hz. Does beating occur? If so, what is the period of beating?

4.8. A mass-spring system whose natural frequency is ω_n is at rest in equilibrium when a harmonic excitation of frequency ω is applied.

(*a*) Set up an integral that represents the work done by the harmonic excitation between $t = 0$ and an arbitrary time t.

(*b*) Consider only the free-vibration contribution to the total response. That is, the velocity is given as $v = A \sin \omega_n t$. Assume $\omega \neq \omega_n$. Calculate the work done by the excitation from 0 to $2\pi/\omega_n$.

(*c*) Assume $\omega = \omega_n$. Calculate the work done by the excitation over one cycle of motion.

(*d*) Use the results of parts (*b*) and (*c*) to provide an explanation in terms of work and energy for the phenomenon of resonance.

4.9. A system of equivalent mass 30 kg has a natural frequency of 120 rad/s and a damping ratio of 0.12 is subject to a harmonic excitation of amplitude 2000 N and frequency 150 rad/s. What is the steady-state amplitude and phase angle of the response?

4.10. A 30-kg block is suspended from a spring of stiffness 3×10^2 N/m and attached to a dashpot of damping coefficient 1200 N · s/m. The block is subject to a harmonic excitation of magnitude 1150 N at a frequency of 450 Hz. What is the block's steady-state vibration amplitude?

4.11. What is the amplitude of steady-state oscillations of the 30-kg block of Fig. P4.11?

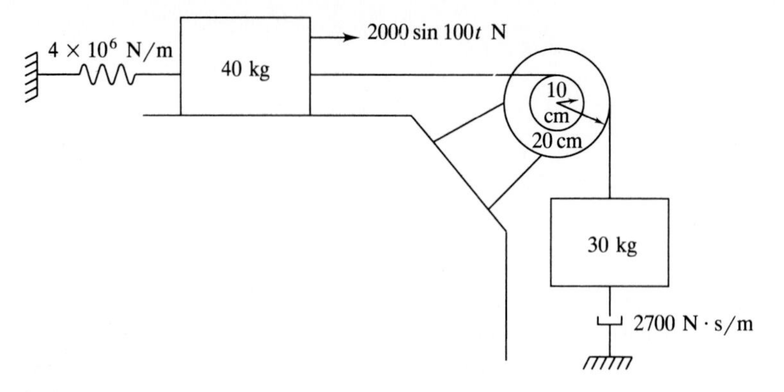

FIGURE P4.11

4.12. If $\omega = 16.5$ rad/s, what is the maximum value of M_0 such that the disk of Fig. P4.12 rolls without slip in the steady state?

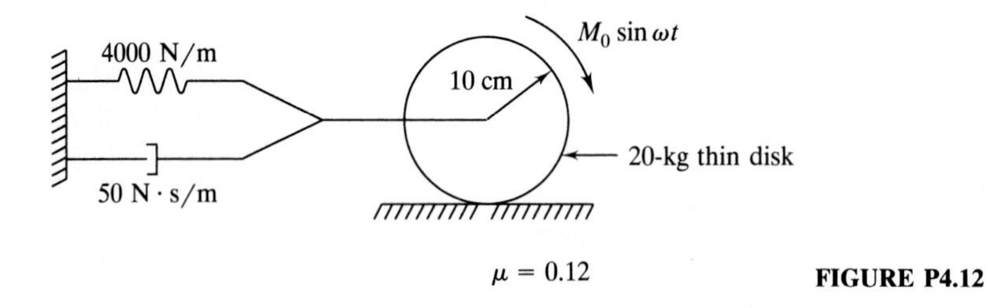

FIGURE P4.12

4.13. If $M_0 = 2.0$ N · m, for what values of ω will the disk of Fig. P4.13 roll without slip in the steady state?

4.14. A machine is attached to the end of a steel ($E = 210 \times 10^9$ N/m^2) cantilever beam of length 60 cm and moment of inertia 1×10^{-6} m^4. The machine produces a harmonic excitation whose magnitude of 1000 N is independent of frequency. A frequency sweep reveals that the maximum steady-state amplitude of vibration of the end of the beam is 1.5 mm and occurs at a frequency of 1400 Hz. What is the equivalent damping ratio and equivalent mass for this system?

4.15. For what values of d will the steady-state amplitude of angular oscillations of the slender rod of Fig. P4.15 be less than 1°?

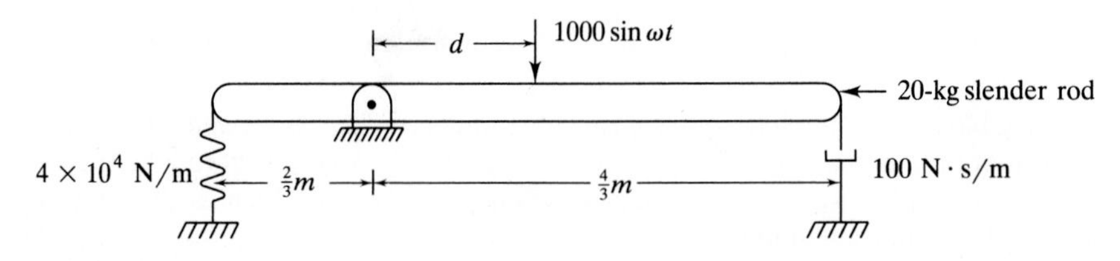

FIGURE P4.15

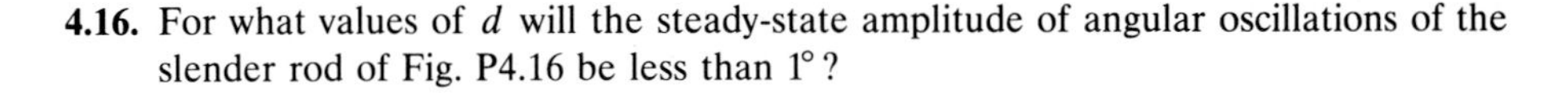

4.16. For what values of d will the steady-state amplitude of angular oscillations of the slender rod of Fig. P4.16 be less than 1°?

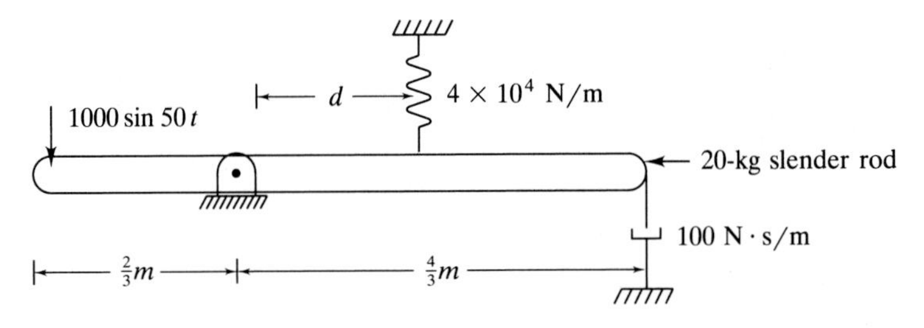

FIGURE P4.16

4.17. A 30-kg mass is mounted to an isolator pad of stiffness 6×10^5 N/m. When subject to a harmonic excitation of magnitude 350 N and frequency 100 rad/s, the phase difference between the excitation and the steady-state response is 24.3°. What is the damping ratio of the isolator pad and its maximum deflection due to this excitation?

4.18. A thin disk of mass 5 kg and radius 10 cm is connected to a torsional damper of coefficient 4.1 N · s · m/rad and a solid circular shaft of radius 10 mm, length 40 cm, and shear modulus 80×10^9 N/m^2. The disk is subject to a harmonic moment of magnitude 250 N · m and frequency 600 Hz. What is the amplitude of the steady-state torsional oscillations?

4.19. When a 10-kg recording device is mounted on an isolator, the isolator deflects 22 mm. When the recorder operates at 35 rad/s, and unbalanced force of 10 N is produced and the system vibrates with an amplitude of 1 mm. What is the damping ratio of the system?

4.20. Derive Eqs. (4.32) and (4.33).

4.21. A 100-kg machine has a 2-kg rotating component. When the machine is mounted on an isolator and its operating speed is very large, the steady-state vibration amplitude is 0.7 mm. How far is the center of mass of the rotating component from its axis of rotation?

4.22. A 120-kg machine with a rotating unbalance of 0.35 kg · m is to be placed at the midspan of a 2.6-m simply supported beam. The beam, to be made of steel ($E = 210 \times 10^9$ N/m^2), will have a uniform rectangular cross section with a height of 5 cm. For what values of the cross-sectional depth will the steady-state amplitude of the midspan be limited to 5 mm for all operating speeds between 50 and 125 rad/s?

4.23. A 1000-kg machine with a rotating unbalance is placed on springs and viscous dampers in parallel. When the operating speed is 20 Hz, the observed steady-state amplitude is 0.08 mm. As the operating speed is increased, the steady-state amplitude increases with 0.25 mm at 40 Hz, and is 0.5 mm for much larger speeds. Determine the equivalent stiffness and damping coefficient of the system.

4.24. A 15-kg motor is placed at the end of the rigid bar of Fig. P4.24. The motor produces a force of $2000 \sin \omega t$ N. What is the steady-state amplitude of the end of the bar when $\omega = 30$ Hz?

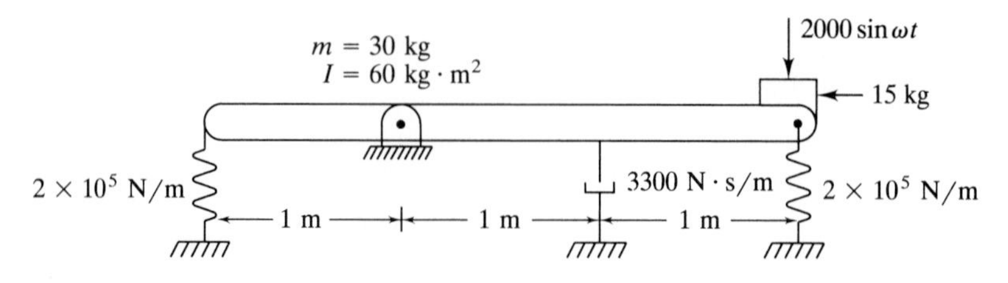

FIGURE P4.24

4.25. A 620-kg fan has a rotating unbalance of 0.25 kg · m. What is the maximum stiffness of the fan's mounting such that the steady-state amplitude is 0.5 mm or less at all operating speeds greater than 100 Hz? Assume a damping ratio of 0.08.

4.26. It is proposed to build a 6-m smokestack on the top of a 60-m factory. The smokestack will be made of steel ($\rho = 7850$ kg/m^3) and will have an inner radius of 40 cm and an outer radius of 45 cm. What is the maximum amplitude of vibration due to vortex shedding? Use a one-degree-of-freedom model for the smokestack with a concentrated mass at the end of the smokestack to account for inertia effects. Assume $\zeta = 0.05$.

4.27. Repeat Prob. 4.26 assuming a damping ratio of 0.11.

4.28. Repeat parts (a) and (b) of Example 4.6 if the light pole is hollow with an inner diameter of 15 cm and an outer diameter of 20 cm.

4.29. Repeat Prob. 4.28 including the inertia effects of the lamp pole ($\rho = 7500$ kg/m^3).

4.30. A factory is using the piping system of Fig. P4.30 to discharge environmentally safe wastewater into a small river. The velocity of the river is estimated as 5.5 m/s. Determine the allowable values of l such that the amplitude of torsional oscillations of the vertical pipe due to vortex shedding is less than 0.1°. Assume the vertical pipe is rigid and rotates about an axis perpendicular to the page through the elbow. The horizontal pipe is restrained from rotation at the river bank. Assume a damping ratio of 0.05.

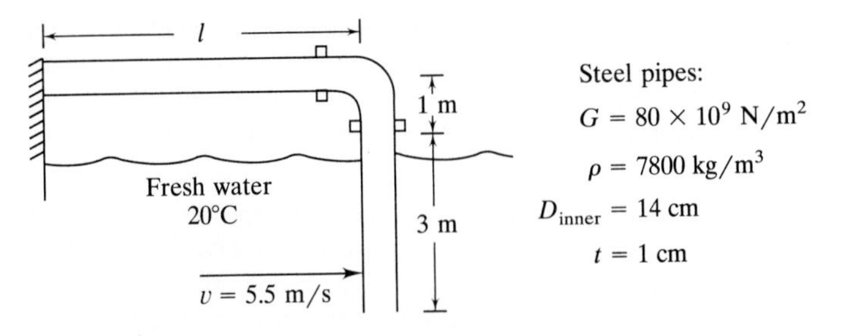

FIGURE P4.30

4.31. A 40-kg mass is attached to a base through a spring of stiffness 2×10^4 N/m in parallel with a dashpot of coefficient 150 N · s/m. The base is given a time-dependent displacement $0.15 \sin 30.1t$ m. Determine the amplitude of the absolute displacement of the mass and the amplitude of the mass's displacement relative to the base.

4.32–4.37. Determine the amplitude of steady-state vibration for the system shown. Use the generalized coordinate identified in the diagram.

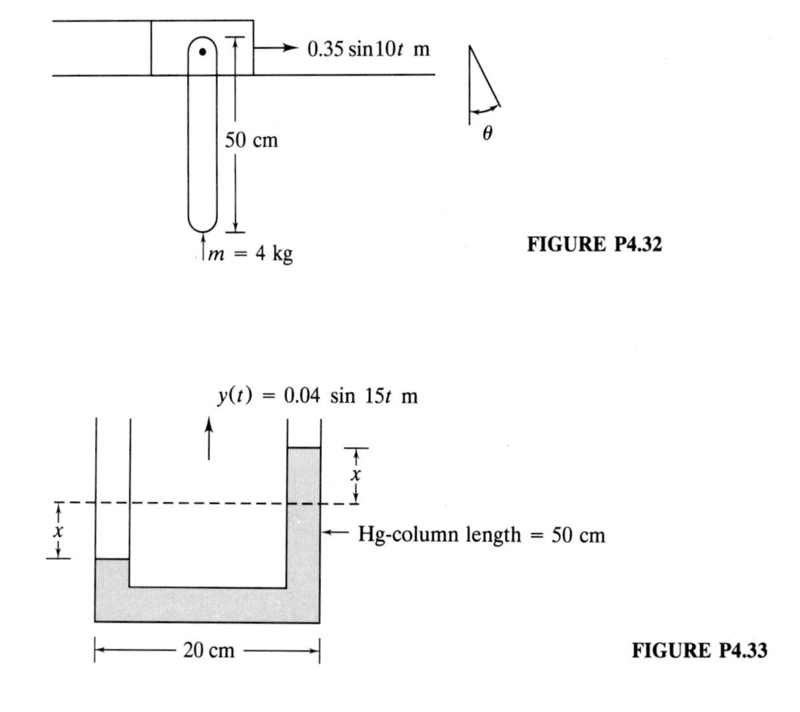

FIGURE P4.32

FIGURE P4.33

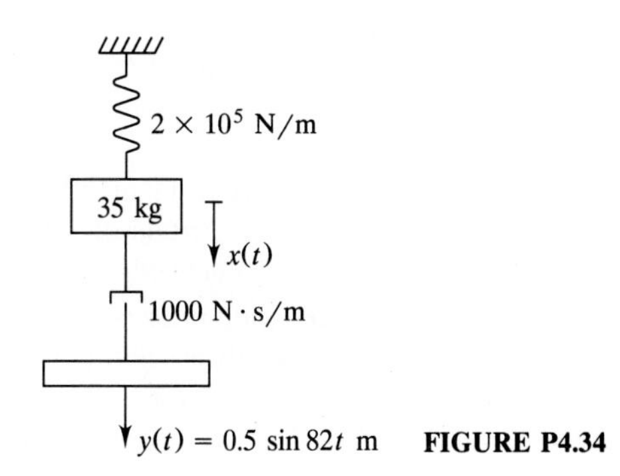

FIGURE P4.34

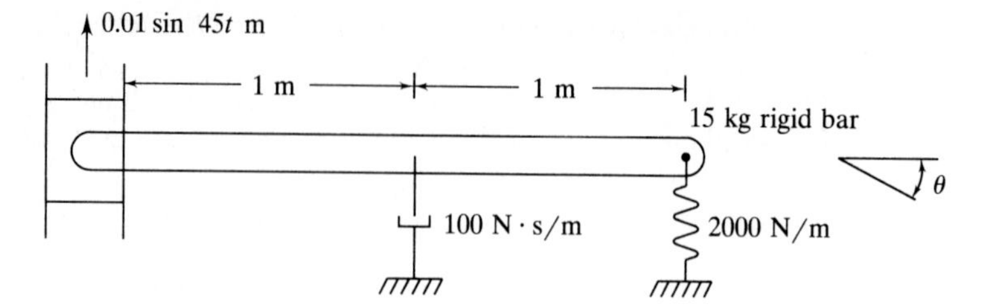

FIGURE P4.35

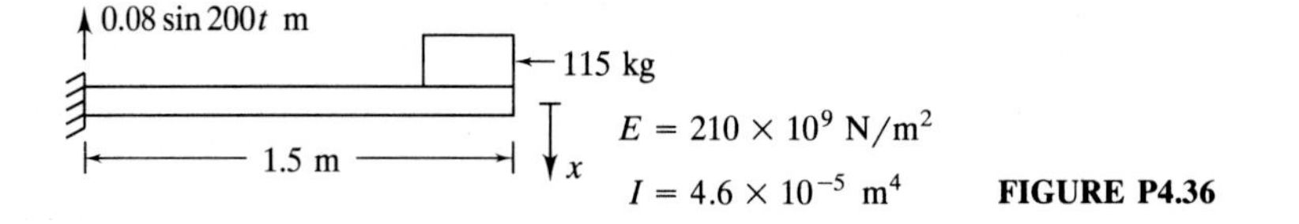

FIGURE P4.36

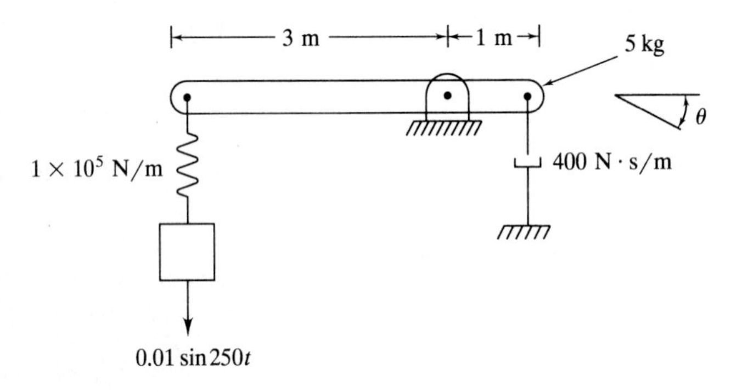

FIGURE P4.37

4.38. The suspension system for a small vehicle is being designed such that passengers are comfortable as the vehicle travels over a sinusoidal road contour similar to that shown in Fig. 4.25. Specify the natural frequency and damping ratio used in a suspension system such that the static deflection is no greater than 1.5 mm and the acceleration amplitude felt by the passengers is no greater than 70 m/s^2 for vehicle speeds between 20 and 40 m/s and road contours with $d < 5$ mm and $l > 2$ m.

4.39. Use of a seismometer of natural frequency 20 Hz and a damping ratio of 0.15 reveals that a system vibrates with a frequency of 100 Hz and an amplitude of 1.3 mm. What is the error in the measurement and what is the actual amplitude?

4.40. Use of an accelerometer of natural frequency 100 Hz and a damping ratio of 0.15 reveals that a system vibrates at a frequency of 20 Hz and has an acceleration amplitude of 14.3 m/s^2. What is the error in the measurement, the actual acceleration amplitude, and the actual displacement amplitude?

4.41. An accelerometer of natural frequency 200 Hz and damping ratio 0.7 is used to measure the vibrations of a system whose actual displacement is

$$x(t) = 1.6 \sin 45.1t \text{ mm}$$

What is the output of the accelerometer?

4.42. What is the minimum natural frequency of an undamped accelerometer that is used to measure the vortex-induced oscillations of the street light of Example 4.6 for wind speeds up to 60 m/s with an error less than 3%?

4.43. What is the minimum natural frequency of an accelerometer of damping ratio 0.7 used to measure the acceleration of the vehicle of Prob. 4.38 with an error of less than 3%?

4.44. A 550-kg industrial sewing machine has a rotating unbalance of 0.24 kg · m. The machine operates at speeds between 2000 and 3000 rpm. The machine is placed on an isolator pad of stiffness 5×10^6 N/m and damping ratio 0.12. What is the maximum natural frequency of an undamped seismometer that can be used to measure the steady-state vibrations at all operating speeds with an error less than 4%? If this seismometer is used, what is its output when the machine is operating at 2500 rpm?

4.45. Show that phase distortion is not a problem when multifrequency measurements are made using a seismometer.

4.46. The system of Fig. P4.46 is subject to the excitation

$$F(t) = 1000 \sin 25.4t + 800 \sin(48t + 0.35) - 300 \sin(100t + 0.21) \text{ N}$$

What is the output, in mm/s^2, of an accelerometer of natural frequency 100 Hz and damping ratio 0.7 placed at A?

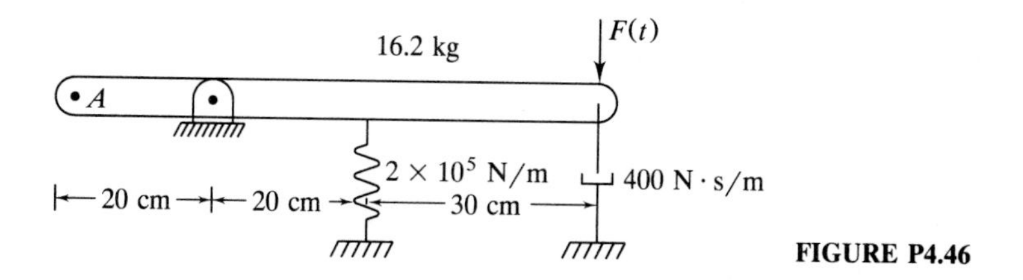

FIGURE P4.46

4.47. What is the output, in mm, of a seismometer with a natural frequency of 2.5 Hz and a damping ratio of 0.05 placed at point A for the system of Prob. 4.46?

4.48. A 20-kg block is connected to a moveable support through a spring of stiffness 1×10^5 N/m in parallel with a viscous damper of coefficient 600 N · s/m. The support is given a harmonic displacement of amplitude 25 mm and frequency 40 rad/s. An accelerometer of natural frequency 25 Hz and damping ratio 0.2 is attached to the block. What is the output of the accelerometer, in mm/s^2?

4.49. An accelerometer has a natural frequency of 80 Hz and a damping coefficient of 8.0 N · s/m. When attached to a vibrating structure, it measures an amplitude of 8.0 m/s^2 and a frequency of 50 Hz. The true acceleration of the structure is 7.5 m/s^2. Determine the mass and spring constant of the accelerometer.

4.50. A 20-kg block is connected to a spring of stiffness 1×10^5 N/m and placed on a surface which makes an angle of 30° with the horizontal. The coefficient of friction between the block and surface is 0.15. A force $300 \sin 80t$ N is applied to the block. What is the amplitude of the resulting steady-state oscillations?

4.51. When the mass-spring system of Prob. 4.50 is placed on a different surface, which also makes an angle of 30° with the horizontal and the same force applied, the steady-state amplitude is measured as 10.6 mm. What is the coefficient of friction between the block and the surface?

4.52. A 40-kg block is connected to a spring of stiffness 1×10^5 N/m and slides on a surface with a coefficient of friction of 0.2. When a harmonic force of frequency 60 rad/s is applied to the block, the resulting amplitude of steady-state vibrations is 3 mm. What is the amplitude of the excitation?

4.53. The 2-kg swing of Example 3.19 is subject to a harmonic moment of $1.5 \sin 5t$ N · m. What is the amplitude of the steady-state oscillation?

4.54. Use the equivalent viscous damping approach to determine the response of a harmonically excited system with both viscous and Coulomb damping.

4.55. The area under the hysteresis curve for a particular helical coiled spring is 0.2 N · m when subject to a 350-N load. The spring has a stiffness 4×10^5 N/m. If a 44-kg block is hung from the spring and subject to an excitation force of $350 \sin 35t$ N, what is the amplitude of the resulting steady-state oscillations?

4.56. An engineer wants to isolate a foundation from the forces produced by a machine with a rotating unbalance. The machine has a wide operating range. The engineer is considering mounting the machine on springs in parallel with a viscous damper such that the system has a damping coefficient of 0.1. The engineer is also considering using a pad with hysteretic damping such that the equivalent viscous damping coefficient of the pad is 0.1 at the lower end of the operating range. Discuss the advantages of using each system. Which isolation system should he choose, assuming all other factors are equal?

4.57. A schematic of a single-cylinder engine mounted on springs and a viscous damper is shown in Fig. P4.57. The crank rotates about O with a constant speed ω. The connecting rod of mass m_r connects the crank and the piston of mass m_p such that the piston moves in a vertical plane. The center of gravity of the crank is at its axis of rotation.

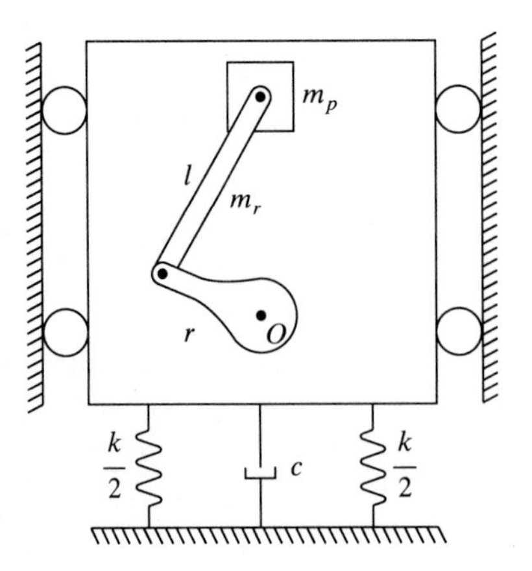

FIGURE P4.57

(a) Derive the differential equation governing the absolute vertical displacement of the engine including the inertia forces of the crank and piston, but ignoring forces due to combustion. Use an exact expression for the inertia forces in terms of m_r, m_p, ω, the crank length r, and the connecting rod length l. Write the differential equation in the form of Eq. (4.3).

(b) Since $F(t)$ is periodic, a Fourier series representation can be used. Set up, but do not evaluate, the integrals required for a Fourier series expansion for $F(t)$.

(c) Assume $r/l \ll 1$. Rearrange $F(t)$ and use a binomial expansion such that

$$F(t) = \sum_{i=1}^{\infty} a_i \left(\frac{r}{l}\right)^i$$

(d) Truncate the preceding series after $i = 3$. Use trigonometric identities to approximate

$$F(t) \approx b_1 \cos \omega t + b_2 \cos 2\omega t + b_3 \cos 3\omega t$$

(e) Find an approximation to the steady-state form of $x(t)$.

4.58. Using the results of Prob. 4.57, determine the maximum steady-state response of a single-cylinder engine with $m_r = 1.5$ kg, $m_p = 1.7$ kg, $r = 5.0$ cm, $l = 15.0$ cm, $\omega = 800$ rpm, $k = 1 \times 10^5$ N/m, $c = 500$ N · s/m, and total mass 7.2 kg.

4.59. An accelerometer of natural frequency 200 rad/s and damping ratio 0.7 is attached to the engine of Prob. 4.58. What is the output of the accelerometer, in mm/s^2?

4.60–4.64. Determine the response of a mass-spring-dashpot system with $m = 100$ kg, $\omega_n = 40$ rad/s and $\zeta = 0.2$ subject to the periodic excitation shown. Approximate the maximum displacement.

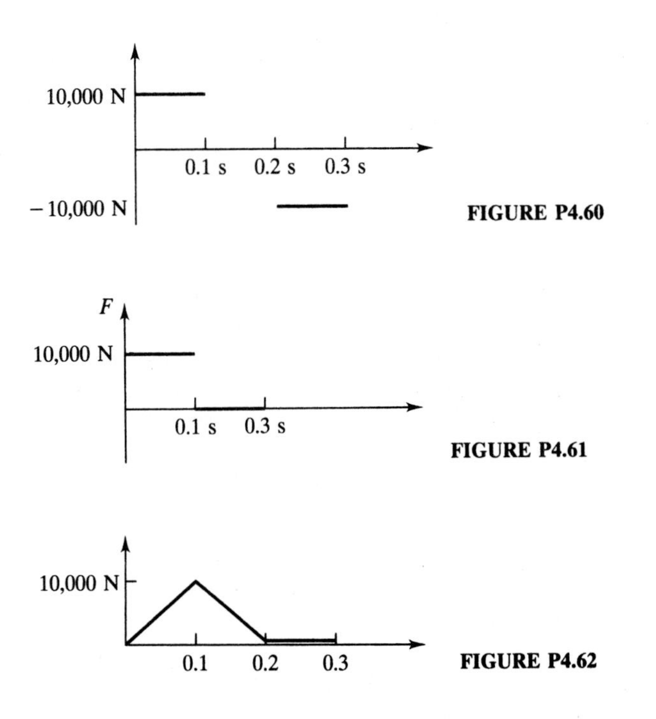

FIGURE P4.60

FIGURE P4.61

FIGURE P4.62

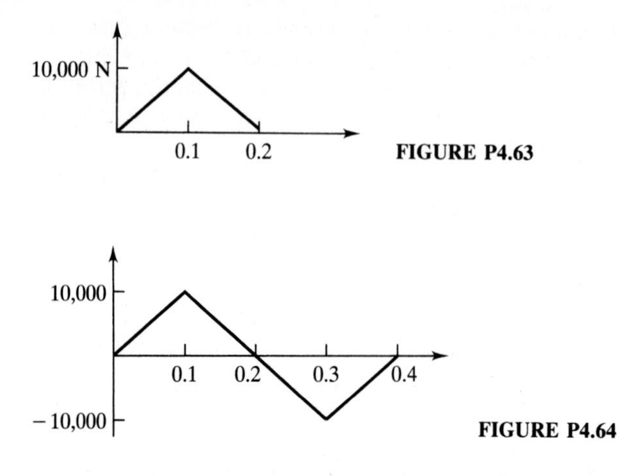

FIGURE P4.63

FIGURE P4.64

4.65. The base of a mass-spring-dashpot system with $\omega_n = 20$ rad/s and $\zeta = 0.1$ is subject to the displacement of Fig. P4.65. Determine the absolute displacement of the mass.

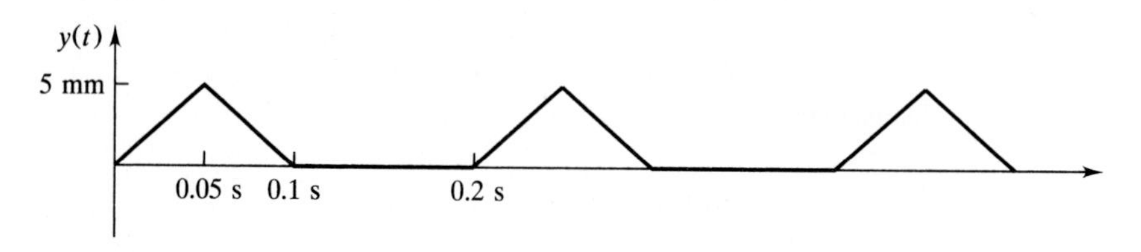

FIGURE P4.65

4.66. The linearized differential equation governing the displacement of an offshore structure composed of a rigid body supported by a spring and viscous damper is

$$m\ddot{x} + 2\zeta m\omega_n\dot{x} + kx = F_I + F_D$$

where m is the mass of the structure plus the added mass of the fluid, F_I is an inertia force, and F_D is the drag force. The velocity of the ocean is periodic due to wave motion and is given by $U \sin \omega t$. For a circular cylinder of radius R and length L, the drag force takes the form

$$F_D = \rho RLU^2 |\sin \omega t| \sin \omega t$$

and the inertia force takes the form

$$= 2\rho\pi R^2 LU\omega \cos \omega t$$

(*a*) Expand the drag force in a Fourier series.
(*b*) Determine the steady-state displacement of the cylinder.
(*c*) Consider a system composed of a 25-cm-radius steel cylinder of length 1 m. The cylinder is attached to a spring such that the natural frequency is 20 rad/s and the damping ratio is 0.05. The ocean has an oscillating flow of amplitude 5 m/s and frequency 1 rad/s. Use a one-degree-of-freedom model to approximate the cylinder's maximum displacement.

The program FORCED.BAS included in the accompanying diskette simulates the forced response of a one-degree-of-freedom system to a single-frequency harmonic excitation. It also provides plots of the magnification factor vs. frequency ratio and a dimensional plot of steady-state amplitude vs. excitation frequency. Use FORCED.BAS to solve Probs. 4.67 through 4.73.

4.67. A 200-kg block is suspended from a spring of stiffness 400,000 N/m and damping coefficient 2600 N · s/m. It is subject to a harmonic excitation of magnitude 200 N at a frequency of 30 rad/s. Use FORCED.BAS to determine the steady-state amplitude and the phase angle.

4.68. The block of Prob. 4.67 is subject to a harmonic excitation of magnitude 200 N. For what frequencies will the steady-state amplitude be less than 1 mm?

4.69. For what excitation frequency will the steady-state amplitude of the block of Prob. 4.67 be a maximum? What is this maximum amplitude?

4.70. If the frequency of excitation of the block of Prob. 4.67 is fixed at 40 rad/s and the excitation magnitude is 200 N, for what values of the damping coefficient c will the steady-state amplitude be less than 2 mm?

4.71. The thin disk of Figure P4.12 is subject to a moment with $M_0 = 2.0$ N · m and $\omega = 18$ rad/s. Use FORCED.BAS to determine the steady-state amplitude.

4.72. Use FORCED.BAS to determine how long it takes the system of Prob. 4.67 to reach steady-state when
(*a*) $c = 2600$ N · s/m
(*b*) $c = 200$ N · s/m
(*c*) $c = 4000$ N · s/m

4.73. A 400-kg machine is placed on an elastic foundation of stiffness 4,000,000 N/m. The damping ratio of the system is unknown, so a forced vibration test is run.
(*a*) Select "random choice of damping ratio" from the main menu. When prompted for a seed number, input the last four digits of your social security number. Then use FORCED.BAS to simulate any forced vibrations test which you think will help determine the value of λ.
(*b*) Repeat part (*a*) using twice the last four digits of your social security number as the seed number.

The program COULOMB.BAS included on the accompanying diskette simulates the free and harmonically forced vibrations of a one-degree-of-freedom system with Coulomb damping. Use COULOMB.BAS to solve Probs. 4.74–4.78.

4.74. A 50-kg block is attached to a spring of stiffness 10,000 N/m. The block is excited by a harmonic excitation of frequency 20 rad/s and magnitude 370 N. The coefficient of friction between the block and the surface is 0.15. Use COULOMB.BAS to determine the maximum displacement of the block.

4.75. Solve Prob. 4.50 using COULOMB.BAS.

4.76. Solve Prob. 4.53 using COULOMB.BAS.

4.77. Use COULOMB.BAS to develop the response of a one-degree-of-freedom system with $r = 1.5$ and $\iota = 0.8$.

4.78. Use COULOMB.BAS to help develop the frequency response curve for a one-degree-of-freedom system with Coulomb damping for
(*a*) $\iota = 0.1$
(*b*) $\iota = 0.4$
(*c*) $\iota = 0.8$

Compare the exact solution computed by COULOMB.BAS to the approximate solution presented in Sec. 4.10.

REFERENCES

1. Amann, O. H., T. VonKarman, and G. B. Woodruff: *The Failure of the Tacoma Narrows Bridge*, Federal Works Agency, Washington, D.C., 1941.
2. Blevins, R. D.: *Flow Induced Vibrations*, 2nd ed., Van Nostrand Reinhold, New York, 1990.
3. Boyce, W. E., and R. C. DiPrima: *An Introduction to Ordinary Differential Equations*, Wiley, New York, 1970.
4. Churchill, R. V.: *Fourier Series and Boundary Value Problems*, McGraw-Hill, New York, 1941.
5. Crandall, S. H.: "The Role of Damping in Vibration Theory," *Journal of Sound and Vibration*, vol. 11, pp. 3–18, 1970.
6. DenHartog, J. P.: "Forced Vibrations with Combined Coulomb and Viscous Friction," *Transactions of the ASME, Applied Mechanics*, vol. 53, pp. 107–115, 1931.
7. DenHartog, J. P.: *Mechanical Vibration*, 4th ed., McGraw-Hill, New York, 1956.
8. Dimarogonas, A. D., and S. Haddad: *Vibrations for Engineers*, Prentice-Hall, Englewood Cliffs, N.J., 1992.
9. Greif, R., B. Schwarz, and N. Weinstock: "Resonant Response of a System with Coulomb Friction," *Journal of Vibration and Acoustics*, vol. 112, pp. 427–428, 1990.
10. Holman, J. P.: *Experimental Methods for Engineers*, 4th ed., McGraw-Hill, New York, 1984.
11. Hundal, M. S.: "Response of a Base Excited System with Coulomb and Viscous Friction," *Journal of Sound and Vibration*, vol. 64, pp. 371–378, 1979.
12. James, M. L., G. M. Smith, J. C. Wolford, and P. W. Whaley: *Vibrations of Mechanical and Structural Systems with Microcomputer Applications*, Harper and Row, New York, 1989.
13. Jones, G. W.: "Unsteady Lift Forces Generated by Vortex Shedding about a Large Stationary, Oscillating Cylinder at High Reynolds Numbers," ASME Symposium on Unsteady Flow, 1968.
14. Lazer, A. C., and P. J. McKenna: "Large Amplitude Oscillations in Suspension Bridges: Some New Connections with Nonlinear Analysis," *SIAM Review*, vol. 32, pp. 537–575, 1990.
15. Kreider, D. L., R. G. Kuller, D. R. Ostberg, and F. W. Perkins: *An Introduction to Linear Analysis*, 2nd ed., Addison-Wesley, Reading, Mass., 1966.
16. Meirovitch, L.: *Elements of Vibration Analysis*, McGraw-Hill, New York, 1975.
17. Rabenstein, A. L.: *Elementary Differential Equations with Linear Algebra*, 3rd ed., Academic Press, New York, 1982.
18. Rao, S. S.: *Mechanical Vibrations*, 2nd ed., Addison-Wesley, Reading, Mass., 1990.
19. Roshko, A.: "On the Development of Turbulent Wakes from Vortex Streets," NACA Report 1191, 1954.
20. Ross, A. D., and D. J. Inman: "A Design Criterion for Avoiding Resonance in Lumped Mass Normal Mode Systems," *Journal of Vibrations, Acoustics, Stress, and Reliability in Design*, vol. 111, pp. 49–52, 1989.

21. Shabana, A. A.: *Theory of Vibration: An Introduction*, vol. 1, Springer-Verlag, New York, 1991.
22. Shabana, A. A.: *Theory of Vibration: Discrete and Continuous Systems*, vol. 2, Springer-Verlag, New York, 1991.
23. Shames, I. H.: *Mechanics of Fluids*, McGraw-Hill, New York, 1992.
24. Simiu, E., and R. H. Scanlan: *Wind Effects on Structures*, 2nd ed., Wiley-Interscience, New York, 1986.
25. Steidel, R. F.: *An Introduction to Mechanical Vibrations*, 3rd ed., Wiley, New York, 1988.
26. Thomson, W. T.: *Theory of Vibrations with Applications*, 3rd ed., Prentice-Hall, Englewood Cliffs, N.J., 1988.
27. Weaver, W., S. P. Timoshenko, and D. H. Young: *Vibration Problems in Engineering*, 5th ed., Wiley-Interscience, New York, 1990.
28. Wylie, C. R., and R. Barrett: *Advanced Engineering Mathematics*, McGraw-Hill, New York, 1972.

CHAPTER 5

TRANSIENT VIBRATIONS OF ONE-DEGREE-OF-FREEDOM SYSTEMS

5.1 INTRODUCTION

5.1.1 Examples of Nonperiodic Excitation

Examples of systems subject to nonperiodic excitations are shown in Figs. 5.1 through 5.3. During start-up, the operating speed of the rotating machine of Fig. 5.1 builds up to its steady-state speed. During this time the machine is subject to a harmonic excitation whose frequency and amplitude vary with time. If the system is designed to provide for vibration isolation, then its natural frequency must be passed when the speed of the rotating component builds up to its final operating speed. Thus large-amplitude oscillations occur during start-up. The maximum response can be approximated using the methods of Chap. 4, ignoring the time dependency of the frequency and amplitude. The methods developed in this chapter can take into account the transient nature of the operating speed to give a better estimate of the maximum response as resonance is passed.

Figure 5.2 shows a sensitive computer system mounted in the nose cone of a rocket. To guard against shocks, the computer system is mounted on a spring-dashpot system. Relative motion between the computer system and the rocket occurs as the rocket is thrust into its orbit. The resulting motion of the

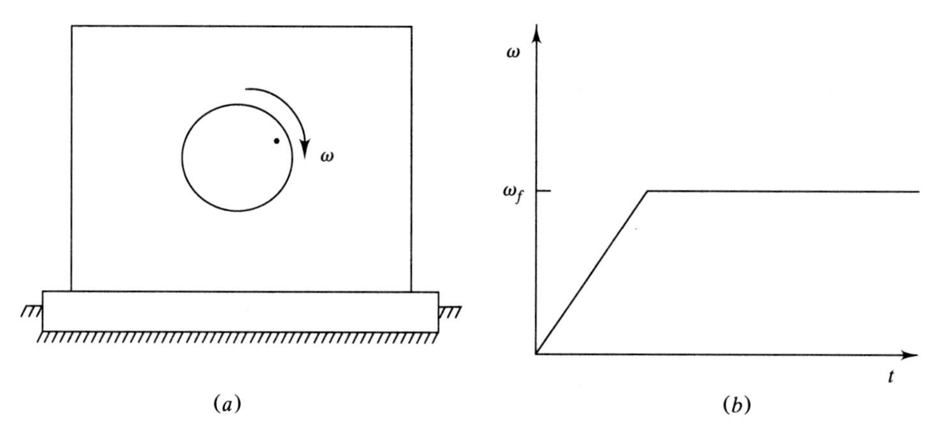

FIGURE 5.1
(*a*) The rotating unbalance provides excitation to the machine as it operates; (*b*) while the speed is building up to its final operating speed, the machine is subject to an excitation that is harmonic in form but its frequency and amplitude change with time.

computer is analyzed as a support motion problem with a nonharmonic base excitation.

A simplified model of a one-story frame structure is shown in Fig. 5.3. The one-degree-of-freedom model is adequate if the girder is assumed infinitely rigid in comparison to the columns. This particular structure houses a chemical laboratory. The structure must be designed to withstand an explosion which could occur due to a laboratory accident. The explosion produces a large excitation force over a small time.

In each of the preceding examples, the short-term transient behavior is more important than the steady state. In the first example the maximum response occurs during buildup to steady state. In the other examples the maximum response occurs shortly after a change occurs. After the excitations are removed, damping causes the systems to return to their equilibrium positions, resulting in trivial steady states.

5.1.2 Methods of Solution

Forced vibrations of one-degree-of-freedom systems are described by the differential equation

$$\ddot{x} + 2\zeta\omega_n\dot{x} + \omega_n^2 x = \frac{\tilde{F}(t)}{\tilde{m}} \tag{5.1}$$

Initial conditions, values of $x(0)$ and $\dot{x}(0)$, complete the problem formulation. Solution of Eq. (5.1) for periodic forms of $\tilde{F}(t)$ is discussed in Chap. 4.

The purpose of this chapter is to analyze the motion of systems undergoing transient vibrations. Equation (5.1) is a second-order linear nonhomoge-

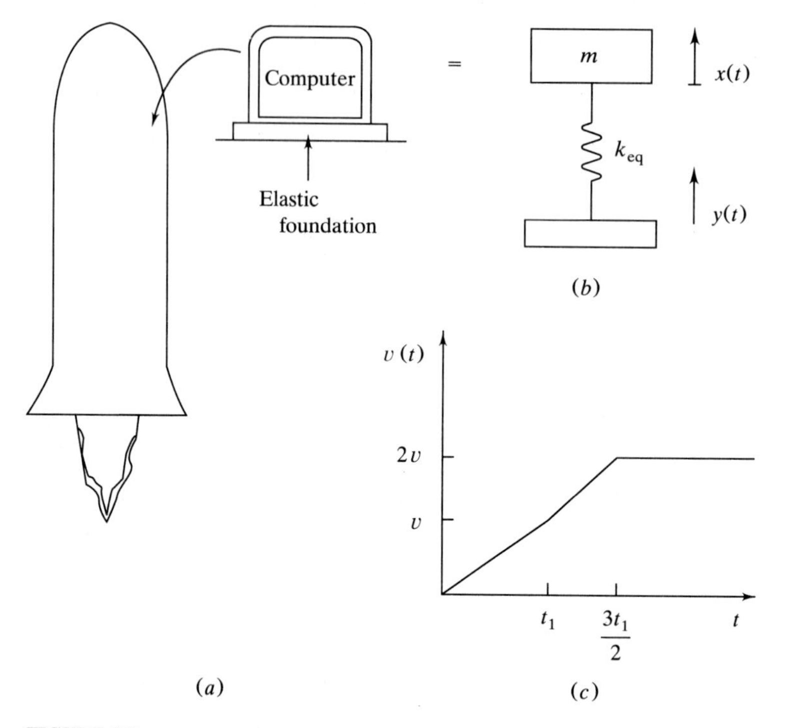

FIGURE 5.2
(a) A sensitive computer system used to control flight is mounted in the nose cone of a rocket; (b) the computer system is modeled as a mass attached through a spring to a moving base; (c) the acceleration of the rocket is constant until a second stage is fired. The velocity then increases linearly at a different rate until the final orbit speed is reached. This produces a nonharmonic excitation to the computer system.

neous ordinary differential equation. For certain forms of $\tilde{F}(t)$, the method of undetermined coefficients, as applied in Chap. 4, can be used to determine the particular solution. The homogeneous solution is added to the particular solution, resulting in a general solution involving two constants of integration. Initial conditions are applied to evaluate the constants of integration. If damping is present the homogeneous solution dies out, leaving the particular solution as a steady-state solution. The method of undetermined coefficients is best suited for harmonic, polynomial, or exponential excitations and not useful for most excitations studied in this chapter.

The initial conditions and the homogeneous solution have an important effect on the short-term transient motion of vibrating systems. For these problems, it is convenient to use a solution method in which the homogeneous solution and particular solution are obtained simultaneously and the initial conditions are incorporated in the solution.

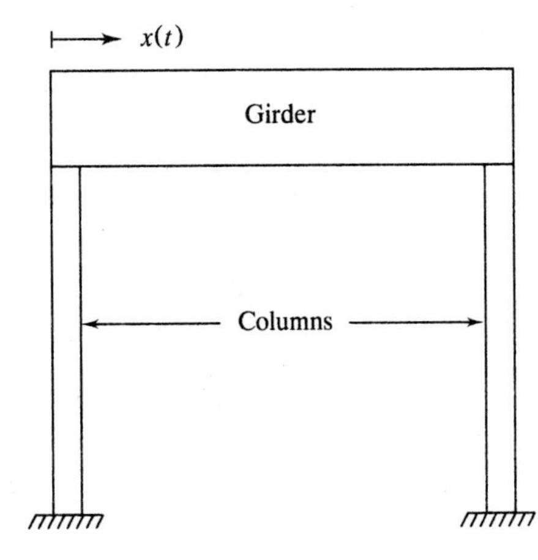

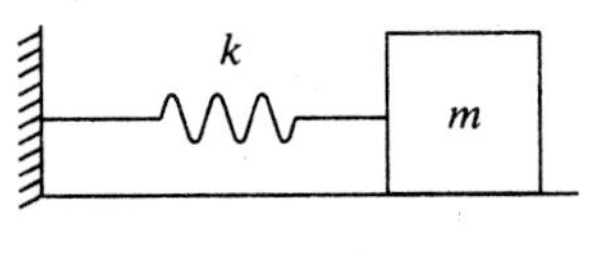

FIGURE 5.3
(a) A one-story chemical laboratory is modeled as a frame structure; (b) the frame structure is modeled as a one-degree-of-freedom mass-spring system.

Many excitations are of short duration. For short-duration responses, the maximum response may occur after the excitation has ceased. Thus it is necessary to develop a solution method which determines the response of a system for all time, even after the excitation is removed. In addition, many excitations change form at discrete times. For these excitations a solution method in which a unified mathematical form of the response is determined is a great convenience.

The primary method of solution presented in this chapter is use of the convolution integral. The convolution integral is derived using the principle of impulse and momentum and linear superposition. It can also be derived by application of the method of variation of parameters. The convolution integral is the most general closed-form solution of Eq. (5.1). The initial conditions are applied in the derivation of the integral, and need not be applied during every application. The convolution integral can be used to generate a unified mathematical response for excitations whose form changes at discrete times. Since it only requires evaluation of an integral, it is easy to apply.

A second method presented in this chapter is the Laplace transform method. Initial conditions are applied during the transform procedure and the Laplace transform can be used to develop a unified mathematical response for excitations whose form changes at discrete times. Use of tables of transforms makes application of the method convenient. The algebraic effort can be less than that using the convolution integral for damped systems, if appropriate transforms are available in a table. However, if the appropriate transforms are not available in a table, determination of the response is difficult.

There are some excitations in which a closed-form solution of Eq. (5.1) does not exist. In these cases, the convolution integral does not have a closed-form evaluation and the Laplace transform method can be used to derive only the response in terms of an integral. In addition, situations exist when the excitation is not known explicitly at all values of time. The excitation may be obtained empirically. In these situations, numerical methods must be used to develop approximations to the response at discrete times. These numerical methods include numerical evaluation of the convolution integral and direct numerical solution of Eq. (5.1).

5.2 DERIVATION OF CONVOLUTION INTEGRAL

5.2.1 Response Due to a Unit Impulse

Consider a one-degree-of-freedom system initially at rest in equilibrium. Let $x(t)$ be a generalized coordinate, representing the displacement of a particle, A. The motion of the system, using x as the generalized coordinate, is modeled by the equivalent system of Fig. 5.4. An impulse of magnitude I is applied at A

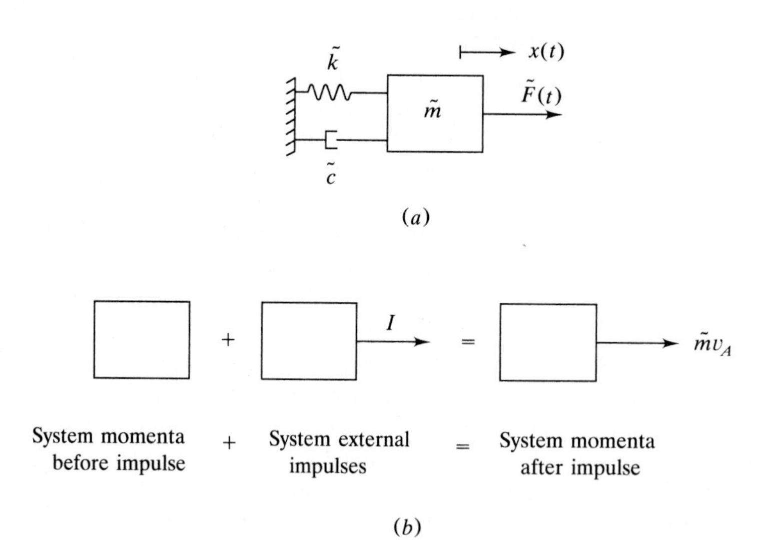

FIGURE 5.4
(*a*) Equivalent system used to model linear one-degree-of-freedom systems undergoing forced vibrations; (*b*) impulse and momentum diagrams used to obtain velocity immediately after impulse is applied.

initiating motion. The principle of impulse and momentum is used to determine the velocity of A immediately after application of the impulse

$$v_A = \frac{I}{\tilde{m}} \tag{5.2}$$

Application of the impulse initiates free vibration of the system. The motion is governed by Eq. (5.1) with $\tilde{F}(t) = 0$ and $x(0) = 0$, $\dot{x}(0) = v_A$. If $\zeta < 1$, the resulting motion of the system is determined using Eq. (3.31) as

$$x(t) = \frac{I}{\tilde{m}\omega_d} e^{-\zeta\omega_n t} \sin \omega_d t \tag{5.3}$$

or

$$x(t) = Ih(t) \tag{5.4}$$

where

$$h(t) = \frac{1}{\tilde{m}\omega_d} e^{-\zeta\omega_n t} \sin \omega_d t \tag{5.5}$$

is the response of the system due to a unit impulse applied at $t = 0$.

5.2.2 Response Due to a General Excitation

Consider a one-degree-of-freedom system subject to an arbitrary external excitation as represented by Fig. 5.5. The time interval from 0 to t is broken into n equal intervals, each of duration $\Delta\tau = t/n$. Define

$$\tau_k = \left(k + \tfrac{1}{2}\right) \Delta\tau$$

If n is large, the effect of applying a force $F(t)$ during the time interval $k\,\Delta\tau$ to $(k+1)\,\Delta\tau$ can be approximated by the effect of applying an impulse of magnitude $F(\tau_k)\,\Delta\tau$ at time τ_k. Thus, as shown in Fig. 5.5, $F(t)$ applied between 0 and t is approximated by applying impulses of appropriate magnitudes at τ_k, $k = 0, 1, \ldots, n-1$.

The response of the system at any time t due solely to an impulse applied at time τ_k, $t > \tau_k$, is obtained by solving

$$\frac{d^2 x_k}{d\xi_k^2} + 2\zeta\omega_n \frac{dx_k}{d\xi_k} + \omega_n^2 x_k = 0 \tag{5.6a}$$

$$x_k(\xi_k = 0) = 0 \tag{5.6b}$$

$$\frac{dx_k}{d\xi_k}(\xi_k = 0) = \frac{F(\tau_k)\,\Delta\tau}{\tilde{m}} \tag{5.6c}$$

where $\xi_k = t - \tau_k$. The solution of Eq. (5.6) is obtained by direct comparison to

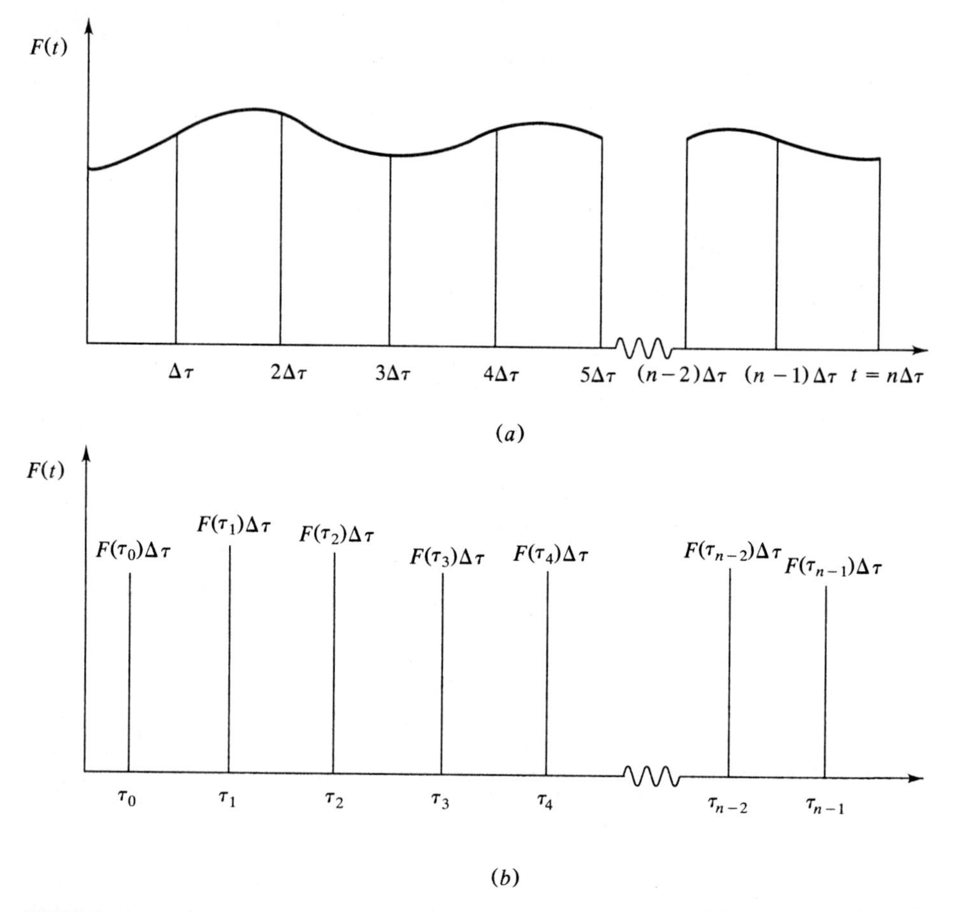

FIGURE 5.5
(*a*) Discretization of interval form 0 to t by n intervals of duration $\Delta\tau = t/n$; (*b*) approximation of $F(t)$ by impulses applied at τ_k, $k = 0, 1, \ldots, n-1$. As $n \to \infty$ the series of impulses exactly duplicates $F(t)$.

the response of a system due to an impulse,

$$x_k(\xi_k) = F(\tau_k)\,\Delta\tau\, h(\xi_k) \tag{5.7}$$

Since Eq. (5.1) is linear, the principle of linear superposition is used to determine the response at time t to all impulses applied before time t,

$$x^n(t) = \sum_{k=0}^{n-1} x_k(\xi_k) \tag{5.8}$$

Since t is broken into discrete intervals, Eq. (5.8) only provides an approximation to the response. The limit as $n \to \infty$ $(\Delta\tau \to 0)$ of the series of impulses applied before time t yields the exact excitation force applied before time t.

Thus the exact solution is obtained by taking the limit of Eq. (5.8) as $n \to \infty$,

$$x(t) = \lim_{n \to \infty} x^n(t)$$

$$= \lim_{n \to \infty} \sum_{k=0}^{n-1} F(\tau_k) h(t - \tau_k) \, \Delta\tau$$

In the limit the sum becomes an integral yielding

$$x(t) = \int_0^t F(\tau) h(t - \tau) \, d\tau \tag{5.9}$$

The integral representation of Eq. (5.9) is called the convolution integral. It can be used to determine the response of a one-degree-of-freedom system initially at rest in equilibrium subject to any form of excitation. The convolution integral solution is valid for all linear systems when $h(t)$ is viewed as the response of the system due to a unit impulse at $t = 0$. The appropriate forms of $h(t)$ for systems whose free vibrations are critically damped and overdamped are, respectively,

$$h(t) = \frac{te^{-\omega_n t}}{\tilde{m}} \qquad \zeta = 1 \tag{5.10}$$

$$h(t) = \frac{e^{-\zeta \omega_n t}}{\tilde{m} \omega_n \sqrt{\zeta^2 - 1}} \sinh\left(\omega_n \sqrt{\zeta^2 - 1}\, t\right) \qquad \zeta > 1 \tag{5.11}$$

The response of a system with a nonzero initial velocity is obtained by adding to the convolution integral of Eq. (5.9) the response of the system due to a unit impulse at $t = 0$, necessary to cause the initial velocity. The response of a system that is not in its equilibrium position at $t = 0$ is obtained through defining a new independent variable,

$$y = x - x(0)$$

The differential equation governing $y(t)$ is

$$\ddot{y} + 2\zeta\omega_n \dot{y} + \omega_n^2 y = -\frac{\tilde{k}}{\tilde{m}} x(0) + \frac{\tilde{F}}{\tilde{m}}$$

The convolution integral is used to obtain

$$y(t) = \int_0^t \left[-\tilde{k}x(0) + \tilde{F}(t)\right] h(t - \tau) \, d\tau$$

The resulting general solution for a system whose free vibrations are underdamped is

$$x(t) = x(0)e^{-\zeta\omega_n t}\cos\omega_d t + \frac{\dot{x}(0) + \zeta\omega_n x(0)}{\omega_d}e^{-\zeta\omega_n t}\sin\omega_d t$$

$$+ \frac{1}{\tilde{m}\omega_d}\int_0^t F(\tau)e^{-\zeta\omega_n(t-\tau)}\sin\omega_d(t-\tau)\,d\tau \tag{5.12}$$

Example 5.1. Find the response of a one-degree-of-freedom mass-spring-dashpot system initially at rest in equilibrium when the force

$$F(t) = F_0 e^{-\alpha t}$$

is applied.

Application of Eq. (5.9) for this particular form of $F(t)$ gives

$$x(t) = \int_0^t \frac{F_0 e^{-\alpha\tau}}{\tilde{m}\omega_d}e^{-\zeta\omega_n(t-\tau)}\sin\omega_d(t-\tau)\,d\tau$$

$$= \frac{F_0}{\tilde{m}\omega_d\left(\omega_n^2 - 2\zeta\omega_n\alpha + \alpha^2\right)}$$

$$\times\left\{e^{-\zeta\omega_n t}\left[(\alpha - \zeta\omega_n)\sin\omega_d t - \omega_d\cos\omega_d t\right] - \omega_d e^{-\alpha t}\right\}$$

Example 5.2. A mass m, falling vertically, strikes a spring and dashpot in parallel with a velocity v. Upon impact, the mass sticks to the assembly, as shown in Fig. 5.6. Determine $x(t)$, the displacement of the resulting mass-spring-dashpot system, measured from the position of the mass as it strikes the assembly. Assume the spring-dashpot assembly is massless.

The differential equation governing $x(t)$ is

$$\ddot{x} + 2\zeta\omega_n\dot{x} + \omega_n^2 x = g$$

Since $x(t)$ is not measured from the assembly's equilibrium position, gravity does not cancel with static deflection and acts as an excitation force in the differential equation.

The initial conditions are

$$x(0) = 0$$

$$\dot{x}(0) = v$$

The convolution integral for a system with nonzero initial conditions, Eq. (5.12), is

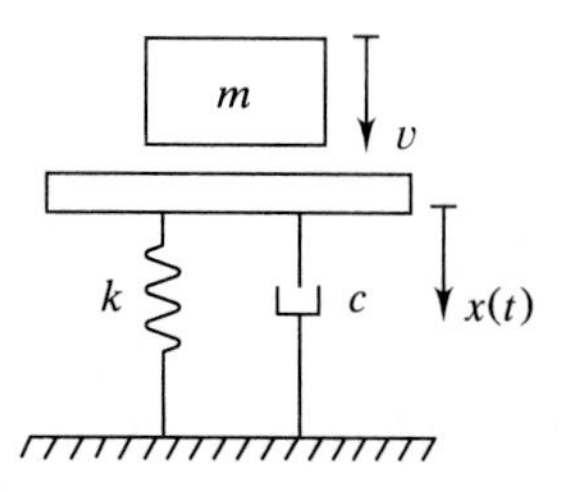

FIGURE 5.6
Spring-dashpot system of Example 5.2. Transient vibrations occur when the mass strikes the structure. The differential equation is nonhomogeneous because the initial position of the system is not the equilibrium position of the mass-spring-dashpot system after impact.

used to determine the response

$$x(t) = \frac{v}{\omega_d} e^{-\zeta\omega_n t} \sin \omega_d t$$

$$+ \int_0^t \frac{g}{\omega_d} e^{-\zeta\omega_n (t-\tau)} \sin \omega_d (t - \tau)\, d\tau$$

$$= \frac{v}{\omega_d} e^{-\zeta\omega_n t} \sin \omega_d t + \frac{g}{\omega_n^2}\left[1 - e^{-\zeta\omega_n t}\left(\frac{\zeta}{\sqrt{1-\zeta^2}} \sin \omega_d t + \cos \omega_d t\right)\right]$$

5.3 EXCITATIONS WHOSE FORMS CHANGE WITH TIME

Many engineering systems are subject to excitations whose forms change with time. The excitations for the systems of Figs. 5.1 through 5.3 each change form at a discrete time. The force provided to the laboratory structure of Fig. 5.3 might be approximated by the force of Fig. 5.7, whose mathematical representation is

$$F(t) = \begin{cases} F_0\left(1 - \dfrac{t}{t_0}\right) & t \le t_0 \\ 0 & t \ge t_0 \end{cases}$$

The convolution integral is used to determine the response of the structure as

$$x(t) = \begin{cases} F_0 \displaystyle\int_0^t \left(1 - \frac{\tau}{t_0}\right) h(t - \tau)\, d\tau & t \le t_0 \\ F_0 \displaystyle\int_0^{t_0} \left(1 - \frac{\tau}{t_0}\right) h(t - \tau)\, d\tau & t \ge t_0 \end{cases}$$

If t_0 is small, the maximum response may occur for $t > t_0$. Thus, in order to determine the response characteristics of the system, both integrals must be evaluated.

In the preceding discussion, the form of the response is different for different time intervals. It is more convenient to have a unified mathematical expression for the excitation and for the response. To this end, the unit step

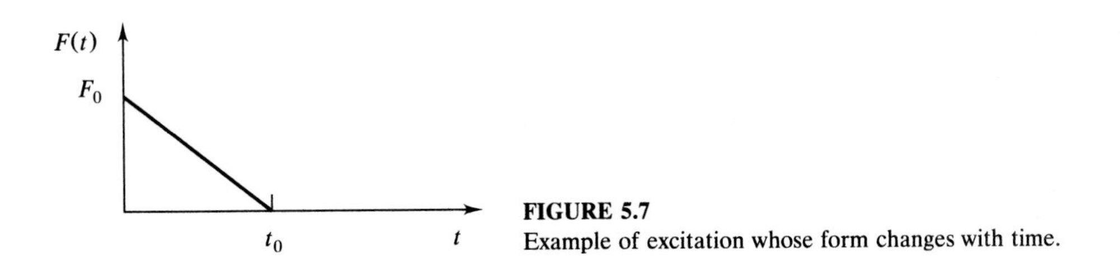

FIGURE 5.7
Example of excitation whose form changes with time.

function, introduced in App. A, is used

$$u(t) = \begin{cases} 0 & t \leq 0 \\ 1 & t > 0 \end{cases} \tag{5.13}$$

Using the unit step function, the excitation of Fig. 5.7 is written as

$$F(t) = F_0\left(1 - \frac{t}{t_0}\right)[u(t) - u(t - t_0)]$$

A single equation representing the response over all time is obtained using the convolution integral

$$x(t) = F_0\int_0^t\left(1 - \frac{\tau}{t_0}\right)[u(\tau) - u(\tau - t_0)]h(t - \tau)\,d\tau$$

Example 5.3. Use the unit step function to write a unified mathematical expression for each of the forces of Fig. 5.8.

Each of the forces of Fig. 5.8 can be written as the sum and/or difference of functions that are nonzero only after a discrete time. The graphical breakdown for

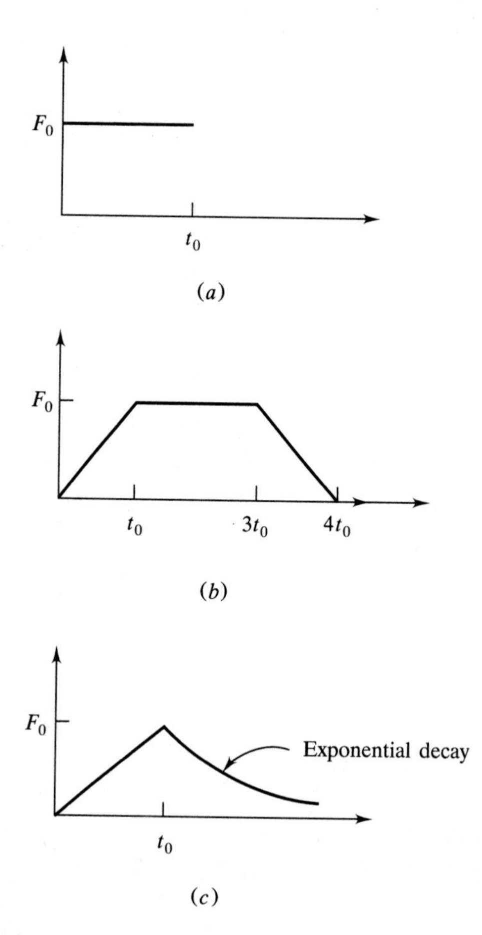

FIGURE 5.8
Excitations of Example 5.3.

each function is shown in Fig. 5.9. The unit step function is used to write a mathematical expression for each term in the forcing functions, leading to

$$(a)\quad F(t) = F_0[u(t) - u(t - t_0)]$$

$$(b)\quad F(t) = \frac{F_0 t}{t_0}[u(t) - u(t - t_0)] + F_0[u(t - t_0) - u(t - 3t_0)]$$

$$+ F_0\left(4 - \frac{t}{t_0}\right)[u(t - 3t_0) - u(t - 4t_0)]$$

$$(c)\quad F(t) = \frac{F_0 t}{t_0}[u(t) - u(t - t_0)] + F_0 e^{-\alpha(t - t_0)}u(t - t_0)$$

The unit impulse function, defined in App. A, is used to represent the excitation when subject to a unit impulse. The response of a system at time t due to a unit impulse applied at time t_0 is $h(t - t_0)u(t - t_0)$.

Many functions found in practice can be written as combinations of impulses, step functions, ramp functions, exponentially decaying functions, and sinusoidal pulses. Many functions which cannot be mathematically defined in terms of these functions are often approximated by these functions for estimation purposes.

Table 5.1 provides the response of an undamped one-degree-of-freedom system to common excitation terms delayed by a time t_0. The responses are derived from the convolution integral making use of the following formula:

$$\int_0^t F(\tau)u(\tau - t_0)\,d\tau = u(t - t_0)\int_{t_0}^t F(\tau)\,d\tau \tag{5.14}$$

Example 5.4. Use the convolution integral to determine the response of an undamped one-degree-of-freedom system subject to the delayed exponential function shown in Table 5.1.

The mathematical representation of the forcing function is

$$F(t) = F_0 e^{-\alpha(t - t_0)}u(t - t_0)$$

The convolution integral, Eq. (5.9), is used to write the solution as

$$x(t) = \frac{F_0}{\tilde{m}\omega_n}\int_0^t e^{-\alpha(\tau - t_0)}u(\tau - t_0)\sin\omega_n(t - \tau)\,d\tau$$

which using Eq. (5.14) is rearranged as

$$x(t) = u(t - t_0)\frac{F_0}{\tilde{m}\omega_n}\int_{t_0}^t e^{-\alpha(\tau - t_0)}\sin\omega_n(t - \tau)\,d\tau$$

$$= u(t - t_0)\frac{F_0}{\tilde{m}\omega_n(\alpha^2 + \omega_n^2)}\left[\omega_n e^{-\alpha(t - t_0)} + \alpha\sin\omega_n(t - t_0)\right.$$

$$\left.-\omega_n\cos\omega_n(t - t_0)\right]$$

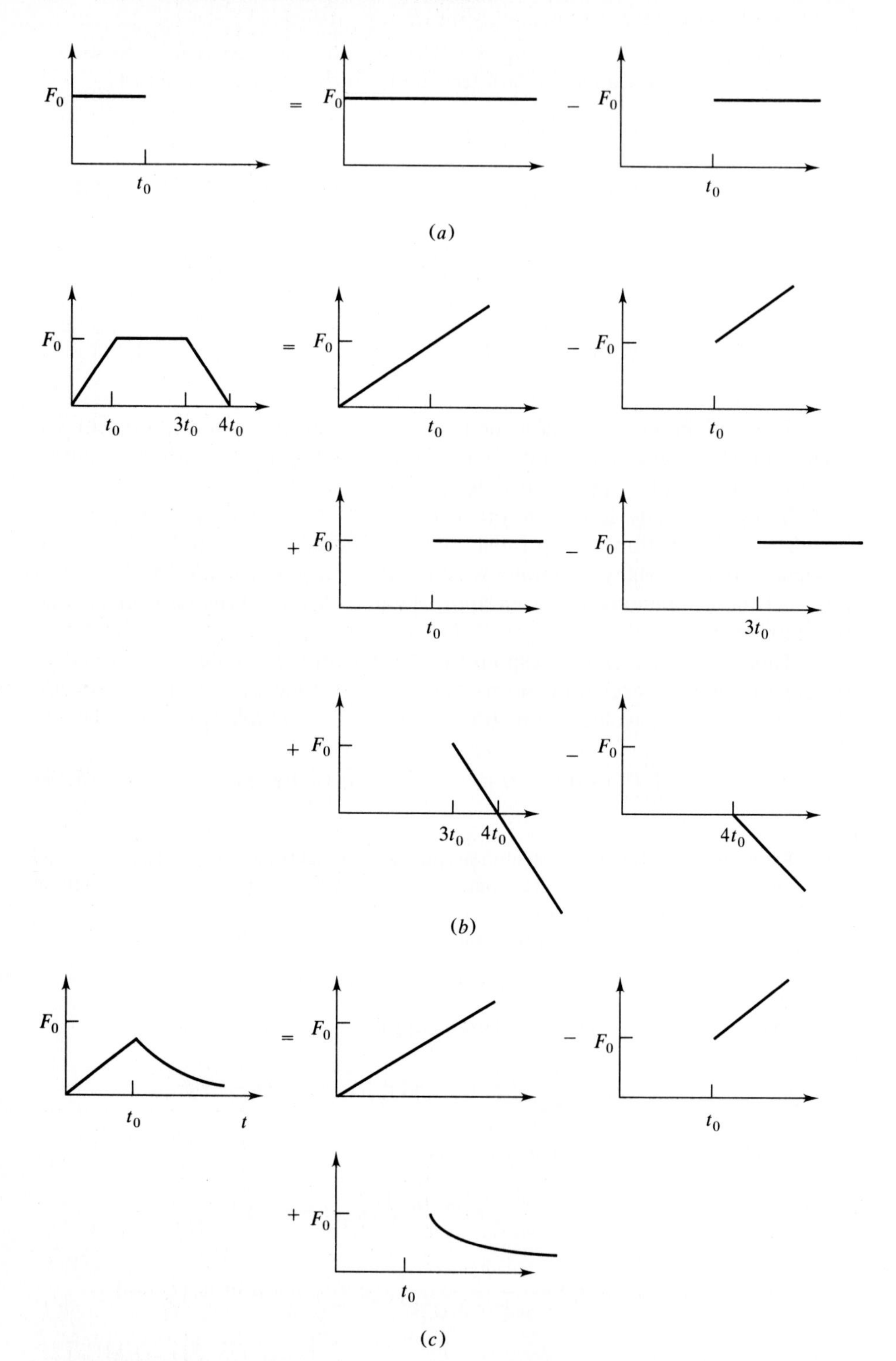

FIGURE 5.9
Graphical breakdown of excitations of Fig. 5.8 into functions that can be written using unit step functions.

TABLE 5.1
Response of an undamped one-degree-of-freedom to common forms of excitation

Delayed impulse
Excitation: $F(t) = A\delta(t - t_0)$
Response: $\tilde{m}\omega_n^2 x(t)/A = \omega_n \sin \omega_n(t - t_0)u(t - t_0)$

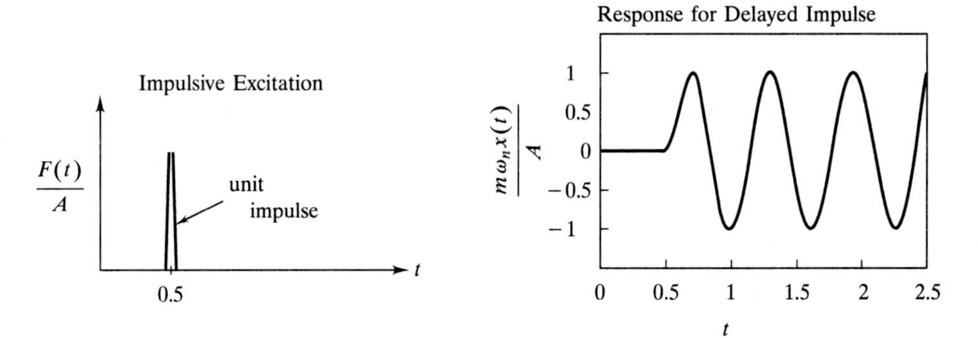

Delayed step function
Excitation: $F(t) = Au(t - t_0)$
Response: $\tilde{m}\omega_n^2 x(t)/A = [1 - \cos \omega_n(t - t_0)]u(t - t_0)$

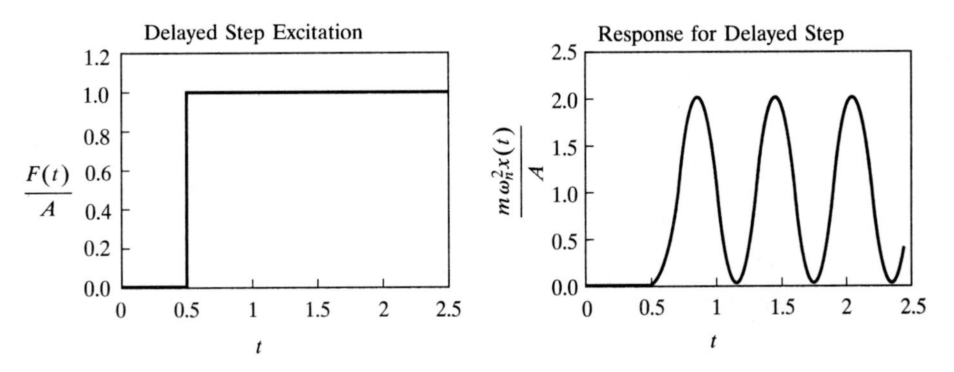

Delayed ramp function
Excitation: $F(t) = (At + B)u(t - t_0)$
Response: $\tilde{m}\omega_n^2 x(t)/A = [t + B/A - (t_0 + B/A)\cos \omega_n(t - t_0)$
$- \sin \omega_n(t - t_0)/\omega_n]u(t - t_0)$

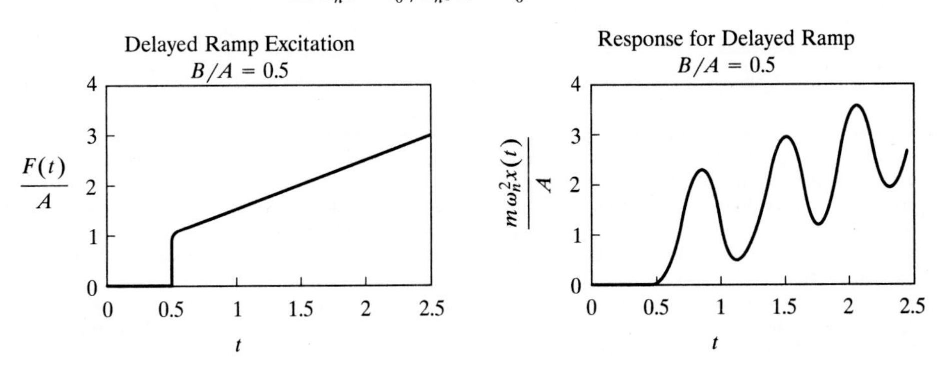

TABLE 5.1
(Continued)

Delayed exponential function
Excitation: $F(t) = Ae^{-\alpha(t-t_0)}u(t - t_0)$
Response: $\tilde{m}\omega_n^2 x(t)/A = [e^{-\alpha(t-t_0)} + \alpha/\omega_n \sin \omega_n(t - t_0)$
$- \cos \omega_n(t - t_0)]/(1 + \alpha^2/\omega_n^2)u(t - t_0)$

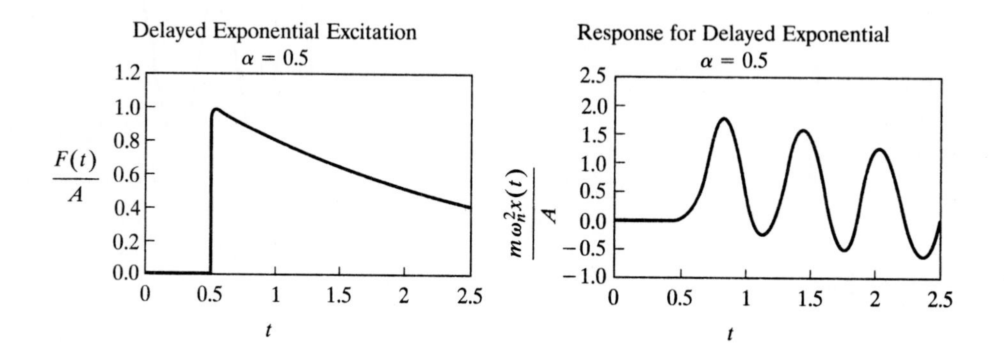

Delayed sine function:
Excitation: $F(t) = A \sin \omega(t - t_0)u(t - t_0)$

Response: $$\frac{\tilde{m}\omega_n^2 x(t)}{A} = \frac{1}{2}\left\{\left(\frac{1}{\omega/\omega_n - 1}\right)[\sin \omega(t - t_0) - \sin \omega_n(t - t_0)] - \left(\frac{1}{\omega/\omega_n + 1}\right)[\sin \omega(t - t_0) + \sin \omega_n(t - t_0)]\right\}u(t - t_0)$$

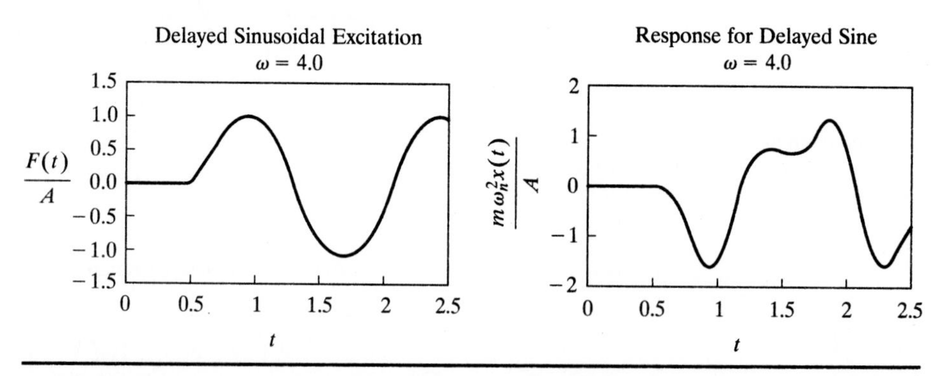

Note: This table provides the response of an undamped one-degree-of-freedom system to common forms of excitation. Many forms of excitation can be written as combinations of the excitations whose system responses are provided in the table. Superposition can be used to determine the response due to these excitations. In other cases excitations can be approximated by combinations of excitations in this table. Then this table and superposition is used to approximate the response of an undamped one-degree-of-freedom system.

The table provides the mathematical form of the excitation and response as well as graphical representations. In all cases values of $\omega_n = 10$ rad/s and $t_0 = 0.5$ s were used to generate the graphs. The values of specific parameters used for specific excitations are given.

Often, excitations are linear combinations of the functions whose responses are presented in Table 5.1 The general form of an excitation that changes form at discrete times $t_1, t_2, \ldots, t_n$ is

$$F(t) = \sum_{i=1}^{n} f_i(t)u(t - t_i) \tag{5.15}$$

Application of the convolution integral to the excitation of Eq. (5.15), using Eq. (5.14), yields

$$x(t) = \sum_{i=1}^{n} u(t - t_i)\int_{t_i}^{t} f_i(\tau)h(t - \tau)\, d\tau \tag{5.16}$$

Equation (5.16) shows that the total response is the sum of the responses due to the individual terms of the excitation. This result is due to the linearity of Eq. (5.1). The effects of any nonzero initial conditions are included with the response due to $f_1(t)$.

Example 5.5. Use Table 5.1 to develop the response of an undamped one-degree-of-freedom system to the triangular pulse of Fig. 5.10.

The triangular pulse can be written as the sum and difference of ramp functions as shown in Fig. 5.10. The response due to the triangular pulse is obtained by adding and subtracting the responses due to each ramp function

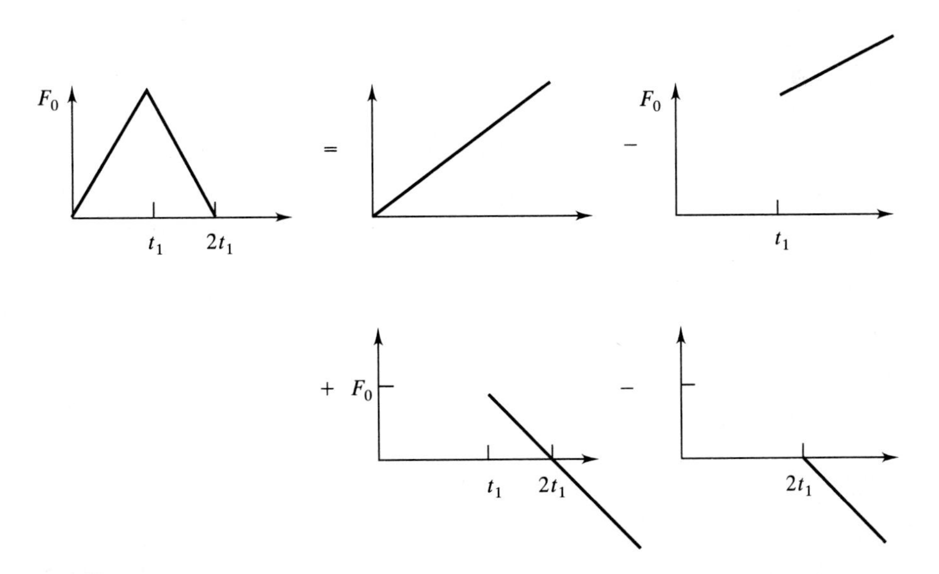

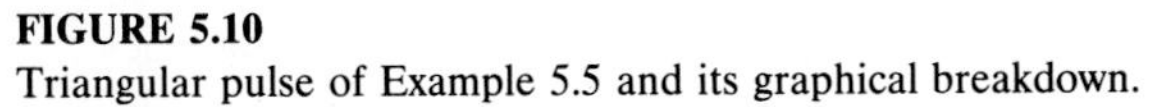

FIGURE 5.10
Triangular pulse of Example 5.5 and its graphical breakdown.

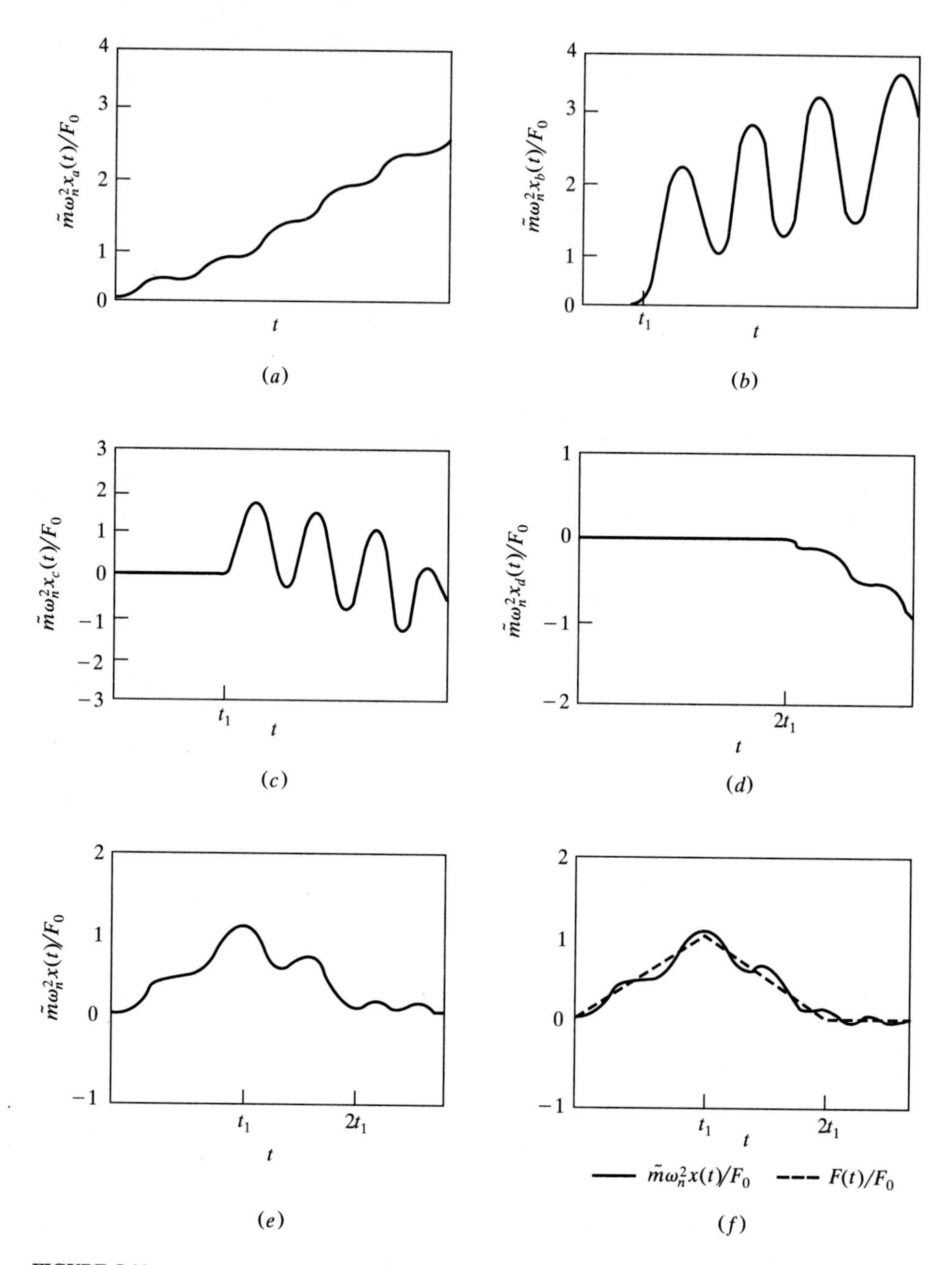

FIGURE 5.11
(a)–(d) Response of a one-degree-of-freedom system due to the component parts of a triangular pulse excitation obtained using Table 5.1; (e) response of a one-degree-of-freedom system due to a triangular pulse excitation obtained using superposition; (f) comparison of triangular pulse and the resulting excitation.

according to

$$x(t) = x_a(t) - x_b(t) + x_c(t) - x_d(t)$$

where the individual responses are determined using Table 5.1.

For $x_a(t)$, the ramp function entry of Table 5.1 is used with $A = F_0/t_1$, $B = 0$, and $t_0 = 0$ leading to

$$x_a(t) = \frac{F_0}{\tilde{m}\omega_n^2}\left[\frac{t}{t_1} - \frac{1}{\omega_n t_1}\sin \omega_n t\right]$$

$x_b(t)$ is determined using the ramp function entry of Table 5.1 with $A = F_0/t_1$, $B = 0$, $t_0 = t_1$. This gives

$$x_b(t) = \frac{F_0}{\tilde{m}\omega_n^2}\left[\frac{t}{t_1} - \cos \omega_n(t - t_1) - \frac{1}{\omega_n t_1}\sin \omega_n(t - t_1)\right]u(t - t_1)$$

For $x_c(t)$, the ramp function entry of Table 5.1 is used with $A = -F_0/t_1$, $B = 2F_0$, and $t_0 = t_1$. This leads to

$$x_c(t) = \frac{F_0}{\tilde{m}\omega_n^2}\left[\left(2 - \frac{t}{t_1}\right) - \cos \omega_n(t - t_1) + \frac{1}{\omega_n t_1}\sin \omega_n(t - t_1)\right]u(t - t_1)$$

$x_d(t)$ is determined using the ramp function entry of Table 5.1 with $A = -F_0/t_1$, $B = 2F_0$, and $t_0 = 2t_1$. This gives

$$x_d(t) = \frac{F_0}{\tilde{m}\omega_n^2}\left[\left(2 - \frac{t}{t_1}\right) + \frac{1}{\omega_n t_1}\sin \omega_n(t - t_1)\right]u(t - 2t_1)$$

Simplifying the resulting expression in each interval of time yields

$$x(t) = \frac{F_0}{\tilde{m}\omega_n^2}\begin{cases} \dfrac{t}{t_1} - \dfrac{1}{\omega_n t_1}\sin \omega_n t & 0 \le t \le t_1 \\ 2 - \dfrac{t}{t_1} + \dfrac{1}{\omega_n t_1}[2\sin \omega_n(t - t_1) - \sin \omega_n t] & t_1 \le t \le 2t_1 \\ \dfrac{1}{\omega_n t_1}[2\sin \omega_n(t - t_1) - \sin \omega_n t - \sin \omega_n(t - 2t_1)] & t > 2t_1 \end{cases}$$

The response of each component part and the total response is shown in Fig. 5.11.

Example 5.6. A frame structure of mass 1200 kg and equivalent stiffness 2×10^6 N/m is subject to a blast whose time-dependent resultant force is given in Fig. 5.12. Assuming the structure is undamped, approximate its maximum displacement.

A reasonable approximation to the blast excitation is shown in Fig. 5.12*b*. The blast force is approximated by a ramp function building up to 5000 N at $t = 0.2$ s, followed by a rectangular pulse of duration 0.8 s. The decay of the force after $t = 1$ s is approximated by an exponential pulse.

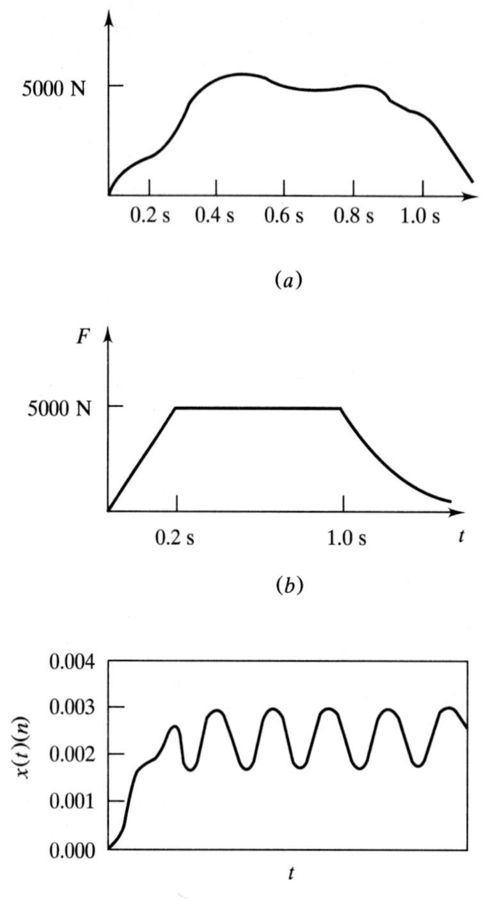

FIGURE 5.12
(*a*) Time-dependent resultant force excitation of Example 5.6; (*b*) approximation of blast force by functions whose response can be obtained by use of Table 5.1 and superposition; (*c*) response of the frame structure due to the approximation of the blast excitation.

The displacement in each time interval is approximated using Table 5.1. The natural frequency of the structure is calculated as

$$\omega_n = \sqrt{\frac{\tilde{k}}{\tilde{m}}} = 40.82 \frac{\text{rad}}{\text{s}}$$

The response for t during $0 \leq t \leq 0.2$ s is determined using the ramp function entry of Table 5.1 with $A = 5000$ N/0.2 s = 25,000 N/s, $B = 0$, and $t_0 = 0$. This leads to

$$x(t) = 0.0125\left(t - \frac{1}{\omega_n}\sin\omega_n t\right)\text{ m} \qquad 0 \leq t \leq 0.2\text{ s}$$

The maximum displacement during this time interval either occurs at a time when $dx/dt = 0$ or at $t = 0.2$ s. To this end

$$\frac{dx}{dt} = 0.0125(1 - \cos \omega_n t)$$

The times at which $dx/dt = 0$ are given by

$$t_m = \frac{2n\pi}{\omega_n} = 0.154n \text{ s} \qquad n = 1, 2, \ldots$$

There is only one time in the interval between 0 and 0.2 s where $dx/dt = 0$. This time corresponds to a maximum of $x(t)$ and implies that $x(t)$ is decreasing at 0.2 s. Hence the maximum response of the structure between $t = 0$ and $t = 0.2$ s is

$$x_{\max} = x(0.153 \text{ s}) = 0.0019 \text{ m}$$

For 0.2 s $\leq t \leq$ 1.0 s, the approximation to the response is obtained by subtracting the response due to a ramp function with $A = 25{,}000$ N, $B = 0$, and $t_0 = 0.2$ s from the previously determined response and adding the response for a rectangular pulse with $F_0 = 5000$ N and $t_0 = 0.2$ s. Thus, using Table 5.1,

$$x(t) = 0.0125\left(t - \frac{1}{\omega_n}\sin \omega_n t\right) - 0.0125\left[t - 0.2\cos \omega_n(t - 0.2) - \frac{1}{\omega_n}\sin \omega_n(t - 0.2)\right] + 0.0025[1 - \cos \omega_n(t - 0.2)] \qquad 0.2 \text{ s} \leq t \leq 1.0 \text{ s}$$

$$= 0.0025 + \frac{0.0125}{\omega_n}[\sin \omega_n(t - 0.2) - \sin \omega_n t]$$

$$= 0.0125[0.2 + 0.04\cos \omega_n(t - 0.1)]$$

The maximum displacement of 0.003 m first occurs during this time interval at $t = 0.253$ s.

The energy level of the excitation decays after $t = 1.0$ s. Since the motion is periodic between 0.2 s and 1.0 s, the decrease in energy input causes a decrease in the response after $t = 1.0$ s. Thus the maximum displacement of the structure due to the blast is approximately 3 mm.

The preceding example has several interesting features. The response due to the ramp function reaches a maximum before the excitation reaches its peak because $t_0 = 0.2 \text{ s} > 2\pi/\omega_n$. However, since the level of energy input increases beyond this time, the absolute maximum is reached after the response becomes periodic. Additionally, the duration of the rectangular pulse is long enough such that a maximum is reached before the magnitude of the blast force begins to decay. If this is not the case, the absolute maximum response could occur during the exponential decay of the blast force. Thus the duration of the ramp and rectangular pulse compared to the natural period are important in determining the maximum response and when it occurs.

5.4 TRANSIENT MOTION DUE TO BASE EXCITATION

Many mechanical systems and structures are subject to base excitation. A rigid wheel traveling along a road contour excites motion of a vehicle through the suspension system. The computer in the nose cone of the rocket of Fig. 5.2 is subject to base excitation as the rocket travels. Earthquakes produce base excitation of structures.

Recall the governing equation for the relative displacement between a mass and its base when the mass is connected to the base through a spring and viscous damper in parallel

$$\ddot{z} + 2\zeta\omega_n\dot{z} + \omega_n^2 z = -\ddot{y} \tag{5.17}$$

where y is the prescribed base motion. If $z(0) = 0$ and $\dot{z}(0) = 0$, the convolution integral is used to solve Eq. (5.17), yielding

$$z(t) = -\tilde{m}\int_0^t \ddot{y}(\tau)h(t-\tau)\,d\tau \tag{5.18}$$

Equation (5.18) is integrated by parts to write the solution in terms of the base velocity

$$z(t) = \tilde{m}\left[\dot{y}(0)h(t) - \int_0^t \dot{y}(\tau)\dot{h}(t-\tau)\,d\tau\right] \tag{5.19}$$

where

$$\dot{h}(t) = -\frac{e^{-\zeta\omega_n t}}{\tilde{m}\sqrt{1-\zeta^2}}\sin\omega_d(t-\chi) \tag{5.20}$$

where

$$\chi = \tan^{-1}\left(\frac{\sqrt{1-\zeta^2}}{\zeta}\right)$$

If the base displacement is known, it can be differentiated to calculate the velocity and Eq. (5.19) can be used to determine the relative displacement. Alternately, the absolute displacement of the base can be attained by solving

$$\ddot{x} + 2\zeta\omega_n\dot{x} + \omega_n^2 x = -2\zeta\omega_n\dot{y} - \omega_n^2 y \tag{5.21}$$

When applied to Eq. (5.21) the convolution integral yields

$$x(t) = -\tilde{m}\int_0^t \left[2\zeta\omega_n\dot{y}(\tau) + \omega_n^2 y(\tau)\right]h(t-\tau)\,d\tau \tag{5.22}$$

Example 5.7. A sensitive computer system is mounted in the nose cone of a rocket, as shown in Fig. 5.2. Thrust is developed in the rocket such that its speed is as shown in Fig. 5.2*b*. The discontinuity in slope occurs when a second stage is fired. Determine the displacement of the computer relative to the rocket.

The unified mathematical expression for the rocket's velocity is

$$v(t) = \frac{vt}{t_1}[1 - u(t - t_1)] + v\left(\frac{2t}{t_1} - 1\right)\left[u(t - t_1) - u\left(t - \frac{3t_1}{2}\right)\right]$$
$$+ 2vu\left(t - \frac{3t_1}{2}\right)$$

Equation (5.19) is used to evaluate the displacement of the computer relative to the rocket, noting that for an undamped system

$$\dot{h}(t) = \frac{1}{\tilde{m}} \cos \omega_n t$$

This leads to

$$z(t) = \frac{1}{\omega_n} \sin \omega_n t\left[\frac{vt}{t_1} + v\left(\frac{t}{t_1} - 1\right)u(t - t_1) + v\left(2 - \frac{2t}{t_1}\right)u\left(t - \frac{3t_1}{2}\right)\right]$$
$$- \int_0^t \left[\frac{v\tau}{t_1} + v\left(\frac{\tau}{t_1} - 1\right) + v\left(3 - \frac{2\tau}{t_1}\right)u\left(\tau - \frac{3t_1}{2}\right)\right]\cos \omega_n(t - \tau)\, d\tau$$

Integration yields

$$z(t) = \frac{v}{\omega_n^2 t_1}\Bigg\{\omega_n t \sin \omega_n t - 1 + \cos \omega_n t + u(t - t_1)[\cos \omega_n(t - t_1) - 1$$
$$+\omega_n(t - t_1)\sin \omega_n t] + u\left(t - \frac{3t_1}{2}\right)\left[2 - 2\cos \omega_n\left(t - \frac{3t_1}{2}\right)\right.$$
$$\left.\left. +\omega_n(3t_1 - 2)\sin \omega_n t\right]\right\}$$

An alternate solution is to differentiate the velocity with respect to time to yield the base acceleration. Then Eq. (5.17) is used to determine the relative displacement. The derivative of the unit step functions is the unit impulse function as shown in App. A. However, since the velocity is continuous, no real impulses are applied. The impulse functions do not contribute mathematically to the solution. Since the system is undamped, Table 5.1 can be used to develop the response when the base acceleration is calculated.

Example 5.8. Determine the transient response of a mass m connected through a spring of stiffness k to a base when the base is subject to the rectangular velocity pulse if Fig. 5.13. Use (*a*) Eq. (5.19) and (*b*) Eq. (5.18).

The mathematical expression for the rectangular velocity pulse is

$$v[u(t) - u(t - t_0)]$$

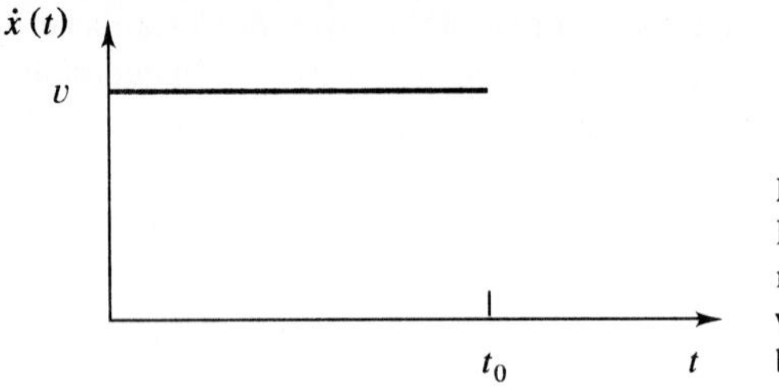

FIGURE 5.13
Base of System of Example 5.8 is subject to a rectangular velocity pulse. The instantaneous velocity changes at $t = 0$ and $t = t_0$ can only be caused by applied impulses.

(a) It is noted in using Eq. (5.19) that $u(0) = 0$. Hence $y(0) = 0$. Substituting the base velocity into Eq. (5.19) yields

$$z(t) = -\int_0^t v[u(\tau) - u(\tau - t_0)]\cos \omega_n(t - \tau)\, d\tau$$

which is integrated to yield

$$z(t) = \frac{v}{\omega_n}[\sin \omega_n(t - t_0)u(t - t_0) - \sin \omega_n t u(t)]$$

(b) The base acceleration is obtained by differentiating the base velocity with respect to time. Noting that the derivative of a unit step function is the unit impulse function gives

$$\ddot{y}(t) = v[\delta(t) - \delta(t - t_0)]$$

The base velocity changes instantaneously at $t = 0$ and $t = t_0$. Instantaneous velocity changes are only the result of applied impulses. This gives rise to unit impulse functions in the mathematical expression for the base acceleration.

Substituting the base acceleration into Eq. (5.18) gives

$$z(t) = -\int_0^t v[\delta(\tau) - \delta(\tau - t_0)]\sin \omega_n(t - \tau)\, d\tau$$

Noting that

$$\int_0^t \delta(\tau - t_0)f(\tau)\, d\tau = f(t_0)u(t - t_0)$$

the relative displacement is

$$z(t) = \frac{v}{\omega_n}[\sin \omega_n(t - t_0)u(t - t_0) - \sin \omega_n t u(t)]$$

5.5 LAPLACE TRANSFORM SOLUTIONS

The Laplace transform method is a convenient method for finding the response of a system due to any excitation. The basic method is to use known properties of the transform to transform an ordinary differential equation into an algebraic equation, using the initial conditions. The algebraic equation is solved to find the transform of the solution. This transform is inverted using properties of the transform and a table of known transform pairs.

The Laplace transform can be used to solve linear ordinary differential equations with constant or polynomial coefficients. The method easily handles excitations whose form changes with time. Such excitations are written in a unified mathematical expression using the unit step functions. The shifting theorems help perform the transform and evaluate the inversions.

The Laplace transform is not as easy to apply as the convolution integral unless one has extensive experience in its use. The main drawback of the method is the difficulty in inverting the transform. A formal inversion theorem, involving contour integration in the complex plane, is available, but is beyond the scope of this text.

The transform pairs and properties used in the following discussion are summarized and explained in App. B.

Let $\bar{x}(s)$ be the Laplace transform of the generalized coordinate for a one-degree-of-freedom system. That is,

$$\bar{x}(s) = \int_0^\infty x(t)e^{-st}\,dt \tag{5.23}$$

Let $\bar{F}(s)$ be the Laplace transform of the known forcing function which, for a specific form of $\tilde{F}(t)$, is calculated from the transform definition, referring to a table of transform pairs, or using basic properties in conjunction with a table.

Taking the Laplace transform of Eq. (5.1) and using linearity of the transform,

$$\mathscr{L}\{\ddot{x}\} + 2\zeta\omega_n\mathscr{L}\{\dot{x}\} + \omega_n^2\bar{x}(s) = \frac{\bar{F}(s)}{\tilde{m}} \tag{5.24}$$

The property for transform of derivatives allows the transform of the differential equation for $x(t)$ into an algebraic equation for $\bar{x}(s)$. Its application to Eq. (5.24) gives

$$s^2\bar{x}(s) - sx(0) - \dot{x}(0) + 2\zeta\omega_n[s\bar{x}(s) - x(0)] + \omega_n^2\bar{x}(s) = \frac{\bar{F}(s)}{\tilde{m}}$$

which rearranges to

$$\bar{x}(s) = \frac{\dfrac{\bar{F}(s)}{\tilde{m}} + (s + 2\zeta\omega_n)x(0) + \dot{x}(0)}{s^2 + 2\zeta\omega_n s + \omega_n^2} \tag{5.25}$$

The definition and linearity of the inverse transform is used to find $x(t)$,

$$x(t) = \frac{1}{\tilde{m}}\mathscr{L}^{-1}\left\{\frac{\bar{F}(s)}{s^2 + 2\zeta\omega_n s + \omega_n^2}\right\} + \mathscr{L}^{-1}\left\{\frac{(s + 2\zeta\omega_n)x(0) + \dot{x}(0)}{s^2 + 2\zeta\omega_n s + \omega_n^2}\right\} \tag{5.26}$$

The inverse transform of each term of Eq. (5.26) depends upon the types of roots in the denominator, which, in turn, depend upon the value of ζ. For a given ζ, the inverse transform of the last term of Eq. (5.26) is directly determined. The inverse transform of the first term is determined only after specifying $\tilde{F}(t)$ and taking its Laplace transform.

If the free vibrations are underdamped, then the denominator has two complex roots. In this case it is convenient to complete the square of the denominator and write it as

$$(s + \zeta\omega_n)^2 + \omega_n^2(1 - \zeta^2)$$

The inverse transform of the last term of Eq. (5.26) is found by applying the first shifting theorem and known transform pairs $B4$ and $B5$,

$$\mathscr{L}^{-1}\left\{\frac{(s + 2\zeta\omega_n)x(0) + \dot{x}(0)}{s^2 + 2\zeta\omega_n s + \omega_n^2}\right\}$$

$$= e^{-\zeta\omega_n t}\left[x(0)\cos\omega_d t + \frac{\dot{x}(0) + \zeta\omega_n x(0)}{\omega_d}\sin\omega_d t\right] \tag{5.27}$$

For systems whose free vibrations are critically damped or overdamped, similar algebraic manipulations and the use of tables of transforms give

$$\mathscr{L}^{-1}\left\{\frac{(s + 2\zeta\omega_n)x(0) + \dot{x}(0)}{s^2 + 2\zeta\omega_n s + \omega_n^2}\right\}$$

$$= \begin{cases} e^{-\omega_n t}\{x(0) + [\omega_n x(0) + \dot{x}(0)]t\} & \zeta = 1 \\ \dfrac{1}{\sqrt{2\zeta^2 - 1}}\{[x(0)\alpha_1 + 2\zeta\omega_n x(0) + \dot{x}(0)]e^{\alpha_1 t} & \\ \quad -[x(0)\alpha_2 + 2\zeta\omega_n x(0) + \dot{x}(0)]e^{-\alpha_2 t}\} & \zeta > 1 \\ \alpha_{1,2} = \omega_n\left(-\zeta \pm \sqrt{\zeta^2 - 1}\right) & \end{cases}$$

The inverse transform of the first term of Eq. (5.26) is found by finding $\bar{F}(s)$ for the particular form of $F(t)$, forming $\bar{F}(s)/(s_2 + s\zeta\omega_n s + \omega_n^2)$, and inverting using algebraic manipulations, transform properties, and a table of known transform pairs.

Example 5.9. A 200-kg machine is to be mounted on an elastic surface of equivalent stiffness 2×10^5 N/m with no damping. During operation, the machine is subject to a constant force of 2000 N for 3 s. Can steady-state vibrations be eliminated without adding damping? If so, what is the maximum deflection of the machine?

The differential equation governing motion of the machine is

$$\ddot{x} + \omega_n^2 x = F_0[u(t) - u(t - 3)]$$

where $F_0 = 2000$ N and $\omega_n = 25.82$ rad/s. The Laplace transform of $F(t)$ is obtained using the second shifting theorem

$$\mathscr{L}\{F_0[u(t) - u(t-3)]\} = \frac{F_0}{s}(1 - e^{-3s})$$

Then from Eq. (5.26) with $x(0) = 0$ and $\dot{x}(0) = 0$,

$$x(t) = \frac{F_0}{\tilde{m}}\mathscr{L}^{-1}\left\{\frac{1 - e^{-3s}}{s(s^2 + \omega_n^2)}\right\}$$

Partial fraction decomposition yields

$$\bar{x}(s) = \frac{F_0}{\tilde{m}\omega_n^2}\left(\frac{1}{s} - \frac{s}{s^2 + \omega_n^2}\right)(1 - e^{-3s})$$

The second shifting theorem is used to help invert the transform

$$x(t) = \frac{F_0}{\tilde{m}\omega_n^2}[1 - \cos\omega_n t - u(t-3)(1 - \cos\omega_n(t-3))]$$

The solution for $t > 3$ s is

$$x(t) = \frac{F_0}{\tilde{m}\omega_n^2}[\cos\omega_n(t-3) - \cos\omega_n t] \qquad t > 3 \text{ s}$$

For no steady-state motion,

$$\cos\omega_n t = \cos\omega_n(t-3)$$

which is satisfied if $3\omega_n = 2n\pi$ for any positive integer n. Thus steady-state vibrations are eliminated by requiring

$$\omega_n = \frac{2n\pi}{3} = 2.09n\ \frac{\text{rad}}{\text{s}}$$

The smallest mass above the mass of the machine for which this is true is $m = 203.5$ kg when $n = 15$. Thus steady-state vibrations are eliminated if 3.5 kg is rigidly added to the machine.

The machine undergoes 15 cycles while the force is applied and motion ceases when the force is removed. The maximum displacement during operation is

$$x_{\max} = \frac{2F_0}{\tilde{m}\omega_n^2} = \frac{2F_0}{k} = 0.02 \text{ m}$$

Example 5.10. Use the Laplace transform method to determine the response of an underdamped one-degree-of-freedom system to the rectangular velocity pulse of Fig. 5.13.

Using the analysis in Example 5.7, the differential equation governing displacement of the mass relative to its base when the base is subject to a rectangular velocity pulse is

$$\ddot{z} + 2\zeta\omega_n z + \omega_n^2 z = -v[\delta(t) - \delta(t - t_0)]$$

Using transform pair $B1$, and assuming $z(0) = 0$ and $\dot{z}(0) = 0$, Eq. (5.25) becomes

$$\bar{x}(s) = \frac{v(1 - e^{-st_0})}{s^2 + 2\zeta\omega_n s + \omega_n^2}$$

The transform is inverted by completing the square in the denominator and using both the first shifting theorem and the second shifting theorem to obtain

$$z(t) = \frac{v}{\omega_n}\left[e^{-\zeta\omega_n t}\sin\omega_n t - e^{-\zeta\omega_n(t-t_0)}\sin\omega_n(t-t_0)u(t-t_0)\right]$$

5.6 SHOCK SPECTRUM

Example 5.6 presents a situation in which the maximum displacement for a system due to a given form of excitation is needed. Similar situations occur frequently in design, but with the system parameters to be chosen such that specified design criteria are satisfied. In many problems the design criteria involve limiting maximum displacements and/or maximum stresses for a given type of excitation. For example, if it is determined that all earthquakes in a given area have similar forms of excitations, only with different levels of severity, then knowledge of the maximum displacement as a function of system parameters is useful in the design of a structure to withstand a certain level of earthquake. The structure's ability to withstand the earthquake depends upon the maximum displacement developed in the structure during the earthquake and the maximum stresses developed. A structure in California along the San Andreas fault will usually be designed to withstand a more severe earthquake than a structure in Ohio. This, of course, depends upon the use of the structure.

Thus it is useful for the designer to know the maximum response of a structure as a function of system parameters. The transmissibility curves presented in Chap. 4 and discussed in more detail in Chap. 11 actually do this for the steady-state response due to harmonic excitations. For a given value of the damping ratio, the transmissibility curve plots the nondimensional ratio of the amplitude of the transmitted force to the maximum amplitude of the excitation force against the nondimensional frequency ratio.

Similar curves are useful for analysis and design of systems that are subject to shock excitations. A shock is a large force applied over a short interval resulting in transient vibration. The maximum response is a function of the type of shock and system parameters.

A shock spectrum is a nondimensional plot of the maximum response of a one-degree-of-freedom system for a specified excitation as a function of a nondimensional time ratio. The vertical axis of the plot is the maximum value of the force developed in the spring divided by the maximum of the excitation force. The horizontal axis is the ratio of a characteristic time for the excitation divided by the natural period. For a shock excitation the characteristic time is usually taken as the duration of the shock. For a periodic excitation the

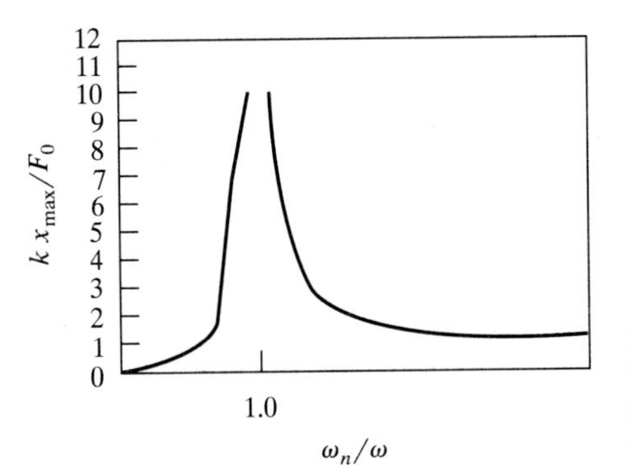

FIGURE 5.14
The response spectrum for a harmonic excitation is related to the frequency response curve presented in Chap. 4.

characteristic time is the period of excitation. The shock spectrum for a harmonic excitation is given in Fig. 5.14.

Shock spectra are often only plotted for undamped systems as damping tends to act favorably to reduce the maximum response. Also, a shock spectrum is very tedious to calculate and plot. Inclusion of damping in the development of a shock spectrum greatly increases the amount of algebra performed. The resulting complexity may obscure the usefulness of the results.

Example 5.11. A one-story frame structure is to be built to house a chemical laboratory. The experiments performed in the laboratory involve highly volatile chemicals and the possibility of explosion is great. It is estimated that the worst explosion will generate a force of 5×10^6 N lasting 0.5 s. The structure is to be designed such that the maximum displacement due to such an explosion is 10 mm. The equivalent mass of the structure is 500,000 kg. Draw the shock spectrum for the structure subject to a rectangular pulse and determine the minimum allowable stiffness.

The laboratory frame structure is modeled as an undamped one-degree-of-freedom system with $x(t)$ representing the displacement at the top of the columns. The excitation is modeled as a rectangular pulse of magnitude $F_0 = 5 \times 10^6$ N and duration $t_0 = 0.5$ s. The response of an undamped one-degree-of-freedom system with zero initial conditions is calculated as

$$x(t) = \frac{F_0}{k}\{1 - \cos \omega_n t - u(t - t_0)[1 - \cos \omega_n(t - t_0)]\}$$

For $t < t_0$ the nondimensional force ratio is

$$\frac{kx}{F_0} = 1 - \cos \omega_n t$$

The preceding function increases until $t = \pi/\omega_n$ when it reaches a maximum value of 2. If $t_0 < \pi/\omega_n$, the maximum nondimensional force ratio in this interval is

$$\frac{kx_{\max}}{F_0} = 1 - \cos \omega_n t_0$$

However, since the response is continuous, the maximum response for $t > \pi/\omega_n$ must be at least this large. For $t > t_0$ the nondimensional force ratio is

$$\frac{kx}{F_0} = \cos \omega_n(t - t_0) - \cos \omega_n t$$

The time at which the maximum occurs for $t > t_0$ is found by setting the time derivative of the preceding function to zero. This yields

$$\omega_n \sin \omega_n t = \omega_n \sin \omega_n(t - t_0)$$

which is solved as

$$t = \frac{1}{2}\left[t_0 + \frac{(2n-1)\pi}{2\omega_n}\right] \qquad n = 1, 2, \ldots$$

Substituting into the nondimensional force ratio and using trigonometric identities to simplify yields

$$\frac{kx_{\max}}{F_0} = 2\sin\frac{\omega_n t_0}{2} \qquad t_0 < \frac{\pi}{\omega_n}$$

In summary,

$$\frac{kx_{\max}}{F_0} = \begin{cases} 2\sin\dfrac{\omega_n t_0}{2} & t_0 < \dfrac{\pi}{\omega_n} \\ 2 & t_0 > \dfrac{\pi}{\omega_n} \end{cases}$$

The shock spectrum is plotted in Fig. 5.15.

Returning to the specific problem, assume first that $\pi/\omega_n > 0.5$ s. The smallest stiffness is obtained by setting $x_{\max} = 0.01$ m, leading to

$$1 \times 10^{-8}k = \sin(3.53 \times 10^{-4}\sqrt{k})$$

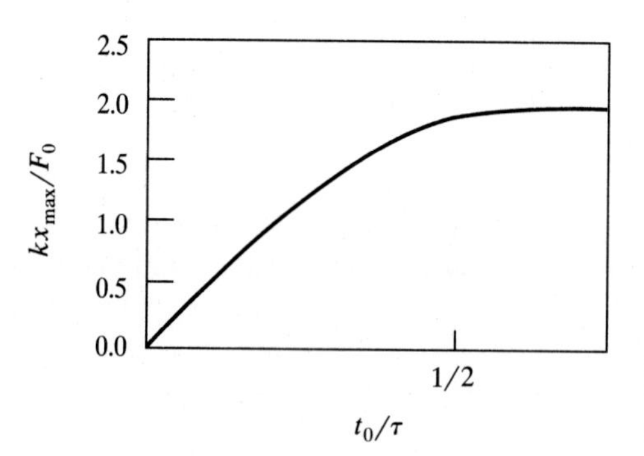

FIGURE 5.15
The shock spectrum for a rectangular pulse excitation.

The smallest solution of the preceding equation is $k = 5.33 \times 10^7$ N/m. Then

$$\frac{\pi}{\omega_n} = \pi\sqrt{\frac{m}{k}} = 0.304 \text{ s} < 0.5 \text{ s}$$

Hence there is no solution for $\pi/\omega_n > 0.5$ s.

Now let $\pi/\omega_n < 0.5$ s. Setting $x_{max} = 0.01$ m leads to

$$k = \frac{2F_0}{x_{max}} = \frac{2(5 \times 10^6 \text{ N})}{0.01 \text{ m}} = 1 \times 10^9 \frac{\text{N}}{\text{m}}$$

Then

$$\frac{\pi}{\omega_n} = 0.070 \text{ s} < 0.5 \text{ s}$$

Hence the maximum deflection will be less than 10 mm if the structure is designed such that $k > 1 \times 10^9$ N/m.

5.7 NUMERICAL METHODS

The convolution integral and Laplace transform methods are easy methods of solving Eq. (5.1) for any excitation. However, closed-form solutions using these methods are limited to cases where the forcing function has an explicit mathematical formulation and closed-form evaluation of the convolution integral is possible. In addition, there are explicitly defined forcing functions such as those proportional to nonintegral powers of time where a closed-form evaluation of the convolution integral or evaluation of the inverse Laplace transform is very difficult. When these situations occur numerical methods must be used to obtain an approximate solution to the differential equation at discrete values of time.

Numerical solutions of forced one-degree-of-freedom vibrations problems are of two classes: numerical evaluation of the convolution integral and direct numerical evaluation of Eq. (5.1).

5.7.1 Numerical Evaluation of Convolution Integral

Many numerical integration techniques are available for evaluation of integrals. Most numerical integration techniques use piecewise defined functions to interpolate the integrand. A closed-form integration of the interpolated integrand is performed. The method described here uses an interpolation for $\tilde{F}(t)$ from which an approximation to the convolution integral is obtained. The discretization of a time interval and possible interpolations to $\tilde{F}(t)$ are shown in Fig. 5.16.

Let $t_1, t_2, \ldots$ be values of time at which an approximate solution is to be obtained. Let $F_1(t), F_2(t), \ldots$ be the interpolating functions such that $F_k(t)$ interpolates $\tilde{F}(t)$ on the interval $t_{k-1} < t < t_k$. Let x_k be the numerical approximation for $x(t_k)$. Also define

$$\Delta_j = t_j - t_{j-1}$$

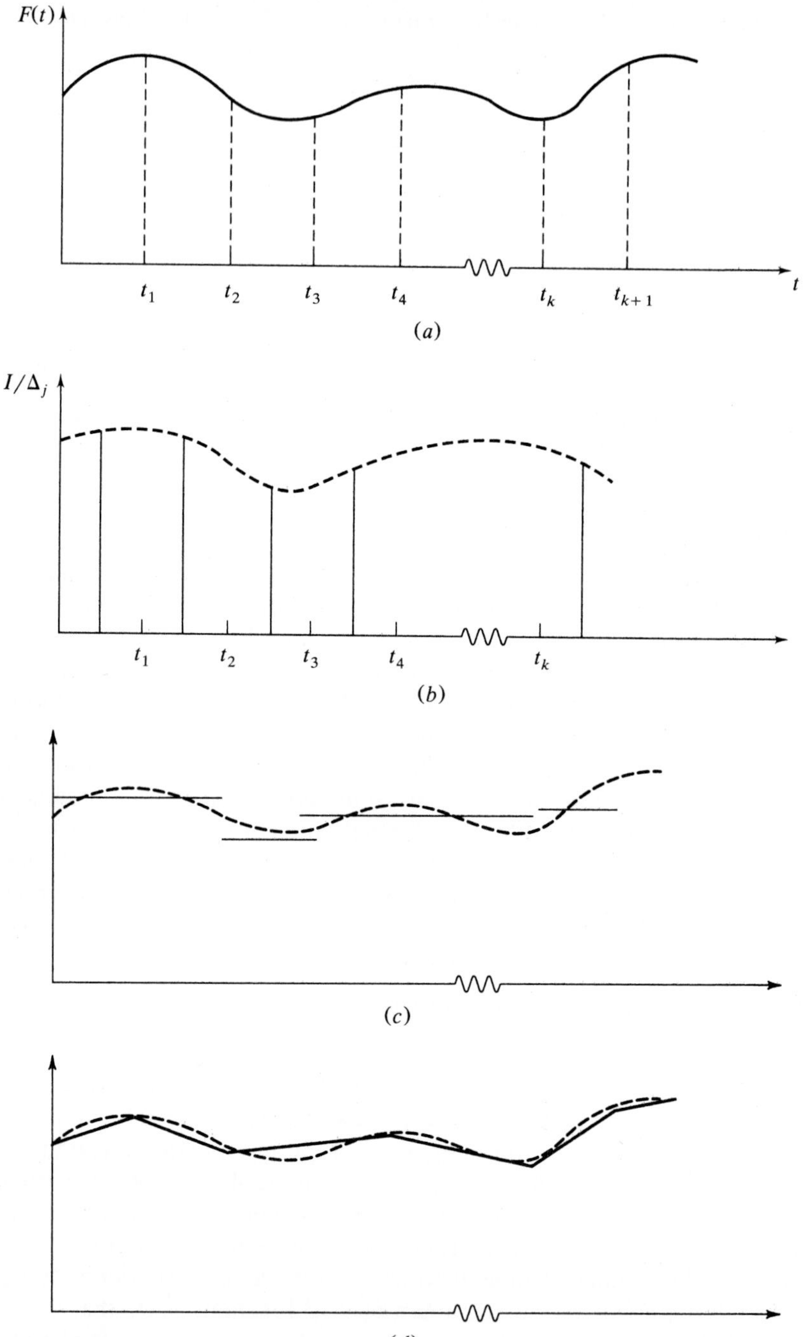

FIGURE 5.16
(a) Discretization of time for numerical integration of convolution integral; (b) $F(t)$ can be interpolated by impulses of magnitude $F(t_j^*)\Delta t_j$ applied at t_j^* for some t_j^*, $t_{j-1} < t_j^* < t_j$, (c) Interpolation of $F(t)$ by piecewise constants. For $t_{j-1} < t < t_j$, $F(t) = F(t_j^*)$ for some t_j^*, $t_{j-1} < t_j^* < t_j$, (d) Interpolation of $F(t)$ by piecewise linear functions.

The convolution integral is used to obtain the response of an underamped one-degree-of-freedom system as

$$x(t) = x(0)e^{-\zeta\omega_n t}\cos\omega_d t + \frac{\dot{x}(0)+\zeta\omega_n x(0)}{\omega_d}e^{-\zeta\omega_n t}\sin\omega_d t + \int_0^t \frac{\tilde{F}(\tau)}{\tilde{m}\omega_d}e^{-\zeta\omega_n(t-\tau)}\sin\omega_d(t-\tau)\,d\tau \tag{5.29}$$

The trigonometric identity for the sine of the difference of angles is used to rewrite Eq. (5.29) as

$$x(t) = e^{-\zeta\omega_n t}\left[x(0)\cos\omega_d t + \frac{\dot{x}(0)+\zeta\omega_n x(0)}{\omega_d}\sin\omega_d t\right] + \frac{1}{\tilde{m}\omega_d}e^{-\zeta\omega_n t}\left[\sin\omega_d t\int_0^t \tilde{F}(\tau)e^{\zeta\omega_n\tau}\cos\omega_d\tau - \cos\omega_d t\int_0^t \tilde{F}(\tau)e^{\zeta\omega_n\tau}\sin\omega_d\tau\,d\tau\right] \tag{5.30}$$

Define

$$G_{1j} = \int_{t_{j-1}}^{t_j} \tilde{F}(\tau)e^{\zeta\omega_n\tau}\cos\omega_d\tau\,d\tau \tag{5.31}$$

and

$$G_{2j} = \int_{t_{j-1}}^{t_j} \tilde{F}(\tau)e^{\zeta\omega_n\tau}\sin\omega_d\tau\,d\tau \tag{5.32}$$

Using the definitions in Eqs. (5.31) and (5.32) in Eq. (5.30) leads to

$$x_k = e^{-\zeta\omega_n t_k}\left[x(0)\cos\omega_d t_k + \frac{\zeta\omega_n x(0)+\dot{x}(0)}{\omega_d}\sin\omega_d t_k\right] + \frac{1}{\tilde{m}\omega_d}e^{-\zeta\omega_n t_k}\left[\sin\omega_d t_k\sum_{j=1}^{k} G_{1j} - \cos\omega_d t_k\sum_{j=1}^{k} G_{2j}\right] \tag{5.33}$$

Interpolating functions are chosen for $\tilde{F}(t)$ such that Eqs. (5.31) and (5.32) have closed-form evaluations when the interpolating function is used in place of $\tilde{F}(t)$. Then Eq. (5.33) is used to calculate approximations to the solution at discrete times.

First, consider the case where $\tilde{F}(t)$ is interpolated by a series of impulses, as illustrated in Fig. 5.16*b*. During the interval between t_{j-1} and t_j, application of $\tilde{F}(t)$ results in an impulse of magnitude

$$I_j = \int_{t_{j-1}}^{t_j} \tilde{F}(\tau)\, d\tau$$

The mean value theorem of integral calculus implies that there exists a t_j^*, $t_{j-1} < t_j^* < t_j$, such that

$$I_j = \tilde{F}(t_j^*)\Delta_j$$

For the sake of interpolation, approximate t_j^* by

$$t_j^* \approx \frac{t_j + t_{j-1}}{2}$$

Thus, on the interval $t_{j-1} < t < t_j$, $F(t)$ is interpolated by an impulse of magnitude I_j applied at the midpoint of the interval. With this choice of interpolation, Eqs. (5.31) and (5.32) are evaluated as

$$G_{1j} = \tilde{F}(t_j^*)\Delta_j e^{\zeta\omega_n t_j^*} \cos \omega_d t_j^* \tag{5.34}$$

$$G_{2j} = \tilde{F}(t_j^*)\Delta_j e^{\zeta\omega_n t_j^*} \sin \omega_d t_j^* \tag{5.35}$$

It is also possible to interpolate $\tilde{F}(t)$ with piecewise constants. Over the interval from t_{j-1} to t_j, the interpolate for $\tilde{F}(t)$ assumes the value of $\tilde{F}(t)$ at the interval's midpoint, as illustrated in Fig. 5.16*c*. Call the value of the interpolate f_j. Then

$$G_{1j} = f_j C_j \tag{5.36}$$

$$G_{2j} = f_j D_j \tag{5.37}$$

where

$$C_j = \frac{1-\zeta^2}{\omega_d}\left[e^{\zeta\omega_n t_j}\left(\sin \omega_d t_j + \frac{\zeta\omega_n}{\omega_d}\cos \omega_d t_j\right) - e^{\zeta\omega_n t_{j-1}}\left(\sin \omega_d t_{j-1} + \frac{\zeta\omega_n}{\omega_d}\cos \omega_d t_{j-1}\right)\right] \tag{5.38}$$

$$D_j = \frac{1-\zeta^2}{\omega_d}\left[e^{\zeta\omega_n t_j}\left(-\cos \omega_d t_j + \frac{\zeta\omega_n}{\omega_d}\sin \omega_d t_j\right) - e^{\zeta\omega_n t_{j-1}}\left(-\cos \omega_d t_{j-1} + \frac{\zeta\omega_n}{\omega_d}\sin \omega_d t_{j-1}\right)\right] \tag{5.39}$$

Finally, consider the case where $\tilde{F}(t)$ is interpolated linearly between t_{j-1} and t_j, as illustrated in Fig. 5.16*d*. Then if $g_j = f(t_j)$,

$$G_{1j} = \frac{1}{\Delta_j}\left[(g_j - g_{j-1})A_j + (g_{j-1}t_j - g_j t_{j-1})C_j\right] \tag{5.40}$$

$$G_{2j} = \frac{1}{\Delta_j}\left[(g_j - g_{j-1})B_j + (g_{j-1}t_j - g_j t_{j-1})D_j\right] \tag{5.41}$$

where C_j and D_j are given by Eqs. (5.38) and (5.39), respectively, and

$$\begin{aligned} A_j = \frac{1-\zeta^2}{\omega_d}\Bigg[& t_j e^{\zeta\omega_n t_j}\left(\sin\omega_d t_j + \frac{\zeta\omega_n}{\omega_d}\cos\omega_d t_j\right) \\ & - t_{j-1}e^{\zeta\omega_n t_{j-1}}\left(\sin\omega_d t_{j-1} + \frac{\zeta\omega_n}{\omega_d}\cos\omega_d t_{j-1}\right) - \left(D_j + \frac{\zeta\omega_n}{\omega_d}C_j\right)\Bigg] \end{aligned} \tag{5.42}$$

$$\begin{aligned} B_j = \frac{1-\zeta^2}{\omega_d}\Bigg[& t_j e^{\zeta\omega_n t_j}\left(\frac{\zeta\omega_n}{\omega_d}\sin\omega_d t_j - \cos\omega_d t_j\right) \\ & - t_{j-1}e^{\zeta\omega_n t_{j-1}}\left(\frac{\zeta\omega_n}{\omega_d}\sin\omega_d t_{j-1} - \cos\omega_d t_{j-1}\right) + \left(C_j - \frac{\zeta\omega_n}{\omega_d}D_j\right)\Bigg] \end{aligned} \tag{5.43}$$

A FORTRAN computer program, CONVOL, using Eq. (5.33) to approximate the solutions of forced-vibration problems, is presented at the end of the chapter. The user has the flexibility to choose piecewise impulses, piecewise constants, or piecewise linear functions as interpolates for $\tilde{F}(t)$. The program uses only a constant Δ_j and requires a user-furnished subprogram for $\tilde{F}(t)$. The BASIC version CONVOL.BAS is included on the accompanying diskette. It requires a user furnished subprogram, written in BASIC, to provide the excitation. Instructions for its use are provided on the diskette.

Other choices for interpolating functions for $\tilde{F}(t)$ are possible. Higher-order piecewise polynomials may be used, as well as interpolates which require more smoothness at each t_j, such as splines. Any form of interpolating function can be chosen as long as Eqs. (5.31) and (5.32) have closed-form evaluations. However, the more complicated the interpolating function, the more algebra is involved in the evaluation of G_{1j} and G_{2j}. The numerical evaluation of the convolution integral also requires more computations for more complicated interpolating functions.

If $\tilde{F}(t)$ is known empirically, any of the methods presented may be used to evaluate the convolution integral. If piecewise impulses or piecewise constants are used, the times where $\tilde{F}(t)$ is known are taken as midpoints of the intervals. If piecewise linear interpolates are used, the times where $\tilde{F}(t)$ is known are taken as the t_j's.

Error analysis of the preceding methods is beyond the scope of this text. Better accuracy for the response is, of course, obtained with better accuracy of the interpolate. Error analysis usually involves comparing the interpolation with a Taylor series expansion to estimate the error in the interpolation. The error is usually expressed as being the order of some power of Δ_j. Bounds on the error in using the convolution integral are obtained. Integration tends to smooth errors.

Determination of the response using these methods requires evaluation of the convolution integral at discrete values of time. Since errors are introduced in the evaluation of G_{1j} and G_{2j}, the more of these terms used in the evaluation, the larger is the error. Hence the error in approximation grows with increasing t. Reduction of error can be achieved by using smaller time intervals, if possible, or by using more accurate interpolates.

5.7.2 Numerical Solution of Eq. (5.1)

An alternative to numerical evaluation of the convolution integral is to approximate the solution of Eq. (5.1) by direct numerical integration. Many methods are available for numerical solution of ordinary differential equations.

Since vibrations of discrete systems are governed by initial value problems, it is best to use a numerical method that is self-starting. That is, previous knowledge of the solution at only one time is required to start the procedure. Finite-difference methods, though widely used in engineering applications, are not self-starting and thus are not the best methods to use for numerical solution of vibration problems.

Best application of self-starting methods requires the rewriting of an nth-order differential equation as n first-order differential equations. This is done for Eq. (5.1) by defining

$$\begin{aligned} y_1(t) &= x(t) \\ y_2(t) &= \dot{x}(t) \end{aligned} \tag{5.44}$$

Thus

$$\dot{y}_1(t) = y_2(t) \tag{5.45a}$$

and from Eq. (5.1)

$$\dot{y}_2(t) = \frac{\tilde{F}}{\tilde{m}} - 2\zeta\omega_n y_2(t) - \omega_n^2 y_1(t) \tag{5.45b}$$

Equations (5.45a) and (5.45b) are two simultaneous linear first-order ordinary differential equations whose numerical solution yields the values of displacement and velocity at discrete times.

In the following let t_i, $i = 1, 2, \ldots,$ be the discrete times at which the solution is obtained and let $y_{1,i}$ and $y_{2,i}$ be the displacements and velocities at these times and define

$$\Delta_j = t_{j+1} - t_j$$

The recurrence relations for the simplest self-starting method, called the Euler method, are obtained from truncating Taylor series expansions for $y_{k,i+1}$ about $y_{k,i}$ after the linear terms. These recurrence relations are

$$\begin{aligned} y_{1,i+1} &= y_{1,i} + (t_{i+1} - t_i) y_{2,i} \\ y_{2,i+1} &= y_{2,i} + (t_{i+1} - t_i)\left[\frac{\tilde{F}(t_i)}{\tilde{m}} - 2\zeta\omega_n y_{2,i} - \omega_n^2 y_{1,i}\right] \end{aligned} \tag{5.46}$$

Given initial values of y_1 and y_2, Eq. (5.46) is used to calculate recursively the displacement and velocity at increasing times. The Euler method is first-order accurate meaning that the error is of the order of Δ_j.

Runge-Kutta methods are more popular than the Euler method because of their better accuracy, while still being easy to use. A Runge-Kutta formula for the solution of the first-order differential equation

$$\dot{y}(t) = f(y, t)$$

is of the form

$$y_{i+1} = y_i + \sum_{j=1}^{n} a_j k_j \tag{5.47}$$

where

$$k_1 = (t_{i+1} - t_i) f(y_i, t_i) \tag{5.48a}$$

$$k_2 = (t_{i+1} - t_i) f(y_1 + q_{1,1}k_1, t_i + p_1) \tag{5.48b}$$

$$k_3 = (t_{i+1} - t_i) f(y_i + q_{2,1}k_1 + q_{2,2}k_2, t_i + p_2) \tag{5.48c}$$

$$\vdots$$

$$k_n = (t_{i+1} - t_i) f(y_i + q_{n-1,1}k_1 + q_{n-2,2}k_2 + \cdots + q_{n-1,n-1}k_{n-1}, t_i + p_{n-1})$$

and the a, q, and p coefficients are chosen by using Taylor series expansions to approximate the differential equation to the desired accuracy.

The error for a fourth-order Runge-Kutta formula is proportional to Δ_j^4. A fourth-order Runge-Kutta formula is

$$y_{i+1} = y_i + \tfrac{1}{6}(k_1 + 2k_2 + 2k_3 + k_4) \tag{5.49}$$

where

$$\begin{aligned} k_1 &= (t_{i+1} - t_i) f(y_i, t_i) \\ k_2 &= (t_{i+1} - t_i) f\left(y_i + \tfrac{1}{2}k_1, \tfrac{1}{2}(t_i + t_{i+1})\right) \\ k_3 &= (t_{i+1} - t_i) f\left(y_i + \tfrac{1}{2}k_2, \tfrac{1}{2}(t_i + t_{i+1})\right) \\ k_4 &= (t_{i+1} - t_i) f(y_i + k_3, t_{i+1}) \end{aligned} \tag{5.50}$$

Equation (5.49) can be used for higher-order differential equations by rewriting them as a system of first-order equations as has been done in Eq. (5.45) for a one-degree-of-freedom system. The result is

$$y_{1,i+1} = y_{1,i} + \tfrac{1}{6}(k_{1,1} + 2k_{1,2} + 2k_{1,3} + k_{1,4}) \tag{5.51a}$$

$$y_{2,i+1} = y_{2,i} + \tfrac{1}{6}(k_{2,1} + 2k_{2,2} + 2k_{2,3} + k_{2,4}) \tag{5.51b}$$

where
$$k_{1,1} = (t_{i+1} - t_i)y_{2,i} \tag{5.52a}$$

$$k_{1,2} = (t_{i+1} - t_i)(y_{2,i} + \tfrac{1}{2}k_{2,1}) \tag{5.52b}$$

$$k_{1,3} = (t_{i+1} - t_i)(y_{2,i} + \tfrac{1}{2}k_{2,2}) \tag{5.52c}$$

$$k_{1,4} = (t_{i+1} - t_i)(y_{2,i} + k_{2,3}) \tag{5.52d}$$

$$k_{2,1} = (t_{i+1} - t_i)\left[\frac{\tilde{F}(t_i)}{\tilde{m}} - 2\zeta\omega_n y_{2,i} - \omega_n^2 y_{1,i}\right] \tag{5.52e}$$

$$k_{2,2} = (t_{i+1} - t_i)\left[\frac{\tilde{F}(\tfrac{1}{2}(t_i + t_{i+1}))}{\tilde{m}} - 2\zeta\omega_n(y_{2,i} + \tfrac{1}{2}k_{2,1}) - \omega_n^2(y_{1,i} + \tfrac{1}{2}k_{1,1})\right] \tag{5.52f}$$

$$k_{2,3} = (t_{i+1} - t_i)\left[\frac{\tilde{F}(\tfrac{1}{2}(t_i + t_{i+1}))}{\tilde{m}} - 2\zeta\omega_n(y_{2,i} + \tfrac{1}{2}k_{2,2}) - \omega_n^2(y_{1,i} + \tfrac{1}{2}k_{1,2})\right] \tag{5.52g}$$

$$k_{2,4} = (t_{i+1} - t_i)\left[\frac{\tilde{F}(t_{i+1})}{\tilde{m}} - 2\zeta\omega_n(y_{2,i} + k_{2,3}) - \omega_n^2(y_{1,i} + k_{1,3})\right] \tag{5.52h}$$

The Runge-Kutta method is often used because it is easy to program for a digital computer. Its most restrictive limitation is that extension of the approximation between two discrete times requires evaluation of the excitation at an intermediate time. If the forcing function is known only at discrete times, evaluation at the appropriate intermediate times is often impossible. In addition, a large number of function evaluations can lead to large computer times.

A FORTRAN computer program, RUNGE, using the fourth-order Runge-Kutta formula of Eq. (5.49) as applied in Eqs. (5.51) and (5.52) to solve for the response of a linear one-degree-of-freedom system, is presented at the end of the chapter. The program requires a user-provided function subprogram for $\tilde{F}(t)$.

Adams' formulas provide more accurate approximations of ordinary differential equations. An open Adams formula requires knowledge of the functions at the two previous time steps to calculate the approximation at the desired time. A closed Adams formula requires knowledge of the function at only the previous time step, but the formula involves the evaluation of the function at the time step of interest. Thus a closed Adams formula requires an iterative solution at each time step. The closed Adams formula is much more accurate than an open formula of the same order. The closed formula is self-starting, whereas the open formula is not self-starting.

A predictor-corrector method is a compromise that uses the closed formula for increased accuracy but uses the open formula to reduce computation time. The open formula is used to "predict" the solution at the desired time, then the closed formula is used to "correct" by using the predicted value as an initial guess. Iterations are not necessary as the first correction is very accurate. Since the open Adams formulas are not self-starting, a self-starting method such as the Runge-Kutta method of the same order is used to calculate the solution at the first time. The predictor-corrector method is used for the remainder of the calculations.

Example 5.12. A one-degree-of-freedom system has a natural frequency of 20 rad/s, a damping ratio of 0.1, and an equivalent mass of 35 kg. The system is subject to the sine pulse excitation of Fig. 5.17. The exact solution of Eq. (5.1), with the excitation of Fig. 5.17, can be calculated from the convolution integral. Compare the exact solution with the numerical solutions obtained using numerical integration of the convolution integral with piecewise impulse interpolation, piecewise constant interpolation, and piecewise linear interpolation. Also compare the exact solution with numerical solutions obtained using the fourth-order Runge-Kutta methods of Eq. (5.49).

The computer programs CONVOL and RUNGE presented at the end of the chapter are used with the function subroutine shown as part of RUNGE. A time increment, $\Delta_j = 0.01$ s is used for all calculations. The ratio of the time step to the natural period is 0.032.

The results are tabulated in Table 5.2 at selected times and show that all methods work well for this example.

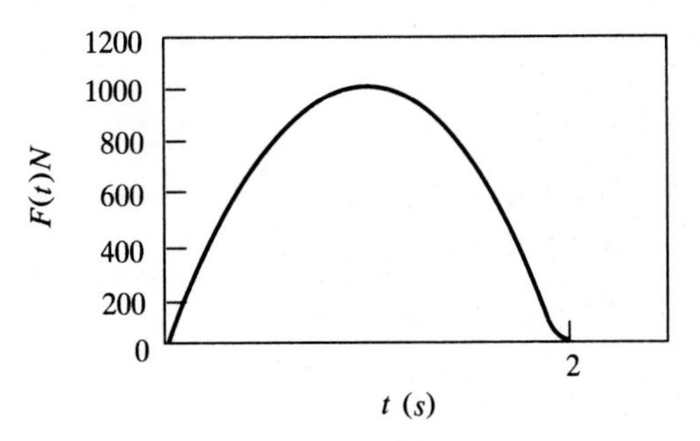

FIGURE 5.17
Sine pulse excitation for Example 5.12.

TABLE 5.2
Numerical results of Example 5.12

	$x(t)$ (m)				
Time (s)	Exact solution	Piecewise impulses	Piecewise constants	Piecewise linear	Runge-Kutta
0.01	0.00007	0.00011	0.00011	0.00007	0.00007
0.05	0.00847	0.00863	0.00861	0.00847	0.00847
0.10	0.05584	0.05609	0.05598	0.05584	0.05584
0.20	0.23385	0.23416	0.23375	0.23384	0.23385
0.30	0.33142	0.33194	0.33138	0.33141	0.33142
0.40	0.38779	0.38552	0.38786	0.38777	0.38779
0.50	0.50669	0.50750	0.50664	0.50667	0.50669
0.60	0.58722	0.58816	0.58718	0.58720	0.58722
0.70	0.62222	0.62333	0.62227	0.62221	0.62222
0.80	0.67998	0.68111	0.67996	0.67996	0.67998
0.90	0.71644	0.71760	0.71641	0.71642	0.71644
1.00	0.71278	0.71398	0.71279	0.71276	0.71278
1.20	0.69200	0.69312	0.69197	0.69198	0.69199
1.40	0.58600	0.58697	0.58600	0.58599	0.58599
1.60	0.43102	0.43174	0.43104	0.43103	0.43102
1.80	0.23419	0.23458	0.23420	0.23421	0.23419
2.00	0.01037	0.01040	0.01040	0.01039	0.01037
2.05	−0.03682	−0.03672	−0.04200	−0.03680	−0.03682
2.10	−0.04477	−0.04090	−0.04464	−0.04477	−0.04477
2.15	−0.01397	−0.01398	−0.01397	−0.01399	−0.01397
2.20	0.02289	0.02282	0.02277	0.02286	0.02288
2.30	0.01476	0.01475	0.01474	0.01477	0.01476
2.40	−0.02518	−0.02513	−0.02509	−0.02517	−0.02518
2.50	0.00689	0.00685	0.00684	0.00687	0.00688
2.60	0.01229	0.01227	0.01226	0.01229	0.01229
2.70	−0.01280	−0.01278	−0.01276	−0.01280	−0.01280
2.80	0.00030	0.00029	0.00029	0.00029	0.00030
2.90	0.00838	0.00837	0.00836	0.00839	0.00839
3.00	−0.00579	−0.00577	−0.00576	−0.00579	−0.00579

Example 5.13. An airfoil, modeled as a one-degree-of-freedom system, is subject to aerodynamic pulse loading. The loading is experimentally determined. The results of measurements are tabulated in Table 5.3. The duration of the pulse is 1.7 s. It is anticipated that while the magnitudes of the loads will change, the airfoil will often be subject to pulses of the same shape. Determine the shock spectrum for this pulse.

Since the value of the excitation is known only at discrete times, numerical approximations are used to generate the shock spectrum. Program CONVOL is used with $x(0) = 0$, $\dot{x}(0) = 0$, and $\zeta = 0$. Piecewise linear functions are used to interpolate $F(t)$. The program was run for different values of ω_n.

The shock spectrum is a plot of the maximum value of kx/F_0 vs. the ratio of the pulse duration to the natural period. For lack of better information, F_0 is taken as 2.38 kN. The ratio kx/F_0 is independent of specific values of k and m, and is dependent only on ω_n and the frequency and shape of the pulse. However, values of the system parameters must be chosen to effect numerical calculations.

TABLE 5.3
Experimentally determined forcing function for Example 5.13

Time (s)	$F(t)$ (kN)	Time (s)	$F(t)$ (kN)
0.0	0.0	0.9	1.84
0.1	1.02	1.0	2.17
0.2	1.58	1.1	2.38
0.3	1.68	1.2	2.16
0.4	1.47	1.3	1.51
0.5	1.21	1.4	1.00
0.6	1.13	1.5	0.61
0.7	1.25	1.6	0.21
0.8	1.61	1.7	0.0

Numerical calculations can be performed by using the same equivalent mass for different natural frequencies and changing the equivalent stiffness according to

$$\tilde{k} = \tilde{m}\omega_n^2$$

The maximum value of x was found for each natural frequency for which the program was run, the values of $kx_{\max}/F_0$ and $2\pi(1.7 \text{ s})/\omega_n$ calculated, and the response spectrum plotted. The result is shown in Fig. 5.18.

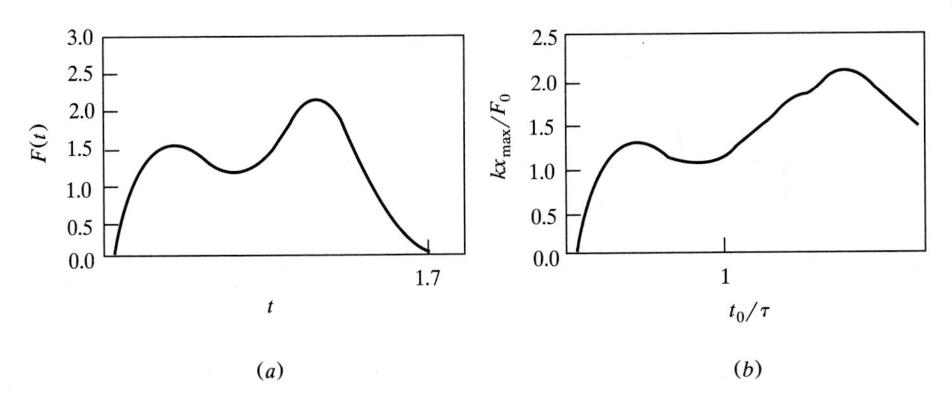

FIGURE 5.18
(a) plot of the forcing function for Example 5.13 using experimental data; (b) shock spectrum determined by numerical methods for Example 5.13.

```
C   PROGRAM CONVOL
C   SOLUTION OF ONE DEGREE OF FREEDOM FORCED VIBRATIONS
C   PROBLEMS USING NUMERICAL EVALUATION OF CONVOLUTION INTEGRAL
C
      REAL IK,JK
      WRITE(6,100)
```

```
  100 FORMAT(16X, 'FORCED VIBRATIONS OF A ONE-DEGREE-OF-FREE'
     $'DOM SYSTEM' / 18X, 'NUMERICAL INTEGRATION OF CONVOLUTION'
     $'INTEGRAL' ///)
C
C   INPUT INITIAL CONDITIONS AND SYSTEM PARAMETERS
C
      WRITE(6,110)
  110 FORMAT(5X, 'INPUT INITIAL CONDITIONS' // 10X, 'INITIAL'
     $'DISPLACEMENT (M) = ')
      READ(5,*) X0
      WRITE(6,120)
  120 FORMAT(10X, 'INITIAL VELOCITY (M / SEC) = ')
      READ(5,*) V0
      WRITE(6,130)
  130 FORMAT( / 5X, 'INPUT SYSTEM PARAMETERS' // 10X, 'DAMPING'
     $'RATIO = ')
      READ(5,*) ZETA
      WRITE(6,140)
  140 FORMAT(10X, 'NATURAL FREQUENCY (RAD / SEC) = ')
      READ(5,*) WN
      WRITE(6,150)
  150 FORMAT(10X, 'EQUIVALENT MASS(KG) = ')
      READ(5,*) EQM
      ZON = ZETA*WN
      WD = WN*SQRT(1.- ZETA*ZETA)
      C1 = 1. / (EQM*WD)
      C2 = (V0 + ZON*X0) / WD
      C3 = (1.- ZETA*ZETA) / WD
      C4 = ZETA*WN / WD
C
C   DECIDE ON TYPE OF INTERPOLATION
C       I1 = 1, PIECEWISE IMPULSE INTERPOLATION
C       I1 = 2, PIECEWISE CONSTANT INTERPOLATION
C       I1 = 3, PIECEWISE LINEAR INTERPOLATION
C
      WRITE(6,160)
  160 FORMAT( / 5X, 'DECIDE ON THE TYPE OF INTERPOLATION' /
               10X, 'IF PIECEWISE IMPULSES ARE TO BE USED ENTER'
     $         '1' /
               10X, 'IF PIECEWISE CONSTANTS ARE TO BE USED ENTER'
     $         '2' /
               10X, 'IF PIECEWISE LINEAR FUNCTIONS ARE TO BE'
     $         'USED ENTER 3')
      READ(5,*)I1
C
C   INPUT TIME INCREMENT AND FINAL VALUE OF TIME
C
      WRITE(6,170)
```

```
  170 FORMAT(5X, 'TIME INCREMENT (SEC) = ')
      READ(5,*)DT
      WRITE(6,180)
  180 FORMAT(5X, 'FINAL TIME (SEC) = ')
      READ(5,*)TF
      WRITE(6,190)
  190 FORMAT(20X, 'TIME (SEC)', 20X, 'DISPLACEMENT (M)')
      K = 1
      TK = 0
      SUM1 = 0.0
      SUM2 = 0.0
      WRITE(6,200)TK,X0
  200 FORMAT(17X,F12.6,21X,F12.6)
      GO TO (1,2,3)I1
C
C   SOLUTION FOR PIECEWISE IMPULSE INTERPOLATION
C
      TK = TK + DT
      TS = TK- DT / 2.
      A = FUN(TS)*EXP(ZON*TS)*DT
      SUM1 = SUM1 + A*COS(WD*TS)
      SUM2 = SUM2 + A*SIN(WD*TS)
      X = EXP(- ZON*TK)*((X0- C1*SUM2)*COS(WD*TK) + (C2 + C1*SUM1)*
     $SIN(WD*TK)
      WRITE(6,200)TK,X
      K = K + 1
      IF(TK.GE.TF)GO TO 4
      GO TO 1
C
C   SOLUTION FOR PIECEWISE CONSTANT INTERPOLATION
C
    2 C3 = (1. - ZETA * ZETA) / WD
      C4 = ZETA * WN / WD
    5 TK1 = TK
      TK = TK + DT
      TS = TK- DT / 2.
      IK = C3*(EXP(ZON*TK)*(C4*SIN(WD*TK)- COS(WD*TK))- EXP(ZON*
     $TK1)*(C4*$SIN(WD*TK1)- COS(WD*TK1)))
      JK = C3*(EXP(ZON*TK)*(SIN(WD*TK) + C4*COS(WD*TK))- EXP(ZON*
     $TK1)*(C4*COS(WD*TK1) + SIN(WD*TK1)))
      A = FUN(TS)
      SUM1 = SUM1 + A*JK
      SUM2 = SUM2 + A*IK
      X = EXP(- ZON*TK)*((X0- C1*SUM2)*COS(WD*TK) + (C2 + C1*SUM1)*
     $SIN(WD*TK))
      WRITE(6,200)TK,X
      K = K + 1
      IF(TK.GE.TF)GO TO 4
      GO TO 5
```

```
C
C   SOLUTION FOR PIECEWISE LINEAR INTERPOLATION
C
    3 C3 = (1. - ZETA * ZETA) / WD
      C4 = ZETA * WN / WD
      C5 = 1./ WN**2
      C7 = 1.- ZETA*ZETA
      C6 = 2.*ZETA*SQRT(C7)
      C8 = 1.- 2.*ZETA*ZETA
    6 TK1 = TK
      TK = TK + DT
      IK = C3*(EXP(ZON*TK)*(C4*SIN(WD*TK)- COS(WD*TK))- EXP(ZON*
     $TK1)*(C4*SIN(WD*TK1)- COS(WD*TK1)))
      JK = C3*(EXP(ZON*TK)*(SIN(WD*TK) + C4*COS(WD*TK))- EXP(ZON*
     $TK1)*(C4*COS(WD*TK1) + SIN(WD*TK1)))
      AK = C3*(TK*EXP(ZON*TK)*(SIN(WD*TK) + C4*COS(WD*TK))- TK1*
     $EXP(ZON*TK1)*(SIN(WD*TK1) + C4*COS(WD*TK1))- IK- C4*JK
      BK = C3*(TK*EXP(ZON*TK)*(C4*SIN(WD*TK)- COS(WD*TK))- TK1*
     $EXP(ZON*TK1)*(C4*SIN(WD*TK1)- COS(WD*TK1)) + JK- C4*IK)
      FK = FUN(TK)
      FK1 = FUN(TK1)
      SUM1 = SUM1- 1./ DT*((FK1- FK)*AK + (FK*TK1- FK1*TK)*JK)
      SUM2 = SUM2- 1./ DT*((FK1- FK)*BK + (FK*TK1- FK1*TK)*IK)
      X = EXP(- ZON*TK)*((XO- C1*SUM2)*COS(WD*TK) + (C2 + C1*SUM1)*
     $SIN(WD*TK))
      WRITE(6,200)TK,X
      K = K + 1
      IF(TK.LE.TF)GO TO 6
    4 STOP
      END

C PROGRAM RUNGE
C FOURTH ORDER RUNGE-KUTTA SOLUTION OF FORCED VIBRATIONS OF
C ONE DEGREE OF FREEDOM SYSTEMS WITH CONSTANT TIME STEP
C
C RUNGE-KUTTA SOLUTION OF EQ. (5.1) FOR ARBITRARY F(t)
C SECOND ORDER DIFFERENTIAL EQUATION IS REWRITTEN AS TWO FIRST
C ORDER DIFFERENTIAL EQUATIONS- EQS. (5.45) WHERE
C          Y1(t) = DISPLACEMENT
C          Y2(t) = VELOCITY
  FORCING FUNCTION IS PROVIDED BY USER AS FUNCTION SUBPROGRAM
  FUN(T)
C
      REAL K11,K12,K13,K14,K21,K22,K23,K24
C INPUT INITIAL CONDITIONS AND SYSTEM PARAMETERS
      WRITE(6,110)
  110 FORMAT(5X, 'INPUT INITIAL CONDITIONS'//10X, 'INITIAL'
     $'DISPLACEMENT (M) = ')
      READ(5,*)Y1
```

```
      WRITE(6,120)
  120 FORMAT(10X, 'INITIAL VELOCITY (M / SEC) = ')
      READ(5,*)Y2
      WRITE(6,130)
  130 FORMAT( / 5X, 'INPUT SYSTEM PARAMETERS' / / 10X, 'NATURAL'
     $'FREQUENCY (RAD / SEC = ')
      READ(5,*)WN
      WRITE(6,140)
  140 FORMAT(10X, 'DAMPING RATIO = ')
      READ(5,*)ZETA
      WRITE(6,150)
  150 FORMAT(10X, 'EQUIVALENT MASS (KG) = ')
      READ(5,*)EQM
      WRITE(6,160)
  160 FORMAT( / 5X, 'TIME STEP (SEC) = ')
      READ(5,*)DT
      WRITE(6,170)
  170 FORMAT( / 5X, 'FINAL TIME (SEC) = ')
      READ(5,*)TF
      WRITE(6,180)
  180 FORMAT(10X, 'TIME (SEC)', 10X, 'DISPLACEMENT (M)', 10X,
     $'VELOCITY (M / SEC)')
      C1 = 2,*ZETA*WN
      C2 = WN*WN
      TIME = 0.0
      WRITE(6,190)TIME,Y1,Y2
  190 FORMAT(8X,F10.4,12X,F12.6,14X,F12.6)
C
C CALCULATION OF K'S ACCORDING TO EQS. (5.52)
C
    1 K11 = DT*Y2
      K21 = DT*(FUN(TIME) / EQM- C1*Y2- C2*Y1)
      K12 = DT*(Y2 + K21 / 2.)
      FS = FUN(TIME + DT / 2.) / EQM
      K22 = DT*(FS- C1*(Y2 + K21 / 2.)- C2*(Y1 + K11 / 2.))
      K13 = DT*(Y2 + K22 / 2.)
      K23 = DT*(FS- C1*(Y2 + K22 / 2.)- C2*(Y1 + K12 / 2.))
      K14 = DT*(Y2 + K23)
      K24 = DT*(FUN(TIME + DT) / EQM- C1*(Y2 + K23)- C2*(Y1 + K13))
C
C CALCULATION OF Y1 AND Y2 ACCORDING TO EQS. (5.51)
C
      Y1 = Y1 + (K11 + 2.*K12 + 2.*K13 + K14) / 6.
      Y2 = Y2 + (K21 + 2.*K22 + 2.*K23 + K24) / 6.
      WRITE(6,190)TIME,Y1,Y2
      TIME = TIME + DT
      IF(TIME.LE.TF)GO TO 1
      STOP
      END
```

```
C USER PROVIDED FUNCTION SUBPROGRAM FUN(T)
C
C EXAMPLE IS FOR TEXT EXAMPLE 5.12
C
      FUNCTION FUN(T)
      FUN = 0.0
      IF(T.GE.2)GO TO 1
      FUN = 10000*SIN(1.5708*T)
    1 RETURN
      END
```

PROBLEMS

5.1. Use the method of variation of parameters to obtain the general solution of Eq. (5.1) and show that it can be written in the form of the convolution integral, Eq. (5.9).

5.2. Use the convolution integral to calculate the response of an underdamped system to a unit step function.

5.3. Let $g(t)$ be the response of an underdamped system to a unit step function and $h(t)$ the response of an underdamped system to the unit impulse function. Show

$$h(t) = \frac{dg}{dt}$$

5.4. Use the convolution integral, Eq. (5.9), and the notation and result of Prob. 5.3 to show that an alternate expression for the response of a system subject to an excitation, $F(t)$, is

$$x(t) = F(0)g(t) + \int_0^t \frac{dF(\tau)}{d\tau} g(t-\tau)\, d\tau$$

5.5. A one-degree-of-freedom undamped system is initially at rest in equilibrium and subject to a time-dependent force, $F(t) = F_0 t e^{-t/2}$. Use the convolution integral to determine the response of the system.

5.6. The mass of Fig. P5.6 has a velocity v when it engages the spring-dashpot mechanism. Let $x(t)$ be the displacement of the mass from its position when the mechanism is engaged. Use the convolution integral to determine $x(t)$. Assume the system is underdamped.

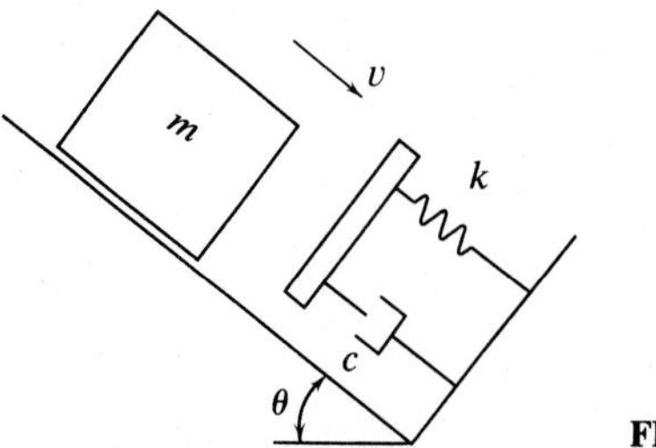

FIGURE P5.6

5.7. Use the convolution integral to determine the response of the system of Fig. P.57.

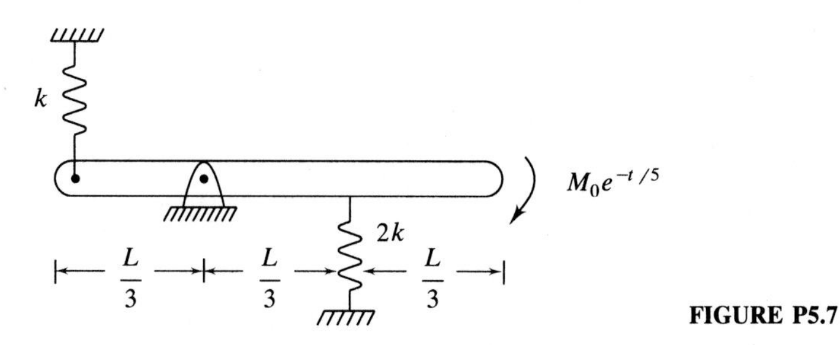

FIGURE P5.7

5.8. A one-degree-of-freedom system of natural frequency ω_n and damping ratio $\zeta < 1$ is subject to a rectangular pulse of magnitude F_0 and duration t_0. Under what conditions on ω_n, ζ, and t_0 will the maximum response occur after the pulse is removed?

5.9–5.15. Use superposition and Table 5.1 to develop the response of a one-degree-of-freedom undamped system of equivalent mass 30 kg and equivalent stiffness 1500 N/m to the excitation shown.

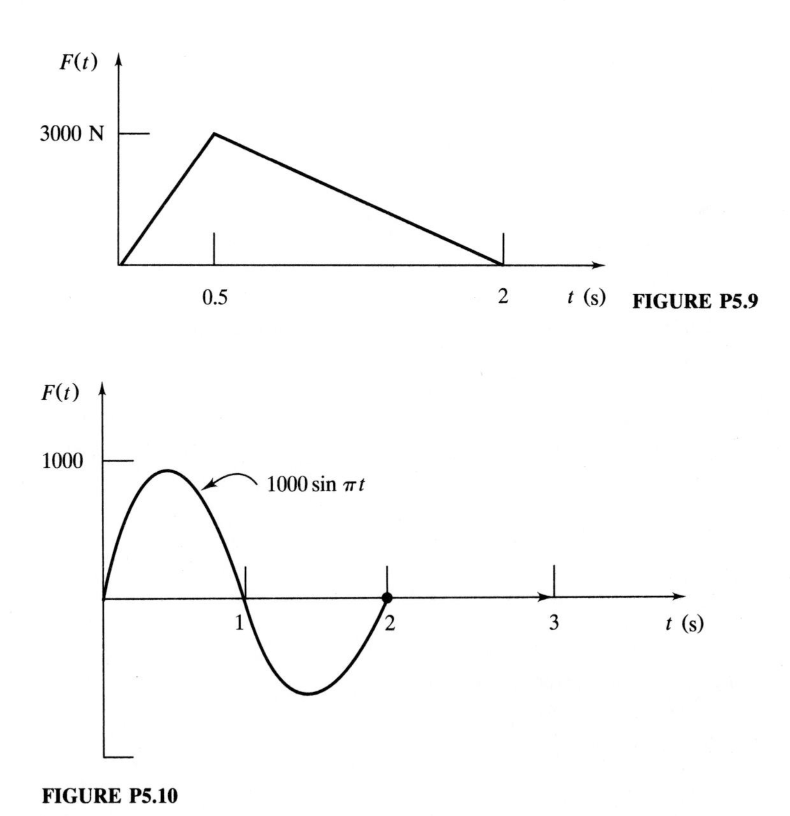

FIGURE P5.9

FIGURE P5.10

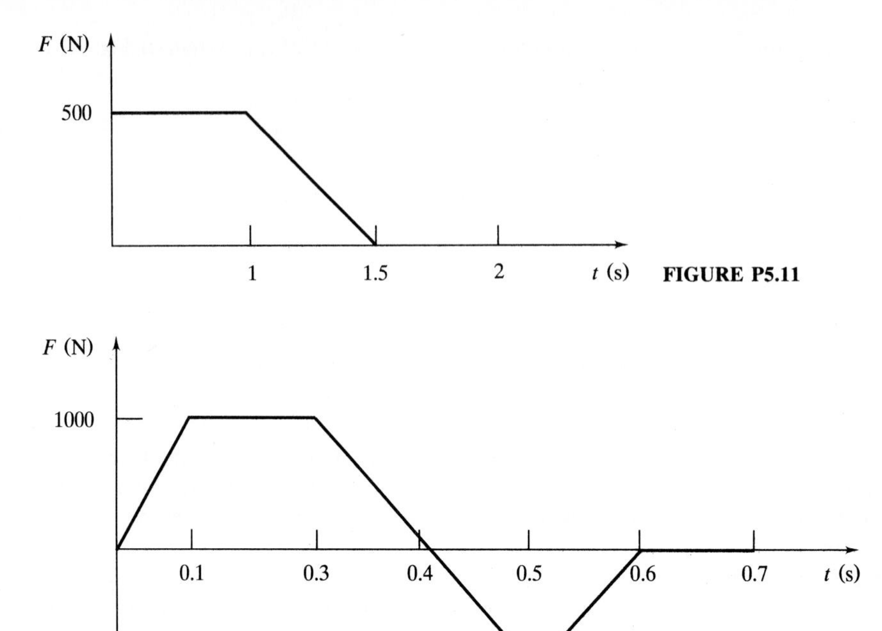

FIGURE P5.11

FIGURE P5.12

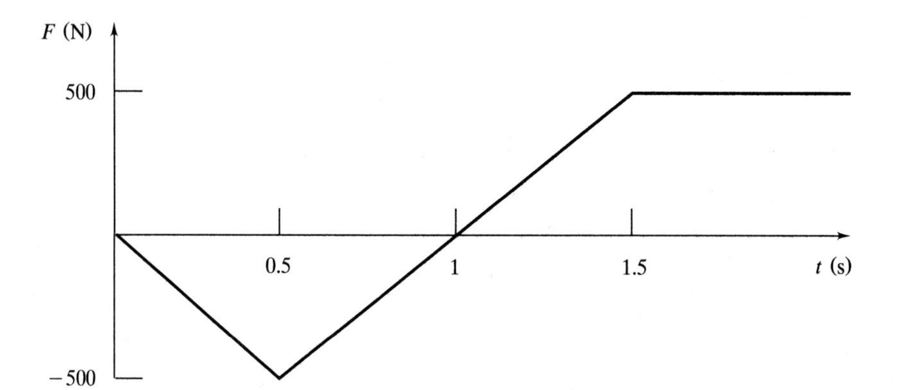

FIGURE P5.13

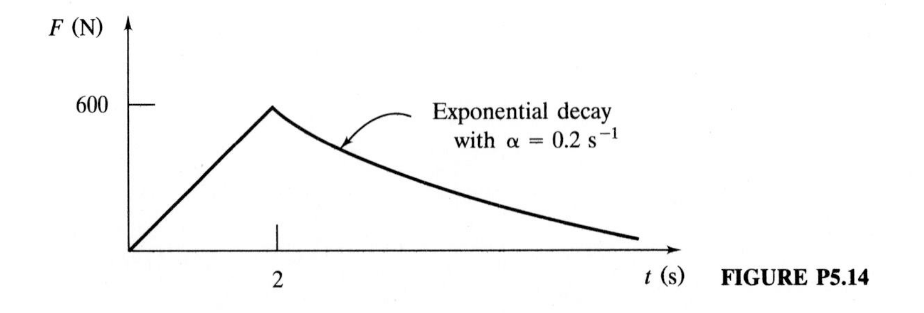

FIGURE P5.14

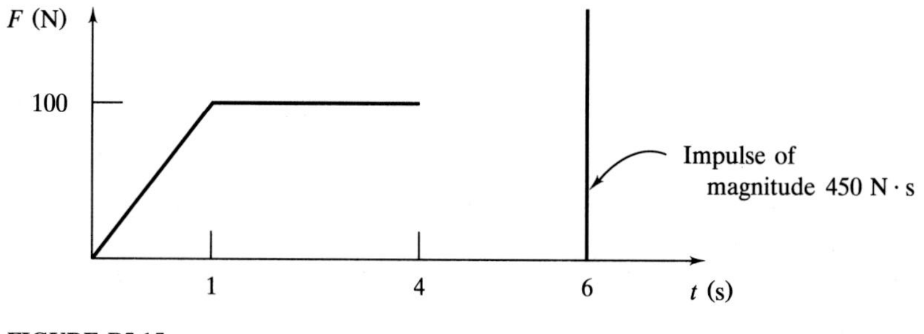

FIGURE P5.15

5.16. A mass-spring system is initially in equilibrium in a cylinder of compressed air at a gauge pressure of 300 kPa. A small hole is punctured in the casing causing the pressure to decrease exponentially. After 5 s the pressure drops to 200 kPa. Determine $x(t)$, the displacement of the piston from its initial position.

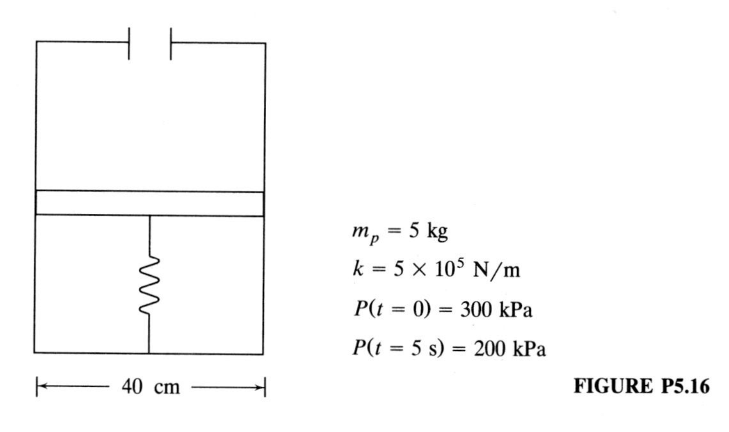

FIGURE P5.16

5.17. A one-story frame structure houses a chemical laboratory. Figure P5.17 shows the results of a model test to predict the transient force to which the structure would be subject if an explosion would occur. The equivalent mass of the structure is 2000

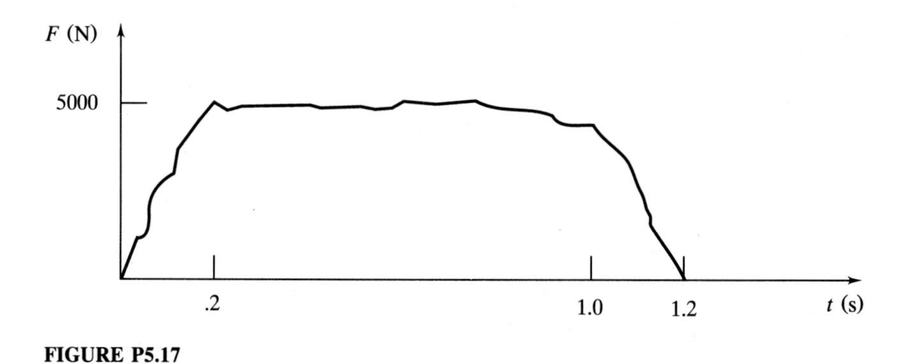

FIGURE P5.17

kg and its equivalent stiffness is 5×10^6 N/m. Approximate the maximum displacement of the structure due to this blast.

5.18. A cart propelled by a water jet slides on a frictionless track. When the cart reaches a certain point, it engages a spring and damper in parallel designed such that the system is critically damped. Determine the steady-state position of the cart, and how long it takes for the system to be within 0.1 mm of steady-state.

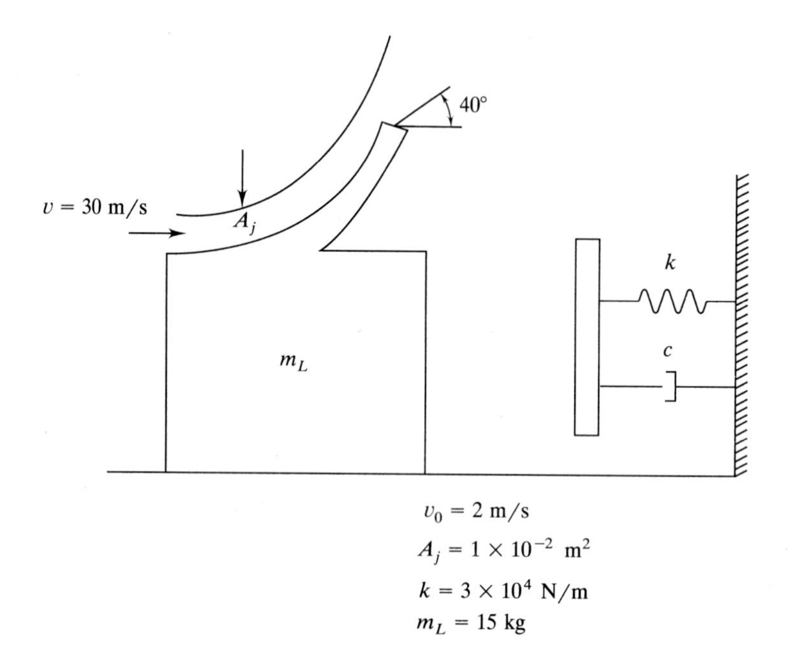

$v_0 = 2$ m/s

$A_j = 1 \times 10^{-2}$ m^2

$k = 3 \times 10^4$ N/m

$m_L = 15$ kg

FIGURE P5.18

5.19. A sensitive computer system of mass 150 kg is to be loaded by crane onto a ship where it will be transported from Baltimore to London by ship. The computer was placed in a heavy-duty crate and packed in a material that exhibits linear behavior

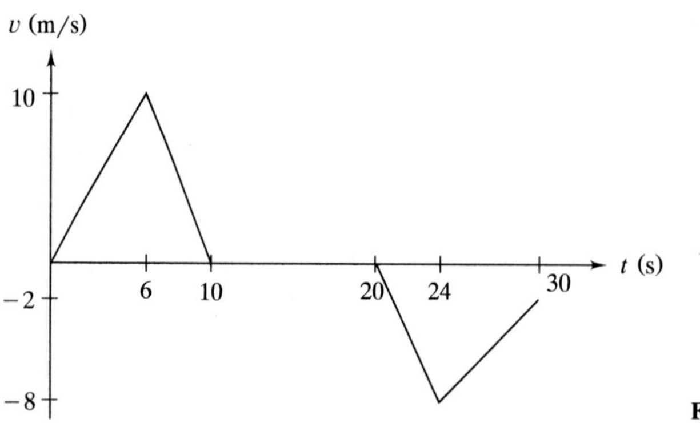

FIGURE P5.19

with a stiffness of 8000 N/m. The packaging was designed to survive rough seas. The estimated vertical velocity of the crate during loading is shown in Fig. P5.19. Note that the velocity of the crate suddenly changes from -2 m/s to 0 when it hits the deck of the ship. What is the maximum displacement of the computer relative to the crate during loading?

5.20. A 500-kg vehicle is traveling at 45 m/s when it encounters a pothole of depth 25 cm and length 1.2 m. The deflection of the suspension spring under the weight of the vehicle is 7.2 mm. The shock absorber has a damping ratio of 0.3. Using a one-degree-of-freedom system to model the vehicle and its suspension system, approximate the maximum vertical acceleration of the vehicle assuming it maintains its horizontal speed as it encounters the pothole.

5.21. The vehicle of Prob. 5.20 is traveling at 40 m/s when it encounters a bump of height 20 cm and length 45 cm. The shape of the bump is approximated by a sinusoidal function. Using a one-degree-of-freedom model, determine the displacement of the vehicle relative to its wheels.

5.22. A machine of mass m is attached to the end of a cantilever beam of length L, elastic modulus E, and moment of inertia I. The base of the beam is given a displacement $w(1 - e^{-t/5})$. Ignoring the inertia of the beam, determine the absolute displacement of the machine.

5.23. A compound pendulum consists of a slender rod of length L and mass m, pinned at one end. The pin support is given a horizontal displacement according to Fig. P5.23. Determine the angular displacement of the pendulum.

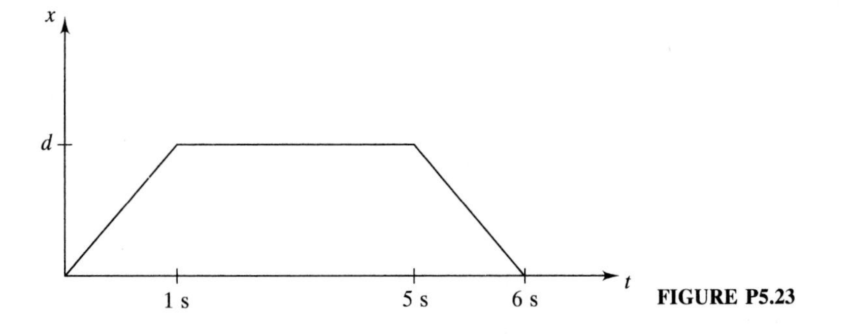

FIGURE P5.23

5.24. A personal computer of mass m is packed inside a box such that the stiffness and damping coefficient of the packing material are k and c, respectively. The package is dropped from a height h and lands on a hard surface without rebound. Determine the maximum displacement of the computer relative to the package.

5.25. Use the Laplace transform method to determine the response of a mass-spring system at rest in equilibrium when subject to an impulse of magnitude I.

5.26. Use the Laplace transform method to find the response of a system at rest in equilibrium when subject to

$$F(t) = F_0 \cos \omega t[1 - u(t - t_0)]$$

for

(a) $\zeta < 1$.

(b) $\zeta = 1$.

(c) $\zeta > 1$.

5.27. Determine the response of an undamped mass-spring system at rest in equilibrium to a symmetric triangular pulse of magnitude F_0 and total duration t_0.

5.28. Use the Laplace transform method to determine the steady-state response of an underdamped mass-spring-dashpot system due to the periodic force of Fig. P5.28.

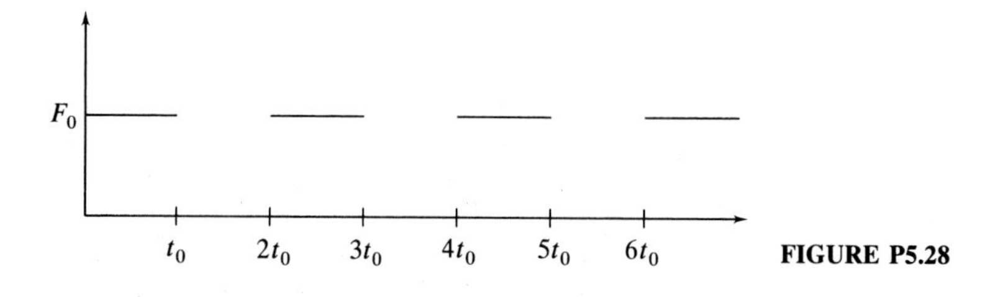

FIGURE P5.28

5.29. Repeat Prob. 5.22 using the Laplace transform method.

5.30. Use the Laplace transform method to derive the response of a one-degree-of-freedom undamped mass-spring system with a harmonic excitation, $F_0 \sin \omega t$, when (a) $\omega \neq \omega_n$ and (b) $\omega = \omega_n$.

5.31. Determine the shock spectrum for a rectangular velocity pulse.

5.32. Determine the shock spectrum for a rectangular acceleration pulse.

5.33. Determine the shock spectrum for a triangular velocity pulse of total duration $2t_0$.

5.34. Is there an approximate $h(t)$ such that the response of a one-degree-of-freedom system with hysteretic damping can be written using the convolution integral of Eq. (5.9)? Why or why not? If so, what is $h(t)$?

5.35. Is there an approximate $h(t)$ such that the response of a one-degree-of-freedom system with Coulomb damping can be written using the convolution integral of Eq. (5.9)? Why or why not? If so, what is $h(t)$?

5.36. A performance test is run on a model of a rocket. The model is connected to an elastic medium of stiffness k and damping coefficient c. The rocket is fired and thrust is developed as exhaust gases leave the nozzle. The thrust is $\dot{m}_e v_e$ where $\dot{m}_e = \rho_e v_e A_e$ is the mass flow rate of the exhaust gases, v_e is the exit velocity, ρ_e is the density of the exit gas, and A_e is the exit area of the nozzle. The velocity of the exhaust gases are predicted in the following table:

t (s)	v_e (m/s)	t (s)	v_e (m/s)	t (s)	v_e (m/s)
0	0	0.35	125.0	0.70	150
0.05	10.0	0.40	144.0	0.75	120
0.10	22.0	0.45	160.0	0.80	95
0.15	40.0	0.50	169.0	0.85	70.0
0.20	61.0	0.55	174.0	0.90	50.0
0.25	89.0	0.60	180.0	0.85	20.0
0.30	100.0	0.65	180.0	1.00	0.0

Burnout is achieved at 1.0 s. The total mass of the rocket and fuel is 3000 kg, the spring stiffness is 40,000 N/m, the damping coefficient is 1000 N · s/m, and

$\rho_e A_e = 0.41$ kg/m. The governing differential equation is

$$M\frac{d^2x}{dt^2} + c\frac{dx}{dt} + kx = \dot{m}_e v_e$$

(*a*) Use the Runge-Kutta method to predict the response of the system. Assume the mass remains constant at 3000 kg and use a linear interpolation to predict the velocity of the exhaust gases between given values.

(*b*) The mass of the rocket is actually a function of time given by

$$M(t) = M_0 - \int_0^t \dot{m}_e \, dt$$

where $M_0 = 3000$ kg. Modify the Runge-Kutta method to account for variable mass. Use a numerical integration scheme to calculate the mass as a function of time. Write a FORTRAN program to determine the displacement as a function of time.

5.37. Derive Eq. (5.34).

5.38. Derive Eq. (5.37).

5.39. Write a FORTRAN program utilizing the Euler method as specified in Eq. (5.46) to determine the response of a one-degree-of-freedom system. Use the program to solve Example 5.10.

The program CONVOL, included on the accompanying diskette, uses numerical integration of the convolution integral to solve Eq. (5.1). Program CONVOL appears on the diskette in two versions. One version includes a library of excitations. The other version requires a user provided BASIC subprogram to provide the excitation. Use CONVOL to help solve Problems 5.40–5.46.

5.40. Use CONVOL.BAS to repeat Example 4.16.

5.41. The system of Fig. P5.38 is initially at rest in equilibrium when the time-dependent moment is applied. Determine its response using

(*a*) Runge-Kutta.

(*b*) Numerical integration of the convolution integral with piecewise linear functions.

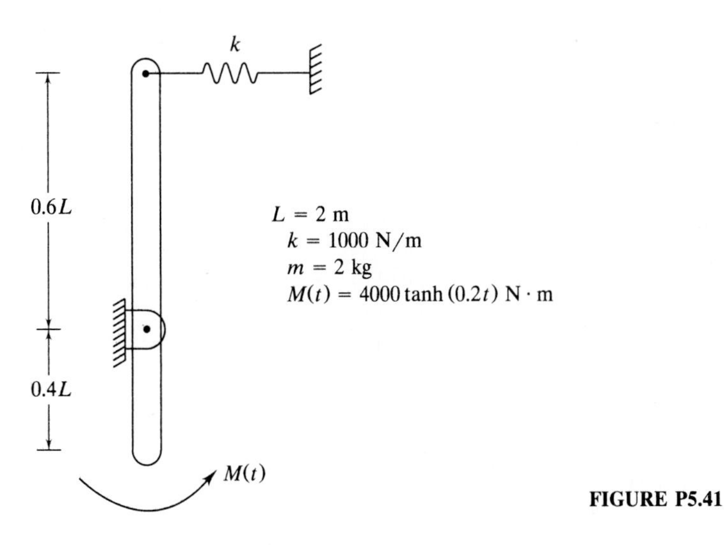

FIGURE P5.41

5.42. A 2000-kg vehicle is traveling along a smooth road at 60 km/h when it encounters the bump of Fig. P5.42*a*. The driver slows down when the wheels first encounter the bump. The velocity of the vehicle is given in Fig. P5.42*b*. The suspension system has a stiffness 4×10^6 N/m and a damping ratio of 0.16. Model the vehicle and its suspension system using one degree of freedom. Use numerical integration of the convolution integral with piecewise linear functions to determine the displacement of the vehicle.

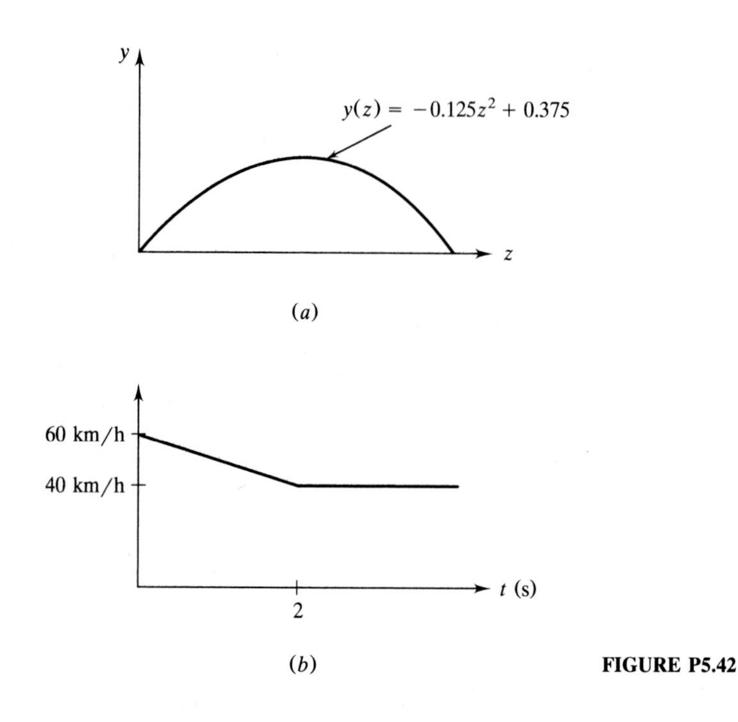

FIGURE P5.42

5.43. Use CONVOL.BAS to determine the response of a block of mass 100 kg suspended from a spring of stiffness 30,000 N/m in parallel with a dashpot of damping coefficient 400 N · s/m subject to a force

$$F(t) = \begin{cases} 500(1 - t^{3/2})N, & 0 \leq t \leq 1 \text{ s} \\ 0, & t > 1 \text{ s} \end{cases}$$

for $t < 2.0$ s.

5.44. A 300-kg machine has a rotating unbalance of 1.5 kg · m. The machine is mounted on an isolator of equivalent stiffness 3.5×10^6 N/m and damping ratio of 0.07. The machine has a steady-state operating speed of 2000 rpm. It takes the machine 1.8 s to build up linearly to its final operating speed.

(*a*) Set up the convolution integral which gives the transient response of the system as it builds up to its steady-state operation.

(*b*) Use piecewise constants to integrate numerically the convolution integral and determine the maximum displacement of the machine during its transient operation.

5.45. Use CONVOL.BAS to determine the response of a block of mass 100 kg suspended from a spring of stiffness 30,000 N/m subject to a force

$$F(t) = 250\, e^{t^2}\ \text{N}$$

for $t < 2.0$ s.

5.46. Use CONVOL to plot the time dependent response of a 100 kg block attached to a spring of stiffness 10,000 N/m in parallel with a viscous damper of damping coefficient 400 N · s/m, subject to a triangular pulse of magnitude 400 N and of total duration 0.45 s. Plot the response between $t = 0$ and $t = 1.5$ s.

The program SPECT.BAS, included in the accompanying diskette, uses Runge-Kutta to develop the shock spectrum for a one-degree-of-freedom-system subject to an excitation provided in a user supplied BASIC subprogram. Use SPECT.BAS to solve problems 5.47–550.

5.47. Use SPECT to develop the shock spectrum for a triangular pulse.

5.48. Develop the shock spectrum for the following excitation:

$$F(t) = \cos\frac{\pi t}{t_0}\left[u(t) - u(t - t_0)\right]$$

5.49. Develop the shock spectrum for the versed sine pulse,

$$F(t) = \begin{cases} \sin^2\left(\dfrac{\pi t}{t_0}\right) & 0 < t < t_0 \\ 0 & t > t_0 \end{cases}$$

5.50. Develop the shock spectrum for the following excitation:

$$F(t) = F_0\left\{u(t) - u(t - t_0) + \left(3 - 2\frac{t}{t_0}\right)\left[u(t - t_0) - u\left(t - \frac{3}{2}t_0\right)\right]\right\}$$

REFERENCES

1. Carnahan, B., H. A. Luther, and J. O. Wilkes: *Applied Numerical Analysis*, Wiley, New York, 1969.
2. Dimarogonas, A. D., and S. Haddad: *Vibrations for Engineers*, Prentice-Hall, Englewood Cliffs, N.J., 1992.
3. James, M. L., G. M. Smith, J. C. Wolford, and P. W. Whaley: *Vibration of Mechanical and Structural Systems with Microcomputer Applications*, Harper and Row, New York, 1989.
4. Ketter, R. L., and S. P. Prawel: *Modern Methods of Engineering Computations*, McGraw-Hill, New York, 1969.
5. Meirovitch, L.: *Elements of Vibration Analysis*, McGraw-Hill, New York, 1975.
6. Rao, S. S.: *Mechanical Vibrations*, 2nd ed., Addison-Wesley, Reading, Mass., 1990.
7. Shabana, A. A.: *Theory of Vibration: An Introduction*, vol. 1, Springer-Verlag, New York, 1991.
8. Shabana, A. A.: *Theory of Vibration: Discrete and Continuous Systems*, vol. 2, Springer-Verlag, New York, 1991.
9. Steidel, R. F.: *An Introduction to Mechanical Vibrations*, 3rd. ed., Wiley, New York, 1988.
10. Thomson, W. T.: *Theory of Vibrations with Applications*, 3rd ed., Prentice-Hall, Englewood Cliffs, N.J., 1988.

CHAPTER

6

MULTI-DEGREE-OF-FREEDOM-SYSTEM DERIVATION OF GOVERNING EQUATIONS

6.1 INTRODUCTION

The analysis of the vibrations of a multi-degree-of-freedom system is significantly more difficult and time consuming than the analysis of the vibrations of a one-degree-of-freedom system. Recall from Chap. 1 that the number of degrees of freedom necessary for the analysis of vibrations of a system is the number of kinematically independent coordinates necessary to specify the motion of every particle contained in the system. The number of degrees of freedom of many mechanical systems is easy to identify. The motion of the rigid bar in Fig. 6.1*a* is analyzed using one degree of freedom. However, when the pin support is removed, as shown in Fig. 6.1*b*, the motion must be analyzed using two degrees of freedom. This is because the kinematic relationship, $x_a = L\theta$, obtained using the small-angle assumption is no longer valid and x_a and θ are kinematically independent. If the cable in the mass-pulley system of Fig. 6.2 is flexible, then the motion of the system is analyzed using two degrees of freedom. The no-slip assumption between the cable and the pulley is still used but kinetics, not kinematics, provides the relationship between the displacement of the mass and the angular displacement of the pulley.

Modeling a structural system with a finite number of degrees of freedom provides approximations to the behavior of the system. All structural systems

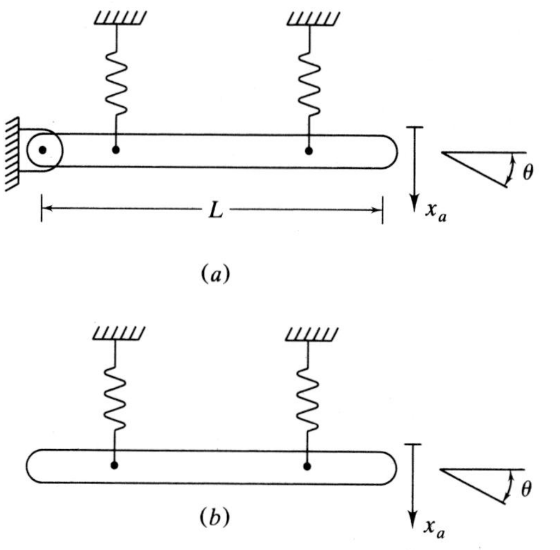

FIGURE 6.1
(*a*) One-degree-of-freedom systems, $x_a \approx L\theta$; (*b*) when pin support is removed, system has two degrees of freedom. No kinematic relation exists between x_a and θ.

are continuous systems with an infinite number of degrees of freedom. The continuous change in internal forces and moments across the length of a beam prevents a kinematic relationship between the displacement of any two particles on the beam's neutral axis. However, the analysis of continuous systems requires the solution of partial differential equations. The analysis of a multi-degree-of-freedom system is easier. Modeling of a structural element using a one-degree-of-freedom system was introduced in Chap. 2. The one-degree-of-freedom model provides only an approximation to the lowest natural frequency, whereas a continuous system possesses an infinite but countable number of free vibration

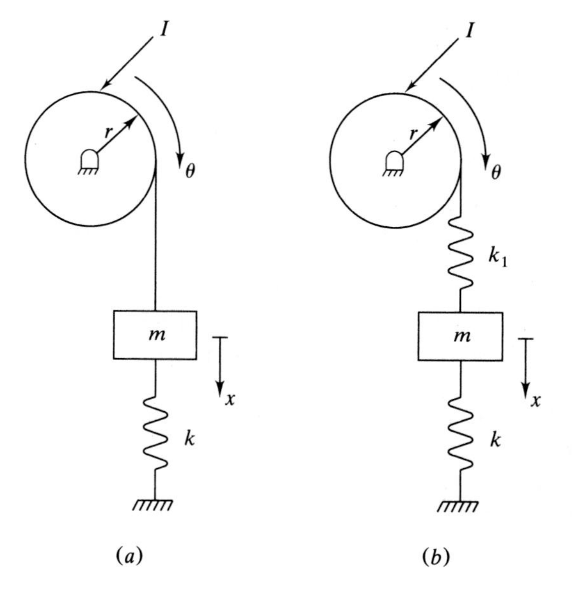

FIGURE 6.2
(*a*) If there is no slip between cable and pulley, $x = r\theta$ and system is modeled using one degree of freedom; (*b*) if cable is elastic, no kinematic relation exists between x and θ.

modes. The use of a multi-degree-of-freedom model yields a better approximation for the lowest natural frequency and also provides approximations for higher frequencies.

This chapter deals with the derivation of the differential equations governing the vibrations of multi-degree-of-freedom systems. Chapter 7 is concerned with the free-vibration response of a multi-degree-of-freedom system, while Chap. 8 is concerned with the forced-vibration response of a multi-degree-of-freedom system.

Two methods of deriving the differential equations are presented: application of Newton's laws to free-body diagrams and energy methods. The application of Newton's laws to free-body diagrams is straightforward, similar to their application for one-degree-of-freedom systems. However, their application can be tedious and the resulting equations may have to be manipulated to be put into a usable form. Energy methods are based upon the use of Lagrange's equation, a general equation derived from energy methods that is used to formulate differential equations for possibly nonlinear systems. A complete understanding of Lagrange's equation and its full use requires knowledge of the calculus of variations and is beyond the scope of this book.

Differential equations for linear systems are summarized in a matrix form. Lagrange's equations are used to show how knowledge of the quadratic forms of potential and kinetic energies for a linear system are used to define the elements of the stiffness and mass matrices. Influence coefficients are developed as an alternative to formally calculating the potential and kinetic energy.

The formulation of the differential equations for discrete models of structural systems is handled differently. The inverse of the stiffness matrix, called the flexibility matrix, is calculated and approximations are used for the mass matrix.

6.2 DERIVATION OF DIFFERENTIAL EQUATIONS USING BASIC PRINCIPLES OF DYNAMICS

Governing differential equations for multi-degree-of-freedom mechanical systems can be derived by applying the basic principles of rigid-body dynamics to the appropriate free-body diagrams. The method of application is very similar to that presented in Chap. 3 for one-degree-of-freedom systems. The method used in this textbook requires two free-body diagrams drawn for each rigid body or system of rigid bodies. One free-body diagram shows the external forces acting on the rigid body and a separate free-body diagram shows the effective forces acting on the body. The effective forces are equivalent to a force equal to the mass times the acceleration of the mass center of the rigid body placed at its mass center, and a couple equal to the centroidal mass moment of inertia of the body times the angular acceleration of the body. Conservation laws are then applied to the free body-diagrams. The following examples illustrate this procedure.

Example 6.1. The blocks in Fig. 6.3 slide on a frictionless surface. Derive the differential equations governing free vibrations using x_1, x_2, and x_3 as generalized coordinates.

The following form of Newton's law:

$$\sum \overset{+}{\vec{F}}_{\text{eff}} = \sum \overset{+}{\vec{F}}_{\text{ext}}$$

is applied to the free-body diagrams of Fig. 6.3*b*. The resulting differential

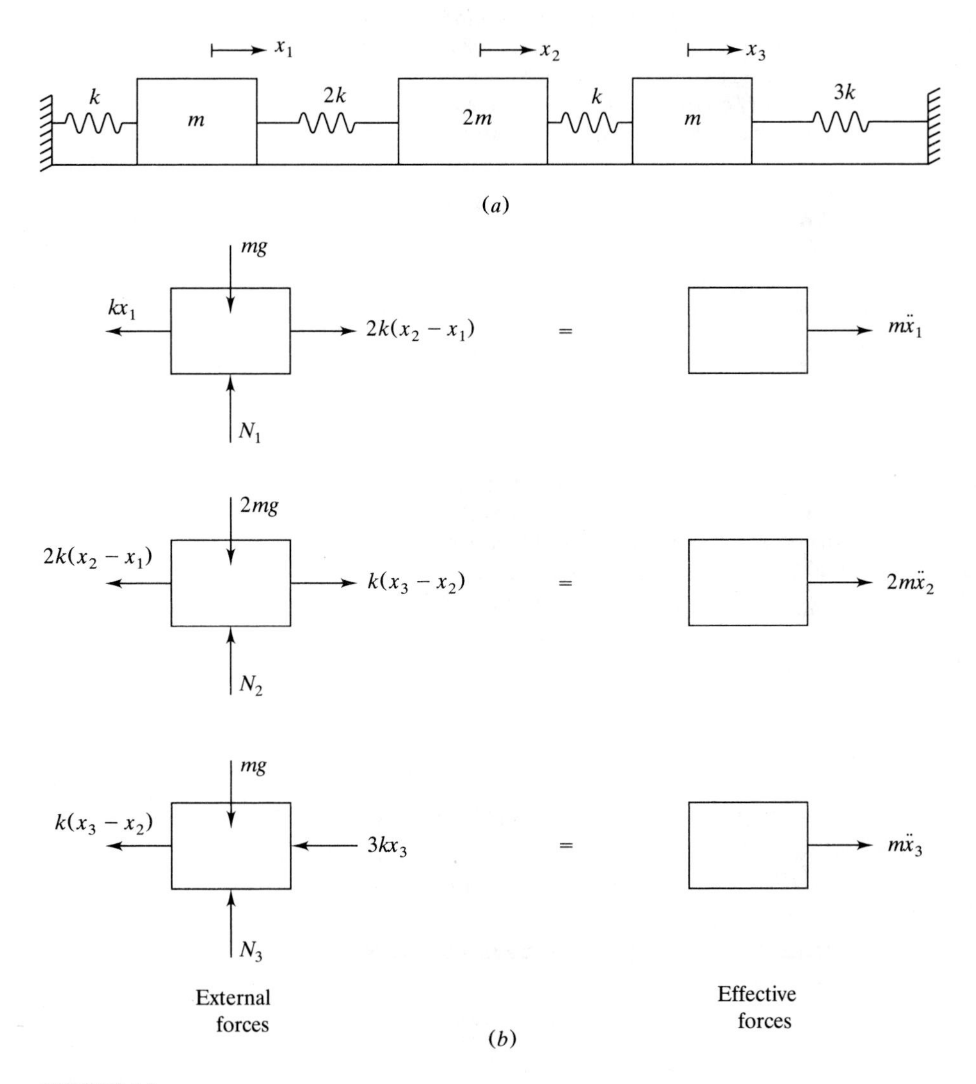

FIGURE 6.3
(*a*) System of Example 6.1; (*b*) free-body diagrams of each block at an arbitrary instant of time.

equations are

$$-kx_1 + 2k(x_2 - x_1) = m\ddot{x}_1$$

$$-2k(x_2 - x_1) + k(x_3 - x_2) = 2m\ddot{x}_2$$

$$-k(x_3 - x_2) - 3kx_3 = m\ddot{x}_3$$

The preceding equations are rearranged and written in matrix form as

$$\begin{bmatrix} m & 0 & 0 \\ 0 & 2m & 0 \\ 0 & 0 & m \end{bmatrix} \begin{bmatrix} \ddot{x}_1 \\ \ddot{x}_2 \\ \ddot{x}_3 \end{bmatrix} + \begin{bmatrix} 3k & -2k & 0 \\ -2k & 3k & -k \\ 0 & -k & 4k \end{bmatrix} \begin{bmatrix} x_1 \\ x_2 \\ x_3 \end{bmatrix} = \begin{bmatrix} 0 \\ 0 \\ 0 \end{bmatrix}$$

Example 6.2. Consider the uniform slender rod of Fig. 6.4. Assume the springs remain vertical, the bar is in equilibrium when it is horizontal, and the small-angle assumption applies. Using x, the displacement of the mass center of the bar, and θ, the clockwise angular displacement of the bar from equilibrium, as the generalized coordinates, derive the differential equations governing the forced vibrations of this two-degree-of-freedom system.

Free-body diagrams of the bar are shown in Fig. 6.4b where θ is exaggerated for illustration. The differential equations are derived using summation of moments

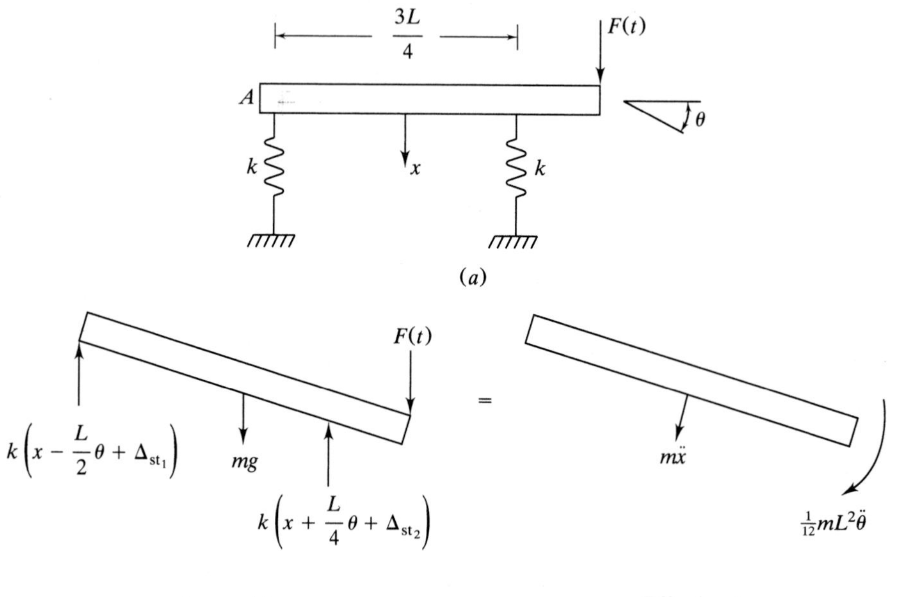

FIGURE 6.4
(a) System of Example 6.2. x and θ are chosen as generalized coordinates; (b) free-body diagrams at an arbitrary instant of time.

with the small-angle assumption

$$\sum \overset{+}{\overleftarrow{M}}_{A_{\text{ext}}} = \sum \overset{+}{\overleftarrow{M}}_{A_{\text{eff}}}$$

$$F(t)L + mg\frac{L}{2} - k\left(x + \frac{L}{4}\theta + \Delta_{s_2}\right)\frac{3L}{4} = m\ddot{x}\frac{L}{2} + \frac{1}{12}mL^2\ddot{\theta}$$

$$\sum \overset{+}{\overleftarrow{M}}_{B_{\text{ext}}} = \sum \overset{+}{\overleftarrow{M}}_{B_{\text{eff}}}$$

$$F(t)\frac{L}{4} + k\left(x - \frac{L}{2}\theta + k\Delta_{s_1}\right)\frac{3L}{4} - mg\frac{L}{4} = -m\ddot{x}\frac{L}{4} + \frac{1}{12}mL^2\ddot{\theta}$$

An analysis of the static-equilibrium position shows that gravity cancels with the static-deflection terms in both equations. Rearranging and writing the equations in matrix form gives

$$\begin{bmatrix} m\frac{L}{2} & m\frac{L^2}{12} \\ -m\frac{L}{4} & m\frac{L^2}{12} \end{bmatrix}\begin{bmatrix} \ddot{x} \\ \ddot{\theta} \end{bmatrix} + \begin{bmatrix} 3k\frac{L}{4} & 3k\frac{L^2}{16} \\ -3k\frac{L}{4} & 3k\frac{L^2}{8} \end{bmatrix}\begin{bmatrix} x \\ \theta \end{bmatrix} = \begin{bmatrix} LF(t) \\ \frac{L}{4}F(t) \end{bmatrix}$$

Example 6.3. The block of Fig. 6.5 slides on a frictionless surface. The slender rod is pinned at the center of the block. Using x, the displacement of the block from equilibrium, and θ, the angular displacement of the bar from the vertical, as generalized coordinates, derive the nonlinear differential equations governing the motion of this system. Linearize the differential equations using the small-angle assumption.

Free-body diagrams of the block-rod assembly and the rod alone are shown in Fig. 6.5*b*. A free-body diagram of the block alone is not analyzed as it includes the unknown pin reactions. These reactions must be known in terms of the generalized coordinates to apply Newton's law to the block. Application of

$$\sum \overset{+}{\vec{F}}_{\text{ext}} = \sum \overset{+}{\vec{F}}_{\text{eff}}$$

to the block-rod assembly yields

$$F(t) - kx = 2m\ddot{x} + m\ddot{x} + m\frac{L}{2}\ddot{\theta}\cos\theta - m\frac{L}{2}\dot{\theta}^2\sin\theta$$

Application of

$$\sum \overset{+}{\circlearrowleft} M_{O_{\text{ext}}} = \sum \overset{+}{\circlearrowleft} M_{O_{\text{eff}}}$$

to the free-body diagram of the rod alone yields

$$M(t) - mg\sin\theta\left(\frac{L}{2}\right) = m\frac{L}{2}\ddot{\theta}\left(\frac{L}{2}\right) + m\frac{L^2}{12}\ddot{\theta} + m\ddot{x}\cos\theta\left(\frac{L}{2}\right)$$

These differential equations are nonlinear due to the presence of the $\sin\theta$, $\cos\theta$, and $\dot{\theta}^2$ terms. The small-angle assumption is used to linearize the differential equations with $\cos\theta \approx 1$, $\sin\theta \approx \theta$, and neglecting $\dot{\theta}^2$ in comparison to linear

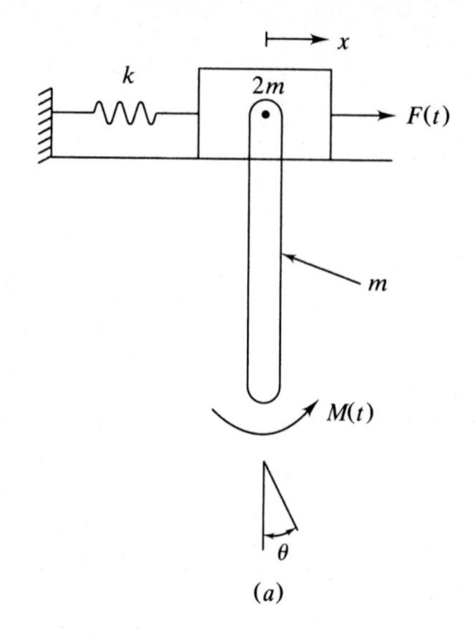

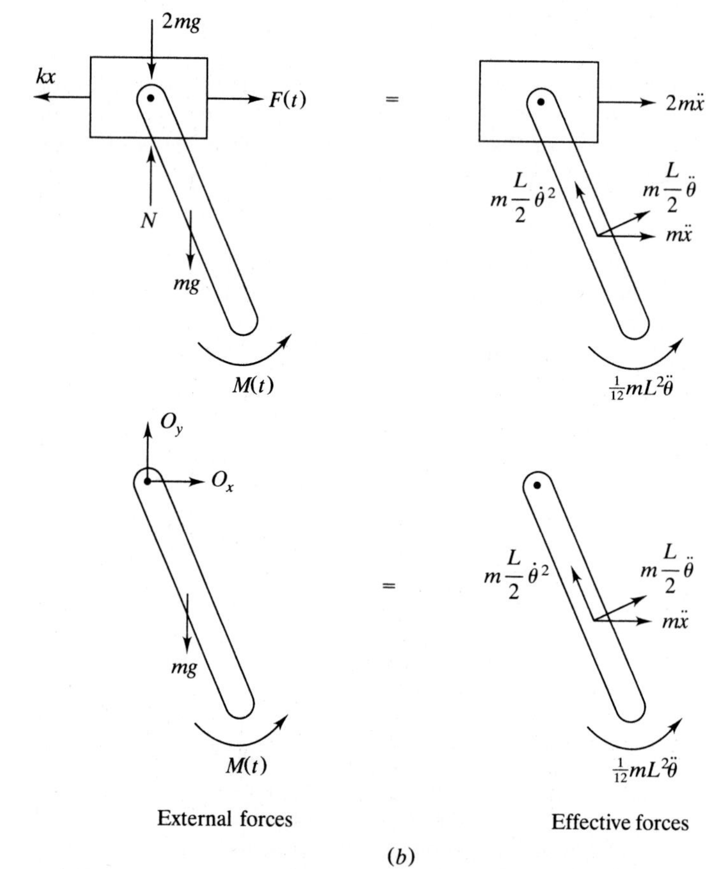

FIGURE 6.5
(*a*) System of Example 6.3; (*b*) free-body diagrams at an arbitrary instant of time.

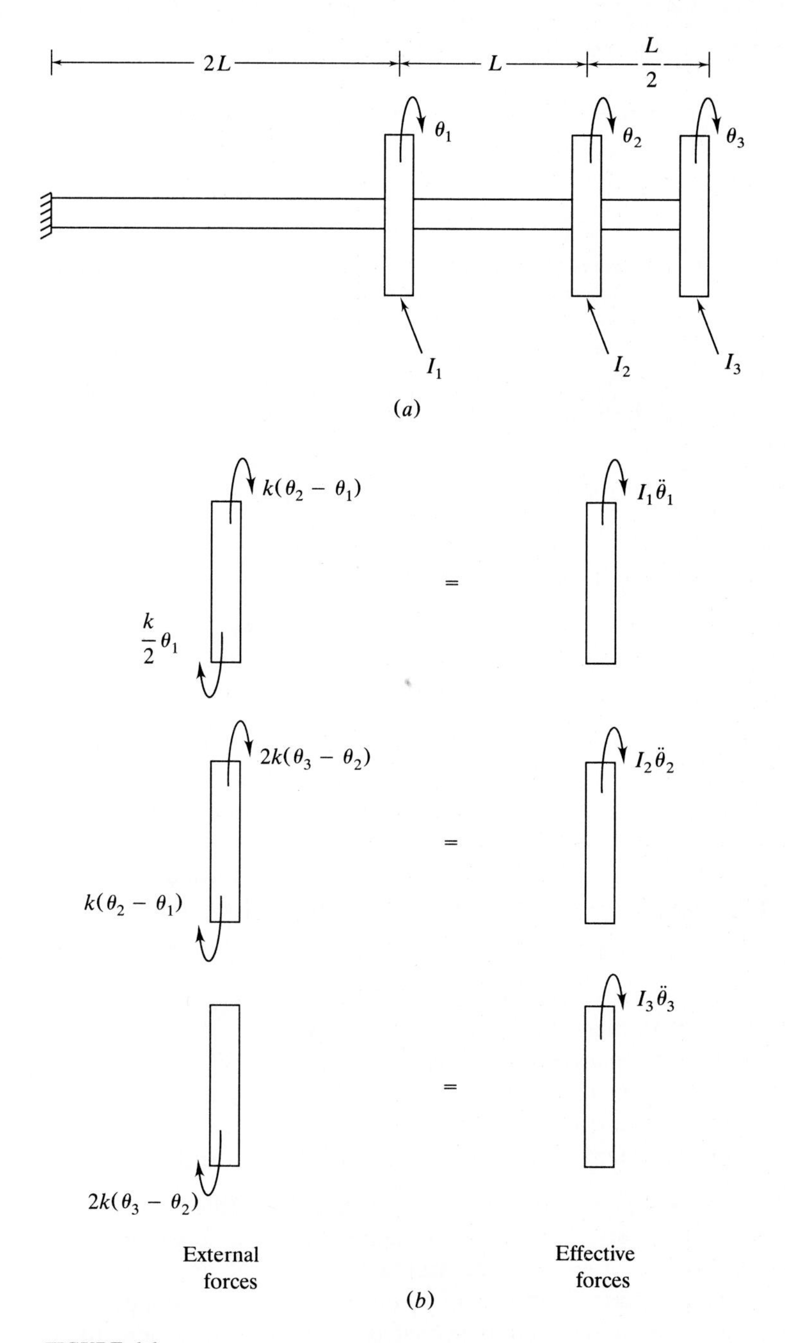

FIGURE 6.6
(*a*) System of Example 6.4; (*b*) free-body diagrams at an arbitrary instant of time.

terms. The resulting equations are written in matrix form as

$$\begin{bmatrix} 3m & m\dfrac{L}{2} \\ m\dfrac{L}{2} & m\dfrac{L^2}{3} \end{bmatrix} \begin{bmatrix} \ddot{x} \\ \ddot{\theta} \end{bmatrix} + \begin{bmatrix} k & 0 \\ 0 & mg\dfrac{L}{2} \end{bmatrix} \begin{bmatrix} x \\ \theta \end{bmatrix} = \begin{bmatrix} F(t) \\ M(t) \end{bmatrix}$$

Example 6.4. A rotor consists of three disks attached to a uniform shaft as shown in Fig. 6.6. Derive the differential equations governing the torsional vibrations using θ_1, θ_2, and θ_3 as generalized coordinates.

Each shaft is replaced by a torsional spring. Free-body diagrams of each disk are shown in Fig. 6.6*b* where $k = JG/L$. The moment equation is applied about the mass center of each disk. The resulting differential equations are written in matrix form as

$$\begin{bmatrix} I_1 & 0 & 0 \\ 0 & I_2 & 0 \\ 0 & 0 & I_3 \end{bmatrix} \begin{bmatrix} \ddot{\theta}_1 \\ \ddot{\theta}_2 \\ \ddot{\theta}_3 \end{bmatrix} + \begin{bmatrix} \dfrac{3k}{2} & -k & 0 \\ -k & 2k & -2k \\ 0 & -2k & 3k \end{bmatrix} \begin{bmatrix} \theta_1 \\ \theta_2 \\ \theta_3 \end{bmatrix} = \begin{bmatrix} 0 \\ 0 \\ 0 \end{bmatrix}$$

6.3 MATRIX FORMULATION OF DIFFERENTIAL EQUATIONS

The differential equations derived in each example in the previous section are coupled; at least one equation in each set involves more than one generalized coordinate. Each equation cannot be solved individually; the equations must be solved as a system. To facilitate the solution, linear systems of differential equations are written using matrices. The general matrix formulation of a system of differential equations for a linear n-degree-of-freedom system is

$$\mathbf{M}\ddot{\mathbf{x}} + \mathbf{C}\dot{\mathbf{x}} + \mathbf{K}\mathbf{x} = \mathbf{F} \tag{6.1}$$

where $\mathbf{M} = n \times n$ mass matrix; its elements are denoted by m_{ij}, that is, m_{ij} is the element in the ith and jth column of $\mathbf{M}$

$\mathbf{C} = n \times n$ damping matrix; its elements are denoted by c_{ij}

$\mathbf{K} = n \times n$ stiffness matrix; its elements are denoted by k_{ij}

$\mathbf{F} = n \times 1$ force vector; its elements are denoted by f_i

$\mathbf{x} = n \times 1$ displacement vector (vector of generalized coordinates)

Each row of the matrix system, Eq. (6.1), represents a differential equation. The basic laws of dynamics are applied to appropriate free-body diagrams for an n-degree-of-freedom linear system, yielding n coupled linear ordinary differential equations. The derived differential equations can be summarized in matrix form and the mass matrix, damping matrix, stiffness matrix, and force vector obtained. Energy methods can be used to calculate the elements of these matrices directly.

If a linear system's stiffness matrix is not diagonal (i.e., at least one off-diagonal term is nonzero), the system is said to be statically coupled. If a system's mass matrix is not diagonal, the system is said to be dynamically coupled. A system may be both statically and dynamically coupled. The system of Example 6.1 is statically coupled and not dynamically coupled. The system of Example 6.3 is dynamically coupled and not statically coupled. The choice of generalized coordinates affects the system's coupling. In addition, one has to be careful in deciding the nature of the coupling when Newton's laws are used to derive the differential equations as evidenced by the following example.

Example 6.5. Discuss the coupling present for the system in Fig. 6.4 of Example 6.2 using

(*a*) x and θ as generalized coordinates
(*b*) x_A, the displacement of point A, and x_B, the displacement of point B, as generalized coordinates
(*c*) x_A and θ as generalized coordinates.

(*a*) It appears when examining the differential equations derived in Example 6.2 that the system is both statically and dynamically coupled when using x and θ as generalized coordinates. Consider, though, a reformulation of the differential equations using summation of forces in the vertical direction and summation of moments about the mass center. The resulting differential equations in matrix form are

$$\begin{bmatrix} m & 0 \\ 0 & m\dfrac{L^2}{12} \end{bmatrix}\begin{bmatrix} \ddot{x} \\ \ddot{\theta} \end{bmatrix} + \begin{bmatrix} 2k & -k\dfrac{L}{4} \\ -k\dfrac{L}{4} & 5k\dfrac{L^2}{16} \end{bmatrix}\begin{bmatrix} x \\ \theta \end{bmatrix} = \begin{bmatrix} F(t) \\ \dfrac{L}{2}F(t) \end{bmatrix}$$

In this formulation the mass matrix is diagonal, suggesting that the system is not dynamically coupled. However, note that the stiffness matrix is still not diagonal but is now symmetric. The equations obtained using this specific application of Newton's laws are not independent with the equations obtained in Example 6.2. Indeed, the equations obtained in Example 6.2 can be algebraically manipulated to obtain the current equations. The differential equations for a linear system can always be algebraically manipulated such that the mass and stiffness matrices are symmetric. It is this formulation which is used to decide upon the coupling of the system for a certain set of generalized coordinates. Thus, using these coordinates, the system is statically coupled, but not dynamically coupled.

(*b*) The free-body diagrams showing the external and effective forces in terms of x_A and x_B are shown in Fig. 6.7. Summation of moments about A and B leads to the following differential equations:

$$\begin{bmatrix} \dfrac{7}{36}m & \dfrac{1}{18}m \\ \dfrac{1}{18}m & \dfrac{4}{9}m \end{bmatrix}\begin{bmatrix} \ddot{x}_A \\ \ddot{x}_B \end{bmatrix} + \begin{bmatrix} 3\dfrac{k}{4} & 0 \\ 0 & 3\dfrac{k}{4} \end{bmatrix}\begin{bmatrix} x_A \\ x_B \end{bmatrix} = \begin{bmatrix} -\dfrac{1}{4}F(t) \\ F(t) \end{bmatrix}$$

In this formulation the mass matrix is symmetric, but not diagonal and the stiffness

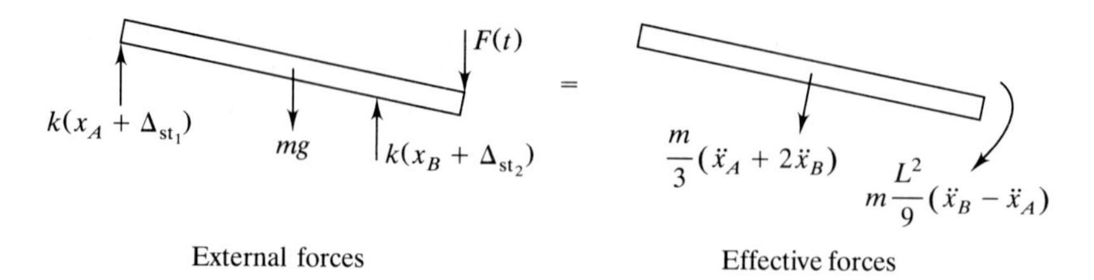

FIGURE 6.7
Free-body diagrams of system of Example 6.1 using x_A and x_B as generalized coordinates.

matrix is diagonal. Thus, using x_A and x_B as generalized coordinates, the system is dynamically coupled but not statically coupled.

(*c*) The free-body diagrams showing the effective and external forces in terms of x_A and θ are shown in Fig. 6.8. Summation of forces in the vertical direction and summation of moments about A leads to the following differential equations:

$$\begin{bmatrix} m & m\dfrac{L}{2} \\ m\dfrac{L}{2} & m\dfrac{L^2}{3} \end{bmatrix}\begin{bmatrix} \ddot{x}_A \\ \ddot{\theta} \end{bmatrix} + \begin{bmatrix} 2k & 3k\dfrac{L}{4} \\ 3k\dfrac{L}{4} & 9k\dfrac{L^2}{16} \end{bmatrix}\begin{bmatrix} x_A \\ \theta \end{bmatrix} = \begin{bmatrix} F(t) \\ LF(t) \end{bmatrix}$$

Both the mass and stiffness matrices are symmetric and nondiagonal, leading to the conclusion that the system is both statically and dynamically coupled when using x_A and θ as generalized coordinates.

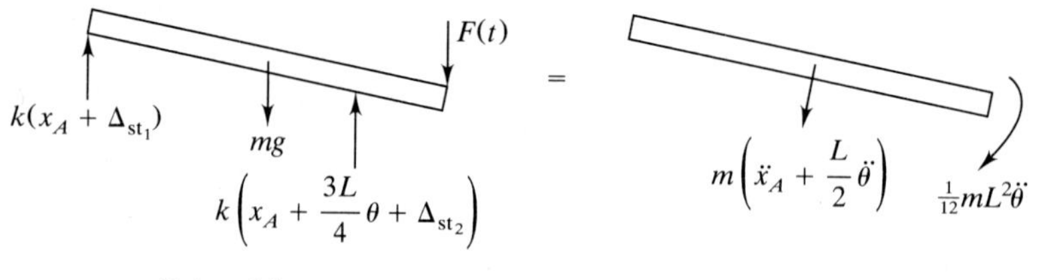

FIGURE 6.8
Free-body diagrams of system of Example 6.1 using x_A and θ as generalized coordinates.

The preceding example also points out a physical means of deciding whether a system is statically and/or dynamically coupled when using a certain set of generalized coordinates. The system is statically coupled if when a static change induced for one generalized coordinate gives rise to a static change in another generalized coordinate. The static change can be introduced by statically applying a concentrated load to the particle whose displacement specifies a generalized coordinate. The system is dynamically coupled if when instantaneous motion occurs for one generalized coordinate then instantaneous motion is induced for another generalized coordinate. Motion corresponding to a generalized coordinate can be instantaneously started by application of an impulse to the particle whose displacement defines the generalized coordinate. If a generalized coordinate represents an angular displacement, then a moment is applied to induce a static displacement and an angular impulse is applied to induce instantaneous motion.

For case (a) a static change in x gives rise to a static change in θ due to the asymmetry of the bar and vice versa. An applied impulse at the mass center of the bar induces only translational motion. Rotational motion is induced only through kinetics. Additionally, an applied angular impulse only leads to instantaneous rotational motion.

For case (b) a static change in x_A does not affect a static change in x_B and vice versa. A static change in x_A causes the bar to rotate statically about point B. However, the system is dynamically coupled using this set of coordinates as an impulse applied at A induces instantaneous motion of A and an instantaneous rotation about the mass center which also causes instantaneous motion of B.

For case (c), as pointed out previously, a static change in x_A cause rotation; hence the system is statically coupled. The system is also dynamically coupled for the same reason as case (b).

6.4 AN INTRODUCTION TO ENERGY METHODS: LAGRANGE'S EQUATIONS

Energy methods are often more useful than application of Newton's laws for deriving differential equations governing vibrations of multi-degree-of-freedom systems. A full understanding of the most convenient energy method first requires knowledge of the calculus of variations. Thus the method and examples of its application are presented without theory.

The lagrangian of a dynamic system is defined as the difference between its kinetic and potential energy at an arbitrary instant of time

$$L = T - V \tag{6.2}$$

The lagrangian is a function of the generalized coordinates and their time derivatives

$$L = L\left(x_1, x_2, \ldots, x_n, \dot{x}_1, \dot{x}_2, \ldots, \dot{x}_n\right) \tag{6.3}$$

The lagrangian is treated as a function of $2n$ variables; the time derivatives of the generalized coordinates are viewed as being independent of the generalized coordinates.

For a conservative system, the dot product of Newton's laws is taken with a virtual displacement defined by any of the generalized coordinates. This energy equation can be manipulated to yield Lagrange's equations

$$\frac{d}{dt}\left(\frac{\partial L}{\partial \dot{x}_i}\right) - \frac{\partial L}{\partial x_1} = 0 \qquad i = 1, 2, \ldots, n \tag{6.4}$$

Equation (6.4) is applied to derive n differential equations for an n-degree-of-freedom conservative system.

Example 6.6. Use Lagrange's equations to derive the differential equations for the system of Example 6.3.

The only external forces acting on the system are the spring force, gravity, and the normal force between the block and the surface on which it slides. All external forces are conservative. This Eq. (6.4) is used with x and θ as generalized coordinates. If the pin support is used as the datum for potential energy calculations, then

$$V = \frac{1}{2}kx^2 - mg\frac{L}{2}\cos\theta$$

Recall that the kinetic energy of a rigid body is

$$T = \tfrac{1}{2}m\bar{v}^2 + \tfrac{1}{2}\bar{I}\omega^2$$

where v is the velocity of the mass center of the body and ω is the angular velocity of the body. The velocity of the mass center of the bar has both a vertical component and a horizontal component. Thus

$$T = \frac{1}{2}(2m)\dot{x}^2 + \frac{1}{2}m\left[\left(\dot{x} + \frac{L}{2}\dot{\theta}\cos\theta\right)^2 + \left(\frac{L}{2}\dot{\theta}\sin\theta\right)^2\right] + \frac{1}{2}\left(\frac{1}{12}mL^2\dot{\theta}^2\right)$$

$$= \frac{1}{2}m\left(3\dot{x}^2 + \dot{x}\dot{\theta}L\cos\theta + \frac{L^2}{3}\dot{\theta}^2\right)$$

and

$$L = \frac{1}{2}m\left(3\dot{x}^2 + \dot{x}\dot{\theta}L\cos\theta + \frac{L^2}{3}\dot{\theta}^2\right) - \frac{1}{2}kx^2 + mg\frac{L}{2}\cos\theta$$

With $x_1 = x$, Eq. (6.4) becomes

$$0 = \frac{d}{dt}\left(\frac{\partial L}{\partial \dot{x}}\right) - \frac{\partial L}{\partial x}$$

which gives

$$0 = \frac{d}{dt}\left(3m\dot{x} + m\frac{L}{2}\dot{\theta}\cos\theta\right) + kx$$

$$= 3m\ddot{x} + m\frac{L}{2}\ddot{\theta}\cos\theta - m\frac{L}{2}\dot{\theta}^2\sin\theta + kx$$

Application of Eq. (6.4) with $x_2 = \theta$ gives

$$0 = \frac{d}{dt}\left(\frac{\partial L}{\partial \dot{\theta}}\right) - \frac{\partial L}{\partial \theta}$$

$$= \frac{d}{dt}\left(m\frac{L}{2}\dot{x}\cos\theta + \frac{L^2}{3}\dot{\theta}\right) + m\frac{L}{2}\dot{x}\dot{\theta}\sin\theta + mg\frac{L}{2}\sin\theta$$

$$= m\frac{L}{2}\ddot{x}\cos\theta + m\frac{L^2}{3}\ddot{\theta} + mg\frac{L}{2}\sin\theta$$

The differential equations derived using Lagrange's equations are identical to the nonlinear equations derived using Newton's laws in Example 6.3.

Lagrange's equations provide a powerful tool to use for deriving complex nonlinear differential equations. However, this text is mostly devoted to linear problems, and the power of Lagrange's equations is not needed. Lagrange's equations can be applied to linear systems to derive differential equations of the form of (6.1).

It can be shown that for an n-degree-of-freedom linear system the potential and kinetic energies must have the quadratic forms

$$V = \frac{1}{2}\sum_{i=1}^{n}\sum_{j=1}^{n} k_{ij}x_i x_j \tag{6.5}$$

$$T = \frac{1}{2}\sum_{i=1}^{n}\sum_{j=1}^{n} m_{ij}\dot{x}_i \dot{x}_j \tag{6.6}$$

The lagrangian for a linear system becomes

$$L = \frac{1}{2}\left[\sum_{i=1}^{n}\sum_{j=1}^{n}\left(m_{ij}\dot{x}_i\dot{x}_j - k_{ij}x_i x_j\right)\right]$$

Application of Eq. (6.4) yields

$$0 = \frac{d}{dt}\left(\frac{\partial L}{\partial \dot{x}_l}\right) - \frac{\partial L}{\partial x_l} \qquad l = 1, 2, \ldots, n$$

$$0 = \frac{1}{2}\sum_{i=1}^{n}\sum_{j=1}^{n}\left\{m_{ij}\frac{d}{dt}\left[\frac{\partial}{\partial \dot{x}_l}(\dot{x}_i\dot{x}_j)\right] + k_{ij}\frac{\partial}{\partial x_l}(x_i x_j)\right\}$$

$$= \frac{1}{2}\sum_{i=1}^{n}\sum_{j=1}^{n}\left\{m_{ij}\frac{d}{dt}\left[\dot{x}_i\frac{\partial \dot{x}_j}{\partial \dot{x}_l} + \dot{x}_j\frac{\partial \dot{x}_i}{\partial \dot{x}_l}\right] + k_{ij}\left(x_i\frac{\partial x_j}{\partial x_l} + x_j\frac{\partial x_i}{\partial x_l}\right)\right\} \tag{6.7}$$

Noting that

$$\frac{\partial x_i}{\partial x_l} = \delta_{il} = \begin{cases} 0 & i \neq l \\ 1 & i = l \end{cases}$$

Eq. (6.7) becomes

$$0 = \frac{1}{2} \sum_{i=1}^{n} \sum_{j=1}^{n} \left[m_{ij} \frac{d}{dt} \left(\dot{x}_i \delta_{jl} + \dot{x}_j \delta_{il} \right) + k_{ij} \left(x_i \delta_{jl} + x_j \delta_{il} \right) \right]$$

The right-hand side of the preceding equation is broken into four terms and the order of summation interchanged on the second and fourth terms. Then, due to the presence of the δ terms, the value of the term on the inner summation is nonzero only for one value of the summation index. Thus the preceding equation can be rewritten using single summations as

$$0 = \frac{1}{2} \left(\sum_{i=1}^{n} m_{il} \ddot{x}_i + \sum_{j=1}^{n} m_{lj} \ddot{x}_j + \sum_{i=1}^{n} k_{il} x_i + \sum_{j=1}^{n} k_{lj} x_j \right)$$

The name of a summation index is arbitrary. Thus these summations are combined, yielding

$$0 = \frac{1}{2} \left[\sum_{i=1}^{n} (m_{il} + m_{li}) \ddot{x}_i + \sum_{i=1}^{n} (k_{il} + k_{li}) x_i \right]$$

Note that in Eq. (6.5) k_{il} and k_{li} both multiply the product $x_i x_l$ and hence must be equal. The same reasoning using Eq. (6.6) leads to $m_{il} = m_{li}$. Thus

$$\sum_{i=1}^{n} m_{li} \ddot{x}_i + \sum_{i=1}^{n} k_{li} x_i = 0 \qquad l = 1, \ldots, n \tag{6.8}$$

Equation (6.8) represents a system of n simultaneous linear differential equations. The matrix formulation of Eq. (6.8) is

$$\mathbf{M\ddot{x}} + \mathbf{Kx} = \mathbf{0} \tag{6.9}$$

Equation (6.9) shows that application of Lagrange's equations for linear conservative systems yields the matrix formulation of Eq. (6.1). It also shows that the elements of the stiffness matrix can be determined directly from knowledge of the potential energy and elements of the mass matrix can be obtained directly from knowledge of the kinetic energy. When energy methods are used the mass and stiffness matrices are symmetric.

The use of Lagrange's equations for nonconservative systems is considered in Sec. 6.8 for calculation of the components of the force vector.

6.5 STIFFNESS INFLUENCE COEFFICIENTS

It is shown in the previous section that the elements of the stiffness matrix for a linear system can be determined directly from the expression for the potential energy in terms of the generalized coordinates. Recall that the change in potential energy between two system configurations is the work done by all conservative forces as the system moves between the two configurations. The work done by conservative forces is independent of path. Hence, to evaluate the potential energy for a specific system configuration of a conservative system, one can look at any possible means in arriving at that configuration.

Consider an n-degree-of-freedom system with generalized coordinates $x_1, x_2, \ldots, x_n$. The system is at rest in a configuration where its potential energy is zero. A system of forces $f_{11}, f_{21}, \ldots, f_{n1}$ is statically applied. The resulting displacements are $x_1 = x_1$, $x_2 = 0$, $x_3 = 0, \ldots, x_n = 0$. The force f_{j1} is applied to the particle whose displacement is given by x_j. If x_j is an angular coordinate, then f_{j1} is an applied moment. The work done by this system of forces is

$$U_{0\to 1} = \tfrac{1}{2} f_{11} x_1$$

Now add a set of forces f_{j2} such that the resulting displacements after application of these forces are $x_1 = x_1$, $x_2 = x_2$, $x_3 = 0$, $x_4 = 0, \ldots, x_n = 0$. The change in work during application of these forces is

$$U_{1\to 2} = \left(f_{21} + \tfrac{1}{2} f_{22}\right) x_2$$

This process is continued. The lth application of forces, f_{jl}, yields displacements of $x_1 = x_1$, $x_2 = x_2, \ldots, x_l = x_l$, $x_{l+1} = 0, \ldots, x_n = 0$. The change in work between the $l-1$st application of forces and the lth application of forces is

$$U_{l-1\to l} = \left(\sum_{i=1}^{l-1} f_{li} + \frac{1}{2} f_{ll}\right) x_l \tag{6.10}$$

After n applications of sets of forces, the displacements of the particles whose displacements define the generalized coordinates are as desired.

Define the set $\{k_{1j}, k_{2j}, \ldots, k_{nj}\}$ as the set of forces required during the jth application to cause a unit displacement for x_j. Since the system is linear, a change in displacement causes a proportional change in forces. Thus

$$f_{ij} = k_{ij} x_j \tag{6.11}$$

Note that the order in which the displacements are prescribed does not matter. Suppose x_2 is prescribed then x_1 is prescribed. The resulting configuration is the same as if x_1 is prescribed then x_2 prescribed. Since all forces are conservative the work done by these forces must be the same in both cases. This implies

$$\tfrac{1}{2} f_{11} x_1 + \left(f_{21} + \tfrac{1}{2} f_{22}\right) x_2 = \tfrac{1}{2} f_{22} x_2 + \left(f_{12} + \tfrac{1}{2} f_{11}\right) x_1$$

which yields

$$f_{21} x_2 = f_{12} x_1$$

Using Eq. (6.11) this gives

$$k_{21} = k_{12}$$

A more general argument is used to show

$$k_{ij} = k_{ji} \qquad i, j = 1, \ldots, n \tag{6.12}$$

Equation (6.12) is called the reciprocity relation.

Substitution of Eq. (6.11) into Eq. (6.10) and summation over l yields the total work done by all conservative forces between the configuration where the system has zero potential energy and an arbitrary system configuration. Reciprocity is used to write the work as

$$U_{\text{total}} = \frac{1}{2} \sum_{i=1}^{n} \sum_{j=1}^{n} k_{ij} x_i x_j$$

Since these forces are applied statically, the potential energy in the final configuration is the total work done by all external forces:

$$V = \frac{1}{2} \sum_{i=1}^{n} \sum_{j=1}^{n} k_{ij} x_i x_j \tag{6.13}$$

which is the same as Eq. (6.5). Thus physical meaning can be assigned to the elements of the stiffness matrix. The column of stiffness matrix elements $\{k_{1j}, k_{2j}, \ldots, k_{nj}\}$ is the set of forces acting on the particles whose displacements are given by x_i, $i = 1, \ldots, n$, to maintain an equilibrium situation with $x_j = 1$ and $x_i = 0$ for $i \neq j$. These coefficients are called stiffness influence coefficients.

Stiffness influence coefficients provide a convenient means to calculate the stiffness matrix for multi-degree-of-freedom systems. Their evaluation simply requires application of the principles of statics. There is no need to evaluate the potential energy function or to apply Newton's laws. Influence coefficients are very powerful for linear systems. The method does not apply for nonlinear systems due to the linearity assumption made in Eq. (6.5).

In summary, the influence coefficient method to find the elements of the stiffness matrix for an n-degree-of-freedom system is as follows:

1. Assign a unit displacement for x_1, maintaining $x_2, x_3, \ldots, x_n$ in their static-equilibrium position. Calculate the system of forces required to maintain this as an equilibrium position. The forces, k_{i1}, are applied at the locations whose displacements define the generalized coordinates. This set of forces yields the first column of the stiffness matrix.
2. Continue this procedure to find all columns of the stiffness matrix. The jth column is found by prescribing $x_j = 1$ and $x_i = 0$, $i \neq j$, and calculating the system of forces necessary to maintain this as an equilibrium position.
3. If x_j is an angular coordinate, then k_{ji} is an applied moment. When calculating the jth column of the stiffness matrix, a unit rotation in radians must be applied to the angle defined by x_j. If the small-angle assumption is necessary to achieve a linear system, it is also used to calculate the stiffness influence coefficients.
4. Reciprocity implies the stiffness matrix must be symmetric: $k_{ij} = k_{ji}$. The symmetry can be used as a check.

Example 6.7. Use the stiffness influence coefficient method to calculate the stiffness matrix for the system of Fig. 6.3 used in Example 6.1.

The first column of the stiffness matrix is obtained by setting $x_1 = 1$, $x_2 = 0$, $x_3 = 0$ and calculating the system of applied forces necessary to maintain this position in equilibrium. Free-body diagrams of the blocks are shown in Fig. 6.9. Setting $\Sigma F = 0$ yields

$$\text{Block } a: \qquad -k - 2k + k_{11} = 0 \Rightarrow k_{11} = 3k$$

$$\text{Block } b: \qquad k + k_{21} = 0 \Rightarrow k_{21} = -2k$$

$$\text{Block } c: \qquad \Rightarrow k_{31} = 0$$

The second column is obtained by setting $x_1 = 0$, $x_2 = 1$, $x_3 = 0$. Summing forces on the free-body diagrams yields

$$\text{Block } a: \qquad 2k + k_{12} = 0 \Rightarrow k_{12} = -2k$$

$$\text{Block } b: \qquad -2k - k + k_{22} = 0 \Rightarrow k_{22} = 3k$$

$$\text{Block } c: \qquad k + k_{32} = 0 \Rightarrow k_{32} = -k$$

The third column is obtained by setting $x_1 = 0$, $x_2 = 0$, $x_3 = 1$. Summing forces

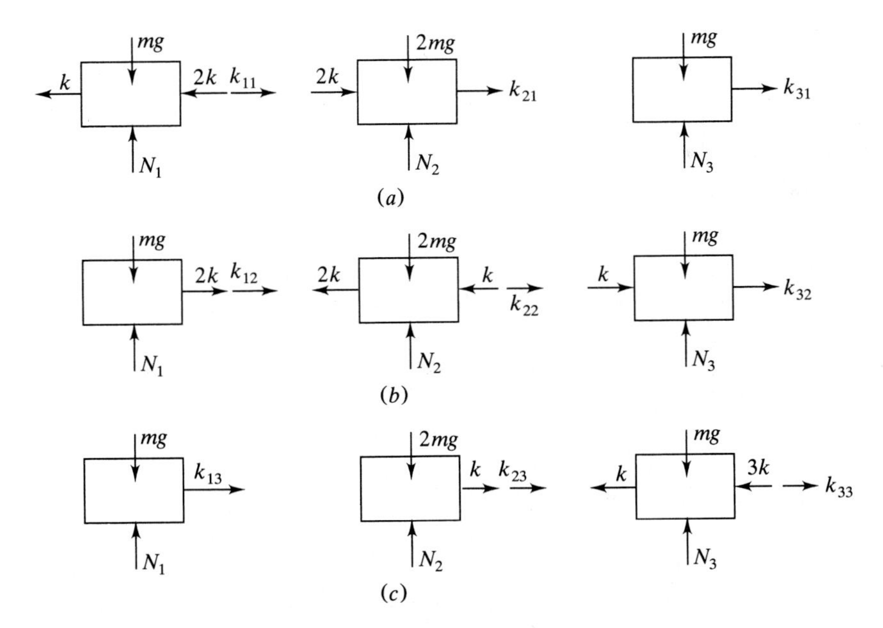

FIGURE 6.9
(*a*) First column of stiffness matrix is calculated by setting $x_1 = 1$, $x_2 = 0$, $x_3 = 0$; (*b*) second column of stiffness matrix is calculated by setting $x_1 = 0$, $x_2 = 1$, $x_3 = 0$; (*c*) third column of stiffness matrix is calculated by setting $x_1 = 0$, $x_2 = 0$, $x_3 = 1$.

on the free-body diagrams yields

$$\text{Block } a: \qquad \Rightarrow k_{13} = 0$$

$$\text{Block } b: \qquad k + k_{23} = 0 \Rightarrow k_{23} = -k$$

$$\text{Block } c: \qquad -k - 3k + k_{33} = 0 \Rightarrow k_{33} = 4k$$

The stiffness matrix is

$$\mathbf{K} = \begin{bmatrix} 3k & -2k & 0 \\ -2k & 3k & -k \\ 0 & -k & 4k \end{bmatrix}$$

Example 6.8. Use the stiffness influence coefficient method to find the stiffness matrix for the system in Fig. 6.10. Use x_A, the downward displacement of block A, x_B, the upward displacement of block B, and θ, the counterclockwise angular rotation of the pulley, as generalized coordinates.

The first column of the stiffness matrix is obtained by setting $x_A = 1$, $x_B = 0$, and $\theta = 0$ and finding the resulting system of forces and moments to maintain this as an equilibrium position. Note that since θ is an angular coordinate k_{31} is a moment. Gravity cancels with static-deflection terms in all equations; thus neither is included.

$$\text{Block } A: \qquad \sum F = 0 \Rightarrow -k + k_{11} = 0 \Rightarrow k_{11} = k$$

$$\text{Block } B: \qquad \sum F = 0 \Rightarrow k_{21} = 0$$

$$\text{Pulley:} \qquad \sum M_O = 0 \Rightarrow k(r) + k_{31} = 0 \Rightarrow k_{31} = -kr$$

The second column is obtained by setting $x_A = 0$, $x_B = 1$, $\theta = 0$. The equations of equilibrium yield

$$\text{Block } A: \qquad \sum F = 0 \Rightarrow k_{12} = 0$$

$$\text{Block } B: \qquad \sum F = 0 \Rightarrow 3k - k_{22} = 0 \Rightarrow k_{22} = 3k$$

$$\text{Pulley:} \qquad \sum M_O = 0 \Rightarrow 3k(2r) + k_{32} = 0 \Rightarrow k_{32} = -6kr$$

The third column is obtained by setting $x_A = 0$, $x_B = 0$, $\theta = 1$. The equations of equilibrium yield

$$\text{Block } A: \qquad \sum F = 0 \Rightarrow kr + k_{13} = 0 \Rightarrow k_{13} = -kr$$

$$\text{Block } B: \qquad \sum F = 0 \Rightarrow 3k(2r) + k_{23} = 0 \Rightarrow k_{23} = -6kr$$

$$\text{Pulley:} \qquad \sum M_O = 0 \Rightarrow -k(r)(r) - 3k(2r)(2r) + k_{33} = 0 \Rightarrow k_{33} = 13kr^2$$

Thus the stiffness matrix for this choice of generalized coordinates is

$$\mathbf{K} = \begin{bmatrix} k & 0 & -kr \\ 0 & 3k & -6kr \\ -kr & -6kr & 13kr^2 \end{bmatrix}$$

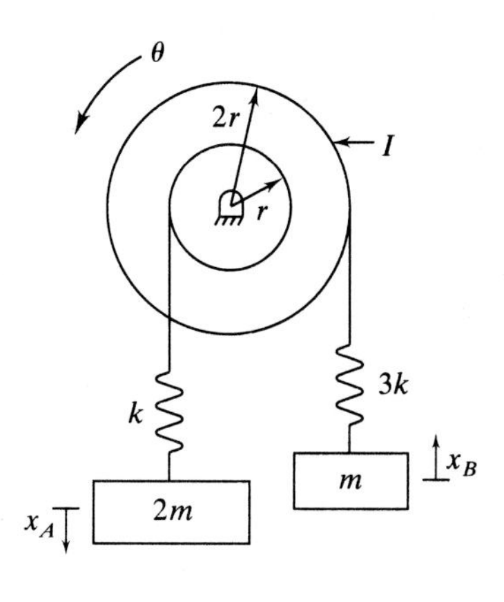

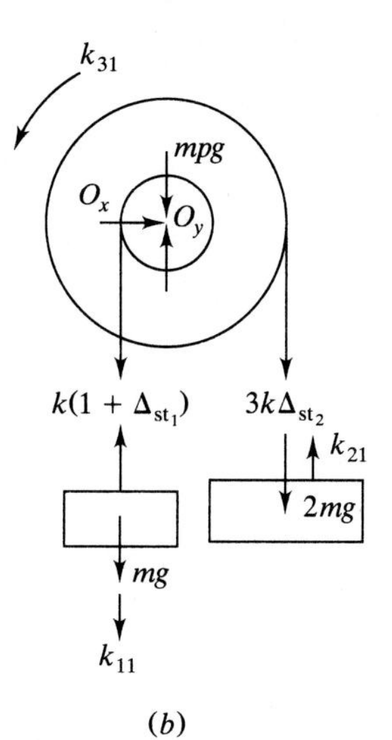

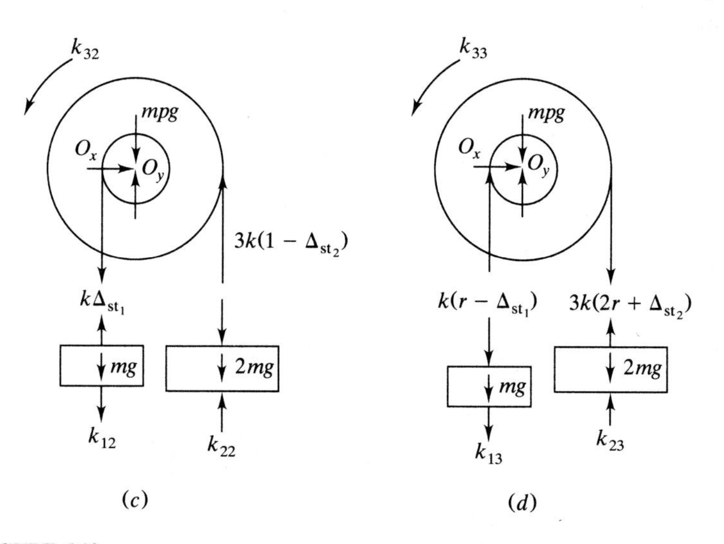

FIGURE 6.10
(*a*) System of Example 6.8; (*b*) first column of stiffness matrix is calculated by setting $x_A = 1$, $x_B = 0$, $\theta = 0$; (*c*) second column of stiffness matrix is calculated by setting $x_A = 0$, $x_B = 1$, $\theta = 0$; (*d*) third column of stiffness matrix is calculated by setting $x_A = 0$, $x_B = 0$, $\theta = 1$.

Example 6.9. Use the influence coefficient method to find the stiffness matrix for the system of Fig. 6.11 using θ_1, the clockwise angular displacement of bar *AB*, and θ_2, the counterclockwise angular displacement of bar *CD*, as generalized coordinates.

The first column of the stiffness matrix is obtained by setting $\theta_1 = 1$ and $\theta_2 = 0$ and finding the moments that must be applied to the bars to maintain this

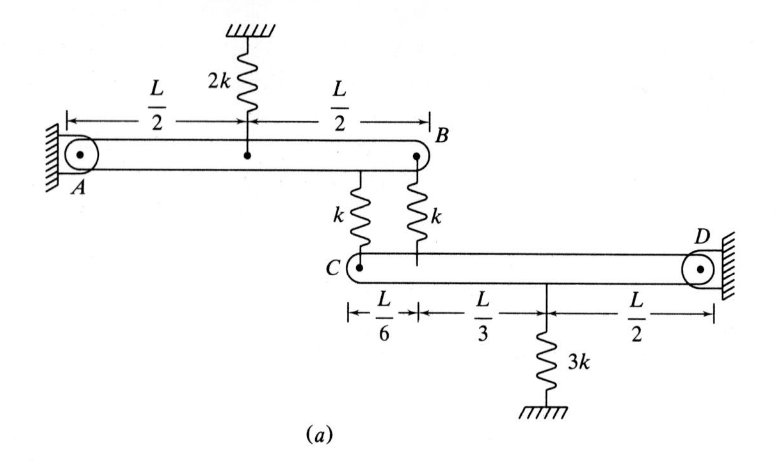

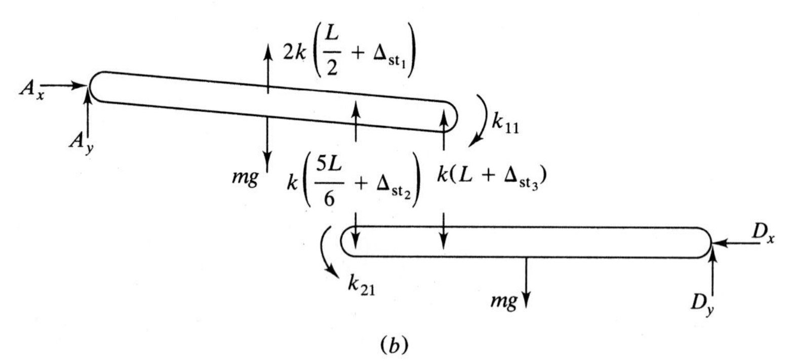

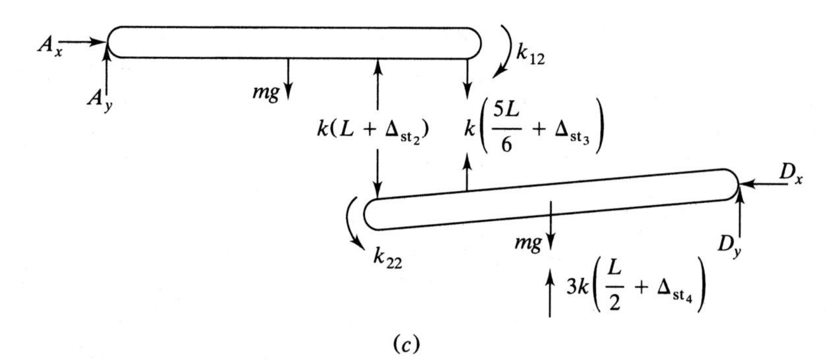

FIGURE 6.11
(a) System of Example 6.9; (b) first column of stiffness matrix is calculated by setting $\theta_1 = 1, \theta_2 = 0$; ($c$) second column of stiffness matrix is calculated by setting $\theta_1 = 0, \theta_2 = 1$.

as an equilibrium position. The small-angle assumption is used. Equilibrium equations are applied to the free-body diagrams of Fig. 6.11*b*. Gravity cancels with the static deflections in the springs.

$$\sum \overset{+}{\circlearrowleft} M_A = 0 = -2k\frac{L}{2}\left(\frac{L}{2}\right) - 5k\frac{L}{6}\left(5\frac{L}{6}\right) - kL(L) + k_{11} \Rightarrow k_{11} = \frac{79}{36}kL^2$$

$$\sum \overset{+}{\circlearrowright} M_D = 0 = 5k\frac{L}{6}(L) + kL\left(5\frac{L}{6}\right) + k_{21} \Rightarrow k_{21} = -5k\frac{L^2}{3}$$

The second column is obtained by setting $\theta_1 = 0$ and $\theta_2 = 1$. The equilibrium equations are applied to the free-body diagrams to yield

$$\sum \overset{+}{\circlearrowleft} M_A = 0 = kL\left(5\frac{L}{6}\right) + 5k\frac{L}{6}(L) + k_{12} \Rightarrow k_{12} = -5k\frac{L^2}{3}$$

$$\sum \overset{+}{\circlearrowright} M_D = 0 = -kL(L) - 5k\frac{L}{6}\left(5\frac{L}{6}\right) - 3k\frac{L}{2}\left(\frac{L}{2}\right) + k_{22} \Rightarrow k_{22} = 22k\frac{L^2}{9}$$

The stiffness matrix is

$$\mathbf{K} = \begin{bmatrix} \frac{79}{36}kL^2 & -5k\frac{L^2}{3} \\ -5k\frac{L^2}{3} & \frac{22}{9}kL^2 \end{bmatrix}$$

Example 6.10. The transverse vibrations of the cantilever beam of Fig. 6.12 are to be approximated by modeling the beam as a two-degree-of-freedom system. The inertia of the beam is modeled by placing discrete masses at the beam's midspan and end. Calculate the stiffness matrix for this two-degree-of-freedom model using the displacements of the midspan and end of the beam as generalized coordinates.

Calculation of the stiffness matrix requires the evaluation of the deflection of the beam due to a concentrated load at the midspan and a concentrated load at the end of the beam. Perhaps the best way of handling the beam deflection problem is to use the method of superposition as shown in Fig. 6.12*b*. The elements of the ith column of the stiffness matrix are calculated from

$$y\left(\frac{L}{2}\right) = k_{1i}y_1\left(\frac{L}{2}\right) + k_{2i}y_2\left(\frac{L}{2}\right)$$

$$y(L) = k_{1i}y_1(L) + k_{2i}y_2(L)$$

where $y(z)$ is the total deflected shape of the beam, $y_1(z)$ is the deflected shape of the beam due to a concentrated unit load at the midspan, and $y_2(z)$ is the deflected shape of the beam due to a concentrated unit load at the end of the

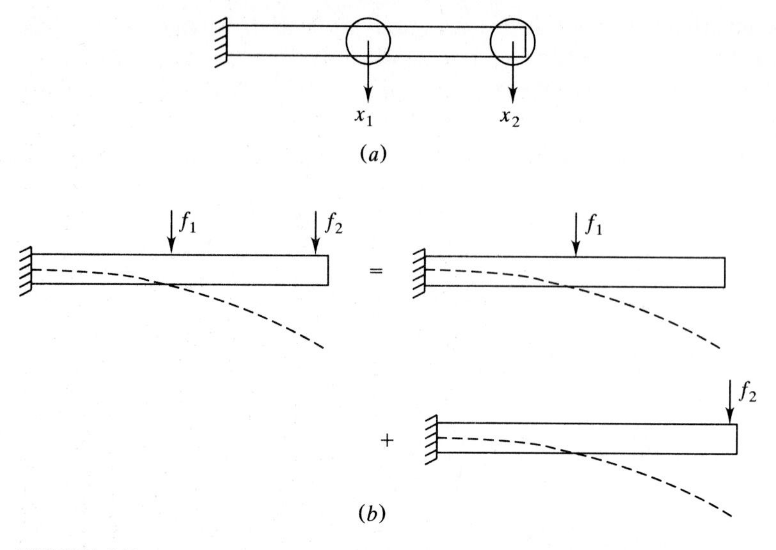

FIGURE 6.12
(*a*) Two-degree-of-freedom model of cantilever beam of Example 6.10; (*b*) superposition method used to calculate stiffness matrix.

beam. Using Table D2 these are evaluated as

$$y_1\left(\frac{L}{2}\right) = \frac{L^3}{24EI} \qquad y_2\left(\frac{L}{2}\right) = \frac{5L^3}{48EI}$$

$$y_1(L) = \frac{5L^3}{48EI} \qquad y_2(L) = \frac{L^3}{3EI}$$

To determine the first column, set $y(L/2) = 1$ and $y(L) = 0$. The equations are solved simultaneously, yielding

$$k_{11} = \frac{768EI}{7L^3} \qquad k_{21} = -\frac{240EI}{7L^3}$$

To determine the second column, set $y(L/2) = 0$ and $y(L) = 1$. The equations are solved simultaneously, yielding

$$k_{12} = -\frac{240EI}{7L^3} \qquad k_{22} = \frac{96EI}{7L^3}$$

6.6 FLEXIBILITY INFLUENCE COEFFICIENTS

Development of the stiffness matrix using stiffness influence coefficients is straightforward. For mechanical systems the calculation of stiffness influence coefficients requires the application of the principles of statics and little algebra. However, as shown in Example 6.10, the calculation of a column of stiffness influence coefficients for a structural system modeled with n degrees of freedom requires the solution of n simultaneous equations. This leads to significant computation time for systems with many degrees of freedom. Flexibility influence coefficients provide a convenient alternative. They are easier to calculate than stiffness influence coefficients for structural systems and their knowledge is sufficient for solution of the free-vibration problem.

If the stiffness matrix, **K**, is nonsingular, then its inverse exists. The flexibility matrix, **A**, is defined by

$$\mathbf{A} = \mathbf{K}^{-1} \tag{6.14}$$

Premultiplying Eq. (6.1) by **A** gives

$$\mathbf{A}\mathbf{M}\ddot{\mathbf{x}} + \mathbf{A}\mathbf{C}\dot{\mathbf{x}} + \mathbf{x} = \mathbf{A}\mathbf{F} \tag{6.15}$$

Equation (6.15) shows that knowledge of **A** instead of **K** is sufficient for solution of the vibration problem.

The elements of **K** are determined using stiffness influence coefficients. Analogously, flexibility influence coefficients can be used to determine **A**. The flexibility influence coefficient a_{ij} is defined as the displacement of the particle whose displacement is represented by x_i when a unit load is applied to the particle whose displacement is represented by x_j and no other loading is applied to the system. If x_j represents an angular coordinate, then a unit moment is applied.

Suppose an arbitrary set of concentrated loads $\{f_1, f_2, \ldots, f_n\}$ is applied statically to an n-degree-of-freedom system. The load f_i is applied to the particle whose displacement is represented by x_i. Using the definition of flexibility influence coefficients, x_j is calculated from

$$x_j = \sum_{i=1}^{n} a_{ji} f_i \tag{6.16}$$

Equation (6.16) is summarized in matrix form as

$$\mathbf{x} = \mathbf{A}\mathbf{f} \tag{6.17}$$

Inverting (6.17) yields

$$\mathbf{f} = \mathbf{A}^{-1}\mathbf{x} = \mathbf{K}\mathbf{x} \tag{6.18}$$

which defines the static relationship between force and displacement. Equation (6.18) shows that the flexibility influence coefficients as defined are the elements of the inverse of the stiffness matrix, defined as the flexibility matrix.

The procedure for determining the flexibility matrix using influence coefficients is as follows:

1. Apply a unit load at the location whose displacement is defined by x_1. The flexibility influence coefficient in the first column, a_{i1}, is the resulting displacement of the particle whose displacement is x_i.

2. Successively apply individual unit loads to particles whose displacements define the remaining generalized coordinates. Calculate columns of flexibility influence coefficients using the principles of statics.
3. If x_l is an angular displacement, then a unit moment is applied to calculate a_{jl}, $j = 1, \ldots, n$. The displacements calculated for a_{li}, $i = 1, \ldots, n$, are angular displacements.
4. Since the stiffness matrix is symmetric, the flexibility matrix must also be symmetric. This condition serves as a check on the analysis.

Example 6.11. Determine the flexibility matrix for the system in Fig. 6.11 of Example 6.9 using flexibility influence coefficients.

The free-body diagrams of Fig. 6.13 show the external forces, in terms of angular displacements, acting on each bar when an arbitrary set of moments is applied. The equations of equilibrium are used to derive equations relating the displacements to the applied forces

$$\text{Bar } AB: \quad \overset{+}{\circlearrowleft} \sum M_A = 0 \Rightarrow m_1 = \frac{79kL^2}{36}\theta_1 - \frac{5kL^2}{3}\theta_2$$

$$\text{Bar } BC: \quad \overset{+}{\circlearrowright} \sum M_D = 0 \Rightarrow m_2 = -\frac{5kL^2}{3}\theta_1 + \frac{22kL^2}{9}\theta_2$$

The first column of the flexibility matrix is obtained by setting $m_1 = 1$, $m_2 = 0$, $\theta_1 = a_{11}$, $\theta_2 = a_{21}$, and solving the resulting equations simultaneously. The second column is obtained by setting $m_1 = 0$, $m_2 = 1$, $\theta_1 = a_{12}$, $\theta_2 = a_{22}$, and solving the resulting simultaneous equations. The flexibility matrix is

$$\mathbf{A} = \begin{bmatrix} \dfrac{396}{419kL^2} & \dfrac{270}{419kL^2} \\ \dfrac{270}{419kL^2} & \dfrac{711}{838kL^2} \end{bmatrix}$$

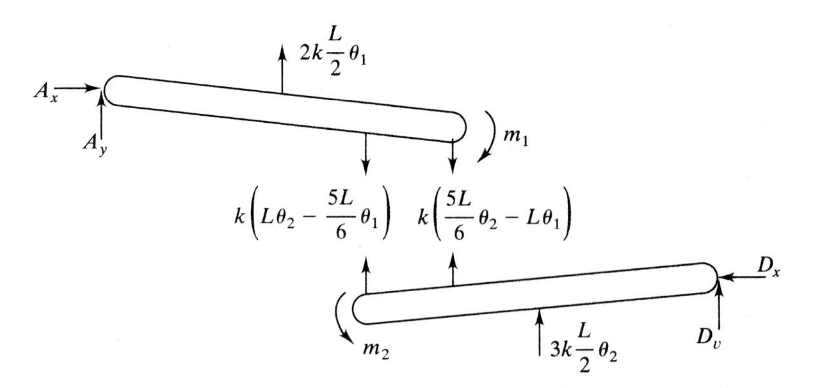

FIGURE 6.13
Free-body diagram of a static position used to calculate flexibility influence coefficients for system of Examples 6.9 and 6.11. For the first column θ_1 and θ_2 are determined by setting $m_1 = 1$, $m_2 = 0$. For the second column θ_1 and θ_2 are determined by setting $m_1 = 0$, $m_2 = 1$.

Example 6.12. Use the flexibility influence coefficient method to find the flexibility matrix for the cantilever beam of Fig. 6.12 and Example 6.10.

The flexibility influence coefficients for this example are determined by evaluating the equations for deflection of a cantilever beam due to concentrated loads. Table D.2 is used to evaluate the deflection due to a concentrated load. For example, a_{12} is the midspan deflection midspan due to an end concentrated load. The flexibility matrix is

$$\mathbf{A} = \begin{bmatrix} \dfrac{L^3}{24EI} & \dfrac{5L^3}{48EI} \\ \dfrac{5L^3}{48EI} & \dfrac{L^3}{3EI} \end{bmatrix}$$

Note that the values of the flexibility influence coefficients were actually calculated during the solution for the stiffness matrix in Example 6.10.

Example 6.13. Two small machines are to be bolted to an overhanging beam as shown in Fig. 6.14. The beam is nonuniform; thus prediction of influence coefficients from strength-of-materials concepts is difficult. Instead, the project engineer performs static measurements. After the first machine is installed, the engineer notes that the deflection directly below the machine is 10 mm and the deflection of the end of the beam is 2 mm. After the second machine is also installed, the deflection of the end of the beam is 0.8 mm.

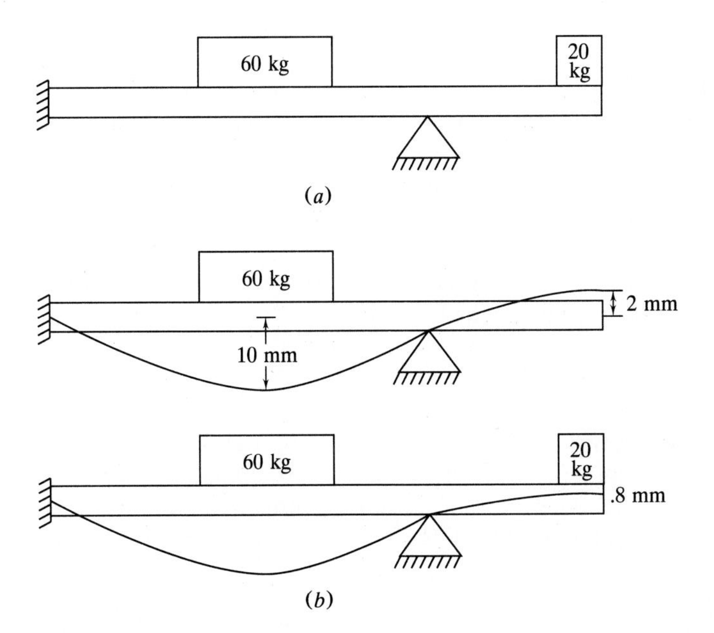

FIGURE 6.14
(*a*) System of Example 6.13; (*b*) as each machine is bolted to the continuous beam, static deflection measurements are made.

(*a*) What is the deflection at the location where the first machine is installed after the second machine is installed?
(*b*) What is the flexibility matrix for this system?

(*a*) Assuming a linear system, the principle of superposition yields the following relationships between the static loads, the influence coefficients, and the deflections:

$$x_1 = a_{11}f_1 + a_{12}f_2$$

$$x_2 = a_{21}f_1 + a_{22}f_2$$

When only the first machine is installed, $f_1 = (60 \text{ kg})(9.81 \text{ m/s}^2) = 588.6$ N, $f_2 = 0$, $x_1 = 0.01$ m, $x_2 = -0.002$ m. Substitution into the preceding equations yields $a_{11} = 1.7 \times 10^{-5}$ m/N, $a_{21} = -3.4 \times 10^{-6}$ m/N. When the second machine is also installed, $f_1 = 588.6$ N, $f_2 = (20 \text{ kg})(9.81 \text{ m/s}^2) = 196.2$ N, and $x_2 = -0.0008$ m. Then the displacement at the location of the first machine when both machines are installed is

$$x_1 = (1.7 \times 10^{-5} \text{ m/N})(588.6 \text{ N}) + (-3.4 \times 10^{-6} \text{ m/N})(196.2 \text{ N}) = 9.3 \text{ mm}$$

(*b*) Noting that $a_{21} = a_{12}$, the second of the preceding equations yields

$$a_{22} = \frac{x_2 - a_{21}f_1}{f_2}$$

$$= \frac{[-0.0008 \text{ m} - (-3.4 \times 10^{-6} \text{ m/N})(588.6 \text{ N})]}{196.2 \text{ N}}$$

$$= 6.1 \times 10^{-6} \frac{\text{m}}{\text{N}}$$

Systems exist in which the stiffness matrix is singular and hence the flexibility matrix does not exist. These systems are called semidefinite or unconstrained. It is shown in Chap. 7 that these systems have a natural frequency of zero and a corresponding mode where the system moves as a rigid body.

The system of Fig. 6.15 has two degrees of freedom and is unconstrained. The stiffness matrix for this system is calculated as

$$\mathbf{K} = \begin{bmatrix} k & -k \\ -k & k \end{bmatrix}$$

The second row of the stiffness matrix is a multiple of the first row which implies that the matrix is singular and a flexibility matrix for this system does not exist. Indeed, when the definition of flexibility influence coefficients is applied in an attempt to calculate the flexibility matrix, as shown in Fig. 6.16, no

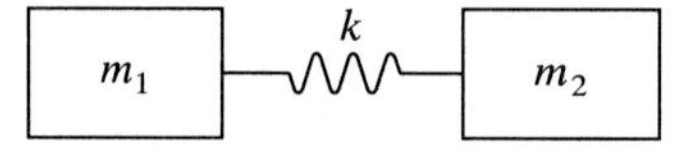

FIGURE 6.15
Unconstrained two-degree-of-freedom system.

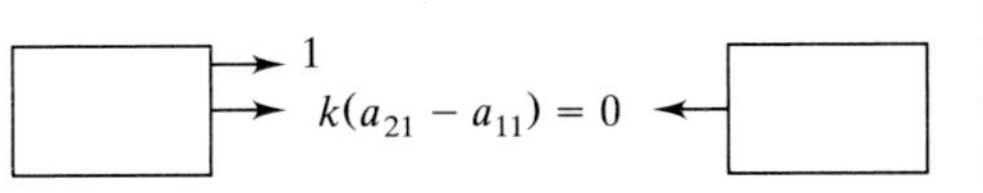

FIGURE 6.16
Free-body diagrams of system of Fig. 6.15 are used to show that the flexibility matrix does not exist.

solution is found. Since the system is unconstrained, when a unit force is applied to either mass, the system cannot remain in equilibrium. Instead, the system will behave as a rigid body with uniform acceleration.

Another example of an unconstrained system is the system of Fig. 6.10 used in Example 6.8. The stiffness matrix for this example is repeated here

$$\mathbf{K} = \begin{bmatrix} k & 0 & -kr \\ 0 & 3k & -6kr \\ -kr & -6kr & 13kr^2 \end{bmatrix}$$

Inspection of this matrix reveals that the first row plus two times the second row is proportional to the third row. Thus the three rows of the stiffness matrix are dependent, which implies that the stiffness matrix is singular, which, in turn, implies that the flexibility matrix does not exist. If, for example, a unit moment were applied to the pulley, then there are no other external forces which develop a moment about the center of the pulley. Hence equilibrium cannot be maintained. If the inertia of the pulley were neglected, then the system only has two degrees of freedom. Knowledge of x_A and x_B would be sufficient to calculate, from kinematics, the displacement of any particle in the system whose inertia is included.

Flexibility influence coefficients can be used to calculate the flexibility matrix. Equation (6.15) shows that knowledge of the flexibility matrix instead of knowledge of the stiffness matrix is sufficient to proceed with solution of the system of differential equations governing the vibrations of a multi-degree-of-freedom system. The choice of whether to determine the stiffness matrix or the flexibility matrix is usually easy.

For structural systems calculation of the flexibility matrix is easier than calculation of the stiffness matrix. For these systems deflection equations from mechanics of solids are used to determine the deflection of a particle due to an applied concentrated load. The deflection equation for the structure is often available in a textbook or handbook (e.g., App. D). Thus calculation of the flexibility matrix is direct, whereas the solution of a system of simultaneous equations is necessary to determine each column of the stiffness matrix. However, calculation of the stiffness matrix is easier than calculation of the flexibility matrix for mechanical systems which are comprised of rigid bodies connected by flexible elements. For these systems application of the equations of static equilibrium to appropriate free-body diagrams is sufficient to calculate the stiffness matrix, while calculation of a column of the flexibility matrix also requires the solution of a system of simultaneous equations.

The stiffness matrix must be calculated for unconstrained systems.

6.7 INERTIA INFLUENCE COEFFICIENTS

In Sec. 6.4 it is shown how an energy formulation is used to derive the differential equations for multi-degree-of-freedom systems of the form of Eq. (6.9). In Sec. 6.5 it is shown how the potential energy and stiffness influence coefficients are used to determine the elements of the stiffness matrix through stiffness influence coefficients. In this section it is shown how the kinetic energy is used to determine the elements of the mass matrix.

Recall the general form of the kinetic energy for a linear n-degree-of-freedom system

$$T = \frac{1}{2} \sum_{i=1}^{n} \sum_{j=1}^{n} m_{ij} \dot{x}_i \dot{x}_j \tag{6.6}$$

where x_i, $i = 1, 2, \ldots, n$, are the generalized coordinates, and m_{ij} are the elements of the $n \times n$ mass matrix. A simple method of calculating the elements of the mass matrix is to develop an expression for the kinetic energy in terms of the time derivatives of the generalized coordinates. The comparison of this expression to Eq. (6.6) yields the elements of the mass matrix. Care must be taken in determining the off-diagonal terms. The mass matrix is symmetric. For any i and j such that $i \neq j$, Eq. (6.6) contains two equal terms: $m_{ij} x_i x_j$ and $m_{ji} x_j x_i$. Thus, when comparing the expression for the kinetic energy of a given system to Eq. (6.6), it must be realized that the term multiplying $x_i x_j$ is actually $2m_{ij}$.

Example 6.14. Develop an expression for the kinetic energy of the two-degree-of-freedom system of Fig. 6.4 used in Examples 6.2 and 6.5. Use x_A and x_B as generalized coordinates. Develop the mass matrix from the kinetic energy.

In terms of these coordinates the velocity of the mass center is

$$\dot{\bar{x}} = \tfrac{1}{3}(\dot{x}_A + 2\dot{x}_B)$$

Assuming small displacements, the angular velocity of the bar is

$$\dot{\theta} = \frac{4}{3L}(\dot{x}_B - \dot{x}_A)$$

The kinetic energy is

$$\begin{aligned} T &= \frac{1}{2} m \left[\frac{1}{3}(\dot{x}_A + 2\dot{x}_B)\right]^2 + \frac{1}{2}\left(\frac{1}{12} m L^2\right)\left[\frac{4}{3L}(\dot{x}_B - \dot{x}_A)\right]^2 \\ &= \frac{1}{2} m \left(\frac{7}{27}\dot{x}_A^2 + \frac{4}{27}\dot{x}_A \dot{x}_B + \frac{16}{27}\dot{x}_B^2\right) \end{aligned}$$

The mass matrix is determined by comparing the preceding expression energy to the general form of the kinetic energy for a linear two-degree-of-freedom system

$$T = \tfrac{1}{2}\left(m_{11}\dot{x}_A^2 + 2m_{12}\dot{x}_A\dot{x}_B + m_{22}\dot{x}_B^2\right)$$

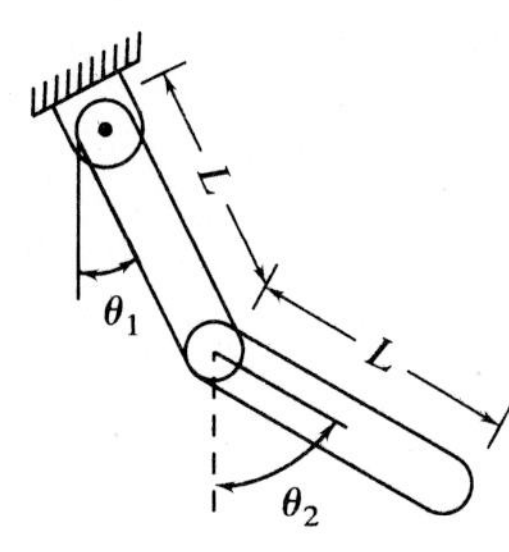

FIGURE 6.17
System of Example 6.15.

The mass matrix is

$$\mathbf{M} = \begin{bmatrix} \dfrac{7m}{27} & \dfrac{2m}{27} \\ \dfrac{2m}{27} & \dfrac{16m}{27} \end{bmatrix}$$

Note that the mass matrix obtained in this example is 4/3 times the mass matrix obtained in Example 6.5(*b*).

Example 6.15. Determine the mass matrix for the system of Fig. 6.17 when θ_1 and θ_2 are used as generalized coordinates. Use the small-angle assumption.

The kinetic energy of bar AB is

$$T_{AB} = \frac{1}{2}m\left(\frac{L}{2}\dot{\theta}_1\right)^2 + \frac{1}{2}\frac{1}{12}mL^2\dot{\theta}_1^2$$

$$= \frac{1}{2}\left(\frac{1}{3}mL^2\right)\dot{\theta}_1^2$$

The velocity of the mass center of bar BC is

$$\bar{\mathbf{v}}_{BC} = \left(L\dot{\theta}_1 \cos\theta_1 + \frac{L}{2}\dot{\theta}_2 \cos\theta_2\right)\mathbf{i} + \left(L\dot{\theta}_1 \sin\theta_1 + \frac{L}{2}\dot{\theta}_2 \sin\theta_2\right)\mathbf{j}$$

After imposing the small-angle approximation, this becomes

$$\bar{\mathbf{v}}_{BC} = \left(L\dot{\theta}_1 + \frac{L}{2}\dot{\theta}_2\right)\mathbf{i}$$

The total kinetic energy of the system is

$$T = \frac{1}{2}\left[\frac{1}{3}mL^2\dot{\theta}_1^2 + \frac{1}{12}mL^2\dot{\theta}_2^2 + m\left(L\dot{\theta}_1 + \frac{L}{2}\dot{\theta}_2\right)^2\right]$$

$$= \frac{1}{2}\left(\frac{4}{3}mL^2\dot{\theta}_1^2 + mL^2\dot{\theta}_1\dot{\theta}_2 + \frac{1}{3}mL^2\dot{\theta}_2^2\right)$$

Comparing the kinetic energy expression to the general form of the kinetic energy yields the mass matrix

$$\mathbf{M} = \begin{bmatrix} \frac{4mL^2}{3} & \frac{mL^2}{2} \\ \frac{mL^2}{2} & \frac{mL^2}{3} \end{bmatrix}$$

The mass matrix for a linear system can also be calculated by developing inertia influence coefficients from an impulse momentum analysis. Consider a linear system initially at rest in equilibrium. Free vibrations occur if the system is given either an initial potential energy or an initial kinetic energy. The stiffness influence coefficients are developed by examining the potential energy developed in the system due to the static application of a system of forces. Inertia influence coefficients are developed by examining the kinetic energy developed in the system due to the application of a system of impulses. An instantaneous change in kinetic energy occurs due to the application of an impulse. If a system is dynamically coupled, then an instantaneous change in the velocity associated with one generalized coordinate may also cause an instantaneous change in the velocities associated with other generalized coordinates.

Consider an n-degree-of-freedom system with generalized coordinates $x_1, x_2, \ldots, x_n$. Assume a system of impulses is applied such that I_i is an impulse applied to the particle whose motion is described by x_i. Motion occurs with possibly nonzero velocities associated with generalized coordinates. These velocities are related to the applied impulses by n applications of the principle of impulse and momentum. For a linear system the resulting equations can be written in the form

$$I_i = \sum_{j=1}^{n} m_{ij}\dot{x}_j \tag{6.19}$$

where the m_{ij} are the inertia influence coefficients. Consider, in particular, a system of applied impulses such that $\dot{x}_l = 1$ and $\dot{x}_j = 0$, $j \neq l$. Then Eq. (6.19) reduces to

$$I_i = m_{il} \tag{6.20}$$

Thus the inertia influence coefficient m_{jl} is one component of a system of impulses that is applied to generate an instantaneous velocity $\dot{x}_l = 1$ with $\dot{x}_i = 0$ for $i \neq l$. In particular, it is the impulse that is applied to the particle whose displacement is represented by x_j. If a system of impulses is applied to a linear system such that the relationship between the applied impulses and their induced velocities is given by Eq. (6.19), then the principle of work-energy can be used to show that the kinetic energy developed by the system is given by Eq. (6.6). Thus the inertia influence coefficients are the elements of the mass matrix.

Frequently, in solving the differential equations governing vibrations of a multi-degree-of-freedom system, the inverse of the mass matrix, $\mathbf{M}^{-1}$, is required. If the system is not dynamically coupled, then the mass matrix is diagonal and its inverse is also diagonal with the reciprocals of the elements of **M** along its diagonal. However, if the system is dynamically coupled, the inverse must be calculated using principles of linear algebra (App. C). Inverse inertia influence coefficients provide an alternative to calculating the mass matrix and then finding its inverse. Suppose a unit impulse is applied to the particle in the system whose displacement is represented by x_l and no other impulses are applied. Then Eq. (6.19) takes the form

$$\delta_{il} = \sum_{j=1}^{n} m_{ij}\dot{x}_j \tag{6.21}$$

The matrix form of Eq. (6.21) is

$$\mathbf{q}_l = \mathbf{M}\dot{\mathbf{x}} \tag{6.22}$$

where $\mathbf{q}_l$ is an $n \times 1$ column vector whose only nonzero element is the lth element, which is unity. The mass matrix is always nonsingular. Inversion of Eq. (6.22) yields

$$\begin{aligned}\dot{\mathbf{x}} &= \mathbf{M}^{-1}\mathbf{q}_l \\ &= \mathbf{M}_l^{-1}\end{aligned} \tag{6.23}$$

where $\mathbf{M}_l^{-1}$ is the lth column of the inverse of the mass matrix. Equation (6.23) implies that the lth column of the inverse of the mass matrix can be calculated by applying a unit impulse to the particle whose displacement is represented by x_l and calculating the induced velocities.

The following summarizes the calculation of inertia influence coefficients and inverse inertia influence coefficients.

Inertia Influence Coefficients

1. Assume that a system of impulses, I_i, $i = 1, \ldots, n$, are applied such that $\dot{x}_1 = 1$, $\dot{x}_2 = 0$, $\dot{x}_3 = 0, \ldots, \dot{x}_n = 0$. Note that I_j is the impulse applied to the particle whose displacement is described by the generalized coordinate x_j. Repeated application of the principle of impulse and momentum allows for the solution of the applied impulses. The inertia influence coefficient $m_{i1} = I_i$.
2. The procedure in step 1 is repeated with $\dot{x}_k = 1$ and all other velocities equal to zero for $k = 2, \ldots, n$.

3. If x_j represents an angular coordinate, then I_j is an angular impulse and x_j is an angular velocity.
4. The mass matrix is symmetric, $m_{ij} = m_{ji}$. This serves as a check on the calculations.

Inverse Inertia Influence Coefficients

1. Apply a unit impulse to the particle whose displacement is represented by x_1 and no other impulses. Calculate the induced velocities. The column of inverse inertia influence coefficients, $n_{j1} = \dot{x}_j$.
2. Repeat the procedure in step 1 by applying unit impulses to particles whose displacements are represented by generalized coordinates. Columns of inverse inertia influence coefficients are solved successively.
3. If x_j represents an angular coordinate, then the applied unit impulse is a unit angular impulse.
4. The inverse inertia matrix is symmetric. This serves as a check in the calculations.

The following example illustrates applications of the procedure to determine inertia influence coefficients.

Example 6.16. Use influence coefficient methods to find the mass matrix and the inverse of the mass matrix for the system of Fig. 6.5 using x_A and θ as generalized coordinates.

Mass matrix. To find the first column of the mass matrix, set $\dot{x}_A = 1$ and $\dot{\theta} = 0$ and solve for the system of impulses necessary to impart this motion. The impulse diagrams for this situation are shown in Fig. 6.18*a*. Application of the principle of linear impulse and linear momentum gives

$$\mathbf{L}_1 + \mathbf{I}_{1 \to 2} = \mathbf{L}_2$$

$$0 + m_{11} = 3m$$

Application of the principle of angular impulse and angular momentum about A gives

$$\mathbf{H}_{A_1} + \mathbf{G}_{1 \to 2} = \mathbf{H}_{A_2}$$

$$0 + m_{21} = m\frac{L}{2}$$

The second column of the mass matrix is found by assigning $\dot{x}_A = 0$, $\dot{\theta} = 1$ and solving for the system of impulses necessary to impart this motion. The impulse momentum diagrams are shown in Fig. 6.18*b*. Application of the principle

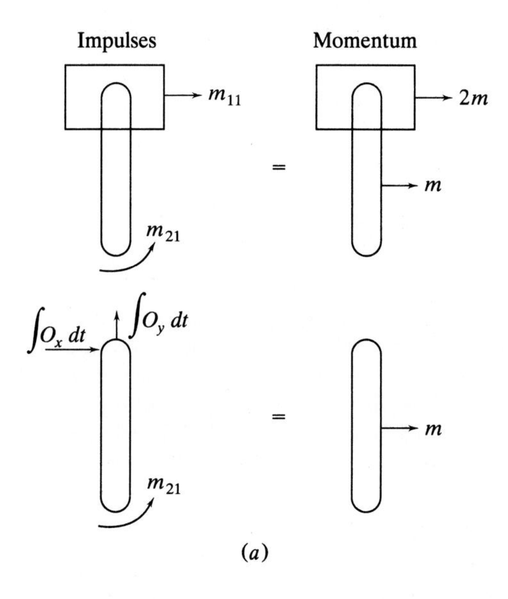

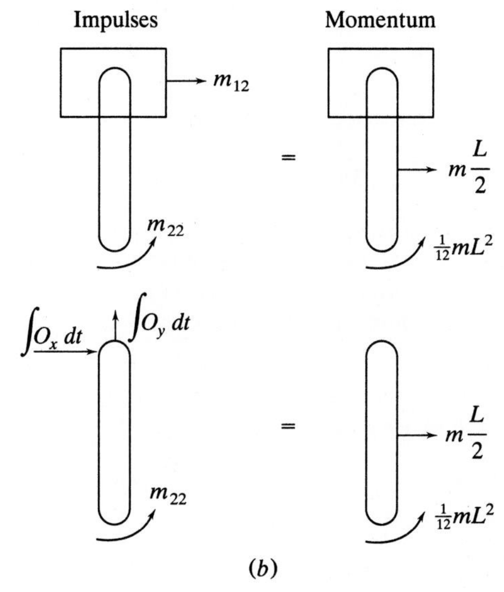

FIGURE 6.18
(*a*) The first column of the mass matrix is obtained by setting $\dot{x}_A = 1$, $\dot{\theta} = 0$; (*b*) The second column of the mass matrix is obtained by setting $\dot{x}_A = 1$, $\dot{\theta} = 0$.

of linear impulse and momentum yields

$$m_{12} = m\frac{L}{2}$$

Application of the principle of angular impulse and angular momentum about A yields

$$m_{22} = m\frac{L}{2}\left(\frac{L}{2}\right) + \frac{1}{12}mL^2 = m\frac{L^2}{3}$$

Thus the mass matrix is

$$\mathbf{M} = \begin{bmatrix} 3m & m\dfrac{L}{2} \\ m\dfrac{L}{2} & m\dfrac{L^2}{3} \end{bmatrix}$$

which agrees with the mass matrix calculated in Example 6.3 using summation of force and moment equations.

Inverse of mass matrix. The first column of the inverse of the mass matrix is obtained by calculating the velocities imparted to the system when a unit impulse is applied to the block. The impulse diagrams for this situation are shown in Fig. 6.19*a*. Applying the principle of linear impulse and linear momentum in the horizontal direction and the principle of angular impulse and angular momentum about point A yields the following equations:

$$1 = 3mn_{11} + m\frac{L}{2}n_{21}$$

$$0 = m\frac{L}{2}n_{11} + m\frac{L^2}{3}n_{21}$$

which yields

$$n_{11} = \frac{4}{9m} \qquad n_{21} = -\frac{2}{3mL}$$

The second column of the inverse of the mass matrix is obtained by calculating the velocities imparted to the system by an applied unit angular impulse. Application of the principle of linear impulse and linear momentum and the principle of angular impulse and angular momentum about point A using the diagrams of Fig. 6.19*b* yields

$$0 = 3mn_{12} + m\frac{L}{2}n_{22}$$

$$1 = m\frac{L}{2}n_{12} + m\frac{L^2}{3}n_{22}$$

which yields

$$n_{12} = -\frac{2}{3mL} \qquad n_{22} = \frac{4}{mL^2}$$

Thus the inverse of the mass matrix is

$$\mathbf{M}^{-1} = \begin{bmatrix} \dfrac{4}{9m} & -\dfrac{2}{3mL} \\ -\dfrac{2}{3mL} & \dfrac{4}{mL^2} \end{bmatrix}$$

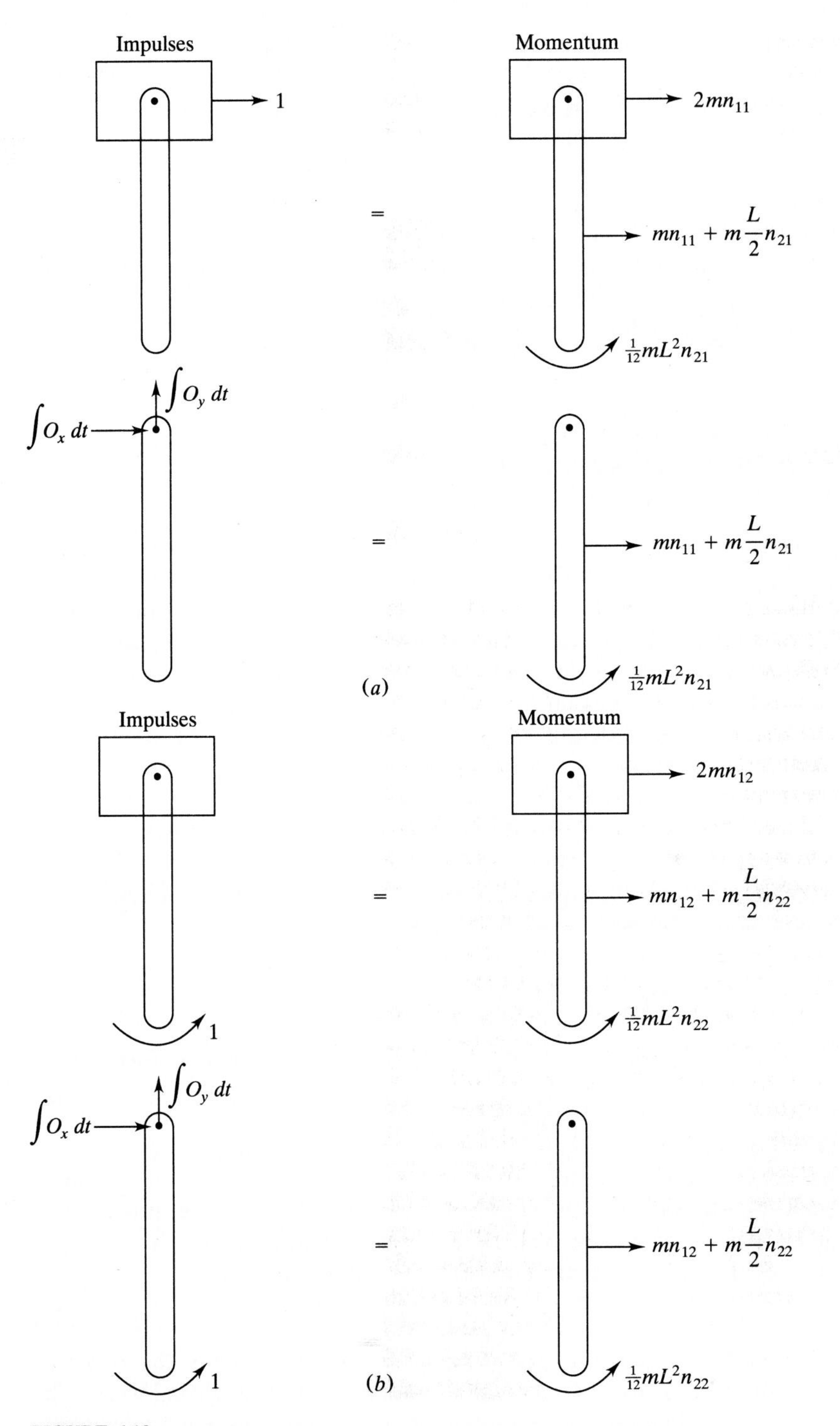

FIGURE 6.19
(a) Impulse-momentum diagrams used to determine first column of inverse mass matrix when unit impulse is applied to block; (b) impulse-momentum diagrams used to determine second column of inverse mass matrix when unit angular impulse is applied to bar.

The mass matrix for a linear system can be developed directly from the expression for the system's kinetic energy or it can be developed using inertia influence coefficients. For most mechanical systems the stiffness matrix is easier to determine than the flexibility matrix. Thus, as shown in Chap. 7, it is necessary to find the inverse of the mass matrix. The inversion is easiest if the mass matrix is diagonal. Thus it is convenient to choose generalized coordinates such that the system is not dynamically coupled. However, if this is difficult or impossible, then the inverse inertia influence coefficients provide a convenient method of determining the inverse of the mass matrix.

Determination of the appropriate form of the mass matrix when using a discrete approximation for a continuous system is considered in Sec. 6.9.

6.8 ENERGY METHODS FOR NONCONSERVATIVE SYSTEMS: GENERALIZED FORCES

Consider an n-degree-of-freedom system with generalized coordinates $x_1, x_2, \ldots, x_n$ acted on by external nonconservative forces. The system is moved through small displacements to a new arbitrary state specified by $x_1 + \delta x_1, x_2 + \delta x_2, \ldots, x_n + \delta x_n$. The changes in displacements are called virtual displacements. The work done by the nonconservative forces as the system moves through the virtual displacements is called the virtual work and is calculated using the usual definition of work done by a force. The virtual work can be written in the form

$$\delta W = \sum_{i=1}^{n} Q_i \, \delta x_i \tag{6.24}$$

The Q_i terms are called generalized forces. It can be shown that Lagrange's equations for nonconservative systems takes the form

$$\frac{d}{dt}\left(\frac{\partial L}{\partial \dot{x}_i}\right) - \frac{\partial L}{\partial x_i} = Q_i \qquad i = 1, \ldots, n \tag{6.25}$$

For a linear nonconservative system, Eq. (6.25) leads to the matrix formulation

$$\mathbf{M}\ddot{\mathbf{x}} + \mathbf{K}\mathbf{x} = \mathbf{Q} \tag{6.26}$$

where $\mathbf{Q}$ is a column vector whose elements are the generalized forces.

The nonconservative force vector can be written as a viscous damping force vector and a vector of other nonconservative forces that are independent of the generalized coordinates. The viscous damping force vector is written as the product of the damping matrix and the velocity vector. Thus

$$\mathbf{Q} = -\mathbf{C}\dot{\mathbf{x}} + \mathbf{F} \tag{6.27}$$

Substitution of Eq. (6.27) into Eq. (6.5) yields Eq. (6.1). Thus the damping matrix and the external force vector can be determined using virtual work. The following example illustrates the derivation of differential equations for a multi-degree-of-freedom system with nonconservative forces.

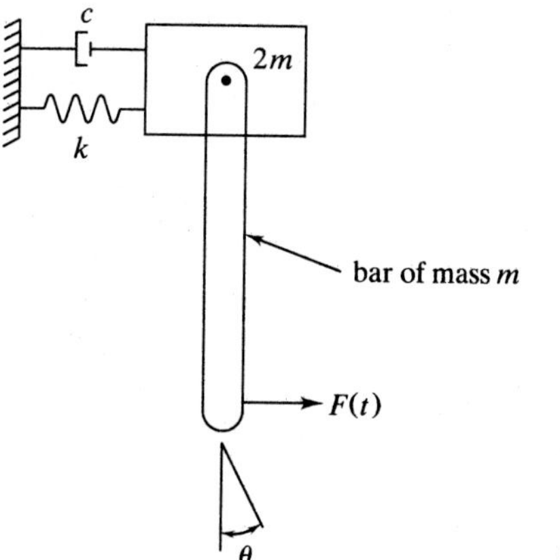

FIGURE 6.20
(a) System of Example 6.17.

Example 6.17. Formulate, in matrix form, the differential equations, for the two-degree-of-freedom system of Fig. 6.20.

A quick inspection of the system shows that it is dynamically, but not statically coupled. The stiffness matrix is easily calculated using stiffness influence coefficients as

$$\mathbf{K} = \begin{bmatrix} k & 0 \\ 0 & mg\dfrac{L}{2} \end{bmatrix}$$

The mass matrix is as calculated using inertia influence coefficients in Ex. 6.16. The force developed in the dashpot and the externally applied $F(t)$ are nonconservative. The work done by these forces as the system moves through a virtual displacement defined by δx and $\delta\theta$ is

$$\delta W = -c\dot{x}\,\delta x + F(t)(\delta x + L\,\delta\theta)$$

from which the generalized forces are determined as

$$Q_1 = -c\dot{x} + F(t)$$

$$Q_2 = LF(t)$$

The resulting differential equations, in matrix form, are

$$\begin{bmatrix} 3m & m\dfrac{L}{2} \\ m\dfrac{L}{2} & m\dfrac{L^2}{3} \end{bmatrix}\begin{bmatrix} \ddot{x} \\ \ddot{\theta} \end{bmatrix} + \begin{bmatrix} c & 0 \\ 0 & 0 \end{bmatrix}\begin{bmatrix} \dot{x} \\ \dot{\theta} \end{bmatrix} + \begin{bmatrix} k & 0 \\ 0 & mg\dfrac{L}{2} \end{bmatrix}\begin{bmatrix} x \\ \theta \end{bmatrix} = \begin{bmatrix} F(t) \\ LF(t) \end{bmatrix}$$

6.9 DISCRETE MODELING OF CONTINUOUS SYSTEMS

Vibrations of continuous systems are governed by partial differential equations. Free vibrations lead to homogeneous partial differential equations, while forced vibrations require the solution of nonhomogeneous partial differential equations.

Analytical solutions to some partial differential equations may not exist, while others are difficult to obtain. Thus approximate and numerical methods are often used to approximate the vibration properties and system response for continuous systems. One such method is to break the continuous system into a finite number of elements, and then apply energy principles to approximate the solution in each element. This powerful method is called the finite-element method and is beyond the scope of the text. A simpler method of approximation is to replace the distributed inertia of the continuous system by a finite number of lumped inertia elements. A point where a lumped mass is placed is called a node. All inertia effects are concentrated at the nodes. The nodes are assumed to be connected by elastic, but massless elements. Generalized coordinates are chosen as the displacements and/or angular displacements of the nodes.

The number of nodes and generalized coordinates used is influenced by the accuracy required and the computational power available. A continuous system has an infinite, but countable, number of natural frequencies. Approximating the continuous system with n degrees of freedom will yield approximations to only the first n natural frequencies. The approximation is better for the lower calculated frequencies and the approximation for each frequency improves with the number of degrees of freedom used. Modal analysis shows that the influence of a free-vibration mode on the forced response of a continuous system decreases as the natural frequency increases. Thus accurate computation of the lower natural frequencies and application of modal analysis using these frequencies will often provide an excellent approximation for the forced response of a continuous system. Often many more degrees of freedom are used than the number of natural frequencies calculated to give accurate approximations for the lower natural frequencies.

Discrete models for several continuous systems are shown in Figs. 6.21 through 6.24. The elastic rod of Fig. 6.21*a* is modeled by n particles located as shown in Fig. 6.21*b*. The n generalized coordinates are the longitudinal displacements of the nodes. The analogous mechanical system is n masses connected by linear springs, as shown in Fig. 6.21*c*. The simply supported beam of Fig. 6.22 is modeled by n particles located along the axis of the beam. The generalized coordinates are the transverse displacements of the nodes. The torsional shaft of Fig. 6.23 is discretized by placing thin disks along the axis of the shaft. The generalized coordinates are the angular displacements of the disks. The four-story frame structure of Fig. 6.24 is modeled with four degrees of freedom assuming the girders are infinitely rigid compared to the columns. The inertia effects are concentrated at the floors. The generalized coordinates are chosen to represent the horizontal displacements of the floors. The one-story frame structure of Fig. 6.25*a* has a flexible girder. Three degrees of freedom can be used to model this structure. The generalized coordinate x_A represents the horizontal displacement of point A. The generalized coordinate x_B represents the horizontal displacement of point B. The generalized coordinate θ represents the slope of the girder at point A. If all corners remain perpendicular, a geometric relationship is developed between the slope of the girder at

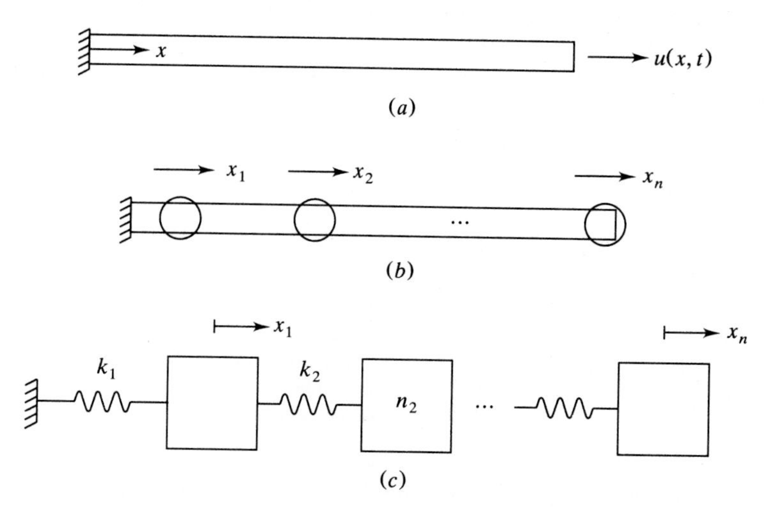

FIGURE 6.21
(a) Longitudinal vibrations of rod are described by $u(x, t)$; (b) discrete model of continuous system. $x_1, \ldots, x_n$ are used as generalized coordinates; (c) analogous mechanical system.

point B and the generalized coordinates; hence a fourth generalized coordinate is not needed.

Discretizing a continuous system using n degrees of freedom results in n simultaneous coupled differential equations. These differential equations are of the form of Eq. (6.1). These equations governing the discrete model are best derived using energy methods.

The flexibility matrix for a discrete model of a continuous system is determined using flexibility influence coefficients. The flexibility influence coefficients are obtained from the static-deflection properties of the structural member.

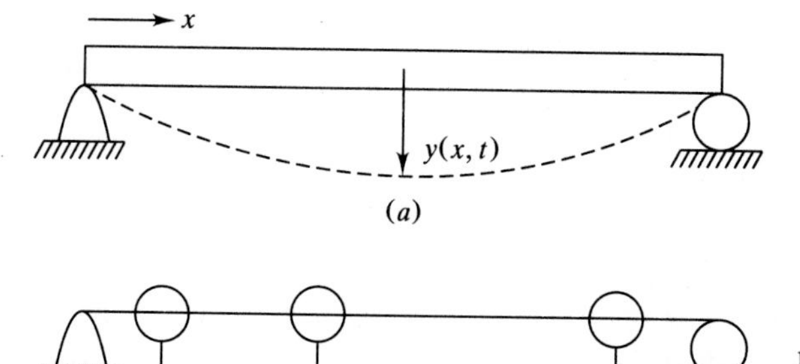

FIGURE 6.22
(a) Transverse vibrations of simply supported beam are described by $y(x, t)$; (b) discrete model of continuous system.

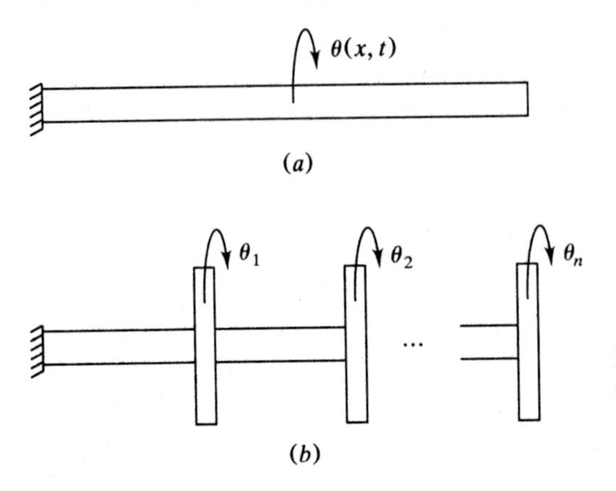

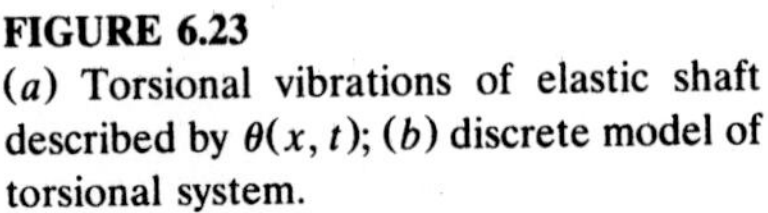

FIGURE 6.23
(*a*) Torsional vibrations of elastic shaft described by $\theta(x,t)$; (*b*) discrete model of torsional system.

The damping matrix for the discrete model of a continuous system is determined by the hysteretic damping characteristics of the structure. However, as shown in Chap. 8, modal damping is usually assumed.

The mass matrix for the discrete model of a continuous system is not as well defined as the flexibility matrix. The flexibility influence coefficient calculations are based upon static-deflection properties of the element, which are known, while the calculations for the elements of the mass matrix should be based upon the dynamic deflection properties, which are not known. If the time-dependent displacement of the system were known and written as a function of the displacement of the nodes,

$$w(x,t) = w(x; x_1 x_2, \ldots, x_n)$$

then the kinetic energy could be calculated from

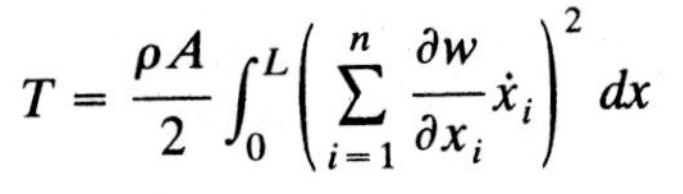

$$T = \frac{\rho A}{2} \int_0^L \left(\sum_{i=1}^{n} \frac{\partial w}{\partial x_i} \dot{x}_i \right)^2 dx$$

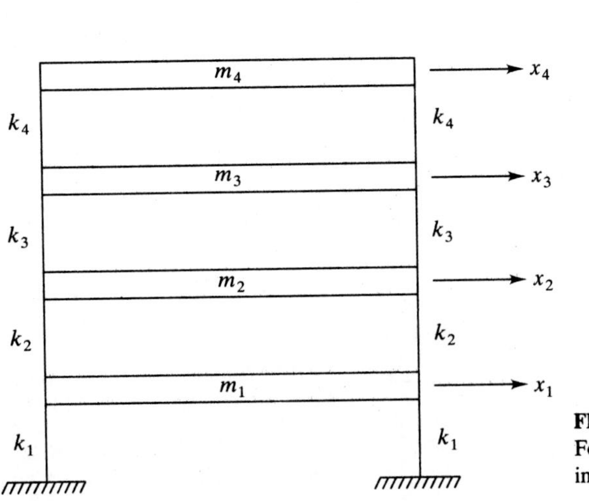

FIGURE 6.24
Four-story structure is modeled using four degrees of freedom.

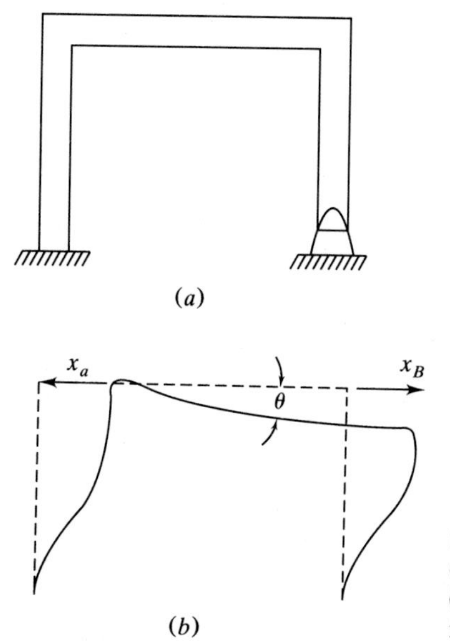

FIGURE 6.25
(*a*) One-story frame structure where girder is not rigid; (*b*) structure can be modeled using three degrees of freedom.

Thus, when the elements of the mass matrix are determined from the kinetic energy, the mass matrix is not diagonal. However, the exact expression for the kinetic energy is not known. Thus any formulation of the mass matrix only adds to the approximations made when the continuous system is modeled with a discrete representation. An approximate expression for the displacement, perhaps the static-deflection relationship, could be used to develop an expression for the kinetic energy and thus lead to a nondiagonal mass matrix. However, the required algebra and calculations are lengthy and tedious. The increased accuracy is usually not worth the effort. If greater accuracy is required, then perhaps a full finite-element method should be used.

It is desirable to specify a diagonal mass matrix, in order to reduce the computations involved in calculation of the elements of the mass matrix and to reduce later computations. If a discretization is used where the mass of the system is simply lumped at nodes, then an obvious approximation to the mass matrix is a diagonal mass matrix with the nodal masses along the diagonal. In problems where an angular displacement of a flexural member is chosen as a generalized coordinate, such as the frame with the flexible girder, the approximation of the mass matrix is not as straightforward. It would be difficult to calculate an equivalent moment of inertia to model the inertia effects of the girder and columns. In these problems an approximation for the displacement is used to calculate the kinetic energy and a nondiagonal mass matrix.

If a diagonal mass matrix is assumed with nodal masses along the diagonal, the choice of the values of the nodal masses affects the accuracies of the natural frequencies, the mode shapes, and the forced response. The inertia effects of a uniform structural element discretized by an n-degree-of-freedom system could be modeled by equally spacing n particles each of mass m/n where m is the total mass of the element. If a one-degree-of-freedom model of a cantilever beam were used with all of the mass lumped at the end of the beam, the approximation for the lowest natural frequency is

$$\omega_n = \sqrt{\frac{3EI}{mL^3}}$$

compared to the exact natural frequency of

$$\omega_1 = 3.515\sqrt{\frac{EI}{mL^3}}$$

The error in the approximation is 51%.

The assumption that all particles used in discretizing a uniform structural system have the same mass implies that all particles in the system contribute equally to the kinetic energy. This is not true. For a flexural member undergoing transverse vibrations, particles located near supports which prevent transverse movement have little kinetic energy compared to particles far away from these supports. When a diagonal mass matrix is used to model the inertia effects of a discretized continuous system, the mass lumped at each node should represent the mass of an identifiable region of the structural element. A good scheme for calculation of the nodal mass is to add half of the mass of the actual system between the node and the node to its left to half of the mass of the system between the node and the node to its right. If the node has no neighbor on one side, but is adjacent to a free end, then all of the mass between the location of the particle and the free end is used. If the particle is adjacent to a support which prevents motion in the direction described by the generalized coordinates, then only half of the mass between the node and the support is used. Thus the kinetic energy of the particles less than halfway between a node and its adjacent support is ignored. When this method is used to model a cantilever beam with a one-degree-of-freedom system, a particle of mass $m/2$ is placed at the end of the beam leading to a natural frequency of

$$\omega_n = \sqrt{\frac{6EI}{mL^3}}$$

which is an error of 30%.

It is noted that if the static deflection were used to approximate the kinetic energy of the beam due to a concentrated unit load applied at its end, a particle of mass $0.2357m$ would be placed at the end of the beam leading to a natural frequency approximation of

$$\omega_n = 3.567\sqrt{\frac{EI}{mL^3}}$$

an error of only 1.5%.

Calculation of the force vector may also require additional approximations. As shown in Sec. 6.8, the force vector is obtained by calculating the generalized forces, which occur when the method of virtual work is used. If a concentrated load is applied at a node, then the generalized force for the node's generalized coordinate is the value of the concentrated load and the generalized forces for all other coordinates are zero. However, if a concentrated load is applied at a location other than a node or the loading is distributed, calculation of the generalized forces requires additional approximations. The dynamic displacement is not available to apply the method of virtual work. In these cases it is suggested that the loading be replaced by a series of concentrated loads, calculated as follows, such that the resulting system is approximately statically equivalent to the applied loading. Static equivalence does not imply dynamic equivalence.

If the applied loading is replaced by a system of concentrated loads, the following method is used. The loading between any two nodes is replaced by a concentrated load at each of the nodes. The two concentrated loads are statically equivalent to the loading between the nodes. The sum of the concentrated loads is the resultant of the loading between the nodes. The moment of the distributed loading about either node is the same as the moment of the two concentrated loads about that point. Thus the total generalized force applied at a node is approximated by the sum of the contribution from the loading between the node and its neighbor to the left and the contribution from the loading between the node and its neighbor to the right. If the node is adjacent to a free end, the contribution to the loading between the node and the free end is the resultant of the loading. If the particle is adjacent to a support that prevents displacement, only the resultant of the loading between the node and the point halfway between the node and the support is used. In this case the work done by particles near supports are ignored in modeling the system, just as these particles' kinetic energy is ignored. The concentrated load is not statically equivalent to the actual loading if the particle is adjacent to a free end or a support.

Example 6.18. Derive the differential equations whose solution approximates the forced response of the cantilever beam of Fig. 6.26. Use four degrees of freedom to discretize the system. The beam is made of a material of elastic modulus E and mass density ρ. It has a cross-sectional area A and moment of inertia I. Neglect damping.

The beam is discretized by lumping its mass in four particles as shown in Fig. 6.25*b*. The nodes are chosen to be equally spaced. The generalized coordinates are the displacements of the nodes. The mass of each particle models the inertia effects of the regions shown in the figure. The loading is replaced by time-dependent concentrated loads at the nodes, as shown in Fig. 6.25*c*.

The flexibility matrix for this discretized system is determined from flexibility influence coefficients, as described in Sec. 6.6. The first column is obtained by placing a unit load at the first node and calculating the resulting deflections at

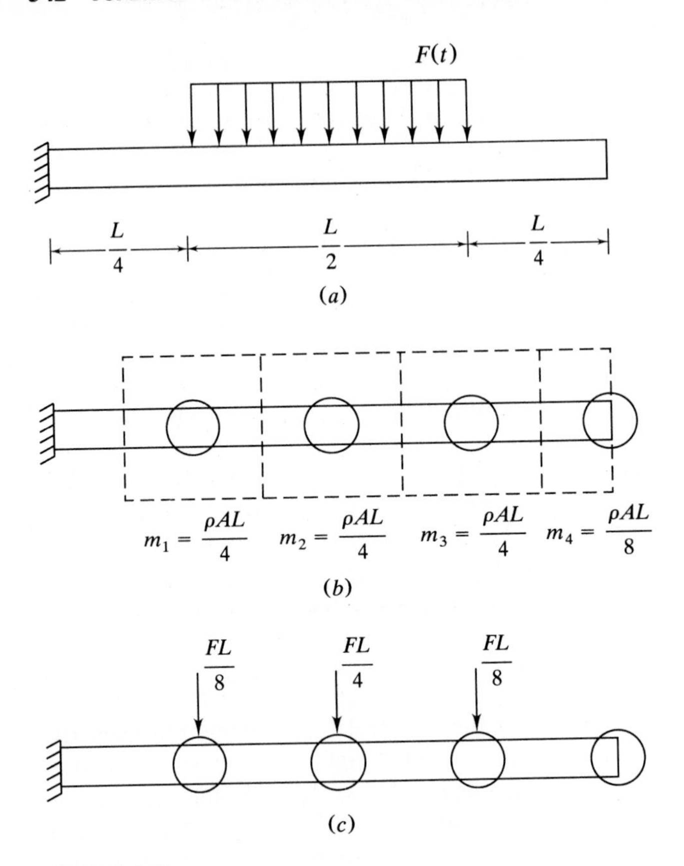

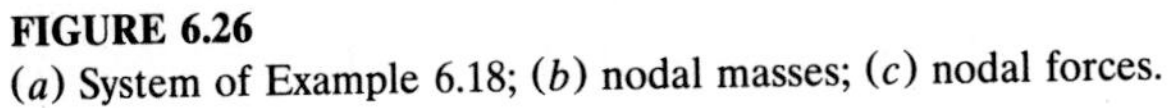

FIGURE 6.26
(a) System of Example 6.18; (b) nodal masses; (c) nodal forces.

each of the nodes. The result is

$$\mathbf{A} = \frac{L^3}{384EI}\begin{bmatrix} 2 & 5 & 8 & 11 \\ 5 & 16 & 28 & 40 \\ 8 & 28 & 54 & 81 \\ 11 & 40 & 81 & 128 \end{bmatrix}$$

The mass matrix is a diagonal matrix with the nodal masses along the diagonal. The force vector is simply the vector of concentrated loads from Fig. 6.25c. Then Eq. (6.15) becomes

$$\frac{\rho AL^3}{1536EI}\begin{bmatrix} 4 & 10 & 16 & 11 \\ 10 & 32 & 56 & 40 \\ 16 & 56 & 108 & 81 \\ 22 & 80 & 162 & 128 \end{bmatrix}\begin{bmatrix} \ddot{x}_1 \\ \ddot{x}_2 \\ \ddot{x}_3 \\ \ddot{x}_4 \end{bmatrix} + \begin{bmatrix} x_1 \\ x_2 \\ x_3 \\ x_4 \end{bmatrix} = \frac{\rho AL^4F(t)}{3072EI}\begin{bmatrix} 18 \\ 65 \\ 118 \\ 172 \end{bmatrix}$$

PROBLEMS

6.1–6.14. Use conservation laws to derive the differential equations governing the motion of the systems shown in Figures P6.1–P6.14 using the indicated coordinates as generalized coordinates. Make any linearizing assumptions and write the differential equations in the form of Eq. (6.1).

6.15–6.28. Use Lagrange's equations to derive the differential equations governing the motion of the systems shown in Figures P6.1–P6.14 using the indicated coordinates as generalized coordinates. Make any linearizing assumptions and write the differential equation in the form of Eq. (6.1). Indicate whether the system is statically coupled, dynamically coupled, both, or neither.

6.29–6.40. Derive the stiffness matrix for the systems shown in Figures P6.1–P6.14 using the indicated generalized coordinates and stiffness influence coefficients.

6.41–6.46. Derive the flexibility matrix for the systems shown in Figures P6.1–P6.14 using the indicated generalized coordinates and flexibility influence coefficients.

6.47–6.52. Derive the mass matrix for the systems shown in Figures P6.1–P6.14 using the indicated generalized coordinates and inertia influence coefficients.

6.53–6.58. Derive the inverse of the mass matrix for the systems shown in Figure P6.1–P6.14 using the indicated generalized coordinates and inverse inertia influence coefficients.

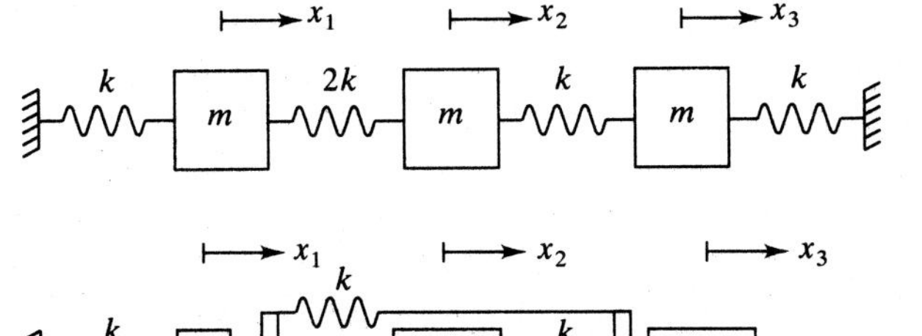

FIGURE P6.1
(Problems 6.1, 6.15, 6.29)

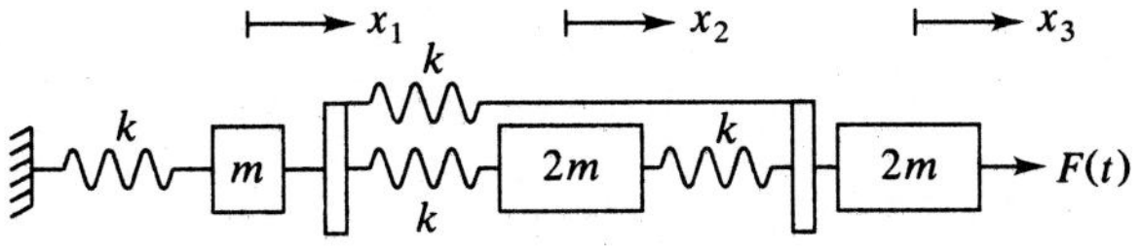

FIGURE P6.2
(Problems 6.2, 6.16, 6.30, 6.41)

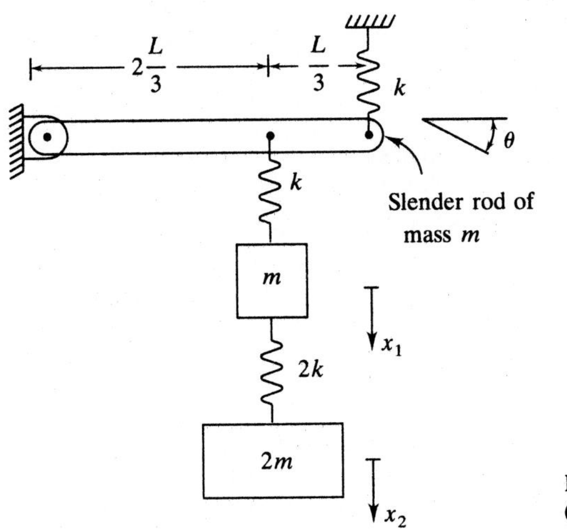

FIGURE P6.3
(Problems 6.3, 6.17, 6.31, 6.47, 6.53)

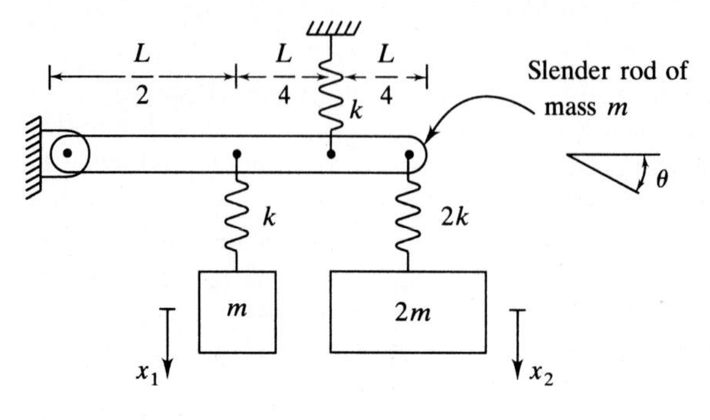

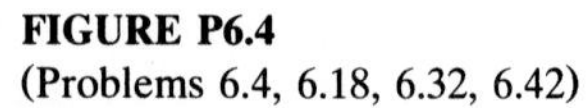
FIGURE P6.4
(Problems 6.4, 6.18, 6.32, 6.42)

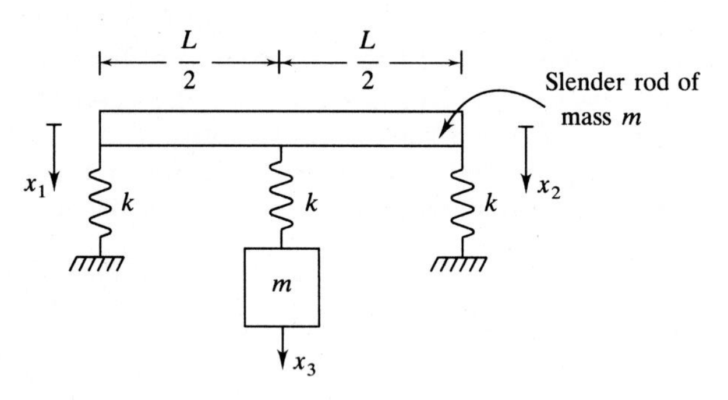

FIGURE P6.5
(Problems 6.5, 6.19, 6.33, 6.48, 6.54)

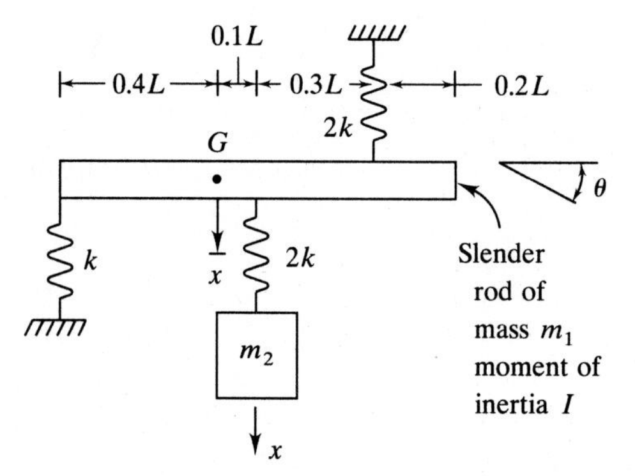

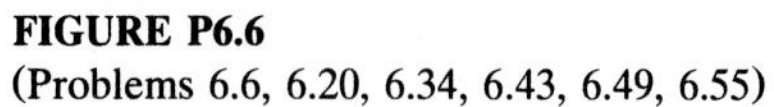
FIGURE P6.6
(Problems 6.6, 6.20, 6.34, 6.43, 6.49, 6.55)

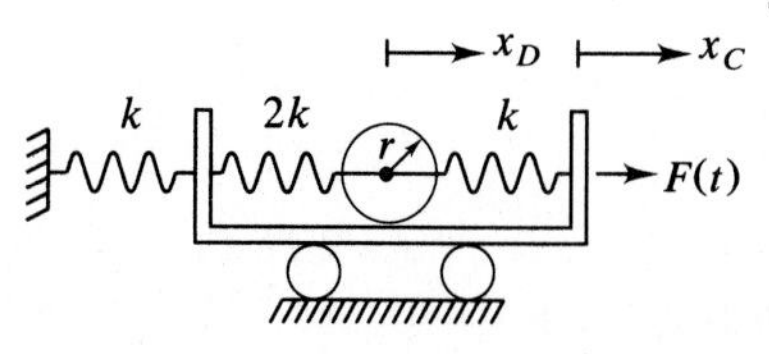

x_C $F(t)$

Thin disk of mass m and radius r rolls without slip relative to center of mass $2m$. x_D is absolute displacement of mass center of disk

FIGURE P6.7
(Problems 6.7, 6.21, 6.35)

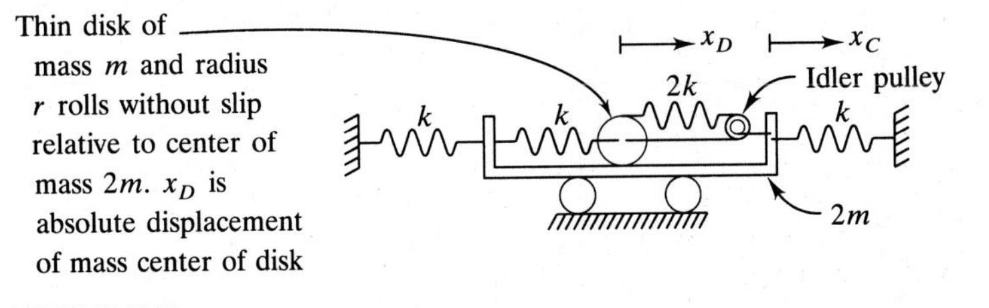

FIGURE P6.8
(Problems 6.8, 6.22, 6.36, 6.50, 6.56)

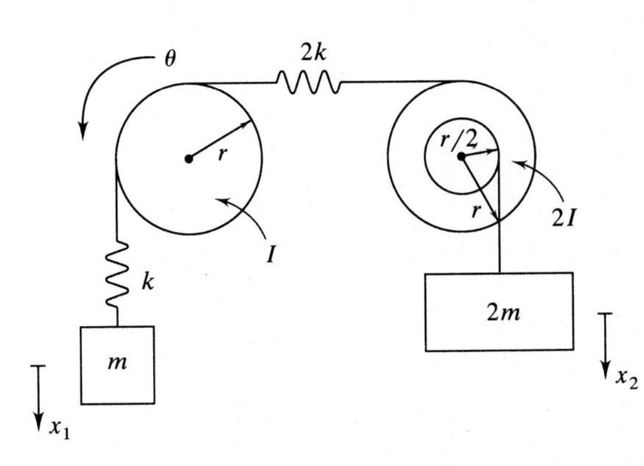

FIGURE P6.9
(Problems 6.9, 6.23, 6.37)

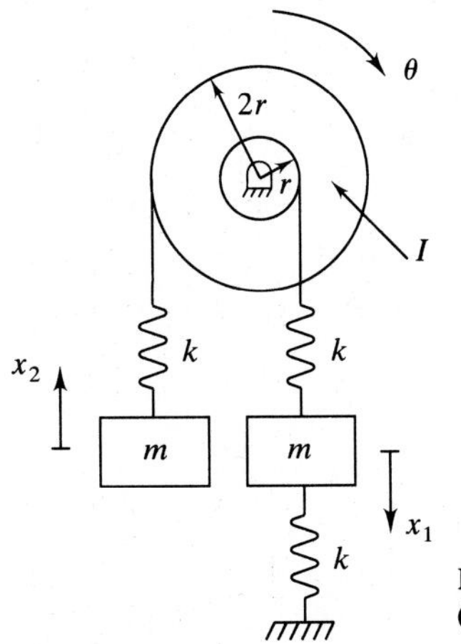

FIGURE P6.10
(Problems 6.10, 6.24, 6.38, 6.44)

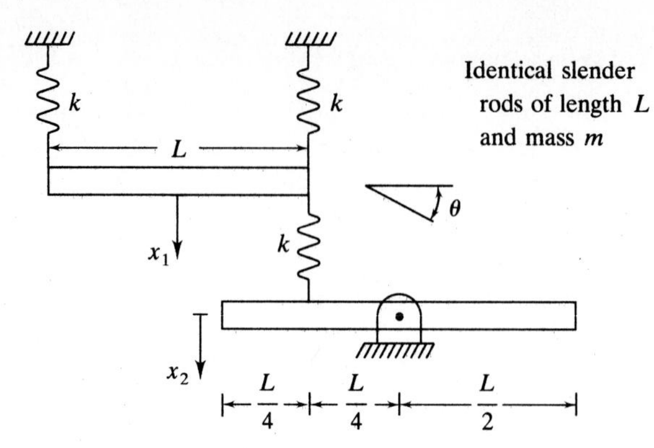

FIGURE P6.11
(Problems 6.11, 6.25, 6.39, 6.45)

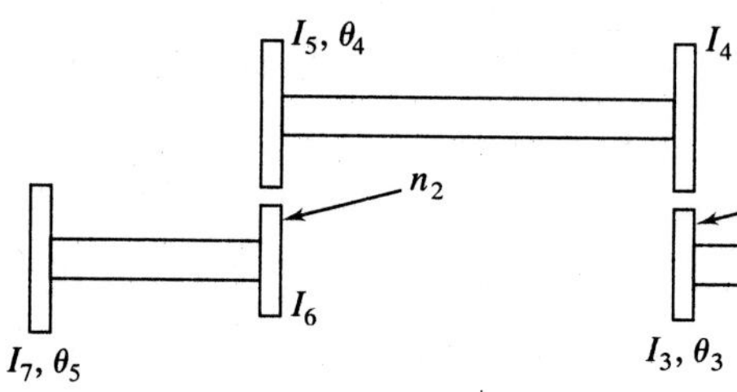

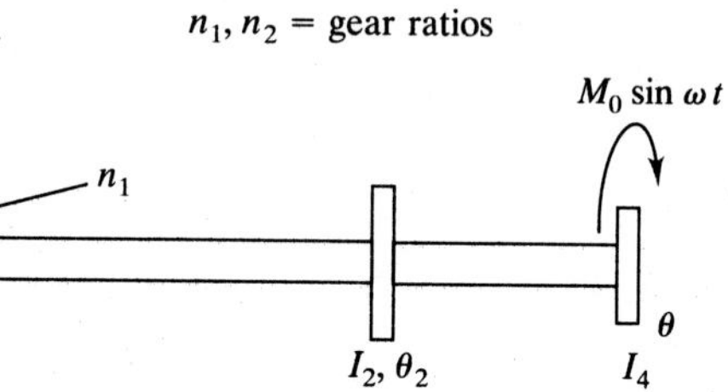

FIGURE P6.12
(Problems 6.26, 6.40, 6.51, 6.57)

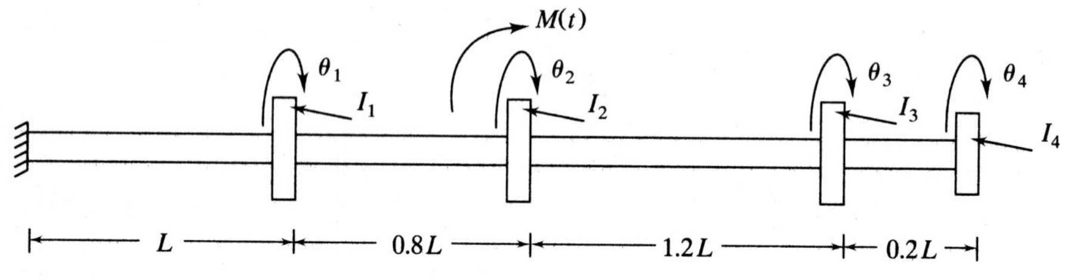

FIGURE P6.13
(Problems 6.27, 6.46)

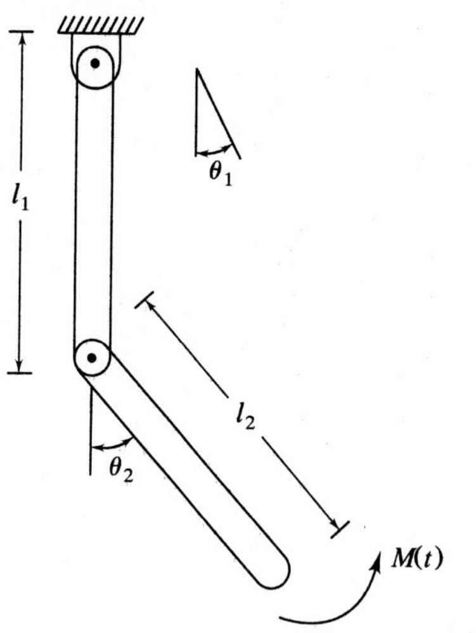

FIGURE P6.14
(Problems 6.14, 6.28, 6.52, 6.58)

6.59–6.62. A rigid bar is attached to two springs, as shown in Fig. P6.59. Since the mass and moment of inertia of the bar are unknown, an engineer runs an experiment. When he applies a 10-N · s vertical impulse to point A, the instantaneous velocity of point A is measured as 4.5 m/s downward and the velocity of B is measured as 0.5 m/s downward. When a 10-N · s impulse is applied to B, the instantaneous velocity of A is measured as 0.5 m/s downward and the velocity of B is measured as 10.5 m/s downward.

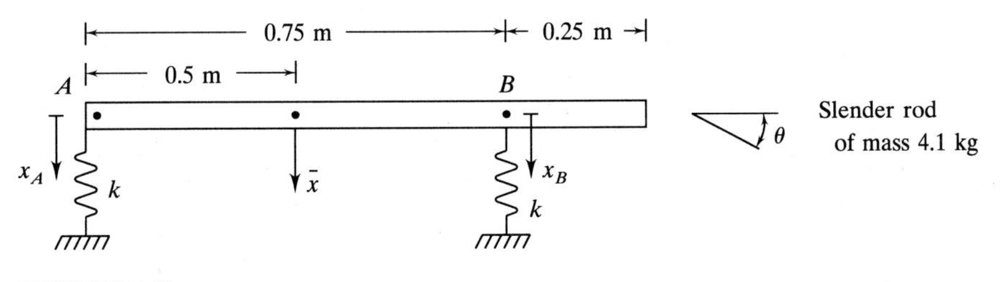

FIGURE P6.59

6.59. Determine the inverse of the mass matrix when x_A and x_B are used as generalized coordinates.

6.60. Determine the mass matrix when x_A and x_B are used as generalized coordinates.

6.61. Determine the mass matrix when x_A and θ are used as generalized coordinates.

6.62. Determine the mass matrix when $\bar{x}$, the displacement of the mass center, and θ are used as generalized coordinates.

6.63–6.66. The beams shown are made of a material of elastic modulus 210×10^9 N/m^2 and have a cross-sectional moment of inertia of 1.3×10^{-5} m^4. Determine the flexibility matrix for the beam when three degrees of freedom are used to analyze the beam's vibrations. Use the displacements of the particles shown as generalized coordinates. Use Table D2 for deflection calculations.

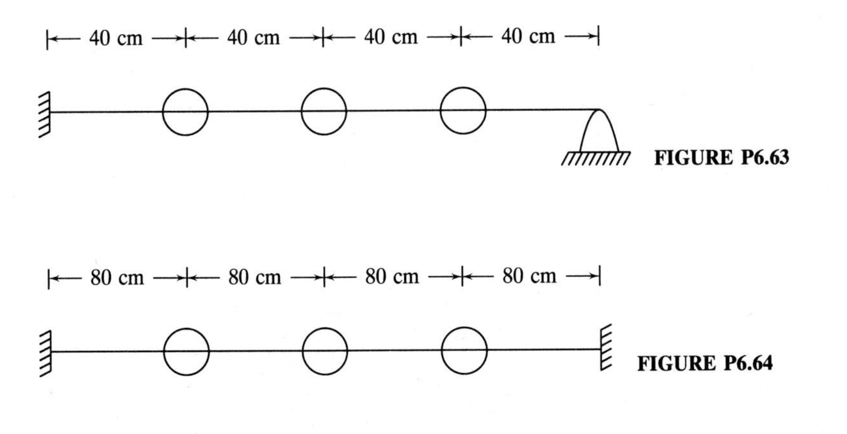

FIGURE P6.63

FIGURE P6.64

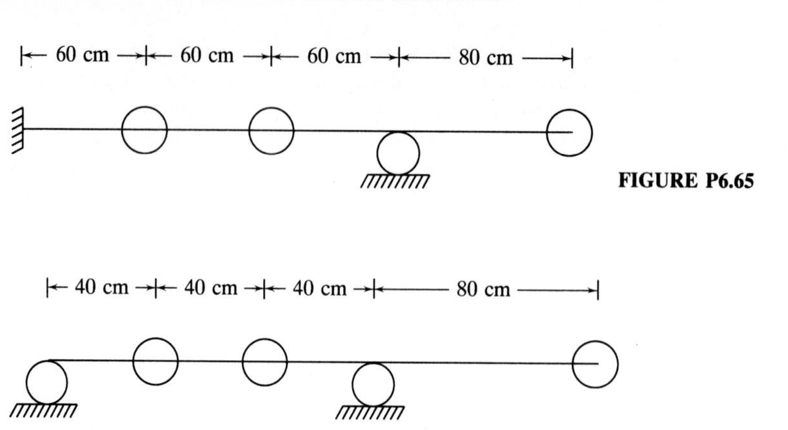

FIGURE P6.65

FIGURE P6.66

6.67. Determine the stiffness matrix for the free-free beam of Fig. P6.67 when three degrees of freedom are used to analyze the beam's vibrations.

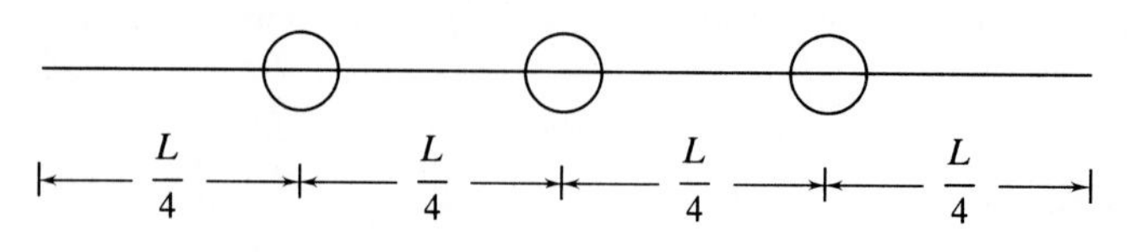

FIGURE P6.67

6.68. Using a two-degree-of-freedom model, derive the differential equations governing forced vibration of the system of Fig. P6.68.

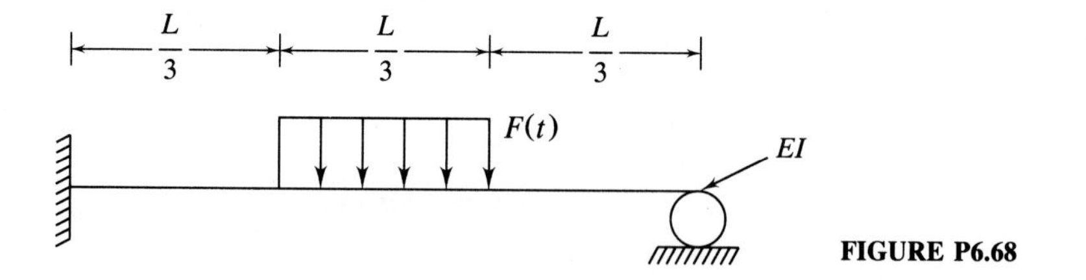

FIGURE P6.68

6.69. Using a three-degree-of-freedom model, derive the differential equations governing forced vibrations of the system of Fig. P6.68.

6.70. A disk of mass m and moment of inertia I is attached to the end of the cantilever beam of Fig. P6.70. The beam is subject to a harmonic force which varies linearly over the beam's span. Let x be the vertical displacement of the disk and θ the slope of the beam at the location where the disk is attached. Derive the differential equations governing forced vibrations of this beam using x and θ as generalized coordinates.

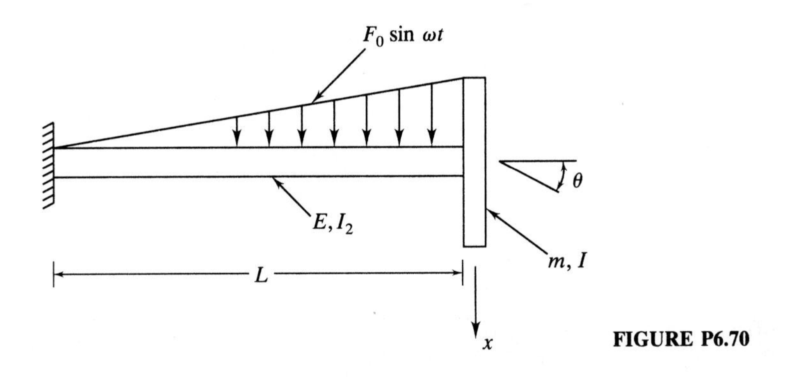

FIGURE P6.70

6.71. A mass m is attached to the end of the tapered circular bar of Fig. P6.71. The bar is made of a material of mass density ρ. A two-degree-of-freedom model is used to model the longitudinal vibrations of the bar. Use x_1, the displacement of the particles in the cross section at the midspan of the bar, and x_2, the displacement of the mass, as generalized coordinates. Derive the differential equations governing free vibrations.

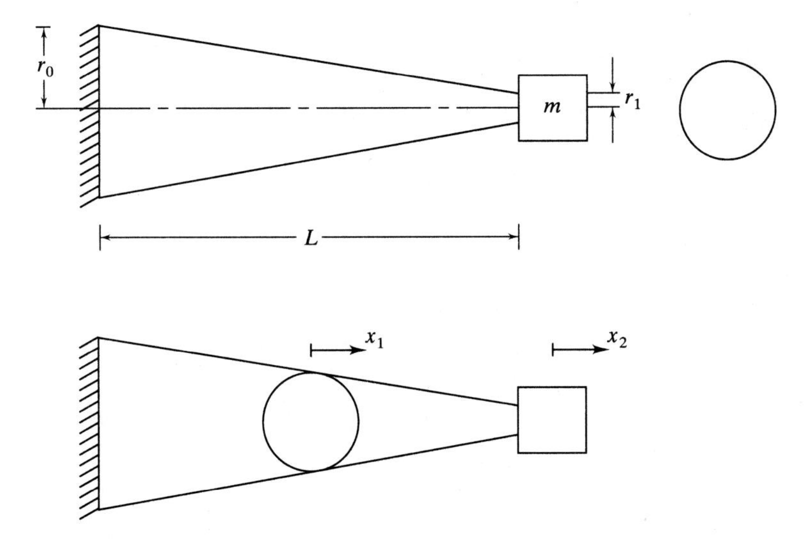

FIGURE P6.71

The program BEAM.BAS, included in the accompanying diskette, determines the flexibility matrix when an n-degree-of-freedom model is used for a beam. Use BEAM.BAS to solve Probs. 6.72–6.75.

6.72. Solve Prob. 6.64 using BEAM.BAS.

6.73. Solve Prob. 6.66 using BEAM.BAS.

6.74. Determine the flexibility matrix when seven degrees of freedom are used to model the beam of Prob. 6.63.

6.75. Model the beam of Prob. 6.66 with six degrees of freedom. The locations of the nodes are $x_1 = 20$ cm, $x_2 = 50$ cm, $x_3 = 75$ cm, $x_4 = 100$ cm, $x_5 = 150$ cm, $x_6 = 200$ cm.

REFERENCES

1. Clough, R. W., and J. Penzien: *Dynamics of Structures*, McGraw-Hill, New York, 1975.
2. Dimarogonas, A. D., and S. Haddad: *Vibrations for Engineers*, Prentice-Hall, Englewood Cliffs, N.J., 1992.
3. Fertis, D. G.: *Dynamics of Vibrations and Structures*, Wiley-Interscience, New York, 1973.
4. Goldstein, H.: *Classical Mechanics*, Addison-Wesley, Reading, Mass., 1950.
5. James, M. L., G. M. Smith, J. C. Wolford, and P. W. Whaley: *Vibration of Mechanical and Structural Systems with Microcomputer Applications*, Harper and Row, New York, 1989.
6. Meirovitch, L.: *Elements of Vibration Analysis*, McGraw-Hill, New York, 1975.
7. Rao, S. S.: *Mechanical Vibrations*, 2nd ed., Addison-Wesley, Reading, Mass., 1990.
8. Shabana, A. A.: *Theory of Vibration: An Introduction*, vol. 1, Springer-Verlag, New York, 1991.
9. Shabana, A. A.: *Theory of Vibration: Discrete and Continuous Systems*, vol. 2, Springer-Verlag, New York, 1991.
10. Steidel, R. F.: *An Introduction to Mechanical Vibrations*, 3rd ed., Wiley, New York, 1988.
11. Thomson, W. T.: *Theory of Vibrations with Applications*, 3rd ed., Prentice-Hall, Englewood Cliffs, N.J., 1988.
12. Weaver, W., S. P. Timoshenko, and D. H. Young: *Vibration Problems in Engineering*, 5th ed., Wiley-Interscience, 1990.

CHAPTER

7

FREE VIBRATIONS OF MULTI-DEGREE-OF-FREEDOM SYSTEMS

7.1 INTRODUCTION

Methods used to determine the vibration response of a multi-degree-of-freedom system differ greatly from those used for a one-degree-of-freedom system. An n-degree-of-freedom system is governed by n coupled differential equations and has n natural frequencies. However, these natural frequencies are not apparent once the differential equations are written in a standard form, as is the case for a one-degree-of-freedom system.

The solution of a system of coupled differential equations can be written as the sum of a homogeneous solution and a particular solution. The homogeneous solution reflects the free-vibration properties of the system and is the focus of this chapter. The particular solution represents the forced response.

Knowledge of the natural frequencies, and thus implicitly the homogeneous solution, is essential for solving the forced-vibration problem for one-degree-of-freedom systems. For example, the integrand in the convolution integral contains a trigonometric term whose frequency is the damped natural frequency of the system. Knowledge of the free-vibration solution is also essential for determining the forced response of multi-degree-of-freedom systems. Application of the modal analysis method of Chap. 8 requires knowledge of all natural frequencies and mode shapes.

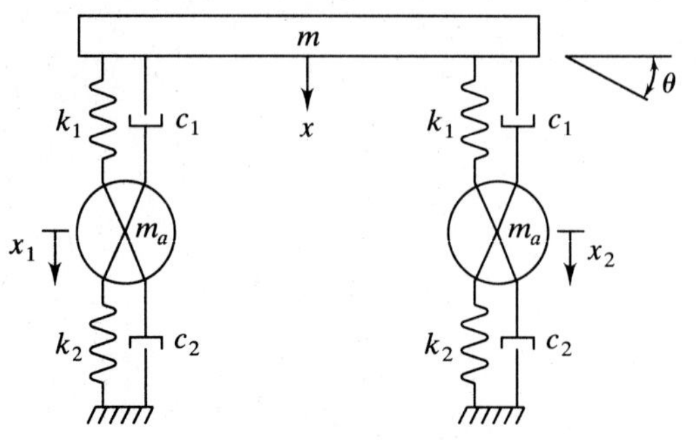

FIGURE 7.1
Four-degree-of-freedom model of automobile suspension system, including elasticity and damping of tires.

When damping is included the free-vibrations response of a one-degree-of-freedom system decays (exponentially for viscous damping, linearly for Coulomb damping) and is usually ignored in comparison to the forced response for large time. Damping also causes the free-vibration response of multi-degree-of-freedom systems to decay. Small viscous damping does not have much effect on the natural frequencies. The general solution for a multi-degree-of-freedom system with damping is very complicated and beyond the scope of this text. For these reasons damping will be limited to systems with only a few degrees of freedom or to special forms of damping such as proportional damping or modal damping.

Examples of multi-degree-of-freedom systems undergoing free vibrations are shown in Figs. 7.1 through 7.3. Figure 7.1 shows a simplified model of an automobile suspension system. The body is modeled as a rigid bar which can translate and rotate about any axis. The body is connected through springs and dampers (shock absorbers) to the axles. The tires also have elasticity and damping. An actual suspension system is much more complicated and can have up to 18 degrees of freedom. If the mass of the axles and wheels are neglected, the system can be modeled using two degrees of freedom. If the inertia of the axles and wheels are included, four degrees of freedom are necessary. The natural frequencies, neglecting damping, are obtained for this system in Sec. 7.10.

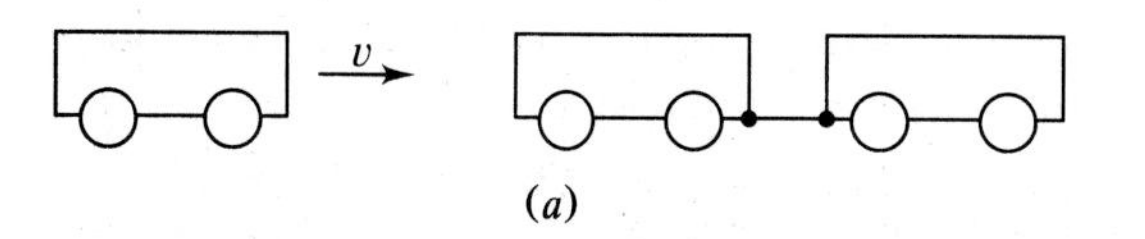

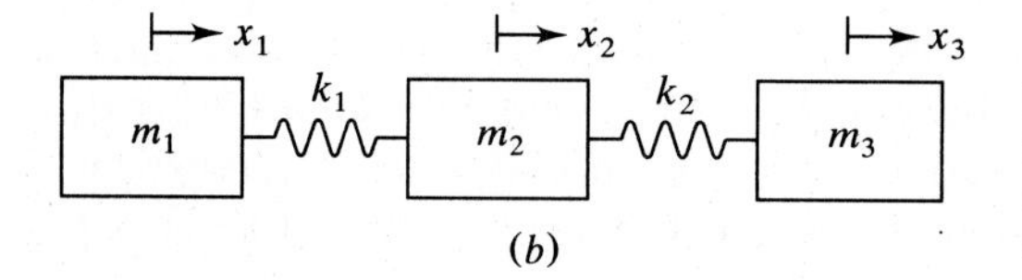

FIGURE 7.2
(*a*) Shunting of railroad cars; (*b*) three-degree-of-freedom model used to model resulting coupled system.

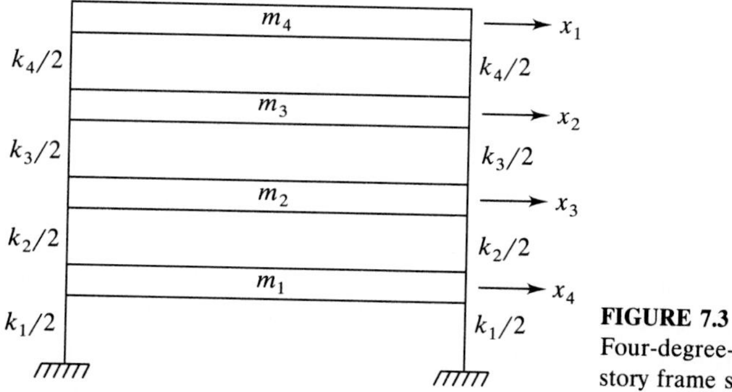

FIGURE 7.3
Four-degree-of-freedom model of four-story frame structure.

The two railroad cars of Fig. 7.2 are connected by elastic couplings. A third car is coupled to these cars as it moves with an initial velocity. The three coupled railroad cars are modeled as a three-degree-of-freedom unrestrained system undergoing free vibrations.

If the girders of the four-story frame structure of Fig. 7.3 are rigid compared to the columns and all inertia concentrated at the floors, the structure is modeled using four degrees of freedom. The modal analysis method of Chap. 8 can be used to determine the response of this structure to an earthquake. However, the natural frequencies and mode shapes must first be known in order to apply modal analysis.

The free-vibration analysis of a multi-degree-of-freedom system is significantly more complex than the free-vibration analysis of a one-degree-of-freedom system. The natural frequencies of an n-degree-of-freedom system are determined by solving a matrix eigenvalue problem involving an $n \times n$ matrix. The analysis of free and forced vibrations of multi-degree-of-freedom systems requires extensive matrix analysis. A reader unfamiliar with matrix algebra and terminology is encouraged to read App. C before proceeding with Chaps. 7 and 8.

7.2 NORMAL-MODE SOLUTION

The general formulation of the differential equations governing free vibrations of a linear undamped n-degree-of-freedom system is

$$\mathbf{M}\ddot{\mathbf{x}} + \mathbf{K}\mathbf{x} = \mathbf{0} \tag{7.1}$$

where $\mathbf{M}$ and $\mathbf{K}$ are the symmetric $n \times n$ mass and stiffness matrices, respectively, and $\mathbf{x}$ is the n-dimensional column vector of generalized coordinates. The symmetry of the mass and stiffness matrices is guaranteed if energy methods are used to formulate the differential equations. If energy methods are not used and

$\mathbf{M}$ and $\mathbf{K}$ are not symmetric, then the system can be algebraically manipulated such that a reformulation involves symmetric matrices. Symmetry is not necessary to solve the free-vibration problem. It is required in order to apply certain numerical methods for calculation of natural frequencies, and is necessary to apply the modal analysis method for solving forced-vibration problems as described in Chap. 8.

Free vibrations of a multi-degree-of-freedom system are initiated due to the presence of an initial potential or kinetic energy. If the system is undamped, there are no dissipative mechanisms and it is expected that the free vibrations described by the solution of Eq. (7.1) are periodic. It is assumed that the vibrations are synchronous in that all dependent variables execute motion with the same time-dependent behavior. Thus, when free vibrations at a single frequency are initiated for a particular system, the ratio of any two dependent variables is independent of time. These assumptions lead to hypothesizing the normal-mode solution of Eq. (7.1) in the form

$$\mathbf{x}(t) = \mathbf{X}e^{i\omega t} \tag{7.2}$$

where ω is the frequency of vibration and $\mathbf{X}$ is an n-dimensional vector of constants, called a mode shape. This hypothesis implies that certain initial conditions lead to a solution of the form of Eq. (7.2) for specific values of ω. The values of ω such that Eq. (7.2) is a solution of Eq. (7.1) are called the natural frequencies. Each natural frequency has at least one corresponding mode shape. Since the differential equations represented by Eq. (7.1) are linear and homogeneous, their general solution is a linear superposition over all possible modes.

Substitution of Eq. (7.2) into Eq. (7.1) leads to

$$(-\omega^2\mathbf{MX} + \mathbf{KX})e^{i\omega t} = \mathbf{0} \tag{7.3}$$

Since $e^{i\omega t} \neq 0$,

$$-\omega^2\mathbf{MX} + \mathbf{KX} = \mathbf{0} \tag{7.4}$$

Recall from Chap. 6 that if the equations are formulated using energy methods, the mass matrix is nonsingular. Thus $\mathbf{M}^{-1}$ exists. Premultiplying Eq. (7.4) by $\mathbf{M}^{-1}$ and rearranging gives

$$(\mathbf{M}^{-1}\mathbf{K} - \omega^2\mathbf{I})\mathbf{X} = 0 \tag{7.5}$$

where $\mathbf{I}$ is the $n \times n$ identity matrix. Equation (7.5) is the matrix representation of a system of n simultaneous linear algebraic equations for the n components of the mode shape vector. The system is homogeneous. Application of Cramer's rule gives the solution for the jth component of $\mathbf{X}$, X_j, as

$$X_j = \frac{0}{\det|\mathbf{M}^{-1}\mathbf{K} - \omega^2\mathbf{I}|} \tag{7.6}$$

Thus the trivial solution ($\mathbf{X} = \mathbf{0}$) is obtained unless

$$\det|\mathbf{M}^{-1}\mathbf{K} - \omega^2\mathbf{I}| = 0 \tag{7.7}$$

Hence, applying the definitions of App. C, ω^2 must be an eigenvalue of $\mathbf{M}^{-1}\mathbf{K}$. The square root of a real positive eigenvalue has two possible values, one positive and one negative. While both are used to develop the general solution, the positive square root is identified as a natural frequency. The corresponding eigenvector is called a mode shape.

7.3 NATURAL FREQUENCIES AND MODE SHAPES

In the previous section it is shown that the natural frequencies are the positive square roots of the eigenvalues of $\mathbf{M}^{-1}\mathbf{K}$. As shown in App. C, the evaluation of Eq. (7.7) leads to an nth-order polynomial equation, called the characteristic equation. Since all elements of the mass and stiffness matrices are real, all coefficients of the characteristic equation are real and thus if complex roots exist, then they exist as complex conjugate pairs. However, it can be shown (see Sec. 7.7) that only real roots of the characteristic equation are obtained when the mass and stiffness matrices are symmetric. Negative eigenvalues of $\mathbf{M}^{-1}\mathbf{K}$ are possible, but lead to imaginary values for the natural frequency. When the negative square root of a negative eigenvalue is multiplied by i to form the exponent in the normal-mode solution, Eq. (7.2), a real positive exponent is developed. This part of the normal-mode solution grows without bound as time increases. Such a system is unstable.

Assume that all eigenvalues of $\mathbf{M}^{-1}\mathbf{K}$ corresponding to symmetric mass and stiffness matrices are nonnegative. Then there exist n real natural frequencies that can be ordered by $\omega_1 \leq \omega_2 \leq \cdots \leq \omega_n$. Each distinct eigenvalue ω_i^2, $i = 1, 2, \ldots, n$, has a corresponding nontrivial eigenvector, $\mathbf{X}_i$, which satisfies

$$\mathbf{M}^{-1}\mathbf{K}\mathbf{X}_i = \omega_i^2\mathbf{X}_i \tag{7.8}$$

This mode shape, $\mathbf{X}_i$, is an n-dimensional column vector of the form

$$\mathbf{X}_i = \begin{bmatrix} X_{i1} \\ X_{i2} \\ \vdots \\ X_{in} \end{bmatrix} \tag{7.9}$$

Since the system of equations represented by Eq. (7.8) is homogeneous, the mode shape is not unique. However, if ω_i^2 is not a repeated root of the characteristic equation, then there is only one linearly independent nontrivial solution of Eq. (7.8). The eigenvector is unique only to an arbitrary multiplicative constant. Normalization schemes exist such that the constant is chosen so the eigenvector satisfies an externally imposed condition. One such scheme is presented in Sec. 7.9.

If ω_i^2 is a repeated root of the characteristic equation of multiplicity r, there are r linearly independent nontrivial solutions of Eq. (7.8). Each of these mode shapes is also unique to a multiplicative constant.

Solution of the eigenvalue-eigenvector problem is an important part of the vibration analysis of multi-degree-of-freedom systems. The quadratic formula is used to find the roots of the characteristic equation for a two-degree-of-freedom system. The natural frequencies of a three-degree-of-freedom system are obtained by finding the roots of a cubic polynomial, which can be done by trial and error or an iterative method. The algebraic complexity of the solution grows exponentially with the number of degrees of freedom. The development of characteristic equation for an n-degree-of-freedom system requires the evaluation of an $n \times n$ determinant. N roots of the resulting nth-order polynomial must be found. The determination of each eigenvector requires the solution of n homogeneous simultaneous algebraic equations. Thus numerical methods which do not require the evaluation of the characteristic equation are used for systems with a large number of degrees of freedom.

Example 7.1. Find the natural frequencies and mode shapes for the slender rod of Example 6.2 and Fig. 6.4. Use x and θ as generalized coordinates.

The differential equations of motion using x and θ as generalized coordinates are obtained in Example 6.2 and are presented as follows for the case of free vibrations

$$\begin{bmatrix} m & 0 \\ 0 & m\dfrac{L^2}{12} \end{bmatrix} \begin{bmatrix} \ddot{x} \\ \ddot{\theta} \end{bmatrix} + \begin{bmatrix} 2k & -k\dfrac{L}{4} \\ -k\dfrac{L}{4} & 5k\dfrac{L^2}{16} \end{bmatrix} \begin{bmatrix} x \\ \theta \end{bmatrix} = \begin{bmatrix} 0 \\ 0 \end{bmatrix}$$

The matrix $\mathbf{M}^{-1}\mathbf{K}$ is calculated as

$$\mathbf{M}^{-1}\mathbf{K} = \begin{bmatrix} \dfrac{1}{m} & 0 \\ 0 & \dfrac{12}{mL^2} \end{bmatrix} \begin{bmatrix} 2k & -k\dfrac{L}{4} \\ -k\dfrac{L}{4} & 5k\dfrac{L^2}{16} \end{bmatrix} = \begin{bmatrix} 8\phi & -\phi L \\ -\phi\dfrac{12}{L} & 15\phi \end{bmatrix}$$

where
$$\phi = \frac{k}{4m}$$

The determinant whose evaluation yields the characteristic equation is

$$\det \begin{bmatrix} 8\phi - \lambda & -\phi L \\ -\phi\dfrac{12}{L} & 15\phi - \lambda \end{bmatrix} = 0$$

The resulting characteristic equation is

$$\beta^2 - 23\beta + 108 = 0$$

where

$$\beta = \frac{\lambda}{\phi}$$

Application of the quadratic formula yields

$$\beta = \frac{23 \pm \sqrt{(23)^2 - 4(108)}}{2} = 16.42, 6.58$$

Since the natural frequencies are the positive square roots of the eigenvalues

$$\omega_1 = 1.28\sqrt{\frac{k}{m}}$$

and

$$\omega_2 = 2.03\sqrt{\frac{k}{m}}$$

The equations from which the mode shapes are obtained are

$$\begin{bmatrix} 8\phi - \lambda_i & -\phi L \\ -\phi\dfrac{12}{L} & 15\phi - \lambda_i \end{bmatrix} \begin{bmatrix} X_{i1} \\ X_{i2} \end{bmatrix} = \begin{bmatrix} 0 \\ 0 \end{bmatrix}$$

Since the determinant of the matrix is zero, these equations are not independent and only one is used. The first equation gives the following relation between the components of each mode shape:

$$X_{i2} = \frac{8\phi - \lambda_i}{\phi L} X_{i1}$$

Arbitrarily choosing $X_{i1} = 1$, the following mode shapes are obtained

$$\mathbf{X}_1 = \begin{bmatrix} 1 \\ \dfrac{1.43}{L} \end{bmatrix} \qquad \mathbf{X}_2 = \begin{bmatrix} 1 \\ -\dfrac{8.42}{L} \end{bmatrix}$$

Graphical representations of the mode shapes are shown in Fig. 7.4. A node, or point of zero velocity, occurs at a distance $0.119L$ to the right of the mass center for the second mode. There are no nodes on the bar for the first mode.

Example 7.2. Resolve Example 7.1 using x_A and x_B as generalized coordinates.

The differential equations using x_A and x_B as the generalized coordinates are

$$\begin{bmatrix} \frac{7}{36}m & \frac{1}{18}m \\ \frac{1}{18}m & \frac{4}{9}m \end{bmatrix} \begin{bmatrix} \ddot{x}_A \\ \ddot{x}_B \end{bmatrix} + \begin{bmatrix} \frac{3}{4}k & 0 \\ 0 & \frac{3}{4}k \end{bmatrix} \begin{bmatrix} x_A \\ x_B \end{bmatrix} = \begin{bmatrix} 0 \\ 0 \end{bmatrix}$$

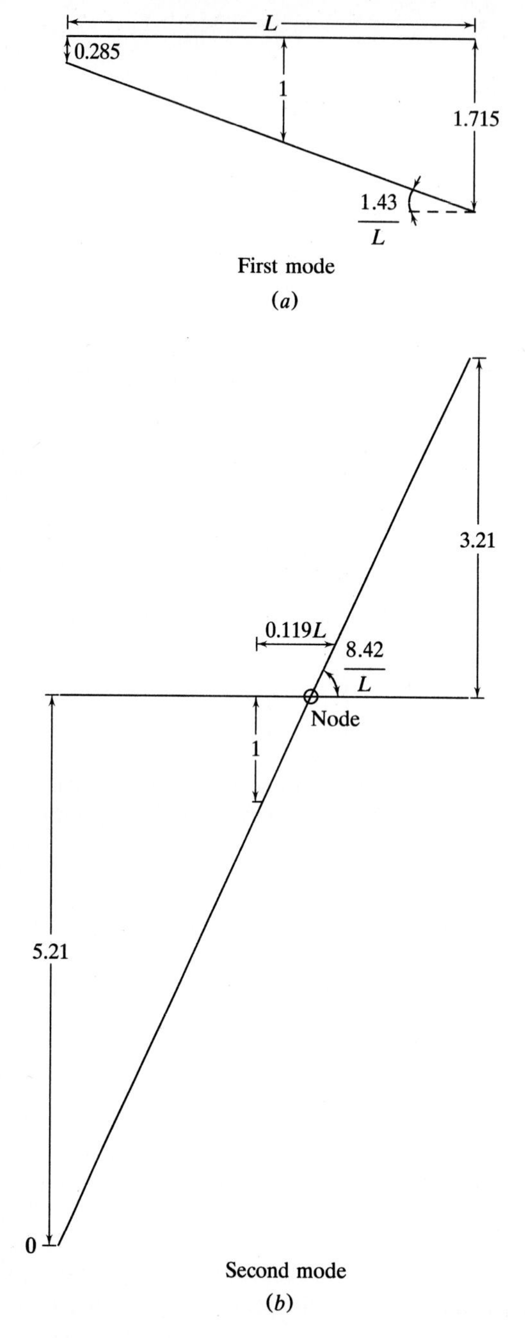

FIGURE 7.4
Mode shapes for Example 7.1. Displacement of mass center is taken as one. (a) $\omega = 1.28\sqrt{k/m}$; (b) $\omega = 2.03\sqrt{k/m}$, a node occurs a distance $0.119L$ to right of mass center. Mode shape represents rigid body rotation about node.

The inverse of the mass matrix, $\mathbf{M}^{-1}$, is calculated using the methods of App. C, leading to

$$\mathbf{M}^{-1}\mathbf{K} = \begin{bmatrix} \dfrac{16}{3m} & -\dfrac{2}{3m} \\ -\dfrac{2}{3m} & \dfrac{7}{3m} \end{bmatrix} \begin{bmatrix} \dfrac{3}{4}k & 0 \\ 0 & \dfrac{3}{4}k \end{bmatrix} = \begin{bmatrix} 16\phi & -2\phi \\ -2\phi & 7\phi \end{bmatrix}$$

where ϕ is as defined in Example 7.1. Setting the determinant of $\mathbf{M}^{-1}\mathbf{K} - \lambda\mathbf{I}$ to 0 yields the characteristic equation

$$\beta^2 - 23\beta + 108 = 0$$

where β is as defined in Example 7.1. Note that this is the same characteristic equation as in Example 7.1 with the same definitions of ϕ and β. Thus the natural

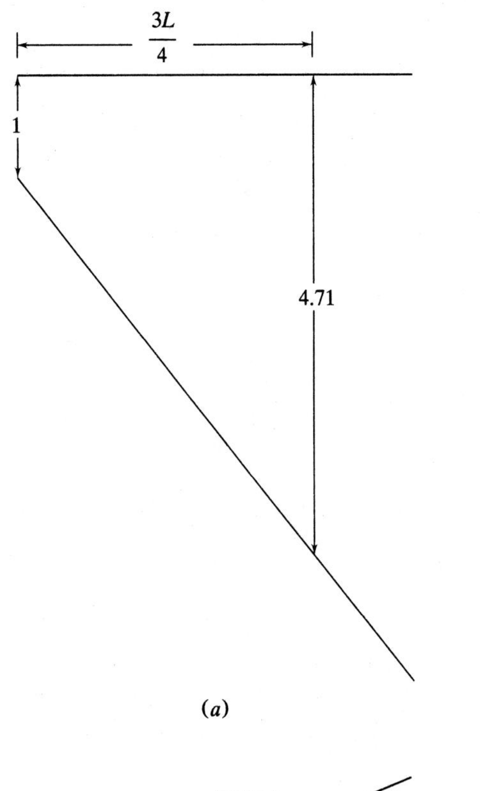

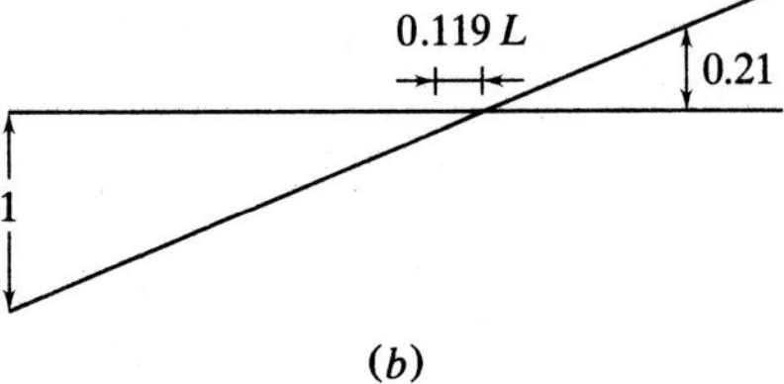

FIGURE 7.5
Mode shapes for Example 7.2. These are identical to mode shapes of Fig. 7.4 except that displacement of left end is taken to be one. (*a*) $\omega = 1.28\sqrt{k/m}$; (*b*) $\omega = 2.03\sqrt{k/m}$.

frequencies are the same. This points out that, as with one-degree-of-freedom systems, natural frequencies are system properties, independent of the choice of generalized coordinates. However, the mode shape vectors are specific to the choice of generalized coordinates. The first of the component equations of $(\mathbf{M}^{-1}\mathbf{K} - \omega_i^2\mathbf{I})\mathbf{X}_i = \mathbf{0}$ gives

$$X_{i2} = \left(8 - \frac{\lambda_i}{2\phi}\right)X_{i1}$$

which, after arbitrarily choosing $X_{i1} = 1$, leads to the mode shape vectors

$$\mathbf{X}_1 = \begin{bmatrix} 1 \\ 4.71 \end{bmatrix} \qquad \mathbf{X}_2 = \begin{bmatrix} 1 \\ -0.21 \end{bmatrix}$$

The graphical representation of these mode shapes is given in Fig. 7.5. Note that, as in Example 7.1, a node occurs a distance of $0.119L$ from the mass center for the second mode.

Example 7.3. A simplified model of an automobile suspension system is shown in Fig. 7.6. The body of the vehicle weights 3000 lb, has a centroidal moment of inertia of 300 slug-ft^2, and is connected to four massless springs. The springs are connected to the axle-wheel assembly. If the inertial of the axle-wheel assembly is ignored, then the elasticity of the tires is included by adding a spring in series with each of the existing springs. The result is the two-degree-of-freedom model with equivalent spring stiffnesses of 2000 lb/ft. Determine the natural frequencies for this two-degree-of-freedom model.

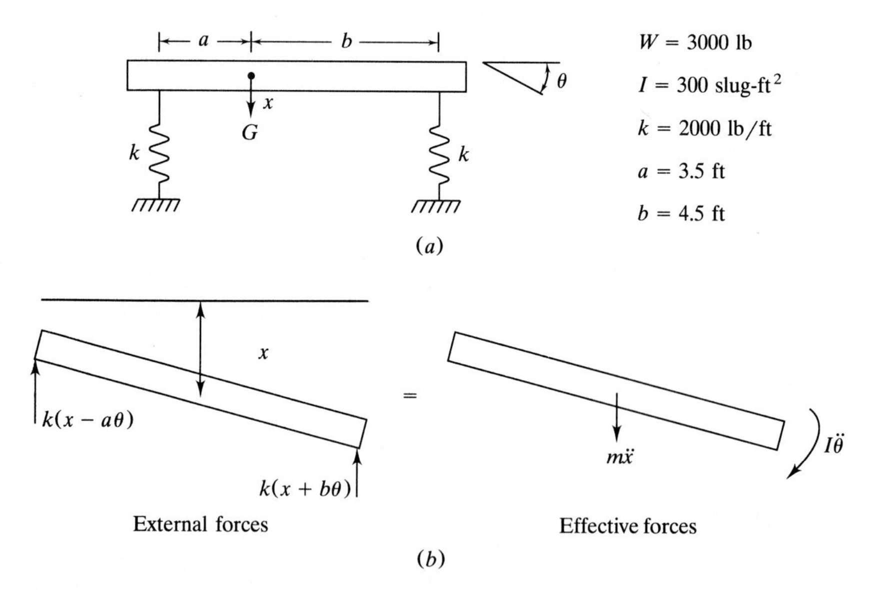

FIGURE 7.6
(a) Two-degree-of-freedom model of automobile suspension system; (b) free-body diagrams used to derive differential equations. Note that static deflections cancel with gravity.

Let x be the displacement of the mass center and θ be the angular rotation of the bar from its horizontal position. The differential equations derived by either summing forces in the vertical direction and summing moments about the mass center or by using energy methods are

$$\begin{bmatrix} m & 0 \\ 0 & I \end{bmatrix}\begin{bmatrix} \ddot{x} \\ \ddot{\theta} \end{bmatrix} + \begin{bmatrix} 2k & (b-a)k \\ (b-a)k & (b^2+a^2)k \end{bmatrix}\begin{bmatrix} x \\ \theta \end{bmatrix} = \begin{bmatrix} 0 \\ 0 \end{bmatrix}$$

The characteristic equation for the eigenvalues of $\mathbf{M}^{-1}\mathbf{K}$ is

$$\lambda^2 - \left(\frac{2}{m} + \frac{a^2+b^2}{I}\right)k\lambda + \frac{(a+b)^2}{Im}k^2 = 0$$

After substituting the given values, this equation becomes

$$\lambda^2 - 259.7\lambda + 9159.1 = 0$$

The quadratic formula is used to determine the eigenvalues as

$$\lambda = 42.1\ \text{s}^{-2}, 217.3\ \text{s}^{-2}$$

The natural frequencies are

$$\omega_1 = 6.49\ \text{rad/s} \qquad \omega_2 = 14.74\ \text{rad/s}$$

Example 7.4. Calculate the natural frequencies and the mode shapes for the three-degree-of-freedom system of Fig. 7.7.

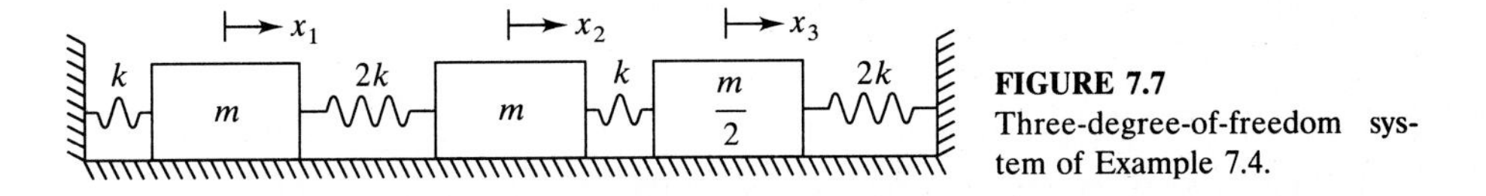

FIGURE 7.7 Three-degree-of-freedom system of Example 7.4.

The differential equations for free vibrations using the displacements of the masses from equilibrium as the generalized coordinates are

$$\begin{bmatrix} m & 0 & 0 \\ 0 & m & 0 \\ 0 & 0 & \frac{m}{2} \end{bmatrix}\begin{bmatrix} \ddot{x}_1 \\ \ddot{x}_2 \\ \ddot{x}_3 \end{bmatrix} + \begin{bmatrix} 3k & -2k & 0 \\ -2k & 3k & -k \\ 0 & -k & 3k \end{bmatrix}\begin{bmatrix} x_1 \\ x_2 \\ x_3 \end{bmatrix} = \begin{bmatrix} 0 \\ 0 \\ 0 \end{bmatrix}$$

Application of Eq. (7.7) gives

$$\det\begin{bmatrix} 3\phi - \lambda & -2\phi & 0 \\ -2\phi & 3\phi - \lambda & -\phi \\ 0 & -2\phi & 6\phi - \lambda \end{bmatrix} = 0$$

where
$$\phi = \frac{k}{m}$$

Expansion of the determinant yields the characteristic equation

$$-\beta^3 + 12\beta^2 - 39\beta + 24 = 0$$

where $\beta = \lambda/\phi$. A plot of the preceding cubic polynomial is given in Fig. 7.8. The

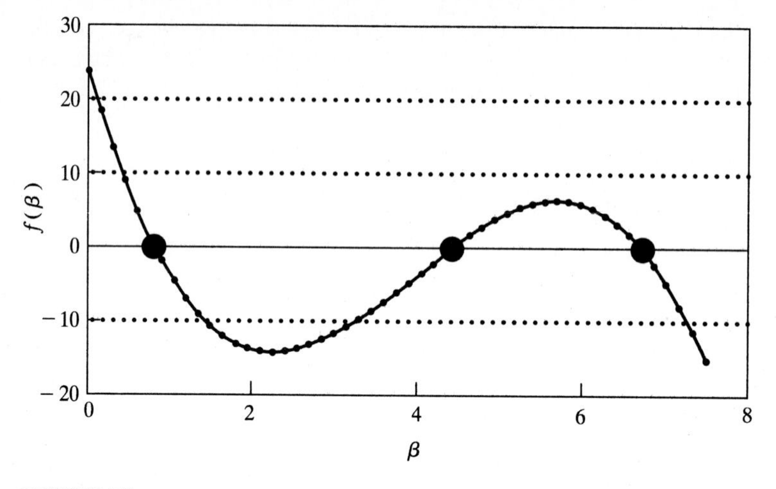

FIGURE 7.8
Plot of characteristic equation of Example 7.4. Roots occur for values of β where curve intersects β axis.

roots of this equation are

$$\beta = 0.798, 4.455, 6.747$$

which leads to the natural frequencies

$$\omega_1 = 0.893\sqrt{\frac{k}{m}} \qquad \omega_2 = 2.110\sqrt{\frac{k}{m}} \qquad \omega_3 = 2.597\sqrt{\frac{k}{m}}$$

The mode shapes are obtained by finding the nontrivial solutions of

$$\begin{bmatrix} 3\phi - \lambda_i & -2\phi & 0 \\ -2\phi & 3\phi - \lambda_i & -\phi \\ 0 & -2\phi & 6\phi - \lambda_i \end{bmatrix} \begin{bmatrix} X_{i1} \\ X_{i2} \\ X_{i3} \end{bmatrix} = \begin{bmatrix} 0 \\ 0 \\ 0 \end{bmatrix}$$

The first equation leads to

$$X_{i1} = \frac{2\phi}{3\phi - \lambda_i} X_{i2}$$

while the third equation leads to

$$X_{i3} = \frac{2\phi}{6\phi - \lambda_i} X_{i2}$$

Arbitrarily choosing $X_{i2} = 1$ leads to the following mode shape vectors:

$$\mathbf{X}_1 = \begin{bmatrix} 0.908 \\ 1 \\ 0.384 \end{bmatrix} \qquad \mathbf{X}_2 = \begin{bmatrix} -1.375 \\ 1 \\ 1.294 \end{bmatrix} \qquad \mathbf{X}_3 = \begin{bmatrix} -0.534 \\ 1 \\ -2.677 \end{bmatrix}$$

The graphical representations of the mode shapes in Fig. 7.9 are made assuming the displacement in each spring is a linear function of position along

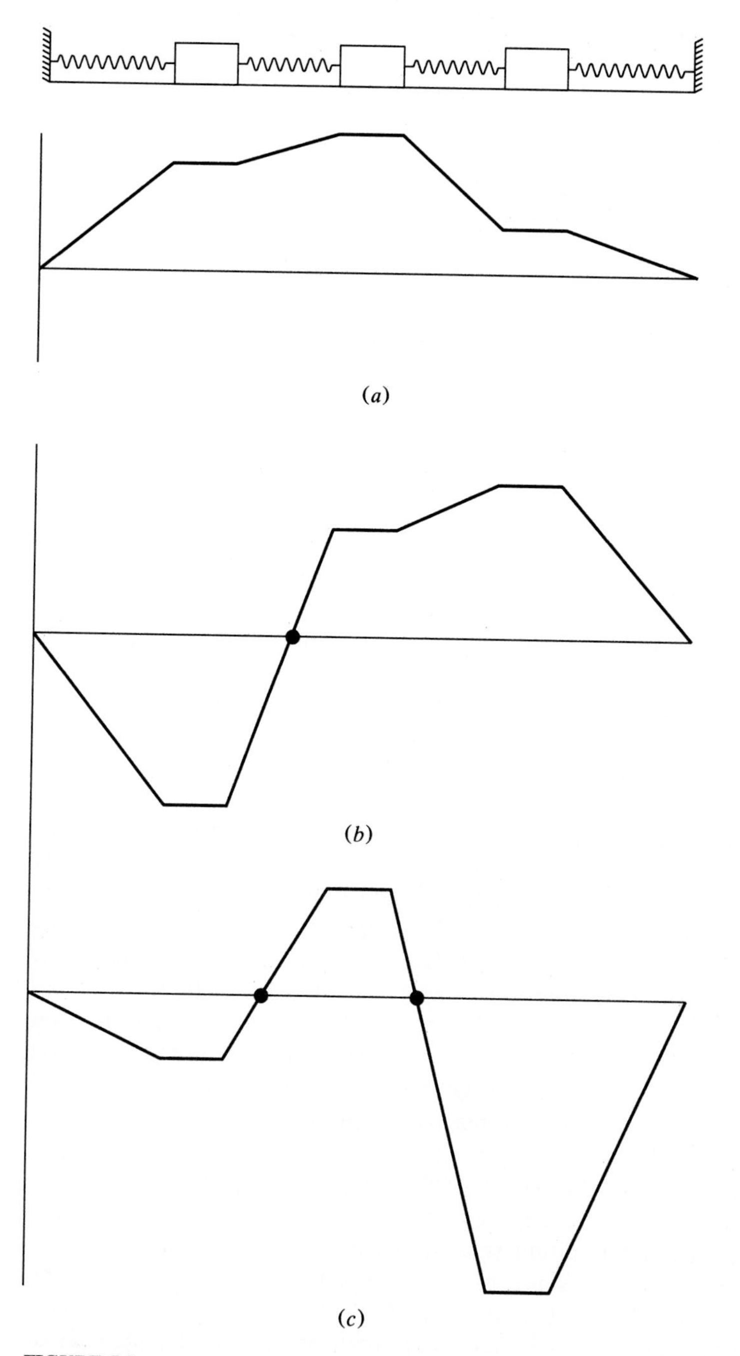

FIGURE 7.9
Mode shapes for Example 7.4. Nodes are shown for second and third modes. (*a*) $\omega = 0.893\sqrt{k/m}$; (*b*) $\omega = 2.110\sqrt{k/m}$; (*c*) $\omega = 2.597\sqrt{k/m}$.

the length of the spring. There are no nodes for the first mode. The second mode has a node in the spring between the first and second mass. The third mode has one node in the spring between the first and second mass and one node in the spring between the second and third masses.

7.4 NATURAL FREQUENCIES USING FLEXIBILITY MATRIX

It is shown in Sec. 6.6 that when the stiffness matrix, **K**, is nonsingular, its inverse is the flexibility matrix, **A**. Premultiplying Eq. (7.4) by **A** leads to

$$(-\omega^2\mathbf{AM} + \mathbf{I})\mathbf{X} = \mathbf{0} \tag{7.10}$$

Dividing by ω^2 gives

$$\left(\mathbf{AM} - \frac{1}{\omega^2}\mathbf{I}\right)\mathbf{X} = \mathbf{0} \tag{7.11}$$

Thus the natural frequencies are the reciprocals of the positive square roots of the eigenvalues of **AM** and the mode shapes are its eigenvectors. The matrix, **AM**, is often called the dynamical matrix.

Natural frequencies of multi-degree-of-freedom systems are calculated as either the square roots of the eigenvalues of $\mathbf{M}^{-1}\mathbf{K}$ or as the reciprocals of the square roots of the eigenvalues of **AM**. The mode shapes are the corresponding eigenvectors of either matrix.

The flexibility matrix is easier to calculate than the stiffness matrix for structural systems. Thus the natural frequencies of structural systems are usually calculated from **AM** as inversion of a matrix is not required.

Influence coefficient methods are used to formulate the stiffness matrix for a mechanical system. If the system is statically coupled, but not dynamically coupled, inversion of the mass matrix is simple and the natural frequencies are calculated using $\mathbf{M}^{-1}\mathbf{K}$. If a mechanical system is dynamically coupled, but not statically coupled, inversion of the stiffness matrix is easier and the natural frequencies are calculated using **AM**. If the system is both statically coupled and dynamically coupled, inversion of the matrix which contains the most zeros is usually easier. Since the natural frequencies are independent of the choice of generalized coordinates, but the nature of the coupling depends on the choice of generalized coordinates, the generalized coordinates should be chosen, if possible, to render the system either statically or dynamically uncoupled.

The flexibility matrix does not exist for unrestrained systems. Thus the natural frequencies must be calculated using $\mathbf{M}^{-1}\mathbf{K}$ for unrestrained systems.

Often numerical methods are used to approximate natural frequencies of systems with a large number of degrees of freedom when not all natural frequencies are required. These methods usually use the dynamical matrix.

Example 7.5. Approximate the two lowest natural frequencies and mode shapes for a fixed-free beam using a two-degree-of-freedom model.

The flexibility and mass matrices for this example are calculated in Secs. 6.6 and 6.8, respectively. They are

$$\mathbf{A} = \begin{bmatrix} \dfrac{L^3}{24EI} & \dfrac{5L^3}{48EI} \\ \dfrac{5L^3}{48EI} & \dfrac{L^3}{3EI} \end{bmatrix} \qquad \mathbf{M} = \begin{bmatrix} \dfrac{\rho AL}{2} & 0 \\ 0 & \dfrac{\rho AL}{4} \end{bmatrix}$$

from which the dynamical matrix is calculated

$$\mathbf{AM} = \begin{bmatrix} 4\phi & 5\phi \\ 10\phi & 16\phi \end{bmatrix}$$

where

$$\phi = \frac{\rho AL^4}{192EI}$$

The characteristic equation of this matrix is

$$\beta^2 - 20\beta + 14 = 0$$

where

$$\beta = \frac{\lambda}{\phi}$$

and λ is an eigenvalue of the dynamical matrix. The roots of the characteristic equation are

$$\beta = 0.73, 19.27$$

which lead to the natural frequencies

$$\omega_1 = \sqrt{\frac{192EI}{19.27\rho AL^4}} = \frac{3.16}{L^2}\sqrt{\frac{EI}{\rho A}}$$

$$\omega_2 = \sqrt{\frac{192EI}{0.73\rho AL^4}} = \frac{16.22}{L^2}\sqrt{\frac{EI}{\rho A}}$$

Solution of the resulting homogeneous equations for the mode shapes leads to the following relation between the components of the mode shape vectors:

$$X_{i2} = -\frac{4\phi - \lambda_i}{5\phi}X_{i1}$$

Arbitrarily choosing $X_{i1} = 1$ leads to

$$\mathbf{X}_1 = \begin{bmatrix} 1 \\ 3.054 \end{bmatrix} \qquad \mathbf{X}_2 = \begin{bmatrix} 1 \\ -0.654 \end{bmatrix}$$

The graphical representations of these mode shapes are approximated in Fig. 7.10. The second mode has a node between the locations where the masses are lumped.

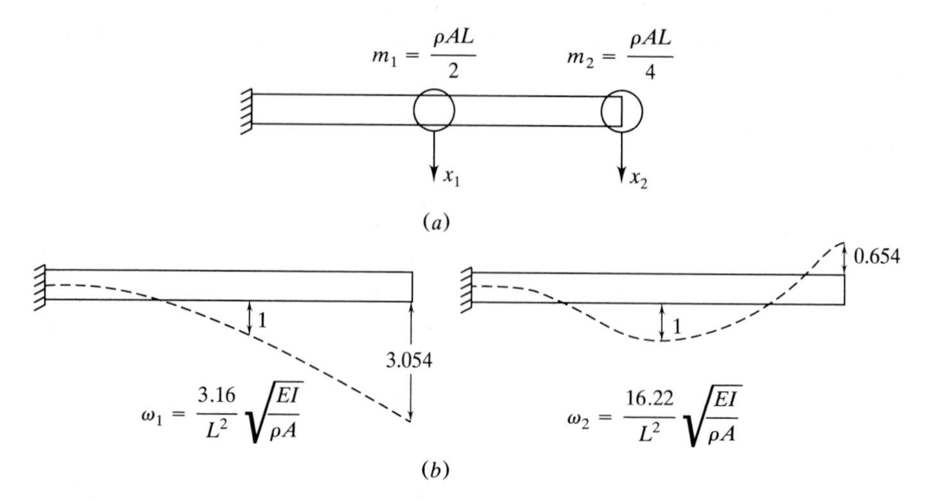

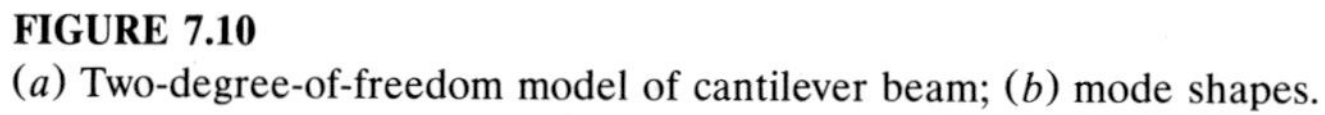

FIGURE 7.10
(*a*) Two-degree-of-freedom model of cantilever beam; (*b*) mode shapes.

7.5 GENERAL SOLUTION

Equation (7.1) represents a homogeneous system of n second-order linear differential equations. The normal-mode assumption, Eq. (7.2), leads to the determination of n natural frequencies. If λ is an eigenvalue of $\mathbf{M}^{-1}\mathbf{K}$, then both $\omega = +\sqrt{\lambda}$ and $\omega = -\sqrt{\lambda}$ satisfy Eq. (7.7) and give rise to the same solution, $\mathbf{X}$, of Eq. (7.5). The functions $e^{i\omega t}$ and $e^{-i\omega t}$ are linearly independent with each other and linearly independent with other functions of the same form with different values of ω. Thus the normal-mode solution generates $2n$ linearly independent solutions of Eq. (7.1). The most general solution of a linear homogeneous problem is a linear combination of all possible solutions. To this end,

$$\mathbf{x}(t) = \sum_{i=1}^{n} \mathbf{X}_i\left(\tilde{C}_{i1}e^{i\omega t} + \tilde{C}_{i2}e^{-i\omega t}\right) \tag{7.12}$$

Using Euler's identity to replace the complex exponential by trigonometric functions and redefining the arbitrary constants gives

$$\mathbf{x}(t) = \sum_{i=1}^{n} \mathbf{X}_i(C_{i1}\cos\omega_i t + C_{i2}\sin\omega_i t) \tag{7.13}$$

Trigonometric identities are used to write Eq. (7.13) in the alternate form

$$\mathbf{x}(t) = \sum_{i=1}^{n} \mathbf{X}_i A_i \sin(\omega_i t - \phi_i) \tag{7.14}$$

Initial conditions must be specified for each dependent variable

$$\mathbf{x}(0) = \begin{bmatrix} x_1(0) \\ x_2(0) \\ \vdots \\ x_n(0) \end{bmatrix} \qquad \dot{\mathbf{x}}(0) = \begin{bmatrix} \dot{x}_1(0) \\ \dot{x}_2(0) \\ \vdots \\ \dot{x}_n(0) \end{bmatrix}$$

Application of the $2n$ initial conditions to Eq. (7.14) yields $2n$ equations to be solved for the $2n$ integration constants.

$$\mathbf{x}(0) = -\sum_{i=1}^{n} \mathbf{X}_i A_i \sin \phi_i \tag{7.15}$$

and

$$\dot{\mathbf{x}}(0) = \sum_{i=1}^{n} \mathbf{X}_i \omega_i A_i \cos \phi_i \tag{7.16}$$

Example 7.6. Consider again the system of Examples 7.1 and 7.2. The springs are designed such that the static-equilibrium position coincides with the horizontal orientation of the bar. Both ends of the bar are displaced a distance δ from the equilibrium position. The bar is held in this position and released. Solve for the time-dependent history of the resulting free vibrations.

Using x and θ as generalized coordinates, the initial conditions are

$$\begin{bmatrix} x(0) \\ \theta(0) \end{bmatrix} = \begin{bmatrix} \delta \\ 0 \end{bmatrix} \qquad \begin{bmatrix} \dot{x}(0) \\ \dot{\theta}(0) \end{bmatrix} = \begin{bmatrix} 0 \\ 0 \end{bmatrix}$$

Applying the initial conditions and using the results of Example 7.1 in Eqs. (7.15) and (7.16) lead to

$$-A_1 \sin \phi_1 - A_2 \sin \phi_2 = \delta$$

$$-1.43\frac{A_1}{L} \sin \phi_1 + 8.42\frac{A_2}{L} \sin \phi_2 = 0$$

$$1.28\sqrt{\frac{k}{m}} A_1 \cos \phi_1 + 2.03\sqrt{\frac{k}{m}} A_2 \cos \phi_2 = 0$$

$$1.83\sqrt{\frac{k}{m}} \frac{A_1}{L} \cos \phi_1 - 17.09\sqrt{\frac{k}{m}} \frac{A_2}{L} \cos \phi_2 = 0$$

The solution of the preceding equations is

$$A_1 = 0.855\delta \qquad \phi_1 = \frac{\pi}{2}$$

$$A_2 = 0.145\delta \qquad \phi_2 = \frac{\pi}{2}$$

which leads to

$$x(t) = \delta\left[0.855\sin\left(1.28\sqrt{\frac{k}{m}} + \frac{\pi}{2}\right) + 0.125\sin\left(2.03\sqrt{\frac{k}{m}} + \frac{\pi}{2}\right)\right]$$

$$\theta(t) = 1.22\frac{\delta}{L}\left[\sin\left(1.28\sqrt{\frac{k}{m}} + \frac{\pi}{2}\right) - \sin\left(2.03\sqrt{\frac{k}{m}} + \frac{\pi}{2}\right)\right]$$

7.6 SPECIAL CASES

Repeated eigenvalues of $\mathbf{M}^{-1}\mathbf{K}$ and $\mathbf{AM}$ occur when the natural frequencies of two distinct modes coincide. It is usually possible to identify the separate modes of vibration. For example, consider the circular cantilever beam of Fig. 7.11. The beam has a thin disk attached at its end. If the disk is vertically displaced and released, the disk undergoes free transverse vibrations. Using a one-degree-of-freedom model, and ignoring inertia effects of the beam, the natural frequency of free transverse vibrations of the disk is

$$\omega_1 = \sqrt{\frac{3EI}{mL^3}}$$

where E is the elastic modulus of the beam, I is the cross-sectional moment of inertia of the beam, L is the length of the beam, and m is the mass of the disk. If the disk is twisted and released, it undergoes free torsional oscillations. Using a one-degree-of-freedom model, and ignoring inertia effects of the beam, the natural frequency of free torsional oscillations is

$$\omega_2 = \sqrt{\frac{JG}{I_d L}}$$

where J is the polar moment of inertia of the cross section of the beam, G is the beam's shear modulus, and I_D is the mass moment of inertia of the disk. These two natural frequencies are equal for a steel shaft when the ratio of the length of the beam to the radius of the disk is 1.40. The two modes of vibration are independent but happen to have the same natural frequency.

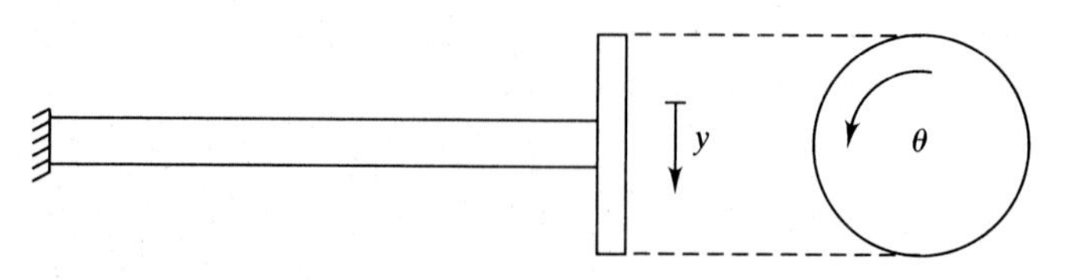

FIGURE 7.11
For certain combination of parameters the natural frequency for transverse vibration coincides with the natural frequency for torsional oscillation.

If ω_i is a natural frequency calculated from an eigenvalue of multiplicity m, then only $n - m$ of the linear algebraic equations from which the mode shape is calculated are independent. Thus m elements of the mode shape can be arbitrarily chosen. The most general mode shape involves m arbitrary constants. Then m linearly independent mode shapes, $\mathbf{X}_i, \mathbf{X}_{i+1}, \ldots, \mathbf{X}_{i+m}$, are specified. The general solution of Eq. (7.1) is still given by Eq. (7.14), but $\omega_i = \omega_{i+1} = \cdots = \omega_{i+m}$.

Example 7.7. The two-degree-of-freedom system of Fig. 7.12 has a natural frequency of $\sqrt{6k/m}$ corresponding to a rotational mode and a natural frequency of $\sqrt{2k/m}$ corresponding to a translational mode. The system is neither statically nor dynamically coupled. A block of mass m is attached to the mass center of the bar through a spring as shown in Fig. 7.13, adding a degree of freedom and leading to static coupling. The differential equations governing free vibration of this vibration of this three-degree-of-freedom system are

$$\begin{bmatrix} m & 0 & 0 \\ 0 & m & 0 \\ 0 & 0 & m\dfrac{L^2}{12} \end{bmatrix} \begin{bmatrix} \ddot{x}_1 \\ \ddot{x}_2 \\ \ddot{\theta} \end{bmatrix} + \begin{bmatrix} 2k + k_1 & -k_1 & 0 \\ -k_1 & k_1 & 0 \\ 0 & 0 & k\dfrac{L^2}{2} \end{bmatrix} \begin{bmatrix} x_1 \\ x_2 \\ \theta \end{bmatrix} = \begin{bmatrix} 0 \\ 0 \\ 0 \end{bmatrix}$$

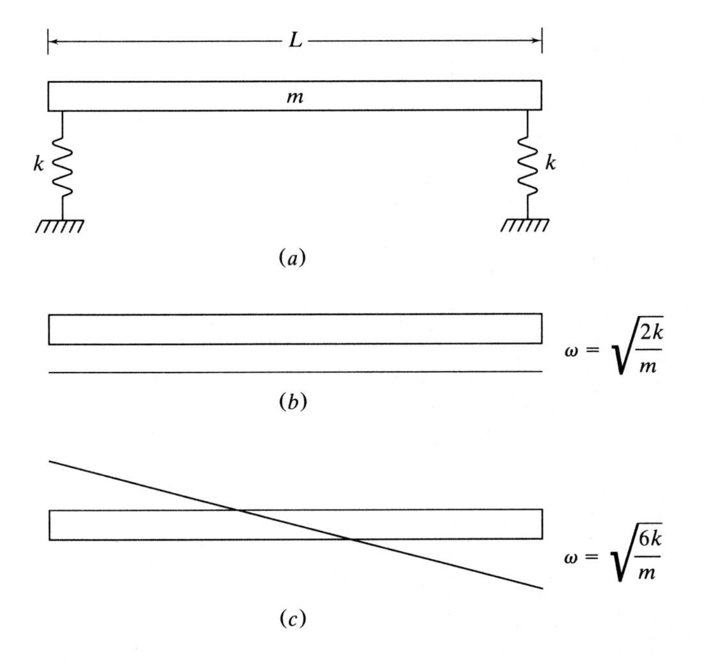

FIGURE 7.12
(a) Original system of Example 7.7; (b) mode shape for translational mode, $\omega = \sqrt{2k/m}$; (c) mode shape for rotational mode $\omega = \sqrt{6k/m}$.

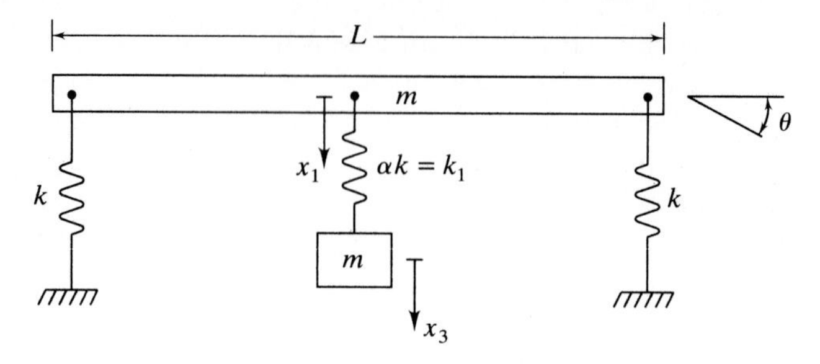

FIGURE 7.13
System of Example 7.7.

The rotational mode is still uncoupled from the other modes. Find a value of k_1 such that another natural frequency of the system coincides with the natural frequency of the rotational mode. Find the mode shapes for all modes.

The determinant leading to the characteristic equation is

$$\det\begin{bmatrix} (2+\alpha)\phi - \lambda & -\alpha\phi & 0 \\ -\alpha\phi & \alpha\phi - \lambda & 0 \\ 0 & 0 & 6-\lambda \end{bmatrix} = 0$$

where
$$\phi = \frac{k}{m}$$

and
$$\alpha = \frac{k_1}{k}$$

The characteristic equation obtained by row expansion of the determinant, using the third row, is

$$(6-\beta)\left[\beta^2 - 2(1+\alpha)\beta + 2\alpha\right] = 0$$

where
$$\beta = \frac{\lambda}{\phi}$$

The roots of the characteristic equation are

$$\beta = 6, 1 + \alpha \pm \sqrt{1+\alpha^2}$$

The $\beta = 6$ root corresponds to the natural frequency of the rotational mode. Requiring one of the other natural frequencies to be equal to the natural frequency of the rotational mode leads to

$$1 + \alpha \pm \sqrt{1+\alpha^2} = 6 \Rightarrow \alpha = \tfrac{12}{5}$$

Then the natural frequencies become

$$\omega_1 = \sqrt{\frac{4k}{5m}} \qquad \omega_2 = \omega_3 = \sqrt{6\frac{k}{m}}$$

The mode shape corresponding to the lowest natural frequency is

$$\mathbf{X}_1 = \begin{bmatrix} 1 \\ 1.5 \\ 0 \end{bmatrix}$$

For $\beta = 6$ the mode shapes are determined from

$$\begin{bmatrix} -1.6\phi & -2.4\phi & 0 \\ -2.4\phi & -3.6\phi & 0 \\ 0 & 0 & 0 \end{bmatrix} \begin{bmatrix} X_{21} \\ X_{22} \\ X_{23} \end{bmatrix} = \begin{bmatrix} 0 \\ 0 \\ 0 \end{bmatrix}$$

The general solution of this system contains two arbitrary constants and can be written as

$$\begin{bmatrix} a \\ -\frac{2}{3}a \\ b \end{bmatrix} = a \begin{bmatrix} 1 \\ -\frac{2}{3} \\ 0 \end{bmatrix} + b \begin{bmatrix} 0 \\ 0 \\ 1 \end{bmatrix}$$

Thus the two linearly independent mode shapes corresponding to $\omega = \sqrt{6k/m}$ are

$$\mathbf{X}_2 = \begin{bmatrix} 1 \\ -\frac{2}{3} \\ 0 \end{bmatrix} \qquad \mathbf{X}_3 = \begin{bmatrix} 0 \\ 0 \\ 1 \end{bmatrix}$$

A second special case occurs when one of the eigenvalues of $\mathbf{M}^{-1}\mathbf{K}$ is zero. The general solution for a system with a zero eigenvalue is

$$x(t) = (C_1 + C_2 t)\mathbf{X}_1 + \sum_{i=2}^{n} A_i \mathbf{X}_i \sin(\omega_i t - \phi_i) \tag{7.17}$$

where C_1, C_2, and A_i are constants determined from application of the initial conditions. The first part of the solution corresponds to a rigid-body motion. The summation term corresponds to oscillatory motion.

A system has a natural frequency of zero only when it is unrestrained. For example, if both masses of the two-degree-of-freedom system of Fig. 7.14 are given the same initial displacement with no initial velocity, they will remain in their displaced positions indefinitely. If the shaft connecting the two flywheels of Fig. 7.15 is rotating at a constant speed, both flywheels will continue to rotate at this speed.

When motion of an unrestrained system occurs, either linear or angular momentum is conserved for the entire system. Application of the principle of conservation of linear momentum or the principle of conservation of angular momentum provides a relationship between the generalized coordinates of the

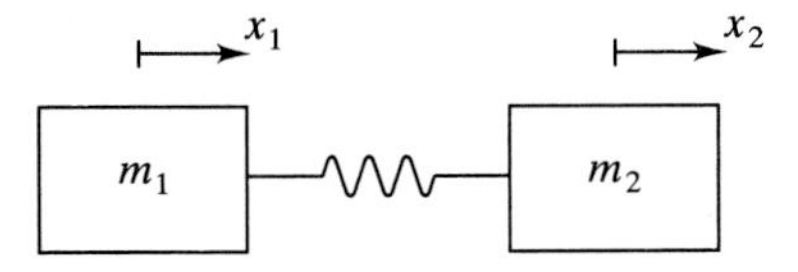

FIGURE 7.14
A two-degree-of-freedom unrestrained system. If both blocks are given the same displacement, they will move as a rigid body. If blocks are given different displacements, free oscillations occur.

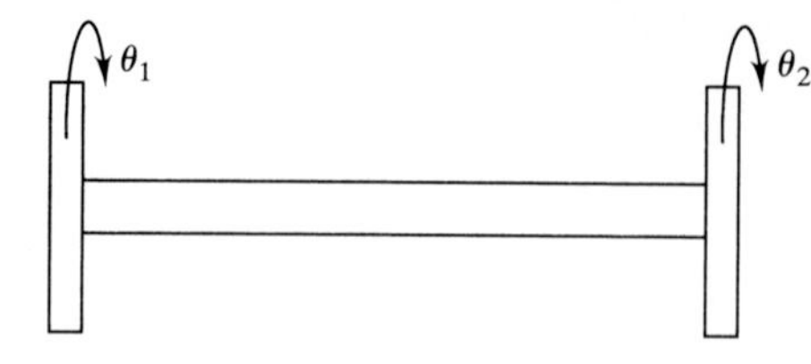

FIGURE 7.15
An unrestrained torsional system.

form

$$\sum_{l=1}^{n} \alpha_l \dot{x}_l = C_1 \tag{7.18}$$

where C_1 is a constant determined from the initial state. Equation (7.18) can be integrated to provide a constraint between the generalized coordinates of the form

$$\sum_{l=1}^{n} \alpha_l x_l = C_1 t + C_2 \tag{7.19}$$

Equation (7.19) could be used to reduce the number of degrees of freedom by one.

Example 7.8. A railroad car of mass 1500 kg is to be coupled to an assembly of two precoupled identical railroad cars. The couplers are elastic connections of stiffness 4.2×10^7 N/m. The single car is rolled toward the other cars with a velocity of 7 m/s. Describe the motion of the three railroad cars after coupling is achieved.

After coupling, the motion of the three railroad cars is modeled using three degrees of freedom, as shown in Fig. 7.16*b*. The differential equations of motion are

$$\begin{bmatrix} m & 0 & 0 \\ 0 & m & 0 \\ 0 & 0 & m \end{bmatrix} \begin{bmatrix} \ddot{x}_1 \\ \ddot{x}_2 \\ \ddot{x}_3 \end{bmatrix} + \begin{bmatrix} k & -k & 0 \\ -k & 2k & -k \\ 0 & -k & k \end{bmatrix} = \begin{bmatrix} 0 \\ 0 \\ 0 \end{bmatrix}$$

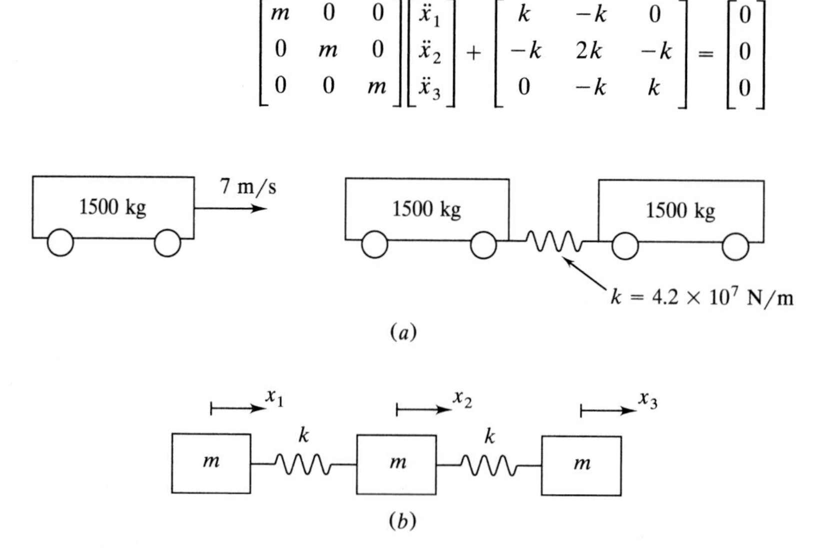

FIGURE 7.16
(*a*) Shunting of railroad cars; (*b*) three-degree-of-freedom model of coupled system.

The initial conditions are

$$\begin{bmatrix} x_1(0) \\ x_2(0) \\ x_3(0) \end{bmatrix} = \begin{bmatrix} 0 \\ 0 \\ 0 \end{bmatrix} \qquad \begin{bmatrix} \dot{x}_1(0) \\ \dot{x}_2(0) \\ \dot{x}_3(0) \end{bmatrix} = \begin{bmatrix} 7\text{ m/s} \\ 0 \\ 0 \end{bmatrix}$$

The natural frequencies are determined from

$$\det \begin{bmatrix} \phi - \lambda & -\phi & 0 \\ -\phi & 2\phi - \lambda & -\phi \\ 0 & -\phi & \phi - \lambda \end{bmatrix} = 0$$

where $\phi = \sqrt{k/m}$. The resulting characteristic equation is solved to yield

$$\omega_1 = 0$$

$$\omega_2 = \sqrt{\frac{k}{m}} = 167.3 \frac{\text{rad}}{\text{s}}$$

$$\omega_3 = \sqrt{\frac{3k}{m}} = 289.8 \frac{\text{rad}}{\text{s}}$$

The corresponding mode shapes are

$$\mathbf{X}_1 = \begin{bmatrix} 1 \\ 1 \\ 1 \end{bmatrix} \qquad X_2 = \begin{bmatrix} 1 \\ 0 \\ -1 \end{bmatrix} \qquad \mathbf{X}_3 = \begin{bmatrix} 1 \\ -2 \\ 1 \end{bmatrix}$$

The general solution of the differential equations is

$$\begin{bmatrix} x_1(t) \\ x_2(t) \\ x_3(t) \end{bmatrix} = (C_1 + C_2 t) \begin{bmatrix} 1 \\ 1 \\ 1 \end{bmatrix} + C_3 \begin{bmatrix} 1 \\ 0 \\ -1 \end{bmatrix} \sin(167.3t + \phi_1)$$

$$+ C_4 \begin{bmatrix} 1 \\ -2 \\ 1 \end{bmatrix} \sin(289.8t + \phi_2)$$

Application of the initial conditions leads to

$$C_1 = \phi_1 = \phi_2 = 0$$

$$C_2 = 2.32\text{ m/s}$$

$$C_3 = 0.021\text{ m}$$

$$C_4 = 0.004\text{ m}$$

The equation expressing conservation of linear momentum of the railroad cars after coupling is achieved is

$$m\dot{x}_1(t) + m\dot{x}_2(t) + m\dot{x}_3(t) = C$$

Example 7.9. Consider the unrestrained three-degree-of-freedom system of Example 6.8 and Fig. 6.10. Let $mr^2/I = 2$. Calculate the natural frequencies and illustrate the development of the constraint from momentum considerations.

The differential equations are

$$\begin{bmatrix} 2m & 0 & 0 \\ 0 & m & 0 \\ 0 & 0 & I \end{bmatrix}\begin{bmatrix} \ddot{x}_A \\ \ddot{x}_B \\ \ddot{\theta} \end{bmatrix} + \begin{bmatrix} k & 0 & -kr \\ 0 & 3k & -6kr \\ -kr & -6kr & 13kr^2 \end{bmatrix}\begin{bmatrix} x_A \\ x_B \\ \theta \end{bmatrix} = \begin{bmatrix} 0 \\ 0 \\ 0 \end{bmatrix}$$

The characteristic equation is developed from

$$\det\begin{bmatrix} \frac{1}{2}\phi - \lambda & 0 & -\frac{r}{2}\phi \\ 0 & 3\phi - \lambda & -6r\phi \\ -\frac{mr}{I}\phi & -\frac{6mr}{I}\phi & \frac{13mr^2}{I}\phi - \lambda \end{bmatrix} = 0$$

where $\phi = \sqrt{k/m}$. The characteristic equation is

$$-\beta^3 + \frac{33}{2}\beta^2 - \frac{31}{2}\beta = 0$$

where $\beta = \lambda/\phi$. The roots of this equation are

$$\beta = 0, 1, 15.5$$

Which lead to natural frequencies of

$$\omega_1 = 0 \qquad \omega_2 = \sqrt{\frac{k}{m}} \qquad \omega_3 = 3.94\sqrt{\frac{k}{m}}$$

Momentum diagrams for this system are given in Fig. 7.17. Application of the principle of conservation of angular momentum about the center of the pulley leads to

$$2mr\dot{x}_A(t) + 2mr\dot{x}_B(t) + I\dot{\theta} = 2mr\dot{x}_A(0) + 2mr\dot{x}_B(0) + I\dot{\theta}(0)$$

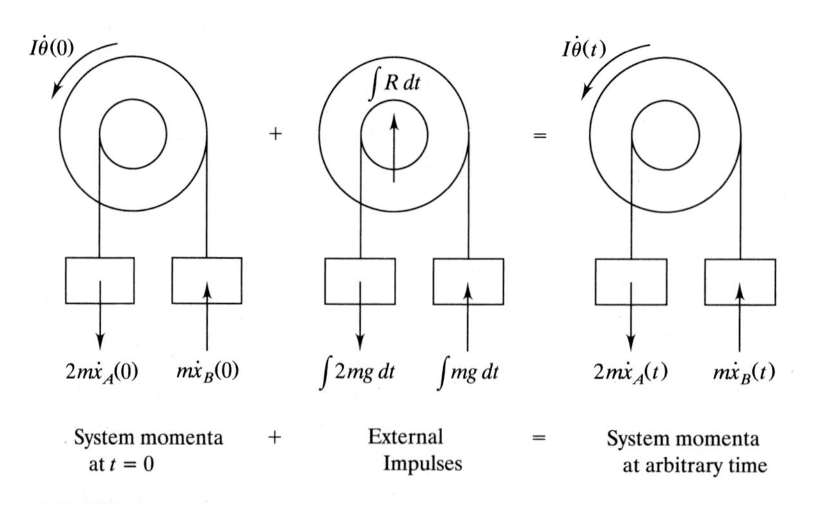

FIGURE 7.17
Angular momentum about center of pulley is conserved for this unrestrained system.

7.7 ENERGY SCALAR PRODUCTS

A scalar product is an operation performed on two vectors such that the result is a scalar. In order for the operation to be termed a scalar product, it must satisfy certain rules as outlined in App. C. When the differential equations governing the motion of a linear n-degree-of-freedom system are formulated using energy methods, the mass and stiffness matrices are symmetric. Then for a stable restrained system the following two operations satisfy all requirements to be called scalar products. Let **y** and **z** be any two n-dimensional vectors, define

$$(\mathbf{y},\mathbf{z})_K = \mathbf{y}^T\mathbf{K}\mathbf{z} \tag{7.20}$$

and

$$(\mathbf{y},\mathbf{z})_M = \mathbf{y}^T\mathbf{M}\mathbf{z} \tag{7.21}$$

The scalar product defined by Eq. (7.20) is called the potential energy scalar product. Let $\mathbf{X}_i$ be the mode shape corresponding to a natural frequency ω_i. If the system response includes only this mode, then from Eq. (7.14)

$$\mathbf{x}(t) = A_i\mathbf{X}_i\sin(\omega_i t - \phi_i) \tag{7.22}$$

From Eq. (7.5) the potential energy is calculated as

$$V = \frac{A_i^2}{2}\sin^2(\omega_i t - \phi_i)\sum_{r=1}^{n}\sum_{s=1}^{n}k_{rs}X_{ir}X_{is}$$

$$= \frac{A_i^2}{2}\sin^2(\omega_i t - \phi_i)(\mathbf{X}_i,\mathbf{X}_i)_K \tag{7.23}$$

Thus, at a given instant of time, the potential energy scalar product of a mode shape with itself is proportional to the potential energy associated with that mode.

The scalar product defined by Eq. (7.21) is called the kinetic energy scalar product. It can be shown using Eqs. (6.6) and (7.21) that

$$T = \frac{A_i^2}{2}\omega_i^2\cos^2(\omega_i t - \phi_i)(\mathbf{X}_i,\mathbf{X}_i)_M \tag{7.24}$$

or that for a linear system the kinetic energy scalar product of a mode shape with itself is proportional to the kinetic energy associated with that mode.

The mass and stiffness matrices for a linear system are guaranteed to be symmetric. In addition, the mass matrix is positive definite. The stiffness matrix for a stable system is positive definite unless it is unrestrained. The stiffness matrix for an unstable system is not positive definite. Thus, from Example C.6 of App. C, Eq. (7.21) defines a valid scalar product for all n-degree-of-freedom systems and Eq. (7.20) defines a valid scalar product for all stable constrained one-degree-of-freedom systems.

The validity of Eqs. (7.20) and (7.21) as being valid scalar products leads to the following useful properties:

1. *Commutivity of scalar products*. For all real n-dimensional vectors $\mathbf{y}$ and $\mathbf{z}$:

$$(\mathbf{y},\mathbf{z})_K = (\mathbf{z},\mathbf{y})_K \tag{7.25a}$$

and

$$(\mathbf{y},\mathbf{z})_M = (\mathbf{z},\mathbf{y})_M \tag{7.25b}$$

2. *Linearity*. For all real n-dimensional vectors $\mathbf{w}$, $\mathbf{y}$, and $\mathbf{z}$ and all real scalars α and β:

$$(\alpha\mathbf{w} + \beta\mathbf{y},\mathbf{z})_K = \alpha(\mathbf{w},\mathbf{z})_K + \beta(\mathbf{y},\mathbf{z})_K \tag{7.26a}$$

and

$$(\alpha\mathbf{w} + \beta\mathbf{y},\mathbf{z})_M = \alpha(\mathbf{w},\mathbf{z})_M + \beta(\mathbf{y},\mathbf{z})_M \tag{7.26b}$$

Two vectors are said to be orthogonal with respect to a scalar product if their scalar product is zero. The n-dimensional vectors $\mathbf{y}$ and $\mathbf{z}$ are orthogonal with respect to the potential energy scalar product if

$$(\mathbf{y},\mathbf{z})_K = 0 \tag{7.27a}$$

The vectors are orthogonal with respect to the kinetic energy scalar product if

$$(\mathbf{y},\mathbf{z})_M = 0 \tag{7.27b}$$

The use of scalar product notation is not essential to analyze and understand free and forced vibrations of multi-degree-of-freedom systems. Writing equations using scalar product notation is usually less confusing than using matrix and vector notation. In addition, since the scalar products have identifiable physical meaning, it may be easier to identify the physical significance of an equation when it is written using scalar product notation. At the very least the energy scalar products can be thought of as shorthand notation for the products defined by Eqs. (7.20) and (7.21). For these reasons the remainder of the discussion in Chap. 7 and the entire discussion in Chap. 8 uses scalar product notation. In addition, a scalar product is developed for use with continuous systems in Chap. 9. Many equations are also written using matrix notation for those not comfortable with scalar product notation.

7.8 PROPERTIES OF NATURAL FREQUENCIES AND MODE SHAPES

Let ω_i and ω_j be distinct natural frequencies of an n-degree-of-freedom system. Let $\mathbf{X}_i$ and $\mathbf{X}_j$ be their respective mode shapes. From Eq. (7.8) the equations satisfied by these natural frequencies and mode shapes are

$$\omega_i^2\mathbf{M}\mathbf{X}_i = \mathbf{K}\mathbf{X}_i \tag{7.28}$$

and

$$\omega_j^2\mathbf{M}\mathbf{X}_j = \mathbf{K}\mathbf{X}_j \tag{7.29}$$

Premultiplying Eq. (7.28) by $\mathbf{X}_j^T$ gives

$$\omega_i^2 \mathbf{X}_j^T \mathbf{M} \mathbf{X}_i = \mathbf{X}_j^T \mathbf{K} \mathbf{X}_i$$

or using scalar product notation

$$\omega_i^2 (\mathbf{X}_j, \mathbf{X}_i)_M = (\mathbf{X}_j, \mathbf{X}_i)_K \tag{7.30}$$

Premultiplying Eq. (7.29) by $\mathbf{X}_i^T$ gives

$$\omega_j^2 (\mathbf{X}_i, \mathbf{X}_j)_M = (\mathbf{X}_i, \mathbf{X}_j)_K \tag{7.31}$$

Subtracting Eq. (7.31) from Eq. (7.30) gives

$$\omega_i^2 (\mathbf{X}_j, \mathbf{X}_i)_M - \omega_j^2 (\mathbf{X}_i, \mathbf{X}_j)_M = (\mathbf{X}_j, \mathbf{X}_i)_K - (\mathbf{X}_i, \mathbf{X}_j)_K \tag{7.32}$$

Using the commutivity of the scalar products, Eq. (7.32) reduces to

$$\left(\omega_i^2 - \omega_j^2\right)(\mathbf{X}_j, \mathbf{X}_i)_M = 0 \tag{7.33}$$

Since $\omega_i \neq \omega_j$,

$$(\mathbf{X}_i, \mathbf{X}_j)_M = 0 \tag{7.34}$$

or mode shapes corresponding to distinct natural frequencies are orthogonal with respect to the kinetic energy scalar product. Then from Eq. (7.30) these mode shapes are also orthogonal with respect to the potential energy scalar product, or

$$(\mathbf{X}_i, \mathbf{X}_j)_K = 0 \tag{7.35}$$

If a system has a zero natural frequency, then it is strictly improper to define a potential energy scalar product. Property 3 required for scalar products is violated. However, it can be shown that the mode shape for the rigid-body mode for an unrestrained system is orthogonal to all other mode shapes for the system.

If an eigenvalue is not distinct, but has a multiplicity $m > 1$, then there are m linearly independent mode shapes corresponding to that eigenvalue. The preceding analysis shows that each of these mode shapes is orthogonal to mode shapes corresponding to different natural frequencies. Independent mode shapes obtained by solving Eq. (7.5) for the same eigenvalue may or may not be mutually orthogonal with respect to the energy scalar products. However, a procedure known as the Gram-Schmidt orthogonalization process can be used to replace these mode shapes with a set of m mutually orthogonal mode shapes. These orthogonalized mode shapes are linearly dependent with the original mode shapes.

Example 7.10. Demonstrate orthogonality of the mode shapes with respect to the kinetic energy scalar product for the system of Example 7.4.

The mass matrix, stiffness matrix, and mode shapes are as given in Example 7.4. Orthogonality with respect to the kinetic energy inner product is as follows:

$$(\mathbf{X}_1, \mathbf{X}_2)_M = \mathbf{X}_1^T \mathbf{M} \mathbf{X}_2$$

$$= [0.908 \quad 1 \quad 0.384] \begin{bmatrix} m & 0 & 0 \\ 0 & m & 0 \\ 0 & 0 & \frac{m}{2} \end{bmatrix} \begin{bmatrix} -1.375 \\ 1 \\ 1.294 \end{bmatrix}$$

$$= [0.908 \quad 1 \quad 0.384] \begin{bmatrix} -1.375m \\ m \\ 0.647m \end{bmatrix}$$

$$= (0.908)(-1.375m) + (1)(m) + (0.384)(0.647m)$$

$$= -0.000052m \approx 0$$

$$(\mathbf{X}_1, \mathbf{X}_3)_M = \mathbf{X}_1^T \mathbf{M} \mathbf{X}_3$$

$$= [0.908 \quad 1 \quad 0.384] \begin{bmatrix} m & 0 & 0 \\ 0 & m & 0 \\ 0 & 0 & \frac{m}{2} \end{bmatrix} \begin{bmatrix} -0.534 \\ 1 \\ -2.677 \end{bmatrix}$$

$$= [0.908 \quad 1 \quad 0.384] \begin{bmatrix} -0.534m \\ m \\ -1.239m \end{bmatrix}$$

$$= (0.908)(-0.534m) + (1)(m) + (0.384)(-1.339m)$$

$$= 0.00095m \approx$$

$$(\mathbf{X}_2, \mathbf{X}_3)_M = \mathbf{X}_2^T \mathbf{M} \mathbf{X}_3$$

$$= [-1.375 \quad 1 \quad 1.294] \begin{bmatrix} m & 0 & 0 \\ 0 & m & 0 \\ 0 & 0 & \frac{m}{2} \end{bmatrix} \begin{bmatrix} -0.534 \\ 1 \\ -2.677 \end{bmatrix}$$

$$= [1.375 \quad 1 \quad 1.294] \begin{bmatrix} -0.535m \\ m \\ -1.338m \end{bmatrix}$$

$$= (-1.375)(-0.534m) + (1)(m) + (1.294)(-1.339m)$$

$$= -0.00159m \approx 0$$

A variation of the preceding argument is used to prove that the eigenvalues are all real. The formal proof of this statement involves the introduction of a complex energy scalar product and is beyond the scope of this text.

The argument can also be used to show that if $\mathbf{M}$ and $\mathbf{K}$ are positive definite then the eigenvalues of $\mathbf{M}^{-1}\mathbf{K}$ are all positive. Let $\mathbf{X}_i = \mathbf{X}_j$ in Eq. (7.30).

$$\omega_i^2 = \frac{(\mathbf{X}_i, \mathbf{X}_i)_K}{(\mathbf{X}_i, \mathbf{X}_i)_M} \tag{7.36}$$

If $\mathbf{M}$ and $\mathbf{K}$ are positive definite, then the energy scalar products are valid scalar products. This implies that both scalar products in the quotient of Eq. (7.36) are positive. Hence

$$\omega_i^2 > 0 \tag{7.37}$$

This, in turn, shows that a system in which both the mass and stiffness matrices are positive definite is stable.

The ratio of Eq. (7.36) is called Rayleigh's quotient. For a given mode it is the ratio of the potential energy to the kinetic energy for that mode.

It is possible to construct n orthogonal, and hence linearly independent, mode shapes for an n-degree-of-freedom system. Thus any n-dimensional vector can be written as a linear combination of these n mode shapes. To this end, if $\mathbf{y}$ is any n-dimensional vector

$$\mathbf{y} = \sum_{i=1}^{n} c_i \mathbf{X}_i \tag{7.38}$$

where c_i, $i = 1, 2, \ldots, n$, are coefficients. Equation (7.38) is known as the expansion theorem. Premultiplying Eq. (7.38) by $\mathbf{X}_j^T\mathbf{M}$ for some j, $1 \le j \le n$. The result is written using scalar product notation as

$$(\mathbf{X}_j, \mathbf{y})_M = \left(\mathbf{X}_j, \sum_{i=1}^{n} c_i \mathbf{X}_i\right)_M \tag{7.39}$$

Interchanging the scalar product operation with the summation and using the property of Eq. (7.25) gives

$$(\mathbf{X}_j, \mathbf{y})_M = \sum_{i=1}^{n} c_i (\mathbf{X}_j, \mathbf{X}_i)_M \tag{7.40}$$

The orthogonality of the mode shapes implies that the only nonzero term in the summation occurs when $i = j$. Then Eq. (7.38) reduces to

$$c_j = \frac{(\mathbf{X}_j, \mathbf{y})_M}{(\mathbf{X}_j, \mathbf{X}_j)_M} \tag{7.41}$$

7.9 NORMALIZED MODE SHAPES

A mode shape corresponding to a specific natural frequency of an n-degree-of-freedom system is unique only to a multiplicative constant. The arbitrariness can be alleviated by requiring the mode shape to satisfy the normalization constraint. A mode shape chosen to satisfy the normalization constraint is called a normalized mode shape. The normalization constraint, itself, is arbitrary. However, all mode shapes are required to satisfy the same normalization constraint. The constraint should be chosen such that subsequent use of the normalized mode shape is convenient.

It is convenient to normalize mode shapes by requiring that the kinetic energy scalar product of a mode shape with itself is equal to one. That is,

$$(\mathbf{X}_i, \mathbf{X}_i)_M = \mathbf{X}_i^T \mathbf{M} \mathbf{X}_i = 1 \tag{7.42}$$

If the mode shape, $\mathbf{X}_i$, is normalized according to Eq. (7.42), then from Rayleigh's quotient, Eq. (7.36),

$$\mathbf{X}_i^T \mathbf{K} \mathbf{X}_i = (\mathbf{X}_i, \mathbf{X}_i)_K = \omega_i^2 \tag{7.43}$$

The orthogonality relations, Eqs. (7.34) and (7.35), the normalization constraint, Eq. (7.42), and the subsequent result of the choice of normalization, Eq. (7.43), are summarized by

$$(\mathbf{X}_i, \mathbf{X}_j)_M = \delta_{ij} \tag{7.44}$$

and

$$(\mathbf{X}_i, \mathbf{X}_j)_K = \omega_i^2 \delta_{ij} \tag{7.45}$$

where δ_{ij} is the Kronecker delta. From this point, mode shapes will be assumed to be normalized by Eq. (7.42).

With the normalization scheme of Eq. (7.42), the expansion theorem, Eqs. (7.38) and (7.41), becomes

$$\mathbf{y} = \sum_{i=1}^{n} (\mathbf{X}_i, \mathbf{y})_M \mathbf{X}_i \tag{7.46}$$

Example 7.11. Expand the vector

$$\mathbf{y} = \begin{bmatrix} 1 \\ 4 \\ -2 \end{bmatrix}$$

using the normalized mode shapes of Example 7.4.

The general mode shapes of Example 7.4 are

$$\mathbf{X}_1 = B_1 \begin{bmatrix} 0.908 \\ 1 \\ 0.384 \end{bmatrix} \qquad \mathbf{X}_2 = B_2 \begin{bmatrix} -1.375 \\ 1 \\ 1.294 \end{bmatrix} \qquad \mathbf{X}_3 = B_3 \begin{bmatrix} -0.534 \\ 1 \\ -2.677 \end{bmatrix}$$

where B_1, B_2, and B_3 are arbitrary constants. The normalization of the first mode shape proceeds as follows

$$1 = (\mathbf{X}_1, \mathbf{X}_1)_M$$

$$= B_1^2 [0.908 \quad 1 \quad 0.384] \begin{bmatrix} m & 0 & 0 \\ 0 & m & 0 \\ 0 & 0 & \frac{m}{2} \end{bmatrix} \begin{bmatrix} 0.908 \\ 1 \\ 0.384 \end{bmatrix}$$

which yields $B_1 = 0.726/\sqrt{m}$ and

$$\mathbf{x}_1 = \frac{1}{\sqrt{m}} \begin{bmatrix} 0.659 \\ 0.726 \\ 0.279 \end{bmatrix}$$

The other mode shapes are normalized in the same manner yielding

$$\mathbf{X}_2 = \frac{1}{\sqrt{m}} \begin{bmatrix} -0.712 \\ 0.518 \\ 0.670 \end{bmatrix} \qquad \mathbf{X}_3 = \frac{1}{\sqrt{m}} \begin{bmatrix} -0.242 \\ 0.453 \\ -1.213 \end{bmatrix}$$

The first coefficient in the expansion is calculated by

$$c_1 = (\mathbf{X}_1, \mathbf{y})_M$$

$$= \frac{1}{\sqrt{m}} [0.659 \quad 0.726 \quad 0.279] \begin{bmatrix} m & 0 & 0 \\ 0 & m & 0 \\ 0 & 0 & \frac{m}{2} \end{bmatrix} \begin{bmatrix} 1 \\ 4 \\ -2 \end{bmatrix}$$

$$= 3.284\sqrt{m}$$

The other coefficients are calculated in a similar manner, yielding $c_2 = 0.690\sqrt{m}$, $c_3 = 2.777\sqrt{m}$. Thus

$$\begin{bmatrix} 1 \\ 4 \\ -2 \end{bmatrix} = 3.284 \begin{bmatrix} 0.659 \\ 0.726 \\ 0.279 \end{bmatrix} + 0.690 \begin{bmatrix} -0.712 \\ 0.518 \\ 0.670 \end{bmatrix} + 2.787 \begin{bmatrix} -0.242 \\ 0.455 \\ -1.213 \end{bmatrix}$$

7.10 DETERMINATION OF NATURAL FREQUENCIES BY MATRIX ITERATION

Determination of the characteristic equation for an n-degree-of-freedom system requires the expansion of an $n \times n$ determinant. The expansion is tedious and time consuming, even on a digital computer. Thus numerical methods have been developed such that natural frequencies and mode shapes can be determined without first determining the characteristic equation.

As shown in Sec. 7.12, good approximations to lower natural frequencies of structural systems often require the system to be modeled using a large number of degrees of freedom. The higher natural frequencies that are predicted from a discrete analysis are inaccurate and not useful for computation. Thus numerical methods have been developed such that natural frequencies and mode shapes are calculated sequentially.

A numerical method that calculates natural frequencies and mode shapes sequentially without using the characteristic equation is the matrix iteration or power iteration method described in this section. A subroutine using matrix iteration to calculate the natural frequencies and mode shapes for an arbitrary n-degree-of-freedom system is provided and used for several examples. A matrix iteration program, MITER, is provided in the accompanying diskette.

Consider an n-degree-of-freedom system with natural frequencies $\omega_1 \leq \omega_2 \leq \cdots \leq \omega_n$ and corresponding normalized mode shapes $\mathbf{X}_1, \mathbf{X}_2, \ldots, \mathbf{X}_n$. Let $\mathbf{u}_1$ be an arbitrary n-dimensional vector. From the expansion theorem, Eq. (7.46), there exist constants $c_1, c_2, \ldots, c_n$ such that

$$\mathbf{u}_1 = \sum_{i=1}^{n} c_i \mathbf{X}_i \tag{7.47}$$

Starting with $\mathbf{u}_1$, define a sequence of n-dimensional vectors $\mathbf{u}_i$, $i = 1, 2, \ldots,$ such that

$$\mathbf{u}_i = \mathbf{AMu}_{i-1} \tag{7.48}$$

Then

$$\begin{aligned} \mathbf{u}_2 &= \mathbf{AMu}_1 \\ &= \mathbf{AM} \sum_{i=1}^{n} c_i \mathbf{X}_i \\ &= \sum_{i=1}^{n} c_i \mathbf{AMX}_i \\ &= \sum_{i=1}^{n} \frac{c_i}{\omega_i^2} \mathbf{X}_i \end{aligned}$$

Then

$$\begin{aligned} \mathbf{u}_3 &= \mathbf{AMu}_2 = \sum_{i=1}^{n} \frac{c_i}{\omega_i^4} \mathbf{X}_i = \mathbf{u}_3 \\ \mathbf{u}_4 &= \mathbf{AMu}_3 = \sum_{i=1}^{n} \frac{c_i}{\omega_i^6} \mathbf{X}_i \\ &\vdots \\ \mathbf{u}_{j+1} &= \mathbf{AMu}_j = \sum_{i=1}^{n} \frac{c_i}{\omega_i^{2j}} \mathbf{X}_i \end{aligned} \tag{7.49}$$

As j increases the first term in the summation of Eq. (7.49) is much larger than the remaining terms. Thus, for large j,

$$\mathbf{u}_j \approx \frac{c_1}{\omega_1^{2(j-1)}}\mathbf{X}_1 \tag{7.50}$$

and

$$\mathbf{u}_{j+1} \approx \frac{c_1}{\omega_1^{2j}}\mathbf{X}_1 \tag{7.51}$$

Thus, for large j, $\mathbf{u}_j$ is proportional to $\mathbf{X}_1$, and the ratio of corresponding elements of $\mathbf{u}_j$ and $\mathbf{u}_{j+1}$ is constant and is equal to the square of the lowest natural frequency. That is,

$$\frac{(\mathbf{u}_j)_k}{(\mathbf{u}_{j+1})_k} = \omega_1^2 \qquad k = 1, 2, \ldots, n \tag{7.52}$$

The preceding iteration process is often called power iteration because

$$\mathbf{u}_j = (\mathbf{AM})^{j-1}\mathbf{u}_1 \tag{7.53}$$

Matrix iteration can also be used to obtain the second and higher natural frequencies and their corresponding mode shapes. Choose $\mathbf{u}_1^2$ such that it is orthogonal to $\mathbf{X}_1$ with respect to the kinetic energy scalar product. Then if the expansion theorem is used to expand $\mathbf{u}_1^2$ in a summation of mode shapes, as in Eq. (7.47), Eq. (7.41) implies $c_1 = 0$ and

$$\mathbf{u}_1^2 = \sum_{i=2}^{n} \frac{c_i}{\omega_i^2}\mathbf{X}_i$$

The iteration process develops as before and leads to

$$\mathbf{u}_j^2 \approx \frac{c_2}{\omega_2^{2(j-1)}}\mathbf{X}_2 \tag{7.54}$$

for large j. This yields both ω_2 and $\mathbf{X}_2$ which is subsequently normalized using Eq. (7.42).

The preceding process is continued until all desired natural frequencies and mode shapes are found. If $\mathbf{u}_1^{k+1}$ is a vector orthogonal to $\mathbf{X}_1, \mathbf{X}_2, \ldots, \mathbf{X}_k$, then the coefficients from the expansion theorem for $\mathbf{u}_1^{k+1}$, $c_1, c_2, \ldots, c_k$, all equal zero. Thus, when matrix iteration is used with this choice for $\mathbf{u}_1^{k+1}$ as an initial guess, the iteration process leads to ω_{k+1} and $\mathbf{X}_{k+1}$.

It is always possible to construct an n-dimensional vector orthogonal to m mutually orthonormal vectors where $m < n$. Let $\mathbf{v}$ be any vector. Then consider

$$\mathbf{u}_1^{k+1} = \mathbf{v} - \sum_{s=1}^{k} (\mathbf{v}, \mathbf{X}_s)_M \mathbf{X}_s \tag{7.55}$$

Properties of scalar products (App. C) are used to write the kinetic energy scalar product of $\mathbf{u}_1^{k+1}$ with the normalized mode shape $\mathbf{X}_j$, $j \leq k$, as

$$\left(\mathbf{u}_1^{k+1}, \mathbf{X}_j\right)_M = \left(\mathbf{v}, \mathbf{X}_j\right)_M - \sum_{s=1}^{k} \left(\mathbf{v}, \mathbf{X}_s\right)_M \left(\mathbf{X}_s, \mathbf{X}_j\right)_M$$

Since the mode shapes are mutually orthonormal with respect to the kinetic energy scalar product, the term being summed is nonzero only for $s = j$. Thus

$$\left(\mathbf{u}_1^{k+1}, \mathbf{X}_j\right)_M = \left(\mathbf{v}, \mathbf{X}_j\right)_M - \left(\mathbf{v}, \mathbf{X}_j\right)_M = 0 \tag{7.56}$$

The initial guess vector used in the iteration to determine ω_1 can be modified to yield an appropriate initial guess on each subsequent iteration sequence by subtracting the kinetic energy scalar product of the original guess with the new normalized mode shape from the previous initial guess.

Matrix iteration using Eq. (7.55) to calculate $\mathbf{u}_1^{k+1}$ from an arbitrary $\mathbf{v}$ converges to $\mathbf{X}_{k+1}$ which must be normalized and leads to ω_{k+1}.

If all natural frequencies and mode shapes are desired, it is not necessary to use matrix iteration to determine the highest natural frequency and its corresponding mode shape. Since $\mathbf{X}_n$ is orthogonal to all modes corresponding to lower natural frequencies and there are at most n mutually orthogonal vectors, the final mode shape is simply obtained by using Eq. (7.55) for $k = n - 1$. The highest natural frequency is obtained from any component equation of

$$\mathbf{AMX}_n = \frac{1}{\omega_n^2} \mathbf{X}_n \tag{7.57}$$

A similar procedure is available using $\mathbf{M}^{-1}\mathbf{K}$, on each iteration step. However, this leads to successive calculation of the natural frequencies and their corresponding mode shapes beginning with the largest natural frequency.

The matrix iteration process as described previously has advantages and disadvantages over other numerical methods of natural frequency calculation. Its advantages are

1. The method is straightforward and is easy to program on a computer.
2. The natural frequency and mode shape for a given mode are calculated simultaneously.
3. The natural frequencies and mode shapes for different modes are calculated separately. For systems with a large number of degrees of freedom, often only a small percentage of free-vibration modes are necessary to find the forced-vibration solution accurately using modal analysis. Other methods calculate all eigenvectors and eigenvalues simultaneously.

The disadvantages of matrix iteration are

1. Matrix iteration only works for symmetric matrices. It cannot be generalized to eigenvalue problems which involve nonsymmetric matrices.

2. The rate of convergence is highly dependent upon the initial guess.
3. If two natural frequencies are close together, the rate of convergence is slow.
4. Matrix iteration may not work for problems with repeated natural frequencies.
5. Matrix iteration will diverge for a zero natural frequency.
6. Since matrix iteration calculates natural frequencies and mode shapes successively, the numerical error propagates, and decreases the accuracy with which each higher natural frequency and mode shape are calculated.

Numerical practicalities must be considered when using matrix iteration. As matrix iteration progresses from an initial trial vector, the components of the iterates grow large and awkward to handle. It is easier to divide each component of the iterate vector by the largest component of the vector. This leads to the largest component of the revised iterate being unity. When this scheme is used and convergence is close to being achieved, the square of the natural frequency will appear in the component location that was unity on the previous iteration.

During the iteration process to approximate higher natural frequencies, the numerical propagation of error leads to the deterioration of the orthogonality conditions imposed by Eq. (7.55). When this occurs the largest natural frequency is no longer suppressed and will eventually dominate the results as in the first iteration. This problem is resolved by reorthogonalizing each newly calculated iterate using Eq. (7.55).

Subroutine ITER uses matrix iteration to calculate the natural frequencies and mode shapes for an n-degree-of-freedom system given the mass matrix and either the stiffness matrix or the flexibility matrix. If the stiffness matrix is used, subroutine ITER calls subroutine INVERSE to calculate the inverse of the mass matrix. Subroutine ITER returns to a calling program a vector of the desired natural frequencies and a matrix of their corresponding normalized mode shapes.

Subroutine ITER and subroutine INVERSE are written to handle up to 10-degree-of-freedom systems. If more degrees of freedom are required, the dimension statements must be revised.

```
C ***************************************************
C                                                   *
C  SUBROUTINE INVERSE CALCULATES THE INVERSE        *
C  OF A MATRIX BY USING GAUSS ELIMINATION           *
C  ON AN AUGMENTED MATRIX                           *
C        A = MATRIX TO BE INVERTED                  *
C        B = INVERSE OF A                           *
C        N = NUMBER OF ROWS IN A                    *
C                                                   *
C ***************************************************
      SUBROUTINE INVERSE(N,A,B)
      DIMENSION A(9,9),B(9,9),C(9,18)
```

```
      DATA C / 162 * 0.0 /
      N2 = 2*N
      N1 = N - 1
C
C  FORMATION OF AUGMENTED MATRIX: A AUGMENTED WITH
C  I, THE NXN IDENTITY
C        MATRIX
C
      DO 1  I = 1,N
      DO 1  J = 1,N
    1 C(I,J) = A(I,J)
      DO 2  I = 1,N
    2 C(I,N + I) = 1.0
C
C  GAUSS ELIMINATION TO FORM LOWER TRIANGULAR MATRIX
C
      DO 3  I = 1,N1
      I1 = I + 1
      DO 3  J = I1,N
      P = C(J,I) / C(I,I)
      DO 3  K = I,N2
    3 C(J,K) = C(J,K)- P*C(I,K)
      DO 4  I = 1,N1
      I2  =  N - I + 1
      I1 = I + 1
C
C  GAUSS ELIMINATION TO FORM DIAGONAL MATRIX
C
      DO 4  J = I1,N
      J2 = N - J + 1
      P = C(J2,I2) / C(I2,I2)
      DO 4  K = J2,N2
    4 C(J2,K) = C(J2,K)- P*C(I2,K)
C
C  DIVISION BY DIAGONAL ELEMENT
C  THIS RESULTS IN I AUGMENTED BY THE INVERSE OF A
C
      DO 5  I = 1,N
      P = C(I,I)
      DO 5  J = 1,N
    5 B(I,J) = C(I,N + J) / P
      RETURN
      END

C  *************************************************
C  *                                               *
C  *   SUBROUTINE ITER(N,L,LK,A,M,U,Q,OM,ITMAX,TOL) *
C  *   PERFORMS MATRIX ITERATION TO CALCULATE      *
C  *   NATURAL FREQUENCIES                         *
C  *   AND MODE SHAPES FOR A MULTI DEGREE OF       *
C  *   SYSTEM FREEDOM                              *
C  *                                               *
C  *************************************************
```

```
C
C  THE FOLLOWING MUST BE PROVIDED BY THE CALLING
C  PROGRAM
C
C  N = DEGREES OF FREEDOM
C  L = NUMBER OF NATURAL FREQUENCIES TO BE CALCULATED
C  LK IS A PARAMETER THAT SPECIFIES WHETHER A
C  STIFFNESS OR
C     FLEXIBILITY FORMULATION IS USED
C          LK = 0 - FLEXIBILITY FORMULATION NATURAL
C              FREQUENCIES ARE EIGENVALUES OF
C              FLEXIBILITY MATRIX TIMES MASS MATRIX
C          LK = 1 - STIFFNESS FORMULATIONS NATURAL
C              FREQUENCIES ARE EIGENVALUES OF
C              INVERSE OF MASS MATRIX TIMES STIFFNESS
C              MATRIX
C  A IS AN NXN MATRIX
C      IF LK = 0, A IS THE FLEXIBILITY MATRIX
C      IF LK = 1, A IS THE STIFFNESS MATRIX
C  M = NXN MASS MATRIX
C  U = NX1 INITIAL GUESS VECTOR
C  ITMAX = MAXIMUM NUMBER OF ITERATIONS BEFORE
C  EXECUTION IS STOPPED
C  TOL = TOLERANCE FOR RESULTS
C        AFTER ADJUSTING THE NEW ITERATE, V, SUCH
C        THAT THE LARGEST ABSOLUTE VALUE OF ITS
C        ELEMENTS IS ONE CONVERGENCE IS
C        ACHIEVED IF
C        ABS((V(J) - U(J)) / (U(J)) < TOL FOR ALL
C        J = 1,...,N
C        WHERE U IS THE PREVIOUS ITERATE
C
C  THE FOLLOWING IS RETURNED TO THE CALLING PROGRAM
C
C  OM = NX1 COLUMN VECTOR OF NATURAL FREQUENCIES
C       IF LK = 0, OM(1) < OM(2) < ... < OM(N)
C       IF LK = 1, OM(1) > OM(2) > ... > OM(N)
C  Q = NXN MATRIX OF NORMALIZED MODE SHAPES
C  ERR IS AN INTEGER ERROR INDICATOR. IF ERR = 1,
C  THE MAXIMUM NUMBER OF ITERATIONS WAS EXCEEDED
C
C      SUBROUTINE ITER(N,L,LK,A,M,U,Q,OM,
C      ITMAX,TOL,ERR)
C      REAL A(9,9),M(9,9),D(9,9),U(9),V(9),W(9),
C      Q(9,9),OM(9)
       REAL MINV(9,9)
       INTEGER ERR
       IF(LK.NE.0)GO TO 101
       ERR = 0
C  CALCULATION OF DYNAMICAL MATRIX
       DO 1 I = 1,N
```

```
      DO 1 J=1,N
      SUM=0.0
      DO 2 K=1,N
    2 SUM=SUM+A(I,K)*M(K,J)
    1 D(I,J)=SUM
      GO TO 102
C   CALCULATION OF MINVERSE K
  101 CALL INVERSE (N,M,MINV)
      DO 103 I=1,N
      DO 103 J=1,N
      SUM=0.0
      DO 104 K=1,N
  104 SUM=SUM+MINV(I,K)*A(K,J)
  103 D(I,J)=SUM
C   INITIALIZATION OF VALUES
  102 KK=1
   12 DO 4 I=1,N
   4  V(I)=U(I)
      JTER=1
C   BEGINNING OF ITERATION LOOP
   18 LL=0
      DO 5 I=1,N
    5 W(I) = V(I)
      DO 6 I=1,N
      V(I) = 0.0
      DO 7 J=1,N
    7 V(I) = V(I) + D(I,J)*W(J)
    6 CONTINUE
C   REORTHOGONALIZATION
      IF(KK.EQ.1) GO TO 37
      KK1=KK-1
      DO 38 KL=1,KK1
      SUM=0.0
      DO 39 I=1,N
      DO 39 J=1,N
   39 SUM=SUM+V(J)*Q(I,KL)*M(I,J)
      DO 40 I=1,N
   40 V(I)=V(I) - SUM*Q(I,KL)
   38 CONTINUE
   37 XM=V(1)
      DO 8 I=2,N
    8 IF (ABS(V(I)).GT.ABS(SM))XM=V(I)
C   CONVERGENCE CHECK
      DO 9 I=1,N
      V(I)=V(I)/XM
      IF(ABS(V(I)).LT.0.0001)GO TO 9
      IF(ABS(1.-V(I)/W(I)).GT.TOL)LL=1
    9 CONTINUE
      IF(LL.EQ.0) GO TO 10
      JTER=JTER+1
      IF(JTER.LE.ITMAX) GO TO 18
      ERR=1
      GO TO 13
   10 OM(KK)=SQRT(ABS(XM))
```

```
      IF(LK.EQ.0)OM(KK) = 1./ OM(KK)
C  NORMALIZATION OF MODE SHAPE
      SUM = 0.0
      DO 20 I = 1,N
      DO 20 J = 1,N
   20 SUM = SUM + M(I,J)*V(I)*V(J)
      do 21 I = 1,N
   21 V(I) = V(I) / SQRT(SUM)
   26 DO 41 I = 1,N
   41 Q(I,KK) = V(I)
      KK = KK + 1
C  ORTHOGONALIZATION SCHEME
      SUM = 0.0
      DO 14 I = 1,N
      DO 14 J = 1,N
   14 SUM = SUM + U(I)*V(J)*M(I,J)
      DO 15 I = 1,N
   15 U(I) = U(I) - SUM*V(I)
      IF(KK.GT.L)GO TO 13
      IF(KK.LT.N)GO TO 12
C  CALCULATION OF LARGEST NATURAL FREQUENCY AND MODE
C  SHAPE
      SUM = 0.0
      DO 16 I = 1,N
   16 SUM = SUM + D(1,I)*U(I)
      OM(N) = SQRT(SUM / U(1))
      IF(LK.EQ.0)OM(N) = 1./ OM(N)
      SUM = 0.0
      DO 42 I = 1,N
      DO 42 J = 1,N
   42 SUM = SUM + M(I,J)*U(I)*U(J)
      DO 43 I = 1,N
   43 Q(I,N) = U(I) / SQRT(SUM)
   13 RETURN
      END
```

Example 7.12. Write a calling program that uses subroutine ITER to calculate the natural frequencies and mode shapes for Example 7.4, with $k = 1000$ N/m and m = 10 kg.

The mass matrix and stiffness matrix are given in Example 7.4. The calling program and the results follow. The results compare favorably with the exact results calculated in Example 7.4 and the normalized mode shapes calculated in Ex. 7.11.

```
C ***********************************************
C                                               *
C        CALLING PROGRAM FOR EXAMPLE 7.12       *
C                                               *
C                                               *
C ***********************************************
```

```
      EXTERNAL INVERSE
      REAL K(9,9),M(9,9),Q(9,9),OM(9),U(9)
      INTEGER ERR
      DATA K / 81* 0.0 /
      DATA M / 81* 0.0 /
      OPEN(7,FILE = 'EX812.RES')
      WRITE(7,110)
110   FORMAT(5X,'SOLUTION OF EXAMPLE 7.12')
      XK = 1000.
      XM = 10.
      M(1,1) = XM
      M(2,2) = XM
      M(3,3) = XM / 2.
      K(1,1) = 3.*XK
      K(1,2) = -2.*XK
      K(2,1) = -2.*XK
      K(2,2) = 3.*XK
      K(2,3) = -XK
      K(3,2) = -XK
      K(3,3) = 3.*XK
      U(1) = 1.0
      U(2) = 0.0
      U(3) = 0.0
      CALL ITER(3,3,1,K,M,U,Q,OM,200,1.E-5,ERR)
      WRITE(7,105)ERR
105   FORMAT(5X,'ERR = ',2X,I2)
      DO 1 I = 1,3
      J = 4 - I
      WRITE(7,100)J,OM(J)
100   FORMAT(///,5X,'THE NATURAL FREQUENCY FOR MODE',
     $I2,2X,' = ',F10.4,X 'RAD / SEC')
  1   WRITE(7,101)Q(1,J),Q(2,J),Q(3,J)
101   FORMAT( / ,5X,'THE CORRESPONDING MODE SHAPE IS',
     $/15X,3(F10.4,2X))
      STOP
      END
```

```
SOLUTION OF EXAMPLE 7.12
ERR =    0

THE NATURAL FREQUENCY FOR MODE 1 = 8.9360RAD / SEC
THE CORRESPONDING MODE SHAPE IS
              0.2085     0.2295     0.0882

THE NATURAL FREQUENCY FOR MODE 2 = 21.1066RAD / SEC
THE CORRESPONDING MODE SHAPE IS
              0.2252    -0.1638    -0.2120
```

```
THE NATURAL FREQUENCY FOR MODE 3 = 25.9742RAD / SEC
THE CORRESPONDING MODE SHAPE IS
                0.0765    -0.1432    0.3838
```

Example 7.13. Use subroutine ITER to calculate the two lowest natural frequencies and mode shapes of the six-story building shown in Fig. 7.18. Assume the girders are infinitely rigid compared to the columns and the inertia effects are all lumped at the floors.

The differential equations of free vibration are

$$\begin{bmatrix} m & 0 & 0 & 0 & 0 & 0 \\ 0 & m & 0 & 0 & 0 & 0 \\ 0 & 0 & m & 0 & 0 & 0 \\ 0 & 0 & 0 & m & 0 & 0 \\ 0 & 0 & 0 & 0 & m & 0 \\ 0 & 0 & 0 & 0 & 0 & m \end{bmatrix} \begin{bmatrix} \ddot{x}_1 \\ \ddot{x}_2 \\ \ddot{x}_3 \\ \ddot{x}_4 \\ \ddot{x}_5 \\ \ddot{x}_6 \end{bmatrix}$$

$$+ \begin{bmatrix} 2k & -k & 0 & 0 & 0 & 0 \\ -k & 2k & -k & 0 & 0 & 0 \\ 0 & -k & 2k & -k & 0 & 0 \\ 0 & 0 & -k & 2k & -k & 0 \\ 0 & 0 & 0 & -k & 2k & -k \\ 0 & 0 & 0 & 0 & -k & k \end{bmatrix} \begin{bmatrix} x_1 \\ x_2 \\ x_3 \\ x_4 \\ x_5 \\ x_6 \end{bmatrix} = \begin{bmatrix} 0 \\ 0 \\ 0 \\ 0 \\ 0 \\ 0 \end{bmatrix}$$

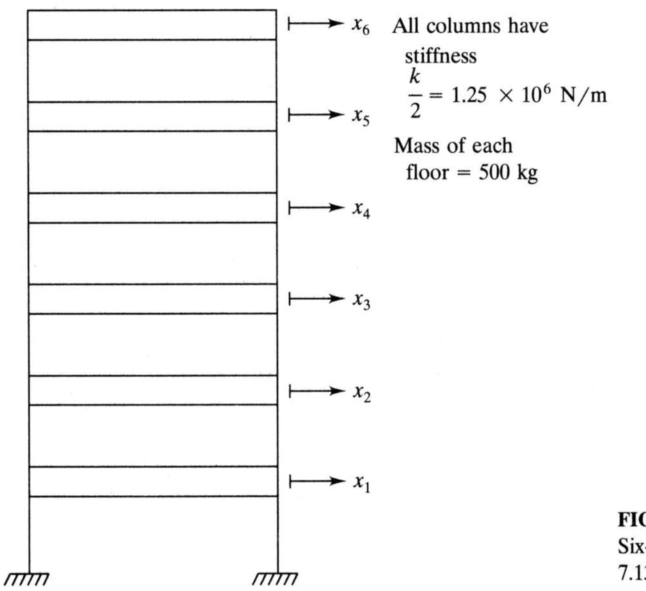

FIGURE 7.18
Six-story building of Example 7.13.

Since only the two lowest modes are required and the stiffness matrix is known, the calling program calls subroutine INVERSE to provide the flexibility matrix. Matrix iteration is then used on **AM**. An alternative is to use matrix iteration on $\mathbf{M}^{-1}\mathbf{K}$ and find all six modes. However, since the highest modes are found first by this procedure, the lowest modes would have the greatest error. The calling program using subroutine ITER and the results for the first two modes follow.

```
C ********************************************
C                                            *
C       CALLING PROGRAM FOR EXAMPLE 7.13     *
C                                            *
C                                            *
C ********************************************
        EXTERNAL INVERSE
        REAL K(9,9),M(9,9),Q(9,9),OM(0),U(9),A(9,9)
        INTEGER ERR
        DATA K / 81* 0.0 /
        DATA M / 81* 0.0 /
        OPEN(7,FILE='EX813.RES')
        WRITE(7,110)
  110   FORMAT(5X,'SOLUTION OF EXAMPLE 7.13')
        XK=2.5E6
        XM=500.
        M(1,1)=XM
        M(2,2)=XM
        M(3,3)=XM
        M(4,4)=XM
        M(5,5)=XM
        M(6,6)=XM
        K(1,1)=2.*XK
        K(1,2)=-XK
        K(2,1)=-XK
        K(2,2)=2.*XK
        K(2,3)=-XK
        K(3,2)=-XK
        K(3,3)=2.*XK
        K(3,4)=-XK
        K(4,3)=-XK
        K(4,4)=2.*XK
        K(4,5)=-XK
        K(5,4)=-XK
        K(5,5)=2.*XK
        K(5,6)=-XK
        K(6,5)=-XK
        K(6,6)=XK
        U(1)=1.0
        U(2)=0.0
        U(3)=0.0
        U(4)=1.0
```

```
      U(5)=-1.0
      U(6)=.21
      CALL INVERSE(6,K,A)
      CALL ITER(6,2,0,A,M,U,Q,OM,200,1.E-5,ERR)
      WRITE(7,105)ERR
105   FORMAT(5X,'ERR=',2X,I2)
      DO 1 I=1,2
  1   WRITE(7,100)I,OM(I)
100   FORMAT(///,5X,'THE NATURAL FREQUENCY FOR MODE',
     $I2,2X,'=',F10.4,X'RAD/SEC')
      STOP
      END
```

```
SOLUTION OF EXAMPLE 7.13
ERR=   0

THE NATURAL FREQUENCY FOR MODE 1=17.0465RAD/SEC

THE NATURAL FREQUENCY FOR MODE 2=50.1487RAD/SEC
```

Example 7.14. Determine the natural frequencies of the automobile suspension system shown in Fig. 7.19. Use a four-degree-of-freedom model, including the inertia of the axle and wheels. The springs of stiffness 1000 lb/ft model the elasticity of the tires.

The generalized coordinates are as shown in Fig. 7.19. These are chosen to illustrate the use of subroutine ITER with a nondiagonal mass matrix. The kinetic energy of the system, using this choice of generalized coordinates, is

$$T(\dot{x}_1, \dot{x}_2, \dot{x}_3, \dot{x}_4) = \frac{1}{2}m\left(\frac{a\dot{x}_2 + b\dot{x}_1}{a+b}\right)^2 + \frac{1}{2}I\left(\frac{\dot{x}_2 - \dot{x}_1}{a+b}\right)^2 + \frac{1}{2}m_a\dot{x}_3^2 + \frac{1}{2}m_a\dot{x}_4^2$$

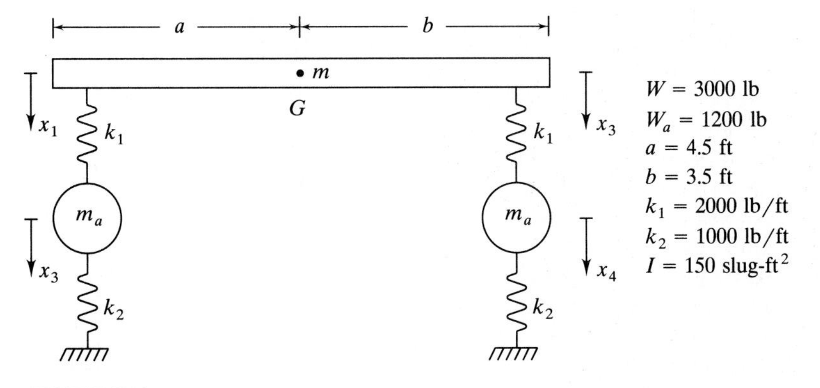

FIGURE 7.19
Four-degree-of-freedom model of automobile suspension system of Example 7.14.

from which the mass matrix is determined as

$$\mathbf{M} = \begin{bmatrix} \frac{mb^2 + I}{(a+b)^2} & \frac{mab - I}{(a+b)^2} & 0 & 0 \\ \frac{mab - I}{(a+b)^2} & \frac{ma^2 + I}{(a+b)^2} & 0 & 0 \\ 0 & 0 & m_a & 0 \\ 0 & 0 & 0 & m_a \end{bmatrix}$$

Stiffness influence coefficients are used to determine the stiffness matrix as

$$\mathbf{K} = \begin{bmatrix} k_1 & 0 & -k_1 & 0 \\ 0 & k_1 & 0 & -k_1 \\ -k_1 & 0 & k_1 + k_2 & 0 \\ 0 & -k_1 & 0 & k_1 + k_2 \end{bmatrix}$$

The calling program and the results follow.

```
C *********************************************
C                                             *
C       CALLING PROGRAM FOR EXAMPLE 7.14      *
C                                             *
C                                             *
C *********************************************
        EXTERNAL INVERSE
        REAL K(9,9),M(9,9),Q(9,9),OM(9),U(9),Z(9,9),I
        INTEGER ERR
        DATA K / 81* 0.0 /
        DATA M / 81* 0.0 /
        OPEN (7,FILE = 'EX714.RES')
        WRITE(7,110)
  110   FORMAT(5X,'SOLUTION FOR EXAMPLE 7.14')
        XK1 = 2000.
        XK2 = 1000.
        A = 4.5
        B = 3.5
        Y2 = (A + B)**2
        I = 150.
        XM = 3000./ 32.2
        XMA = 1200./ 32.2
        M(1,1) = (XM*B*B + I) / Y2
        M(1,2) = (XM*A*B - I) / Y2.
        M(2,1) = M(1,2)
        M(2,2) = (XM*A*A + I) / Y2
        M(3,3) = XMA
        M(4,4) = XMA
```

```
      K(1,1) = XK1
      K(1,3) = -XK1
      K(2,2) = XK1
      K(2,4) = -XK1
      K(3,1) = -XK1
      K(3,3) = XK1 + XK2
      K(4,2) = -XK1
      K(4,4) = XK1 + XK2
      U(1) = 0.0
      U(2) = 0.0
      U(3) = 00.0
      U(4) = 1.0
      CALL ITER(4,4,1,K,M,U,Q,OM,500,1.E-7,ERR)
      WRITE(7,102)ERR
  102 FORMAT(5X,'ERR = ',2X,I2)
      DO 1 L = 1,4
      J = 5 - L
      WRITE(7,100)L,OM(J)
  102 FORMAT(///,5X,'THE NATURAL FREQUENCY FOR MODE',
     $I2,2X,' = ',F10.4,X 'RAD / SEC')
    1 WRITE(7,101)Q(1,J),Q(2,J),Q(3,J),Q(4,J)
  101 FORMAT(/,5X,'THE CORRESPONDING MODE SHAPE IS',
     $/15X,4(F10.4,2X))
      STOP
      END
```

```
SOLUTION FOR EXAMPLE 7.14
ERR =   0

THE NATURAL FREQUENCY FOR MODE 1 = 3.1705RAD / SEC
THE CORRESPONDING MODE SHAPE IS
               0.0726     0.0960     0.0553     0.0731

THE NATURAL FREQUENCY FOR MODE 2 = 4.8713RAD / SEC
THE CORRESPONDING MODE SHAPE IS
               0.1295    -0.0980     0.0819    -0.0926

THE NATURAL FREQUENCY FOR MODE 3 = 10.6136RAD / SEC
THE CORRESPONDING MODE SHAPE IS
              -0.0491    -0.0649     0.0819     0.1083

THE NATURAL FREQUENCY FOR MODE 4 = 22.1548RAD / SEC
THE CORRESPONDING MODE SHAPE IS
               0.3482    -0.2634    -0.0455     0.0344
```

7.11 RAYLEIGH'S QUOTIENT

Consider a situation where the free vibrations of a one-degree-of-freedom system are generated such that only one mode is present. The frequency of the mode is ω and its mode shape is $\mathbf{X}$. The maximum potential energy associated with this mode of vibration is determined from Eq. (7.23) as

$$V_{max} = \tfrac{1}{2}(\mathbf{X}, \mathbf{X})_K \tag{7.58}$$

The maximum kinetic energy associated with this mode is determined from Eq. (7.24) as

$$T_{max} = \tfrac{1}{2}\omega^2(\mathbf{X}, \mathbf{X})_M \tag{7.59}$$

For a conservative system, where a continual process of transfer of kinetic and potential energy occurs without dissipation, the maximum potential energy equals the maximum kinetic energy. Thus from Eqs. (7.58) and (7.59)

$$\omega^2(\mathbf{X}, \mathbf{X})_M = (\mathbf{X}, \mathbf{X})_K$$

or

$$\omega^2 = \frac{(\mathbf{X}, \mathbf{X})_K}{(\mathbf{X}, \mathbf{X})_M} \tag{7.60}$$

For a general n-dimensional vector $\mathbf{X}$, not necessarily a mode shape, Eq. (7.60) is generalized to

$$R(\mathbf{X}) = \frac{(\mathbf{X}, \mathbf{X})_K}{(\mathbf{X}, \mathbf{X})_M} \tag{7.61}$$

The scalar function defined in Eq. (7.61) is called Rayleigh's quotient. If $\mathbf{X}$ is a mode shape of the linear n degree of freedom whose stiffness and mass matrices are $\mathbf{K}$ and $\mathbf{M}$, respectively, then $R(\mathbf{X})$ takes on the value of the natural frequency associated with that mode. If $\mathbf{X}$ is not a mode shape, then $R(\mathbf{X})$ takes on some other value.

Rayleigh's quotient can be useful in determining an upper bound on the lowest natural frequency. In some cases it can be used to attain a good approximation to the lowest natural frequency.

From the expansion theorem, an arbitrary vector $\mathbf{X}$ can be written as a linear combination of the normalized mode shapes

$$\mathbf{X} = \sum_{i=1}^{n} c_i \mathbf{X}_i \tag{7.62}$$

Substituting Eq. (7.62) in Rayleigh's quotient, using properties of the scalar products and orthonormality of the mode shapes leads to

$$R(\mathbf{X}) = \frac{\sum_{i=1}^{n} c_i^2 \omega_i^2}{\sum_{i=1}^{n} c_i^2} \tag{7.63}$$

Stationary values of $R(\mathbf{X})$ occur when

$$\frac{\partial R}{\partial c_1} = \frac{\partial R}{\partial c_2} = \cdots = \frac{\partial R}{\partial c_n} = 0 \tag{7.64}$$

The n solutions of Eq. (7.64) are summarized by $c_i = \delta_{ij}$ for $j = 1, \ldots, n$. That is, Rayleigh's quotient is stationary only when **X** is an eigenvector. It is also possible to show that these stationary values are minimums. Hence ω_1^2 is the minimum value of Rayleigh's quotient.

The preceding result implies that an upper bound and perhaps an approximation for the lowest natural frequency can be obtained using Rayleigh's quotient. Rayleigh's quotient can be calculated for several trial vectors. The lowest natural frequency can be no greater than the square root of the smallest value obtained. The closer a trial vector is to the actual mode shape, the closer the value of Rayleigh's quotient is to the square of the lowest natural frequency.

Example 7.15. Use Rayleigh's quotient to obtain an approximation to the lowest natural frequency of the system of Example 7.4. Use the trial vectors

$$\mathbf{X} = \begin{bmatrix} 1 \\ 1 \\ 0.5 \end{bmatrix} \qquad \mathbf{Y} = \begin{bmatrix} 1 \\ -1 \\ 1 \end{bmatrix} \qquad \mathbf{Z} = \begin{bmatrix} 1 \\ 2 \\ -1 \end{bmatrix}$$

Calculating Rayleigh's quotient

$$R(\mathbf{X}) = \frac{[1 \quad 1 \quad 0.5]\begin{bmatrix} 3k & -2k & 0 \\ -2k & 3k & -k \\ 0 & -k & 3k \end{bmatrix}\begin{bmatrix} 1 \\ 1 \\ 0.5 \end{bmatrix}}{[1 \quad 1 \quad 0.5]\begin{bmatrix} m & 0 & 0 \\ 0 & m & 0 \\ 0 & 0 & \dfrac{m}{2} \end{bmatrix}\begin{bmatrix} 1 \\ 1 \\ 0.5 \end{bmatrix}}$$

$$= 0.823\frac{k}{m}$$

Similar calculations yield

$$R(\mathbf{Y}) = 6.0\frac{k}{m}$$

$$R(\mathbf{Z}) = 2.55\frac{k}{m}$$

From the preceding equations an upper bound on the lowest natural frequency is

$$\omega_1 < 0.907\sqrt{\frac{k}{m}}$$

From Example 7.4 the lowest natural frequency for this system is $0.893\sqrt{k/m}$.

7.12 DISCRETE APPROXIMATIONS FOR CONTINUOUS SYSTEMS

As discussed in Sec. 6.9, discrete modeling of a continuous system leads to approximation of the system's lowest natural frequencies. While the approximation of the natural frequencies may be easier than exact determination from a

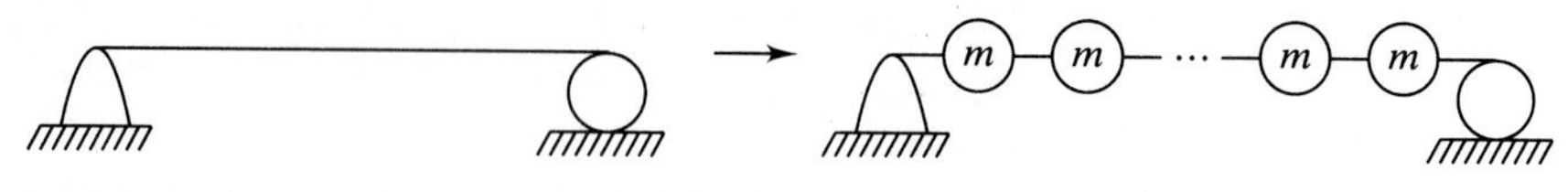

FIGURE 7.20
Simply supported beam modeled with n equal discrete masses.

continuous system analysis, their approximation still requires solution of a matrix eigenvalue problem. The matrix iteration method introduced in Sec. 7.10 is convenient for determining the natural frequencies of the approximate discrete system. It is used in this section to analyze the discrete approximations for several continuous systems.

A simply supported beam is modeled using n discrete masses as shown in Fig. 7.20. The masses are of equal value

$$m = \frac{\rho AL}{\beta} \tag{7.65}$$

where ρ is the beam's mass density, A is its constant cross-sectional area, L is the total length of the beam, and β is a parameter dependent upon the method of discretization. If the sum of the discrete masses equals the total mass of the beam, $\beta = n$. If the mass of each particle represents the mass of a region surrounding the particle and the particle is at the center of the region, then $\beta = n + 1$.

Flexibility influence coefficients yield the elements of the flexibility matrix. These elements can be written in the form

$$a_{i4} = \frac{EI}{L^3}\alpha_{ij} \tag{7.66}$$

where E is the beam's elastic modulus, I is its cross-sectional moment of inertia, and

$$\alpha_{ij} = \frac{6(n+1)^3}{ij\left[(n+1)^2 - i^2 - j^2\right]} \tag{7.67}$$

The differential equations governing the free vibrations of the approximate system are written as

$$\ddot{\mathbf{x}} + \phi^2\mathbf{A}\mathbf{x} = \mathbf{0} \tag{7.68}$$

where

$$\phi = \sqrt{\frac{\beta EI}{\rho AL^4}} \tag{7.69}$$

Nondimensional variables are introduced

$$x_i^* = x_i/L \qquad t^* = \phi t \tag{7.70}$$

TABLE 7.1
Mode number

ω	1	2	3	4	5	6	7
$n = 2$	5.6922	22.046	—	—	—	—	—
$n = 3$	4.9333	19.596	41.607	—	—	—	—
$n = 4$	4.4133	17.637	39.988	64.202	—	—	—
$n = 5$	4.0290	16.100	36.000	62.356	89.194	—	—
$n = 6$	3.7302	14.913	33.456	58.826	88.776	116.19	—
$n = 7$	3.4894	13.954	31.348	55.427	85.221	117.68	145.52

Nondimensional frequencies for simply supported beam.

leading to nondimensional differential equations of the form

$$\ddot{\mathbf{x}}^* + \mathbf{A}\mathbf{x}^* = \mathbf{0} \tag{7.71}$$

where a dot now represents differentiation with respect to t^*. The nondimensional natural frequencies, ω^*, are the square roots of the eigenvalues of **A**. The dimensional natural frequencies are given by

$$\omega = \phi\omega^* \tag{7.72}$$

A FORTRAN program using subroutine ITER is used to determine the nondimensional natural frequencies. The number of degrees of freedom is an input parameter. The program is run for $n = 2$ through $n = 7$ and the results summarized in Table 7.1. The dimensional frequencies are calculated using $\beta = n$ and $\beta = n + 1$ and given in Tables 7.2 and 7.3. The exact natural frequencies, calculated by the methods of Chap. 9, are given for comparison.

Tables 7.2 and 7.3 show that the exact natural frequencies are greater than the approximate natural frequencies for all modes. It also shows that using

TABLE 7.2
Mode number

ω	1	2	3	4	5	6	7
Exact	9.8696	39.478	88.826	157.91	246.74	355.31	483.61
$n = 2$	9.8591	38.184	—	—	—	—	—
$n = 3$	9.8666	39.192	83.214	—	—	—	—
$n = 4$	9.8685	39.381	87.179	143.56	—	—	—
$n = 5$	9.8691	39.437	88.182	152.74	218.48	—	—
$n = 6$	9.8693	39.457	88.523	155.64	234.88	307.40	—
$n = 7$	9.8694	39.467	88.664	156.77	241.04	332.85	411.60

$$\omega = \dot{\omega}\sqrt{\frac{EI}{\rho AL^4}}$$

where ω is the dimensional natural frequency. Dimensional frequency versus degrees of freedom for a simply supported beam with $\beta = n + 1$.

TABLE 7.3
Mode shape

ω	1	2	3	4	5	6	7
Exact	9.8696	39.478	88.826	157.91	246.74	355.31	483.61
$n = 2$	8.0499	31.177	—	—	—	—	—
$n = 3$	8.5447	33.941	72.065	—	—	—	—
$n = 4$	8.8267	35.223	77.973	128.40	—	—	—
$n = 5$	9.0092	36.000	80.499	139.43	199.44	—	—
$n = 6$	9.1372	33.820	81.956	144.09	217.46	284.60	—
$n = 7$	9.2320	36.918	82.938	146.64	225.47	311.35	295.93

$$\omega = \hat{\omega}\sqrt{\frac{EI}{\rho AL^4}}$$

where ω is the natural frequency of a simply supported beam. The approximate solution was calculated using $\beta = n$.

$\beta = n + 1$ provides the better approximation. For an arbitrary mode, say the jth mode, a sequence

$$\omega_{j,j}, \omega_{j,j+1}, \omega_{j,j+2}, \ldots$$

is defined where $\omega_{j,k}$ is the approximation to ω_j when k degrees of freedom are used to model the continuous system. This sequence converges to the exact value of ω_j.

A discrete model of a fixed-free beam is shown in Fig. 7.21. This continuous system is discretized by n equally spaced particles. In view of the results for the simply supported beam, the values of the discrete masses are determined by identifying a portion of the beam such that the inertia effects of that portion are modeled by the discrete mass. This leads to

$$m_i = \begin{cases} \dfrac{\rho AL}{n} & i < n \\ \dfrac{\rho AL}{2n} & i = n \end{cases} \tag{7.73}$$

Flexibility influence coefficients are used to determine the flexibility matrix. A calling program is written using subroutine ITER to determine the natural frequencies, with the results summarized in Table 7.4.

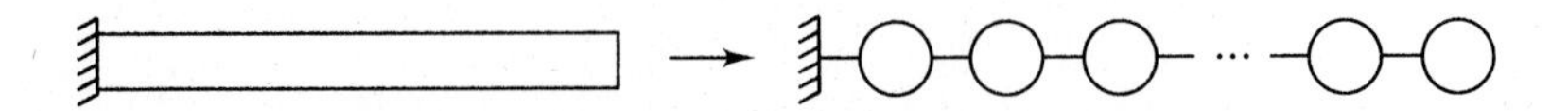

FIGURE 7.21
Fixed-free beam discretized with n equally spaced particles.

TABLE 7.4
Mode number

ω	1	2	3	4	5	6	7
Exact	3.5160	22.034	61.698	120.90	199.86	298.56	416.99
$n = 2$	3.1562	16.258	—	—	—	—	—
$n = 3$	3.3457	18.886	47.029	—	—	—	—
$n = 4$	3.4180	20.090	53.202	92.727	—	—	—
$n = 5$	3.4527	20.734	55.952	104.44	153.01	—	—
$n = 6$	3.4717	21.109	57.557	109.51	172.08	227.65	—
$n = 7$	3.4834	21.345	58.584	112.40	180.65	255.42	316.36

$$\omega = \dot{\omega}\sqrt{\frac{EI}{\rho A L^4}}$$

ω is a natural frequency of a fixed-free beam.

Example 7.16. The short single span bridge of Fig. 7.22 is modeled as a simply supported beam for vibration analysis. The bridge is 18 ft long, weighs 900 lb/ft, and has a rigidity of 6.5×10^9 lb · in.^{2}. Use a discrete model with six degrees of freedom to approximate the three lowest natural frequencies of transverse vibrations of the bridge.

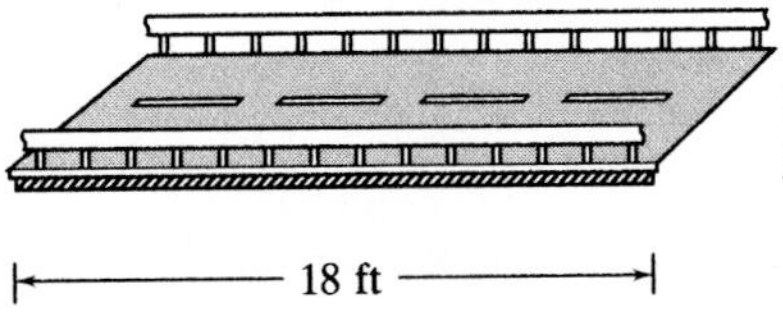

$W = 900$ lb/ft
$EI = 6.5 \times 10^9$ lb · in^2
18 ft

FIGURE 7.22
Bridge modeled as simply supported beam.

The nondimensional square root factor of Table 7.2 is calculated as

$$\sqrt{\frac{EI}{\rho A L^4}} = \sqrt{\frac{(6.5 \times 10^9 \text{ lb} \cdot \text{in}^2)(1 \text{ ft}^2/144 \text{ in.}^2)}{(900 \text{ lb/ft}/32.2 \text{ ft/s}^2)(18 \text{ ft})^4}} = 3.922$$

The three lowest natural frequencies are calculated using the entries from Table 7.2 for $n = 6$:

$$\omega_1 = 3.922(9.8693) = 38.71 \text{ rad/s}$$

$$\omega_2 = 3.922(39.457) = 154.8 \text{ rad/s}$$

$$\omega_3 = 3.922(88.523) = 347.2 \text{ rad/s}$$

PROBLEMS

7.1. Determine the natural frequencies and mode shapes for the system of Fig. P7.1. Graphically illustrate the mode shapes.

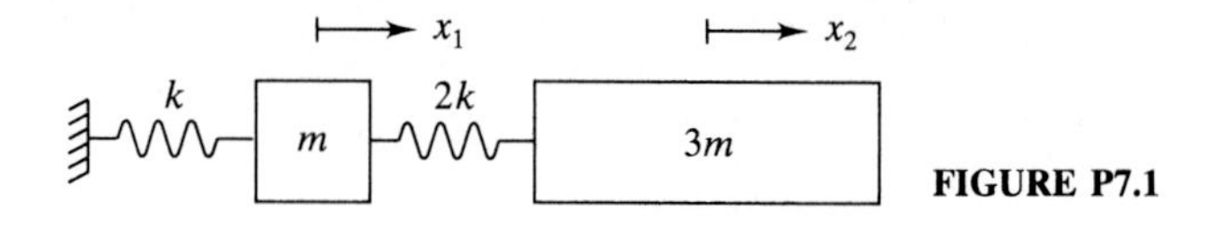

FIGURE P7.1

7.2. Determine the natural frequencies and mode shapes for the system of Fig. P7.2. Graphically illustrate the mode shapes and identify any nodes.

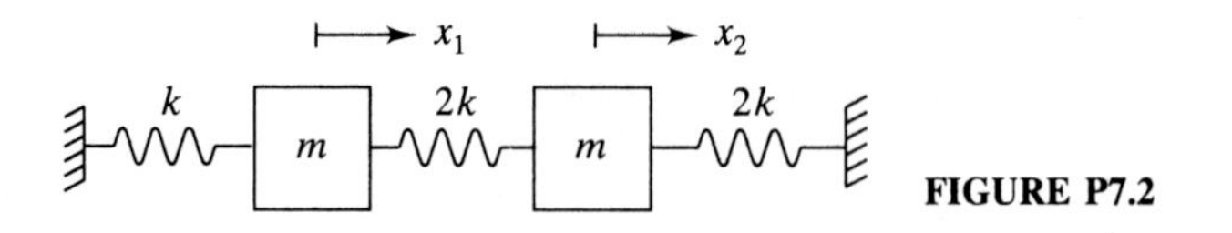

FIGURE P7.2

7.3. The bar of Fig. P7.3 has a mass of 1.5 kg, its center of gravity is 1.3 m from its left end, and it has a centroidal moment of inertia of 0.6 kg · m^2. Determine the natural frequencies and mode shapes of the system. For each mode of free vibration, identify any nodes.

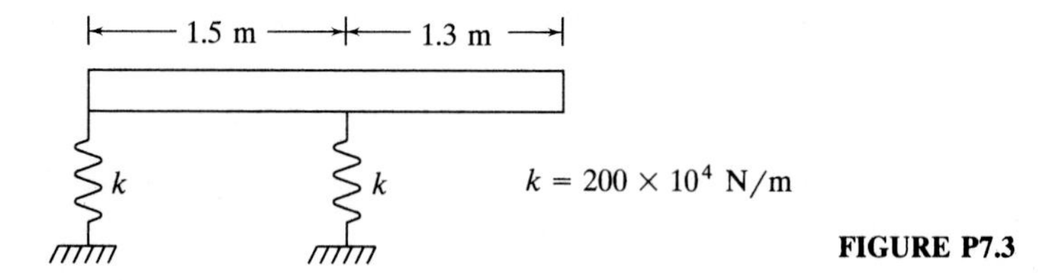

FIGURE P7.3

7.4. Determine the natural frequencies and mode shapes of the system of Fig. P7.4.

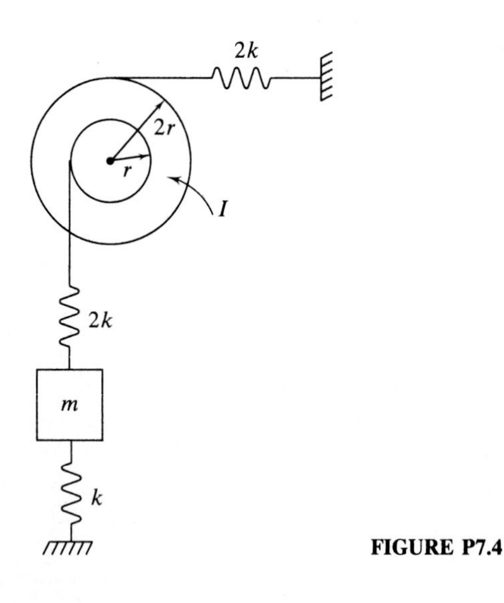

FIGURE P7.4

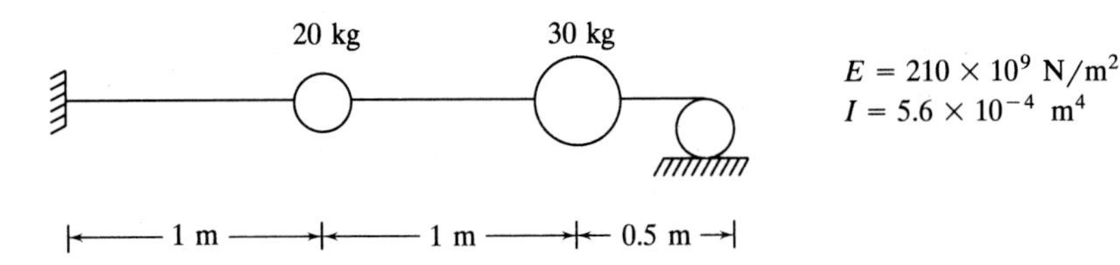

FIGURE P7.5

7.5. Two particles are placed on a massless beam as shown in Fig. P7.5. Determine the natural frequencies of the system.

7.6. The natural frequencies of the system of Fig. P7.6 are measured as 5.0 rad/s and 7.8 rad/s. When free vibrations occur at the lower natural frequency only, the two blocks always move in the same direction and the displacement of the block of mass m_2 is 0.5 of the displacement of the block of mass m_1. When free vibrations occur at the larger natural frequency only, the two blocks move in opposite

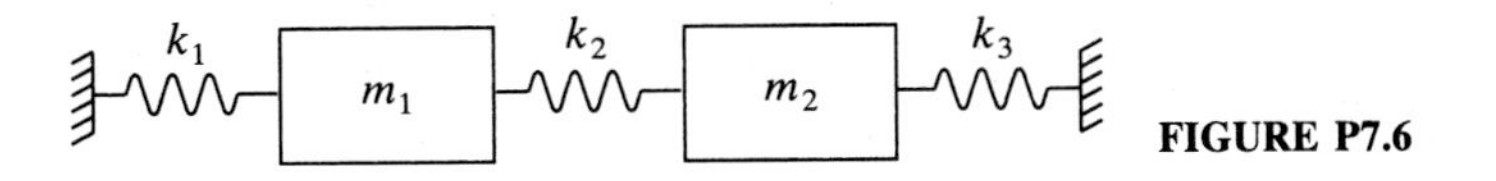

FIGURE P7.6

directions and the displacement of the block of mass m_2 is 1.3 times the displacement of the block of mass m_1. If $m_1 = 4$ kg determine m_2, k_1, k_2, and k_3.

7.7. Determine the natural frequencies and mode shapes for the system of Fig. P6.3 if $k = 3.4 \times 10^5$ N/m, $L = 1.5$ m, and $m = 4.6$ kg.

7.8. Determine the natural frequencies and mode shapes for the system of Fig. P6.6 if $k = 2500$ N/m, $m_1 = 2.4$ kg, $m_2 = 1.6$ kg, $I = 0.65$ kg · m^2, and $L = 1$ m.

7.9. Determine the natural frequencies and mode shapes for the system of Fig. P6.9 if $k = 10{,}000$ N/m, $m = 3$ kg, and $I = 0.6$ kg · m^2, and $r = 80$ cm.

7.10. Determine the natural frequencies of the system of Fig. 6.13 if each of the shafts are of 20 mm radius, made of a material with a shear modulus $G = 90 \times 10^9$ N/m^2, $L = 1$ m, $I_1 = 1.4$ kg · m^2, $I_2 = 1.1$ kg · m^2, $I_3 = 0.85$ kg · m^2, and $I_4 =$ 0.5 kg · m^2.

7.11. The beam of Fig. P6.65 is made of a material of elastic modulus $E = 200 \times 10^9$ N/m^2 and a cross-sectional centroidal moment of inertia of 1.4×10^{-5} m^4, and a total mass of 300 kg. Approximate the lowest natural frequency and its corresponding mode shape using the three-degree-of-freedom model shown.

7.12. A 1500-kg compressor is mounted on springs of stiffness 4×10^5 N/m at the middle of a floor in an industrial plant, which can be modeled as a fixed-fixed beam of length 10 m, elastic modulus 210×10^9 N/m^2, mass 7500 kg, and a cross-sectional moment of inertia of 2.3×10^{-4} m^4. The beam rests on soil whose equivalent stiffness at the midspan of the beam is 6×10^6 N/m. Use the two-degree-of-freedom model of Fig. P7.12 to approximate the natural frequencies of the system.

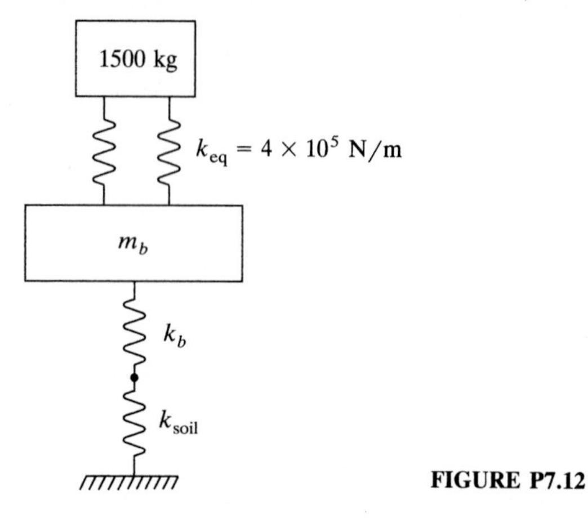

FIGURE P7.12

7.13. The 400-kg lathe of Fig. P7.13 has a centroidal moment of inertia of 15 kg · m^2. It is mounted on equal springs of stiffness 3×10^5 N/m. Determine the lathe's natural frequencies.

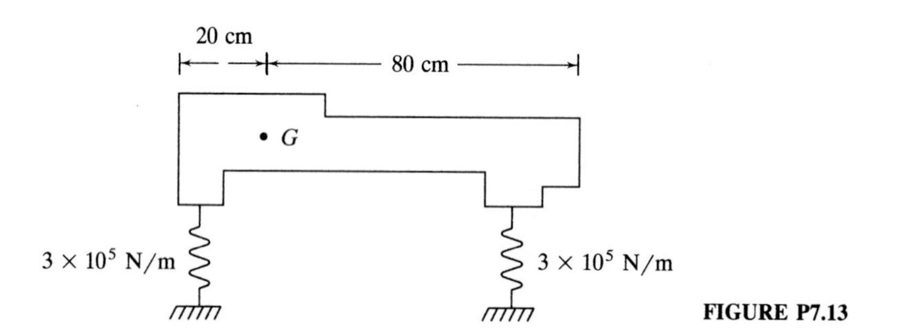

FIGURE P7.13

7.14. The spindle of a lathe is modeled as a beam supported by two elastic bearings. Use a two-degree-of-freedom model to calculate the natural frequencies of a 2.3-kg spindle of elastic modulus 200×10^9 N/m^2, centroidal moment of inertia 1.4×10^{-6} m^4, and length 2.5 m. The bearings are placed at the ends of the spindle and each has an equivalent stiffness of 1.5×10^5 N/m.

7.15. A robot arm is 60 cm long, made of a material of elastic modulus 200×10^9 N/m^2, and has the cross section of Fig. P7.15. The total mass of the arm is 850 g. A tool of mass 1.0 kg is attached to the end of the arm. Assume one end of the arm is pinned while the end with the tool is free. Use a three-degree-of-freedom model to determine the arm's natural frequencies.

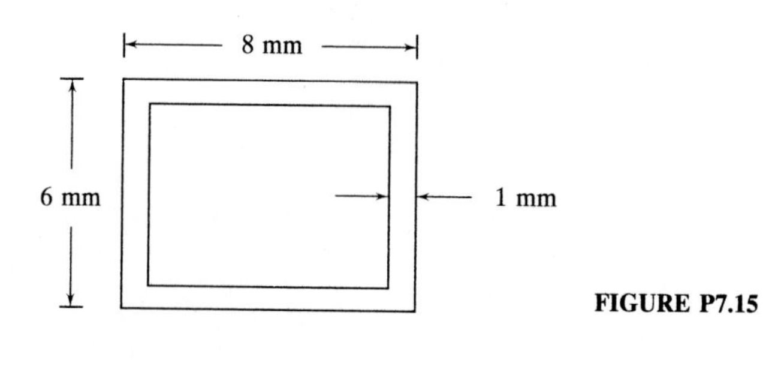

FIGURE P7.15

7.16. Free vibrations of the system of Fig. P7.16 are initiated by moving the block of mass $3m$ a distance δ to the right while holding the block of mass m in its equilibrium position and releasing the system from rest. Mathematically describe the resulting motion.

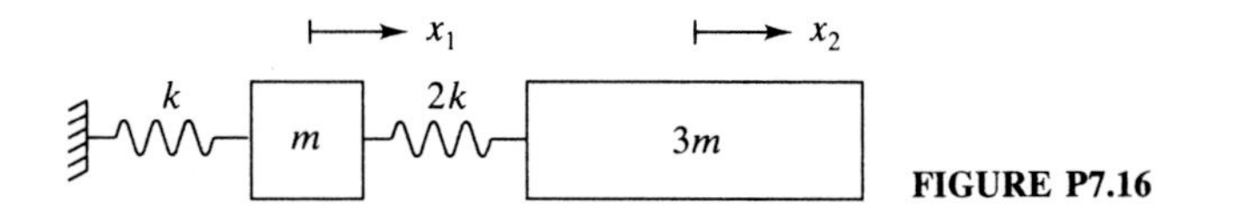

FIGURE P7.16

7.17. Determine a set of initial conditions for the system of Fig. P7.16 such that free vibrations occur only at the system's lowest natural frequency.

7.18. Free vibrations of the system of Fig. P7.18 are initiated by applying a vertical impulse of 1.5 N · s to the left end of the bar when the bar is in equilibrium. Mathematically describe the resulting motion of the right end of the bar.

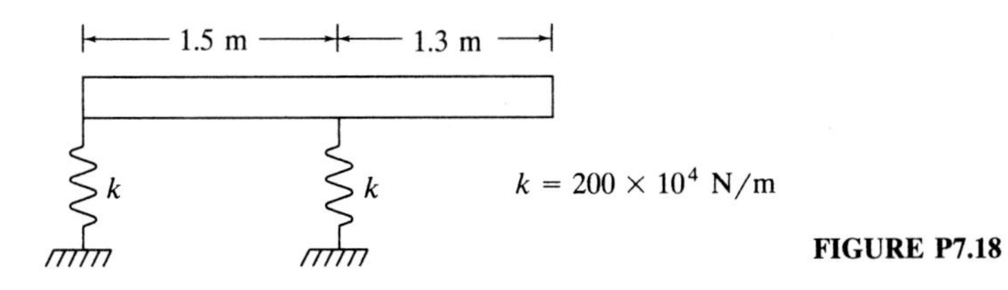

FIGURE P7.18

7.19. Specify a necessary condition between the parameters defining the system of Fig. P7.19 such that the system is stable.

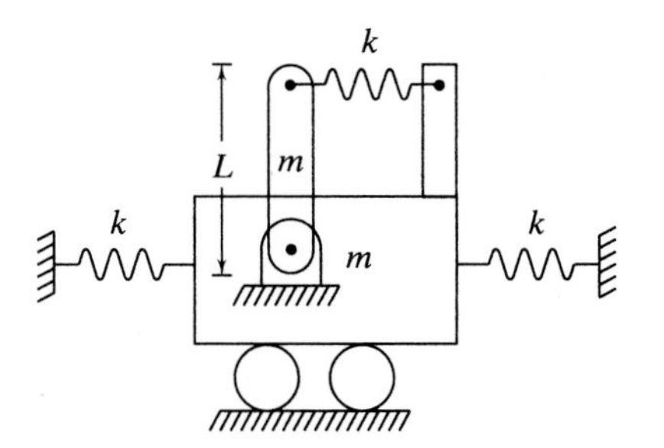

FIGURE P7.19

7.20. A pipe extends from a wall as shown in Fig. P7.20. The pipe is supported at A to prevent transverse displacement, but not to prevent rotation. Under what condi-

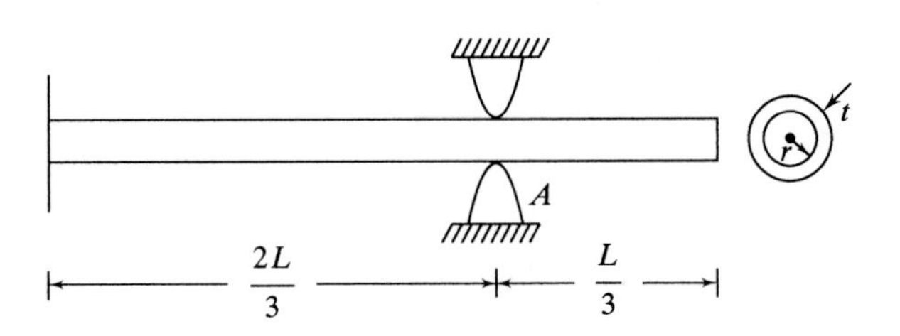

FIGURE P7.20

tions will the unfilled pipe's lowest natural frequency of transverse vibrations coincide with its frequency of free torsional vibrations?

7.21. A 30,000-kg locomotive is coupled to a fully loaded 20,000-kg boxcar and moving at 6.5 m/s. The assembly is coupled to a stationary and empty 5000-kg cattle car. The stiffness of each of the couplings is 5.7×10^5 N/m.
(*a*) What is the velocity of the three cars immediately after coupling?
(*b*) What are the natural frequencies of the three-car assembly?

7.22. The three-car assembly of Prob. 7.21 is moving at 3 m/s when it comes in contact with a bumper of stiffness 4×10^4 N/m. What is the maximum deflection of the bumper?

7.23. Show that the kinetic energy scalar product of a mode shape with itself is proportional to the kinetic energy associated with that mode.

7.24. Show that symmetry of the mass matrix guarantees commutivity of the kinetic energy scalar product.

7.25. The mode shape for the lowest mode of a two-degree-of-freedom system is

$$\mathbf{X}_1 = \begin{bmatrix} 1.0 \\ 1.5 \end{bmatrix}$$

The mass matrix for the system is

$$\mathbf{m} = \begin{bmatrix} 5.4 & 1.2 \\ 1.2 & 3.8 \end{bmatrix}$$

Determine the mode shape corresponding to the system's higher natural frequency.

7.26. Demonstrate orthogonality of the mode shapes of the system of Prob. 7.1.

7.27. Demonstrate the orthogonality of the mode shapes of the system of Prob. 7.3.

7.28. Expand the vector

$$\begin{bmatrix} 1.2 \\ -3.4 \end{bmatrix}$$

in terms of the mode shapes of the system of Prob. 7.1.

7.29. Use matrix iteration to calculate the natural frequencies and normalized mode shapes for the system of Fig. P7.29.

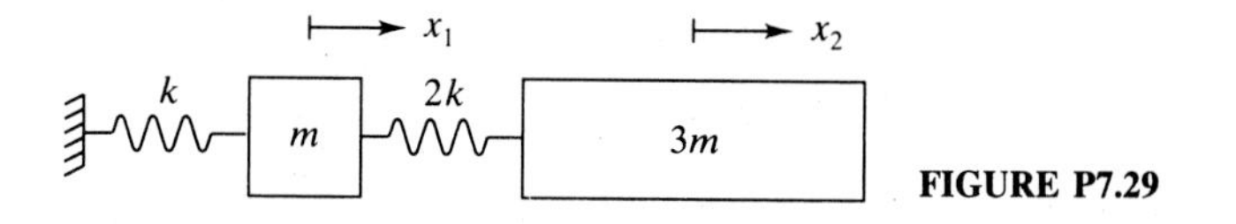

FIGURE P7.29

7.30. Use matrix iteration to calculate the natural frequencies and normalized mode shapes for the system of Fig. P7.30.

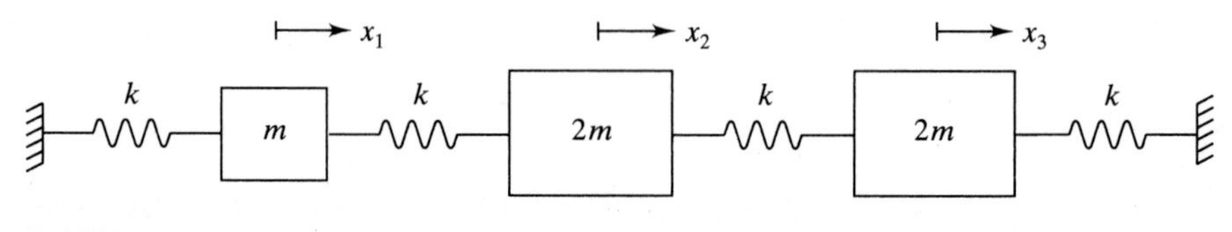

FIGURE P7.30

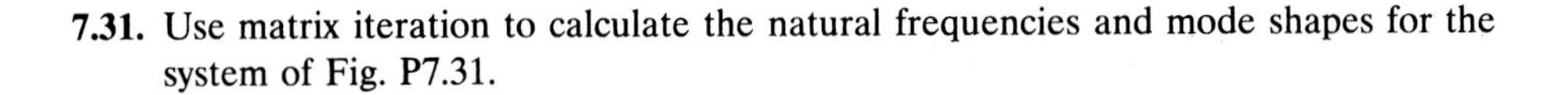

7.31. Use matrix iteration to calculate the natural frequencies and mode shapes for the system of Fig. P7.31.

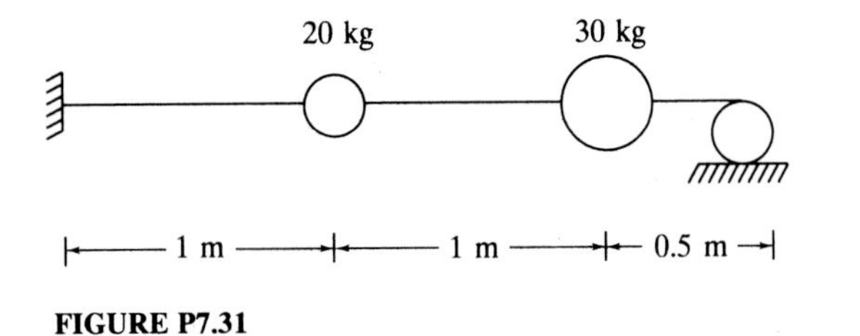

FIGURE P7.31

7.32. Write a FORTRAN program using subroutine ITER to determine the natural frequencies and mode shapes for the frame structure of Fig. P7.32.

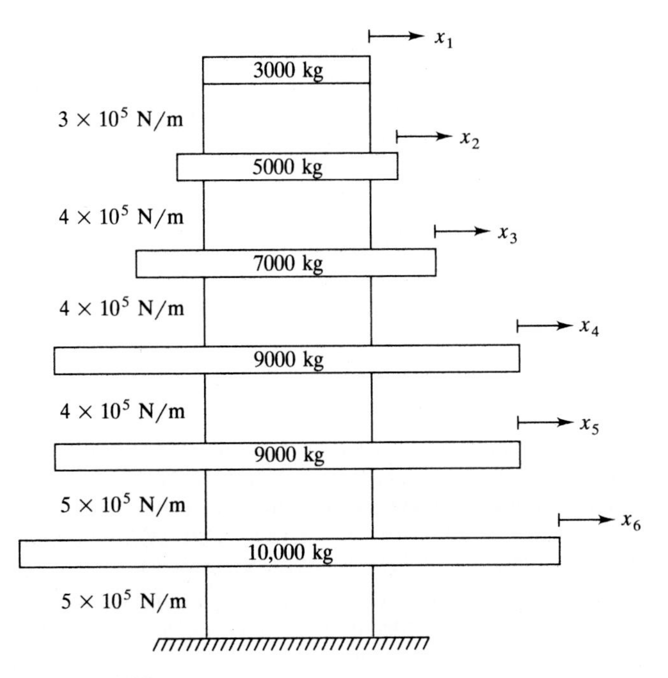

FIGURE P7.32

7.33. Write a FORTRAN program using subroutine ITER to determine the natural frequencies and mode shapes for the torsional system of Fig. P7.33.

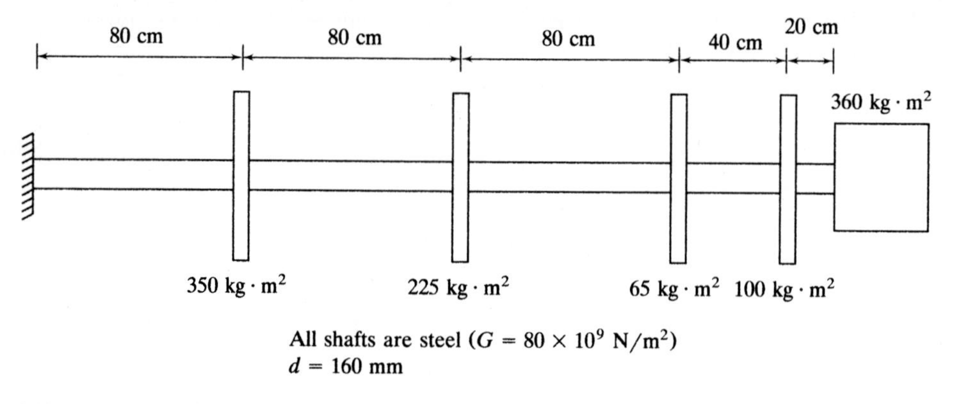

FIGURE P7.33

7.34. For what values of m will the lowest natural frequency of the system of Fig. P7.34 not be in the range of 100–350 rad/s? Use three degrees of freedom to model the beam. Use subroutine ITER to perform the natural frequency calculations.

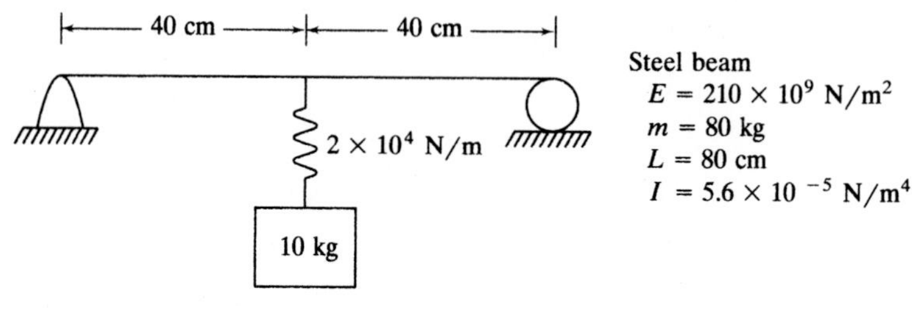

FIGURE P7.34

7.35. The 300-kg tail section of a helicopter is modeled as a cantilever beam of length 3.6 m with a concentrated mass of 30 kg from the tail blades at its end. Use four degrees of freedom to model the tail section. Assume the tail section is uniform, with $E = 210 \times 10^9$ N/m^2 and $I = 5.3 \times 10^{-3}$ m^4. Write a FORTRAN program using subroutine ITER to calculate its two lowest natural frequencies.

7.36. Find the natural frequencies of the automobile suspension system of Example 7.14 using the following numerical values: $W = 4000$ lb, $W_a = 450$ lb, $a = 3$ ft, $b = 2$ ft, $k_1 = 2500$ lb/ft, $k_2 = 1000$ lb/ft, $I = 100$ slug-ft^2.

7.37. Derive Eq. (7.63).

7.38. Show that Eq. (7.64) implies that $R(\mathbf{X})$ is stationary if and only if $\mathbf{X}$ is a mode shape.

7.39. Use Rayleigh's quotient to find an upper bound to the lowest natural frequency of the system of Prob. 7.30. Use at least three trial vectors.

7.40. Use Rayleigh's quotient to find an upper bound to the lowest natural frequency of the system of Prob. 7.31.

7.41. Use Rayleigh's quotient to determine an upper bound to the lowest eigenvalue of the system of Example 7.13.

7.42. If the first mode shape is known, then Rayleigh's quotient can be used to determine an upper bound to the second lowest natural frequencies by using trial vectors orthogonal to the first mode shape. Illustrate this procedure on the system of Example 7.4.

7.43. Rayleigh's quotient can be used to simultaneously approximate more than one frequency. Let $\mathbf{Y}_1, \mathbf{Y}_2, \ldots, \mathbf{Y}_k$ be k linearly independent vectors. Assume a trial solution of

$$\mathbf{X} = \sum_{i=1}^{k} c_i \mathbf{Y}_i$$

Substitute the preceding expansion into Rayleigh's quotient. Recognizing that Rayleigh's quotient is a minimum if $\mathbf{X}$ is a mode shape, choose c_i such that $\mathbf{X}$ is as close to a mode shape as possible, by setting

$$\frac{\partial R}{\partial c_1} = \frac{\partial R}{\partial c_2} = \cdots = \frac{\partial R}{\partial c_k} = 0$$

Develop a procedure to approximate k lowest natural frequencies and mode shapes.

7.44. Illustrate the method of Prob. 7.43 by approximating the two lowest natural frequencies of the system of Example 7.13 using

$$\mathbf{Y}_1 = \begin{bmatrix} 1 \\ 1 \\ 1 \\ 1 \\ 0 \\ 1 \end{bmatrix} \qquad \mathbf{Y}_2 = \begin{bmatrix} 1 \\ -1 \\ -1 \\ 0 \\ -1 \\ 0 \end{bmatrix}$$

7.45. Write a FORTRAN program to determine the natural frequencies and mode shapes of a fixed-fixed beam when n degrees of freedom are used to model the beam. Run the program for $n = 2$, 3, 4, and 5.

7.46. Write a FORTRAN program to determine the natural frequencies and mode shapes of a beam fixed at one end and attached to a spring of stiffness k at the other end when n degrees of freedom are used to model the beam. The program should provide the results in terms of the nondimensional parameter $\beta = kL^3/EI$. Run the program for $n = 2$, 3, 4, and 5 with $\beta = 0.2$ and 0.4.

7.47. Write a FORTRAN program to determine the natural frequencies and mode shapes of a beam fixed at one end, free at its other end, but with a mass m attached to its midspan. The program should provide the results in terms of the nondimensional parameter β, the ratio of the attached mass to the mass of the beam. Run the program with $\beta = 0.2$ and 0.8 for $n = 2, 3, 4$, and 5.

The program MITER.BAS included on the accompanying diskette uses matrix iteration to determine the natural frequencies of a n-degree-of-freedom system. Use MITER.BAS to solve Probs. 7.48 through 7.54.

7.48. Use five degrees of freedom to model the fixed-free beam with an overhang of App. D. Let L be the total length of the beam and let x_1 be the location of the support, measured from the fixed support. Approximate the five lowest natural frequencies of the beam (in nondimensional form) if
(*a*) $x_1/L = 0.4$ (place two nodal points between the supports)
(*b*) $x_1/L = 0.6$ (place three nodal points between the supports)
(*c*) $x_1/L = 0.8$ (place four nodal points between the supports)
The flexibility matrix for each case can be generated using BEAM.BAS

7.49. Use five degrees of freedom to approximate the five lowest natural frequencies of a 10 m 200 kg beam fixed at one end, free at its other end, but with a pin support 6 m from the fixed end. A 75 kg machine is attached to the beam's free end. Take $E = 210 \times 10^9 \text{ N/m}^2$, $I = 3.24 \times 10^{-6} \text{ m}^4$.

7.50. Use MITER.BAS to solve Prob. 7.32.

7.51. Use MITER.BAS to solve Prob. 7.30.

7.52. Use MITER.BAS to solve Prob. 7.33.

7.53. Attempt to run MITER.BAS for the system of Ex. 7.7 when the natural frequencies for two distinct modes coincide.

7.54. Resolve Ex. 7.14 if $k_1 = 2000$ N/m and $k_2 = 1800$ N/m.

REFERENCES

1. Carnahan, B., H. A. Luther, and J. O. Wilkes: *Applied Numerical Analysis*, Wiley, New York, 1969.
2. Dimarogonas, A. D., and S. Haddad: *Vibrations for Engineers*, Prentice-Hall, Englewood Cliffs, N.J., 1992.
3. James, M. L., G. M. Smith, J. C. Wolford, and P. W. Whaley: *Vibration of Mechanical and Structural Systems with Microcomputer Applications*, Harper and Row, New York, 1989.
4. Ketter, R. L., and S. P. Prawel: *Modern Methods of Engineering Computations*, McGraw-Hill, New York, 1969.
5. Kreider, D. L., R. O. Kuller, D. R. Ostberg, and F. W. Perkins: *An Introduction to Linear Analysis*, 2nd ed., Addison-Wesley, Reading, Mass., 1966.
6. Meirovitch, L.: *Elements of Vibration Analysis*, McGraw-Hill, New York, 1975.
7. Rabenstein, A. L.: *Elementary Differential Equations with Linear Algebra*, 3rd ed., Academic Press, New York, 1982.
8. Rao, S. S.: *Mechanical Vibrations*, 2nd ed., Addison-Wesley, Reading, Mass., 1990.
9. Shabana, A. A.: *Theory of Vibrations: Discrete and Continuous Systems*, vol. 1, Springer-Verlag, New York, 1991.
10. Steidel, R. F.: *An Introduction to Mechanical Vibrations*, 3rd ed., Wiley, New York, 1988.
11. Thomson, W. T.: *Theory of Vibrations with Applications*, 3rd ed., Prentice-Hall, Englewood Cliffs, N.J., 1988.
12. Weaver, W., S. P. Timoshenko, and D. H. Young: *Vibration Problems in Engineering*, 5th ed., Wiley-Interscience, New York, 1990.
13. Wilkinson, J. H.: *The Algebraic Eigenvalue Problem*, Clarendon Press, Oxford, 1965.

CHAPTER 8

FORCED VIBRATIONS OF MULTI-DEGREE-OF-FREEDOM SYSTEMS

8.1 INTRODUCTION

Multi-degree-of-freedom systems are subject to the same types of excitation as one-degree-of-freedom systems. The street light fixture of Fig. 8.1 is similar to that considered in Example 4.6. If the moment of inertia of the light fixture is large, a two-degree-of-freedom model provides more accurate approximations to the natural frequencies. If the excitation is due to vortex shedding, as described in Sec. 4.6, then over a range of Reynolds numbers the excitation is harmonic. Forced vibrations of multi-degree-of-systems subject to harmonic excitation is considered in Sec. 8.2 and the street light problem is resolved using a two-degree-of-freedom model.

The natural frequencies of the four-degree-of-freedom model of the automobile suspension system of Fig. 8.2 are calculated in Example 7.14. As the vehicle travels horizontally, the suspension system is subject to a transient excitation dependent on the road contour. The forced response of this system due to a general road contour is considered in Sec. 8.7.

The forced response of a multi-degree-of-freedom system, as for a one-degree-of-freedom system, is the sum of a homogeneous solution and a particular solution. The homogeneous solution is dependent upon system properties, while the particular solution is the response due to the particular form of the

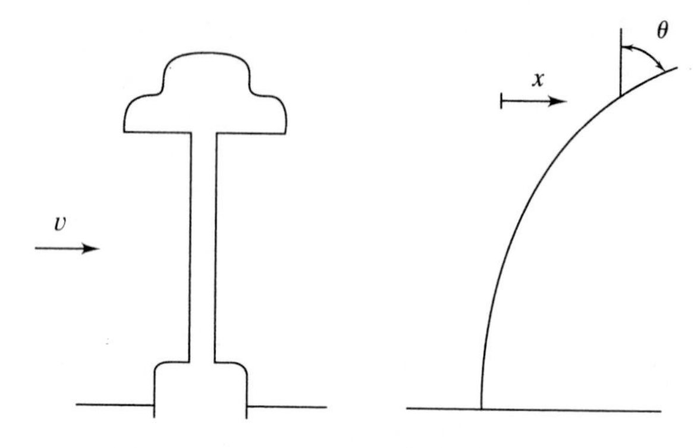

FIGURE 8.1
Wind at constant velocity causes vortex shedding from light pole leading to a harmonic excitation. If moment of inertia of the light fixture is large, a two-degree-of-freedom model may be appropriate to model the response due to vortex shedding.

excitation. The free-vibration response is usually ignored for systems whose long-term behavior is important, such as the steady-state response of a system due to a harmonic excitation. The free-vibration solution is important for systems where the short-term behavior is important, such as a system subject to a shock excitation.

Several methods are available to determine the particular solution for a multi-degree-of-freedom system. The method of undetermined coefficients can be applied for systems subject to harmonic excitations. However, due to the algebraic complexity, its usefulness is restricted to systems with only a few degrees of freedom. The Laplace transform method can be applied to determine system properties, but its usefulness is limited because its application requires the solution of a system of simultaneous equations whose coefficients are functions of the transform variable.

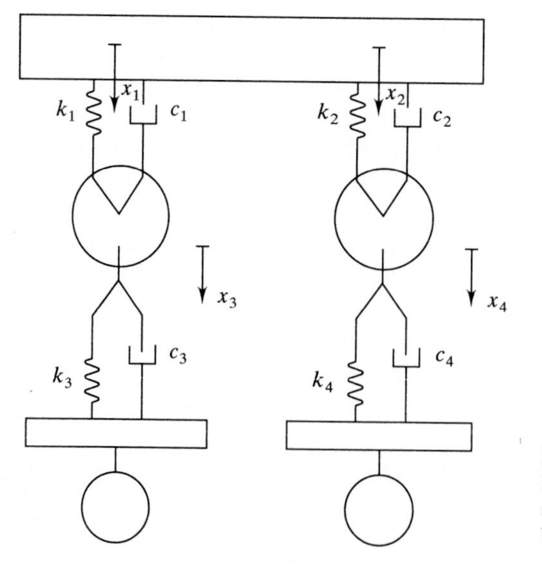

FIGURE 8.2
Four-degree-of-freedom model of vehicle suspension system. Excitation is provided by road contour.

The most useful method for determining the forced-vibration response of a linear multi-degree-of-freedom system is modal analysis. The orthogonality conditions between the mode shapes determined from the free-vibration analysis are used to define a transformation between the generalized coordinates and a new set of coordinates, called the principal coordinates. When the principal coordinates are used as the dependent variables, the differential equations are uncoupled. The resulting uncoupled differential equations are solved by standard techniques (i.e., Laplace transforms, undetermined coefficients, convolution integral). The inverse transformation between the generalized coordinates and principal coordinates is used to find the solution in terms of the original generalized coordinates.

Modal analysis is very powerful and can be applied to all linear undamped multi-degree-of-freedom systems. It can also be applied to systems subject to proportional damping, as shown in Sec. 8.6.

Often, the uncoupled differential equations cannot be solved in closed form. The convolution integral can be used to develop a solution for each principal coordinate in terms of a quadrature. Each integral can be evaluated numerically using the methods of Chapter 5. A numerical method such as Runge-Kutta can be used to solve the uncoupled differential equations. A general program for modal analysis using matrix iteration to first find the required natural frequencies and mode shapes, and then Runge-Kutta to solve the uncoupled differential equations is presented in Sec. 8.7.

8.2 HARMONIC EXCITATIONS

The response of a multi-degree-of-freedom system due to a harmonic excitation is the sum of the homogeneous solution and the particular solution. Even if damping is not included, the homogeneous solution is often ignored. In a real situation, damping is present, causing the homogeneous solution to decay with time. The long-time or steady-state solution is only the particular solution.

The method of undetermined coefficients can be adapted to find the particular solution for a multi-degree-of-freedom system subject to a harmonic excitation. The method of undetermined coefficients can be used for damped or undamped systems. Its application for an n-degree-of-freedom system requires the solution of at least one set of n simultaneous equations. Thus the method of undetermined coefficients is efficient for systems with only a few degrees of freedom.

Consider an n-degree-of-freedom linear system subject to a harmonic excitation at a single frequency. The governing differential equations are of the form

$$\mathbf{M}\ddot{\mathbf{x}} + \mathbf{C}\dot{\mathbf{x}} + \mathbf{K}\mathbf{x} = \mathbf{R}\sin\omega t + \mathbf{S}\cos\omega t \tag{8.1}$$

where $\mathbf{R}$ and $\mathbf{S}$ are n-dimensional column vectors of constants. The method of undetermined coefficients is used and a particular solution of the form

$$\mathbf{x}(t) = \mathbf{U}\sin\omega t + \mathbf{V}\cos\omega t \tag{8.2}$$

is assumed where **U** and **V** are n-dimensional vectors of undetermined coefficients. Substitution of Eq. (8.2) into Eq. (8.1) and equating multipliers of like trigonometric terms leads to

$$(-\omega^2\mathbf{M} + \mathbf{K})\mathbf{U} - \omega\mathbf{C}\mathbf{V} = \mathbf{R} \tag{8.3}$$

and

$$\omega\mathbf{C}\mathbf{U} + (-\omega^2\mathbf{M} + \mathbf{K}) = \mathbf{S} \tag{8.4}$$

Equations (8.3) and (8.4) represent $2n$ simultaneous equations for the $2n$ unknown coefficients. If the system is undamped, $\mathbf{C} = \mathbf{0}$, and Eqs. (8.3) and (8.4) become

$$(-\omega^2\mathbf{M} + \mathbf{K})\mathbf{U} = \mathbf{R} \tag{8.5}$$

and

$$(-\omega^2\mathbf{M} + \mathbf{K})\mathbf{V} = \mathbf{S} \tag{8.6}$$

Equations (8.5) and (8.6) are uncoupled. The left-hand sides of both equations are the same, but the right-hand sides are different. Unique solutions of Eqs. (8.5) and (8.6) exist.

$$\det\{-\omega^2\mathbf{M} + \mathbf{K}\} = 0 \tag{8.7}$$

Equation (8.7) is satisfied only when the excitation frequency coincides with any of the system's natural frequencies. When this occurs, use of Eq. (8.2) is inappropriate. The response grows linearly with time, producing a resonance condition.

The method of linear superposition is used to obtain the response of a multi-degree-of-freedom system to multifrequency excitations. A particular solution is obtained corresponding to each frequency. The total solution is the sum of the particular solutions. For an n-degree-of-freedom damped system subject to a harmonic excitation involving k distinct frequencies, the undetermined coefficients are determined by solving k sets of $2n$ simultaneous equations.

If the excitation is periodic, but not harmonic, a Fourier series is used to represent the excitation as an infinite series of harmonic terms. The response is obtained for each harmonic and the method of superposition is used to determine the response as an infinite series.

Example 8.1. Determine the response of the two-degree-of-freedom system shown in Fig. 8.3. Formulate the equations for the steady-state amplitudes in nondimensional form.

The differential equations governing the behavior of the system of Fig. 8.3 are

$$m_1\ddot{x}_1 + (k_1 + k_2)x_1 - k_2x_2 = F\sin\omega t$$

$$m_2\ddot{x}_2 - k_2x_1 + k_2x_2 = 0$$

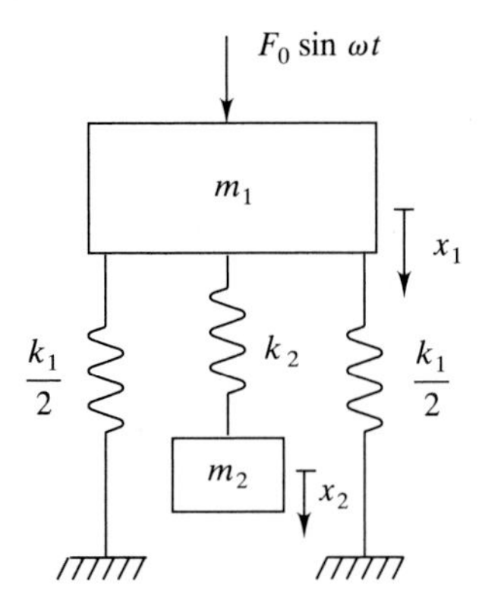

FIGURE 8.3
System of Example 8.1. Response is determined using undetermined coefficients.

The particular solution is determined using Eq. (8.2). Since the system is undamped and $\mathbf{S} = 0$, then $\mathbf{V} = \mathbf{0}$, and

$$\begin{bmatrix} x_1(t) \\ x_2(t) \end{bmatrix} = \begin{bmatrix} U_1 \\ U_2 \end{bmatrix} \sin \omega t$$

Equation (8.3) becomes

$$\begin{bmatrix} -\omega^2 m_1 + k_1 + k_2 & -k_2 \\ -k_2 & -\omega^2 m_2 + k_2 \end{bmatrix} \begin{bmatrix} U_1 \\ U_2 \end{bmatrix} = \begin{bmatrix} F \\ 0 \end{bmatrix}$$

The equations are solved simultaneously, yielding

$$U_1 = \frac{(-\omega^2 m_2 + k_2)F}{D(\omega)}$$

$$U_2 = \frac{k_2 F}{D(\omega)}$$

where $$D(\omega) = m_1 m_2 \omega^4 - (k_2 m_1 + k_1 m_2 + k_2 m_2)\omega^2 + k_1 k_2$$

is the determinant of the coefficient matrix. The roots of the equation $D(\omega) = 0$ are the natural frequencies.

The steady-state amplitudes are each functions of six parameters:

$$U_1 = f(m_1, m_2, k_1, k_2, F, \omega)$$

$$U_2 = g(m_1, m_2, k_1, k_2, F, \omega)$$

The dimensions of the six parameters involve mass, length, and time. Thus the Buckingham Π theorem implies that nondimensional relationships between the steady-state amplitudes and the independent parameters involve four nondimensional parameters. Nondimensional forms of the dimensional equations are obtained by multiplying each equation by k_1/F. The resulting nondimensional relationships are

$$M_1 = \frac{1 - r_2^2}{r_1^2 r_2^2 - r_2^2 - (1 + \mu) r_1^2 + 1}$$

and
$$M_2 = \frac{1}{r_1^2 r_2^2 - r_2^2 - (1+\mu)r_1^2 + 1}$$

where
$$M_1 = \frac{k_1 U_1}{F} \qquad M_2 = \frac{k_1 U_2}{F}$$

$$r_1 = \omega\sqrt{\frac{m_1}{k_1}} \qquad r_2 = \omega\sqrt{\frac{m_2}{k_2}}$$

$$\mu = \frac{m_2}{m_1}$$

It is noted from these equations that the steady-state amplitude of the mass m_1 is zero when $r_2 = 1$. Thus when correctly tuned the second mass-spring system absorbs the steady-state motion of the first mass. This is the concept of the dynamic vibration absorber, which is considered in more detail in Chap. 11.

Example 8.2. Reconsider the street light of Example 4.6, shown again in Fig. 8.4. The moment of inertia of the light fixture about the vertical axis is 14.5 kg · m^2. In view of the large moment of inertia, the use of a two-degree-of-freedom model is suggested such that rotational effects are included. The horizontal displacement of the fixture and the angular rotation of the fixture, measured from the vertical, are selected as generalized coordinates.

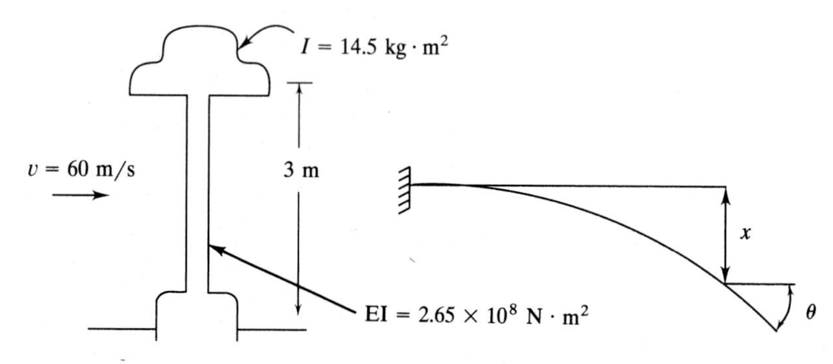

FIGURE 8.4
If moment of inertia of light fixture is significant, two-degree-of-freedom model is used.

For a wind speed of 60 m/s, approximate the maximum displacement of the light fixture and the maximum normal stress in the light pole. Assume no damping and neglect the inertia of the light pole.

The structure is modeled as a cantilever beam with a rigid mass with a substantial moment of inertia attached at its end. The first column of the flexibility matrix is determined by applying a unit load at the location of the light fixture and determining the resulting horizontal displacement and angular rotation of the beam at this point. The second column of the flexibility matrix is obtained by applying a unit moment at the location of the light fixture and determining the resulting horizontal displacement and angular rotation. The resulting flexibility matrix is

$$\mathbf{A} = \frac{L}{EI_b}\begin{bmatrix} \frac{L^2}{3} & \frac{L}{2} \\ \frac{L}{2} & 1 \end{bmatrix} = 10^{-8}\begin{bmatrix} 3.39\ \frac{\text{m}}{\text{N}} & 1.70\ \frac{\text{m}}{\text{N}\cdot\text{m}} \\ 1.70\ \frac{\text{rad}}{\text{N}} & 1.13\ \frac{\text{rad}}{\text{N}\cdot\text{m}} \end{bmatrix}$$

The mass matrix is diagonal with the mass and moment of inertia of the fixture along its diagonal

$$\mathbf{M} = \begin{bmatrix} 60\ \text{kg} & 0 \\ 0 & 14.5\ \text{kg}\cdot\text{m}^2 \end{bmatrix}$$

The vortex shedding results in a uniform force per unit length along the span of the light pole

$$\frac{F}{L} = \frac{F_0}{L}\sin\omega t$$

where $$\omega = \frac{0.4\pi v}{D} = \frac{0.4\pi(60\ \text{m/s})}{0.2\ \text{m}} = 377.0\ \frac{\text{rad}}{\text{s}}$$

and

$$\frac{F_0}{L} = 0.317\rho D^3\omega^2 = 0.317\left(1.204\ \frac{\text{kg}}{\text{m}^3}\right)(0.2\ \text{m})^3\left(377.0\ \frac{\text{rad}}{\text{s}}\right)^2 = 434.0\ \frac{\text{N}}{\text{m}}$$

The force vector is approximated by replacing the uniform force per unit length distribution by a concentrated load and moment that form a system statically equivalent to the uniform loading

$$\mathbf{F}(t) = \begin{bmatrix} \frac{F_0}{L}L \\ \frac{F_0}{L}\frac{L^2}{2} \end{bmatrix}\sin\omega t = \begin{bmatrix} 1300\ \text{N} \\ 1950\ \text{N}\cdot\text{m} \end{bmatrix}\sin\omega t$$

The differential equations modeling the two-degree-of-freedom system are

$$10^{-8}\begin{bmatrix} 3.39 & 1.70 \\ 1.70 & 1.13 \end{bmatrix}\begin{bmatrix} 60 & 0 \\ 0 & 14.5 \end{bmatrix}\begin{bmatrix} \ddot{x} \\ \ddot{\theta} \end{bmatrix} + \begin{bmatrix} x \\ \theta \end{bmatrix} = 10^{-8}\begin{bmatrix} 3.39 & 1.70 \\ 1.70 & 1.13 \end{bmatrix}\begin{bmatrix} 1300 \\ 1950 \end{bmatrix}\sin 377.0t$$

or $$10^{-6}\begin{bmatrix} 2.03 & 0.247 \\ 1.02 & 0.164 \end{bmatrix}\begin{bmatrix} \ddot{x} \\ \ddot{\theta} \end{bmatrix} + \begin{bmatrix} x \\ \theta \end{bmatrix} = 10^{-5}\begin{bmatrix} 6.61 \\ 4.41 \end{bmatrix}\sin 377.0t$$

The particular solution is assumed as

$$\begin{bmatrix} x(t) \\ \theta(t) \end{bmatrix} = \begin{bmatrix} X \\ \Theta \end{bmatrix}\sin 377.5t$$

Substitution into the differential equations leads to simultaneous algebraic equations for the steady-state amplitudes

$$\begin{bmatrix} 0.711 & -0.035 \\ -0.145 & 0.770 \end{bmatrix}\begin{bmatrix} X \\ \Theta \end{bmatrix} = 10^{-4}\begin{bmatrix} 6.61 \\ 4.41 \end{bmatrix}$$

The solution is

$$\begin{bmatrix} X \\ \Theta \end{bmatrix} = 10^{-4}\begin{bmatrix} -9.66 \text{ m} \\ 7.55 \text{ rad} \end{bmatrix}$$

Free-body diagrams showing external and effective forces acting on the light pole at an arbitrary instant of time are shown in Fig. 8.5. The maximum stress occurs at the light pole support when the maximum excitation occurs. Summing

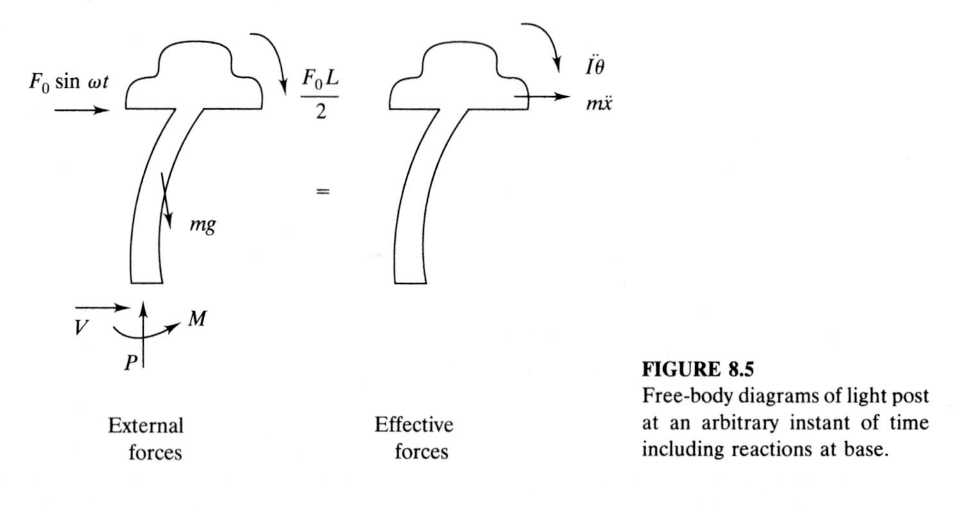

FIGURE 8.5
Free-body diagrams of light post at an arbitrary instant of time including reactions at base.

moments about the support yields

$$\begin{aligned} M_{\max} &= F_0 L + F_0 \frac{L}{2} + m\omega^2 XL + I\omega^2 \Theta \\ &= (1300 \text{ N})(3 \text{ m}) + 1950 \text{ N}\cdot\text{m} + 60 \text{ kg}(377.0 \text{ rad/s})^2 (9.66 \times 10^{-4} \text{ m})(3 \text{ m}) \\ &\quad + (14.5 \text{ kg}\cdot\text{m}^2)(377.0 \text{ rad/s})^2 (7.55 \times 10^{-4} \text{ rad}) \\ &= 15600 \text{ N}\cdot\text{m} \end{aligned}$$

Assuming the pole is made of an elastic material and the stress is less than the proportional limit, the maximum bending stress is calculated using the elastic flexure formula

$$\sigma_{\max} = \frac{M_{\max} r}{I_b} = 1.24 \times 10^6 \, \frac{\text{N}}{\text{m}^2}$$

8.3 MODAL MATRIX

The modal matrix **P** for an n-degree-of-freedom system is the $n \times n$ matrix whose columns are the normalized mode shapes. Let

$$\mathbf{D} = \mathbf{P}^T \mathbf{M} \mathbf{P} \tag{8.8}$$

Let d_{ij} represent the element in the ith row and jth column of $\mathbf{D}$. Then

$$\begin{aligned} d_{ij} &= \sum_{r=1}^{n} (\mathbf{P}^T)_{ir}(\mathbf{MP})_{rj} \\ &= \sum_{r=1}^{n} X_{ri} \sum_{s=1}^{n} M_{rs} P_{sj} \\ &= \sum_{r=1}^{n} \sum_{s=1}^{n} X_{ri} M_{rs} X_{sj} \\ &= \mathbf{X}_i^T \mathbf{M} \mathbf{X}_j \end{aligned}$$

Using the definition of the kinetic energy scalar product, Eq. (7.21), the preceding equation becomes

$$d_{ij} = (\mathbf{X}_i, \mathbf{X}_j)_M$$

Since the modal matrix is defined using normalized mode shapes, Eq. (7.44) implies

$$d_{ij} = \delta_{ij}$$

Thus $\mathbf{D}$ is the $n \times n$ identity matrix and Eq. (8.8) becomes

$$\mathbf{P}^T \mathbf{M} \mathbf{P} = \mathbf{I} \tag{8.9}$$

In a similar fashion it is shown that

$$\mathbf{P}^T \mathbf{K} \mathbf{P} = \mathbf{\Omega} \tag{8.10}$$

where $\mathbf{\Omega}$ is an $n \times n$ diagonal matrix with the squares of the natural frequencies along the diagonal. That is,

$$\mathbf{\Omega} = \begin{bmatrix} \omega_1^2 & 0 & 0 & \cdots & 0 \\ 0 & \omega_2^2 & 0 & \cdots & 0 \\ 0 & 0 & \omega_3^2 & \cdots & 0 \\ \vdots & \vdots & \vdots & \ddots & \vdots \\ 0 & 0 & 0 & \cdots & \omega_n^2 \end{bmatrix} \tag{8.11}$$

8.4 PRINCIPAL COORDINATES: DECOUPLED EQUATIONS

The principal coordinates for a multi-degree-of-freedom system are a set of coordinates related to the chosen generalized coordinates by the linear transformation

$$\mathbf{x} = \mathbf{P}\mathbf{p} \tag{8.12}$$

The columns of the modal matrix are the normalized mode shapes corresponding to the chosen generalized coordinates. These are shown to be mutually orthogonal with respect to the energy scalar products in Sec. 7.8. Thus they are also linearly independent. Hence the modal matrix is nonsingular, and its inverse exists. Premultiplying Eq. (8.12) by $\mathbf{P}^{-1}$ leads to

$$\mathbf{p} = \mathbf{P}^{-1}\mathbf{x} \tag{8.13}$$

The general matrix form of the differential equation for forced vibrations of an undamped multi-degree-of-freedom linear system is

$$\mathbf{M}\ddot{\mathbf{x}} + \mathbf{K}\mathbf{x} = \mathbf{F}(t) \tag{8.14}$$

These differential equations are written using the principal coordinates as dependent variables by substituting Eq. (8.12) into Eq. (8.14), leading to

$$\mathbf{M}\mathbf{P}\ddot{\mathbf{p}} + \mathbf{K}\mathbf{p} = \mathbf{F}(t) \tag{8.15}$$

Premultiplying Eq. (8.15) by $\mathbf{P}^T$ yields

$$\mathbf{P}^T\mathbf{M}\mathbf{P}\ddot{\mathbf{p}} + \mathbf{P}^T\mathbf{K}\mathbf{P}\mathbf{p} = \mathbf{G}(t) \tag{8.16}$$

where

$$\mathbf{G}(t) = \mathbf{P}^T\mathbf{F}(t) \tag{8.17}$$

Equations (8.9) and (8.10) are used in Eq. (8.17), giving

$$\ddot{\mathbf{p}} + \mathbf{\Omega}\mathbf{p} = \mathbf{G}(t) \tag{8.18}$$

The n equations represented by Eq. (8.18) are

$$\begin{aligned} \ddot{p}_1 + \omega_1^2 p_1 &= G_1(t) \\ \ddot{p}_2 + \omega_2^2 p_2 &= G_2(t) \\ \ddot{p}_3 + \omega_3^2 p_3 &= G_3(t) \\ &\vdots \\ \ddot{p}_n + \omega_n^2 p &= G_n(t) \end{aligned} \tag{8.19}$$

Thus, when the principal coordinates are used as dependent variables, the governing differential equations are uncoupled. Each equation in Eq. (8.19) can be solved by any appropriate method. Equation (8.12) is then used to determine the solution in terms of the chosen set of generalized coordinates.

If the initial displacements and velocities of every particle in the system are zero, the use of the convolution integral yields the following solution for the principal coordinates:

$$p_i(t) = \frac{1}{\omega_i}\int_0^t G_i(\tau)\sin\omega_i(t-\tau)\,d\tau \qquad i = 1,\ldots,n \tag{8.20}$$

While the choice of generalized coordinates is arbitrary, the definition of the principal coordinates is unique. The modal matrix $\mathbf{P}$ is dependent upon the choice of generalized coordinates. However, Eq. (8.13) yields the same physical definition of the principal coordinates when applied for different sets of generalized coordinates.

Example 8.3. Calculate the principal coordinates for the system of Example 7.1.

Using x and θ as generalized coordinates, the mode shapes are calculated in Example 7.1 as

$$\mathbf{X}_1 = \begin{bmatrix} 1 \\ \dfrac{1.43}{L} \end{bmatrix} \qquad \mathbf{X}_2 = \begin{bmatrix} 1 \\ -\dfrac{8.42}{L} \end{bmatrix}$$

Each mode shape is normalized by dividing by the kinetic energy scalar product of the mode shape with itself. This leads to

$$\mathbf{X}_1 = \frac{1}{\sqrt{m}} \begin{bmatrix} 0.925 \\ \dfrac{1.33}{L} \end{bmatrix} \qquad \mathbf{X}_2 = \frac{1}{\sqrt{m}} \begin{bmatrix} 0.411 \\ -\dfrac{3.461}{L} \end{bmatrix}$$

The modal matrix is the matrix whose columns are the normalized mode shapes

$$\mathbf{P} = \frac{1}{\sqrt{m}} \begin{bmatrix} 0.925 & 0.411 \\ \dfrac{1.33}{L} & -\dfrac{3.46}{L} \end{bmatrix}$$

The inverse of $\mathbf{P}$ is calculated using the methods of App. C as

$$\mathbf{P}^{-1} = \sqrt{m} \begin{bmatrix} 0.924 & 0.110L \\ 0.353 & -0.247L \end{bmatrix}$$

Then from Eq. (8.13)

$$\begin{bmatrix} p_1 \\ p_2 \end{bmatrix} = \sqrt{m} \begin{bmatrix} 0.924 & 0.110L \\ 0.353 & -0.247L \end{bmatrix} \begin{bmatrix} x \\ \theta \end{bmatrix}$$

The preceding equations can be rewritten as

$$p_1(t) = 0.924\sqrt{m}\,(x + 0.119L\theta)$$

and

$$p_2(t) = 0.353\sqrt{m}\,(x - 0.700L\theta)$$

The preceding equations show that p_1 is proportional to the displacement of a particle a distance $0.119L$ to the right of the mass center. This particle is identified in Example 7.1 as a node for the second mode. The principal coordinate p_2 is proportional to the displacement of a particle $0.700L$ to the left of the mass center. However, this particle does not exist on the rigid bar.

The preceding discussion shows that vibrations at the lowest natural frequency of this system are rigid-body oscillations about the node for the second mode. Vibrations at the higher natural frequency are rigid-body oscillations about a point lying off the bar.

Equation (8.12) shows that the generalized coordinates are linear combinations of the principal coordinates. The generalized coordinates for a linear system are chosen such that the displacement of any particle in the system is a linear combination of the generalized coordinates. Thus the displacement of any particle in the system is a linear combination of the principal coordinates. This implies that if a particle is a node for the higher mode of a two-degree-of-freedom system, then p_1 is proportional to the displacement of that particle. If a particle is a node for the second mode of a three-degree-of-freedom system, then a linear combination of the first and third principal coordinates represents the displacement of that point. Nothing can be inferred about the physical interpretation of either principal coordinate.

8.5 MODAL ANALYSIS: A SUMMARY

The procedure used to uncouple and solve the differential equations for forced vibrations of a linear n-degree-of-freedom undamped system is called modal analysis. Application of modal analysis for a linear system only requires symmetry of the stiffness matrix and the mass matrix. These symmetries are guaranteed if energy methods are used to derive the governing differential equations.

The steps used in analyzing forced vibrations of a linear multi-degree-of-freedom system using modal analysis are as follows:

1. A set of generalized coordinates is chosen.
2. The differential equations governing the forced vibrations are derived using the generalized coordinates as dependent variables. The use of energy methods is recommended to guarantee symmetry of the mass and stiffness matrices. The result is a set of n differential equations, summarized in matrix form by Eq. (8.14).
3. The natural frequencies and mode shapes are obtained. The natural frequencies can be obtained as the square roots of the eigenvalues of $\mathbf{M}^{-1}\mathbf{K}$ or as the reciprocals of the square roots of the eigenvalues of $\mathbf{AM}$. The mode shapes are the eigenvectors of either matrix.
4. The mode shapes are normalized by requiring that the kinetic energy scalar product of a mode shape with itself be one.
5. The modal matrix, $\mathbf{P}$, is developed as the matrix whose columns are the normalized mode shapes. The column vector $\mathbf{G}(t)$ is formed from Eq. (8.17).
6. The differential equations for the principal coordinates are written as Eq. (8.18). Each component equation is solved by standard solution methods for ordinary differential equations.
7. The solution in terms of the chosen generalized coordinates is obtained using Eq. (8.12).

Application of modal analysis is straightforward as evidenced by the following examples.

Example 8.4. The third block of the three-degree-of-freedom system of Example 7.12 is subject to the time-dependent external force of Fig. 8.6. No external forces are applied to the other blocks. Use modal analysis to determine the system response of the system if all blocks are at rest in equilibrium at $t = 0$.

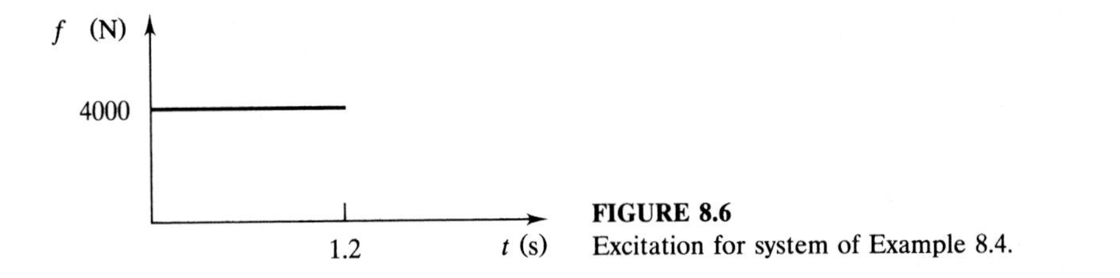

FIGURE 8.6
Excitation for system of Example 8.4.

Steps 1 to 3 of the modal analysis procedure as outlined previously are performed in Examples 7.4 and 7.12 except for calculation of the force vector which is

$$\mathbf{F}(t) = \begin{bmatrix} 0 \\ 0 \\ f(t) \end{bmatrix}$$

where from Fig. 8.6

$$f(t) = 4000 \text{ N}[1 - u(t - 1.2 \text{ s})]$$

The natural frequencies obtained in Example 7.12 are

$$\omega_1 = 8.936 \text{ rad/s} \qquad \omega_2 = 21.107 \text{ rad/s} \qquad \omega_3 = 25.974 \text{ rad/s}$$

and the modal matrix is

$$\mathbf{P} = \begin{bmatrix} 0.2085 & 0.2252 & 0.0765 \\ 0.2295 & -0.1638 & -0.1432 \\ 0.0882 & -0.2120 & 0.3838 \end{bmatrix} (\text{kg})^{-1/2}$$

The vector $\mathbf{G}(t)$ is then calculated using Eq. (8.17)

$$\mathbf{G}(t) = \mathbf{P}^T\mathbf{F} = \begin{bmatrix} 0.0882 \\ -0.2120 \\ 0.3838 \end{bmatrix} f(t)$$

The differential equations satisfied by the principal coordinates are written using Eq. (8.18)

$$\ddot{p}_1 + 79.852p_1 = 352.8[1 - u(t - 1.2)]$$

$$\ddot{p}_2 + 445.5p_2 = -848.0[1 - u(t - 1.2)]$$

$$\ddot{p}_3 + 674.6p_3 = 1535.2[1 - u(t - 1.2)]$$

The convolution integral is used to solve for p_1 as

$$p_1(t) = \frac{1}{8.936}\int_0^t 352.8[1 - u(\tau - 1.2)]\sin 8.936(t - \tau)\, d\tau$$

$$= 4.418\{\cos 8.936t - 1 + u(t - 1.2)[1 - \cos 8.936(t - 1.2)]\}$$

The convolution integral is also used to solve for p_2 and p_3, yielding

$$p_2(t) = -1.903\{\cos 21.107t - 1 + u(t - 1.2)[1 - \cos 21.107(t - 1.2)]\}$$

$$p_3(t) = 2.276\{\cos 25.974t - 1 + u(t - 1.2)[1 - \cos 25.974(t - 1.2)]\}$$

The solution in terms of the original generalized coordinates is obtained using Eq. (8.12)

$$\begin{bmatrix} x_1 \\ x_2 \\ x_3 \end{bmatrix} = \begin{bmatrix} 0.2085 & 0.2252 & 0.0765 \\ 0.2295 & -0.1638 & -0.1432 \\ 0.0882 & -0.2120 & 0.3838 \end{bmatrix} \begin{bmatrix} p_1(t) \\ p_2(t) \\ p_3(t) \end{bmatrix}$$

which leads to

$$x_1(t) = 0.921h_1(t) - 0.429h_2(t) + 0.174h_3(t)$$
$$x_2(t) = 1.014h_1(t) + 0.312h_2(t) - 0.326h_3(t)$$
$$x_3(t) = 0.390h_1(t) + 0.403h_2(t) + 0.874h_3(t)$$

where $h_1(t) = \cos 8.936t - 1 + u(t - 1.2)[1 - \cos 8.936(t - 1.2)]$

$$h_2(t) = \cos 21.107t - 1 + u(t - 1.2)[1 - \cos 21.107(t - 1.2)]$$
$$h_3(t) = \cos 25.974t - 1 + u(t - 1.2)[1 - \cos 25.974(t - 1.2)]$$

Example 8.5. A machine of mass 150 kg is placed as shown on the simply supported beam of Fig. 8.7. The machine has a rotating unbalance of 0.965 kg · m and operates at 1250 rpm. The beam has a total mass of 280 kg, a cross-sectional moment of inertia of 1.2×10^{-4} m^4, a length of 3 m, and an elastic modulus of 210×10^9 N/m^2. Model the beam with three degrees of freedom and use modal analysis to predict the steady-state amplitude of displacement for the point where the machine is attached.

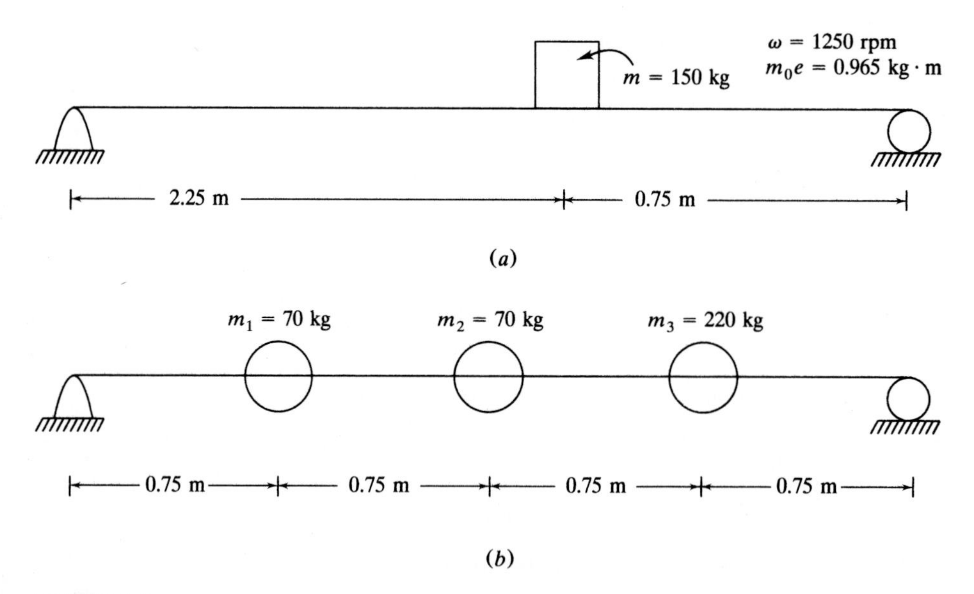

FIGURE 8.7
(*a*) Machine with rotating unbalance is attached to pinned-pinned beam; (*b*) three-degree-of-freedom model of beam.

The beam is modeled by three lumped masses as shown in Fig. 8.7*b*. Using the results of Sec. 7.12, the inertia of the beam is modeledusing three particles of mass 70 kg. Thus the mass matrix is

$$\mathbf{M} = \begin{bmatrix} 70 & 0 & 0 \\ 0 & 70 & 0 \\ 0 & 0 & 220 \end{bmatrix} \text{kg}$$

Flexibility influence coefficients are used to determine the flexibility matrix as

$$\mathbf{A} = 10^{-9} \begin{bmatrix} 12.53 & 15.33 & 9.75 \\ 15.33 & 22.29 & 15.33 \\ 9.75 & 15.33 & 12.53 \end{bmatrix} \frac{\text{m}}{\text{N}}$$

The governing differential equations are

$$\mathbf{AM\ddot{x}} + \mathbf{x} = \mathbf{AF}$$

where

$$\mathbf{F}(t) = \begin{bmatrix} 0 \\ 0 \\ 16500 \sin 130.9t \end{bmatrix} \text{N}$$

Matrix iteration is used to determine the natural frequencies and mode shapes. The program calling SUBROUTINE ITER and its output is

```
      REAL A(9,9),M(9,9),U(9),Q(9,9),OM(9)
      DATA M / 81 * 0.0 /
      DATA A / 81 *0.0 /
      OPEN(7,FILE='EX96.RES')
      M(1,1)=70.
      M(2,2)=70.
      M(3,3)=220.
      A(1,1)=12.53E-9
      A(1,2)=15.33E-9
      A(1,3)=9.75E-9
      A(2,1)=A(1,2)
      A(2,2)=22.29E-9
      A(2,3)=A(1,2)
      A(3,1)=A(1,3)
      A(3,2)=A(2,3)
      A(3,3)=A(1,1)
      U(1)=1.0
      U(2)=0.0
      U(3)=-0.4
      CALL ITER(3,3,0,A,M,U,Q,OM,200,1E-5)
      WRITE(7,100)
100   FORMAT(5X,'NATURAL FREQUENCIES AND MODE SHAPES FOR'
     $'EXAMPLE 8.5' / )
      DO 1 I=1,3
      WRITE(7,101)I,OM(I)
101   FORMAT(5X,'THE NATURAL FREQUENCY FOR MODE NUMBER',
     $2X,I2,2X,'IS', XE14.7,'RAD / SEC')
    1 WRITE(7,102)Q(1,I),Q(2,I),Q(3,I)
```

```
102 FORMAT(5X,'THE CORRESPONDING MODE SHAPE IS'
   $/5X,3(F12.6,3X)/)
    STOP
    END
```

```
NATURAL FREQUENCIES AND MODE SHAPES FOR EXAMPLE 8.5

THE NATURAL FREQUENCY FOR MODE NUMBER 1 IS 0.4558454E+03
  RAD/SEC
THE CORRESPONDING MODE SHAPE IS
     0.045310        0.066603        0.049807

THE NATURAL FREQUENCY FOR MODE NUMBER 2 IS 0.1736550E+04
  RAD/SEC
THE CORRESPONDING MODE SHAPE IS
     0.085087        0.039972        -0.041636

THE NATURAL FREQUENCY FOR MODE NUMBER 3 IS 0.4474021E+04
  RAD/SEC
THE CORRESPONDING MODE SHAPE IS
     0.070661       -0.090840         0.018198
```

Using these results, the modal matrix is written as

$$\mathbf{P} = \begin{bmatrix} 0.0453 & 0.0851 & 0.0707 \\ 0.0666 & 0.400 & -0.0908 \\ 0.0498 & -0.0416 & 0.0182 \end{bmatrix} (\text{kg})^{-1/2}$$

The vector $\mathbf{G}(t)$ is calculated as

$$\mathbf{G} = \mathbf{P}^T\mathbf{F} = \begin{bmatrix} 0.0453 & 0.0666 & 0.0498 \\ 0.0851 & 0.4000 & -0.0416 \\ 0.0707 & -0.0908 & 0.0182 \end{bmatrix} \begin{bmatrix} 0 \\ 0 \\ 16500 \sin 130.9t \end{bmatrix}$$

$$= \begin{bmatrix} 823.2 \\ -687.6 \\ 300.8 \end{bmatrix} \sin 130.9t \text{ N } (\text{kg})^{-1/2}$$

The differential equations for the principal coordinates are written using Eq. (8.18)

$$\ddot{p}_1 + (455.8)^2 p_1 = 823.2 \sin 130.9t$$

$$\ddot{p}_2 + (1736.5)^2 p_2 = -687.6 \sin 130.9t$$

$$\ddot{p}_3 + (4474)^2 p_3 = 300.8 \sin 130.9t$$

The steady-state solution for each principal coordinate is obtained using Eq. (4.8)

$$\begin{bmatrix} p_1 \\ p_2 \\ p_3 \end{bmatrix} = 10^{-5} \begin{bmatrix} 432.0 \\ -22.93 \\ 1.504 \end{bmatrix} \sin 130.9t \ (\text{kg})^{1/2}$$

Equation (8.12) is used to determine $x_3(t)$ as

$$x_3(t) = 0.0498p_1(t) - 0.0416p_2(t) + 0.0182p_3(t)$$

$$= 2.25 \times 10^{-4} \sin 130.9t \text{ m}$$

Thus the maximum steady-state displacement of the point on the beam where the machine is placed is 0.225 mm.

8.6 PROPORTIONAL DAMPING

The modal analysis procedure described in Sec. 8.5 will not, in general, uncouple the differential equations for multi-degree-of-freedom systems with viscous or hysteretic damping. If the damping matrix is a linear combination of powers of the mass and stiffness matrices,

$$\mathbf{C} = \alpha \mathbf{K}^r + \beta \mathbf{M}^s \tag{8.21}$$

where α and β are real values and r and s are integers, the damping is said to be proportional. Without loss of generality, r and s are taken to be one.

The differential equations for a linear n-degree-of-freedom system with proportional viscous damping are

$$\mathbf{M}\ddot{\mathbf{x}} + (\alpha \mathbf{K} + \beta \mathbf{M})\dot{\mathbf{x}} + \mathbf{K}\mathbf{x} = \mathbf{F}(t) \tag{8.22}$$

The undamped system has n natural frequencies and normalized mode shapes. The nonsingular modal matrix $\mathbf{P}$ exists. Principal coordinates for the undamped system are defined by Eq. (8.12). Using the principal coordinates as dependent variables in Eq. (8.22) and premultiplying by $\mathbf{P}^T$ leads to

$$\mathbf{P}^T\mathbf{M}\mathbf{P}\ddot{\mathbf{q}} + (\alpha \mathbf{P}^T\mathbf{K}\mathbf{P} + \beta \mathbf{P}^T\mathbf{M}\mathbf{P})\dot{\mathbf{q}} + \mathbf{P}^T\mathbf{K}\mathbf{P}\mathbf{q} = \mathbf{P}^T\mathbf{F} \tag{8.23}$$

Equations (8.9) and (8.10) allow Eq. (8.23) to be rewritten as

$$\ddot{\mathbf{q}} + (\alpha \boldsymbol{\Omega} + \beta \mathbf{I})\dot{\mathbf{q}} + \boldsymbol{\Omega}\mathbf{q} = \mathbf{G} \tag{8.24}$$

The differential equations represented by Eq. (8.24) are uncoupled. The ith equation represented by the set of equations is

$$\ddot{q}_i + (\alpha\omega_i^2 + \beta)q_i + \omega_i^2 = g_i(t) \qquad i = 1, 2 \ldots, n \tag{8.25}$$

Equation (8.25) is written in a form analogous to the governing differential equation for the forced vibrations of a one-degree-of-freedom system, Eq. (4.3),

$$\ddot{q}_i + 2\zeta_i\omega_i\dot{q}_i + \omega_i^2 q_i = g_i(t) \tag{8.26}$$

where ζ_i is called the modal damping coefficient for mode i:

$$\zeta_i = \frac{1}{2}\left(\alpha\omega_i + \frac{\beta}{\omega_i}\right) \tag{8.27}$$

For $\zeta_i < 1$ application of the convolution integral, Eq. (5.10), to solve Eq. (8.26) yields

$$q_i(t) = \frac{1}{\omega_i\sqrt{1-\zeta_i^2}}\int_0^t e^{-\zeta_i\omega_i(t-\tau)}\sin \omega_i\sqrt{1-\zeta_i^2}\,(t-\tau)\,g_i(\tau)\,d\tau \tag{8.28}$$

Equation (8.25) shows that the principal coordinates used to uncouple the differential equations for an undamped system also uncouple the differential equations when the system is subject to proportional damping. If viscous damping is present, but not proportional, the principal coordinates for the undamped system do not, in general, uncouple the differential equations for the damped system. A normal-mode solution for free vibrations yields complex natural frequencies. The free- and forced-vibration solution for these systems is beyond the scope of this text.

The modal damping ratio, defined for each mode by Eq. (8.27), is dependent on the modal frequency. When the damping matrix is proportional to the stiffness matrix, the modal damping ratio is proportional to the modal frequency and thus increases with increasing frequency. When the damping matrix is proportional to the mass matrix, the damping ratio is inversely proportional to the modal frequency and thus decreases with increasing modal frequency.

Damping in structural systems is mostly hysteretic and hard to quantify. Lacking a better model, proportional damping is often assumed. The modal damping ratios are usually determined experimentally. The equivalent damping ratio for a harmonically excited one-degree-of-freedom system with hysteretic damping is proportional to the natural frequency, and inversely proportional to the excitation frequency. This model fits proportional damping where the damping matrix is proportional to the stiffness matrix. In these cases the higher modes are damped more than the lower modes. The natural frequencies in stiff structural systems are usually greatly separated. The effect of the higher modes in the total response is less than the modes with lower natural frequencies, as evidenced in Ex. 8.5. For these reasons, damping ratios are often only specified for the lower modes.

The total response for the generalized coordinate x_i obtained using modal analysis is written as

$$x_i(t) = \sum_{j=1}^{n} p_{ij} q_j(t) \tag{8.29}$$

If proportional damping is assumed, the higher modes are damped more than the lower modes and have a lesser effect on the overall solution. Modes with higher damping ratios die out more quickly when the system is subject to any short-term or shock excitation. If the system is subject to a harmonic excitation, the modes with higher frequencies have lesser effect because their amplitudes are inversely proportional to the square of their frequencies. Thus fewer modes can be calculated without losing significant accuracy. Hence, in practice, Eq. (8.29) is often replaced by

$$x_i(t) = \sum_{j=1}^{m} p_{ij} q_j(t) \tag{8.30}$$

for some $m < n$. Equation (8.30) is often used in situations where the mode shapes are determined experimentally and an experimental modal analysis method is used to determine the response of a system.

Example 8.6. The three-degree-of-freedom system of Example 8.4 is modified by the addition of dashpots as shown in Fig. 8.8. Determine the forced response of the damped system.

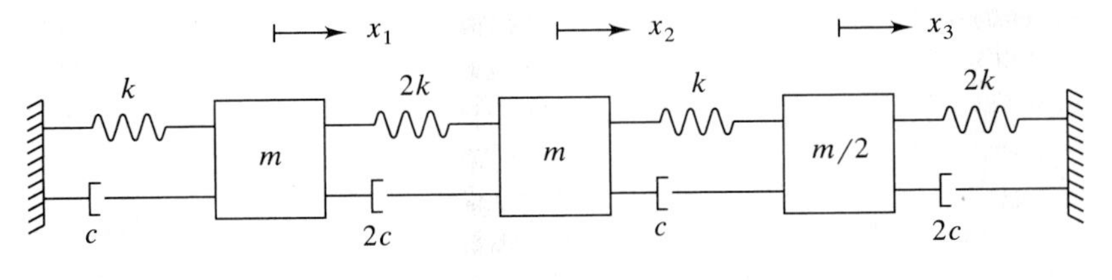

FIGURE 8.8
System of Example 8.6 has viscous damping with the damping matrix proportional to the stiffness matrix.

The damping matrix is

$$\mathbf{C} = \begin{bmatrix} 3c & -2c & 0 \\ -2c & 3c & -c \\ 0 & -c & 3c \end{bmatrix}$$

and is proportional to the stiffness matrix with

$$\alpha = \frac{c}{k} = \frac{40 \text{ N} \cdot \text{s/m}}{1000 \text{ N/m}} = 0.04 \text{ s}$$

Thus the modal damping ratios are given by

$$\zeta_1 = \frac{\alpha}{2}\omega_1 = 0.178 \qquad \zeta_2 = \frac{\alpha}{2}\omega_2 = 0.422 \qquad \zeta_3 = \frac{\alpha}{2}\omega_3 = 0.520$$

All modes are underdamped. The differential equations governing the principal coordinates are

$$\ddot{q}_1 + 1.60\dot{q}_1 + 79.85q_1 = 0.0882f(t)$$

$$\ddot{q}_2 + 8.91\dot{q}_2 + 445.5q_2 = -0.2120f(t)$$

$$\ddot{q}_3 + 13.49\dot{q}_3 + 674.6q_3 = 0.3838f(t)$$

The solution for the principal coordinates is obtained from the convolution integral. It is noted that

$$\int_0^t [1 - u(\tau - 1.2)]e^{-\zeta\omega_n(t-\tau)} \sin \omega_d(t - \tau)\, d\tau$$

$$= -\frac{1-\zeta^2}{\omega_d}\left[1 - e^{-\zeta\omega_n t}\left[\cos \omega_d t + \frac{\zeta}{\sqrt{1-\zeta^2}} \sin \omega_d t\right]\right.$$

$$-u(t - 1.2)\left\{1 - e^{-\zeta\omega_n(t-1.2)}\left[\cos \omega_d(t - 1.2)\right.\right.$$

$$\left.\left.\left. + \frac{\zeta}{\sqrt{1-\zeta^2}} \sin \omega_d(t - 1.2)\right]\right\}\right]$$

The resulting solution for the principal coordinates is

$$p_1(t) = -38.9\big[1 - e^{-1.60t}(\cos 8.79t + 0.181\sin 8.79t)$$
$$-u(t-1.2)\{1 - 6.77e^{-1.60t}[\cos 8.79(t-1.2)$$
$$+0.181\sin 8.79(t-1.2)]\}\big]$$

$$p_2(t) = 36.4\big[1 - e^{-8.91t}(\cos 19.14t + 0.465\sin 19.14t)$$
$$-u(t-1.2)\{1 - e^{-8.91(t-1.2)}[\cos 19.14(t-1.2)$$
$$+0.465\sin 19.14(t-1.2)]\}\big]$$

$$p_3(t) = -50.5\big[1 - e^{-13.48t}(\cos 22.19t + 0.608\sin 22.19t)$$
$$-u(t-1.2)\{1 - e^{-13.48(t-1.2)}[\cos 22.19(t-1.2)$$
$$+0.608\sin 22.19(t-1.2)]\}\big]$$

The solution in terms of the original generalized coordinates is calculated using Eq. (8.12).

8.7 MODAL ANALYSIS COMPUTER SOLUTIONS

The FORTRAN program MODAL provides a modal analysis solution for a multi-degree-of-freedom system with proportional damping. Use of MODAL requires three user-provided subprograms.

Subroutine PARM provides parameters to the main program. The user has the choice of directly specifying the parameters in subroutine PARM or providing input statements to read the parameters into the program. Subroutine PARM provides the main program with the number of degrees of freedom, the number of modes to use in the analysis, the mass matrix, either the stiffness matrix or the flexibility matrix, the modal damping ratios, and the initial conditions. Subroutine PARM can also be used to generate initial output and plotter setups, if desired.

Function FUN specifies the force vector as a function of time.

Subroutine OUT is called to output the results. The small time step used for the Runge-Kutta solution to obtain a desired accuracy may provide information at more times than desired. The times at which the results are to be printed or plotted can be selected in subroutine OUT.

The program immediately calls subroutine PARM to provide necessary parameters. It then calls subroutine ITER (Sec. 7.10) to calculate the desired natural frequencies and mode shapes. If any initial condition is nonzero, the modal matrix is inverted by calling subroutine INVERSE (Sec. 7.10) and the initial conditions for the principal coordinates calculated. If any initial condition is nonzero, the number of calculated modes must equal the number of degrees of freedom.

The initial value of time is set to zero. MODAL uses Runge-Kutta to numerically solve the uncoupled differential equations for the principal coordinates. After each time step the generalized coordinates are calculated by premultiplying the vector of principal coordinates by the modal matrix. Subroutine OUT is called. The time is incremented by a time step specified by subroutine PARM. The Runge-Kutta solution continues until the time exceeds a specified final time.

```
C ************************************************
C                                                *
C              MODAL ANALYSIS PROGRAM            *
C                                                *
C ************************************************
C
C THIS PROGRAM PERFORMS A MODAL ANALYSIS
C     SOLUTION FOR A MULTI DEGREE
C     OF FREEDOM SYSTEM WITH MODAL DAMPING.
C     ITS USE REQUIRES THREE
C     USER SUPPLIED SUBPROGRAMS
C
C     SUBROUTINE PARM IS A USER SUPPLIED
C     SUBROUTINE WHICH PROVIDES INPUT
C     DATA TO THE MAIN PROGRAM AND CAN BE USED
C     TO WRITE TITLES
C
C     SUBROUTINE PARM(N,L,LL,L1,A,M,ZETA,
C     U,X,XDOT,DT,TF)
C
C     N = DEGREES OF FREEDOM
C     L = NUMBER OF MODES TO USE IN ANALYSIS
C     LL = 0 IF FLEXIBILITY FORMULATION IS USED
C     LL = 1 IF STIFFNESS FORMULATION IS USED
C     L1 = 0 IF ALL INITIAL CONDITIONS ARE ZERO
C     L1 = 1 IF ANY INITIAL CONDITION IS NON ZERO
C     A = N X N FLEXIBILITY MATRIX IF LL = 0
C     A = N X N STIFFNESS MATRIX IF LL = 1
C     M = N X N MASS MATRIX
C     ZETA = N X 1 MODAL DAMPING VECTOR
C     U = N X 1 INITIAL GUESS VECTOR FOR MATRIX
C     ITERATION
C     X = N X 1 VECTOR OF INITIAL DISPLACEMENTS
C     (NOT REQUIRED IF L1 =0)
C     XDOT = N X 1 VECTOR OF INITIAL VELOCITIES
C     (NOT REQUIRED IF L1 =0)
C     DT = TIME INCREMENT FOR RUNGE - KUTTA SOLUTION
C     TF = FINAL TIME
C
C     FUN IS A USER PROVIDED FUNCTION SUBPROGRAM
```

```
C       WHICH PROVIDES THE
C       TIME DEPENDENT FORCE VECTOR, FOR THE
C       FORMULATION USING THE
C       ORIGINAL GENERALIZED COORDINATES, FOR USE
C       IN MODAL ANALYSIS
C
C       FUNCTION FUN(T,J)
C
C       THIS FUNCTION SUBPROGRAM RETURNS TO THE
C       CALLING ROUTINE THE
C       JTH ELEMENT OF THE FORCE VECTOR FOR
C       A GIVEN TIME T
C
C       J = NUMBER OF COMPONENT REQUIRED
C       T = TIME
C       FUN = JTH COMPONENT OF FORCE VECTOR
C
C       SUBROUTINE OUT OUTPUTS RESULTS AT EACH
C       TIME STEP
C
C       SUBROUTINE OUT(N,T,X,XDOT,JQ)
C       N = DEGREES OF FREEDOM
C       T = TIME
C       X = N X 1 VECTOR OF DISPLACEMENTS
C       (GENERALIZED COORDINATES)
C       XDOT = N X 1 VECTOR OF VELOCITIES
C       JQ = INTERNAL COUNTER USED TO DECIDE
C       WHETHER TO PRINT RESULTS OF
C            TIME STEP
C
        EXTERNAL FUN
        EXTERNAL INVERSE
        REAL A(9,9),M(9,9),Q(9,9),OM(9,9),X(9),XDOT(9)
        REAL ZETA(9),P(9),PDOT(9),U(9),QINV(9,9)
        REAL K11,K12,K13,K14,K21,K22,K23,K24
        INTEGER ERR
        COMMON Q,I,N
C
C INITIALIZING ALL MATRICES AND VECTORS TO ZERO.
C       THUS USER ONLY HAS TO
C       PROVIDE NONZERO COMPONENTS OF THESE
C       QUANTITIES.
C
        DATA A / 81 * 0.0 /
        DATA M / 81 * 0.0 /
        DATA ZETA / 9 * 0.0 /
        DATA X / 9 * 0.0 /
        DATA XDOT / 9 * 0.0 /
        DATA P / 9 * 0.0 /
```

```
      DATA PDOT/ 9 * 0.0/
      OPEN(7,FILE='EX93.DOC')
C
C DATA INPUT
C
      CALL PARM(N,L,LL,L1,A,M,ZETA,U,X,XDOT,DT,TF)
C
C NATURAL FREQUENCY AND MODAL MATRIX CALCULATIONS
C
      CALL ITER(N,L,LL,A,M,U,Q,OM,100,1.E-5,ERR)
      WRITE(7,201)ERR
      WRITE(7,111)
  111 FORMAL(//,5X,'CALCULATED NATURAL FREQUENCIES AND MODE'
     $'SHAPES'/)
      DO 7 I=1,L
      WRITE(7,112)I,OM(I)
  112 FORMAT(5X,'MODE NUMBER',I2,5X,'NATURAL FREQUENCY'
     $'=',F12.6)
      WRITE(7,113)
  113 FORMAT(5X,'MODE SHAPE')
      DO 8 J=1,N
    8 WRITE(7,135)Q(J,I)
  135 FORMAT(5X,F12.6)
    7 CONTINUE
  201 FORMAT(5X,'ERR=' ,2X,I2)
      CALL OUT(N,T,X,XDOT,JQ)
C
C     CALCULATION OF INITIAL CONDITIONS FOR
C     PRINCIPAL COORDINATES
C
      IF(L1.EQ.0)GO TO 25
      CALL INVERSE(N,Q,QINV)
      DO 221 I=1,N
      SUM1=0.0
      SUM2=0.0
      DO 222=J=1,N
      SUM1=SUM1+QINV(I,J)*X(J)
  222 SUM2=SUM2+QINV(I,J)*XDOT(J)
      P(I)=SUM1
  221 PDOT(I)=SUM2
C
C RUNGE - KUTTA SOLUTION TO UNCOUPLED
C DIFFERENTIAL EQUATIONS
C
   25 T=0.0
   12 DO 9 I=1,L
      C1=2.*ZETA(I)*OM(I)
      C2=OM(I)*OM(I)
      Y1=P(I)
```

```
      Y2 = PDOT(I)
      K11 = DT*Y2
      K21 = DT*(F(T)- C1*Y2- C2*Y1)
      K12 = DT*(Y2 + K21 / 2.)
      FS = F(T + DT / 2.)
      K22 = DT*(FS- C1*(Y2 + K21 / 2.)- C2*(Y1 + K11 / 2.))
      K13 = DT*(Y2 + K22 / 2.)
      K23 = DT*(FS- C1*(Y2 + K22 / 2.)- C2*(Y1 + K12 / 2.))
      K14 = DT*(Y2 + K23)
      K24 = DT*(F(T + DT)- C1*(Y2 + K23)- C2*(Y1 + K13))
      P(I) = Y1 + (K11 + 2.*K12 + 2.*K13 + K14) / 6.
    9 PDOT(I) = Y2 + (K21 + 2.*K22 + 2.*K23 + K24) / 6.
C
C CALCULATION OF ORIGINAL GENERALIZED COORDINATES
C
      DO 10 I = 1,N
      SUM1 = 0.0
      SUM2 = 0.0
      DO 11 J = 1,L
      SUM1 = SUM1 + Q(I,J)*P(J)
   11 SUM2 = SUM2 + Q(I,J)*PDOT(J)
      X(I) = SUM1
   10 XDOT(I) = SUM2
      T = T + DT
      CALL OUT(N,T,X,XDOT,JQ)
      IF(T.LE.TF) GO TO 12
      STOP
      END
C
C FUNCTION F(T) PROVIDES THE APPROPRIATE FUNCTION
C     FOR THE RUNGE- KUTTA
C     SOLUTION. IT CALLS A USER SUPPLIED FUNCTION
      SUBPROGRAM FUN(T)
C     WHICH PROVIDES THE FORCE VECTOR FOR
      THE GENERALIZED COORDINATES.
C
      FUNCTION F(T)
      REAL Q(9,9)
      COMMON Q,I,N
      F = 0.0
      DO 1 J = 1,N
      F = F + Q(J,1)*FUN(T,J)
    1 CONTINUE
      RETURN
      END
```

MODAL is used to solve the following problems.

Example 8.7.

The vehicle which has the suspension system of Fig. 8.9 encounters the bump of Fig. 8.10. The shape of the bump is approximated as a half-wave sinusoid. The vehicle is traveling at a constant horizontal speed of 40 ft/s. The suspension system is in equilibrium when the front axle encounters the pothole. Use MODAL to determine the response of the suspension system for 1 s after the front axle encounters the pothole assuming (a) no damping and (b) proportional damping with $\alpha = 0.04$.

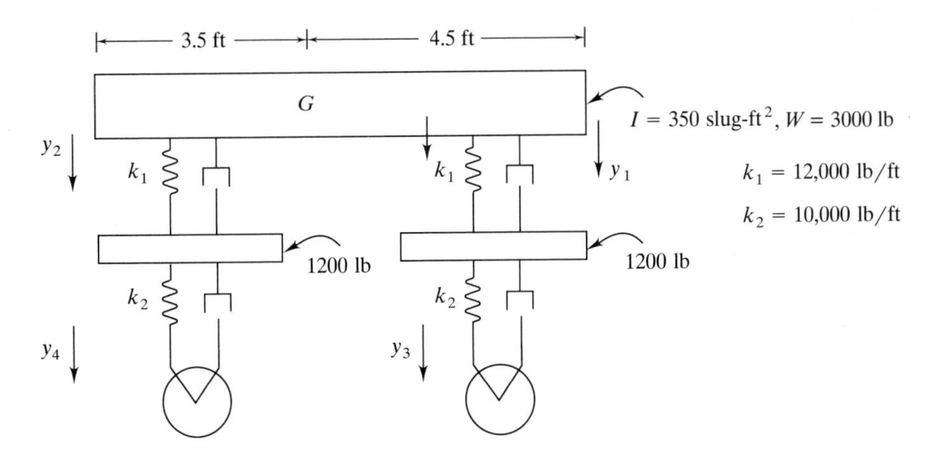

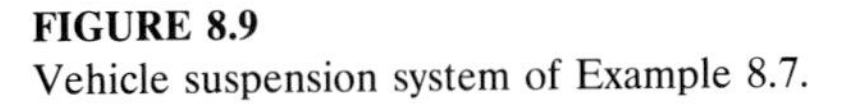

FIGURE 8.9
Vehicle suspension system of Example 8.7.

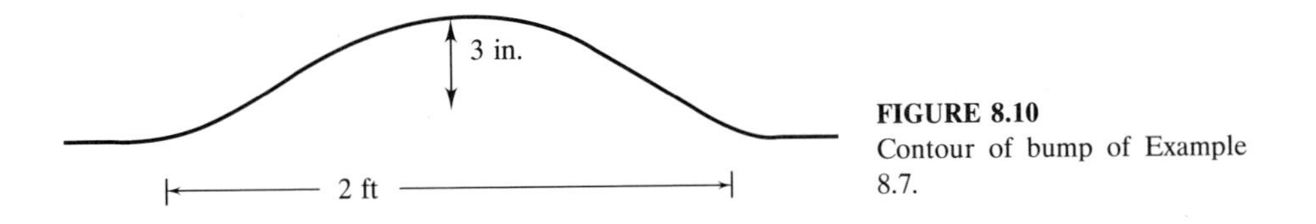

FIGURE 8.10
Contour of bump of Example 8.7.

The differential equations for a similar suspension system are given in Example 7.14. For force vector for the forced-vibration problem is dependent upon the road contour.

Let $y(z)$ denote the road contour where z is the horizontal coordinate measured from the leading edge of the pothole. The front axle encounters the pothole immediately, while the rear axle does not encounter the pothole until $t = 0.2$ s. The principle of virtual work is used to determine the force vector as

$$\mathbf{F}(t) = \begin{bmatrix} 0 \\ 0 \\ k_2 y_3(t) \\ k_2 y_4(t) \end{bmatrix}$$

where $y_3(t) = 0.25 \sin 200\pi(t - 0.2)[u(t - 0.2) - u(t - 0.25)]$

$$y_4(t) = 0.25 \sin 200\pi t[1 - u(1 - 0.05)]$$

The time step must be chosen carefully. The smallest natural period for this example is subsequently determined as 0.16 s. However, one axle traverses the pothole in only 0.05 s. Thus a time step of 0.0025 s is chosen. This gives a nondimensional $\Delta t = 0.0025 \text{ s}/0.05 \text{ s} = 0.05$. Thus the error is expected to be $O(0.05^4) = O(6.25 \times 10^{-6})$.

The user-provided subprograms, subroutine PARM, subroutine OUT, and function FUN, for this example are provided here. Due to the large amount of data, it has been elected only to print every eighth time step.

The results for part (a), the undamped system, are given in the computer output and plotted in Fig. 8.11.

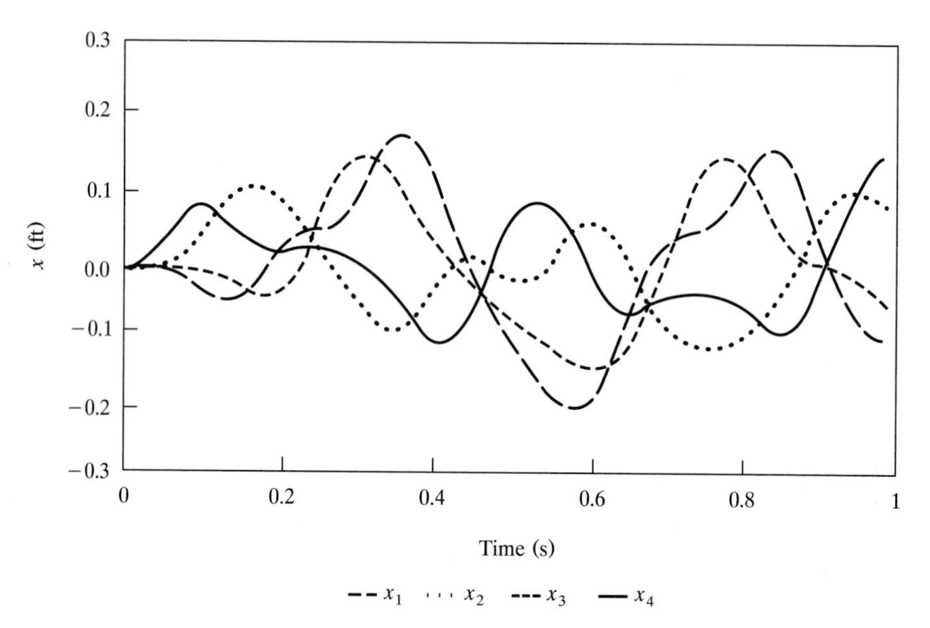

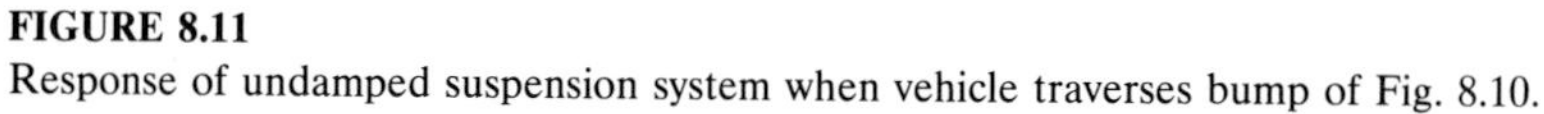

FIGURE 8.11
Response of undamped suspension system when vehicle traverses bump of Fig. 8.10.

```
FUNCTION FUN(T,J)
Y(Z,A,X) = A*SIN(3.1415*Z / X)
FUN = 0.0
IF(J.LE.2)RETURN
XL = 8.0
XK2 = 10000.
X = 2.0
A = 0.25
V = 40.0
```

```
      IF(J.EQ.3)GO TO 1
      Z = V*T
      IF(Z.GT.X)RETURN
      FUN = Y(Z,A,X)*XK2
      RETURN
    1 Z = V*T- XL
      IF(Z.LT.0.0.OR.Z.GT.X)RETURN
      FUN = XK2*Y(Z,A,X)
      RETURN
      END

      SUBROUTINE OUT(N,T,X,XDOT,J)
      REAL X(9),XDOT(9)
      IF(T.GT.1.E-5) GO TO 1
      WRITE(7,101)
      J = 8
  101 FORMAT(7X,'TIME',9X,'X(1)',12X,'X(2)',12X,'X(3)',
     $12X,'X(4)',/ )
    1 IF(J.NE.8)GO TO 2
      WRITE(7,100)T,X(1),X(2),X(3),X(4)
  100 FORMAT(2X,E14.7,2X,E14.7,2X,E14.7,2X,E14.7,
     $2X,F10.4)
    2 J = J + 1
      IF(J.EQ.9)J = 1
      RETURN
      END

C
C USER PROVIDED SUBROUTINE TO PROVIDE INPUT
C       DATA FOR MODAL ANALYSIS
C       PROGRAM FOR EXAMPLE 8.7
C
        SUBROUTINE PARM(N,L,LL,L1,A,M,ZETA,U,
        X,XDOT,DT,TF)
        REAL
  A(9,9),M(9,9),ZETA(9),U(9),X(9),XDOT(9)
        REAL I
        N = 4
        L = 4
        LL = 1
        L1 = 0
        V = 4.5
        W = 3.5
        Y2 = (V + W)**2
        XK1 = 12000.
        XK2 = 10000.
        XM = 3000 / 32.2
        XMA = 1200 / 32.2
        I = 350.
```

```
      M(1,1) = (XM*W*W + I) / Y2
      M(1,2) = (XM*V*W- I) / Y2
      M(2,1) = M(1,2)
      M(2,2) = (XM*V*V + I) / Y2
      M(3,3) = XMA
      M(4,4) = XMA
      A(1,1) = XK1
      A(1,3) = -XK1
      A(2,2) = XK1
      A(2,4) = -XK1
      A(3,1) = -XK1
      A(3,3) = XK1 + XK2
      A(4,2) = -XK1
      A(4,4) = XK1 + XK2
      U(1) = 0.0
      U(2) = 0.0
      U(3) = 0.0
      U(4) = 1.0
      DT = .0025
      TF = 1.0
      RETURN
      END
```

```
ERR = 0
CALCULATED NATURAL FREQUENCIES AND MODE SHAPES
MODE NUMBER 1    NATURAL FREQUENCY = 38.870876
MODE SHAPE
    0.207552
   -0.149569
   -0.072595
    0.052315
MODE NUMBER 2    NATURAL FREQUENCY = 27.437481
MODE SHAPE
   -0.041988
   -0.058263
    0.083207
    0.115464
MODE NUMBER 3    NATURAL FREQUENCY = 14.099642
MODE SHAPE
    0.135355
   -0.097542
    0.111317
   -0.080219
MODE NUMBER 4    NATURAL FREQUENCY = 9.486482
MODE SHAPE
    0.073679
    0.102242
    0.047417
    0.065799
```

TIME	X(1)	X(2)	X(3)	X(4)
0.0000	0.0000000E+00	0.0000000E+00	0.0000000E+00	0.0000000E+00
0.0200	0.4367402E-04	0.5857967E-04	-0.1358749E-06	0.5129081E-02
0.0400	-0.1177300E-02	0.1604953E-02	-0.1568382E-04	0.3095235E-01
0.0600	-0.6662305E-02	0.9375351E-02	-0.2232087E-03	0.6407723E-01
0.0800	-0.1852984E-01	0.2756829E-01	-0.1291477E-02	0.8435199E-01
0.1000	-0.3367131E-01	0.5490262E-01	-0.4438083E-02	0.8867919E-01
0.1200	-0.4472169E-01	0.8464593E-01	-0.1076590E-01	0.7943539E-01
0.1400	-0.4413170E-01	0.1078449E+00	-0.2012236E-01	0.6241910E-01
0.1600	-0.2877999E-01	0.1175768E+00	-0.3023213E-01	0.4442427E-01
0.1800	-0.2367542E-02	0.1119665E+00	-0.3685607E-01	0.3082598E-01
0.2000	0.2594995E-01	0.9447814E-01	-0.3522519E-01	0.2410230E-01
0.2200	0.4617492E-01	0.7150199E-01	-0.1712230E-01	0.2367180E-01
0.2400	0.5464557E-01	0.4787342E-01	0.3250390E-01	0.2678694E-01
0.2600	0.5977404E-01	0.2302885E-01	0.9566589E-01	0.2973447E-01
0.2800	0.7492538E-01	-0.5899191E-02	0.1450258E+00	0.2866031E-01
0.3000	0.1063902E+00	-0.3827339E-01	0.1704515E+00	0.2016583E-01
0.3200	0.1484721E+00	-0.6806269E-01	0.1699679E+00	0.2194833E-02
0.3400	0.1859148E+00	-0.8612484E-01	0.1488344E+00	-0.2475371E-01
0.3600	0.2016876E+00	-0.8521175E-01	-0.1164858E+00	-0.5664161E-01
0.3800	0.1857911E+00	-0.6475368E-01	0.8229409E-01	-0.8596382E-01
0.4000	0.1402597E+00	-0.3240657E-01	0.5219045E-01	-0.1034312E+00
0.4200	0.7775050E-01	-0.1090615E-02	0.2753323E-01	-0.1010558E+00
0.4400	0.1474710E-01	0.1716466E-01	0.6317402E-02	-0.7556432E-01
0.4600	-0.3662675E-01	0.1731375E-01	-0.1436954E-01	-0.3060860E-01
0.4800	-0.7355220E-01	0.3928641E-02	-0.3619634E-01	0.2347199E-01
0.5000	-0.1016206E+00	-0.1123346E-01	-0.5883564E-01	0.7259498E-01
0.5200	-0.1289769E+00	-0.1557571E-01	-0.8068130E-01	0.103610E+00
0.5400	-0.1592735E+00	-0.2583102E-02	-0.1001713E+00	0.1088611E+00
0.5600	-0.1878429E+00	0.2439316E-01	-0.1165686E+00	0.8873526E-01
0.5800	-0.2035172E+00	0.5353598E-01	-0.1295621E+00	0.5117016E-01
0.6000	-0.1951035E+00	0.7040368E-01	-0.1379995E+00	0.8385473E-02
0.6200	-0.1586726E+00	0.6509526E-01	-0.1388210E+00	-0.2769113E-01
0.6400	-0.1012162E+00	0.3697385E-01	-0.1273115E+00	-0.4938415E-01
0.6600	-0.3825693E-01	-0.5406060E-02	-0.9904933E-01	-0.5512879E-01
0.6800	0.1350377E-01	-0.4886951E-01	-0.5280964E-01	-0.4880303E-01
0.7000	0.4446765E-01	-0.8199642E-01	0.7133879E-02	-0.3723596E-01
0.7200	0.5665353E-01	-0.9979346E-01	0.7083583E-01	-0.2723557E-01
0.7400	0.6178484E-01	-0.1043268E+00	0.1252157E+00	-0.2338859E-01
0.7600	0.7390667E-01	-0.1014097E+00	0.1584861E+00	-0.2722288E-01
0.7800	0.1006221E+00	-0.9569825E-01	0.1645587E+00	-0.3753385E-01
0.8000	0.1378923E+00	-0.8733784E-01	0.1453577E+00	-0.5121440E-01
0.8200	0.1714063E+00	-0.7220655E-01	0.1098785E+00	-0.6399538E-01
0.8400	0.1838733E+00	-0.4550101E-01	0.7037835E-01	-0.7095309E-01
0.8600	0.1642739E+00	-0.6294127E-02	0.3745586E-01	-0.6709282E-01
0.8800	0.1140082E+00	0.3996969E-01	0.1627332E-01	-0.4841585E-01
0.9000	0.4660304E-01	0.8288277E-01	0.5534904E-02	-0.1351743E-01
0.9200	-0.1878224E-01	0.1118658E+00	-0.5762497E-03	0.3482608E-01
0.9400	-0.6641559E-01	0.1216925E+00	-0.8731103E-02	0.8902480E-01
0.9600	-0.9066684E-01	0.1154773E+00	-0.2299657E-01	0.1378054E+00
0.9800	-0.9682805E-01	0.1031611E+00	-0.431917E-01	0.1691982E+00
1.0000	-0.9600431E-01	0.9605998E-01	-0.6554167E-01	0.1745037E+00

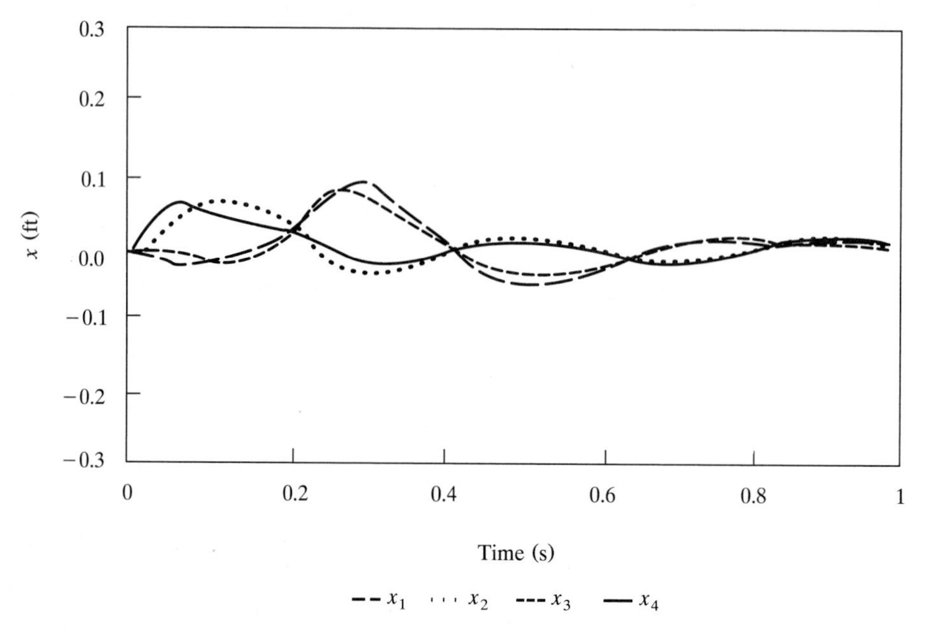

FIGURE 8.12
Response of suspension system designed with proportional damping when vehicle encounters bump of Fig. 8.10.

(*b*) Since proportional damping is assumed, a dashpot whose damping coefficient is proportional to the stiffness is placed in parallel to each spring. For $\alpha = 0.04$, $c_2 = 400\ \mathrm{N \cdot s/m}$. The force vector for the damped system is

$$\mathbf{F} = \begin{bmatrix} 0 \\ 0 \\ k_2 y_3(t) + c_2 \dot{y}_3(t) \\ k_2 y_4(t) + c_2 \dot{y}_4(t) \end{bmatrix}$$

Subroutine PARM is modified by adding the damping ratios for each mode. Since the stiffness matrix is used, matrix iteration gives the largest natural frequency first. Thus, when providing parameters for the main program, ZETA(1) corresponds to the mode with the highest frequency. The results for the damped response are plotted in Fig. 8.12.

Example 8.8. The punch press of Example 4.16 operates on the floor of an industrial plant. The machine is isolated from the floor by an isolator of equivalent stiffness $2 \times 10^6\ \mathrm{N/m}$. The floor is modeled as two identical parallel fixed-fixed beams of length 15 m, elastic modulus $210 \times 10^9\ \mathrm{N/m}$, mass density $7850\ \mathrm{kg/m^3}$, and moment of inertia $3 \times 10^{-4}\ \mathrm{m^4}$. The press is placed at the midspan of the beam. Determine the maximum deflection of the floor. Use five degrees of freedom to model the floor and an additional degree of freedom for the machine, as illustrated in Fig. 8.13. Use only two modes in the modal summation of modal analysis.

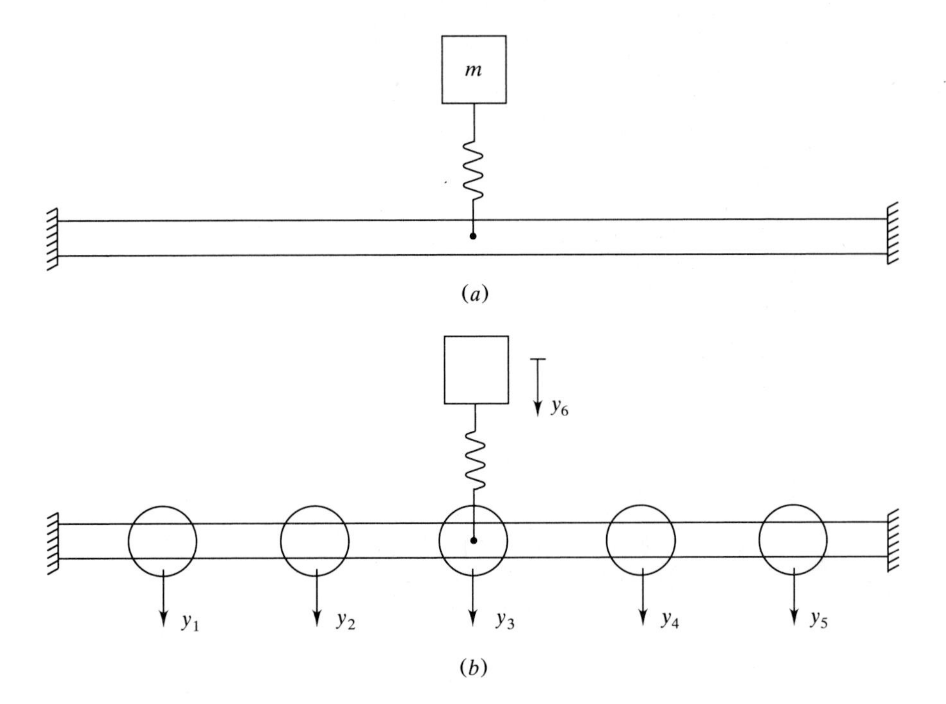

FIGURE 8.13
(*a*) Punch press is mounted on isolator and set on floor in an industrial plant; (*b*) Punch press and floor is modeled with six degrees of freedom.

Flexibility influence coefficients are used to derive the flexibility matrix as

$$\begin{bmatrix} 898.a & 1498.a & 1373.a & 878.a & 357.a & 1373.a \\ 1498.a & 3652.a & 3848.a & 2459.a & 878.a & 3848.a \\ 1373.a & 3848.a & 5205.a & 3848.a & 1373.a & 5205.a \\ 878.a & 2459.a & 3848.a & 3652.a & 1498.a & 3848.a \\ 357.a & 878.a & 1373.a & 1498.a & 898.a & 1373.a \\ 1373.a & 3848.a & 5205.a & 3848.a & 1373.a & 590 \times 10^{-9} \end{bmatrix}$$

where
$$a = \frac{10^{-6}L^3}{EI}$$
$$= 2.69 \times 10^{-10}\ \frac{\text{m}}{\text{N}}$$

The subroutines PARM and FUN for this problem are given in the following program, as well as the output from the matrix iteration part of the modal analysis program. The modal analysis computer solution gives a maximum displacement of 10.0 mm.

```
C
C USER PROVIDED SUBROUTINE TO PROVIDE INPUT DATA
C FOR MODAL ANALYSIS
```

```
C        PROGRAM FOR EXAMPLE 8.8
C
      SUBROUTINE PARM(N,L,LL,L1,A,M,ZETA,U,X,XDOT,DT,TF)
      REAL A(9,9),M(9,9),ZETA(9),U(9),X(9),
C     XDOT(9)
      N = 6
      L = 3
      LL = 0
      L1 = 0
      M(1,1) = 1780.
      M(2,2) = 1780.
      M(3,3) = 1780.
      M(4,4) = 1780.
      M(5,5) = 1780.
      M(6,6) = 500.
      CD = 2.68E-10
      A(1,1) = 898.*CD
      A(1,2) = 1498.*CD
      A(1,3) = 1373.*CD
      A(1,4) = 878.*CD
      A(1,5) = 357.*CD
      A(1,6) = 1373.*CD
      A(2,1) = A(1,2)
      A(2,2) = 3652.*CD
      A(2,3) = 3848.*CD
      A(2,4) = 2459.*CD
      A(2,5) = 878.*CD
      A(2,6) = 3848.*CD
      A(3,1) = A(1,3)
      A(3,2) = A(2,3)
      A(3,3) = 5205.*CD
      A(3,4) = 3848.*CD
      A(3,5) = 1373.*CD
      A(3,6) = 5205.*CD
      A(4,1) = A(1,4)
      A(4,2) = A(2,4)
      A(4,3) = A(3,4)
      A(4,4) = 3652.*CD
      A(4,5) = 1498.*CD
      A(4,6) = 3848.*CD
      A(5,1) = A(1,5)
      A(5,2) = A(2,5)
      A(5,3) = A(3,5)
      A(5,4) = A(4,5)
      A(5,5) = 898.*CD
      A(5,6) = 1373.*CD
      A(6,1) = A(1,6)
      A(6,2) = A(2,6)
      A(6,3) = A(3,6)
```

```
      A(6,4) = A(4,6)
      A(6,5) = A(5,6)
      A(6,6) = 590.E-9
      U(1) = 1.0
      U(2) = 0.5
      U(3) = 0.3
      U(4) = 1.0
      U(5) = -1.
      U(6) = 0.
      DT = .0025
      TF = 3.0
      RETURN
      END

  FUNCTION FUN(T,J)
  IF(J.EQ.6) GO TO 1
  FUN = 0.0
  RETURN
1 N = T / 0.5
  TSTAR = T- .5*N
  IF(TSTAR.LT.0.2) GO TO 2
  FUN = 0.
  RETURN
2 FUN = 4000.
  RETURN
  END
```

```
CALCULATED NATURAL FREQUENCIES AND MODE SHAPES
MODE NUMBER 1    NATURAL FREQUENCY = 12.541233
MODE SHAPE
    0.004334
    0.011310
    0.014663
    0.011310
    0.004334
    0.013790
MODE NUMBER 2    NATURAL FREQUENCY = 36.569687
MODE SHAPE
   -0.008665
   -0.014346
    0.000000
    0.014346
    0.008665
    0.000000
```

The accompanying diskette contains two BASIC versions of the modal analysis program. MODAL.EXE provides a modal analysis solution to the four-degree-of-freedom model of the vehicle suspension system of Example 8.7.

MODAL2.BAS allows the user to use modal analysis for a general n-degree-of-freedom system. The user must supply a BASIC subroutine for the excitation.

8.8 LAPLACE TRANSFORM SOLUTIONS

Let $\bar{\mathbf{x}}(s)$ be the vector of Laplace transforms of generalized coordinates for an n-degree-of-freedom system. Taking the Laplace transform of Eq. (8.1), using linearity of the transform and the property of transforms of derivatives, gives

$$(s^2\mathbf{M} + s\mathbf{C} + \mathbf{K})\bar{\mathbf{x}}(s) = \bar{\mathbf{F}}(s) + (s\mathbf{M} + \mathbf{C})\mathbf{x}(0) + \mathbf{M}\dot{\mathbf{x}}(0) \tag{8.31}$$

where $\mathbf{F}(s)$ is the vector of Laplace transforms of $\mathbf{F}(t)$. If $\mathbf{x}(0) = \mathbf{0}$ and $\dot{\mathbf{x}}(0) = \mathbf{0}$, Eq. (8.31) becomes

$$\mathbf{z}(s)\bar{\mathbf{x}}(s) = \bar{\mathbf{F}}(s) \tag{8.32}$$

where

$$\mathbf{Z}(s) = s^2\mathbf{M} + s\mathbf{C} + \mathbf{K} \tag{8.33}$$

is called the impedance matrix. Premultiplying Eq. (8.32) by $\mathbf{Z}^{-1}$ yields

$$\bar{\mathbf{x}}(s) = \mathbf{Z}^{-1}(s)\bar{\mathbf{F}}(s) \tag{8.34}$$

Using the methods of App. C, the inverse of the impedance matrix is written as

$$\mathbf{Z}^{-1}(s) = \frac{1}{\det(\mathbf{Z}(s))}\mathbf{H}(s) \tag{8.35}$$

where the components of $\mathbf{H}(s)$ are polynomials in s of order $n - 1$ or less. The determinant of the impedance matrix is a polynomial in s of order $2n$, called the characteristic polynomial, and is used to determine the free-vibration characteristics of the system. The roots of this polynomial occur in complex conjugate pairs,

$$s_j = s_{r_j} + is_{i_j} \qquad j = 1, 2, \dots, n \tag{8.36}$$

leading to the following factorization:

$$\det(\mathbf{Z}(s)) = \prod_{j=1}^{n}\left(s^2 - 2s_{r_j}s + s_{r_j}^2 + s_{i_j}^2\right) \tag{8.37}$$

Partial fraction decomposition is used to develop

$$\bar{x}_k(s) = A_k(s) + \sum_{j=1}^{n}\frac{\alpha_{kj}s + \beta_{kj}}{s^2 + 2s_{r_j}s + s_{r_j}^2 + s_{i_j}^2} \qquad k = 1, \dots, n \tag{8.38}$$

Inversion of Eq. (8.38) leads to

$$x_k(t) = x_{k_p}(t) + \sum_{j=1}^{n} e^{s_{r_j}t}\left[C_{kj}\sin\left(\sqrt{s_{r_j}^2 + s_{i_j}^2}\,t\right) + D_{kj}\cos\left(\sqrt{s_{r_j}^2 + s_{i_j}^2}\,t\right)\right] \tag{8.39}$$

where $$x_{k_p}(t) = \mathscr{L}^{-1}\{A_k(s)\}$$

If the real part of all roots of the characteristic polynomial are negative, all terms in the summation in Eq. (8.39) decay with time. The steady state has only a contribution from the particular solution. If any root has a positive real part, then its corresponding term in the summation in Eq. (8.39) grows exponentially and the system is unstable.

Example 8.9. Determine the steady-state amplitudes for the two-degree-of-freedom system of Fig. 8.14.

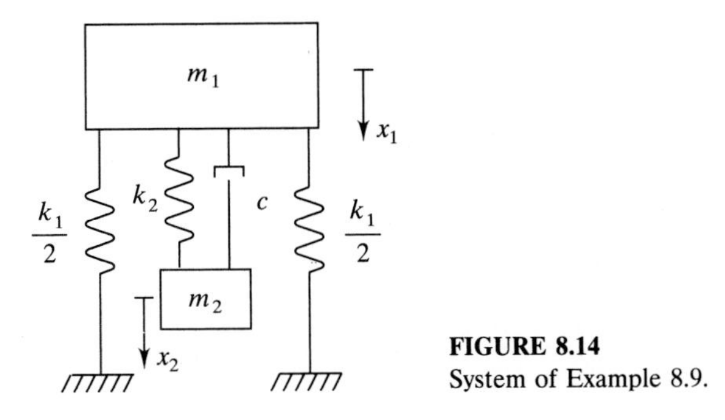

FIGURE 8.14
System of Example 8.9.

The matrix form of the governing differential equations for the system of Fig. 8.14 is

$$\begin{bmatrix} m_1 & 0 \\ 0 & m_2 \end{bmatrix}\begin{bmatrix} \ddot{x}_1 \\ \ddot{x}_2 \end{bmatrix} + \begin{bmatrix} c & -c \\ -c & c \end{bmatrix}\begin{bmatrix} \dot{x}_1 \\ \dot{x}_2 \end{bmatrix} + \begin{bmatrix} k_1 + k_2 & -k_2 \\ -k_2 & k_2 \end{bmatrix}\begin{bmatrix} x_1 \\ x_2 \end{bmatrix} = \begin{bmatrix} F_0 \sin \omega t \\ 0 \end{bmatrix}$$

The system impedance matrix is

$$\mathbf{Z}(s) = \begin{bmatrix} m_1 s^2 + cs + k_1 + k_2 & -cs - k_2 \\ -cs - k_2 & m_2 s^2 + cs + k_2 \end{bmatrix}$$

and its determinant is

$$\begin{aligned} D(s) &= \left(m_1 s^2 + cs + k_1 + k_2\right)\left(m_2 s^2 + cs + k_2\right) - \left(cs + k_2\right)^2 \\ &= m_1 m_2 s^4 + (m_1 + m_2)cs^3 + (m_1 k_2 + k_1 m_2 + m_2 k_2)s^2 + k_1 cs + k_1 k_2 \end{aligned}$$

Cramer's rule is used to solve for $\bar{\mathbf{x}}(s)$, leading to

$$\bar{x}_1(s) = \frac{F_0 \omega\left(m_2 s^2 + cs + k_2\right)}{(s^2 + \omega^2)D(s)}$$

and $$\bar{x}_2(s) = \frac{F_0 \omega(cs + k)}{(s^2 + \omega^2)D(s)}$$

A partial fraction decomposition of $x_1(s)$ allows it to be written as

$$\bar{x}_1(s) = \frac{U_1 s + V_1}{s^2 + \omega^2} + \frac{Q(s)}{D(s)}$$

where the constants U_1 and V_1 and the function $Q(s)$ are to be determined. Inversion of the term with $D(s)$ in the denominator leads to the transient part of the solution represented by the summation in Eq. (8.39). The steady-state solution for $x_1(t)$ is obtained by solving for U_1 and V_1 and inverting only the first term in the preceding equation. Multiplication by the common denominator leads to

$$(U_1 s + V_1)D(s) + Q(s)(s^2 + \omega^2) = F_0\omega\left(m_2 s^2 + cs + k_2\right)$$

Since the preceding equation is valid for all complex s, set $s = i\omega$. Then

$$(i\omega U_1 + V_1)D(i\omega) = F_0\omega\left(-m_2\omega^2 + ic\omega + k_2\right)$$

or
$$i\omega U_1 + V_1 = \frac{F_0\omega\left(-m_2\omega^2 + ic\omega + k_2\right)\overline{D}(i\omega)}{D(i\omega)\overline{D}(i\omega)}$$

The constant V_1 is equal to the real part of the right-hand side of the preceding equation, while ωU_1 is equal to the imaginary part. Performing this algebra leads to

$$U_1 = \frac{F_0\left[c\omega M(\omega) - (k_2 - m_2\omega^2)N(\omega)\right]}{M^2(\omega) + N^2(\omega)}$$

and
$$V_1 = \frac{F_0\omega\left[(k_2 - m_2\omega^2)M(\omega) + c\omega N(\omega)\right]}{M^2(\omega) + N^2(\omega)}$$

where
$$M(\omega) = m_1 m_2\omega^2 - (m_1 k_2 + k_1 m_2 + m_2 k_2)\omega^2 + k_1 k_2$$

and
$$N(\omega) = -(m_1 + m_2)c\omega^3 + k_1 c\omega$$

Inversion of the transform yields the steady-state solution for $x_1(t)$ as

$$x_1(t) = U_1 \cos \omega t + \frac{V_1}{\omega} \sin \omega t$$

An alternate form for the preceding equation is

$$x_1(t) = X_1 \sin(\omega t - \phi_1)$$

where
$$X_1 = \sqrt{U_1^2 + \left(\frac{V_1}{\omega}\right)^2}$$

$$= \sqrt{\frac{(c\omega)^2 + (k_2 - m_2\omega^2)^2}{M^2 + N^2}}$$

and
$$\phi_1 = \frac{\omega U_1}{V_1}$$

A similar procedure is applied to solve for the steady-state solution for $x_2(t)$, yielding

$$x_2(t) = U_2 \cos \omega t + \frac{V_2}{\omega} \sin \omega t$$

$$= X_2 \sin(\omega t - \phi_2)$$

where
$$U_2 = \frac{F_0[k_2 M(\omega) - c\omega N(\omega)]}{M^2(\omega) + N^2(\omega)}$$

$$V_2 = \frac{F_0\omega[c\omega M(\omega) + k_2 N(\omega)]}{M^2(\omega) + N^2(\omega)}$$

and
$$X_2 = \sqrt{\frac{k_2^2 + (c\omega^2)^2}{M^2 + N^2}}$$

The preceding problem could have been solved using the method of Sec. 8.2. However, its application requires the solution of four simultaneous equations. The Laplace transform method has the advantage that $x_1(t)$ is obtained separately from $x_2(t)$, and that the solution of only two simultaneous equations is necessary. In addition, the Laplace transform method can provide the transient solution and the solution for nonharmonic excitations. This problem could not have been solved by modal analysis because the damping matrix is not of the form corresponding to proportional damping.

The system in this example is an undamped one-degree-of-freedom system with a damped vibration absorber attached. This system is discussed in more detail in Sec. 11.7.

8.9 SUMMARY

Three methods for the determination of the response of a multi-degree-of-freedom-system to an excitation are considered: matrix analysis of the response due to harmonic excitations using the method of undetermined coefficients; modal analysis for undamped systems and systems subject to proportional damping; and Laplace transform methods. These are not the only methods available. Other methods include a modal analysis for systems with general viscous damping and direct numerical solution of Eq. (8.1). The methods considered in this chapter are compared in the following discussion.

Modal analysis is the premier method for determining the response of multi-degree-of-freedom linear systems. It applies only to undamped systems and systems with proportional damping. However, proportional damping is often assumed because it usually provides an adequate, if not good, model to the actual damping. Modal analysis can be used to analyze systems whose properties are known or can be used experimentally. Modal analysis allows use of more degrees of freedom in the model than modes used in determining the response. If a discrete analysis of a continuous system is being performed, the higher natural frequencies and mode shapes are inaccurate and inclusion of these modes is in the total response is inappropriate. In addition, these modes are usually highly damped so that their effect on the total solution is small. If the excitation is such that a closed-form solution is impossible or the excitation is given empirically, numerical integration of the convolution integral or Runge-Kutta methods can be used to determine the response of each mode.

The use of modal analysis is preferable over direct numerical integration of Eq. (8.1) because all modes must be used with the direct numerical integration.

The method of Sec. 8.2 applies only for harmonic excitations. Most long-term excitations are harmonic or at least periodic. The method is best applied for systems with only two or three degrees of freedom. If the system has more degrees of freedom, the algebra required to determine its response is lengthy and tedious. The method can be applied to systems with any form of viscous damping. That is where its advantage lies over modal analysis.

The Laplace transform method is also best used for only two- and three-degree-of-freedom systems. The algebra for systems with more degrees of freedom becomes overbearing. The determinant of the impedance matrix must be evaluated as a polynomial in s. The Laplace transform method can be used with any form of excitation and with any form of viscous damping. It is often used to determine the stability of a system and used for systems with controls applications.

In summary, for systems with only two or three degrees of freedom, the method of Sec. 8.2 for harmonic excitations and the Laplace transform method for more general applications are appropriate and used with less algebra than modal analysis. Modal analysis is the most powerful method and can be used for undamped systems or systems with proportional damping. If the system has another form of damping, a more complex version of modal analysis can be derived.

PROBLEMS

8.1. Determine the steady-state amplitudes of each of the blocks of Fig. P8.1.

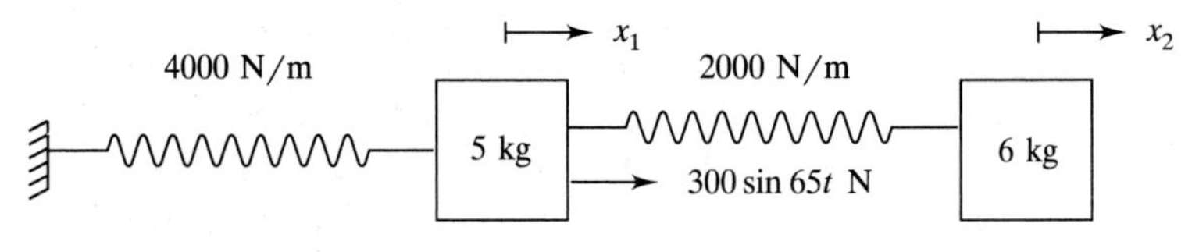

FIGURE P8.1

8.2. Determine the steady-state amplitudes of each of the blocks of Fig. P8.2. Write the frequency response equations in terms of the nondimensional parameters introduced in Example 8.1.

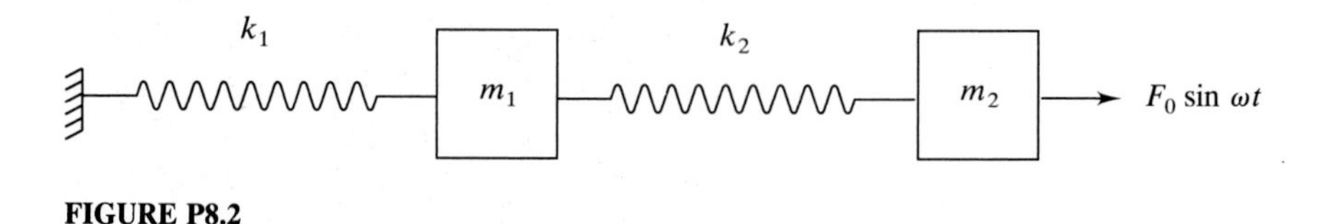

FIGURE P8.2

8.3. Determine the steady-state amplitudes of each of the blocks of Fig. P8.3. Write the frequency response equations in terms of the nondimensional parameters introduced in Example 8.1 and $\zeta = c/(2\sqrt{m_1 k_1})$.

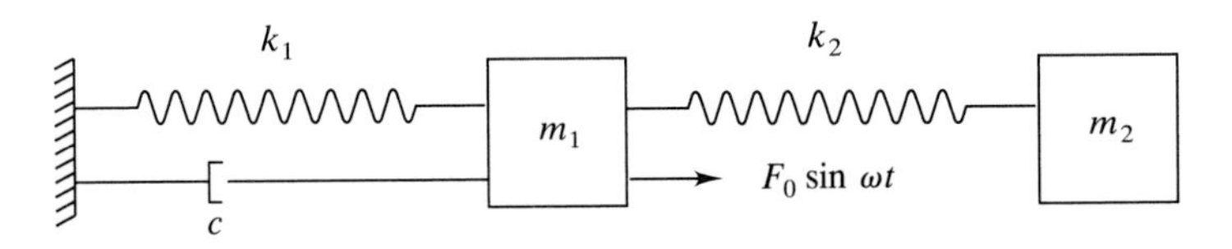

FIGURE P8.3

8.4. For what values of ω will the steady-state amplitude of the block of mass m_2 of Fig. P8.3 be maximized.

8.5. Determine the steady-state amplitude of vibration of the mass center of the bar of Fig. P7.3 when its right end is subject to a moment $M = 4000 \sin 105t \text{ N} \cdot \text{m}$.

8.6. A machine of mass m is attached to the end of a cantilever beam of elastic modulus E, moment of inertia I, length L, and total mass m_b. The machine has a rotating unbalance of magnitude $m_0 e$ and operates at a speed ω. Using two degrees of freedom to model the vibrations of the beam, determine the steady-state amplitude of the location where the machine is placed.

8.7. A machine of mass m is attached to the midspan of a uniform pinned-pinned beam of elastic modulus E, moment of inertia I, length L, and total mass m_b. The machine has a rotating unbalance $m_0 e$ and operates at a speed ω. Using three degrees of freedom to model the vibrations of the beam, determine the steady-state amplitude of the midspan.

8.8. The street light pole of Example 8.2 is made of steel ($\rho = 7850 \text{ kg/m}^3$). Use three degrees of freedom to model the pole. The generalized coordinates represent the deflection of the midspan of the pole, the deflection of the pole where the light is attached, and the rotation of the pole where the light is attached. Neglect the rotational inertia of the pole. Determine the steady-state amplitude of the light fixture.

8.9. Derive Eq. (8.10).

8.10. Find the relations between the principal coordinates and the indicated generalized coordinates of Figure P8.1

8.11. Determine the relations between the principal coordinates and the displacements of the nodal points when two degrees of freedom are used to model a uniform cantilever beam.

8.12. An alternative method to derive the uncoupled differential equations satisfied by the principal coordinates is to make use of the expansion theorem, Eq. (7.36). According to the expansion theorem, the displacement vector at each instant of time can be expanded using the normalized mode shapes

$$\mathbf{x} = \sum_{i=1}^{n} c_i(t)\mathbf{X}_i$$

(*a*) Substitute the preceding equation into Eq. (8.14).

(*b*) Take the standard scalar product of the resulting equation with $\mathbf{X}_j$ (premultiply by $\mathbf{X}_j^T$), for an arbitrary j, and use the orthogonality properties, Eqs. (7.44) and (7.45).

(*c*) Show that the differential equations defining each $c_i(t)$ are uncoupled and that c_i is equivalent to the ith principal coordinate.

8.13. A machine of mass m is attached to the end of a cantilever beam of length L, elastic modulus E, moment of inertia I and total mass m_b. The machine has a rotating unbalance of magnitude $m_0 e$ and operates at a speed ω. Use modal analysis to determine the steady-state amplitude of vibration of the end of the beam if two degrees of freedom are used to model the beam.

8.14. The system of Fig. P8.14 is at rest in equilibrium when the 5-kg block is subject to a 300-N force lasting 0.2 s. Determine the time-dependent response of each block.

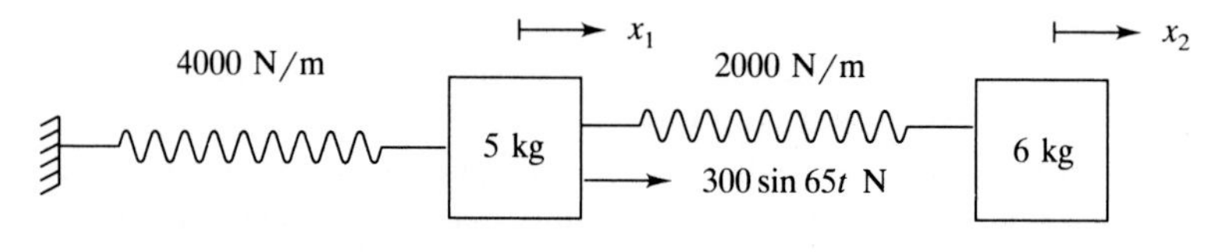

FIGURE P8.14

8.15. Operation of the machine of Prob. 8.11 produces the force of Fig. P8.15 applied to the midspan of the beam. Use a two-degree-of-freedom model for the beam vibrations and determine the time-dependent deflection of the midspan.

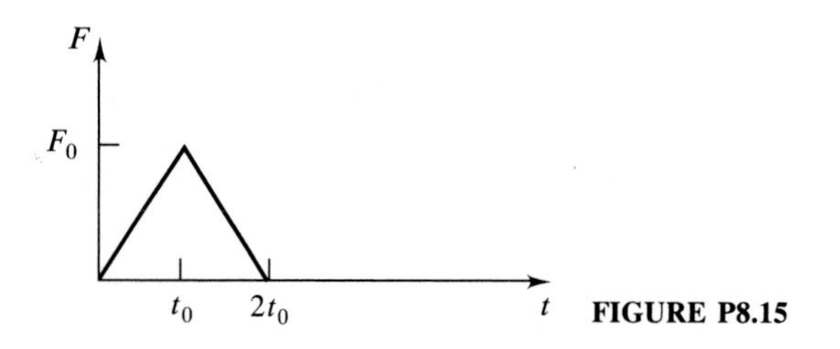

FIGURE P8.15

8.16. The mass center of the bar of Prob. 7.8 is subject to the time-dependent force of Fig. P8.16. Determine the time-dependent response of the attached block.

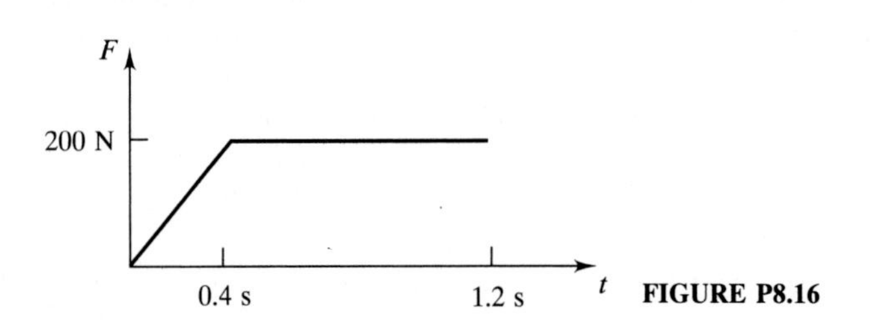

FIGURE P8.16

8.17. The helicopter tail section of Prob. 7.35 is subject to a harmonic excitation at its free end caused by a rotating unbalance in the tail blades. The rotating unbalance has a magnitude of 0.28 kg · m at a frequency of 210 rad/s. Use a four-degree-of-freedom model for the tail section, but use the two modes with the lowest natural frequencies in the modal analysis to predict the maximum displacement of the tail.

8.18. The helicopter of Probs. 7.35 and 8.17 is accelerating upward with an acceleration history shown in Fig. P8.18. Using the same conditions as in Prob. 8.17, determine the maximum displacement of the tail section relative to the helicopter including the rotating unbalance.

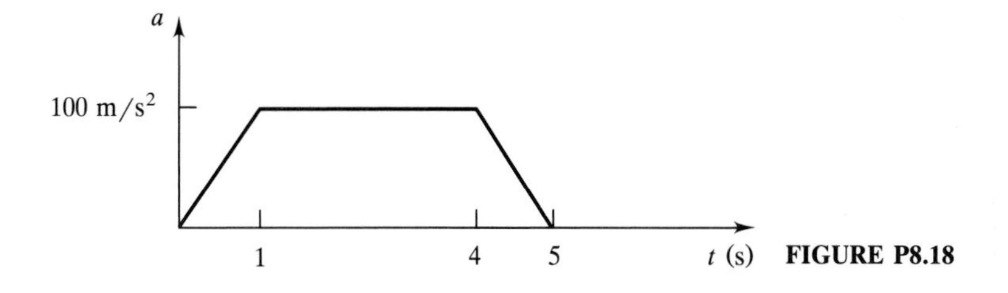

FIGURE P8.18

8.19. The motor at the right end of the shaft of Fig. P7.33 is subject to the periodic torque of Fig. P8.19. Use modal analysis to determine the maximum steady-state displacement of each disk.

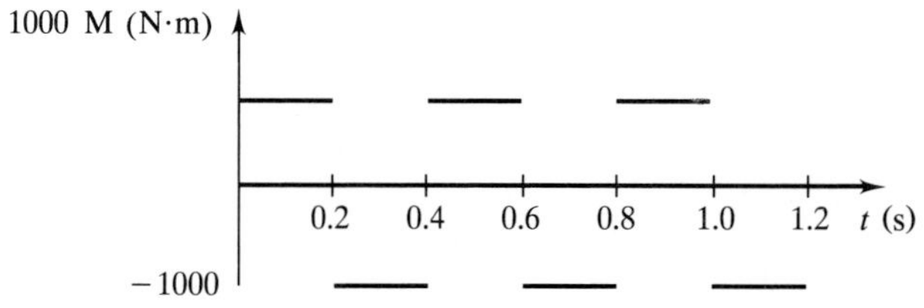

1000 M (N·m)

0.2 0.4 0.6 0.8 1.0 1.2 t (s)

−1000

FIGURE P8.19

8.20. The vehicle of Example 8.7 is traveling on a concrete roadway whose expansion joints are 16 ft apart. The expansion joint is 0.0025 ft wide. When a wheel crosses the joint, it is subject to a 40-lb force during the time it takes the wheel to cross the joint. Use MODAL to determine the response of the vehicle if it travels at a constant horizontal speed of 40 ft/s.

8.21. The vehicle of Example 8.7 encounters a pothole whose contour is given in Fig. P8.21. The vehicle is traveling at a horizontal speed of 80 ft/s when its front wheels first encounter the pothole, but then begins to decelerate at 3 ft/s^2. Determine the response of the suspension system.

8.22. Repeat Prob. 8.21 if the system has proportional damping with $\alpha = 0.02$.

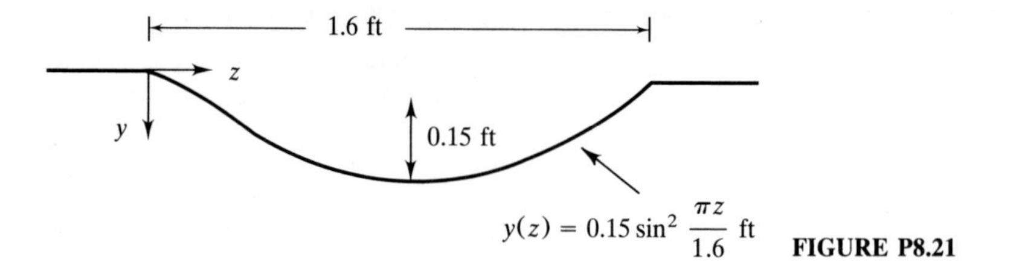

FIGURE P8.21

8.23. Use the FORTRAN program MODAL to determine the response of the system of Fig. P8.23.

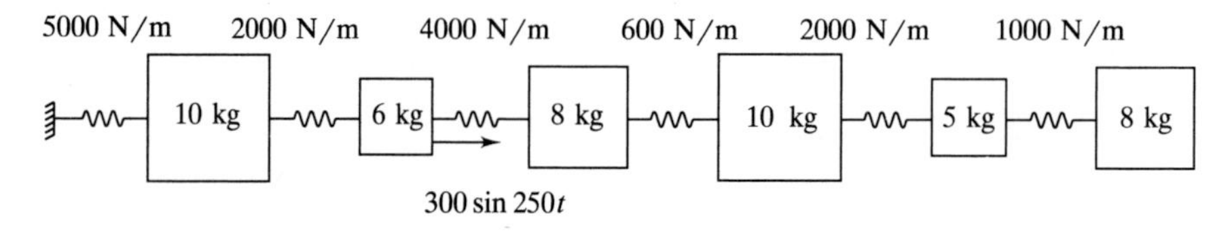

FIGURE P8.23

8.24. As a result of an explosion in the lowest floor of the frame structure of Fig. P7.32, the bottom floor is thus subject to a horizontal force of 8000 N lasting 1.2 s. Determine the response of the structure.

8.25. The frame structure of Fig. P7.32 is subject to a horizontal ground acceleration $0.95 \sin 310t \text{ m/s}^2$. Determine the response of the top floor of the structure using the first four modes.

8.26. The end of the tapered bar of Fig. P8.26 is subject to the time-dependent force shown. Use four degrees of freedom to model the bar, but only include two modes

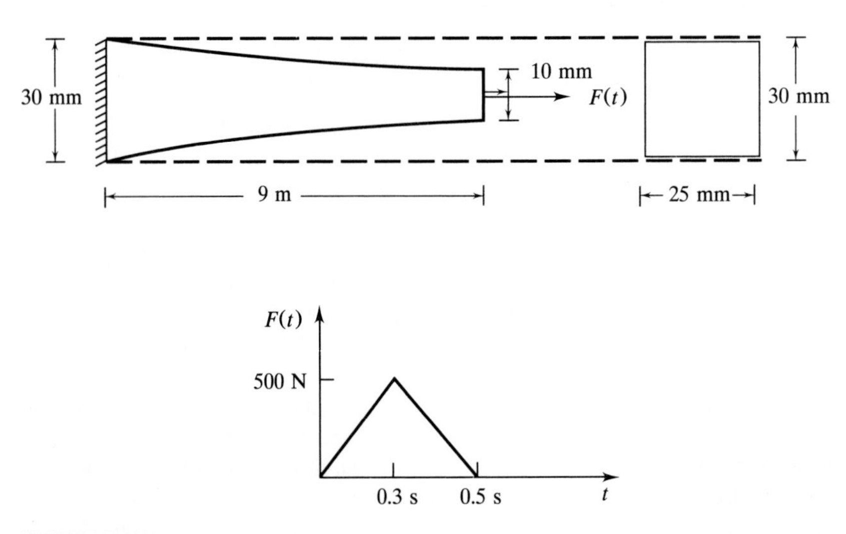

FIGURE P8.26

when using modal analysis to determine the time-dependent lateral displacement of the end of the bar.

8.27. A mass m is placed at the end of a cantilever beam of elastic modulus E, moment of inertia I, length L, and total mass m_b. The beam is subject to a harmonic excitation $F = F_0 \sin \omega$ a distance l from its fixed end. Write the subroutines PARM, FUN, and OUT for use in MODAL to calculate the response of the end of the beam when n degrees of freedom are used to model the beam and m modes are used in the modal analysis.

8.28. Repeat Prob. 8.12 when the system is subject to proportional damping of the form of Eq. (8.21).

8.29. Use only the first three modes in the modal summation to determine the response of the system of Example 7.13 to a rectangular pulse of magnitude 20,000 N and duration 1.2 s applied to the third floor, with $k = 5 \times 10^5$ N/m.

8.30. A 200-kg machine with a rotating unbalance of magnitude 1.5 kg · m operates at 1200 rpm on the second floor of the structure of Example 7.13. Using only three modes in the modal summation and assuming proportional damping with $\alpha = 0.12$, determine the steady-state amplitudes of vibration of each floor.

8.31. A three-story frame structure has girders infinitely rigid compared to its columns. All columns are steel ($E = 200 \times 10^9$ N/m^2, $\rho = 7850$ kg/m^3, $\sigma_y = 200 \times 10^6$ N/m^2) and are circular with radius 12 cm and length 10 m. Each girder has a mass of 1000 kg. The force due to a blast occurring on the third floor of the structure is approximated as a sinusoidal pulse of duration 0.6 s. What is the largest magnitude blast the structure can sustain without the yield stress being exceeded?

8.32. A machine of mass 125 kg is placed on the second floor of the three-story structure of Prob. 8.31. The machine has a rotating unbalance of 1.6 kg and operates at a speed of 100 Hz. What is the largest stress in the columns due to the steady-state vibrations?

8.33. Repeat Prob. 8.32 above assuming proportional damping with $\alpha = 0.01$.

8.34. Repeat Prob. 8.32 assuming that the damping ratios for the three modes are $\zeta_1 = 0.05$, $\zeta_2 = 0.15$, and $\zeta_3 = 0.30$.

8.35. The system of Fig. P8.35 is at rest in equilibrium when the block of mass m_2 is subject to a rectangular pulse of duration t_0 and magnitude F_0. Use the Laplace transform method to determine the response of each of the blocks.

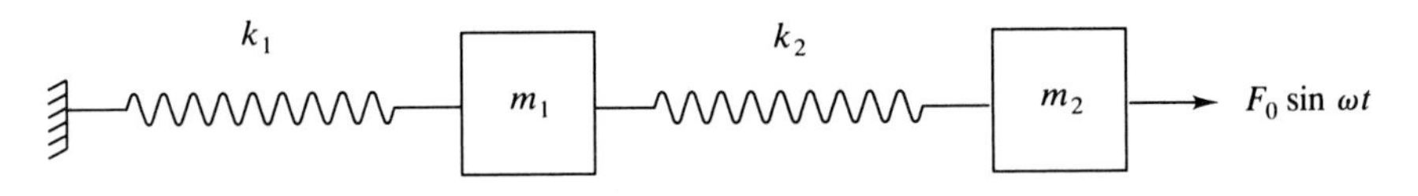

FIGURE P8.35

8.36. The system of Fig. P8.36 is at rest in equilibrium when the block of mass m_1 is subject to an impulse of magnitude I. Use the Laplace transform method to determine the response of each of the blocks.

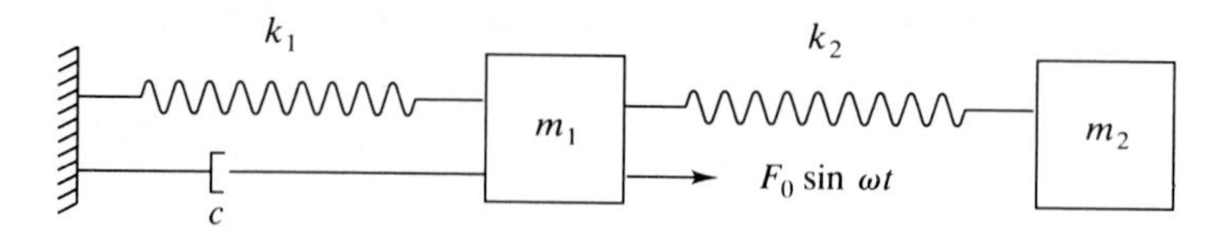

FIGURE P8.36

8.37. Use the Laplace transform method to solve Prob. 8.3.

The programs MODAL.EXE and MODAL2.BAS included on the accompanying diskette use modal analysis to find the forced response of an n-degree-of-freedom system. Use MODAL1 to solve Probs. 8.38 and 8.39. Use MODAL2 to solve Probs. 8.40–8.44.

8.38. Solve Example 8.7 when $v = 80$ ft/sec and $\alpha = 0.02$ sec^{-1}.

8.39. The vehicle of Ex. 8.7 encounters a pothole of depth 0.55 ft and length 1.6 ft when it is traveling at 83 ft/s. Determine response for 1.05 s after the front wheels encounter the pothole.

8.40. The machine of Prob. 7.49 has a rotating unbalance of 1.045 kg · m and operates at 400 rad/sec. Use MODAL2.BAS to determine the maximum displacement of the end of the beam.

8.41. Resolve Ex. 8.8 if the punching operation takes 0.8 s with 0.2 s between punches.

8.42. The tail blades of the helicopter of Prob. 7.35 operate at 800 rpm. Determine the maximum displacement in the tail section if one of the blades falls off, resulting in a rotating unbalance of 1.5 kg · m.

8.43. A blast occurs on the third floor of the structure of Prob. 7.32. The blast force is approximately 10,000 N, lasting 0.45 s. Determine the maximum displacement of the third floor.

8.44. The motor of Prob. 7.33 provides a torque of $100 \sin 50t$ N · m to the system. Determine the response of each rotor.

REFERENCES

1. Caughey, T. K., and M. E. O'Kelly: "Classical Normal Modes in Damped Linear Dynamic Systems," *Journal of Applied Mechanics*, vol. 32, pp. 583–588, 1965.
2. Dimarogonas, A. D., and S. Haddad: *Vibrations for Engineers*, Prentice-Hall, Englewood Cliffs, N.J., 1992.
3. James, M. L., G. M. Smith, J. C. Wolford, and P. W. Whaley: *Vibrations of Mechanical and Structural Systems with Microcomputer Applications*, Harper and Row, New York, 1989.
4. Ketter, R. L., and S. P. Prawel: *Modern Methods of Engineering Computations*, McGraw-Hill, New York, 1969.
5. Kreider, D. L., R. O. Kuller, D. R. Ostberg, and F. W. Perkins: *An Introduction to Linear Analysis*, 2nd ed., Addison-Wesley, Reading, Mass., 1966.
6. Meirovitch, L.: *Elements of Vibration Analysis*, McGraw-Hill, New York, 1975.
7. Rabenstein, A. L.: *Elementary Differential Equations with Linear Algebra*, 3rd ed., Academic Press, New York, 1982.
8. Rao, S. S.: *Mechanical Vibrations*, 2nd ed., Addison-Wesley, Reading, Mass., 1990.

9. Ross, A. D., and D. J. Inman: "A Design Criterion for Avoiding Resonance in Lumped Mass Normal Mode Systems," *Journal of Vibrations, Acoustics, Stress, and Reliability in Design*, vol. 111, pp. 49–52, 1989.
10. Shabana, A. A.: *Theory of Vibrations: Discrete and Continuous Systems*, vol. 2, Springer-Verlag, New York, 1991.
11. Steidel, R. F.: *An Introduction to Mechanical Vibrations*, 3rd ed., Wiley, New York, 1988.
12. Thomson, W. T.: *Theory of Vibrations with Applications*, 3rd ed., Prentice-Hall, Englewood Cliffs, N.J., 1988.
13. Weaver, W., S. P. Timoshenko, and D. H. Young: *Vibration Problems in Engineering*, 5th ed., Wiley-Interscience, New York, 1990.
14. Wylie, C. R., and R. Barrett: *Advanced Engineering Mathematics*, McGraw-Hill, New York, 1982.

CHAPTER 9

VIBRATIONS OF CONTINUOUS SYSTEMS

9.1 INTRODUCTION

All solid objects are made of deformable materials. Often a solid is assumed to be rigid. This allows for simpler modeling and leads to information about essential vibrational characteristics. The validity of a rigid-body assumption in modeling the vibrations of a system depends upon many factors such as geometry and frequency range. For example, consider a machine mounted on springs and operating in an industrial plant. The floor of the industrial plant is often assumed to be rigid and the vibrations of the machine considered by analyzing a one-degree-of-freedom system. However, if the forces developed in the springs are large, then since the floor is really deformable, vibrations are excited in the floor and perhaps the entire structure. In this case the vibrations of the machine are coupled to the structural vibrations.

Examples of continuous systems are shown in Fig. 9.1. All structural elements such as beams, columns, and plates are continuous systems. This includes the suspended piping system of Fig. 9.1*a*, simply supported at locations along its length. Vibrations of the pipeline are excited by the fluid flowing through the pipe, the operation of pumps, or structural vibrations. The vibrations are analyzed by considering a continuous beam with simple supports.

All elements of the frame structure of Fig. 9.1*b* are continuous structural elements. Often the columns of a frame structure are much more flexible than the girders, and the girders are considered rigid resulting in the model shown.

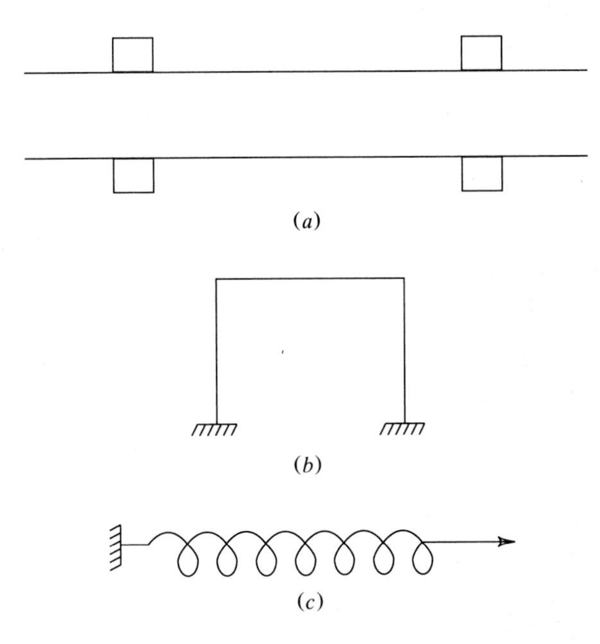

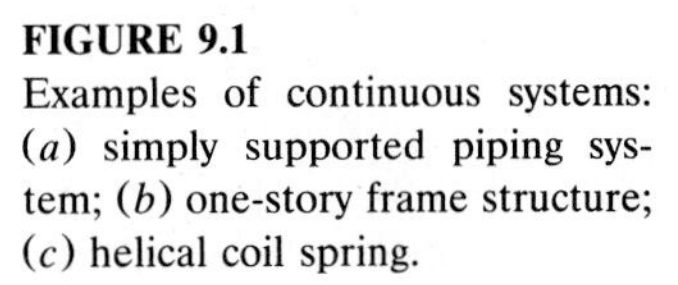

FIGURE 9.1
Examples of continuous systems: (*a*) simply supported piping system; (*b*) one-story frame structure; (*c*) helical coil spring.

The spring of Fig. 9.1*c* is a simple continuous system. As one end of the spring is moved relative to the other, a compression wave is generated and travels throughout the spring. If the excitation frequency is near the frequency of the compression waves, a phenomenon called surge develops. Surge can be a problem in mechanical systems where one end of a spring is given a harmonic displacement.

The vibrations of a rigid body attached to a continuous system are approximated using one degree of freedom in Chaps. 2 through 5. The inertia effects of a continuous element are approximated by adding a particle of a calculated equivalent mass at the location of the rigid body. Multi-degree-of-freedom approximations are considered in Chaps. 6 through 8.

The ordinary differential equations obtained using a discrete model of the continuous system are easier to solve than the governing partial differential equation. Thus discrete approximations are often used, but have limitations. A continuous system has an infinite, but countable, number of natural frequencies and corresponding mode shapes. A discrete approximation only predicts a finite number of modes. Often a large number of degrees of freedom are needed to attain accurate approximations for higher natural frequencies. Consider, for example, the cantilever beam of Fig. 9.2 with a concentrated mass at its end. Figure 9.3 shows the nondimensional lowest natural frequency as a function of β, the ratio of the concentrated mass to the mass of the beam. Figure 9.3 shows natural frequencies calculated using up to six degrees of freedom, including a one-degree-of-freedom approximation.

An introduction to the theory of continuous systems is presented in this chapter. Exact solutions to the governing partial differential equation using

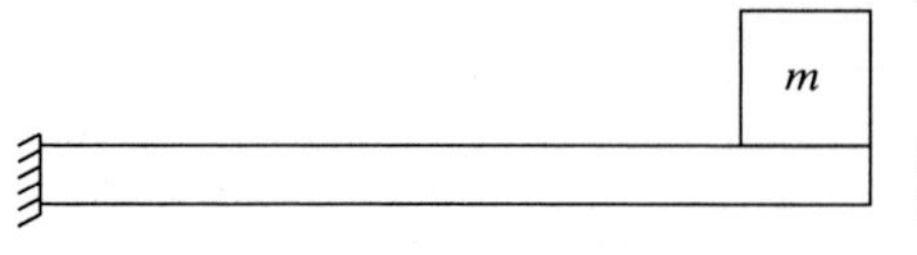

FIGURE 9.2
Fixed-free beam with end mass.

separation of variables are presented. These methods are generally limited to linear continuous systems with uniform geometry. The natural frequencies and mode shapes for systems with nonuniformities, such as the variable-area beam of Fig. 9.4, are often determined using approximate or numerical methods. However, development and use of these methods require knowledge of vibrations of uniform systems. The Rayleigh-Ritz method uses prespecified shape functions to determine natural frequency approximations. It is based upon energy methods as is Rayleigh's quotient introduced in Sec. 7.11 for discrete systems.

The methods used in this chapter are analogous to those used for multi-degree-of-freedom systems. The separation-of-variables method used to determine the natural frequencies is analogous to the normal-mode solution of Eq. (7.2). The method used for the analysis of forced vibrations is a direct result of the expansion theorem and is directly analogous to modal analysis. The approximate methods presented are based upon energy methods. Indeed, similar notation using energy scalar products can be used. The continuous functions

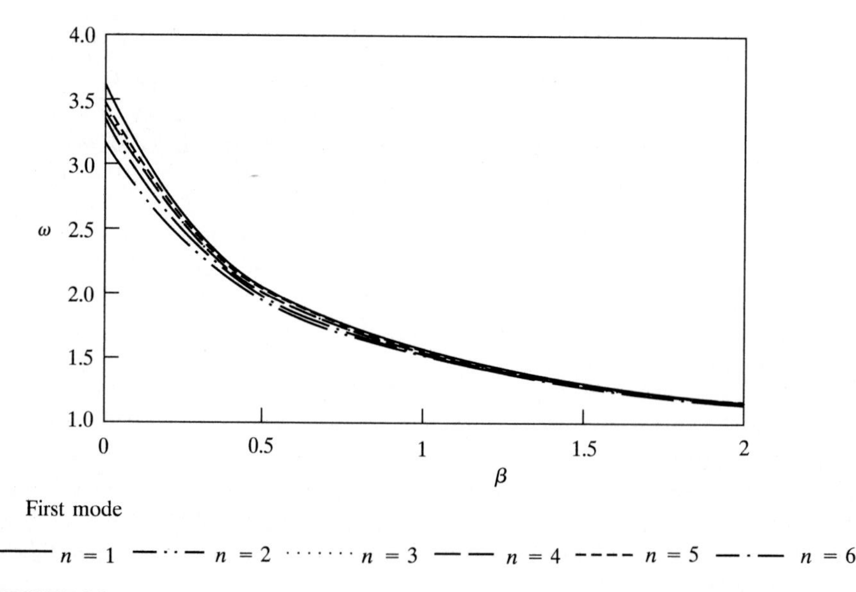

FIGURE 9.3
As the ratio of the concentrated mass to the mass of the beam grows larger, the approximation for the lowest natural frequency using a discrete model with n degrees of freedom improves.

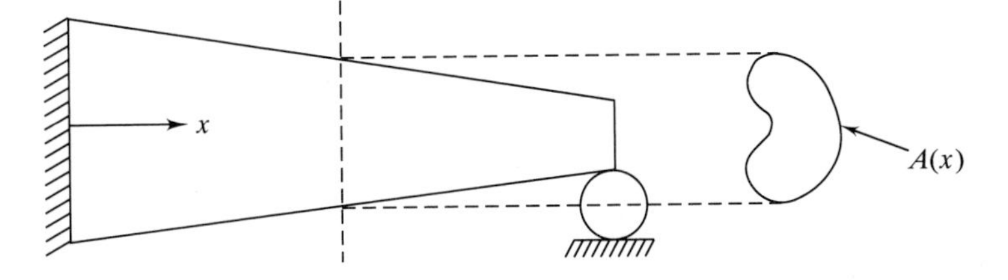

FIGURE 9.4
Natural frequencies and mode shapes for nonuniform structural elements such as a variable area beam can be approximated using the Rayleigh-Ritz method.

used in the analysis of continuous systems are analogous to the column vectors of generalized coordinates used for discrete systems. Energy scalar products are defined for continuous systems using definite integrals.

9.2 GENERAL METHOD

This section presents an outline of an exact closed-form method for analyzing vibrations of continuous systems. The method is applied to analyze the torsional oscillations of a circular shaft and the transverse vibrations of a beam in Secs. 9.3 and 9.4, respectively. This chapter is only intended as an introduction to vibrations of continuous systems. Thus it is assumed that the dependent displacement is a function of only one spatial variable and time, all material properties are constant, and all geometries are uniform.

The analysis procedure is broken into three parts: problem formulation, free-vibration analysis, and forced-vibration analysis. The mathematical theory underlying the analysis of vibrations of continuous systems is developed using an infinite-dimensional vector space, while the mathematical foundation for a multi-degree-of-freedom system is developed using a finite-dimensional vector space. Many of the concepts developed for finite-dimensional spaces have direct extension to infinite-dimensional spaces.

Part I: Problem Formulation

1. An independent spatial variable is chosen, call it x. This independent spatial variable represents the distance of a particle from a reference position when the system is in its equilibrium position. A continuous system has an infinite number of degrees of freedom and hence an infinite number of generalized coordinates are required. These are chosen as the displacement of the particles in the system. They can be summarized by a single dependent variable, $w(x, t)$.
2. Free-body diagrams of an arbitrary differential element are drawn at an arbitrary instant of time. One free-body diagram shows the external forces

acting on the element. Another free-body diagram shows the effective forces for that element. The external forces include forces on the surface of the element that are resultants of stress distributions acting on these surfaces.

3. The appropriate form of Newton's law is applied to the free-body diagram. Appropriate kinematic conditions and constitutive equations are applied to derive a partial differential equation governing $w(x, t)$.
4. Appropriate boundary conditions, dependent upon the end supports of the structural member, are formulated.
5. Appropriate initial conditions are formulated.
6. An optional step is to nondimensionalize the governing equation and boundary conditions by introducing nondimensional forms of the independent and dependent variables. This leads to the formulation of dimensionless parameters which are important in the physical understanding of the results. Assume for the remainder of this discussion that nondimensional variables are introduced and all variables referred to are nondimensional. Also assume that the nondimensional spatial variable x ranges from 0 to 1.

   The governing equations and boundary conditions can also be derived using energy methods. Kinetic and potential energy scalar products directly analogous to those formed for multi-degree-of-freedom systems can be defined.

Part II: Free-Vibration Solution

A true free-vibration problem is one where $w(x, 0)$ or $\partial w/\partial t(x, 0)$ are nonzero and the partial differential equation and all boundary conditions are homogeneous. The initial potential or kinetic energy drives the vibrations, during which no external forces are applied.

As for multi-degree-of-freedom systems, the free-vibration problem is considered to determine the system's natural frequencies and mode shapes. The method presented to solve free vibrations problems for continuous systems is called separation of variables. Application of this method requires that the partial differential equation be of an appropriate form, called separable, and the boundary conditions be homogeneous. If the boundary conditions are nonhomogeneous, their homogeneous form is used to determine the natural frequencies and mode shapes. Problems with nonhomogeneous boundary conditions are actually forced-vibration problems. The governing partial differential equations for torsional vibrations of a uniform shaft, longitudinal vibrations of a uniform elastic bar, and transverse vibrations of a uniform beam are all separable.

1. The dependent variable is assumed to be a product of functions of the independent variables,

$$w(x, t) = X(x)T(t) \tag{9.1}$$

   Equation (9.1) is substituted into the governing partial differential equation.

If the governing partial differential equation is separable then the resulting equation can be written in the form of $L_x X(x) = L_t T(t)$ where L_x and L_t are linear ordinary differential operators. Note that the left hand side of this equation is a function of x only and the right hand side is a function of t only. Since x and t are independent, this can only be true if both sides are equal to the same constant, call it $-\lambda$. The above argument is called the separation argument. Its application leads to ordinary differential equations for $X(x)$ and $T(t)$, both in terms of λ, called the separation constant.

2. Equation (9.1) is applied to the boundary conditions to obtain homogeneous boundary conditions for $X(x)$.
3. If the system is undamped, the solution for $T(t)$ is harmonic. It becomes obvious that the natural frequencies are related to the separation constant and the mode shapes are related to $X(x)$.
4. The problem for $X(x)$ is a homogeneous ordinary differential equation with homogeneous boundary conditions. This is called a differential eigenvalue problem. A nontrivial solution is available only for certain values of the separation constant. Standard solution techniques for ordinary differential equations are applied to determine $X(x)$ in terms of arbitrary constants of integration.
5. Application of the boundary conditions leads to a solvability condition of the form $f(\lambda) = 0$. Nontrivial solutions of the eigenvalue problem exist only for values of λ such that $f(\lambda) = 0$. This results in an infinite, but countable, number of solutions, $\lambda_1 < \lambda_2 < \cdots < \lambda_k < \cdots$. Corresponding to each λ_k, there is a $X_k(x)$, which is unique only to a multiplicative constant.

   If only the natural frequencies and mode shapes are necessary, the solution ends here.
6. An energy scalar product, (X_i, X_j), is defined such that (X_i, X_j) is proportional to the kinetic energy of the ith mode at any instant. It can be shown that for systems governed by the wave equation (torsional vibrations of shafts, longitudinal vibrations of bars) and for uniform beam vibrations, mode shapes for distinct modes are mutually orthogonal with respect to this energy scalar product. For a continuous system, in the absence of discrete masses, the appropriate scalar product is

$$(X_i, X_j) = \int_0^1 X_i(x) X_j(x)\, dx \tag{9.2}$$

   If the system has discrete masses, additional terms are added to the integral of eq. (9.2) to account for the kinetic energy of the discrete masses. In any case the mode shapes are normalized by requiring

$$(X_i, X_i) = 1 \tag{9.3}$$

7. If the mode shapes are normalized with respect to a scalar product for which they are also mutually orthogonal, then an expansion theorem exists which

states that any continuous function, $f(x)$, can be expanded in a series of the mode shapes as

$$f(x) = \sum_{k=1}^{\infty} (f, X_k) X_k \tag{9.4}$$

The expansion converges to $f(x)$ at all x except perhaps at $x = 0$ and $x = 1$.

If a forced-vibration solution is required, the expansion theorem of Eq. (9.4) is noted and the solution proceeds to step 1 of the forced response. If a free-vibration solution is required, the solution continues as follows.

8. The general solution is formed by taking a linear combination over all modes.

$$w(x,t) = \sum_{k=1}^{\infty} X_k(x) T_k(t) \tag{9.5}$$

Two arbitrary constants for each mode are present from the solution for $T_k(t)$. These constants are determined from application of initial conditions. Often the functions involved in the initial conditions must be expanded using the expansion theorem, Eq. (9.4). For example, if $w(x,0)$ is nonzero and is equal to $f(x)$, then $f(x)$ is expanded using Eq. (9.4) and compared to $w(x,0)$ obtained from Eq. (9.5), in terms of arbitrary constants. The linear independence of each $X_k(x)$ is used to determine the constants.

Part III: Forced-Vibration Solution

As for discrete systems, there are several methods available to determine the forced response of continuous systems. As for discrete systems, these include application of the method of undetermined coefficients for harmonic excitations, the Laplace transform method, and modal analysis. Again, as for discrete systems, modal analysis is the most powerful and most often used.

A modal analysis procedure can be developed for forced vibrations of continuous systems. Let $f(x,t)$ represent the nondimensional nonhomogeneous term arising in the partial differential equation due to the external forcing. Nonhomogeneous terms can also occur in the boundary conditions.

1. The expansion theorem, Eq. (9.4) is used to expand $f(x,t)$ as

$$f(x,t) = \sum_{k=1}^{\infty} C_k(t) X_k(x) \tag{9.6}$$

where

$$C_k(t) = (f(x,t), X_k(x)) \tag{9.7}$$

2. The expansion theorem is also used to expand

$$w(x,t) = \sum_{k=1}^{\infty} p_k(t) X_k(x) \tag{9.8}$$

where the $p_k(t)$ are called the principal coordinates for the continuous system. Equations (9.6) and (9.8) are substituted into the governing partial differential equation.

3. The scalar product of the resulting partial differential equation is taken with $X_j(x)$ for an arbitrary j. For a problem whose appropriate scalar product is given by Eq. (9.2), this is equivalent to multiplying the equation by $X_j(x)$ and integrating from 0 to 1. Application of the orthogonality condition leads to uncoupled differential equations for the principal coordinates.
4. The uncoupled differential equations are solved to determine each $p_k(t)$.

9.3 TORSIONAL OSCILLATIONS OF A CIRCULAR SHAFT

9.3.1 Problem Formulation

The circular shaft of Fig. 9.5 is made of a material of mass density ρ and shear modulus G and has a length L, cross-sectional area A, and polar moment of inertia J. Let x be the coordinate along the axis of the shaft, measured from its left end. The shaft is subject to a time-dependent torque per unit length, $T(x, t)$. Let $\theta(x, t)$ measure the resulting torsional oscillations where θ is chosen positive clockwise.

Figure 9.6 shows free-body diagrams of a differential element of the shaft at an arbitrary instant of time. The element is of infinitesimal thickness dx and its left face is a distance x from the left end of the shaft.

The free-body diagram of the external forces shows the time-dependent torque loading as well as the internal resisting torques developed in the cross sections. The internal resisting torques are the resultant moments of the shear stress distributions. If $T_r(x, t)$ is the resisting torque acting on the left face of the element, then a Taylor series expansion truncated after the linear terms

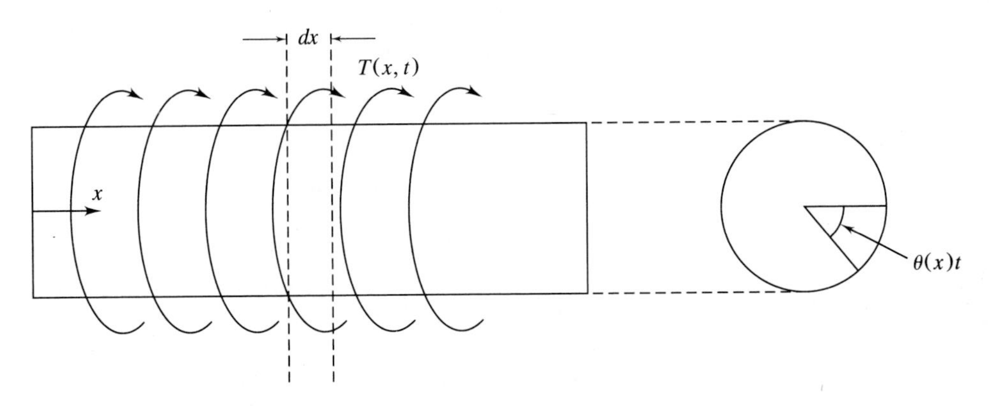

FIGURE 9.5

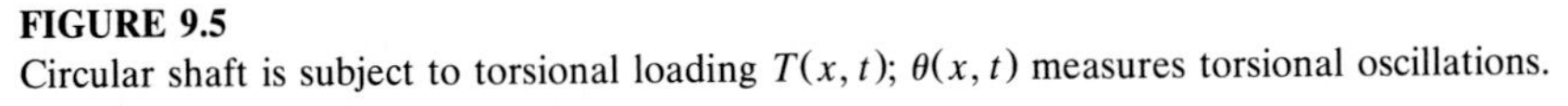
Circular shaft is subject to torsional loading $T(x, t)$; $\theta(x, t)$ measures torsional oscillations.

gives

$$T_r(x + dx, t) = T_r(x, t) + \frac{\partial T_r(x, t)}{\partial t} dx \tag{9.9}$$

The directions of the torques shown on the free-body diagram are consistent with the choice of θ positive clockwise.

Since the disk is infinitesimal, the angular acceleration is assumed constant across the thickness. Thus the free-body diagram of the effective forces simply shows of a moment equal to the mass moment of inertia of the disk times its angular acceleration.

Summation of moments about the mass center of the disk

$$\left(\sum M\right)_{\text{ext}} = \left(\sum M\right)_{\text{eff}}$$

gives

$$T(x, t)\,dx - T_r(x, t) + T_r(x, t) + \frac{\partial T_r(x, t)}{\partial x} dx = \rho J\,dx \frac{\partial^2 \theta(x, t)}{\partial t^2}$$

or

$$T(x, t) + \frac{\partial T_r(x, t)}{\partial x} = \rho J \frac{\partial^2 \theta}{\partial t^2} \tag{9.10}$$

From mechanics of materials,

$$T_r(x, t) = JG \frac{\partial \theta(x, t)}{\partial x} \tag{9.11}$$

which when substituted in Eq. (9.10) leads to

$$T(x, t) + JG \frac{\partial^2 \theta}{\partial x^2} = \rho J \frac{\partial^2 \theta}{\partial t^2} \tag{9.12}$$

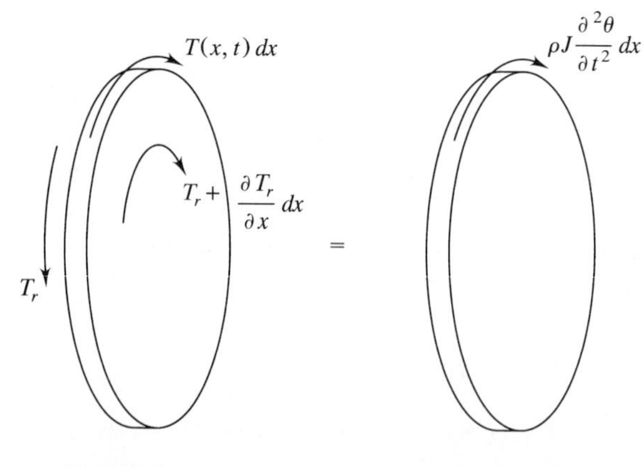

FIGURE 9.6
Free-body diagram of differential element of shaft at arbitrary instant of time.

TABLE 9.1
Boundary conditions for torsional oscillations of a circular shaft

End condition	Boundary condition	Remarks
Fixed, $x = 0$ or $x = 1$	$\theta = 0$	
Free, $x = 0$ or $x = 1$	$\dfrac{\partial \theta}{\partial x} = 0$	
Torsional spring, $x = 0$	$\dfrac{\partial \theta}{\partial x} = \beta\theta$	$\beta = \dfrac{k_t L}{JG}$
Torsional spring, $x = 1$	$\dfrac{\partial \theta}{\partial x} = -\beta\theta$	$\beta = \dfrac{k_t L}{JG}$
Torsional damper, $x = 0$	$\dfrac{\partial \theta}{\partial x} = \beta \dfrac{\partial \theta}{\partial t}$	$\beta = c_t \sqrt{\dfrac{J}{\rho G}}$
Torsional damper, $x = 1$	$\dfrac{\partial \theta}{\partial x} = -\beta \dfrac{\partial \theta}{\partial t}$	$\beta = c_t \sqrt{\dfrac{J}{\rho G}}$
Attached disk, $x = 0$	$\dfrac{\partial \theta}{\partial x} = \beta \dfrac{\partial^2 \theta}{\partial t^2}$	$\beta = \dfrac{I_D}{\rho JL}$
Attached disk, $x = 1$	$\dfrac{\partial \theta}{\partial x} = -\beta \dfrac{\partial^2 \theta}{\partial t^2}$	$\beta = \dfrac{I_D}{\rho JL}$

The following nondimensional variables are introduced:

$$x^* = \frac{x}{L} \qquad t^* = \sqrt{\frac{G}{\rho}}\frac{t}{L} \tag{9.13}$$

and

$$T^*(x^*, t^*) = \frac{T(x,t)}{T_m} \tag{9.14}$$

where T_m is the maximum value of T. Introduction of Eqs. (9.13) and (9.14) in Eq. (9.12) leads to

$$\left(\frac{L^2 T_m}{JG}\right) T(x,t) + \frac{\partial^2 \theta}{\partial x^2} = \frac{\partial^2 \theta}{\partial t^2} \tag{9.15}$$

Table 9.1 provides nondimensional boundary conditions for different types of shaft ends. The problem formulation is completed by specifying appropriate initial conditions of the form

$$\theta(x,0) = g_1(x) \tag{9.16}$$

and

$$\frac{\partial \theta(x,0)}{\partial t} = g_2(x) \tag{9.17}$$

TABLE 9.2
Physical problems governed by the wave equation

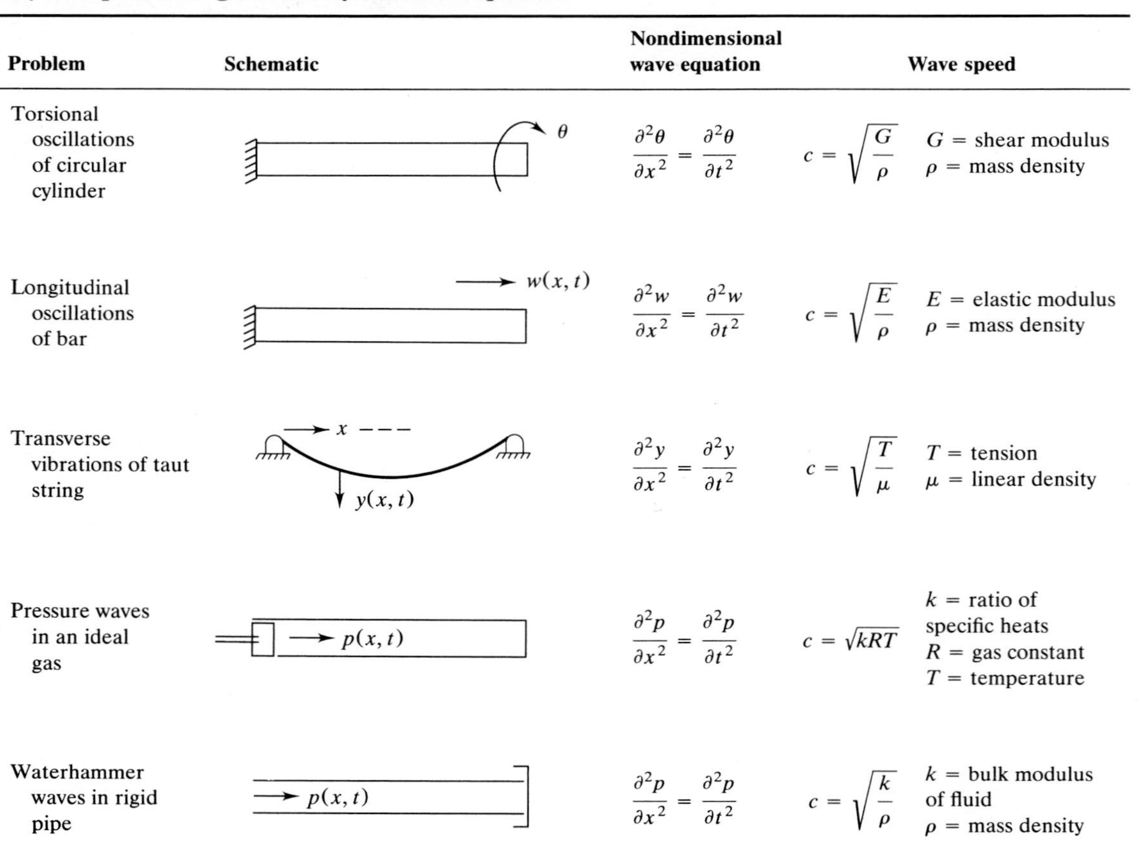

Problem	Schematic	Nondimensional wave equation	Wave speed	
Torsional oscillations of circular cylinder	θ	$\frac{\partial^2\theta}{\partial x^2} = \frac{\partial^2\theta}{\partial t^2}$	$c = \sqrt{\frac{G}{\rho}}$	G = shear modulus ρ = mass density
Longitudinal oscillations of bar	$w(x, t)$	$\frac{\partial^2 w}{\partial x^2} = \frac{\partial^2 w}{\partial t^2}$	$c = \sqrt{\frac{E}{\rho}}$	E = elastic modulus ρ = mass density
Transverse vibrations of taut string	x $y(x, t)$	$\frac{\partial^2 y}{\partial x^2} = \frac{\partial^2 y}{\partial t^2}$	$c = \sqrt{\frac{T}{\mu}}$	T = tension μ = linear density
Pressure waves in an ideal gas	$p(x, t)$	$\frac{\partial^2 p}{\partial x^2} = \frac{\partial^2 p}{\partial t^2}$	$c = \sqrt{kRT}$	k = ratio of specific heats R = gas constant T = temperature
Waterhammer waves in rigid pipe	$p(x, t)$	$\frac{\partial^2 p}{\partial x^2} = \frac{\partial^2 p}{\partial t^2}$	$c = \sqrt{\frac{k}{\rho}}$	k = bulk modulus of fluid ρ = mass density

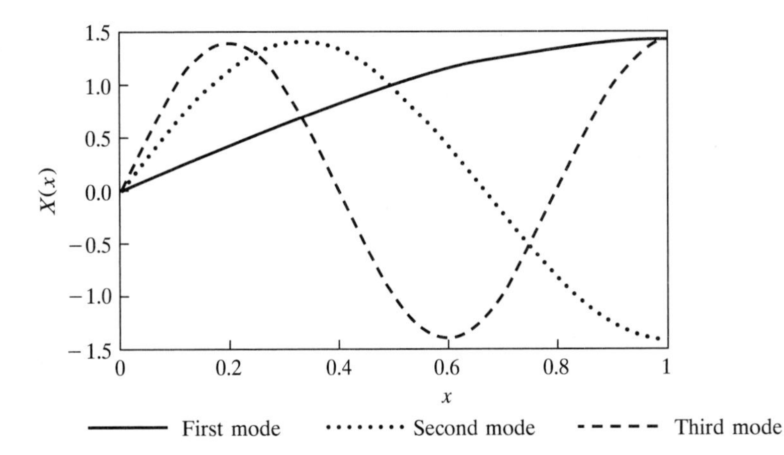

FIGURE 9.7
First three normalized mode shapes of Example 9.1.

Eq. (9.2) as follows:

$$\begin{aligned}\left(X_k(x), X_j(x)\right) &= \int_0^1 D_j D_k \sin(2k-1)\frac{\pi}{2}x \sin(2j-1)\frac{\pi}{2}x\,dx \\ &= \frac{D_j D_k}{\pi}\left[\frac{1}{j-k}\sin(j-k)\pi - \frac{1}{j+k+1}\sin(j+k+1)\pi\right] \\ &= 0\end{aligned}$$

The mode shapes are normalized by requiring

$$\begin{aligned}1 &= (X_k, X_k) \\ &= \int_0^1 D_k^2 \sin^2(2k-1)\frac{\pi}{2}x\,dx \\ &= \frac{D_k^2}{2}\end{aligned}$$

which yields

$$X_k(x) = \sqrt{2}\sin(2k-1)\frac{\pi}{2}x \tag{9.36}$$

The first three normalized mode shapes are shown in Fig. 9.7.

The general solution to the free-vibration problem is

$$\theta(x,t) = \sum_{k=1}^{\infty} \sqrt{2}\sin(2k-1)\frac{\pi}{2}x\left[A_k\cos(2k-1)\frac{\pi}{2}t + B_k\sin(2k-1)\frac{\pi}{2}t\right] \tag{9.37}$$

Application of the initial condition, Eq. (9.23), yields $B_k = 0$. Application of Eq.

(9.22) then gives

$$\gamma x = \sum_{k=1}^{\infty} A_k \sqrt{2} \sin(2k-1)\frac{\pi}{2}x \tag{9.38}$$

The expansion theorem, Eq. (9.4), is used to expand

$$\gamma x = \sum_{k=1}^{\infty} (\gamma x, X_k) X_k \tag{9.39}$$

where

$$\begin{aligned}(\gamma x, X_k) &= \int_0^1 \gamma x \sqrt{2} \sin(2k-1)\frac{\pi}{2}x\,dx \\ &= \frac{4\gamma\sqrt{2}}{\pi^2(2k-1)^2} \sin(2k-1)\frac{\pi}{2} \\ &= \frac{4\gamma\sqrt{2}}{\pi^2(2k-1)^2}(-1)^{k+1}\end{aligned}$$

Comparison of Eqs. (9.38) and (9.39) yields

$$A_k = (\gamma x, X_k) = \frac{4\gamma\sqrt{2}(-1)^{k+1}}{\pi^2(2k-1)^2} \tag{9.40}$$

Equation (9.37) becomes

$$\theta(x,t) = \frac{8\gamma}{\pi^2} \sum_{k=1}^{\infty} (-1)^{k+1} \frac{1}{(2k-1)^2} \sin(2k-1)\frac{\pi}{2}x \cos(2k-1)\frac{\pi}{2}t \tag{9.41}$$

The time-dependent angles of twist at four locations along the axis of the shaft, obtained by numerical evaluation of Eq. (9.41), are plotted in Fig. 9.8.

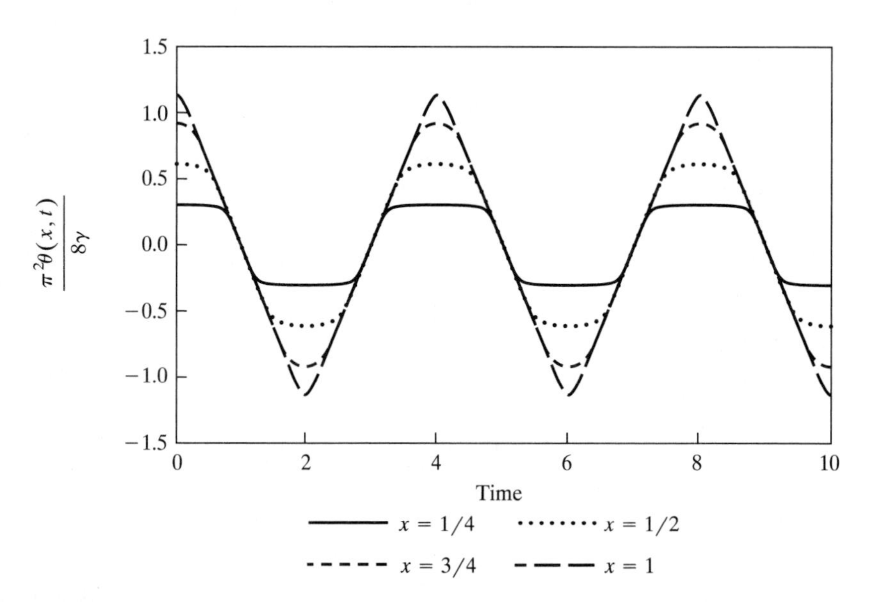

FIGURE 9.8
Time dependent torsional oscillations of circular shaft fixed at $x = 0$, free at $x = 1$.

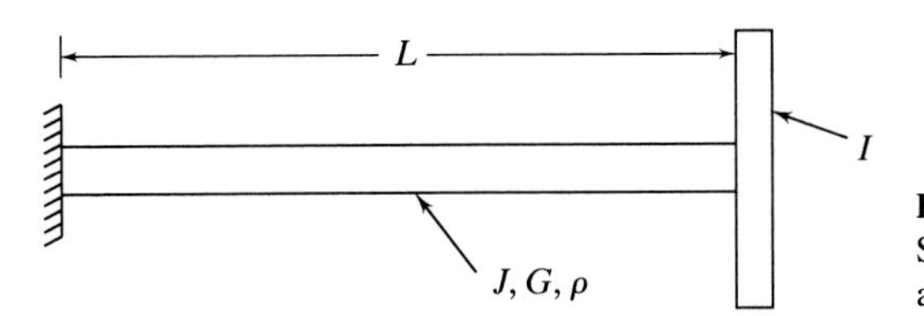

FIGURE 9.9
Shaft has disk of moment of inertia I attached at its free end.

Example 9.2. The circular shaft of Fig. 9.9 is fixed at $x = 0$ and has a thin disk of mass moment of inertia I attached at $x = 1$. Determine the natural frequencies for this system, identify the orthogonality condition satisfied by the mode shapes, and determine the normalized mode shapes.

The partial differential equation governing this system is Eq. (9.18). It is subject to Eq. (9.20) and from Table 9.1

$$\frac{\partial \theta(1,t)}{\partial x} = -\beta \frac{\partial^2 \theta(1,t)}{\partial t^2} \tag{9.42}$$

where
$$\beta = \frac{I}{\rho J L}$$

The separation-of-variables assumption of Eq. (9.24) leads to Eq. (9.27) subject to Eq. (9.30) and

$$\frac{dX(1)}{dx} = \beta \lambda X(1) \tag{9.43}$$

The solution satisfying Eqs. (9.24) and (9.27) is

$$X(x) = D \sin \sqrt{\lambda}\, x \tag{9.44}$$

Application of Eq. (9.43) to Eq. (9.44) yields

$$\sqrt{\lambda} \cos \sqrt{\lambda} = \beta \lambda \sin \sqrt{\lambda} \tag{9.45}$$

or
$$\tan \sqrt{\lambda} = \frac{1}{\beta \sqrt{\lambda}} \tag{9.46}$$

A graphical solution of the transcendental equation, Eq. (9.46), is shown in Fig. 9.10. The values of λ where the curves $\tan \sqrt{\lambda}$ and $1/\beta\sqrt{\lambda}$ intersect are the solutions of Eq. (9.46), and are the values of the separation constant for which nontrivial solutions for $X(x)$ occur. Figure 9.10 shows that there are infinite, but countable, values of λ where these curves intersect. Figure 9.10 also shows that for large k, λ_k approaches $(k\pi)^2$.

The natural frequencies are the square roots of the separation constants. Figure 9.11 shows the first four natural frequencies as a function of β. The first four mode shapes are plotted in Figure 9.12 for $\beta = 0.4$.

Let λ_k be the kth solution of Eq. (9.46) and $X_k(x)$ be its corresponding mode shape, given by Eq. (9.44). Then calculations show that for $j \neq k$,

$$\int_0^1 X_k(x) X_j(x)\, dx = -\beta X_j(1) X_k(1) \tag{9.47}$$

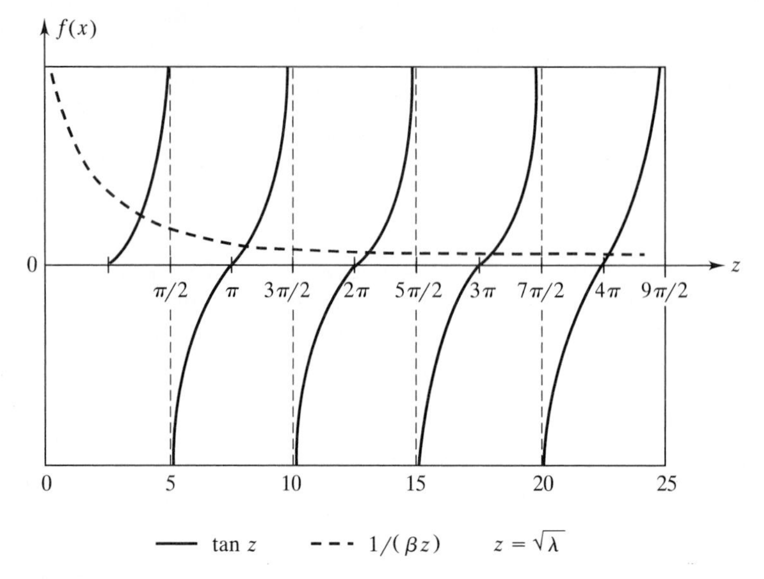

FIGURE 9.10
Graphical solution of transcendental equation $\tan\sqrt{\lambda} = 1/\beta\sqrt{\lambda}$ to determine natural frequencies of system of Example 9.2. Values of separation constants correspond to points of intersection of two curves.

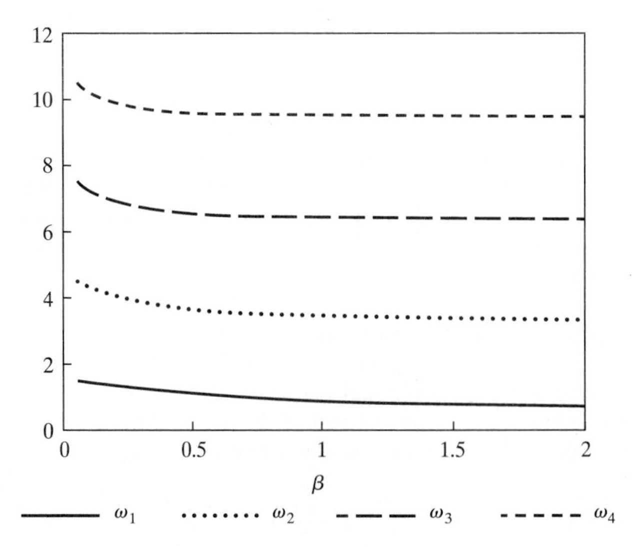

FIGURE 9.11
Nondimensional natural frequencies of Example 9.2 as function of nondimensional parameter β obtained by solving Eq. (9.46).

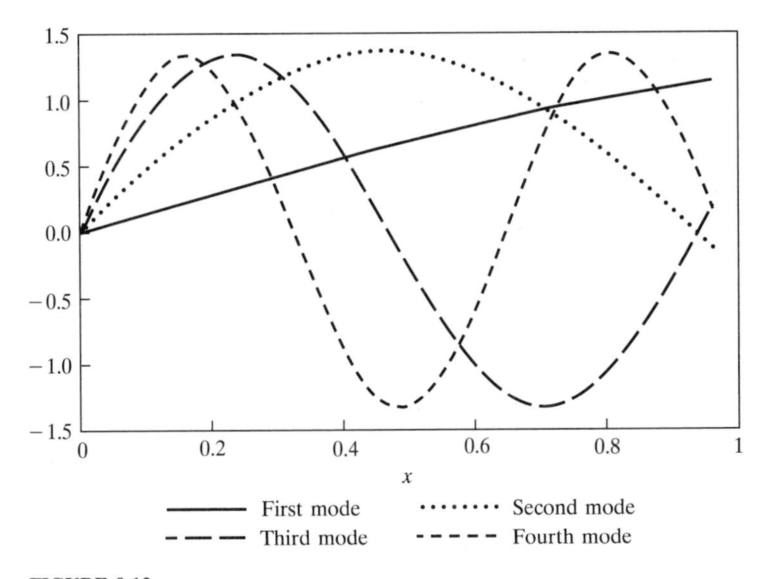

FIGURE 9.12
Mode shapes of Example 9.2 with $\beta = 0.4$

If the scalar product of f and g is defined by

$$(f, g) = \int_0^1 f(x)g(x)\,dx + \beta f(1)g(1) \tag{9.48}$$

then

$$(X_j, X_k) = 0 \tag{9.49}$$

and Eq. (9.48) defines the energy scalar product with which the mode shapes are orthogonal.

Normalization of the mode shape requires

$$1 = (X_k, X_k) = \int_0^1 D_k^2 \sin^2\sqrt{\lambda_k}\,x\,dx + D_k^2\beta\sin^2\sqrt{\lambda_k}$$

$$= D_k^2\left[\int_0^1 \frac{1}{2}\left(1 - \cos 2\sqrt{\lambda_k}\,x\right)dx + \beta\sin^2\sqrt{\lambda_k}\right]$$

$$= D_k^2\left[\frac{1}{2}\left(1 - \frac{1}{2\sqrt{\lambda_k}}\sin 2\sqrt{\lambda_k}\right) + \beta\sin^2\sqrt{\lambda_k}\right]$$

Using the trigonometric identity

$$\sin 2\sqrt{\lambda_k} = 2\sin\sqrt{\lambda_k}\cos\sqrt{\lambda_k}$$

and replacing $\cos\sqrt{\lambda_k}$ from Eq. (9.45) leads to

$$D_k = \sqrt{2}\left(1 + \beta\sin^2\sqrt{\lambda_k}\right)^{-1/2}$$

where λ_k is the kth real solution of Eq. (9.46).

9.3.3 Forced Vibrations

The application of undetermined coefficients for harmonic excitations is illustrated in the following example. Modal analysis is illustrated with examples in Sec. 9.4. Application of the Laplace transform method is beyond the scope of this book.

Example 9.3. The thin disk of Example 9.2 is subject to a harmonic torque,

$$T(t) = T_0 \sin \omega t \tag{9.50}$$

Determine the steady-state response of the system.

The torsional oscillations, in terms of nondimensional variables, are governed by Eq. (9.18) with

$$\theta(0, t) = 0 \tag{9.51}$$

and

$$\frac{\partial \theta}{\partial x}(1, t) = -\beta \frac{\partial^2 \theta}{\partial t^2}(1, t) + \frac{T_0 L}{JG} \sin \tilde{\omega} t \tag{9.52}$$

where

$$\tilde{\omega} = L\sqrt{\frac{\rho}{G}}\,\omega \tag{9.53}$$

Since the external excitation is harmonic, the steady-state response is assumed as

$$\theta(x, t) = u(x) \sin \tilde{\omega} t \tag{9.54}$$

Substituting Eq. (9.54) into Eq. (9.18) leads to

$$\frac{d^2 u}{dx^2} \sin \tilde{\omega} t = -\tilde{\omega}^2 u \sin \tilde{\omega} t$$

or

$$\frac{d^2 u}{dx^2} + \tilde{\omega}^2 u = 0 \tag{9.55}$$

Substituting Eq. (9.53) into the boundary conditions, Eqs. (9.52) and (9.53), leads to

$$u(0) = 0 \tag{9.56}$$

and

$$\frac{du}{dx}(1) - \beta \tilde{\omega}^2 u(1) = \frac{T_0 L}{JG} \tag{9.57}$$

The solution of Eq. (9.55) subject to Eqs. (9.56) and (9.57) is

$$u(x) = \frac{T_0 L}{(\tilde{\omega} \cos \tilde{\omega} - \beta \tilde{\omega}^2 \sin \tilde{\omega}) JG} \sin \tilde{\omega} x \tag{9.58}$$

Note that if $\tilde{\omega}$ is equal to any of the system's natural frequencies, the denominator vanishes. The assumed form of the solution, Eq. (9.53), must be modified to account for this resonance condition.

The steady-state solution is given by Eq. (9.54) where $u(x)$ is given in Eq. (9.58). The total solution is the steady-state solution plus the homogeneous solution which is a summation over all free-vibration modes. Initial conditions can then be applied to determine the constants in the linear combination.

9.4 TRANSVERSE BEAM VIBRATIONS

9.4.1 Problem Formulation

The uniform beam of Fig. 9.13 is made of a material of mass density ρ and elastic modulus E, and has a length L, cross-sectional area A, and centroidal moment of inertia I. Let x be a coordinate along the neutral axis of the beam, measured from its left end. The beam has an external load per unit length, $f(x, t)$. Let $w(x, t)$ be the transverse deflection of the beam, measured from its equilibrium position.

Free-body diagrams of an arbitrary differential element of the beam at an arbitrary instant of time are shown in Fig. 9.14. The element is a slice of the beam of thickness dx and its left face is a distance x from the beam's left end. The external forces shown are the external loading, the internal bending moment which is the resultant moment of the normal stress distribution, and the internal shear force, which is the resultant of the shear stress distribution. It is assumed that the resultant of the normal stress distribution is zero. The effective force is the element mass times its acceleration. The element's longitudinal acceleration and angular acceleration are small in comparison to other effects and are thus ignored.

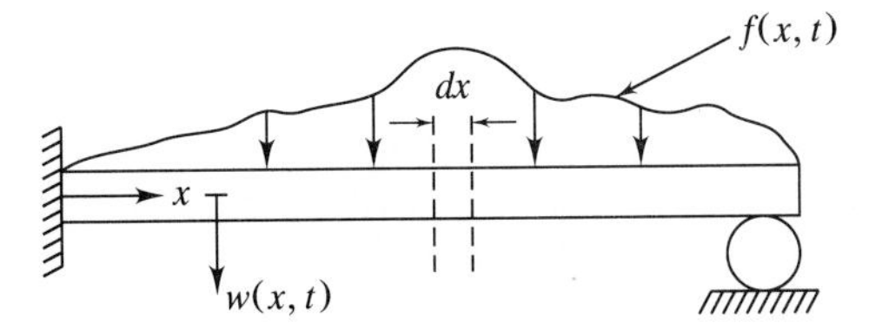

FIGURE 9.13
$w(x, t)$ is the transverse deflection of the beam measured from its static equilibrium position.

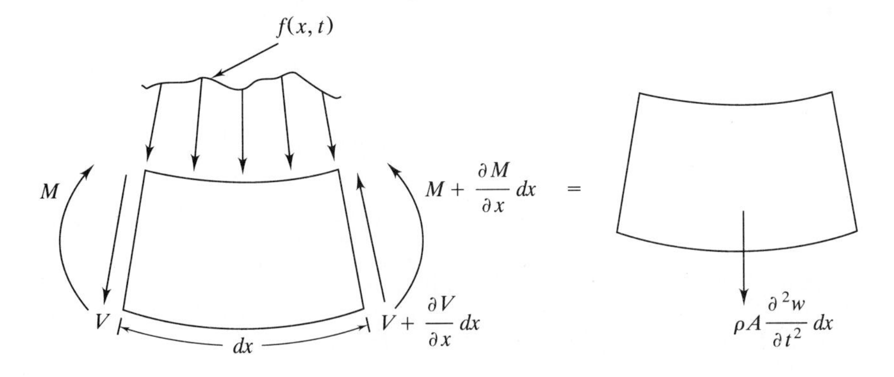

FIGURE 9.14
Free-body diagrams of differential beam element at an arbitrary time.

Summing forces in the vertical direction

$$\left(\sum \overset{\downarrow}{F}\right)_{\text{ext}} = \left(\sum \overset{\downarrow}{F}\right)_{\text{eff}}$$

$$V - \left(V + \frac{\partial V}{\partial x}\,dx\right) + \int_x^{x+dx} f(\zeta, t)\,d\zeta = \rho A \frac{\partial^2 w}{dt^2}\,dx \tag{9.59}$$

The mean value theorem implies that there is an $\tilde{x}$, $x < \tilde{x} < x + dx$, such that

$$\int_x^{x+dx} f(\zeta, t)\,d\zeta = f(\tilde{x}, t)\,dx$$

Since dx is infinitesimal, $\tilde{x} \approx x$. Equation (9.59) becomes

$$f(x,t) - \frac{\partial V}{\partial x} = \rho A \frac{\partial^2 w}{\partial x^2} \tag{9.60}$$

Summing moments about the neutral axis at the left face of the element

$$\left(\sum M_O\right)_{\text{ext}} = \left(\sum M_O\right)_{\text{eff}}$$

$$M - \left(M + \frac{\partial M}{\partial x}\,dx\right) + V\,dx - \int_x^{x+dx} (\zeta - x) f(\zeta, t)\,d\zeta$$

$$= \rho A \frac{\partial^2 w}{\partial t^2}\,dx\left(\frac{dx}{2}\right) \tag{9.61}$$

Since dx is infinitesimal, terms of order dx^2 are negligible compared to terms of order dx. When the mean value theorem is used on the integral, it becomes apparent that this term is of order dx^2. Thus Eq. (9.61) simplifies to

$$V = \frac{\partial M}{\partial x} \tag{9.62}$$

From mechanics of materials

$$M = EI \frac{\partial^2 w}{\partial x^2} \tag{9.63}$$

Substitution of Eqs. (9.62) and (9.63) into Eq. (9.60) leads to

$$\rho A \frac{\partial^2 w}{\partial t^2} + EI \frac{\partial^4 w}{\partial x^4} = f(x,t) \tag{9.64}$$

Equation (9.64) is nondimensionalized by introducing

$$x^* = \frac{x}{L} \qquad t^* = t\sqrt{\frac{EI}{\rho A L^4}}$$
$$w^* = \frac{w}{L} \qquad f^* = \frac{f}{f_m} \tag{9.65}$$

TABLE 9.3
Boundary conditions for transverse vibrations of a beam

End condition	Boundary condition A	Boundary condition B	Remarks
Free, $x = 0$ or $x = 1$	$\frac{\partial^2 w}{\partial x^2} = 0$	$\frac{\partial^3 w}{\partial x^3} = 0$	
Pinned, $x = 0$ or $x = 1$	$w = 0$	$\frac{\partial^2 w}{\partial x^2} = 0$	
Fixed, $x = 0$ or $x = 1$	$w = 0$	$\frac{\partial w}{\partial x} = 0$	
Linear spring, $x = 0$	$\frac{\partial^2 w}{\partial x^2} = 0$	$\frac{\partial^3 w}{\partial x^3} = -\beta w$	$\beta = \frac{kL^3}{EI}$
Linear spring, $x = 1$	$\frac{\partial^2 w}{\partial x^2} = 0$	$\frac{\partial^3 w}{\partial x^3} = \beta w$	$\beta = \frac{kL^3}{EI}$
Viscous damper, $x = 0$	$\frac{\partial^2 w}{\partial x^2} = 0$	$\frac{\partial^3 w}{\partial x^3} = -\beta \frac{\partial w}{\partial t}$	$\beta = \frac{cL}{\sqrt{\rho EIA}}$
Viscous damper, $x = 1$	$\frac{\partial^2 w}{\partial x^2} = 0$	$\frac{\partial^3 w}{\partial x^3} = \beta \frac{\partial w}{\partial t}$	$\beta = \frac{cL}{\sqrt{\rho EIA}}$
Attached mass, $x = 0$	$\frac{\partial^2 w}{\partial x^2} = 0$	$\frac{\partial^3 w}{\partial x^3} = -\beta \frac{\partial^2 w}{\partial t^2}$	$\beta = \frac{m}{\rho AL}$
Attached mass, $x = 1$	$\frac{\partial^2 w}{\partial x^2} = 0$	$\frac{\partial^3 w}{\partial x^3} = \beta \frac{\partial^2 w}{\partial t^2}$	$\beta = \frac{m}{\rho AL}$
Attached inertia element, $x = 0$	$\frac{\partial^2 w}{\partial x^2} = -\beta \frac{\partial^3 w}{\partial x \, \partial t^2}$	$\frac{\partial^3 w}{\partial x^3} = 0$	$\beta = \frac{J}{\rho AL^3}$
Attached inertia element, $x = 1$	$\frac{\partial^2 w}{\partial x^2} = \beta \frac{\partial^3 w}{\partial x \, \partial t^2}$	$\frac{\partial^3 w}{\partial x^3} = 0$	$\beta = \frac{J}{\rho AL^3}$

where f_m is the maximum value of f. The resulting nondimensional form of Eq. (9.64) is

$$\frac{\partial^2 w}{\partial t^2} + \frac{\partial^4 w}{\partial x^4} = \frac{f_m L^3}{EI} f(x, t) \tag{9.66}$$

Four boundary conditions, two at $x = 0$ and two at $x = 1$, must be specified. The forms of the boundary conditions are dependent on the type of end supports. Nondimensional boundary conditions associated with different support conditions are given in Table 9.3.

TABLE 9.4
Natural frequencies and mode shapes for beams

End conditions $X = 0$ $X = 1$	Characteristic equation	Five lowest natural frequencies $\omega_k = \sqrt{\lambda_k}$	Mode shape	Scalar product $(X_j(x), X_k(x))$
Fixed-fixed	$\cos \lambda^{1/4} \cosh \lambda^{1/4} = 1$	$\omega_1 = 22.37$ $\omega_2 = 61.66$ $\omega_3 = 120.9$ $\omega_4 = 199.9$ $\omega_5 = 298.6$	$C_k\left[\cosh \lambda_k^{1/4}x - \cos \lambda_k^{1/4}x\right.$ $\left.-\alpha_k\left(\sinh \lambda_k^{1/4}x - \sin \lambda_k^{1/4}x\right)\right]$ $\alpha_k = \dfrac{\cosh \lambda_k^{1/4} - \cos \lambda_k^{1/4}}{\sinh \lambda_k^{1/4} - \sin \lambda_k^{1/4}}$	$\int_0^1 X_j(x) X_k(x)\,dx$
Pinned-pinned	$\sin \lambda^{1/4} = 0$	$\omega_1 = 9.870$ $\omega_2 = 39.48$ $\omega_3 = 88.83$ $\omega_4 = 157.9$ $\omega_5 = 246.7$	$C_k \sin \lambda_k^{1/4}x$	$\int_0^1 X_j(x) X_k(x)\,dx$
Fixed-free	$\cos \lambda^{1/4} \cosh {}^{1/4} = -1$	$\omega_1 = 3.51$ $\omega_2 = 22.03$ $\omega_3 = 61.70$ $\omega_4 = 120.9$ $\omega_5 = 199.9$	$C_k\left[\cosh \lambda_k^{1/4}x - \cos \lambda_k^{1/4}x\right.$ $\left.-\alpha_k\left(\sinh \lambda_k^{1/4}x - \sin \lambda_k^{1/4}x\right)\right]$ $\alpha_k = \dfrac{\cos \lambda_k^{1/4} + \cosh \lambda_k^{1/4}}{\sin \lambda_k^{1/4} + \sinh \lambda_k^{1/4}}$	$\int_0^1 X_j(x) X_k(x)\,dx$
Free-free	$\cosh \lambda^{1/4} \cos \lambda^{1/4} = 1$	$\omega_1 = 0$ $\omega_2 = 22.37$ $\omega_3 = 61.66$ $\omega_4 = 120.9$ $\omega_5 = 199.9$	$1, \sqrt{3x}\ (k = 1)$ $C_k\left[\cosh \lambda_k^{1/4}x + \cos \lambda_k^{1/4}x\right.$ $\left.+\alpha_k\left(\sinh \lambda_k^{1/4}x + \sin \lambda_k^{1/4}x\right)\right]$ $\alpha_k = \dfrac{\cosh \lambda_k^{1/4} - \cos \lambda_k^{1/4}}{\sin \lambda_k^{1/4} - \sinh \lambda_k^{1/4}}$	$\int_0^1 X_j(x) X_k(x)\,dx$
Fixed-linear spring	$\lambda^{3/4}(\cosh \lambda^{1/4} \cos \lambda^{1/4} + 1)$ $-\beta(\cos \lambda^{1/4} \sinh \lambda^{1/4}$ $-\cosh \lambda^{1/4} \sin \lambda^{1/4})$ $= 0$	For $\beta = 0.25$ $\omega_1 = 3.65$ $\omega_2 = 22.08$ $\omega_3 = 61.70$ $\omega_4 = 120.9$ $\omega_5 = 199.9$	$C_k\left[\cos \lambda_k^{1/4}x - \cosh \lambda_k^{1/4}x\right.$ $\left.-\alpha_k\left(\sin \lambda_k^{1/4}x - \sinh \lambda_k^{1/4}x\right)\right]$ $\alpha_k = \dfrac{\cos \lambda_k^{1/4} + \cosh \lambda_k^{1/4}}{\sin \lambda_k^{1/4} + \sinh \lambda_k^{1/4}}$	$\int_0^1 X_j(x) X_k(x)\,dx$

Pinned-linear spring	$\cot \lambda^{1/4} - \coth \lambda^{1/4} = -\dfrac{2\beta}{\lambda^{3/4}}$	For $\beta = 0.25$ $\omega_1 = 0.8636$ $\omega_2 = 15.41$ $\omega_3 = 49.47$ $\omega_4 = 104.25$ $\omega_5 = 178.27$	$C_k[\sin \lambda_k^{1/4} x + \dfrac{\sin \lambda_k^{1/4}}{\sinh \lambda_k^{1/4}} \sinh \lambda_k^{1/4} x]$	$\int_0^1 X_j(x) X_k(x)\,dx$
Fixed-attached mass	$\lambda^{1/4}(\cos \lambda^{1/4} \cosh \lambda^{1/4} + 1) + \beta(\cos \lambda^{1/4} \sinh \lambda^{1/4} - \cosh \lambda^{1/4} \sin \lambda^{1/4}) = 0$	For $\beta = 0.25$ $\omega_1 = 3.047$ $\omega_2 = 21.54$ $\omega_3 = 61.21$ $\omega_4 = 120.4$ $\omega_5 = 199.4$	$C_k\left[\cos \lambda_k^{1/4} x - \cosh \lambda_k^{1/4} x + \alpha_k\left(\sinh \lambda_k^{1/4} x - \sin \lambda_k^{1/4} x\right)\right]$ $\alpha_k = \dfrac{\cos \lambda_k^{1/4} + \cosh \lambda_k^{1/4}}{\sin \lambda_k^{1/4} + \sinh \lambda_k^{1/4}}$	$\int_0^1 X_j(x) X_k(x)\,dx + \beta X_j(1) X_k(1)$
Pinned-free	$\tan \lambda^{1/4} = \tanh \lambda^{1/4}$	$\omega_1 = 0$ $\omega_2 = 15.42$ $\omega_3 = 49.96$ $\omega_4 = 104.2$ $\omega_5 = 178.3$	$\sqrt{3}\,x \quad (k = 1)$ $C_k\left[\sin \lambda_k^{1/4} x + \dfrac{\sin \lambda_k^{1/4}}{\sinh \lambda_k^{1/4}} \sinh \lambda_k^{1/4} x\right]$ $(k > 1)$	$\int_0^1 X_j(x) X_k(x)\,dx$
Fixed-pinned	$\tan \lambda^{1/4} = \tanh \lambda^{1/4}$	$\omega_1 = 15.42$ $\omega_2 = 49.96$ $\omega_3 = 104.2$ $\omega_4 = 178.3$ $\omega_5 = 272.0$	$C_k\left[\cos \lambda_k^{1/4} x - \cosh \lambda_k^{1/4} x - \alpha_k\left(\sin \lambda_k^{1/4} x - \sinh \lambda_k^{1/4} x\right)\right]$ $\alpha_k = \dfrac{\cos \lambda_k^{1/4} - \cosh \lambda_k^{1/4}}{\sin \lambda_k^{1/4} - \sinh \lambda_k^{1/4}}$	$\int_0^1 X_j(x) X_k(x)\,dx$
Fixed-attached inertia element	$\cos \lambda^{1/4} \cosh \lambda^{1/4} + \beta(\sin \lambda^{1/4} \cosh \lambda^{1/4} + \cos \lambda^{1/4} \sinh \lambda^{1/4}) = -1$	For $\beta = 0.25$ $\omega_1 = 4.425$ $\omega_2 = 27.28$ $\omega_3 = 71.41$ $\omega_4 = 135.4$ $\omega_5 = 219.2$	$C_k\left[\cos \lambda_k^{1/4} x - \cosh \lambda_k^{1/4} x + \alpha_k\left(\sin \lambda_k^{1/4} x - \sinh \lambda_k^{1/4} x\right)\right]$ $\alpha_k = \dfrac{\sin \lambda_k^{1/4} - \sinh \lambda_k^{1/4}}{\cos \lambda_k^{1/4} + \cosh \lambda_k^{1/4}}$	$\int_0^1 X_j(x) X_k(x)\,dx + \beta X_j(1) X_k(1)$

The dimensional natural frequencies are obtained by multiplying the given nondimensional natural frequencies by $\sqrt{EI/\rho A L^4}$; for a given beam β is as defined in Table 9.3.

The formulation of the mathematical problem is completed by specifying two initial conditions.

Equation (9.66) is the governing nondimensional partial differential equation for forced vibrations of a beam assuming no axial load, longitudinal effects are negligible, rotary inertia and transverse shear are negligible, and other standard assumptions of beam theory from mechanics of materials apply.

9.4.2 Free Vibrations

When the product solution

$$w(x,t) = X(x)T(t) \tag{9.67}$$

is substituted into Eq. (9.66) with $f = 0$, the result is

$$\frac{1}{T(t)}\frac{d^2T}{dt^2} = -\frac{1}{X(x)}\frac{d^4X}{dx^4} \tag{9.68}$$

The usual separation argument is used to set both sides of Eq. (9.68) equal to the same constant, say $-\lambda$. This leads to

$$\frac{d^2T}{dt^2} + \lambda T = 0 \tag{9.69}$$

and

$$\frac{d^4X}{dx^4} - \lambda X = 0 \tag{9.70}$$

The solution of Eq. (9.69) is

$$T(t) = A\cos\sqrt{\lambda}\,t + B\sin\sqrt{\lambda}\,t \tag{9.71}$$

from which it is obvious that the natural frequencies are the square roots of the separation constants. The general solution of Eq. (9.70) is

$$X(x) = C_1\cos\lambda^{1/4}x + C_2\sin\lambda^{1/4}x + C_3\cosh\lambda^{1/4}x + C_4\sinh\lambda^{1/4}x \tag{9.72}$$

The solvability condition is determined by applying the homogeneous boundary conditions to Eq. (9.72). Table 9.4 summarizes the solvability conditions for different types of end supports, provides the first five nondimensional natural frequencies for each entry, their corresponding mode shapes, and specifies the scalar product for which the mode shapes are orthogonal.

Free-free and pinned-free beams are unrestrained and thus their lowest natural frequency is zero corresponding to a rigid-body mode. The fixed-pinned beam has the same characteristic equations as the pinned-free beam, and $\lambda = 0$ is a solution of this equation. However, $\lambda = 0$ leads to a trivial mode shape for the fixed-pinned beam.

Example 9.4. Determine the natural frequencies and normalized mode shapes for a simply supported beam.

The boundary conditions for a simply supported beam are

$$w(0,t) = 0 \qquad \frac{\partial^2 w(0,t)}{\partial x^2} = 0$$

and

$$w(1,t) = 0 \qquad \frac{\partial^2 w(1,t)}{\partial x^2} = 0$$

which when applied to Eq. (9.72) gives

$$0 = C_1 + C_3$$

$$0 = -\sqrt{\lambda}\, C_1 - \sqrt{\lambda}\, C_3$$

$$0 = C_1 \cos \lambda^{1/4} + C_2 \sin \lambda^{1/4} + C_3 \cosh \lambda^{1/4} + C_4 \sinh \lambda^{1/4}$$

$$0 = -\sqrt{\lambda}\, C_1 \cos \lambda^{1/4} - \sqrt{\lambda}\, C_2 \sin \lambda^{1/4} - \sqrt{\lambda}\, C_3 \cosh \lambda^{1/4} - \sqrt{\lambda}\, C_4 \sinh \lambda^{1/4}$$

The first two of the preceding equations imply $C_1 = C_3 = 0$. Then the last two equations are multiples of one another. They have a nontrivial solution if and only if

$$\sin \lambda^{1/4} = 0$$

or

$$\lambda_k = (k\pi)^4 \qquad k = 1, 2, \ldots$$

For these values of λ, $C_4 = 0$ and C_2 remains arbitrary, leading to the mode shape

$$X_k(x) = C_k \sin k\pi x$$

The mode shapes are orthogonal with respect to the scalar product of Eq. (9.2), as evidenced by

$$\int_0^1 C_k C_j \sin k\pi x \sin j\pi x \, dx = 0 \qquad k \neq j$$

Normalization with respect to this scalar product yields $C_k = \sqrt{2}$.

Example 9.5. Determine the first four natural frequencies for the beam of Fig. 9.15.

From Table 9.3, the appropriate boundary conditions are

$$w(0,t) = 0 \qquad \frac{\partial w(0,t)}{\partial x} = 0$$

and

$$\frac{\partial^2 w(1,t)}{\partial^2 x} = 0 \qquad \frac{\partial^3 w(1,t)}{\partial x^3} = \beta w(1,t)$$

where

$$\beta = \frac{kL^3}{EI} = \frac{(2 \times 10^6 \text{ N/m})(1 \text{ m})^3}{(210 \times 10^9 \text{ N/m}^2)(5 \times 10^{-5} \text{ m}^4)} = 0.190$$

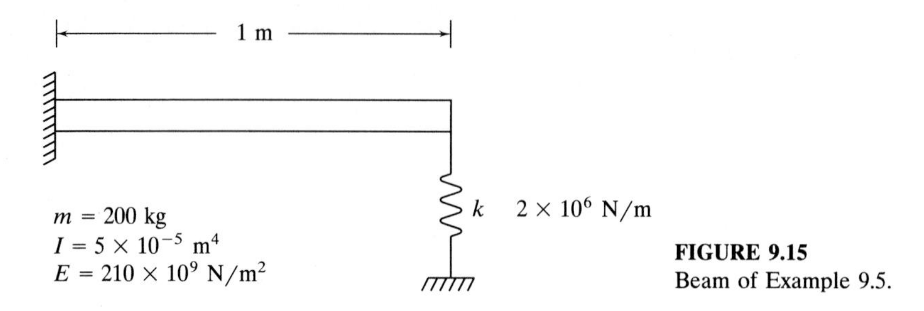

FIGURE 9.15
Beam of Example 9.5.

Application of the boundary conditions to Eq. (9.72) gives

$$C_1 + C_3 = 0$$

$$C_2 + C_4 = 0$$

$$-C_1 \cos\lambda^{1/4} - C_2 \sin\lambda^{1/4} + C_3 \cosh\lambda^{1/4} + C_4 \sinh\lambda^{1/4} = 0$$

$$(\lambda^{3/4} \sin\lambda^{1/4} - \beta\cos\lambda^{1/4})C_1 + (-\lambda^{3/4}\cos\lambda^{1/4} - \beta\sin\lambda^{1/4})C_2$$
$$+(\lambda^{3/4}\sinh\lambda^{1/4} - \beta\cosh\lambda^{1/4})C_3 + (\lambda^{3/4}\cosh\lambda^{1/4} - \beta\sinh\lambda^{1/4})C_4 = 0$$

which leads to the solvability condition

$$\lambda^{3/4}(1 + \cos\lambda^{1/4}\cosh\lambda^{1/4}) = -\beta(\cosh\lambda^{1/4}\sin\lambda^{1/4} - \cos\lambda^{1/4}\sinh\lambda^{1/4})$$

For $\beta = 0.190$ the first four roots of this equation are

$$\lambda = 13.10, 486.2, 380.7, 14161.6, \ldots$$

The nondimensional natural frequencies are the square roots of the values of λ that solve the characteristic equation. The dimensional natural frequencies are obtained by noting the relationship between the dimensional time and the nondimensional time and its application to Eq. (9.65),

$$\omega = \sqrt{\lambda\frac{EI}{\rho AL^4}} = 81.0\sqrt{\lambda}$$

The first four natural frequencies for this beam are

$$\omega_1 = 293.2 \text{ rad/s} \qquad \omega_2 = 1786 \text{ rad/s}$$

$$\omega_3 = 4998 \text{ rad/s} \qquad \omega_4 = 9793 \text{ rad/s}$$

The program CFREQ.BAS, included on the accompanying diskette, calculates the natural frequencies, mode shapes, and normalization constants for all beams listed in Table 9.4. CFREQ could be used to solve Example 9.5.

9.4.3 Forced Vibrations

The modal analysis method, described in Sec. 9.2, for analyzing the forced vibrations of a continuous system is applied to the following examples.

Example 9.6. The simply supported beam of Fig. 9.16 is subject to a harmonic excitation over half of its span. Determine the beam's steady-state response.

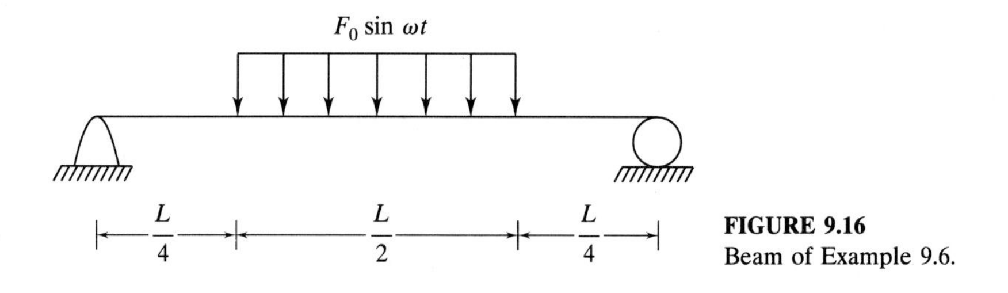

FIGURE 9.16
Beam of Example 9.6.

The nondimensional force per unit length is

$$f(x,t) = \sin \tilde{\omega} t\left[u\left(x - \tfrac{1}{4}\right) - u\left(x - \tfrac{3}{4}\right)\right]$$

where
$$\tilde{\omega} = \omega\sqrt{\frac{\rho A L^4}{EI}}$$

The expansion theorem is used to expand $f(x,t)$ in terms of the normalized mode shapes of the corresponding free-vibration problem, which are determined in Example 9.4. The expansion coefficients are determined using Eq. (9.6), with the scalar product defined by Eq. (9.2),

$$\begin{aligned}
C_k &= \int_0^1 f(x,t)\sqrt{2}\sin k\pi x\,dx \\
&= \sqrt{2}\sin \tilde{\omega} t \int_{1/4}^{3/4} \sin k\pi x\,dx \\
&= \frac{\sqrt{2}}{k\pi}\sin \tilde{\omega} t\left(\cos k\frac{\pi}{4} - \cos k\frac{3\pi}{4}\right) \\
&= \frac{2}{k\pi}\sin \tilde{\omega} t \begin{cases} 0 & k = 2,4,6,\ldots \\ 1 & k = 1,7,9,15,17,23,\ldots \\ -1 & k = 3,5,11,13,19,21,\ldots \end{cases} \\
&= \frac{2}{k\pi} a_k \sin \tilde{\omega} t
\end{aligned}$$

The displacement is expanded as

$$w(x,t) = \sum_{k=1}^{\infty} \sqrt{2}\sin k\pi x p_k(t)$$

Substituting for w and f in Eq. (9.66) leads to

$$\sum_{k=1}^{\infty}\left[\ddot{p}_k + (k\pi)^4 p_k\right]\sqrt{2}\sin k\pi x = \Lambda \sum_{k=1}^{\infty} C_k(t)\sqrt{2}\sin k\pi x$$

where
$$\Lambda = \frac{F_0 L^3}{EI}$$

The preceding equation is multiplied by $\sqrt{2}\sin j\pi x$ for an arbitrary j and integrated from 0 to 1. This is equivalent to taking the scalar product of both sides of the equation with $X_j(x)$. The orthogonality condition, Eq. (9.6), is used such that each

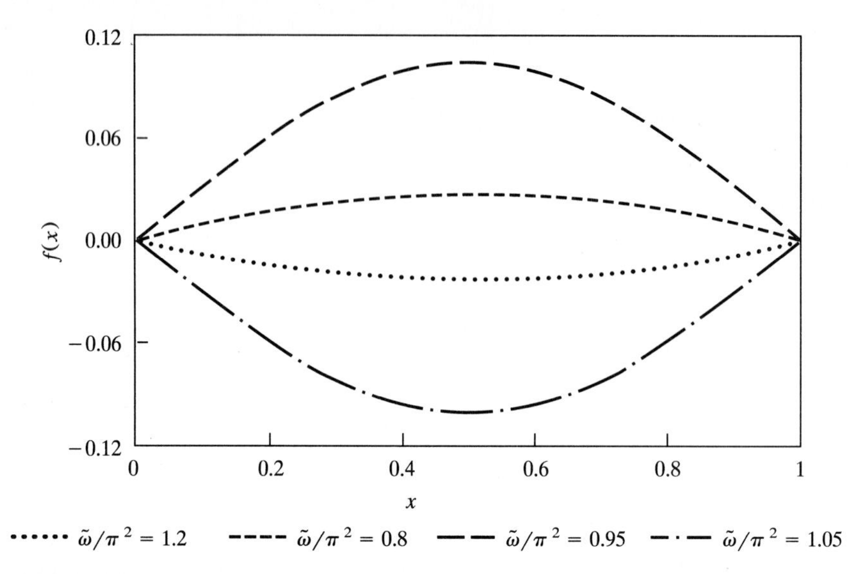

FIGURE 9.17
Steady-state response for Example 9.6.

sum collapses to a single term, yielding

$$\ddot{p}_j + (j\pi)^4 p_j = \Lambda C_j, \qquad j = 1, 2, \ldots$$

whose steady-state solution is

$$p_j(t) = \frac{\Lambda}{(j\pi)^4 - \tilde{\omega}^2} \frac{2}{j\pi} a_j \sin \tilde{\omega} t$$

The steady-state response of the beam is

$$w(x,t) = \frac{2\sqrt{2}\,\Lambda}{\pi} \sin \tilde{\omega} t \left[\frac{1}{\pi^4 - \tilde{\omega}^2} \sin \pi x - \frac{1}{3(81\pi^4 - \tilde{\omega}^2)} \sin 3\pi x - \frac{1}{5(625\pi^4 - \tilde{\omega}^2)} \sin 5\pi x + \frac{1}{7(1501\pi^4 - \tilde{\omega}^2)} \sin 7\pi x + \cdots \right]$$

$$= \frac{2\sqrt{2}\,\Lambda}{\pi} f(x) \sin \tilde{\omega} t$$

The function $f(x)$ is shown in Fig. 9.17 for several values of $\tilde{\omega}$. Note that when $\tilde{\omega}$ is close to π^2 the steady-state amplitude is large at the midspan.

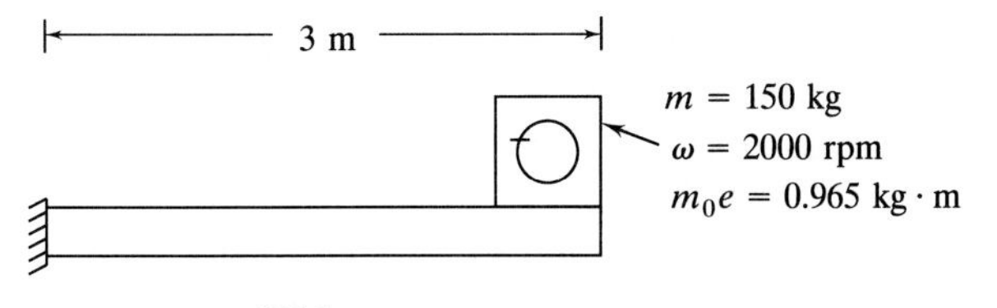

FIGURE 9.18
Cantilever beam of Example 9.7 has machine with rotating unbalance at its end.

Example 9.7. A machine of mass 150 kg is attached to the end of the cantilever beam of Fig. 9.18. The machine operates at 2000 rpm and has a rotating unbalance of 0.965 kg · m. What is the steady-state amplitude of vibration of the end of the beam?

The nondimensional formulation of the governing mathematical problem is

$$\frac{\partial^4 w}{\partial x^4} + \frac{\partial^2 w}{\partial t^2} = 0$$

subject to

$$w(0, t) = 0$$

$$\frac{\partial w(0, t)}{\partial x} = 0$$

$$\frac{\partial^2 w(1, t)}{\partial x^2} = 0$$

$$\frac{\partial^3 w(1, t)}{\partial x^3} = \beta \frac{\partial^2 w(1, t)}{\partial t^2} + \alpha \sin \tilde{\omega} t$$

where
$$\tilde{\omega} = \omega \sqrt{\frac{\rho A L^4}{EI}}$$

$$= 209.4 \frac{\text{rad}}{\text{s}} \sqrt{\frac{(280 \text{ kg})(3 \text{ m})^3}{(210 \times 10^9 \text{ N/m}^2)(1.2 \times 10^{-4} \text{ m}^4)}}$$

$$= 3.63$$

$$\beta = \frac{m}{\rho A L} = \frac{150 \text{ kg}}{280 \text{ kg}} = 0.536$$

and
$$\alpha = \frac{m_0 e \omega^2 L^2}{EI}$$

$$= \frac{(0.965 \text{ kg} \cdot \text{m})(209.4 \text{ rad/s})^2 (3 \text{ m})^2}{(210 \times 10^9 \text{ N/m}^2)(1.2 \times 10^{-4} \text{ m}^4)}$$

$$= 0.015$$

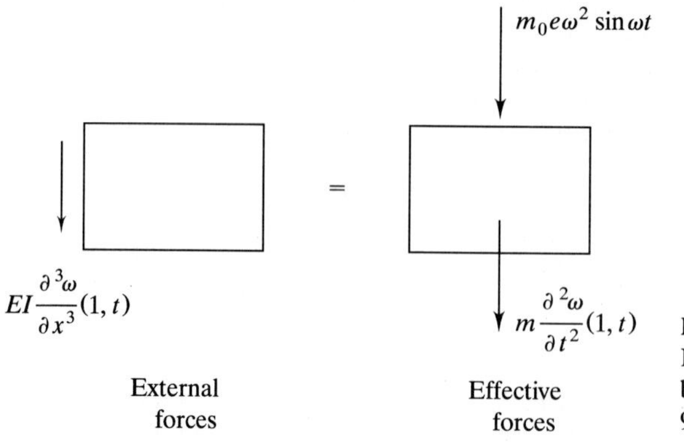

FIGURE 9.19
Free-body diagrams used to derive boundary condition for Example 9.7.

The last boundary condition is developed by applying Newton's law to the machine as shown in Fig. 9.19. The problem is nonhomogeneous due to this boundary condition. From Table 9.4 the characteristic equation for the homogeneous problem of a beam with a concentrated mass at its end is

$$\lambda^{1/4}(1 + \cos\lambda^{1/4}\cosh\lambda^{1/4}) + \beta(\cos\lambda^{1/4}\sinh\lambda^{1/4} - \cosh\lambda^{1/4}\sin\lambda^{1/4}) = 0$$

The corresponding mode shapes for the homogeneous problem are

$$X_k(x) = C_k\left[\cos\lambda^{1/4}x - \cosh\lambda^{1/4}x + \frac{\cos\lambda^{1/4} + \cosh\lambda^{1/4}}{\sin\lambda^{1/4} + \sinh\lambda^{1/4}}(\sinh\lambda^{1/4}x - \sin\lambda^{1/4}x)\right]$$

where C_k is chosen to normalize the mode shape with respect to the scalar product defined by

$$(X_j(x), X_k(x)) = \int_0^1 X_j(x)X_k(x)\,dx + \beta X_j(1)X_k(1)$$

The first six nondimensional natural frequencies and normalization constants are given in Table 9.5. The entries in Table 9.5 are generated using CFREQ.BAS on the accompanying diskette.

The expansion theorem implies that the solution of the nonhomogeneous problem can be expanded in a series of normalized mode shapes. To this end,

$$w(x,t) = \sum_{k=1}^{\infty} p_k(t)X_k(x)$$

Substituting for $w(x,t)$ into the governing partial differential equation, multiplying by $X_j(x)$ for an arbitrary j, and integrating from 0 to 1 leads to

$$\sum_{k=1}^{\infty}(\ddot{p}_k + \lambda p_k)\int_0^1 X_j(x)X_k(X)\,dx = 0$$

TABLE 9.5
Free vibration properties for Example 9.7

λ_k	Nondimensional natural frequency	Natural frequency ω_k (rad / s)	C_k
6.71	2.59	149.55	0.715
443.5	21.06	1216.0	0.617
3682.1	60.68	3483.0	0.593
14371.2	119.88	6922.0	0.584
39533.3	198.83	11480.0	0.582
88513.2	297.51	17178.0	0.434

The mutual orthonormality of the mode shapes implies

$$\int_0^1 X_j(x)X_k(x)\,dx = \delta_{ij} - \beta X_j(1)X_k(1)$$

Use of this orthogonality condition leads to

$$\ddot{p}_j + \lambda_j p_j = \sum_{k=1}^{\infty} (\ddot{p}_k + \lambda_k p_k)\beta X_j(1)X_k(1)$$

Substituting for $w(x,t)$ from the expansion theorem in the nonhomogeneous boundary condition leads to

$$\sum_{k=1}^{\infty} \frac{d^3X_k(1)}{dx^3} p_k(t) = \alpha \sin\tilde{\omega}t + \beta \sum_{k=1}^{\infty} X_k(1)\ddot{p}_k(t)$$

The mode shapes satisfy the boundary conditions for the nonhomogeneous problem. Thus

$$\frac{d^3X(1)}{dx^3} = -\lambda_k \beta X_k(1)$$

which when used in the preceding equation gives

$$\sum_{k=1}^{\infty} (\ddot{p}_k + \lambda_k p_k)X_k(1) = \alpha \sin\tilde{\omega}t$$

and which when substituted into the previously derived differential equations for the principal coordinates uncouples these equations and gives

$$\ddot{p}_j + \lambda_j p_j = \alpha X_j(1)\sin\tilde{\omega}t \qquad j = 1,2,\ldots$$

The steady-state solution for each of the principal coordinates is now easily obtained and the expansion theorem is used to write the steady-state solution as

$$w(x,t) = \left[\sum_{k=1}^{\infty} \frac{\alpha X_k(1)}{\lambda_k - \tilde{\omega}^2} X_k(x)\right] \sin\tilde{\omega}t$$

The nondimensional steady-state amplitude of the end of the beam is

$$\alpha \sum_{k=1}^{\infty} \frac{X_k^2(1)}{\lambda_k - \tilde{\omega}^2} = 8.4 \times 10^{-4}$$

The dimensional amplitude is obtained using Eq. (9.54) as $8.4 \times 10^{-4}(3 \text{ m}) = 2.52$ mm.

The program CMODA.BAS, included on the accompanying diskette, uses modal analysis to determine the steady-state response of any of the beams of Table 9.4 due to an excitation of the form $f(x, t) = g(x)\sin \omega t$. CMODA.BAS could be used to solve Examples 9.6 and 9.7.

9.5 ENERGY METHODS

Consider a differential element of the shaft of Fig. 9.5. Assuming elastic behavior throughout, a shear stress distribution is developed across the cross section of the shaft according to

$$\tau = \frac{Tr}{J} \tag{9.73}$$

where $T(x, t)$ is the resisting torque in the cross section where r is the distance from the center of the shaft to a point in its cross section. The total strain energy in the element is

$$dV = \frac{1}{2G} \int \tau^2 \, dA \, dx \tag{9.74}$$

Substitution of Eqs. (9.73) and (9.11) into Eq. (9.74) leads to

$$dV = \frac{G}{2} \left(\frac{\partial \theta}{\partial x} \right)^2 \int r^2 \, dA \, dx \tag{9.75}$$

Noting that $J = \int r^2 \, dA$ and integrating over the entire length of the shaft, the total strain energy becomes

$$V = \frac{1}{2} \int_0^L JG \left(\frac{\partial \theta}{\partial x} \right)^2 dx \tag{9.76}$$

The kinetic energy of the differential element is

$$dT = \frac{1}{2} \rho J \left(\frac{\partial \theta}{\partial t} \right)^2 dx \tag{9.77}$$

where ρ is the mass density of the shaft's material. The total kinetic energy of the shaft is

$$T = \frac{1}{2} \int_0^L \rho J \left(\frac{\partial \theta}{\partial t} \right)^2 dx \tag{9.78}$$

For a conservative system the maximum potential energy is equal to the maximum kinetic energy. Thus if the free oscillations of the shaft are described by

$$\theta(x,t) = u(x)\sin \omega t \tag{9.79}$$

then

$$\omega^2 = \frac{\int_0^L JG\left(\frac{du}{dx}\right)^2 dx}{\int_0^L \rho J u^2\, dx} \tag{9.80}$$

Introducing the nondimensional variables of Eq. (9.65) into Eq. (9.80) and assuming the shaft is uniform leads to

$$\tilde{\omega}^2 = \frac{\int_0^1 \left(\frac{du}{dx}\right)^2 dx}{\int_0^1 u^2\, dx} \tag{9.81}$$

where

$$\tilde{\omega} = L\sqrt{\frac{\rho}{G}}\,\omega \tag{9.82}$$

For any function $w(x)$ which satisfies the boundary conditions specified for the shaft, define

$$R(w) = \frac{\int_0^L JG\left(\frac{dw}{dx}\right)^2 dx}{\int_0^L \rho J w^2\, dx} \tag{9.83}$$

$R(w)$ is Rayleigh's quotient for this continuous system. If $w(x)$ is a mode shape, then $R(w)$ is equal to the square of the natural frequency of that mode. If w is not a mode shape, then $R(w)$ is a scalar function of w. As for discrete systems, $R(w)$ is a minimum when w is a mode shape. Hence Rayleigh's quotient can be used to approximate the lowest natural frequency for the continuous system.

Example 9.8. Use Rayleigh's quotient to approximate the lowest natural frequency of the tapered circular shaft of Fig. 9.20.

The polar moment of inertia varies over the length of the shaft as

$$J(x) = \frac{\pi}{2}(0.2 - 0.05x)^4$$

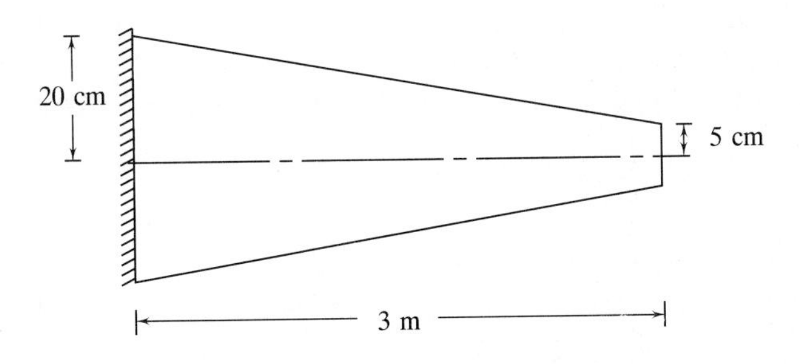

FIGURE 9.20
System of Example 9.8.

A trial function which satisfies the boundary conditions $w(0) = 0$ and $dw/dx(3\text{ m}) = 0$ is

$$w(x) = \sin \frac{\pi}{6} x$$

An upper bound and approximation on the lowest natural frequency is

$$R(w) = \frac{80 \times 10^9 \dfrac{\text{N}}{\text{m}^2} \dfrac{\pi}{2} \displaystyle\int_0^3 (0.2 - 0.05x)^4 \left(\frac{\pi}{6}\right)^2 \cos^2 \frac{\pi}{6} x \, dx}{7850 \dfrac{\text{kg}}{\text{m}^3} \dfrac{\pi}{2} \displaystyle\int_0^3 (0.2 - 0.05x)^4 \sin^2 \frac{\pi}{6} x \, dx}$$

$$\omega_1 \leq [R(w)]^{1/2} = 3767 \text{ rad/s}$$

Rayleigh's quotient can be generalized as the ratio of a potential energy scalar to a kinetic energy scalar product, where the energy products are defined by integrals perhaps with additional terms to account for discrete masses or springs.

$$R(w) = \frac{(w, w)_V}{(w, w)_T} \tag{9.84}$$

Rayleigh's quotient can be applied to any continuous system. Table 9.6 gives the appropriate form of the scalar products for several continuous systems.

A method based upon Rayleigh's quotient, called the Rayleigh-Ritz method, can be used to approximate a finite number of the lowest natural frequencies of a continuous system. Let $u_1(x), u_2(x), \ldots, u_n(x)$ be n linearly independent functions, each of which satisfies the boundary conditions for a specific continuous system. An approximation to the free-vibration response of the continuous system is assumed as

$$w(x) = \sum_{i=1}^{n} c_i u_i(x) \tag{9.85}$$

Equation (9.85) is substituted into Rayleigh's quotient and rewritten as

$$R(w)(w, w)_T = (w, w)_V \tag{9.86}$$

Since $R(w)$ is stationary near a mode shape, a good approximation to the natural frequencies and mode shapes is obtained by setting

$$\frac{\partial R}{\partial c_1} = \frac{\partial R}{\partial c_2} = \cdots = \frac{\partial R}{\partial c_n} = 0 \tag{9.87}$$

Differentiating Eq. (9.86) with respect to c_k for any $k = 1, 2, \ldots, n$ and using Eq. (9.87) gives

$$R(w) \frac{\partial (w, w)_T}{\partial c_k} = \frac{\partial (w, w)_V}{\partial c_k} \tag{9.88}$$

TABLE 9.6
Scalar products for Rayleigh-Ritz method

Structural element	Case	$(u,v)_T$	$(u,v)_V$
Torsional shaft	No added disks or springs	$\int_0^L \rho J u(x) v(x)\, dx$	$\int_0^L GJ \frac{du}{dx}\frac{dv}{dx}\, dx$
	Added disk at $x = \tilde{x}$	$\int_0^L \rho J u(x) v(x)\, dx + I_D u(\tilde{x}) v(\tilde{x})$	$\int_0^L GJ \frac{du}{dx}\frac{dv}{dx}\, dx$
	Torsional spring at $x = \tilde{x}$	$\int_0^L \rho J u(x) v(x)\, dx$	$\int_0^L GJ \frac{du}{dx}\frac{dv}{dx}\, dx + k_t \frac{du(\tilde{x})}{dx}\frac{dv(\tilde{x})}{dx}$
Longitudinal bar	No added masses or springs	$\int_0^L \rho A u(x) v(x)\, dx$	$\int_0^L EA \frac{du}{dx}\frac{dv}{dx}\, dx$
	Added mass at $x = \tilde{x}$	$\int_0^L \rho A u(x) v(x)\, dx + m u(\tilde{x}) v(\tilde{x})$	$\int_0^L EA \frac{du}{dx}\frac{dv}{dx}\, dx$
	Spring at $x = \tilde{x}$	$\int_0^L \rho A u(x) v(x)\, dx$	$\int_0^L EA \frac{du}{dx}\frac{dv}{dx}\, dx + k \frac{du(\tilde{x})}{dx}\frac{dv(\tilde{x})}{dx}$
Beam	No added masses, disks, or springs	$\int_0^L \rho A u(x) v(x)\, dx$	$\int_0^L EI \frac{d^2u}{dx^2}\frac{d^2v}{dx^2}\, dx$
	Added mass at $x = \tilde{x}$	$\int_0^L \rho A u(x) v(x)\, dx + m u(\tilde{x}) v(\tilde{x})$	$\int_0^L EI \frac{d^2u}{dx^2}\frac{d^2v}{dx^2}\, dx$
	Added spring at $x = \tilde{x}$	$\int_0^L \rho A u(x) v(x)\, dx$	$\int_0^L EI \frac{d^2u}{dx^2}\frac{d^2v}{dx^2}\, dx + k \frac{du(\tilde{x})}{dx}\frac{dv(\tilde{x})}{dx}$
	Added disk (I_D) at $x = \tilde{x}$	$\int_0^L \rho A u(x) v(x)\, dx + I_D \frac{du(\tilde{x})}{dx}\frac{dv(\tilde{x})}{dx}$	$\int_0^L EI \frac{d^2u}{dx^2}\frac{d^2v}{dx^2}\, dx$

Developing Eq. (9.88) for each $k = 1, 2, \ldots, n$ leads to n linear homogeneous equations to solve for $c_1, c_2, \ldots, c_n$ in terms of the parameter $R(w)$. Since the equations are homogeneous, a nontrivial solution is available if and only if the determinant is set equal to zero, yielding an nth-order polynomial equation for $R(w)$. The roots of the polynomial are the squares of the approximations to the lowest natural frequencies. Approximations for the mode shapes can be obtained by returning to the homogeneous equations. The method is illustrated in the following example.

Example 9.9. Use the Rayleigh-Ritz method to approximate the two lowest natural frequencies of Example 9.1.

Two polynomials which satisfy the boundary conditions of Example 9.1 are

$$u_1(x) = 2x - x^2$$

$$u_2(x) = 3x - x^3$$

An approximation to the mode shape is developed as

$$w(x) = c_1(2x - x^2) + c_2(3x - x^3)$$

Calculation of the energy scalar products gives

$$(w, w)_T = \int_0^1 \left[c_1(2x - x^2) + c_2(3x - x^3)\right]^2 dx$$

$$= \frac{8}{15}c_1^2 + \frac{61}{30}c_1 c_2 + \frac{204}{105}c_2^2$$

$$(w, w)_V = \int_0^1 \left[c_1(2 - 2x) + c_2(3 - 3x^2)\right]^2 dx$$

$$= \frac{4}{3}c_1^2 + 5c_1 c_2 + \frac{24}{5}c_2^2$$

Application of Eq. (9.88) leads to

$$\left(\frac{8}{3} - \frac{16}{15}R\right)c_1 + \left(5 - \frac{61}{30}R\right)c_2 = 0$$

$$\left(5 - \frac{61}{30}R\right)c_1 + \left(\frac{48}{5} - \frac{136}{35}R\right)c_2 = 0$$

A nontrivial solution of the preceding equations is obtained if and only if

$$\det\begin{bmatrix} \frac{8}{3} - \frac{16}{15}R & 5 - \frac{61}{30}R \\ 5 - \frac{61}{30}R & \frac{48}{5} - \frac{136}{35}R \end{bmatrix} = 0$$

Evaluation of the determinant leads to

$$9.24R^2 - 241.0R + 538.0 = 0$$

whose roots are

$$R = 2.467, 23.610$$

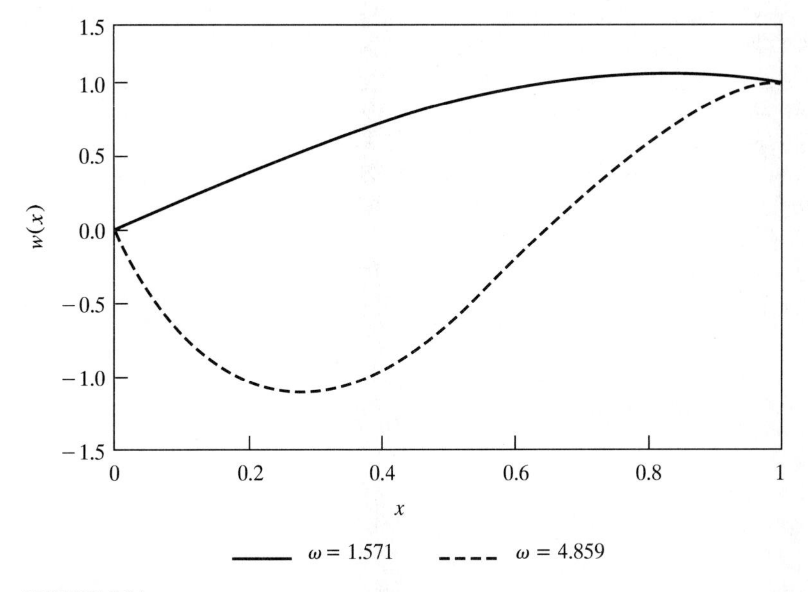

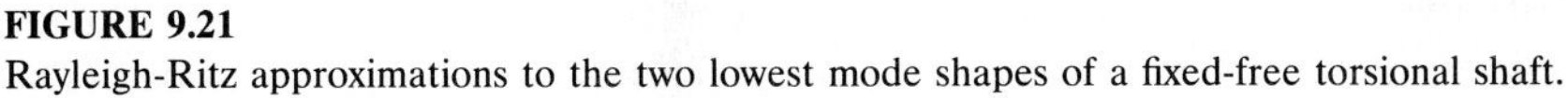

FIGURE 9.21
Rayleigh-Ritz approximations to the two lowest mode shapes of a fixed-free torsional shaft.

The natural frequency approximations are

$$\omega_1 \approx 1.571 \qquad \omega_2 \approx 4.859$$

The approximation to the lowest natural frequency is excellent. The approximation to the second natural frequency is also very good, being only 3.3% higher than the exact value.

The mode shape approximations are obtained by solving for c_2 in terms of c_1 for each R and then substituting into the expression for $w(x)$ with c_1 remaining arbitrary. This leads to

$$w_1(x) = 7.58x - x^2 - 1.86x^3$$

$$w_2(x) = 0.4295x - x^2 + 0.5235x^3$$

The approximate mode shapes plotted in Fig. 9.21 have been normalized such that $w_i(1) = 1$. These compare favorably to the first two mode shapes for a fixed-free torsional shaft plotted in Fig. 9.7.

The finite-element method is a generalized application of the Rayleigh-Ritz method. A continuous system is broken into a number of finite elements. Interpolation functions are defined piecewise on each element with appropriate smoothness at the interelement boundaries. The interpolation functions for elements adjacent to boundaries satisfy boundary conditions. The discretization of a fixed-free bar undergoing longitudinal vibrations is shown in Fig. 9.22. The

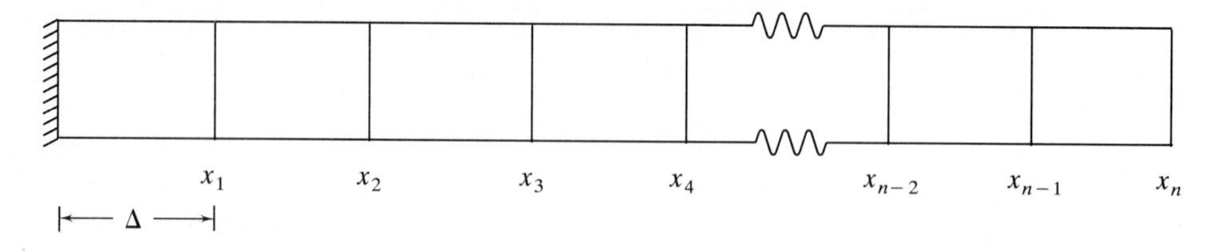

FIGURE 9.22
Discretization of bar undergoing longitudinal vibrations.

bar is broken up into n elements of equal length $\Delta = 1/n$. Figure 9.23 shows a piecewise linear interpolation of the displacement. The total displacement can be written as

$$w(x) = \sum_{i=1}^{n} \phi_i(x)$$

where the piecewise interpolation functions are defined to satisfy the boundary conditions and continuity at the interelement boundaries, called the nodes. The interpolating functions can be written as

$$\phi_0(x) = \frac{w_1 x}{\Delta}[1 - u(x - \Delta)]$$

$$\phi_j(x) = \frac{1}{\Delta}\left[(w_{j+1} - w_j)x + (j+1)\Delta u_j - j\Delta u_{j+1}\right]$$

$$\times\left[u(x - j\Delta) - u(x - (j+1)\Delta)\right] \qquad 2 \le j \le n-1$$

$$\phi_n(x) = w_n u(x - (n-1)\Delta)$$

where w_j is the displacement at the jth node.

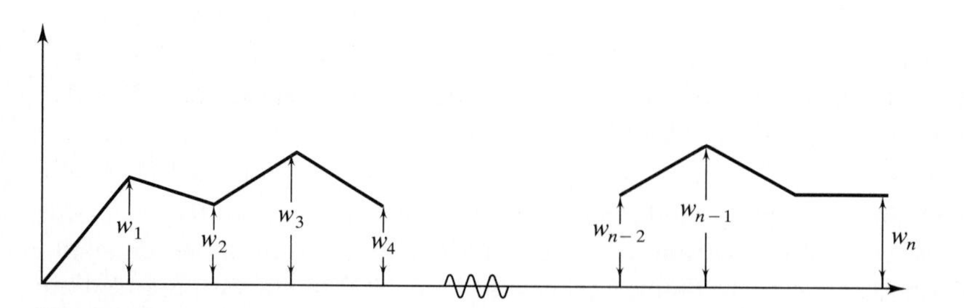

FIGURE 9.23
Piecewise linear interpolation of displacement that satisfies $w(0) = 0$, $dw/dx(0) = 0$. The finite-element method is a generalization of Rayleigh-Ritz using these interpolating functions.

The Rayleigh-Ritz method is applied and used to generate a system of equations used to determine the natural frequencies. Rayleigh's quotient is minimized with respect to the nodal displacements.

In more complicated problems, the discrete stiffness matrix and mass matrix are formulated. The elements of the stiffness matrix are potential energy scalar products of the interpolation functions, while the elements of the mass matrix are kinetic energy scalar products of the interpolation functions. The natural frequencies are the square roots of the eigenvalues of the inverse of the mass matrix times the stiffness matrix or the reciprocals of the square roots of the eigenvalues of the flexibility matrix times the mass matrix. Often finite-element codes use matrix iteration to determine the natural frequencies.

PROBLEMS

9.1. Calculate the speed of longitudinal waves in a 3-m-long steel bar ($E = 210 \times 10^9$ N/m^2, $\rho = 7850$ kg/m^3) of a circular cross section of 20 mm radius.

9.2. Calculate the three lowest torsional natural frequencies of a solid 20-cm-radius steel shaft ($\rho = 7500$ kg/m^3, $G = 80 \times 10^9$ N/m^2) of length 1.5 m that is fixed at one end and free at its other end.

9.3. A 5000-N · m torque is statically applied to the free end of the shaft of Prob. 9.2 and suddenly removed. Plot the time-dependent angular displacement of the free end.

9.4. A 5000-N · m torque is statically applied to the midspan of the shaft of Prob. 9.2 and suddenly removed. Determine an expression for the time-dependent angular displacement of the free end of the shaft.

9.5. A steel shaft ($\rho = 7850$ kg/m^3, $G = 85 \times 10^9$ N/m^2) of inner radius 30 mm and outer radius 50 mm and length 1.0 m is fixed at both ends. Determine the three lowest natural frequencies of the shaft.

9.6. A 10,000-N · m torque is applied to the midspan of the shaft of Prob. 9.5 and suddenly removed. Determine the time-dependent angular displacement of the midspan of the shaft.

9.7. A motor of mass moment of inertia 85 kg · m^2 is attached to the end of the shaft of Prob. 9.2. Determine the three lowest natural frequencies of the shaft and motor assembly. Compare the lowest natural frequency to that obtained by making a one-degree-of-freedom model and approximating the inertia effects of the shaft.

9.8. Show the orthogonality of the two lowest mode shapes of the system in Prob. 9.7.

9.9. Operation of the motor attached to the shaft of Prob. 9.7 produces a harmonic torque of amplitude 2000 N · m at a frequency of 110 Hz. Determine the steady-state angular displacement of the end of the shaft.

9.10. A 20-cm-diameter, 2-m-long steel shaft ($\rho = 7600$ kg/m^3, $G = 80 \times 10^9$ N/m^2) has rotors of mass moment of inertia 110 kg · m^2 and 65 kg · m^2 attached to its ends. Determine the three lowest natural frequencies of the shaft. Compare the lowest nonzero natural frequency to that obtained by using a two-degree-of-freedom model, ignoring the inertia of the shaft.

9.11. Determine an expression for the natural frequencies of the shaft of Fig. P9.11.

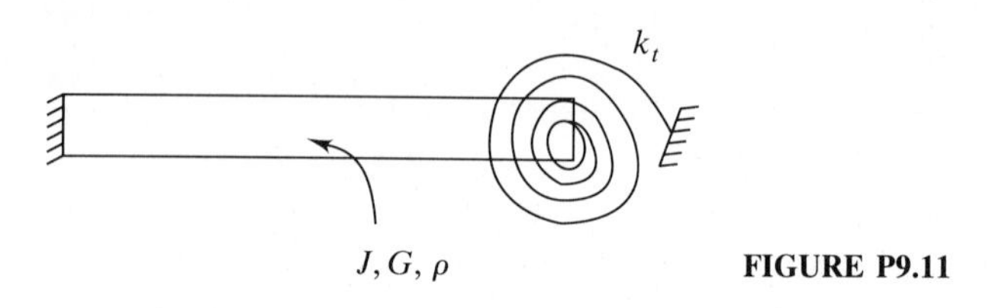

FIGURE P9.11

9.12. An oil well drilling tool is modeled as a bit attached to the end of a long shaft, unrestrained from rotation at its fixed end.

(*a*) Determine the equation defining the natural frequencies of the drilling tool.

(*b*) For a particular operation, the shaft ($\rho = 7500 \text{ kg/m}^3$, $G = 80 \times 10^9 \text{ N/m}^2$) is 20 m long with a 20-cm diameter. The tool operates at a speed of 400 rad/s. What are the limits on the moment of inertia of the drill bit such that the two lowest non-zero natural frequencies of the tool are not within 20% of the operating speed?

9.13. The shaft of Prob. 9.2 is at rest in equilibrium when the time-dependent moment of Fig. P9.13 is applied to the end of the shaft. Determine the time-dependent form of the resulting torsional oscillations.

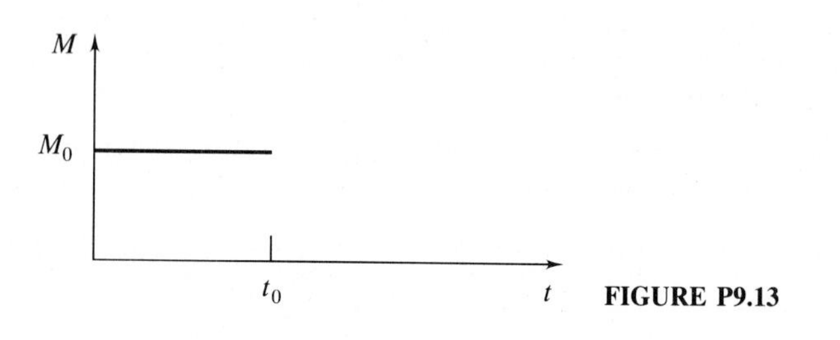

FIGURE P9.13

9.14. The shaft of Prob. 9.2 is at rest in equilibrium when it is subject to the uniform time-dependent torque loading per unit length of Fig. P9.14. Determine the time-dependent form of the resulting torsional oscillations.

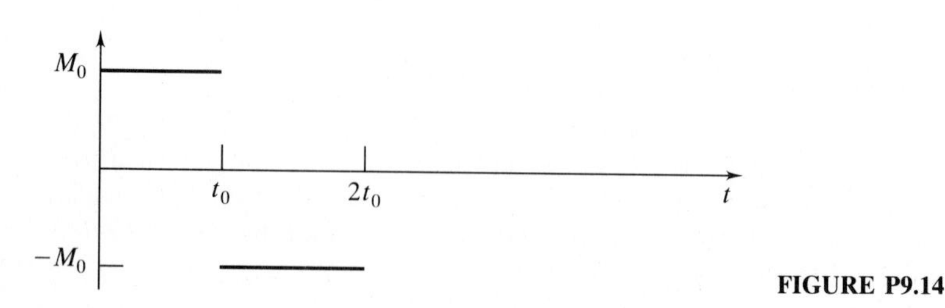

FIGURE P9.14

9.15. The elastic bar of Fig. P9.15 is undergoing longitudinal vibrations. Let $u(x, t)$ be the time-dependent displacement of a particle along the centroidal axis of the bar, initially a distance x from the left support.

(*a*) Draw free-body diagrams showing the external and effective forces acting on a differential element of thickness dx, a distance x from the left end of the bar at an arbitrary instant of time.

(*b*) Show that the governing partial differential equation is

$$E\frac{\partial^2 u}{\partial x^2} = \rho\frac{\partial^2 u}{\partial t^2}$$

(*c*) Introduce nondimensional variables to derive a nondimensional partial differential equation.

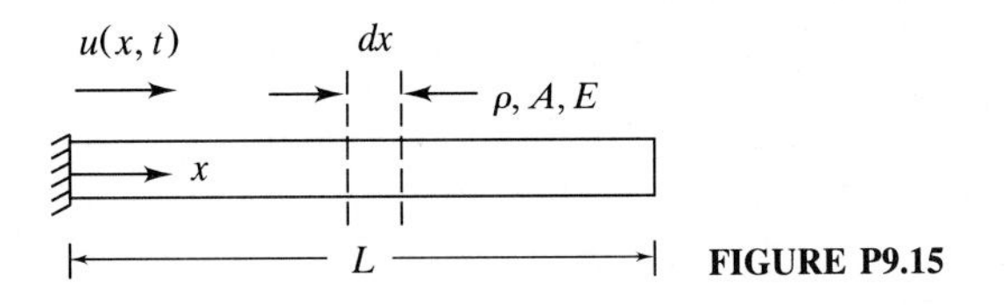

FIGURE P9.15

9.16. Using the results of Prob. 9.15, determine the natural frequencies of longitudinal vibrations of a bar fixed at one end and free at the other.

9.17. Show the orthogonality of mode shapes of longitudinal vibration of a bar fixed at one end and free at its other end.

9.18. Draw frequency response curves for the response of the disk at the end of the shaft in Example 9.3. Plot the curves for $\beta = 0.5$, $\beta = 2$, and $\beta = 20.0$.

9.19. Determine the steady-state response of a circular shaft subject to a uniform torque per unit length $T_0 \sin \omega t$ applied over its entire length.

9.20. Determine the steady-state response of a uniform bar subject to a force $F_0 \sin \omega t$ applied at its end.

9.21. Determine the steady-state response of the system of Fig. P9.21.

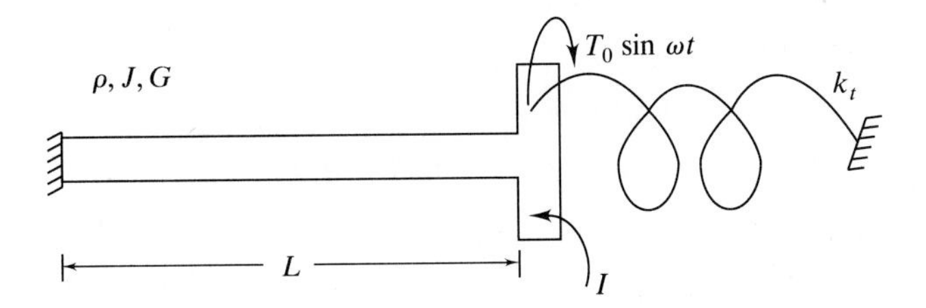

FIGURE P9.21

9.22. Propeller blades totaling 1200 kg with a total mass moment of inertia of 155 $kg \cdot m^2$ are attached to a solid circular shaft ($\rho = 5000\ kg/m^3$, $G = 60 \times 10^9\ N/m^2$, $E = 140 \times 10^9\ N/m^2$) of radius 40 cm and length 20 m. The other end of the shaft is fixed in an ocean liner. Determine
(*a*) The lowest natural frequency for torsional oscillations of the propeller.
(*b*) The lowest natural frequency for longitudinal motion of the propeller.

9.23. A pipe used to convey fluid is cantilevered from a wall. The steel pipe ($\rho = 7500\ kg/m^3$, $G = 80 \times 10^9\ N/m^2$, $E = 200 \times 10^9\ N/m^2$) has an inner radius of 20 cm, a thickness of 1 cm, and a length of 4.6 m. For an empty pipe determine
(*a*) The five lowest natural frequencies for torsional oscillation.
(*b*) The five lowest natural frequencies for longitudinal vibration.
(*c*) The five lowest natural frequencies for transverse motion.

The program CREQ.BAS on the accompanying diskette calculates the natural frequencies and mode shapes for each of the uniform beams of Table 9.4. The program CMODA.BAS determines the forced response of each of the beams of Table 9.4 due to a harmonic excitation. These programs can be used to provide solutions to most of the problems between Probs. 9.24 and 9.46.

9.24–9.27. Each of the beams has $\rho = 8000\ kg/m^3$, $E = 200 \times 10^9\ N/m^2$, $I = 4 \times 10^{-5}\ m^4$, $A = 1.2 \times 10^{-2}\ m^2$, $L = 1.4$ m. Use Table 9.4 or CFREQ.BAS to calculate the beam's three lowest natural frequencies of transverse vibration.

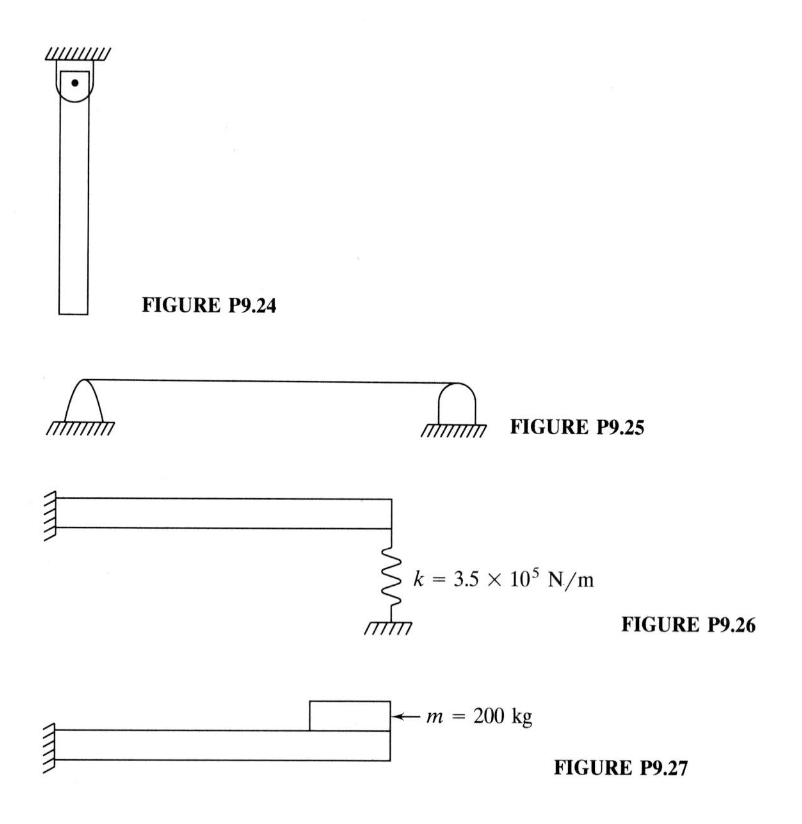

FIGURE P9.24

FIGURE P9.25

FIGURE P9.26

FIGURE P9.27

9.28. Verify the characteristic equation given in Table 9.4 for a pinned-free beam.

9.29. Verify the characteristic equation given in Table 9.4 for a fixed-fixed beam.

9.30. The characteristic equations given in Table 9.4 for a free-free beam and a fixed-fixed beam are identical. Explain both mathematically and physically why the lowest natural frequency for a free-free beam is zero, but not for a fixed-fixed beam.

9.31. Verify the orthogonality of the eigenfunctions given in Table 9.4 for a pinned-free beam.

9.32. Verify the orthogonality of the eigenfunctions given in Table 9.4 for a fixed-attached mass beam.

9.33–9.37. Determine the time-dependent displacement for the beam shown.

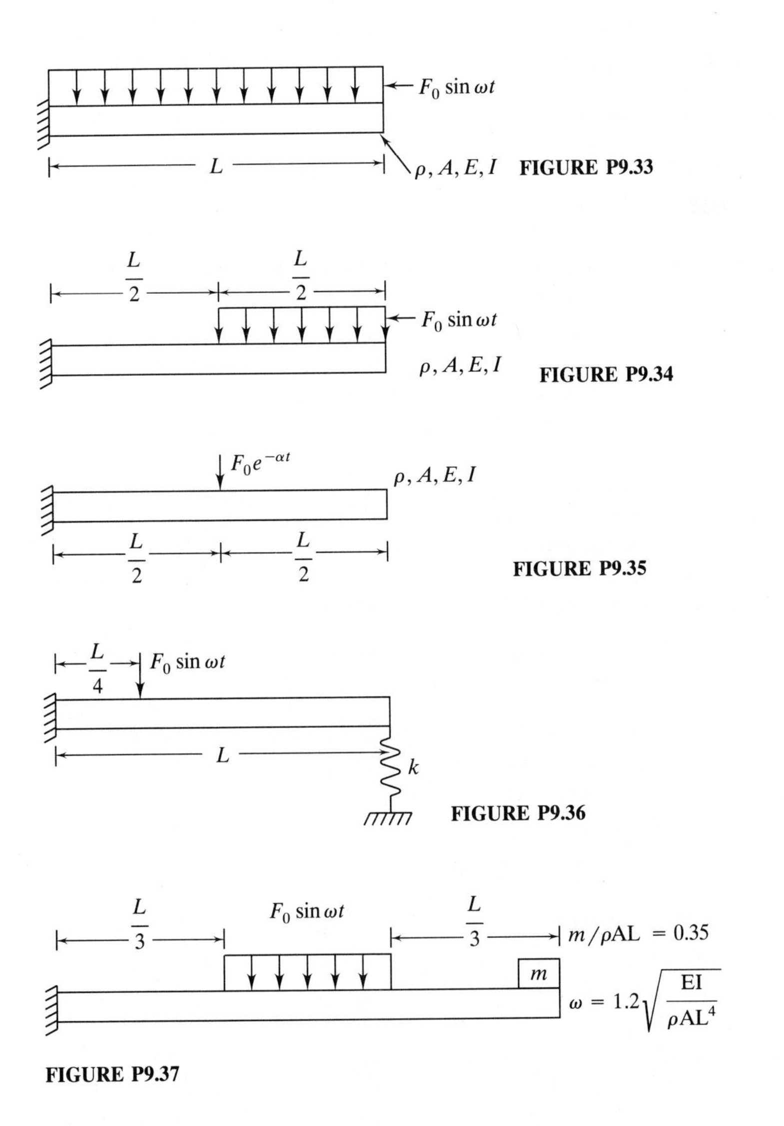

FIGURE P9.33

FIGURE P9.34

FIGURE P9.35

FIGURE P9.36

FIGURE P9.37

9.38. A root manipulator is 60 cm long, made of steel ($E = 210 \times 10^9$ N/m^2, $\rho = 7500$ kg/m^3), and has the cross section of Fig. P9.38. One end of the manipulator is fixed and a 1-kg mass is attached to its opposite end. Determine the three lowest natural frequencies for transverse vibration of the manipulator.

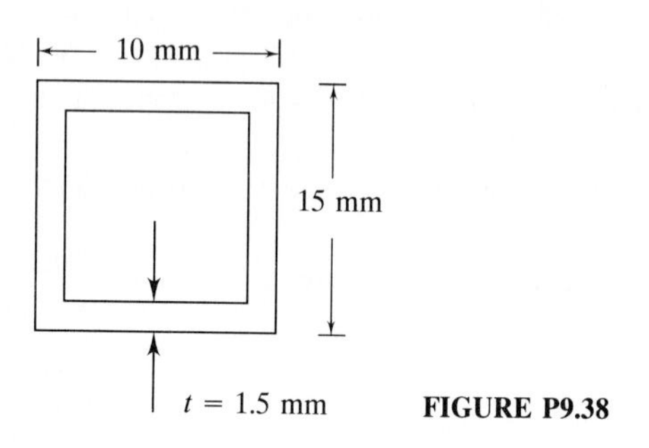

FIGURE P9.38

Problems 9.39 to 9.41 refer to the problem of vortex shedding from a street light fixture considered in Examples 4.4 and 8.2.

9.39. Model the light structure as a cantilever beam with an attached mass. Determine the five lowest natural frequencies of the structure.

9.40. Assume the vortex shedding produces a uniform force per unit length of $F_0 \sin \omega t$ applied over the upper three-fourths of the structure. For a wind speed of 60 mph, calculate the maximum displacement of the light fixture.

9.41. For a wind speed of 20 mph, calculate the maximum displacement of the light fixture.

9.42. A portion of a natural gas pipeline is submerged in the ocean as shown in Fig. P9.42. The pipeline is pinned at two vertical locations and subject to an oscillating flow due to ocean waves. The pipeline is modeled as a simply supported beam with a force per unit length due to an added mass effect and drag. For small oscillations the forced per unit length is assumed to be uniform and approximated as

$$F(t) = 4\rho r L \omega U_m \cos \omega t + \rho r\left(\overline{U}^2 + \tfrac{1}{2}U_m^2 + 2\overline{U}U_m \sin \omega t - \tfrac{1}{2}U_m^2 \cos 2\omega t\right)$$

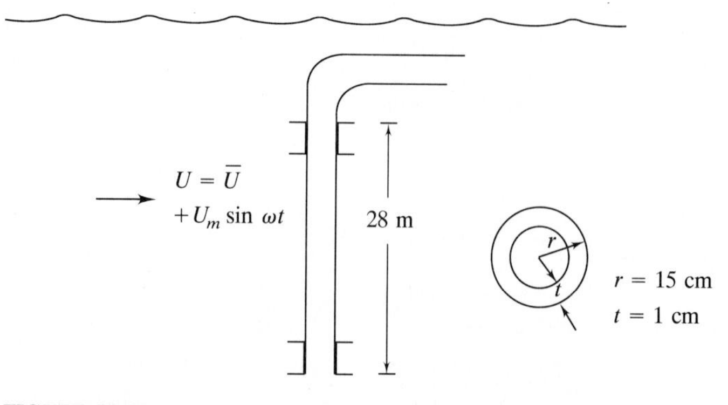

FIGURE P9.42

where ρ is the mass density of the liquid, ω is the frequency of the oscillating flow, $\overline{U}$ is the mean value of the flow velocity, and U_m is the amplitude of the flow oscillations. For the pipeline in ocean water exposed to a tide with a frequency of 0.5 Hz with $\overline{U} = 1.5$ m/s and $U_m = 2.0$ m/s, calculate the maximum deflection of the center of the pipeline.

9.43. The steam pipe of Fig. P9.43 is suspended from the ceiling in an industrial plant. A heavy machine with a rotating unbalance is placed on the floor above the machine causing vibrations of the ceiling. If the frequency of the oscillations is 150 Hz and the amplitude of displacement of the pipe's left support is 0.5 mm and the amplitude of displacement of the pipe's right support is 0.8 mm, determine the maximum deflection of the center of the pipe.

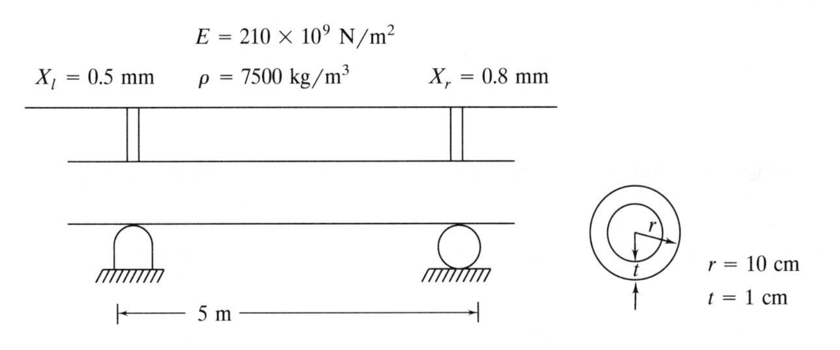

FIGURE P9.43

9.44. A simplified model of the rocket of Fig. P9.44 is a free-free beam.
(*a*) Calculate the five lowest natural frequencies for longitudinal vibration.
(*b*) Calculate the five lowest natural frequencies for transverse vibration.

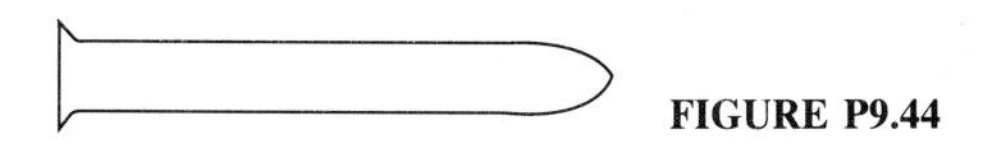

FIGURE P9.44

9.45. Longitudinal vibrations are initiated in the rocket of Fig. P9.44 when thrust is developed. Determine the Laplace transform of the transient response $\overline{u}(x, s)$ when the thrust of Figure P9.45 is developed. Do not invert the transform.

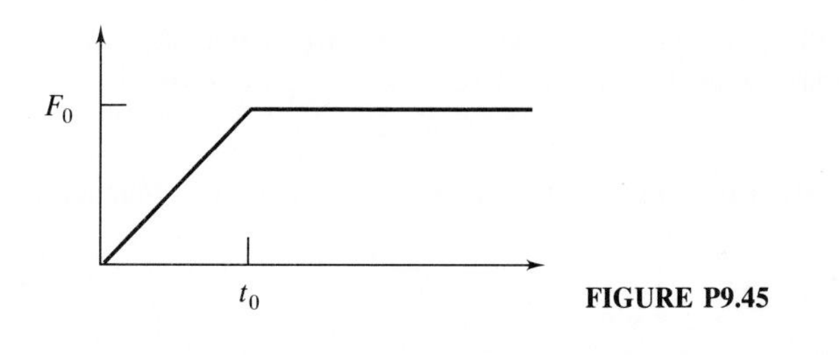

FIGURE P9.45

9.46. Determine the response of a cantilever beam when the fixed support is subject to a displacement $w(t) = A \sin \omega t$. Use the Laplace transform method and determine $\bar{y}(x, s)$. Do not invert.

9.47. The tail rotor blades of a helicopter have a rotating unbalance of magnitude 0.5 kg · m and operate at a speed of 1200 rpm. Modeling the tail section as a cantilever beam of length 3.5 m with $EI = 3.1 \times 10^6$ N · m^2, determine the steady-state response of the tail section.

9.48. Repeat Prob. 9.47 when the helicopter has a constant vertical acceleration of 12.1 m/s^2. Use the Laplace transform method to determine $\bar{y}(x, s)$. Do not invert.

9.49. Show that the differential equation governing free vibration of a uniform beam subject to a constant axial load, P, is

$$EI\frac{\partial^4 w}{\partial x^4} - P\frac{\partial^2 w}{\partial x^2} + \rho A\frac{\partial w}{\partial t^2} = 0$$

9.50. Nondimensionalize the differential equation of Prob. 9.49 by introduction of Eq. (9.65).

9.51. Determine the frequency equation for a simply supported beam subject to an axial load.

9.52. Determine the frequency equation for a fixed-free beam subject to an axial load.

9.53. A fixed-fixed beam is made of a material with a coefficient of thermal expansion α. After installed, the temperature is decreased by ΔT. Determine the beam's frequency equation.

9.54. Show orthogonality of the mode shapes for a simply supported beam subject to an axial load.

9.55. Discuss orthogonality of mode shapes for a fixed-free beam subject to an axial load.

9.56. Use Rayleigh's quotient to approximate the lowest natural frequency of a torsional shaft fixed at both ends.

9.57. Use Rayleigh's quotient to approximate the lowest natural frequency of a torsional shaft with a disk of mass moment of inertia I placed at its midspan. The shaft is fixed at both ends.

9.58. Use Rayleigh's quotient to approximate the lowest natural frequency of a fixed-free beam.

9.59. Use Rayleigh's quotient to approximate the lowest natural frequency of a simply supported beam with a mass m at its midspan. Use $w(x) = \sin \pi x/L$ as the trial function.

9.60. Use the Rayleigh-Ritz method to approximate the two lowest natural frequencies of a fixed-free beam.

9.61. Use the Rayleigh-Ritz method to approximate the two lowest natural frequencies of the system of Fig. P9.61.

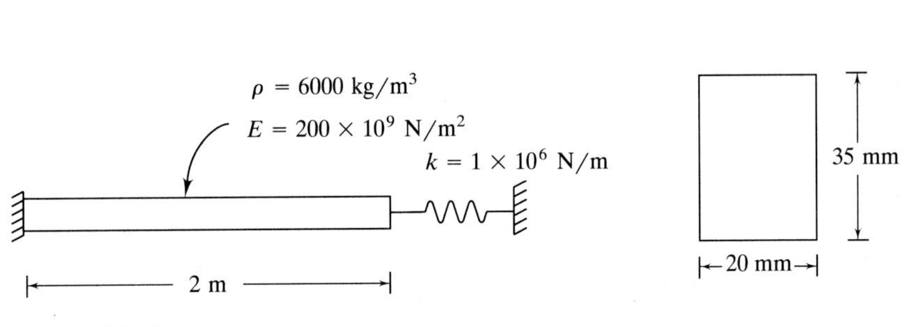

FIGURE P9.61

9.62. Use the Rayleigh-Ritz method to approximate the two lowest natural frequencies for the system of Fig. P9.62.

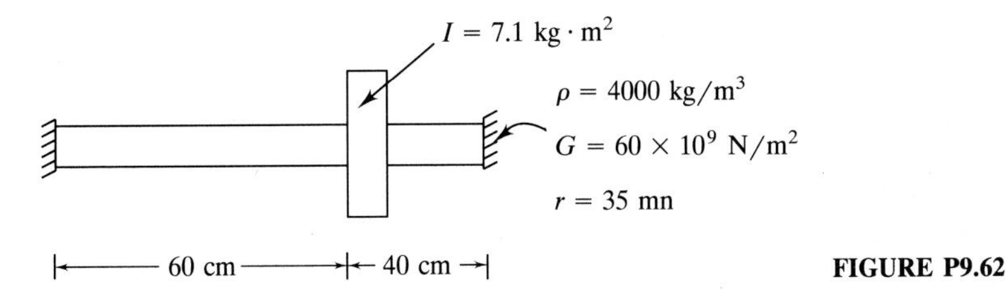

FIGURE P9.62

REFERENCES

1. Dimarogonas, A. D., and S. Haddad: *Vibrations for Engineers*, Prentice-Hall, Englewood Cliffs, N.J., 1992.
2. Dumir, P. C.: "Similarities of Vibration of Discrete and Continuous Systems," *International Journal of Mechanical Engineering Education*, vol. 16, pp. 71–78, 1988.
3. James, M. L., G. M. Smith, J. C. Wolford, and P. W. Whaley: *Vibration of Mechanical and Structural Systems with Microcomputer Applications*, Harper and Row, New York, 1989.
4. Kreider, D. L., R. O. Kuller, D. R. Ostberg, and F. W. Perkins: *An Introduction to Linear Analysis*, 2nd ed., Addison-Wesley, Reading, Mass., 1966.
5. Meirovitch, L.: *Analytical Methods in Vibrations*, Macmillan, New York, 1967.
6. Meirovitch, L.: *Elements of Vibrations Analysis*, McGraw-Hill, New York, 1975.
7. Rao, S. S.: *Mechanical Vibrations*, 2nd ed., Addison-Wesley, Reading, Mass., 1990.
8. Shabana, A. A.: *Theory of Vibrations: Discrete and Continuous Systems*, vol. 2, Springer-Verlag, New York, 1991.
9. Thomson, W. T.: *Theory of Vibrations with Applications*, 3rd ed., Prentice-Hall, Englewood Cliffs, N.J., 1988.
10. Weaver, W., S. P. Timoshenko, and D. H. Young: *Vibration Problems in Engineering*, 5th ed., Wiley-Interscience, New York, 1990.

CHAPTER 10

NONLINEAR VIBRATIONS

10.1 INTRODUCTION

All physical systems are inherently nonlinear. Often assumptions and approximations are made such that the mathematical problem governing the behavior of the system is linear. This is done for an obvious reason; the solution of a linear problem is much easier than the solution of a nonlinear problem. Often, the results obtained using the linear approximation are sufficient for engineering work. Except for the discussions of free and forced oscillations when Coulomb damping is present, this text has thus far considered only linear systems.

Nonlinear systems are rich in phenomena not present in linear systems. These phenomena can be important even when the nonlinearities are small. Nonlinear behavior is not completely understood and is the focus of much current research. Examples of recent research in problems of nonlinear vibrations are given in the following discussion.

Tung and Shaw (1988) used a nonlinear system to model the impact hammer of a dot matrix printer. Plaut and Hsieh (1987) and Rokini and Berger (1988) analyzed the nonlinear vibrations of a two-link mechanism connected to a nonlinear spring. Lazer and McKenna (1990) and McKenna and Walter (1987) included nonlinear effects in the modeling of suspension bridges and hypothesize nonlinear contributions toward the famous Tacoma Narrows Bridge collapse.

Nonlinear systems are much more difficult to analyze than linear systems because the principle of linear superposition is not valid for nonlinear systems.

Among the ramifications of the absence of a superposition principle are

1. The homogeneous solution of a second-order nonlinear differential equation is not a linear combination of two linearly independent solutions.
2. The general solution of a nonlinear differential equation cannot be written as the sum of a homogeneous solution and a particular solution, which is independent of initial conditions. The forced response of a nonlinear system cannot be separated from its free-vibration response.
3. The method of superposition cannot be used to add the forced responses due to a combination of excitations. The nonlinearity causes the responses to interact.
4. Since the convolution integral is derived using linear superposition, it does not apply to nonlinear systems. There is no equivalent of the convolution integral for nonlinear systems.
5. The Laplace transform is not successful in solving nonlinear differential equations.

The focus of this chapter is on the qualitative analysis of nonlinear systems. Quantitative results are presented to show how the nonlinearities act to produce nonlinear phenomena.

10.2 SOURCES OF NONLINEARITY

Let $x_1, x_2, \dots, x_n$ be the generalized coordinates for a conservative n-degree-of-freedom system. The kinetic energy of the system is a function of the generalized coordinates and their derivatives

$$T = T\left(x_1, x_2, \dots, x_n, \dot{x}_1, \dot{x}_2, \dots, \dot{x}_n\right) \tag{10.1}$$

The potential energy of the system is a function of the generalized coordinates

$$V = V(x_1, x_2, \dots, x_n) \tag{10.2}$$

If the system is linear, then its kinetic energy is independent of the generalized coordinates and a quadratic function of their derivatives. A conservative system is nonlinear if either the kinetic or potential energy cannot be written in a quadratic form, as in Eqs. (6.5) and (6.6).

The kinetic energy function contains terms other than quadratic terms when the inertia properties of the system are dependent on the generalized coordinates or due to kinematic relations between the generalized coordinates. Nonlinear terms due to the latter are called geometric nonlinearities.

Terms other than quadratic terms appear in the potential energy function due to geometric nonlinearities or nonlinear force-displacement relations in flexible elements. Nonlinear terms due to the latter are called material nonlinearities.

Example 10.1. Derive the governing differential equation for the simple pendulum of Fig. 10.1.

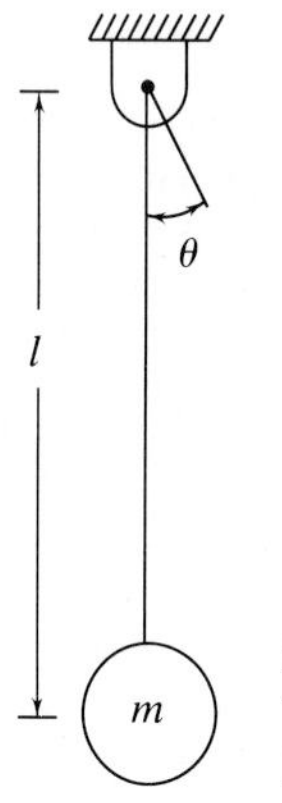

FIGURE 10.1
The differential equation governing oscillations of the simple pendulum of Example 10.1 is nonlinear.

The kinetic energy function for the pendulum is

$$T = \tfrac{1}{2}m(l\dot{\theta})^2$$

Using the plane of the support as the datum,

$$V = -mgl\cos\theta$$

The kinetic energy function is quadratic, but the potential energy function is not. The nonquadratic term in the potential energy function is due to the geometric relationship between the instantaneous position of the particle and the datum.

Lagrange's equation, Eq. (6.4), is applied

$$L = T - V$$

$$\frac{d}{dt}\left(\frac{\partial L}{\partial \dot{\theta}}\right) - \frac{\partial L}{\partial \theta} = 0$$

giving

$$\ddot{\theta} + \frac{g}{l}\sin\theta = 0.$$

The nonlinear term in the differential equation of Example 10.1 is a transcendental function of the dependent variable. Approximate solutions to such equations are made easier by replacing the transcendental function by its Taylor series expansion. For the equation of Example 10.1, this leads to

$$\ddot{\theta} + \frac{g}{l}\left(\theta - \frac{\theta^3}{6} + \frac{\theta^5}{120} - \cdots\right)$$

Approximations can be made by assuming θ is small. A linear approximation is obtained by ignoring all but the linear terms. The simplest nonlinear approximation is obtained by keeping only the largest nonlinear term. Since this term is proportional to the cube of the dependent variable, the nonlinearity is called a cubic nonlinearity.

Example 10.2. Derive the differential equations governing the motion of the system of Fig. 10.2.

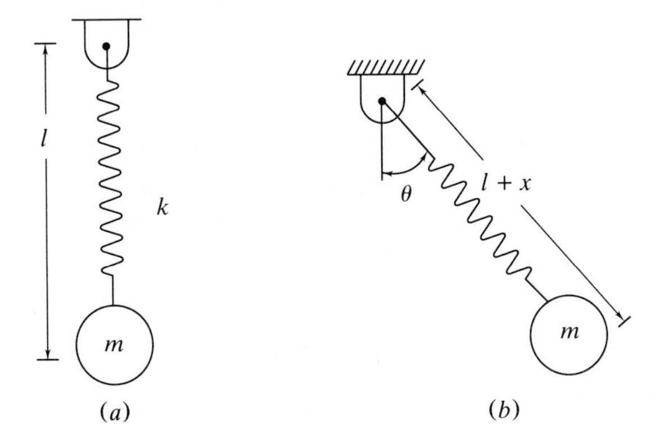

FIGURE 10.2
(*a*) The "swinging spring" in equilibrium; (*b*) the oscillations of the swinging spring are described by coupled nonlinear differential equations. The coupling occurs only in the nonlinear terms. A linear approximation predicts the extensional mode is uncoupled from the swinging mode.

Let x, the change in length of the spring from its length when the system is in equilibrium with a length l, and θ be the generalized coordinates. The system's kinetic energy function is

$$T = \tfrac{1}{2}m\left[\dot{x}^2 + (l + x)^2\dot{\theta}^2\right]$$

Assuming the spring is linear and using the plane of the support as the datum, the system's potential energy function is

$$V = \frac{1}{2}k\left(x + \frac{mg}{k}\right)^2 - mg(l + x)\cos\theta$$

Application of Lagrange's equations leads to

$$m\ddot{x} + kx - m(l + x)\dot{\theta}^2 + mg(1 - \cos\theta) = 0$$

and
$$m(l + x)^2\ddot{\theta} + m(l + x)g\sin\theta + 2m(l + x)\dot{x}\dot{\theta} = 0$$

If x and θ are assumed small, Taylor series expansions used for the transcendental functions, and only linear terms retained, the differential equations of Example 10.2 become

$$m\ddot{x} + kx = 0$$

$$\ddot{\theta} + \frac{g}{l}\theta = 0$$

Thus a linear approximation predicts two uncoupled modes, a spring mode with

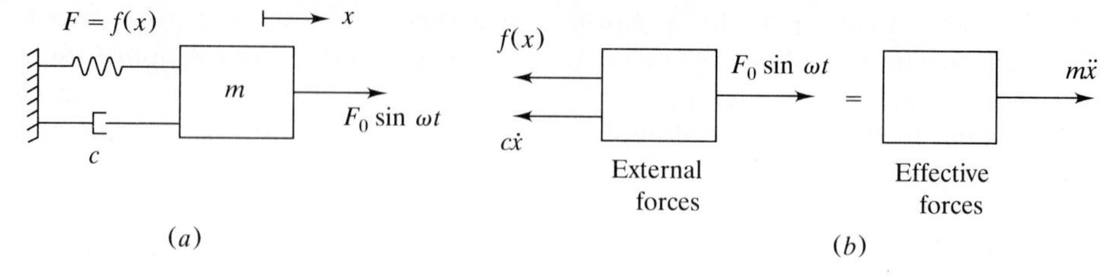

FIGURE 10.3
(*a*) Model system for one-degree-of-freedom system with a nonlinear elastic element, viscous damping and harmonic excitation; (*b*) free-body diagrams used to derive Eq. (10.4). Nonlinear terms are due to a material nonlinearity.

a natural frequency of $\sqrt{k/m}$ and a pendulum mode with a natural frequency of $\sqrt{g/l}$. Coupling occurs only in the nonlinear terms. If only the largest nonlinear terms are retained, the governing differential equations become

$$m\ddot{x} + kx - ml\dot{\theta}^2 + \frac{mg}{2}\theta^2 = 0$$

$$l\ddot{\theta} + g\theta + \frac{g}{l}\theta x + 2\dot{x}\dot{\theta} = 0$$

Since the largest nonlinear terms involve quadratic products of the generalized coordinates and their derivatives, the nonlinearities are termed quadratic.

Note that l is not the unstretched length of the spring, but its length when the system is in static equilibrium, $l = l_0 + mg/k$. Hence the effect of gravity causing a static spring force does not cancel with the static spring force in a nonlinear differential equation. Both must be included in the potential energy formulation.

A material nonlinearity occurs when a flexible component has a nonlinear constitutive equation. The system of Fig. 10.3 is used to model most one-degree-of-freedom systems with viscous damping and harmonic excitation. If the spring has a force-displacement relation of the form

$$F = f(x) \tag{10.3}$$

where f is a nonlinear function of x, then the governing differential equation is nonlinear,

$$m\ddot{x} + c\dot{x} + f(x) = F_0 \sin \omega t \tag{10.4}$$

If the spring is unstretched when it is unloaded, then a Taylor series expansion is used to expand $f(x)$ about $x = 0$. If the spring has the same properties in compression as in tension, only odd powers of x appear in the expansion,

$$m\ddot{x} + c\dot{x} + k_1 x + k_3 x^3 + \cdots = F_0 \sin \omega t \tag{10.5}$$

The coefficients in the Taylor series expansion should decrease as the power increases. Thus the expansion is usually truncated after the cubic term, leading to

$$\ddot{x} + 2\zeta\omega_n\dot{x} + \omega_n^2 x + \alpha\omega_n^2 x^3 = \frac{F_0}{m}\sin\omega t \tag{10.6}$$

where ω_n is the natural frequency of the corresponding linear system, ζ is the damping ratio for the linear system, and

$$\alpha = \frac{k_3}{k_1} \tag{10.7}$$

A spring for which α is positive is called a hardening spring. A spring for which α is negative is called a softening spring.

Equation (10.6) is called Duffing's equation. Duffing's equation is nondimensionalized by introducing

$$x^* = \frac{x}{\Delta} \qquad t^* = \omega_n t \tag{10.8}$$

where

$$\Delta = \frac{mg}{k_1}$$

is the static deflection of a linear spring of stiffness k_1. Substituting Eq. (10.8) into Eq. (10.6), rearranging, and dropping the * from the nondimensional variables leads to

$$\ddot{x} + 2\zeta\dot{x} + x + \epsilon x^3 = \Lambda\sin rt \tag{10.9}$$

where

$$r = \frac{\omega}{\omega_n}$$

$$\Lambda = \frac{F_0}{m\omega_n^2\Delta}$$

and

$$\epsilon = \alpha\Delta^2$$

It is shown in Chap. 3 that the presence of some forms of damping causes nonlinear terms in the differential equation. If the damping force is a function of the velocity,

$$F_d = g(\dot{x})$$

then for Coulomb damping

$$g(\dot{x}) = \mu mg\frac{\dot{x}}{|\dot{x}|}$$

and for aerodynamic drag

$$g(\dot{x}) = c\dot{x}^2$$

The general form of the differential equation for a system subject to a harmonic excitation with nonlinear damping and a nonlinear flexible element is

$$m\ddot{x} + g(\dot{x}) + f(x) = F_0 \sin \omega t \tag{10.10}$$

Nonlinear terms can arise in differential equations due to an external excitation, as in the following example.

Example 10.3. The U-tube manometer of Fig. 10.4 rotates about an axis other than its centroidal axis with an angular velocity $\omega(t)$. The liquid is incompressible with a mass density ρ, the column has a total length l, and the tube has a cross-sectional area A. If the rotational speed is greater than a critical speed, then all of the fluid is drained from the left leg. Assume the column of liquid moves in the manometer as a rigid body and let $h(t)$ represent the instantaneous height of the column in the right leg. The potential energy function for this system is

$$V = \tfrac{1}{2}\rho g A h^2$$

The system's kinetic energy function is

$$T = \tfrac{1}{2}\rho A l \dot{h}^2 + \tfrac{1}{2}\rho A b^2 h \omega^2 + \int_0^b \rho A r^2 \omega^2 \, dr + \int_0^{l-b-h} \rho A r^2 \omega^2 \, dr$$

Neglecting viscous friction, Lagrange's equation is applied to derive

$$l\ddot{h}_2 + g h_2 + \frac{\omega^2}{2}(l - b - h_2)^2 = \frac{\omega^2 b^2}{2}$$

The preceding equation has a quadratic nonlinearity which is the result of the externally imposed rotation. If the speed of rotation is time dependent, the differential equation has variable coefficients and the system is said to parametrically excited.

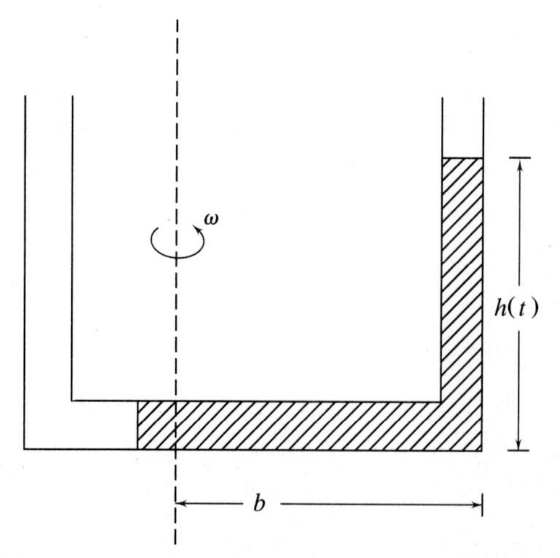

FIGURE 10.4
The oscillations of a column of liquid in a U-tube manometer rotating about a noncentroidal axis, when the angular velocity is large enough to drain fluid from the left leg, are governed by a nonlinear differential equation.

10.3 QUALITATIVE ANALYSIS OF NONLINEAR SYSTEMS

Qualitative analysis of nonlinear systems is of importance since exact analytical solutions are often not available. Qualitative analysis is used to predict general features of the motion including stability and long-time behavior.

The most useful tool for qualitative analysis of a nonlinear system is the state plane, a graphical time history of the relationship between two variables. The state plane for a one-degree-of-freedom system is a family of curves showing the history of the relation between velocity and displacement. The curves in the state plane are called trajectories. Attractors are points or curves to which the trajectories eventually approach.

Example 10.4. Draw the state plane for the unforced Duffing's equation with no damping for a hardening spring.

Let $v = x$. Then

$$\ddot{x} = \frac{dv}{dt} = \frac{dv}{dx}\frac{dx}{dt} = v\frac{dv}{dx}$$

Duffing's equation, Eq. (10.6), becomes

$$v\frac{dv}{dx} = -x - \epsilon x^3$$

Integrating both sides with respect to x gives

$$\tfrac{1}{2}v^2 = C - \tfrac{1}{2}x^2 - \tfrac{1}{4}\epsilon x^4$$

where C is the constant of integration, dependent on initial conditions. The state plane for $\epsilon = 1/2$ is shown in Fig. 10.5. Different trajectories correspond to different values of C.

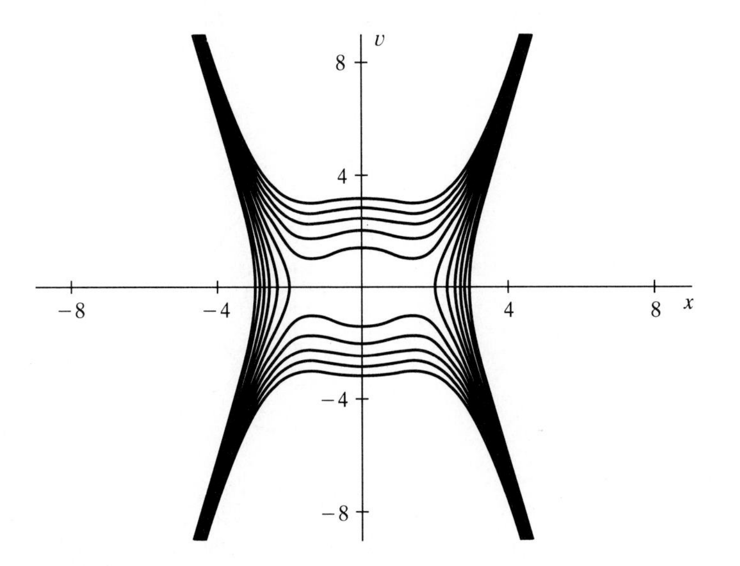

FIGURE 10.5
State plane for unforced undamped Duffing's equation.

The system of Fig. 10.3 is in equilibrium when its velocity is zero and the sum of the spring force and damping force is zero. For a linear system, this occurs only when $v = 0$ and $x = 0$. A nonlinear system may have more than one equilibrium point. An equilibrium point is stable if trajectories approach the equilibrium point for large time. An equilibrium point is unstable if trajectories diverge from the equilibrium point for large time.

The equilibrium points for a system governed by Eq. (10.10) are $v = 0$ and the values of x such that $f = 0$. The stability of an equilibrium point is determined by analyzing the trajectories in the vicinity of the equilibrium point. Let

$$x = x_0 + \Delta x \tag{10.11}$$

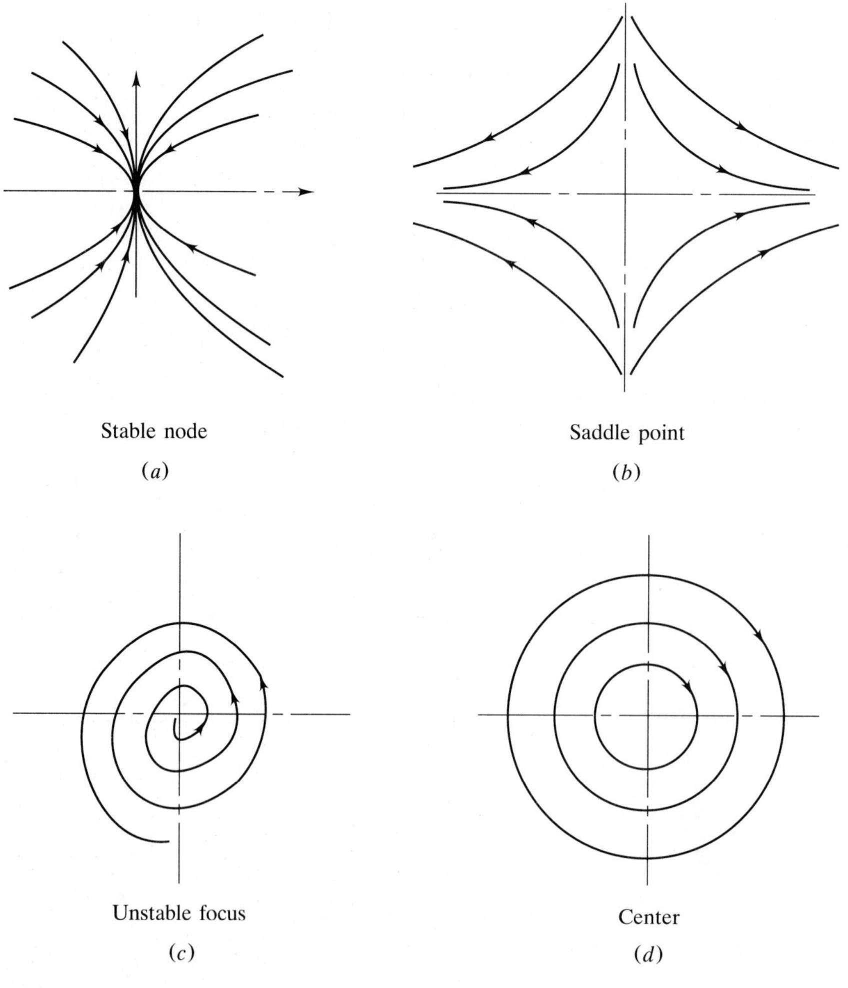

FIGURE 10.6
State planes in the vicinity of equilibrium points.

be a point in the phase plane in the vicinity of the equilibrium point, x_0. Substituting Eq. (10.11) into Eq. (10.10) with $F_0 = 0$ leads to

$$\Delta \ddot{x} + g(\Delta \dot{x}) + f(x_0 + \Delta x) = 0$$

Expanding f and g about $x = x_0$ and $\dot{x} = 0$, respectively, and keeping only linear terms gives

$$\Delta \ddot{x} + \frac{dg(0)}{d\dot{x}} \Delta \dot{x} + \frac{df(x_0)}{dx} \Delta x = 0 \tag{10.12}$$

The general solution of Eq. (10.12) is

$$\Delta x = Ae^{\beta_1 t} + Be^{\beta_2 t} \tag{10.13}$$

If either β_1 or β_2 have a positive real part, then the equilibrium point is unstable.

If β_1 and β_2 are real and have the same sign, the equilibrium point is called a node. If β_1 and β_2 are real and have different signs, the equilibrium point is called a saddle point, and is, by definition, unstable. If β_1 and β_2 are complex conjugates, the equilibrium point is called a focus. A special case of a focus occurs when β_1 and β_2 are purely imaginary, in which case the equilibrium point is called a center. Sketches of state planes in the vicinity of a node, saddle point, focus, and center are given in Fig. 10.6.

Example 10.5. Determine the equilibrium points and their nature for the damped unforced Duffing's equation.

The equilibrium points are the values of x such that

$$x + \epsilon x^3 = 0$$

For a hardening spring, the only equilibrium point for Duffing's equation is $x = 0$. For a softening spring, the system has the additional equilibrium points

$$x_0 = \pm \sqrt{\frac{1}{-\epsilon}}$$

The nature of the equilibrium point corresponding to $x_0 = 0$ is investigated by assuming $x = \Delta x$, which leads to

$$\beta_{1,2} = -\zeta \pm \sqrt{\zeta^2 - 1}$$

Hence the equilibrium point $x = 0$ is a stable node if $\zeta \geq 1$, and is a stable focus if $\zeta < 1$.

For a softening spring, the natures of the additional equilibrium points are determined using

$$x = \pm \sqrt{-\frac{1}{\epsilon}} + \Delta x$$

Substituting into Duffing's equation and linearizing leads to

$$\Delta \ddot{x} + 2\zeta \Delta \dot{x} - 2 \Delta x = 0$$

and

$$\beta = -\zeta \pm \sqrt{\zeta^2 + 2}$$

Since the two values of β are real with opposite signs, these equilibrium points are saddle points and thus, by their very nature, unstable.

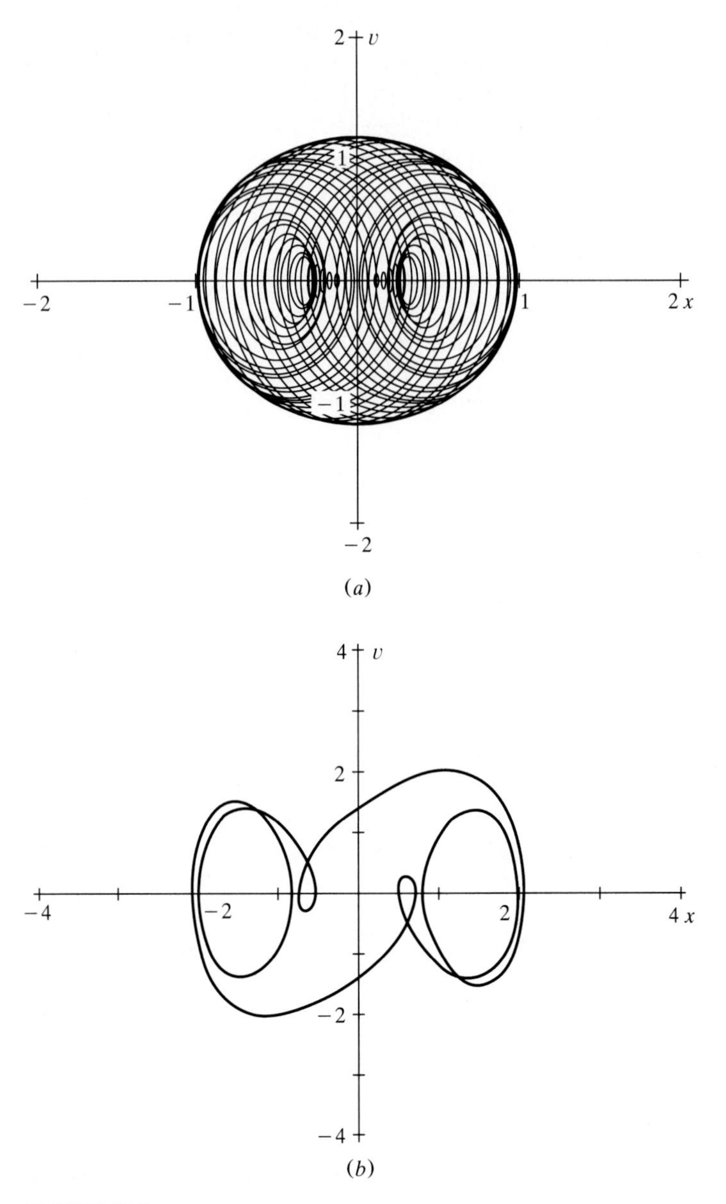

FIGURE 10.7
Examples of state planes for (*a*) forced, undamped Duffing's equation; (*b*) forced, damped Duffing's equation.

The phase plane for a system subject to a forced excitation is usually difficult to determine solely using analytical methods. Often, these phase planes must be drawn using graphical methods or numerical results. Figure 10.7 shows several phase planes corresponding to the forced Duffing's equation.

10.4 QUANTITATIVE METHODS OF ANALYSIS

Exact solutions to nonlinear vibration problems exist only for a few special free-vibration problems. Exact solutions for nonlinear forced-vibration problems are almost nonexistent. Consider, Eq. (10.10) with $F_0 = 0$. Let $v = \dot{x}$. Then, using the chain rule for differentiation, as in Example 10.4, Eq. (10.10) can be written as

$$v\frac{dv}{dx} + g(v) + f(x) = 0 \tag{10.14}$$

For certain forms of $g(v)$ and $f(x)$, Eq. (10.14) can be integrated yielding $v(x)$, which, in turn, can be integrated yielding $t(x)$.

Consider an undamped system, $g(v) = 0$. Integrating Eq. (10.14) with respect to x and using $x = x_0$ and $v = 0$ when $t = 0$ yields

$$v(x) = \left[2\int_{x}^{x_0} f(\eta)\,d\eta\right]^{1/2} \tag{10.15}$$

Rearranging and integrating with respect to x gives

$$t = \int_{x_0}^{x} \frac{d\lambda}{\left[2\int_{\lambda}^{x_0} f(\eta)\,d\eta\right]^{1/2}} \tag{10.16}$$

Since Eq. (10.16) gives t as a function of x, it is not useful for computing the time history of motion, but can be used for frequency calculations. For many forms of $f(x)$, closed-form evaluation of the integral does not exist, and numerical integration is used. Care must be taken when evaluating Eq. (10.16) numerically because the integrand is singular for $\lambda = 0$.

Since exact solutions are not often available, numerical solutions are used. Self-starting methods such as Runge-Kutta are convenient for numerical solution of nonlinear equations.

The general form of the equations for a nonlinear n-degree-of-freedom system is

$$\begin{aligned} \ddot{x}_1 &= h_1(\mathbf{x}, \dot{\mathbf{x}}, t) \\ \ddot{x}_2 &= h_2(\mathbf{x}, \dot{\mathbf{x}}, t) \\ &\vdots \\ \ddot{x}_n &= h_n(\mathbf{x}, \dot{\mathbf{x}}, t) \end{aligned} \tag{10.17}$$

Let $\mathbf{v} = \dot{\mathbf{x}}$ and $\mathbf{x}$ be independent n-dimensional vectors. Equation (10.17) can be rewritten as two systems of first-order equations

$$\frac{dx_1}{dt} = v_1 \qquad \frac{dv_1}{dt} = h_1(\mathbf{x}, \mathbf{v}, t)$$

$$\frac{dx_2}{dt} = v_2 \qquad \frac{dv_2}{dt} = h_2(\mathbf{x}, \mathbf{v}, t)$$

$$\vdots \qquad\qquad \vdots$$

$$\frac{dx_n}{dt} = v_n \qquad \frac{dv_n}{dt} = h_n(\mathbf{x}, \mathbf{v}, t)$$

The program DUFF.BAS, included on the accompanying diskette, uses a fourth-order Runge-Kutta scheme to integrate Duffing's equation, eq. (10.9). The program provides the time history of the response as well as the state plane for any choices of system parameters.

Analytical solutions are preferable to numerical solutions because they can be used to predict trends, analyze the effects of parameters, and develop qualitative results. Thus approximate analytical methods are often used to approximate the solution of nonlinear problems.

If the magnitude of the nonlinear term is small or the amplitude of motion is small, then a perturbation method can be used to develop an approximate solution. Let ϵ be a small nondimensional parameter, $\epsilon \ll 1$. It may be a measure of the amplitude or a measure of the nonlinearity. For a one-degree-of-freedom system, the generalized coordinate is expanded in a series of powers of ϵ,

$$x(t) = x_0(t) + \epsilon x_1(t) + \epsilon^2 x_2(t) + \cdots \tag{10.18}$$

Equation (10.18) is substituted into the governing differential equation. Coefficients of like powers of ϵ are collected and set to zero independently. The result is a set of linear differential equations that are successively solved for $x_i(t)$, $i = 1, 2, \ldots$.

The series of Eq. (10.18) is convergent. However, it converges slowly and thus a finite number of terms are inadequate to represent the solution for all t. When only a few terms are included, nonperiodic terms appear which cause the solution to be unbounded for large t. The terms which produce these nonuniformities are called secular terms. Since it is impossible to include an infinite number of terms in the evaluation, the secular terms must be removed. A variety of perturbation methods have been developed to remove secular terms. These include the method of strained parameters, the method of renormalization, the method of multiple scales, and the method of averaging. The application of these methods to nonlinear oscillation problems is beyond the scope of this book, but an exhaustive treatment is found in Nayfeh and Mook. The method of renormalization is illustrated in Sec. 10.5.

10.5 FREE VIBRATIONS OF ONE-DEGREE-OF-FREEDOM SYSTEMS

The free vibrations of a conservative system are periodic. If the spring in the system of Fig. 10.3 has the same properties in compression as in tension, then each period of motion can be broken into four parts, each of which takes the same amount of time. If the mass is displaced a distance x_0 from equilibrium and released from rest, the period of the resulting motion can be calculated using Eq. (10.16) as four times the time it takes the mass to go from its initial position to $x = 0$,

$$T = \frac{4}{\sqrt{2}} \int_{x_0}^{0} \frac{d\lambda}{\left[\int_{\lambda}^{x_0} f(\eta)\, d\eta\right]^{1/2}} \tag{10.19}$$

Equation (10.19) shows that, in contrast to a linear system, the period and the corresponding natural frequency for a nonlinear system depend upon the initial conditions.

Example 10.6. A mass, attached to a softening spring with a cubic nonlinearity, is displaced a nondimensional distance x_0 from equilibrium and released from rest. Determine the period of the resulting oscillations as a function of ϵ and x_0.

Using the notation of Sec. 10.2 and Eqs. (10.5) through (10.9), the nondimensional force developed in a softening spring is

$$f(x) = x - \epsilon x^3 \qquad \epsilon = \alpha \Delta^2$$

Thus the nondimensional period is determined using Eq. (10.19),

$$T = \frac{4}{\sqrt{2}} \int_{x_0}^{0} \frac{d\lambda}{\left[\int_{\lambda}^{x_0} (\eta - \epsilon\eta^3)\, d\eta\right]^{1/2}}$$

where x_0 is the nondimensional initial displacement. The dimensional period is the nondimensional period divided by the linear natural frequency. Proceeding with the algebra,

$$T = \frac{4}{\sqrt{2}} \int_{x_0}^{0} \frac{d\lambda}{\left[\dfrac{x_0^2}{2} - \epsilon\dfrac{x_0^4}{4} - \dfrac{\lambda^2}{2} + \epsilon\dfrac{\lambda^4}{4}\right]^{1/2}}$$

$$= \frac{4\sqrt{2}}{x_0\sqrt{\epsilon}} \int_{0}^{1} \frac{d\phi}{\sqrt{\dfrac{2}{\epsilon x_0^2} - 1 - \dfrac{2}{\epsilon x_0^2}\phi^2 + \phi^4}}$$

$$= \frac{4\sqrt{2}}{\sqrt{2 - \epsilon x_0^2}} \int_{0}^{1} \frac{d\phi}{(1 - \phi^2)(1 - k^2\phi^2)}$$

$$= \frac{4\sqrt{2}}{\sqrt{2 - \epsilon x_0^2}} F(k, \pi/2)$$

where $F(k, \pi/2)$ is the complete elliptic integral of the first kind of argument k, where

$$k = \sqrt{\frac{2 - \epsilon x_0^2}{\epsilon x_0^2}}$$

A table of elliptic integrals, such as in Abramowitz and Stegun, is used to generate Fig. 10.8.

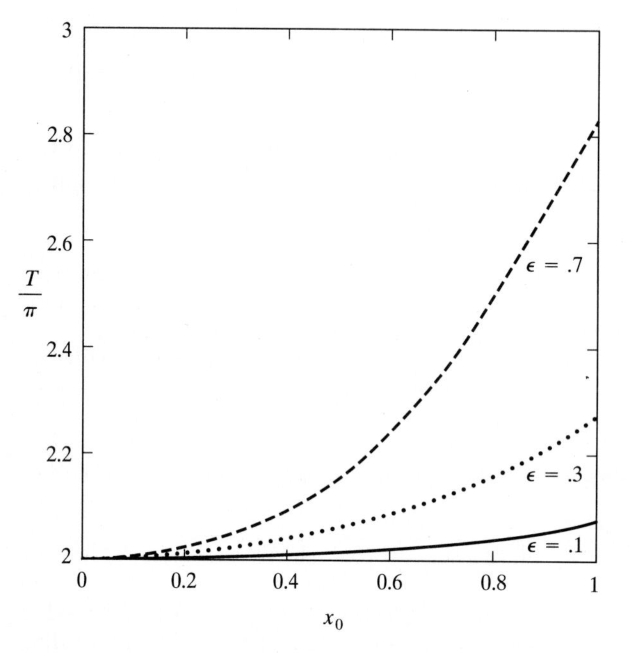

FIGURE 10.8
Period of Duffing's equation as a function of ϵ generated from exact solution.

When the integral of Eq. (10.19) cannot be evaluated in closed form, numerical integration must be used. However, the integrand is singular at $\lambda = x_0$. Let δ be a small nondimensional value. Then for the system of Ex. 10.6,

$$T = \frac{4\sqrt{2}}{x_0\sqrt{\epsilon}}\left[\int_0^{1-\delta} \frac{d\phi}{\sqrt{\dfrac{2}{\epsilon x_0^2} - 1 - \dfrac{2}{\epsilon x_0^2}\phi^2 + \phi^4}} + \int_{1-\delta}^{1} \frac{d\phi}{\sqrt{\dfrac{2}{\epsilon x_0^2} - 1 - \dfrac{2}{\epsilon x_0^2}\phi^2 + \phi^4}}\right]$$

The first integral is evaluated using numerical integration. The integrand of the second integral is expanded using the binomial theorem, and the resulting expansion is integrated term by term. The expansion is truncated such that desired accuracy is achieved.

Perturbation methods can be applied to approximate the period of a nonlinear system. When the straightforward expansion, Eq. (10.18), is substituted into the unforced, undamped Duffing's equation, the following results:

$$\ddot{x}_0 + x_0 + \epsilon\left(\ddot{x}_1 + x_1 + x_0^3\right) + \epsilon^2\left(\ddot{x}_2 + x_2 + 3x_0^2 x_1\right) + \cdots = 0 \quad (10.20)$$

Coefficients of powers of ϵ are set to zero independently, leading to a set of hierarchical equations

$$\begin{aligned} \ddot{x}_0 + x_0 &= 0 \\ \ddot{x}_1 + x_1 &= -x_0^3 \\ \ddot{x}_2 + x_2 &= -3x_0^2 x_1 \\ &\vdots \end{aligned} \quad (10.21)$$

The solution for x_0 is

$$x_0 = A\sin(t + \phi) \quad (10.22)$$

where A and ϕ are determined using initial conditions. Substitution of Eq. (10.22) into the second of Eqs. (10.21) and use of trigonometric identities lead to

$$\ddot{x}_1 + x_1 = -\frac{A^3}{4}[3\sin(t + \phi) - \sin 3(t + \phi)] \quad (10.23)$$

The particular solution of Eq. (10.23) is

$$x_1(t) = \frac{A^3}{8}t\cos(t + \phi) - \frac{A^3}{32}\sin 3(t + \phi) \quad (10.24)$$

and the resulting two-term approximation for $x(t)$ is

$$x(t) = A\sin(t + \phi) + \epsilon\left[\frac{3}{8}A^3 t\cos(t + \phi) - \frac{A^3}{32}\sin 3(t + \phi)\right] + \cdots \quad (10.25)$$

Unfortunately, the expansion of Eq. (10.25) is not periodic and grows without bound. Indeed, when t is as large as $1/\epsilon$, the second term in the expansion is as large as the first term, rendering it invalid.

The problem with the straightforward expansion is that it cannot account for the variation of the period with initial conditions, as mandated by the exact solution. The method of renormalization is used to take this variation into account and render the two-term straightforward expansion uniform. A new time scale is introduced according to

$$t = w(1 + \epsilon\lambda_1 + \epsilon^2\lambda_2 + \cdots) \tag{10.26}$$

Equation (10.25) is rewritten using w as the independent variable

$$x = A\sin(w + \epsilon\lambda_1 w + \cdots + \phi)$$
$$+ \epsilon\left[\frac{3}{8}A^3(w + \epsilon\lambda_1 w + \cdots)\cos(w + \epsilon\lambda_1 w + \cdots + \phi)\right.$$
$$\left. - \frac{A^3}{32}\sin 3(w + \epsilon\lambda_1 w + \cdots + \phi)\right] + \cdots \tag{10.27}$$

Taylor series expansions are used to expand the trigonometric functions and coefficients of powers of ϵ are recollected, leading to

$$x = A\sin(w + \phi) + \epsilon\left[A\lambda_1 w\cos(w + \phi)\right.$$
$$\left. + \frac{3}{8}A^3 w\cos(w + \phi) - \frac{A^3}{32}\sin 3(w + \phi)\right] + \cdots \tag{10.28}$$

The secular term is removed from Eq. (10.28) by choosing

$$\lambda_1 = -\tfrac{3}{8}A^2 \tag{10.29}$$

leading to

$$x = A\sin(w + \phi) - \epsilon\frac{A^3}{32}\sin 3(w + \phi) + \cdots \tag{10.30}$$

where $$t = w(1 - \epsilon\tfrac{3}{8}A^2 + \cdots) \tag{10.31}$$

The binomial expansion is used to invert Eq. (10.31)

$$w = t(1 + \epsilon\tfrac{3}{8}A^2 + \cdots) \tag{10.32}$$

Thus the nondimensional frequency for a system with a hardening cubic spring is increased by $3/8\epsilon A^2$, while it is decreased by the same amount for a softening spring. The amplitude is determined by application of the initial conditions. If $x(0) = \delta$ and $\dot{x}(0) = 0$, then

$$\phi = \frac{\pi}{2}$$

$$\delta = A - \epsilon\frac{A^3}{32}$$

A natural frequency approximation can be obtained to greater accuracy by calculating higher-order terms in the expansion for x, and choosing the λ_i from Eq. (10.31) to eliminate secular terms.

For damped systems, the damping term is often small enough to be ordered with the nonlinearity. To this end, define

$$\zeta = 2\epsilon\mu \tag{10.33}$$

where μ is of order 1. When the straightforward expansion is used in the damped unforced version of Duffing's equation, the following equations result defining x_0 and x_1:

$$\begin{aligned} \ddot{x}_0 + x_0 &= 0 \\ \ddot{x}_1 + x_1 &= -x_0^3 - 2\mu\dot{x}_0 \end{aligned} \tag{10.34}$$

In order to use the method of renormalization for damped systems, the solutions of eqs. (10.34) are written using complex exponentials

$$x_0 = A\cos(t + \phi) = \tfrac{1}{2}A[e^{i(t+\phi)} + e^{-i(t+\phi)}] \tag{10.35}$$

When Eq. (10.31) is used to remove secular terms from the two-term expansion,

$$\lambda_1 = -\frac{3}{8}A^2 - i\frac{\mu}{2} \tag{10.36}$$

and the resulting two-term uniformly valid expansion is

$$x = Ae^{-\zeta t}\sin\left[\left(1 + \epsilon\tfrac{3}{8}A^2\right)t + \phi\right] \tag{10.37}$$

Thus, when secular terms are removed through x_1, damping has no effect on the natural period. The exponential decay, comparable to that of a linear system, is apparent.

In summary, the natural frequency of a nonlinear system is dependent upon its initial conditions. The straightforward perturbation expansion and the method of renormalization can be used to determine an approximation to the natural frequency when the nonlinearity is small or when the amplitude is small. Small viscous damping has a similar effect on free vibrations of nonlinear systems as on free vibrations of linear systems, causing an exponential decay of amplitude.

10.6 FORCED VIBRATIONS OF ONE-DEGREE-OF-FREEDOM SYSTEMS WITH CUBIC NONLINEARITIES

Consider the damped Duffing's equation subject to a two-frequency excitation,

$$\ddot{x} + 2\mu\epsilon\dot{x} + x + \epsilon x^3 = F_1\sin r_1 t + F_2\sin r_2 t \qquad r_1 \neq r_2 \tag{10.38}$$

Use of the straightforward expansion, Eq. (10.18), produces the following two-term approximation to the solution of Eq. (10.38):

$$
\begin{aligned}
x = {} & A \sin(t + \phi) + F_1 M_1 \sin r_1 t + F_2 M_2 \sin r_2 t \\
& + \epsilon \Bigg\{ -\mu A t \sin(t + \phi) - \left(\frac{3}{8} A^3 + \frac{3}{4} A F_1^2 M_1^2 + \frac{3}{4} A F_2^2 M_2^2 \right) t \cos(t + \phi) \\
& - \frac{2\mu F_1 M_1 r_1}{1 - r_1^2} \cos r_1 t - \frac{2\mu F_2 M_2 r_2}{1 - r_2^2} \cos r_2 t \\
& + \frac{3\left(2A^2 F_1 M_1 + F_1^3 M_1^3 + 2F_1 F_2^2 M_1 M_2^2\right)}{4\left(1 - r_1^2\right)} \sin r_1 t \\
& + \frac{3\left(2A^2 F_2 M_2 + F_2^3 M_2^3 + 2F_1^2 F_2 M_1^2 M_2\right)}{4\left(1 - r_2^2\right)} \sin r_2 t \\
& + \frac{A^3}{32} \sin 3(t + \phi) - \frac{3A^2 F_1 M_1}{4\left[1 - (2 + r_1)^2\right]} \sin[(2 + r_1)t + 2\phi] \\
& + \frac{3A^2 F_1 M_1}{4\left[1 - (2 - r_1)^2\right]} \sin[(2 - r_1)t + 2\phi] \\
& - \frac{3AF_1^2 M_1^2}{4\left[1 - (1 + 2r_1)^2\right]} \sin[(1 + 2r_1)t + \phi] \\
& + \frac{3AF_1^2 M_1^2}{4\left[1 - (1 - 2r_1)^2\right]} \sin[(1 - 2r_1)t + \phi] \\
& - \frac{3AF_2 M_2}{4\left[1 - (2 + r_2)^2\right]} \sin[(2 + r_2) + 2\phi] \\
& + \frac{3A^2 F_2 M_2}{4\left[1 - (2 - r_2)^2\right]} \sin[(2 - r_2) + 2\phi] \\
& - \frac{3AF_2^2 M_2^2}{4\left[1 - (1 + 2r_2)^2\right]} \sin[(1 + 2r_2)t + \phi] \\
& + \frac{3AF_2^2 M_2^2}{4\left[1 - (1 - 2r_2)^2\right]} \sin[(1 - 2r_2)t + \phi] \\
& - \frac{3F_1^2 F_2 M_1^2 M_2}{4\left[1 - (2r_1 + r_2)^2\right]} \sin(2r_1 + r_2)t
\end{aligned}
$$

$$
+\frac{3F_1^2F_2M_1^2M_2}{4\left[1-(2r_1-r_2)^2\right]}\sin(2r_1-r_2)t
$$

$$
-\frac{3F_1F_2^2M_1M_2^2}{4\left[1-(2r_2+r_1)^2\right]}\sin(2r_2+r_1)t
$$

$$
+\frac{3F_1F_2^2M_1M_2^2}{4\left[1-(2r_2-r_1)^2\right]}\sin(2r_2-r_1)t
$$

$$
-\frac{F_1^3M_1^3}{4(1-9r_1^2)}\sin 3r_1t-\frac{F_2^3M_2^3}{4(1-9r_2^3)}\sin 3r_2t
$$

$$
+\frac{3AF_1F_2M_1M_2}{2\left[1-(r_1-r_2+1)^2\right]}\sin[r_1-r_2+1)t+\phi]
$$

$$
-\frac{3AF_1F_2M_1M_2}{2\left[1-(r_1-r_2-1)^2\right]}\sin[(r_2-r_1-1)t-\phi]
$$

$$
-\frac{3AF-1F-2M_1M_2}{2\left[1-(r_1+r_2+1)^2\right]}\sin[(r_1+r_2+1)t+\phi]
$$

$$
\left.+\frac{3AF_1F_2M_1M_2}{2\left[1-(r_1+r_2-1)^2\right]}\sin[(r_1+r_2-1)t-\phi]\right\}\cdots \tag{10.39}
$$

where
$$
M_i=\frac{1}{1-r_i^2}
$$

The expansion of Eq. (10.39) is nonuniform due to the secular terms arising from the free-vibration solution. Additional nonuniformities occur when the values of r_1 and r_2 are such that the denominators of other terms are very small. Examination of Eq. (10.39) suggests that an exhaustive study of the frequency response of a one-degree-of-freedom system with a cubic nonlinearity requires the following cases be considered:

1. No resonances.
2. $r_1 = 1$ or $r_2 = 1$, primary resonance.
3. $r_1 = 1/3$ or $r_2 = 1/3$, superharmonic resonance.
4. $r_1 = 3$ or $r_2 = 3$, subharmonic resonance.
5. $2r_2 + r_1 = 1$, $2r_1 - r_2 = \pm 1$, $2r_2 - r_1 = \pm 1$, $r_1 - r_2 + 1 = -1$, $r_1 - r_2 - 1 = \pm 1$, or $r_1 + r_2 - 1 = 1$, combination resonances.
6. Conditions when two resonances occur simultaneously. For example, when $r_1 = 1/3$ and $r_2 = 5/3$, both superharmonic and combination resonances occur.

A resonance condition occurs when the free-vibration contribution to the solution does not decay with time. The steady-state solution has a contribution from the free vibrations as well as the forced steady-state response. For a linear system the free-vibration response is periodic with a frequency equal to the natural frequency and the forced response due to a harmonic excitation is periodic with a frequency equal to the excitation frequency. For a linear system only the primary resonance can occur, when the excitation frequency is near the natural frequency.

For a system with a cubic nonlinearity, Eq. (10.28) shows that the free-vibration response includes a periodic term whose frequency is three times the linear natural frequency. Thus oscillations at this frequency are sustained in the absence of an external excitation. Any additional energy input may lead to growth of the free oscillations and thus producing the subharmonic resonance.

The forced response of a system with a cubic nonlinearity to a harmonic excitation includes a periodic term whose frequency is three times the excitation frequency. Thus, when the excitation frequency is one-third of the natural frequency, this term tends to excite the free vibrations and causes the free-vibration term to be sustained, even in the presence of small damping. This produces the superharmonic resonance.

When a system with a cubic nonlinearity is subject to a multifrequency excitation, the forced response includes periodic terms at frequencies that are combinations of the excitation frequencies. When this combination of frequencies is close to the natural frequency, free oscillations are sustained and a combination resonance exists.

The straightforward expansion is nonuniform for all r_1 and r_2, even when no resonance conditions exist. The method of renormalization can be used to render the two-term expansion uniform, but it can only be used to predict periodic responses, and cannot provide information about the stability of equilibrium points. Possibly the best method for obtaining uniform expansions to approximate the solution of nonlinear forced-vibration problems is the method of multiple scales. The results provided in the following discussion can be obtained using the method of multiple scales. Since its application is beyond the scope of this text, the discussion focuses on qualitative behavior. More detail is available in Nayfeh and Mook.

1. *No resonances.* For most values of r_1 and r_2, no resonance conditions exist. However, the expansion of Eq. (10.39) is still nonuniform. When secular terms are removed, the solution is the sum of the free-vibration response and the forced response. The free vibrations decay exponentially, but the frequency of free vibrations depends upon the initial conditions and the amplitudes and frequencies of the excitation.

2. *Primary resonance.* A primary resonance occurs when an excitation frequency is near the system's linear natural frequency, corresponding to the nondimensional frequency being near 1. When the amplitude of the excitation is of order 1, the straightforward perturbation expansion predicts an infinite amplitude response, even in the presence of small damping. When the amplitude of the excitation is the same order as the nonlinearity and the damping, secular terms occur in x_1.

The frequency response in the vicinity of the primary resonance is studied by introducing a detuning parameter, defined by

$$r_1 = 1 + \epsilon\sigma \tag{10.40}$$

The amplitude and phase of the resulting motion vary with time, but possible steady states can be identified. The following approximate equations can be derived for the steady-state amplitude and the steady-state phase angle:

$$4A^2\left[\mu^2 + \left(\sigma - \tfrac{3}{8}A^2\right)^2\right] = \hat{F}_1^2 \tag{10.41}$$

$$\phi = -\tan^{-1}\left(\frac{\mu}{\sigma - \frac{3}{8}A^2}\right) \tag{10.42}$$

where
$$\hat{F}_1 = \frac{F_1}{\epsilon}$$

Equations (10.41) and (10.42) are plotted in Figs. 10.9 and 10.10. Note from these figures that there is a frequency range where three possible steady-state amplitudes and phases exist for a single frequency. This leads to an interesting phenomenon, peculiar to nonlinear systems, called the jump phenomenon. Imagine that the amplitude of excitation is fixed, but its frequency is slowly increased, starting slightly below the linear natural frequency. As the

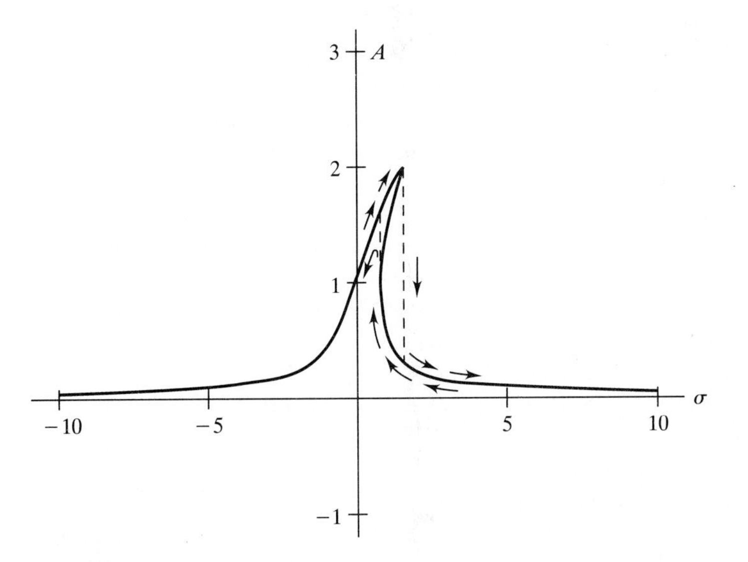

FIGURE 10.9
Frequency response curve for primary resonance of Duffing's equation illustrates the jump phenomenon ($\mu = 0.25$, $\hat{F}_1 = 1$).

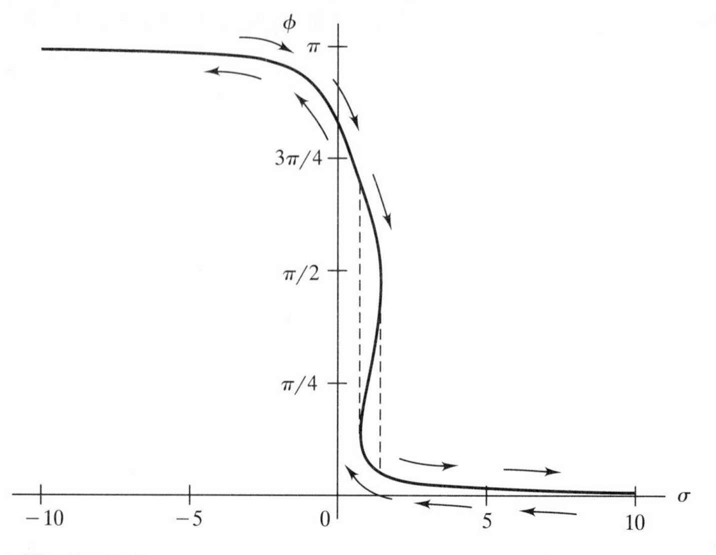

FIGURE 10.10
Phase vs. frequency curve for primary resonance of Duffing's equation also shows a jump phenomenon ($\mu = 0.25, \hat{F}_1 = 1$).

frequency is increased the steady-state amplitude follows the upper branch of the frequency response curve, until the point of vertical tangency is reached. When the frequency is increased beyond this critical value, the only possible steady-state amplitude is finitely lower than the amplitude at the critical frequency, and the amplitude will "jump" to this lower value. Now if the frequency is decreased from this value, the steady-state amplitude will follow the lower branch of the frequency response curve, until the point of vertical tangency is reached, when it will "jump" to the upper branch.

A state plane showing the relation between the amplitude and phase can be plotted for Duffing's equation with a primary resonance for parameters where the triple valuedness exists. Two equilibrium points are stable foci corresponding to the points on the upper and lower branches of the frequency response curve. A third equilibrium point is a saddle point corresponding to the intermediate amplitude between the points of vertical tangency. Since this equilibrium point is unstable, it can never be physically attained. The initial conditions dictate which of the two stable foci corresponds to the steady-state solution.

3. *Superharmonic resonance.* When either r_1 or r_2 is near $1/3$, the free-oscillation term does not decay exponentially. The steady-state response then consists of the forced response whose period is three times that of the linear natural period plus the free response, whose frequency is adjusted to three times that of the excitation. Thus the total response is periodic with the period equal to that of the excitation.

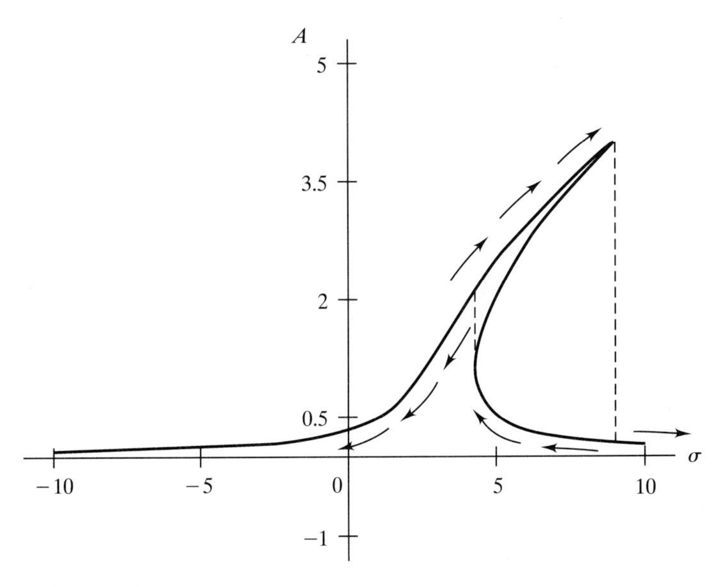

FIGURE 10.11
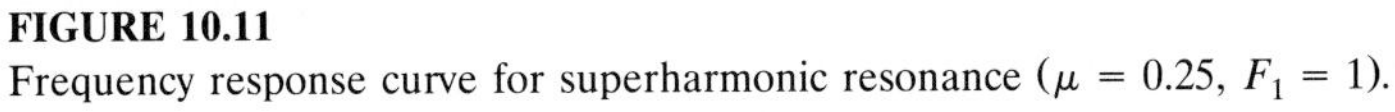
Frequency response curve for superharmonic resonance ($\mu = 0.25$, $F_1 = 1$).

Introduction of a detuning parameter when r_i is near 1/3, $3r_i = 1 + \epsilon\sigma$, leads to the frequency response equation

$$\left[\mu^2 + \left(\sigma - 3F_i^2 - \tfrac{3}{8}A^2\right)^2\right]A^2 = F_i^6 \tag{10.43}$$

which is cubic in A^2 and hence has three solutions. For a certain frequency range, three real solutions exist. The triple valuedness of the amplitude leads to a jump phenomenon similar to that for the primary resonance, as shown in Fig. 10.11.

4. *Subharmonic resonance*. When an excitation frequency is near three times the linear natural frequency, a subharmonic resonance may occur. The frequency response curve when r_i is near 3, $r_i = 3 + \epsilon\sigma$ is given by

$$\left[9\mu^2 + \left(\sigma - 9F_i^2 - \frac{9}{8}A^2\right)^2\right]A^2 = \frac{81}{16}F_i^2A^4 \tag{10.44}$$

Equation (10.44) has the trivial solution, $A = 0$, and two solutions obtained as roots of a quadratic equation in A^2. The quadratic equation yields real solutions for A if and only if the parameters satisfy the following inequality:

$$\frac{\sigma}{\mu} - \sqrt{\left(\frac{\sigma}{\mu}\right)^2 - 63} \le \frac{63F_1^2M_1^2}{4\mu} \le \frac{\sigma}{\mu} + \sqrt{\left(\frac{\sigma}{\mu}\right)^2 - 63} \tag{10.45}$$

When nontrivial solutions exist, one corresponds to a stable focus and one corresponds to a saddle point. The initial conditions determine whether the

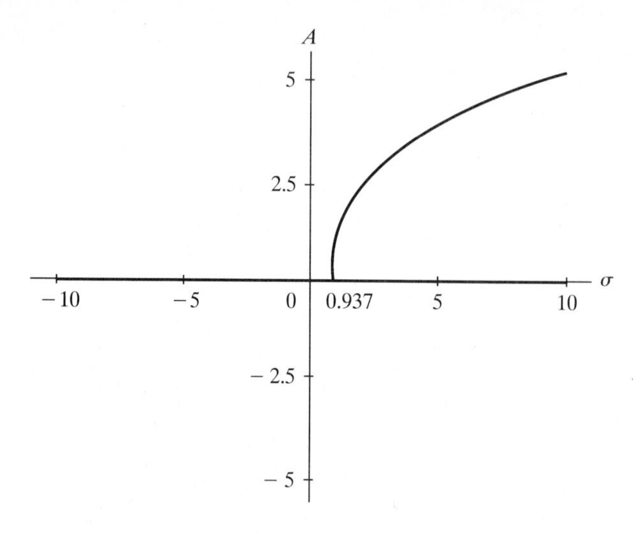

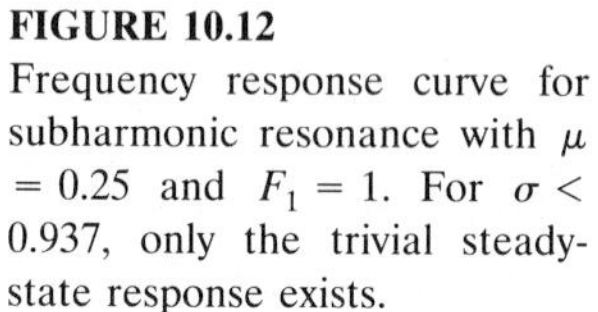
FIGURE 10.12
Frequency response curve for subharmonic resonance with $\mu = 0.25$ and $F_1 = 1$. For $\sigma < 0.937$, only the trivial steady-state response exists.

steady-state contribution from the free-oscillation term is trivial or approaches the stable focus.

Thus, if Eq. (10.45) is satisfied and the initial conditions are appropriate, the free-vibration term will not decay, but will exist with an adjusted frequency of one-third of that of the excitation. The total response is periodic with the period equal to that of the excitation. The frequency response curve is illustrated in Fig. 10.12.

5. *Combination resonances.* Combination resonances are unique to nonlinear systems and occur because of the nonlinear interaction of the particular solutions from x_0 when x_1 is calculated. When a combination resonance is present, a nontrivial free-vibration solution exists. The nonlinearity tunes the free-vibration response to the appropriate combination of frequencies.

The jump phenomenon does occur for cases of combination resonance.

6. *Simultaneous resonances.* Simultaneous resonances occur when two resonance conditions occur simultaneously. A detuning parameter is introduced for each resonance condition. Analysis of the steady state is much more complicated. For some simultaneous resonances, as many as seven equilibrium points exist in the state plane for the same frequency.

10.7 MULTI-DEGREE-OF-FREEDOM SYSTEMS

Nonlinear multi-degree-of-freedom systems exhibit behaviors which are not present for linear systems. It is instructive to consider free and forced vibrations of systems with quadratic nonlinearities and systems with cubic nonlinearities. Let $p_1, p_2, \ldots, p_n$ be the principal coordinates for a linearized system with natural frequencies $\omega_1 < \omega_2 < \cdots < \omega_n$, respectively. Principal coordinates that uncouple a linear system do not uncouple the system when nonlinearities are considered. The differential equations for the principal coordina coupled through nonlinear terms. For example, the free vibrations undamped two-degree-of-freedom system with quadratic nonlinearities a erned by

$$\begin{aligned} \ddot{p}_1 + \omega_1^2 p_1 + \alpha_1 p_1^2 + \alpha_2 p_1 p_2 + \alpha_3 p_2^2 &= 0 \\ \ddot{p}_2 + \omega_2^2 p_2 + \beta_1 p_1^2 + \beta_2 p_1 p_2 + \beta_3 p_2^2 &= 0 \end{aligned} \tag{10.46}$$

10.7.1 Free Vibrations

The free-vibration response of a system with quadratic nonlinearities includes periodic terms with frequencies of $\omega_1 + \omega_2$, $\omega_1 - \omega_2$, $2\omega_1$, and $2\omega_2$. If $\omega_2 \approx 2\omega_1$, then the nonlinearity acts as if it excites the system with a harmonic excitation of frequencies ω_1 and ω_2, producing a self-sustaining free oscillation, called an internal resonance.

In the absence of the internal resonance, and in the presence of damping, the free oscillations of both modes decay exponentially, and are to first approximation independent. When an internal resonance is present, free oscillations are sustained, even when damping is present and causes coupling between the two modes. Even if only one mode is initially excited, the internal resonance excites the other mode as well. Energy is continually exchanged between the two modes.

An internal resonance occurs in a two-degree-of-freedom system with cubic nonlinearities when $\omega_2 \approx 3\omega_1$.

Example 10.7. Reconsider the spring pendulum of Example 10.2. The spring has a stiffness 1×10^3 N/m and an unstretched length of 0.5 m. For what values of m will an internal resonance occur?

Since l is the length of the spring when the system is in equilibrium,

$$l = 0.5m + \frac{mg}{k}$$

Since the approximate linear system is uncoupled when x and θ are used as generalized coordinates, these are also the principal coordinates and the linear natural frequencies are

$$\omega_1 = \sqrt{\frac{g}{0.5 + \dfrac{mg}{k}}} \qquad \omega_2 = \sqrt{\frac{k}{m}}$$

Setting $\omega_2 = 2\omega_1$ gives $m = 12.74$ kg.

10.7.2 Forced Vibrations

The free oscillations are self-sustaining in multi-degree-of-freedom systems subject to harmonic excitations when the frequency of excitation is near certain values. A primary resonance occurs if the excitation frequency is near any of the system's natural frequencies. Subharmonic and superharmonic resonances occur as for one-degree-of-freedom systems. Other secondary resonances occur when the excitation frequency is near a certain combination of natural frequencies.

For a system with quadratic nonlinearities, these resonances occur when the excitation frequency is near the sum or difference of two natural frequencies. Combination resonances occur for multifrequency excitations. Simultaneous resonance conditions can also exist.

A complete summary of the phenomena present in nonlinear multi-degree-of-freedom systems is too extensive. The jump phenomenon occurs for certain types of resonances. A saturation phenomenon exists for certain resonances, where the amplitude of one mode cannot grow beyond a certain value and any additional increase in excitation will only cause growth of the other modes. For a linear system an increase in excitation amplitude will cause an increase in the amplitude of response for all modes. Quenching can also occur in certain systems with simultaneous resonances where introduction of the second resonance causes the total response to decrease.

A saturation phenomenon can also occur for systems with quadratic nonlinearities. The amplitude of a specific mode may build up as the amplitude of excitation is increased. When the excitation amplitude reaches a certain value, the mode may become saturated; its amplitude of response remains constant as the excitation amplitude is further increased. The amplitudes of the other modes will continue to grow with the excitation amplitude.

In addition to primary resonances, subharmonic resonances, and superharmonic resonances, combination resonances occur in a two-degree-of- freedom system with cubic nonlinearities when one of the following conditions are met:

$$\begin{aligned} \Omega &\approx 2\omega_1 \pm \omega_2 \\ \Omega &\approx 2\omega_2 \pm \omega_1 \\ \Omega &\approx \tfrac{1}{2}(\omega_2 \pm \omega_1) \end{aligned} \tag{10.47}$$

where Ω is the excitation frequency

10.8 CONTINUOUS SYSTEMS

The nonlinear dimensionless partial differential equation governing transverse vibrations of a uniform beam of length L and radius of gyration r, subject to a transverse load per unit length $F(x, t)$, is

$$\left(\frac{r}{L}\right)^2\left(\frac{\partial^2 w}{\partial t^2} + \frac{\partial^4 w}{\partial x^4}\right) = \frac{1}{2}\int_0^1 \left(\frac{\partial w}{\partial x}\right)^2 dx \frac{\partial^2 w}{\partial x^2} + F(x, t) \tag{10.48}$$

The nonlinear term is a result of the midplane stretching and is often ignored.

Let $\omega_1, \omega_2, \ldots$ be the natural frequencies of the linearized system and $\phi_1, \phi_2, \ldots$ be their corresponding normalized mode shapes such that

$$\left(\phi_i(x), \phi_j(x)\right) = \delta_{ij}$$

for an appropriate scalar product.

The expansion theorem is used to develop an approximation to the solution of Eq. (10.48) as

$$w(x,t) = \epsilon \sum_{i=1}^{\infty} p_i(t)\phi_i(x) \tag{10.49}$$

where $\epsilon \ll 1$ is a small dimensionless amplitude. Substituting Eq. (10.49) into Eq. (10.48), taking the scalar product with respect to $\phi_j(x)$ for an arbitrary j, and using algebra and mode shape orthonormality lead to

$$\ddot{p}_j + \omega_j^2 p_j = \epsilon\left(\frac{L}{r}\right)^2\left[\frac{1}{2}\sum_{k=1}^{\infty}\sum_{l=1}^{\infty}\sum_{m=1}^{\infty}\left(\phi_j, \frac{\partial^2\phi_k}{\partial x^2}\right)\int_0^1 \frac{\partial\phi_l}{\partial x}\frac{\partial\phi_m}{\partial x}\,dx\, p_k p_l p_m\right]$$
$$+\left(F(x,t), \phi_j(x)\right) \tag{10.50}$$

The preceding procedure is similar to the modal analysis method of Chap. 9, except that the resulting set of ordinary differential equations are still coupled through the nonlinear terms. The nonlinear terms, due to the midplane stretching, are cubic nonlinearities. If the excitation is harmonic with a frequency Ω, then using the results of Sec. 10.7, the following resonances can occur:

1. Internal resonances occur if $\omega_i \approx 3\omega_j$, or $\omega \approx 2\omega_j + \omega_k$ for any i, j, and k. From Table 9.4, for a fixed-pinned beam, $\omega_3 = 2\omega_1 + 2.30$, and for a fixed-fixed beam $\omega_5 = 2\omega_3 + \omega_2 - 4.86$. Internal resonances occur in each of these beams. It is noted that for a pinned-pinned beam $\omega_3 = 2\omega_2 + \omega_1$. However, the coefficient multiplying $p_2^2 p_1$ in Eq. (10.50) is zero for a pinned pinned beam.
2. Primary resonance occurs if $\Omega \approx \omega_i$ for any i.
3. Superharmonic resonance occurs if $\Omega = \omega_i/3$ for any i.
4. Subharmonic resonance occurs if $\Omega = 3\omega_i$ for any i.
5. Combination resonances occur if $\Omega \approx 2\omega_i \pm \omega_j$, $\Omega \approx \omega_i \pm \omega_j \pm \omega_k$, or $\Omega \approx (\omega_i \pm \omega_j)/2$ for any i, j, and k.

10.9 CHAOS

Recent research in nonlinear phenomena has led to the development of a new branch of physics called chaos. The term *chaos* is not well defined but refers to the seemingly random response of a nonlinear system due to deterministic excitation. Chaos occurs when a periodic excitation leads to a nonperiodic response. Chaos occurs when slightly different initial conditions lead to divergent responses.

Chaos has been observed and predicted in nonlinear systems in such diverse fields as physics, medicine, economics, and meteorology. Chaos occurs in mechanical systems, electrical systems, and chemical systems. Researchers observed that chicken pox epidemics are periodic while measles epidemics are chaotic. Others have used chaos to model stock market fluctuations. Many researchers hope that chaos may unlock some of the mysteries of fluid turbulence.

Chaotic motion has been observed in many mechanical systems. Tung and Shaw observed chaotic motions from their nonlinear model of an impact print hammer for a dot matrix printer. Chaotic vibrations for systems modeled by Duffing's equation are well documented as are chaotic motions of a forced pendulum.

Analytical tools have been developed to identify and classify chaotic behavior. These tools can be applied to analytical solutions for vibrating systems as well as experimental observations. Some are described in the following discussion.

1. *State space.* Observation of the state space can indicate whether a system is chaotic. A chaotic motion will have trajectories that do not repeat, when viewed in the phase plane. The trajectories will fill a region of the phase plane without ever repeating. However, viewing of the state plane is by itself insufficient to speculate that a motion is chaotic. An example of a chaotic motion as viewed in a state plane is shown in Fig. 10.13.

2. *Poincaré sections.* A Poincaré section is a graph of the phase plane response taken or sampled only at fixed intervals of time. If the response is periodic and the time interval is equal to the period, then the Poincaré section is only a point, as the same response is obtained on each sampling. If the

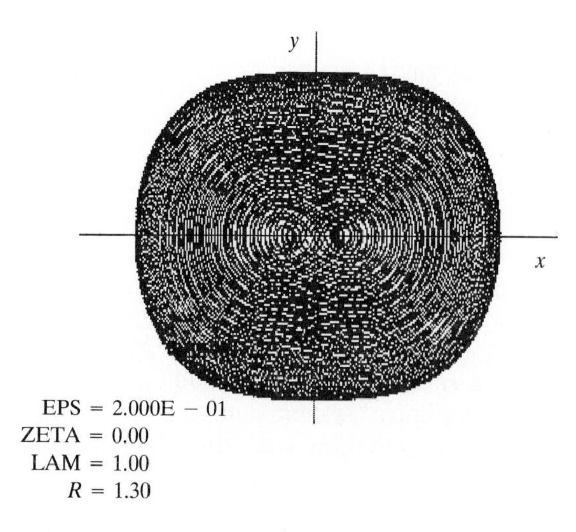

FIGURE 10.13
State plane for an apparently chaotic motion generated from DUFF.BAS.

response is periodic and the time interval is less than the period, but commensurate with the period, the Poincaré section is a finite number of points.

The Poincaré section of a nonlinear system with a quadratic subharmonic resonance, sampled at the period of excitation should have two points. The presence of the subharmonic resonance doubles the period of response. If a system subject to a periodic excitation is sampled at intervals equal to the period of excitation, and the Poincaré section is a seemingly random collection of points, the response can be guessed to be chaotic.

The program DUFF.BAS, included on the accompanying diskette, can develop Poincaré sections for responses of Duffing's equations. The Poincaré sections of Figure 10.14 are generated using DUFF.BAS.

3. *Fourier transforms*. The Fourier transform of a nonperiodic continuous function is an extension of the Fourier series defined for periodic functions. The Fourier transform is obtained from the Fourier series by allowing the period to become infinite. The resulting Fourier transform of $f(t)$ is defined as

$$\bar{f}(\omega) = \frac{1}{2\pi}\int_{-\infty}^{\infty} f(t)e^{-i\omega t}\,dt \tag{10.51}$$

The transform function, $\bar{f}(\omega)$, is a function of the transform variable, ω. If the Fourier transform of a periodic function is taken, then $\bar{f}(\omega) = 0$ unless ω is a multiple of the function's fundamental frequency.

The Fourier transform decomposes a function into its harmonic components. The strength of a component is given by the magnitude of $\bar{f}(\omega)$. The values of ω which have significant nonzero values of $\bar{f}(\omega)$ are called the spectrum of the function. If the Fourier transform of the response of a nonlinear system due to a periodic response is a continuous spectrum, then the response is chaotic.

For computational purposes the Fourier transform is replaced by the fast Fourier transform. If $f(t)$ is known at k times, $t_1, t_2, \ldots, t_k$, then the discrete fast Fourier transform is given by

$$\bar{f}(j) = \sum_{l=1}^{k} f(t_i)e^{-2\pi i(l-1)(j-1)/k} \tag{10.52}$$

Examples of Fourier transforms are given in Fig. 10.15.

4. *Bifurcation diagrams*. Bifurcation diagrams can be used to identify one route to chaos. The steady-state amplitude (or phase) of a nonlinear system as a function of a system parameter is plotted as the parameter is slowly changed. For a nonlinear system the steady-state solution may split at a certain value of the parameter and two possible steady states exist for greater values of the parameter. A bifurcation is said to occur for the value of the parameter where the split occurs. The bifurcation is often the result of the sudden presence of a subharmonic resonance. When this occurs the period of motion doubles. As the parameter is increased additional bifurcations may occur, where the period again doubles. If the system is chaotic, as the parameter increases, bifurcations

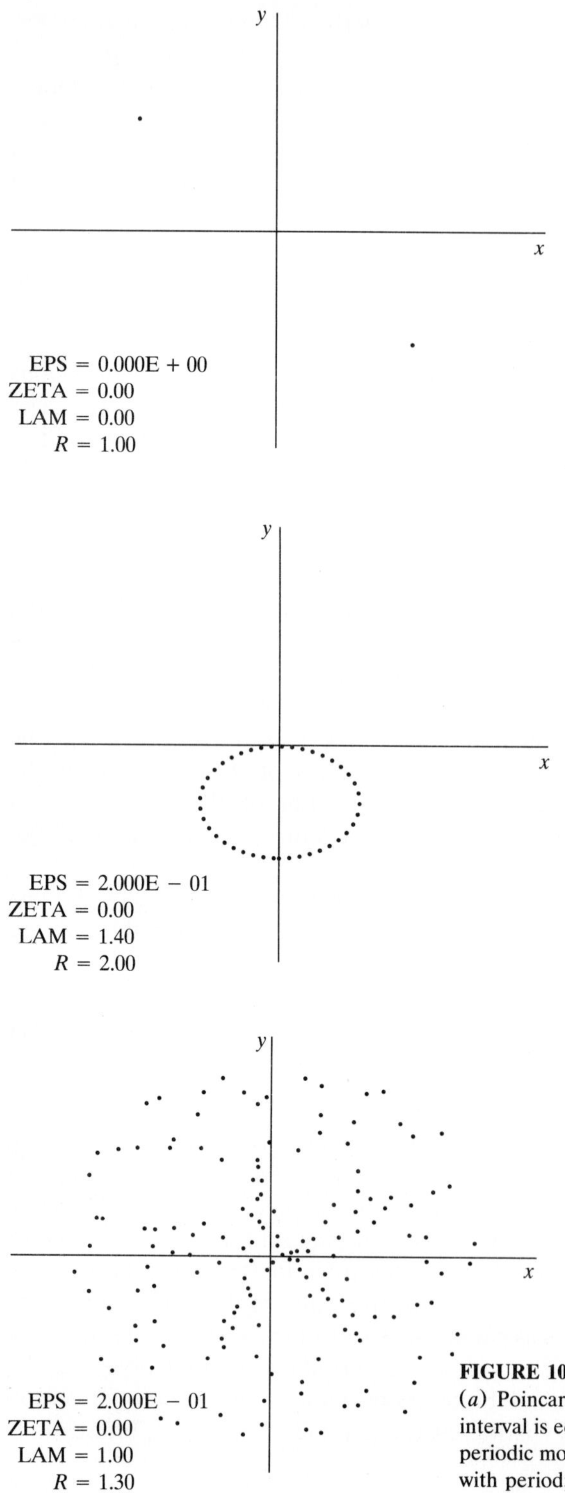

FIGURE 10.14
(a) Poincaré section for periodic motion when sampling interval is equal to half the period; (b) Poincaré section for periodic motion when sampling interval is incommensurate with period; (c) Poincaré section for a chaotic motion.

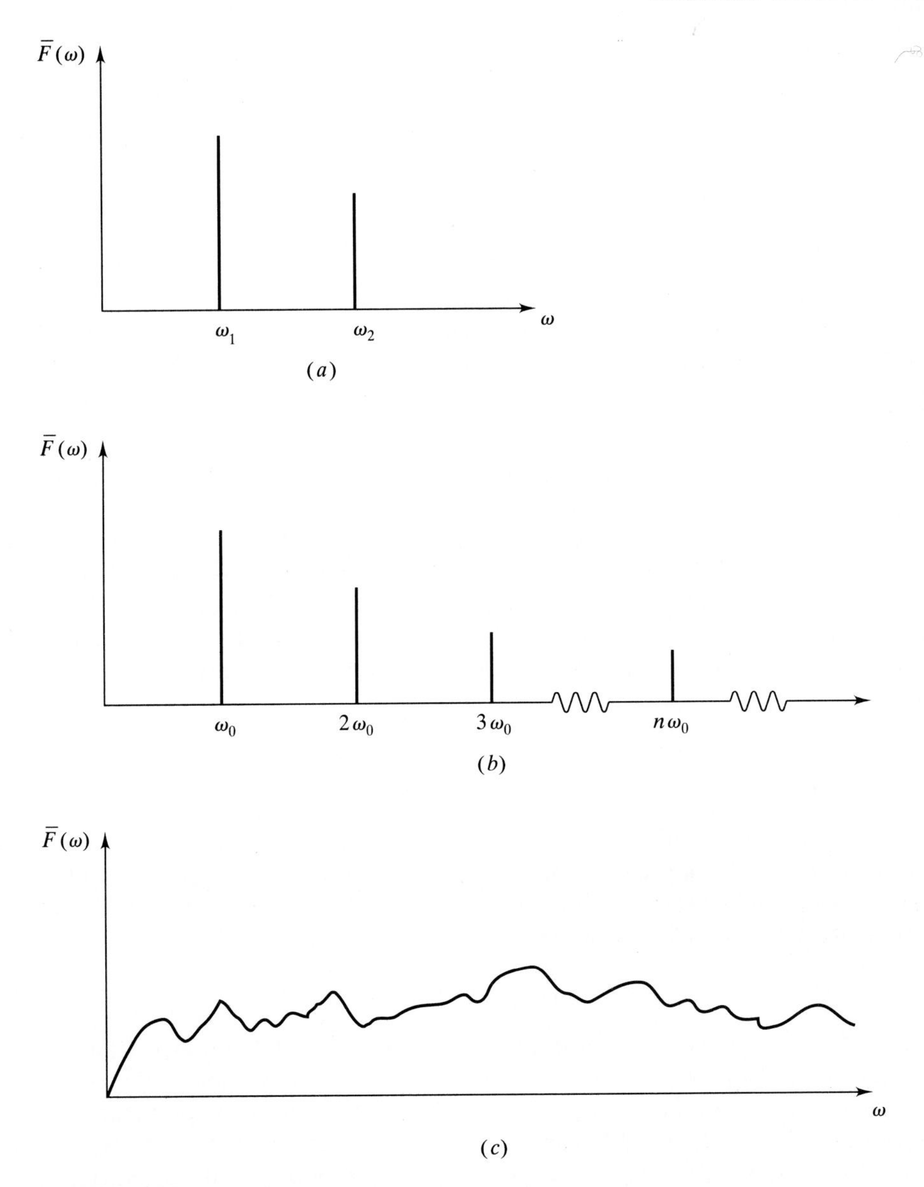

FIGURE 10.15
(*a*) Fourier transform of a periodic function with two distinct frequencies; (*b*) Fourier transform of a periodic function of fundamental frequency ω_0; (*c*) Fourier transform of a chaotic response.

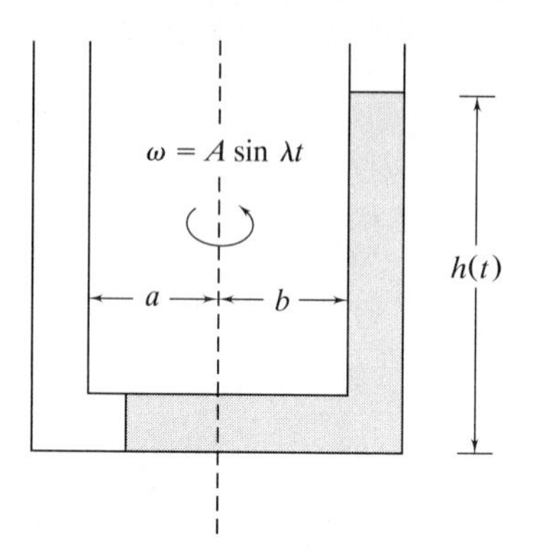

FIGURE 10.16
For certain values of λ and ϵ the motion of the column of liquid in the U tube manometer can be chaotic when the manometer rotates about a non-centroidal axis.

and period doubling occur more rapidly. The chaotic response bounces between amplitudes and has no discernible period. The plot of steady-state amplitudes (or phases) as the parameter increases becomes a blur. It is often the case that as the parameter is increased much further, the motion again becomes periodic.

While chaotic motion is characterized by its unpredictable nature, it has some universal features. Feigenbaum showed that as the number of bifurcations increases the values of the parameter, call it A, for which the bifurcations occur are given by

$$A_n - A_{n-1} = 4.669\ldots(A_{n+1} - A_n) \tag{10.53}$$

There are many routes to chaos. The one described here applies to systems undergoing nonlinear oscillations subject to a harmonic excitation and is illustrated by the rotating U-tube manometer of Example 10.3 and Fig. 10.16. The manometer is rotated about a vertical axis other than its centroidal axis. The rotational speed of the manometer varies as

$$\omega(t) = A \sin \lambda t \tag{10.54}$$

where A is large enough to cause the fluid to be completely drained from the left leg during an initial transient period. The system is subject to viscous damping from the interaction of the fluid with the wall of the manometer.

The behavior of a nonlinear system is heavily influenced by the system parameters. This is evidenced by the state planes of Figs. 10.17 and 10.18. Figure 10.17 shows the state planes for two slightly different values of the frequency for the same amplitude. A steady state is evident for the motion of Fig. 10.17*a*, while the motion of Fig. 10.17*b* appears chaotic. Chaos is also induced by small amplitude changes for the same frequency as shown in Fig. 10.18*a*.

A bifurcation diagram for the parameter A is shown in Fig. 10.19. The frequency of excitation is fixed as its amplitude varies. For $A < 3.33$ the steady-state motion is periodic. The stationary response is periodic of frequency 2λ and a certain amplitude.

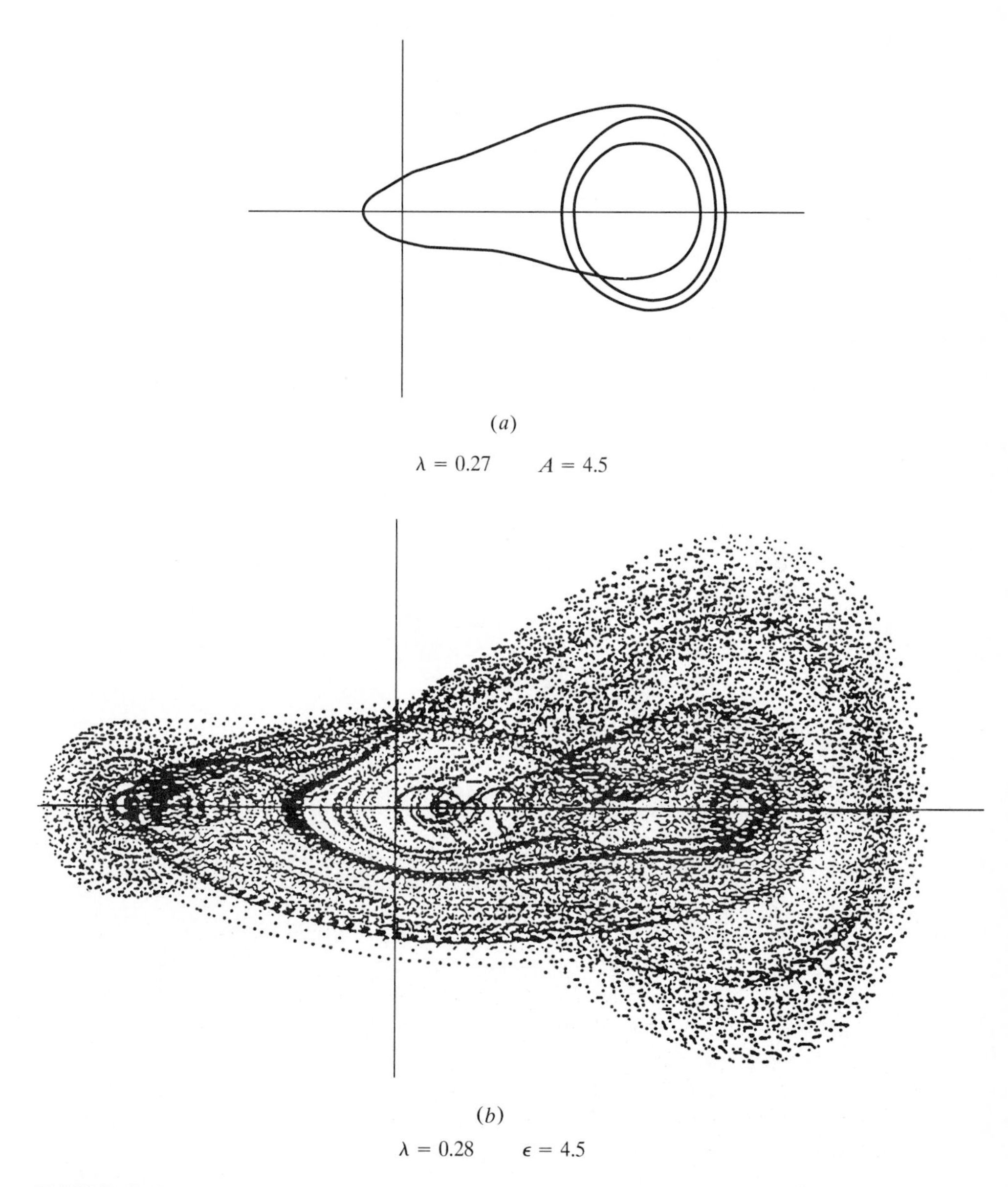

FIGURE 10.17
These state planes show that a small frequency change can cause a change from a periodic response to a chaotic response.

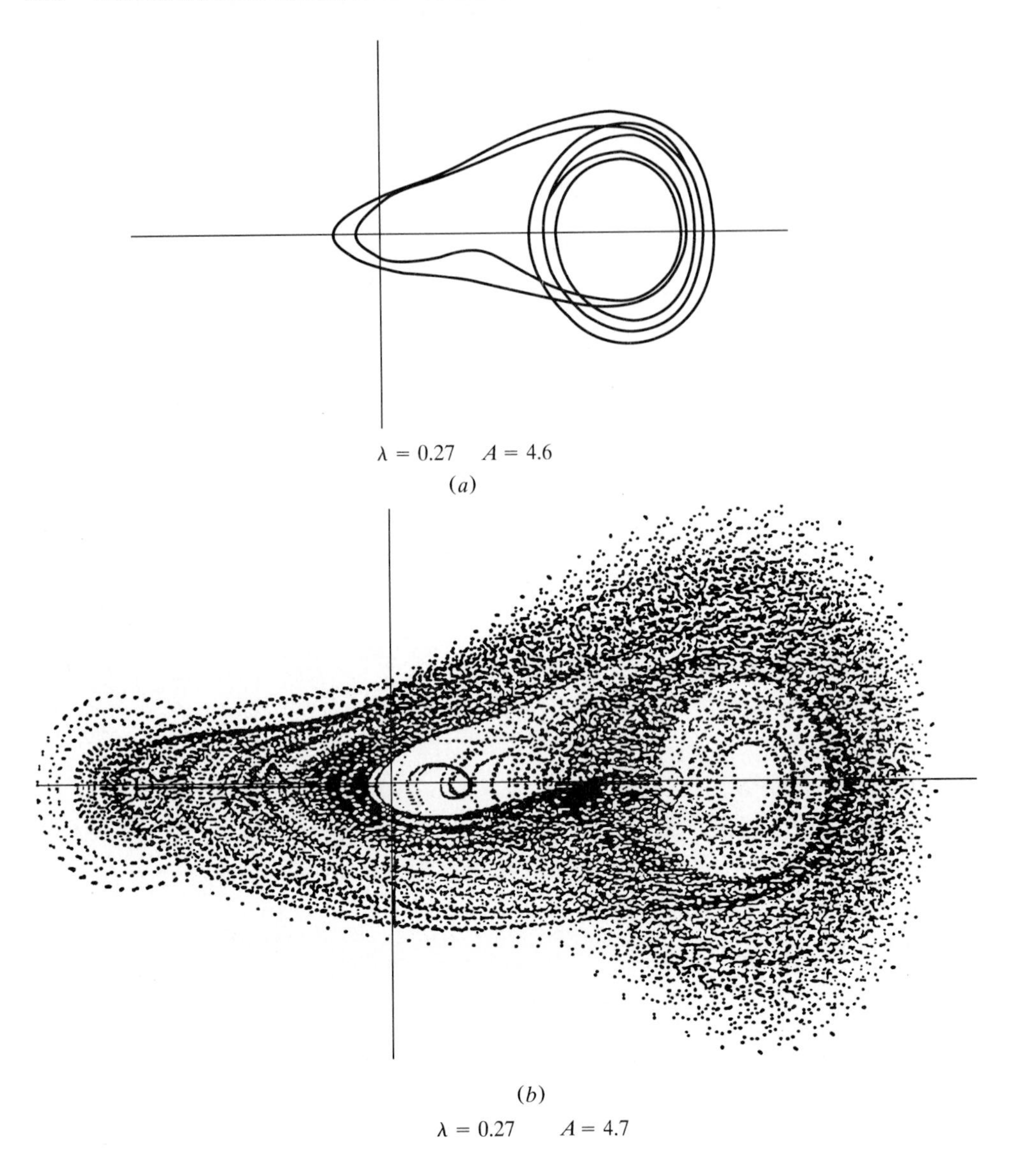

FIGURE 10.18
These state planes show that a small amplitude change can cause a change from a periodic response to a chaotic response.

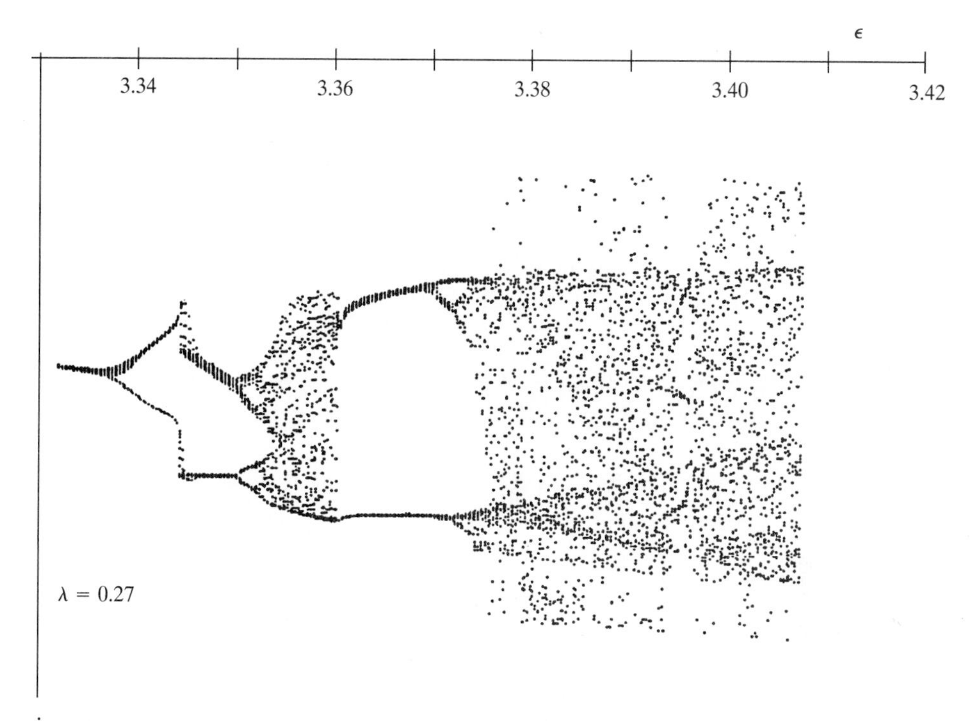

FIGURE 10.19
Bifurcation diagram for rotating manometer. First bifurcation occurs near $A = 3.34$. As A increases chaos develops. Motion is not chaotic for a range of A, then the process to chaos begins again.

For $A \approx 3.33$ the parameters change such that a subharmonic resonance becomes present. A bifurcation is said to occur. The presence of the subharmonic resonance means that the steady-state response is the sum of a free-vibration term and a forced-vibration term and that the period of motion is doubled. Two amplitudes are evident in the stationary oscillations.

For $A \approx 3.35$ another bifurcation occurs. A higher-order subharmonic resonance is induced. The response has a period of four times the original period and is made up of four distinct amplitudes.

As A increases bifurcations occur more rapidly with the period doubling with each bifurcation. Eventually, the response is chaotic. The chaotic response shown in Fig. 10.20 bounces between amplitudes and has no discernible period.

For $A \approx 3.36$ the motion ceases to be chaotic and returns to the doubled period. However, bifurcations begin to occur again at $A \approx 3.37$.

The process described previously is called period doubling through a subharmonic cascade.

Chaos is the subject of much current research. It is hoped that studying chaos can lead to the better understanding of nonlinear systems like turbulent

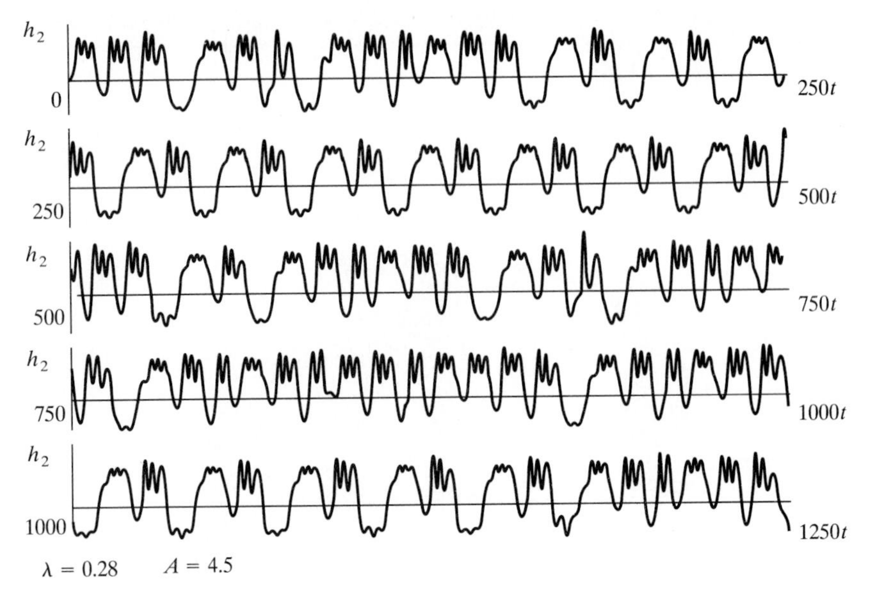

FIGURE 10.20
The time history of motion for these parameters has no discernable period.

fluid flows, the flow and pumping of blood through a human heart, weather patterns, and nonlinear vibrations.

PROBLEMS

10.1. The free-vibration response of a block hanging from a linear spring is the same as that of the block attached to the same spring, but sliding on a frictionless surface. Is the response the same if the spring has a force-displacement relation given by
(a) $F = k_1x + k_3x^3$.
(b) $F = k_1x + k_2x^2$.
(c) $F = k_1x,\ x < x_0$.
$F = k_2x,\ x > x_0$.

10.2. The system of Fig. P10.2 is one of the few for which an exact solution is available. Its solution is obtained in a manner analogous to that of free vibrations with Coulomb damping described in Sec. 3.6. The block is displaced a distance $x_0 > \delta$ to the right from equilibrium and released. Determine the period of the resulting oscillations.

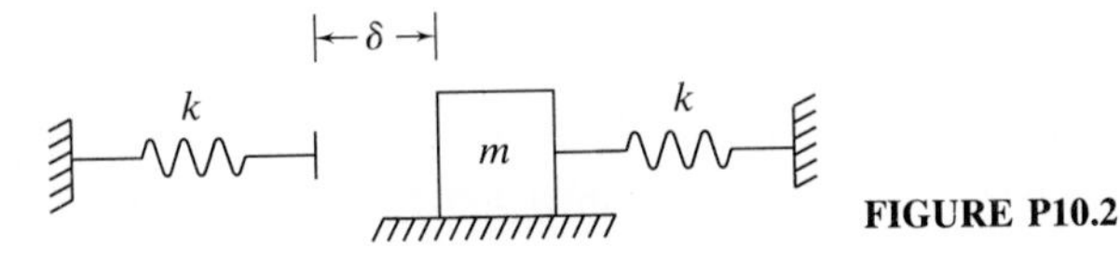

FIGURE P10.2

10.3. The block in Fig. P10.3 is not attached to the springs. Determine the period of the resulting oscillations if the block is displaced a distance x_0 to the right from equilibrium and released.

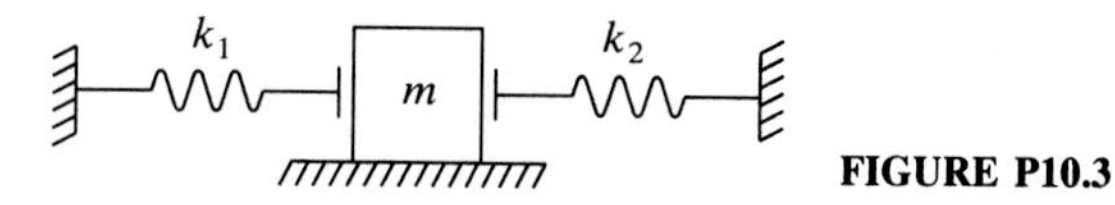

FIGURE P10.3

10.4–10.7. Without making linearizing assumptions, use Lagrange's equations to derive the nonlinear differential equation(s) governing the motion of the system shown. Use the generalized coordinates indicated.

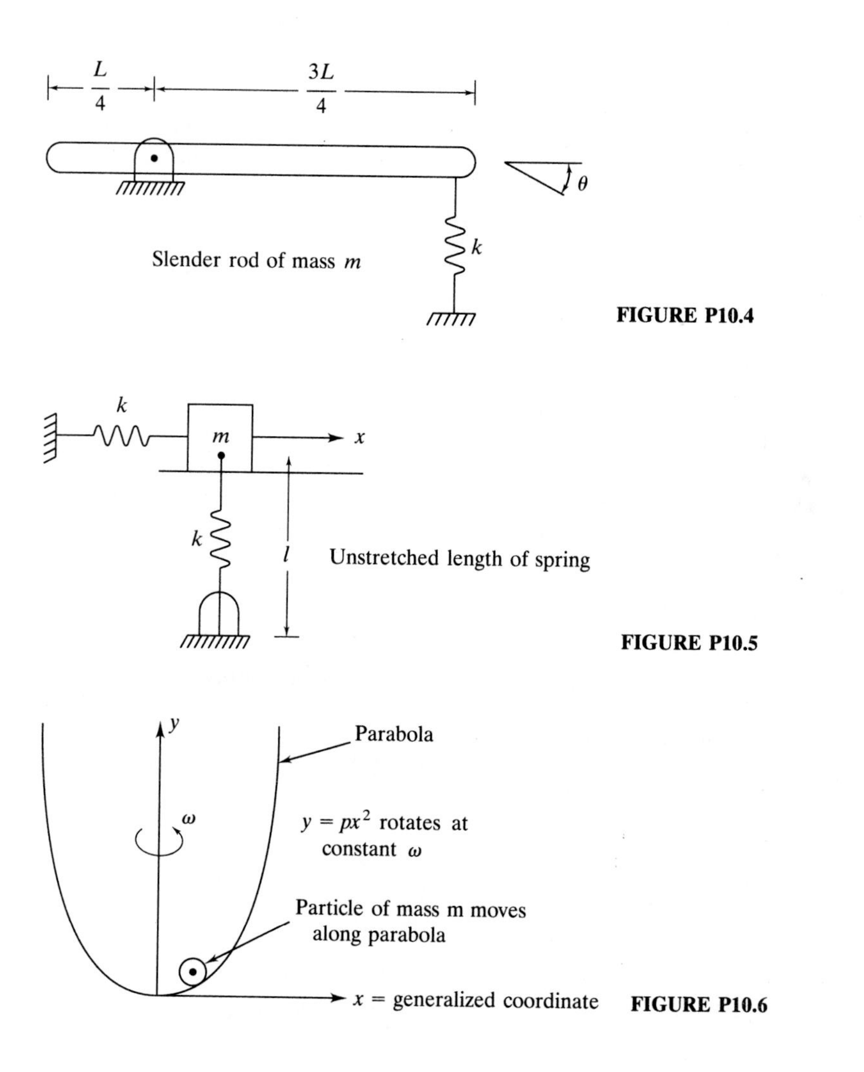

FIGURE P10.4

FIGURE P10.5

FIGURE P10.6

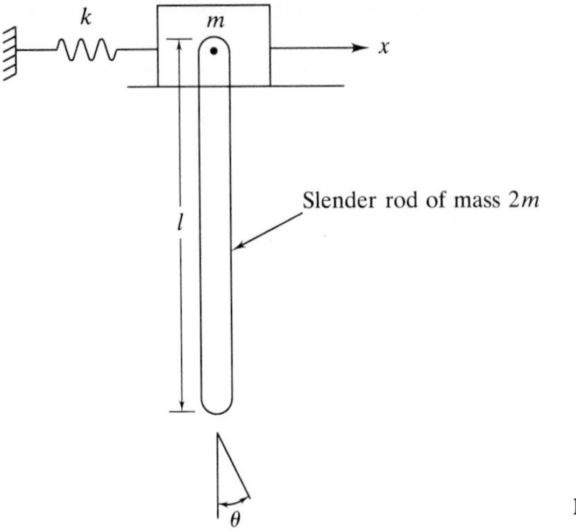

FIGURE P10.7

10.8. A wedge of specific weight γ floats stably on the free surface of a fluid of specific weight γ_w. The wedge is given a vertical displacement, δ, from this equilibrium position.

(a) Derive the differential equation governing the resulting free oscillations of the wedge. Neglect viscous effects and the added mass of the fluid.

(b) What is the equation of the trajectory in the phase plane which describes the resulting motion. Sketch the trajectory.

(c) Assume δ is small and use the method of renormalization to determine a two-term approximation for the frequency-amplitude relationship.

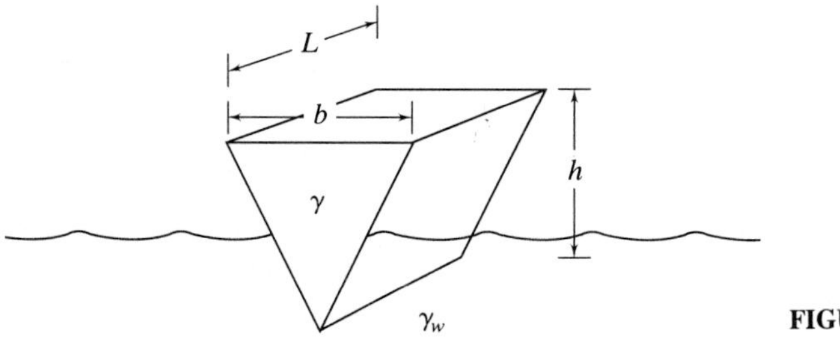

FIGURE P10.8

10.9. Repeat Prob. 10.8 for the inverted cone of Fig. P10.9.

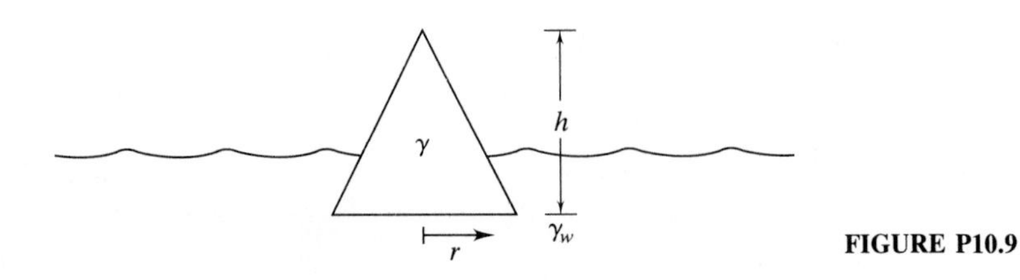

FIGURE P10.9

10.10. Determine the equation defining the state plane for the system of Fig. P10.6. Sketch trajectories in the phase plane when
(a) $p = 1.5\ \text{m}^{-1}$, $\omega = 5$ rad/s.
(b) $p = 1.0\ \text{m}^{-1}$, $\omega = 5$ rad/s.
(c) $p = 5.097\ \text{m}^{-1}$, $\omega = 10$ rad/s.

10.11. Plot the trajectory in the state plane corresponding to the motion of a mass attached to a linear spring free to slide on a surface with Coulomb damping when the mass is displaced from equilibrium and released from rest.

10.12. Determine the equilibrium points and their type for the differential equation

$$\ddot{x} + 2\zeta\dot{x} - x + \epsilon x^3 = 0$$

10.13. Determine the equilibrium points and their type for the differential equation

$$\ddot{x} + 2\zeta\dot{x} - x - \epsilon x^3 = 0$$

10.14. Determine the equilibrium points and their type for the differential equation

$$\ddot{x} + 2\zeta\dot{x} + x + \epsilon x^2 = 0$$

10.15. Determine the equilibrium points and their type for the differential equation

$$\ddot{x} + 2\zeta\dot{x} + x - \epsilon x^2 = 0$$

10.16. The equation of motion for the free oscillations of a pendulum subject to quadratic damping is

$$\ddot{\theta} + 2\zeta\dot{\theta}^2 + \sin\theta = 0$$

(a) Determine an exact equation defining the state plane.
(b) Determine the equilibrium points and their type.

10.17. Determine the period of oscillation of a mass attached to a hardening spring with a cubic nonlinearity.

10.18. Determine an integral expression for the period of oscillation of the system of Fig. P10.6.

10.19. Use the method of renormalization to determine a two-term approximation for the frequency-amplitude relation for the system of Fig. P10.4. If the bar is rotated 4° from equilibrium and released, what is the period for $L = 4$ m, $k = 1000$ N/m, and $m = 10$ kg?

10.20. A 25-kg mass is attached to a hardening spring with $k_1 = 1000$ N/m and $k_3 = 4{,}000\ \text{N/m}^3$. The mass is displaced 15 mm from equilibrium and released from rest. What is the period of the ensuing oscillations?

10.21. Suppose the mass of Prob. 10.20 is subject to an impulse which imparts a velocity of 3.1 m/s to the mass when the mass is in equilibrium. What is the period of the ensuing oscillations?

10.22. Suppose the mass of Prob. 10.20 is attached to the same spring when a 50-N force is statically applied and suddenly removed. What is the period of the ensuing oscillations?

10.23. Use the method of renormalization to determine a two-term frequency-amplitude relationship for the particle on the rotating parabola of Fig. P10.6, assuming the amplitude is small.

10.24. Use the method of renormalization to determine a two-term frequency-amplitude relationship for a block of mass m attached to a spring with a quadratic nonlinearity. When nondimensionalized the differential equation governing free vibrations of the system is

$$\ddot{x} + \omega^2 x + \epsilon x^2 = 0 \qquad \epsilon \ll 1$$

Problems 10.25 to 10.31 refer to the system of Fig. P10.25.

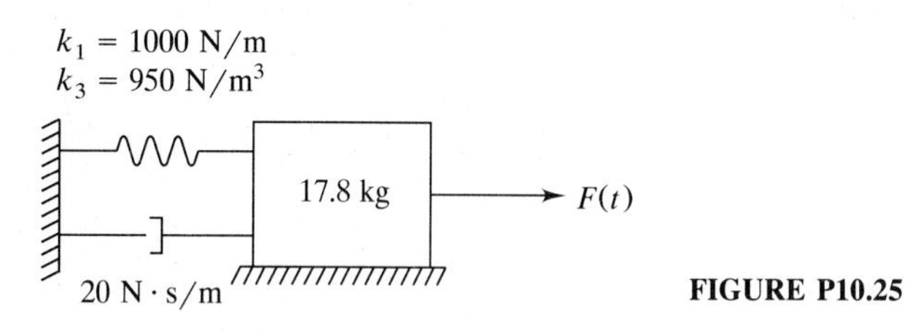

FIGURE P10.25

10.25. If $F(t) = F_0 \sin \omega t$, what values of ω will lead to the presence of
(*a*) A primary resonance.
(*b*) A superharmonic resonance.
(*c*) A subharmonic resonance.

10.26. When $F(t) = 5 \sin 8t$ N a primary resonance condition occurs. Determine the amplitude of the forced response.

10.27. When $F(t) = 150 \sin 2.5t$ N a superharmonic resonance condition occurs. Determine the amplitude of the forced response.

10.28. If $F(t) = F_0 \sin \omega t$ N, for what value of ω will a jump in amplitude occur when ω is increased slightly beyond this value when
(*a*) $F_0 = 5$ N and a primary resonance occurs.
(*b*) $F_0 = 150$ N and a superharmonic resonance occurs.

10.29. If $F(t) = 25 \sin 22t$ N, will a nontrivial subharmonic response exist?

10.30. If $F(t) = 30 \sin 15t + 25 \sin \omega t$ N, what values of ω lead to a combination resonance?

10.31. If $F(t) = 30 \sin 2.5t + 25 \sin \omega t$ N, what values of ω lead to simultaneous resonances?

Problems 10.32 to 10.35 refer to the system of Fig. P10.32. The spring of stiffness k_2 is a linear spring.

10.32. If $m_2 = 10$ kg, for what values of k_2 will internal resonances exist?

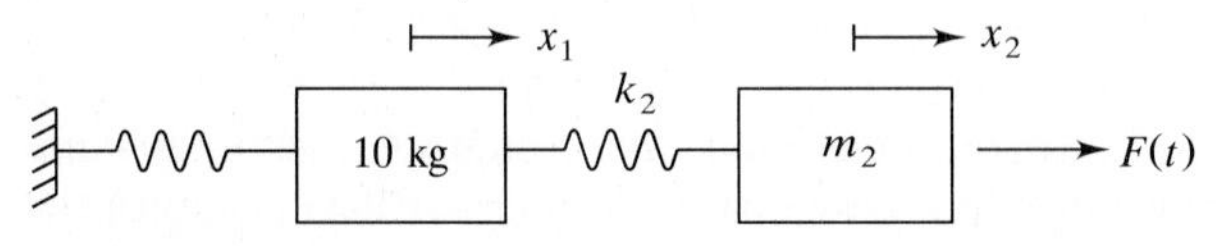

FIGURE P10.32

10.33. For what values of m_2 are internal resonances possible? If an internal resonance is possible, in terms of m_2, for what values of k_2 will they exist?

10.34. Consider the system with $m_2 = 10$ kg and $k_2 = 2000$ N/m. The right mass is displaced 10 mm from equilibrium while the left mass is held in place. The system is released from rest from this configuration.
(a) Determine the natural frequencies, mode shapes, and principal coordinates for the linearized system.
(b) Write the nonlinear differential equations governing the system using the principal coordinates of the linearized system as dependent variables.

10.35. If $m_2 = 10$ kg, $k_2 = 1000$ N/m, and $F(t) = 150 \sin \omega t$ N, for what values of ω will the following resonances exist?
(a) Primary resonance.
(b) Superharmonic resonance.
(c) Subharmonic resonance.
(d) Combination resonance.

10.36. Consider the system of Fig. P10.36.
(a) Derive the nonlinear differential equations governing the motion of the system using the generalized coordinates shown.
(b) Expand trigonometric functions of the generalized coordinates using Taylor series expansions. Rewrite the differential equations keeping only quadratic and cubic nonlinearities.
(c) For what values of l in terms of the other parameters will an internal resonance exist?
(d) In the absence of an internal resonance, for what values of ω will resonance conditions exist?

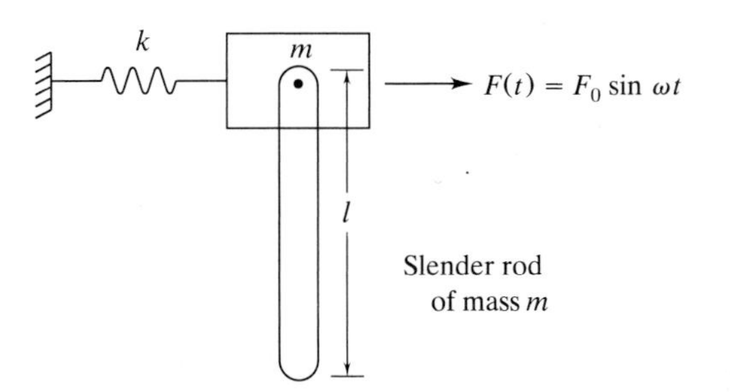

FIGURE P10.36

10.37. Show that the coefficient multiplying $p_2^2 p_1$ for a pinned-pinned beam is zero in Eq. (10.50).

10.38. A fixed-free rectangular steel beam ($\rho = 7850$ kg/m^3, $E = 210 \times 10^9$ N/m^2) of length 1 m, base 2 cm, and height 5 cm is subject to a single-frequency harmonic excitation. List all excitation frequencies that should be avoided to avoid all primary, secondary, and combination resonances involving the three lowest modes.

10.39. If the beam of Prob. 10.38 is fixed-fixed, which of the following excitation frequencies should be avoided and why?
(*a*) 180 rad/s.
(*b*) 1530 rad/s.
(*c*) 2200 rad/s.
(*d*) 7940 rad/s.

The program DUFF.EXE on the accompanying diskette provides a numerical solution to Duffing's equation, Eq. (10.9). Use DUFF.EXE to solve Probs. 10.40 through 10.46.

10.40. Run DUFF.EXE to determine the time history of motion with $\zeta = 0$, $\epsilon = 0.1$, $r = 3$, and
(*a*) $\Lambda = 0.05$.
(*b*) $\Lambda = 0.3$.
(*c*) $\Lambda = 2.5$.
Can you identify the presence of any subharmonic or superharmonic resonances?

10.41. Run DUFF.EXE to determine the time history of motion with $\zeta = 0.05$, $\epsilon = 0.1$, $r = 1/3$, and
(*a*) $\Lambda = 0.05$.
(*b*) $\Lambda = 0.3$.
(*c*) $\Lambda = 2.5$.
Can you identify the presence of any subharmonic or superharmonic resonances?

10.42. Plot the state plane of Duffing's equation with
(*a*) $\zeta = 0$, $\epsilon = 1.0$, $\Lambda = 0$, $x_0 = 0.1$, $\dot{x}_0 = 0.2$.
(*b*) $\zeta = 0.1$, $\epsilon = 0.5$, $r = 1.5$, $\Lambda = 1.2$.
(*c*) $\zeta = 0.2$, $\epsilon = 1.3$, $r = 0.98$, $\Lambda = 1.2$.

10.43. Use DUFF.EXE to help speculate whether the motion is chaotic for each of the following cases:
(*a*) $\zeta = 0$, $\epsilon = 0.1$, $r = 1.2$, $\Lambda = 1.4$.
(*b*) $\zeta = 0$, $\epsilon = 1.2$, $r = 2$, $\Lambda = 2.5$.
(*c*) $\zeta = 0.2$, $\epsilon = 0.5$, $r = 1.1$, $\Lambda = 0.5$.
(*d*) $\zeta = 0.0$, $\epsilon = 2.6$, $r = 1.3$, $\Lambda = 5.0$.
State your reasons for your speculation.

REFERENCES

1. Abramowitz, M., and I. Stegun: *Tables of Mathematical Functions*, Dover, New York, 1964.
2. Baker, G. L., and J. P. Gollub: *Chaotic Dynamics—An Introduction*, Cambridge University Press, Cambridge, 1990.
3. Dimarogonas, A. D., and S. Haddad: *Vibrations for Engineers*, Prentice-Hall, Englewood Cliffs, N.J., 1992.
4. Feigenbaum, M.: "Quantitative Universality for a Class of Nonlinear Transformations," *Journal of Statistical Physics*, vol. 19, pp. 25–52, 1978.
5. Gleick, J.: *Chaos*, Viking, New York, 1987.
6. Kelly, S. G.: "Nonlinear Phenomena in a Column of Liquid in a Rotating Manometer," *SIAM Review*, vol. 32, pp. 652–659, 1990.
7. Lazer, A. C., and P. J. McKenna: "Large Amplitude Oscillations in Suspension Bridges: Some New Connections with Nonlinear Analysis," *SIAM Review*, vol. 32, pp. 537–575, 1990.

8. McKenna, P. J., and W. Walter: "Nonlinear Oscillations in a Suspension Bridge," *Archives for Rational Mechanics and Analysis*, vol. 98, pp. 167–177, 1987.
9. Moon, F. C.: *Chaotic Vibrations, an Introduction for Applied Scientists and Engineers*, Wiley, New York, 1987.
10. Narayanan, S., and K. Jayaraman: "Chaotic Vibration in a Nonlinear Oscillator with Coulomb Damping," *Journal of Sound and Vibration*, vol. 146, pp. 17–31, 1988.
11. Nayfeh, A. H.: *An Introduction to Perturbation Methods*, Wiley-Interscience, New York, 1981.
12. Nayfeh, A. H., and D. T. Mook: *Nonlinear Oscillations*, Wiley-Interscience, New York, 1979.
13. Plaut, R. H., and J. C. Hsieh: "Chaos in a Mechanism with Time Delays Under Parametric and External Excitation," *Journal of Sound and Vibration*, vol. 114, pp. 73–90, 1987.
14. Rao, S. S.: *Mechanical Vibrations*, 2nd ed., Addison-Wesley, Reading, Mass., 1990.
15. Rokini, M. A., and B. S. Berger: "Chaotic Motion of a Two Link Mechanism," *Journal of Sound and Vibration*, vol. 147, pp. 349–357, 1988.
16. Thomson, W. T.: *Theory of Vibrations with Application*, 3rd ed., Prentice-Hall, Englewood Cliffs, N.J., 1988.
17. Tung, P. C., and S. W. Shaw: "The Dynamics of an Impact Print Hammer," *Journal of Vibrations, Acoustics, Stress, and Reliability in Design*, vol. 110, pp. 193–200, 1988.

CHAPTER 11

VIBRATION CONTROL

11.1 INTRODUCTION

The reciprocating machine of Fig. 11.1 is mounted on a rigid foundation. During operation the foundation is subject to a load consisting of a static component equal to the weight of the machine plus a harmonic component due to the inertia of any rotating unbalance.

The forge hammer of Fig. 11.2 is rigidly mounted to its foundation. During operation a large weight is dropped onto an anvil. The impact of the weight causes a large impulsive force to be transmitted to the foundation.

The pump of Fig. 11.3 supplies water to a pipeline. The operating speed of the pump is near a natural frequency of the pipeline, resulting in resonant vibrations.

In each of the preceding examples, an intolerable situation exists. In the first two examples, since the machines are rigidly mounted to their foundations, the foundation feels the full effect of the excitation. If an elastic element is placed between the machine and the foundation, the magnitude of the transmitted force can be changed. The system is said to be isolated if the magnitude of the transmitted force is less than the excitation force. The presence of the elastic element induces vibrations of the machine. A vibration isolator should be designed to reduce the transmitted force to a tolerable level, while keeping the machine vibrations at an acceptable level.

In the pipeline example, excessive vibrations occur due to the resonant operating condition. Vibrations can be reduced to an acceptable level by modifying the system characteristics. One technique is to add a vibration

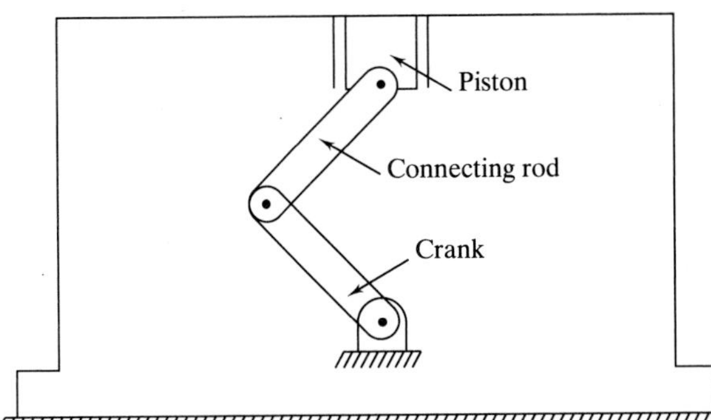

FIGURE 11.1
Schematic of a single cylinder reciprocating machine. During operation, inertia forces lead to harmonic excitation to foundation.

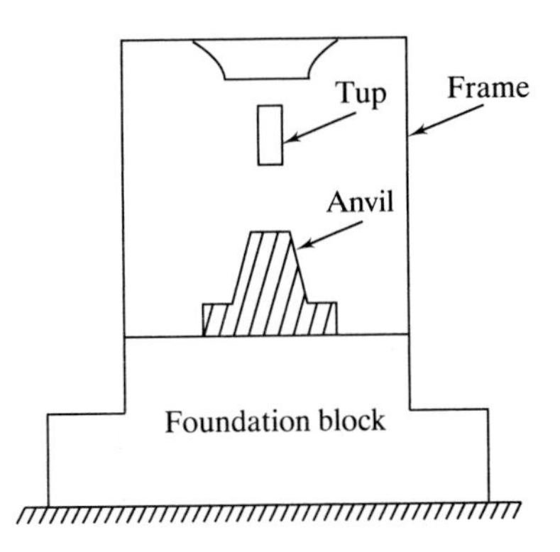

FIGURE 11.2
Schematic of a forge hammer. When hammer impacts anvil, a large impulsive force is produced.

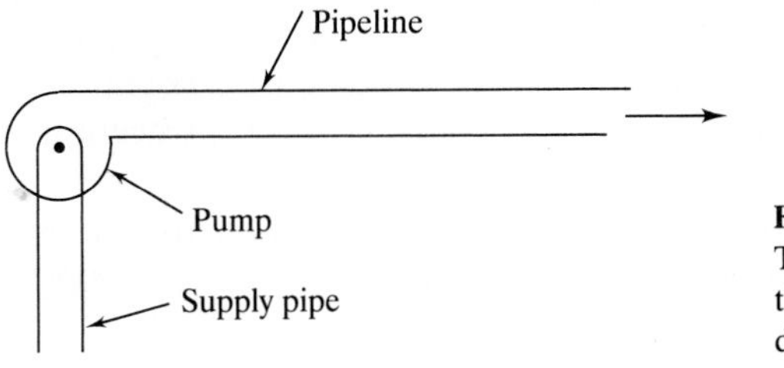

FIGURE 11.3
The operating speed of the pump is close to the lateral frequency of the pipeline, producing resonant vibrations.

absorber, which consists of a mass connected to the pipeline through an elastic element.

The reduction of unwanted vibrations in a mechanical or structural system is known as vibration control. The principles of vibration control are presented in this chapter. Vibration literature abounds with application of these theories to specific systems.

The general theory for one-degree-of-freedom systems is considered in this chapter as well as the extension to some multi-degree-of-freedom systems. Since vibration control often involves modifying parameters to minimize forces and displacements, all algebra is performed in terms of system parameters. The algebra is often intractable for multi-degree-of-freedom and continuous systems, and usually precludes the development of a general theory.

11.2 BASIC CONCEPTS OF VIBRATION ISOLATION

Problems in vibration isolation are of two classes.

1. Foundations and mountings are protected against large forces produced in equipment. Examples of this class of problems include protecting mountings from large inertia forces generated by reciprocating machines and protecting a foundation from a large impulsive force produced by a drop hammer.

A model one-degree-of-freedom system is shown in Fig. 11.4. The differential equation governing the displacement of the mass is

$$\ddot{x} + 2\zeta\omega_n\dot{x} + \omega_n^2 x = \frac{F(t)}{m} \tag{11.1}$$

where $\omega = \sqrt{k/m}$ is the system's natural frequency and $\zeta = c/(2m\omega_n)$ is the system's damping ratio. The force transmitted to the foundation through the spring and viscous damper is

$$F_T(t) = kx + c\dot{x} \tag{11.2}$$

2. Equipment is protected against motion of its foundation. Examples of this class of problems include protecting sensitive computer equipment from harmonic structural vibrations caused by other equipment or protecting a computer on a ship from sudden motions due to waves and rough seas.

A model one-degree-of-freedom system for this class of problems is shown in Fig. 11.5. The differential equation governing the relative displacement between the mass and its foundation is

$$\ddot{z} + 2\zeta\omega_n\dot{z} + \omega_n^2 z = -\ddot{y} \tag{11.3}$$

where $z = x - y$. The acceleration transmitted to the mass due to the motion of

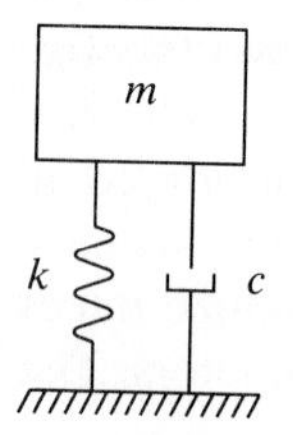

FIGURE 11.4
One-degree-of-freedom model of mass isolated from foundation through isolator.

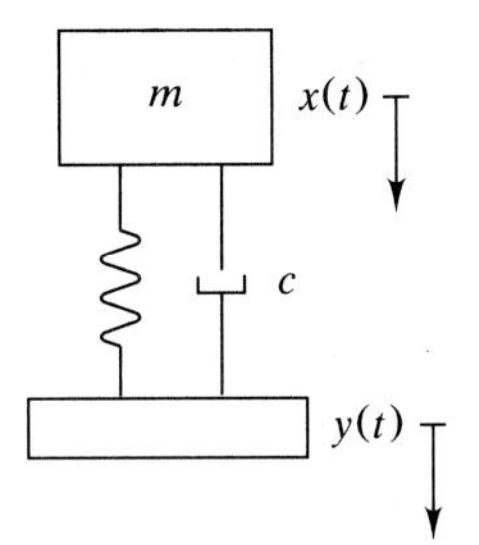

FIGURE 11.5
One-degree-of-freedom model of system protected from motion of base by isolator.

the base is given by

$$\begin{aligned}\ddot{x} &= \ddot{z} + \ddot{y} \\ &= -\left(2\zeta\omega_n z + \omega_n^2 z\right) \\ &= -(kz + c\dot{z})/m \\ &= -\frac{F_T}{m}\end{aligned} \tag{11.4}$$

where F_T is the total force developed in the spring and viscous damper.

For this class of problems the important quantities are the relative displacement of the mass and its absolute acceleration. Consider the case of sensitive equipment mounted in a rocket. After launch, the absolute displacement of the equipment is irrelevant. The displacement of the equipment relative to the rocket may be of interest due to space limitations. The absolute acceleration felt by the equipment may influence its operation.

The preceding discussion shows that the two classes of problems are equivalent. The base acceleration is analogous to the excitation force. The displacement results from class 1 are used to analyze the relative displacement for class 2. The results for transmitted force from class 1 are used to analyze absolute acceleration for class 2.

Often the force transmitted between the mass and the foundation is the primary concern. A large force can lead to large stresses in the foundation. However, space limitations often require that displacement considerations be considered in vibration control. For a given excitation Eq. (11.1) is solved and Eq. (11.2) is used to determine the time-dependent transmitted force. If $F(t)$ is harmonic, the steady-state amplitudes are constant and the important considerations. If $F(t)$ is impulsive, the short-time behavior is important. In either case, vibration control involves minimizing the maximum transmitted force, $F_{T_{\max}}$, the maximum displacement $x_{\max}$, or both.

Let F_0 be a characteristic magnitude and let t_0 be a characteristic time of the excitation. The maximum transmitted force and displacement are dependent upon these characteristic values and system properties

$$F_{T_{max}} = f_1(m, \zeta, \omega_n, F_0, t_0) \tag{11.5}$$

$$X_{max} = f_2(m, \zeta, \omega_n, F_0, t_0) \tag{11.6}$$

Equations (11.5) and (11.6) can be rearranged in nondimensional form

$$\frac{F_{T_{max}}}{F_0} = \bar{f}_1(\zeta, \omega_n t_0) \tag{11.7}$$

$$\frac{m\omega_n^2 X_{max}}{F_0} = \bar{f}_2(\zeta, \omega_n t_0) \tag{11.8}$$

If the excitation is harmonic the characteristic time is taken as the inverse of the excitation frequency and the frequency ratio $r = \omega/\omega_n$ is used in place of $\omega_n t_0$.

Vibration control involves choosing values of the dimensional parameters to minimize the maximum transmitted force and/or displacement. General principles of vibration isolation are obtained by studying the equations in their nondimensional form.

The simplest means of vibration control is to eliminate or modify the external energy source responsible for exciting the vibrations. Unbalanced rotors can be balanced. However, the cost of perfect balancing is usually too high. The harmonic excitation due to the inertia forces in single-cylinder reciprocating engines cannot be eliminated, but balancing is possible for multi-cylinder engines. Environmental conditions such as wave motion, wind speed, and earthquake motion are beyond control.

If the excitation is beyond control, then vibration control requires the design or modification of a system to mitigate the effects of the excitation. The focus of this chapter is the vibration control of one-degree-of-freedom systems.

11.3 VIBRATION ISOLATION THEORY

When the machine of Fig. 11.1 is bolted to the floor in an industrial plant, the floor is subject to a static load equal to the weight of the machine plus a repeated harmonic loading due to the rotating unbalance. If its magnitude is large enough, the repeated loading can lead to fatigue damage to the bolts and the structure and cause unwanted noise.

The magnitude of the repeating component of the transmitted force can be reduced by isolating the floor from the machine by placing the machine on an elastic foundation. The elastic foundation is modeled as springs in parallel with a viscous damper, as shown in Fig. 11.4. If

$$x(t) = X\sin(\omega t - \phi) \tag{11.9}$$

is due to a harmonic excitation

$$F(t) = F_0 \sin \omega t \tag{11.10}$$

then Eq. (11.2) gives

$$F_T(t) = |F_T|\sin(\omega t - \lambda) \tag{11.11}$$

where

$$|F_T| = F_0 T(r, \zeta) \tag{11.12}$$

where $T(r, \zeta)$ is defined in Eq. (4.66) as

$$T(r,\zeta) = \sqrt{\frac{1 + (2\zeta r)^2}{(1 - r^2)^2 + (2\zeta r)^2}} \tag{4.63}$$

The nondimensional ratio of Eq. (4.66) is called the transmissibility ratio. If the transmissibility ratio is less than one, the magnitude of the repeated force transmitted to the floor is less than the magnitude of the excitation force. When this occurs, the vibrations are said to be isolated. Figure 11.6 shows the transmissibility ratio as a function of frequency ratio for different values of the damping ratio. Isolation occurs only when $r > \sqrt{2}$. When isolation occurs, increased damping increases the transmissibility ratio. However, viscous damping is still necessary to limit the amplitude of vibration as the system passes through resonance.

The theory of vibration isolation is used to design isolation systems to protect structures from damage caused by moving components in pumps, turbines, compressors, presses, reciprocating engines, sewing machines, and other industrial equipment. It can also be used to protect equipment from

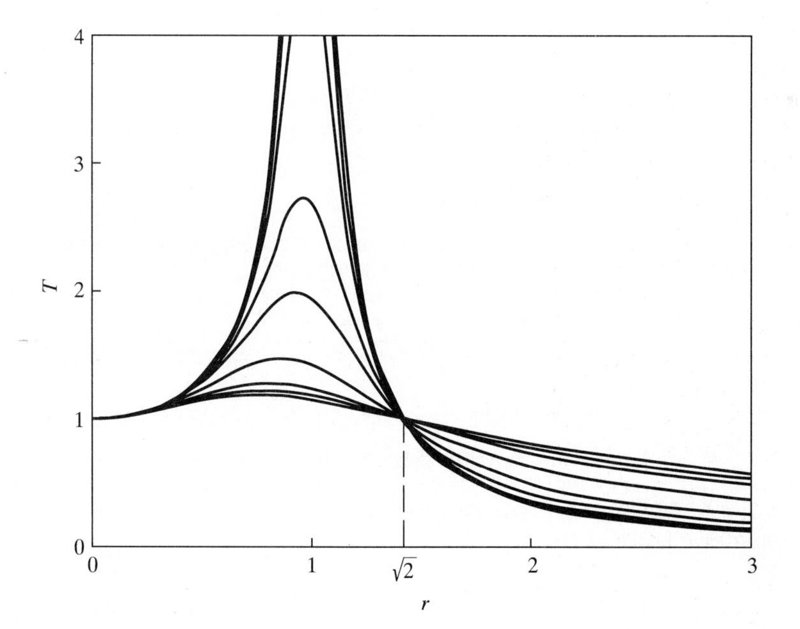

FIGURE 11.6
T vs. r for different values of ζ.

unwanted structural vibrations. Equation (4.64) gives the absolute displacement of a mass connected by springs and a viscous damper in parallel to a base which executes harmonic motion. It can be rearranged as

$$\frac{\omega^2 X}{\omega^2 Y} = T(r, \zeta) \tag{11.13}$$

The left side of Eq. (11.13) is the ratio of the maximum steady-state acceleration of the mass to the maximum acceleration of its base.

Example 11.1. A flow-monitoring device of mass 10 kg is to be installed to monitor the flow of a gas in a manufacturing process. Due to the operation of pumps and compressors, the floor of the plant vibrates with an amplitude of 4 mm at a frequency of 2500 rpm. Effective operation of the flow-monitoring device requires that its acceleration amplitude be limited to $5g$. What is the equivalent stiffness of an isolator with a damping ratio of 0.05 to limit the transmitted acceleration to an acceptable level? What is the maximum displacement of the flow-monitoring device and what is the maximum deformation of the isolator?

The acceleration amplitude of the floor is

$$\omega^2 Y = \left(2500 \frac{\text{rev}}{\text{min}} 2\pi \frac{\text{rad}}{\text{rev}} 1 \frac{\text{min}}{60 \text{ s}}\right)^2 (0.004 \text{ m}) = 274.1 \frac{\text{m}}{\text{s}^2} = 27.95g$$

The maximum allowable transmissibility ratio is

$$T_{\text{max}} = \frac{\omega^2 X}{\omega^2 Y} = \frac{5g}{27.95g} = 0.179$$

From Eq. (4.66) with $\zeta = 0.05$,

$$0.179 = \sqrt{\frac{1 + 0.01r^2}{1 - 1.99r^2 + r^4}}$$

Solution of the preceding equation gives the minimum frequency ratio for which vibrations are sufficiently isolated. It yields $r > 2.60$. Thus

$$\omega_n < \frac{\omega}{2.60} = 100.7 \frac{\text{rad}}{\text{s}}$$

The maximum stiffness of the isolator is

$$k = m\omega_n^2 = 1.01 \times 10^5 \text{ N/m}$$

When $T = 0.179$, Eq. (4.64) is used to calculate the steady-state amplitude of the flow-monitoring device as

$$X = YT = (0.004 \text{ m})(0.179) = 0.72 \text{ mm}$$

Since the isolator is placed between the floor and the flow-monitoring device, its deformation is equal to the relative displacement between the floor and the device.

The steady-state amplitude of the relative displacement is calculated using Eq. (4.61)

$$Z = \Lambda Y = \frac{r^2 Y}{\sqrt{(1 - r^2)^2 + (2\zeta r)^2}} = 4.69 \text{ mm}$$

When the magnitude of the excitation force is independent of the excitation frequency, if sufficient isolation is achieved at a certain speed, then greater isolation is achieved at higher speeds. The transmitted force can theoretically be made as small as desired by making the natural frequency of the machine-isolator system small enough.

A special case occurs when the amplitude of the excitation force is proportional to the square of the excitation frequency, as for the harmonic excitation due to a rotating unbalance. Since the maximum allowable force transmitted to the foundation is independent of the frequency of excitation, the percentage of isolation required varies with the frequency. When the excitation is due to a rotating unbalance, Eq. (11.12) becomes

$$\frac{F_T}{m_0 e \omega^2} = T(r, \zeta) \tag{11.14}$$

or

$$\frac{F_T}{m_0 e \omega_n^2} = r^2 T(r, \zeta) = R(r, \zeta) \tag{11.15}$$

The nondimensional function $R(r, \zeta)$ is defined as

$$R(r, \zeta) = r^2 \sqrt{\frac{1 + (2\zeta r)^2}{(1 - r^2)^2 + (2\zeta r)^2}} \tag{11.16}$$

$R(r, \zeta)$ is plotted in Fig. 11.7. The following is noted about its behavior

1. $R(r, \zeta)$ is asymptotic to the line $f(r) = 2\zeta r$ for large r. That is,

$$\lim_{r \to \infty} R(r, \zeta) = 2\zeta r \tag{11.17}$$

2. For $\zeta < 0.354$, $R(r, \zeta)$ increases with increasing r, from 0 at $r = 0$ and reaches a maximum value. R then decreases and reaches a relative minimum. As r increases from the value where the minimum occurs, R grows without bound and approaches the asymptotic limit given by Eq. (11.17). The values of r where the maximum and relative minimum occur are obtained by setting $dR/dr = 0$, yielding

$$1 + (8\zeta^2 - 1)r^2 + 8\zeta^2(2\zeta^2 - 1)r^4 + 2\zeta^2 r^6 = 0 \tag{11.18}$$

Equation (11.18) is a cubic polynomial in r^2. It has three roots. One root is the value of r where the maximum occurs, another is the value of r where

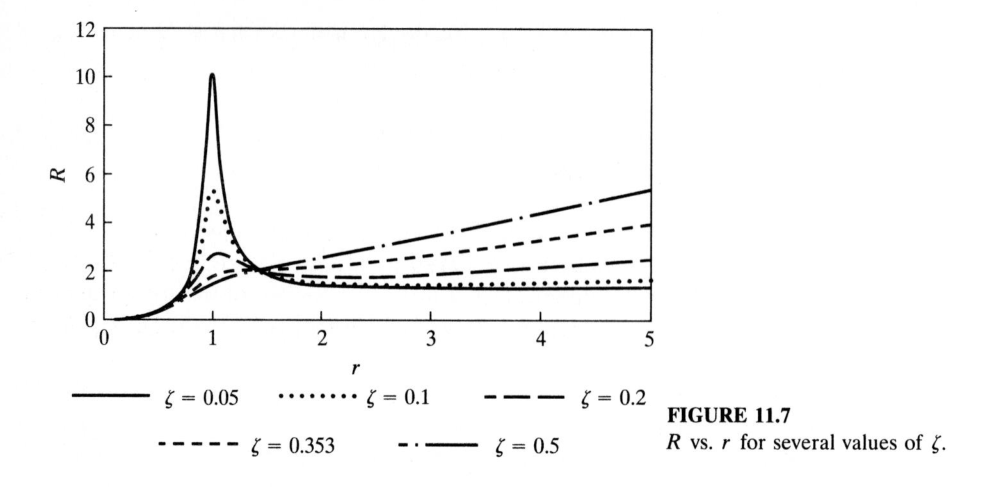

FIGURE 11.7
R vs. r for several values of ζ.

the relative minimum occurs, and one root is negative and irrelevant. Figure 11.8 shows the value of r for which the minimum occurs as a function of ζ. Figure 11.9 shows the corresponding value of R at its relative minimum.

3. $R = 2$ for $r = \sqrt{2}$ for all values of ζ.
4. Equation (11.18) has a double root of $r = \sqrt{2}$ for $\zeta = \sqrt{2}/4 = 0.354$. The maximum and minimum coalesce for this value of ζ. For $\zeta = 0.354$, $R = 2$ at $r = \sqrt{2}$ is an inflection point.
5. For $\zeta > \sqrt{2}/4$, Eq. (11.18) has no positive roots. Thus R does not reach a maximum, but grows without bound from $R = 0$ at $r = 0$.

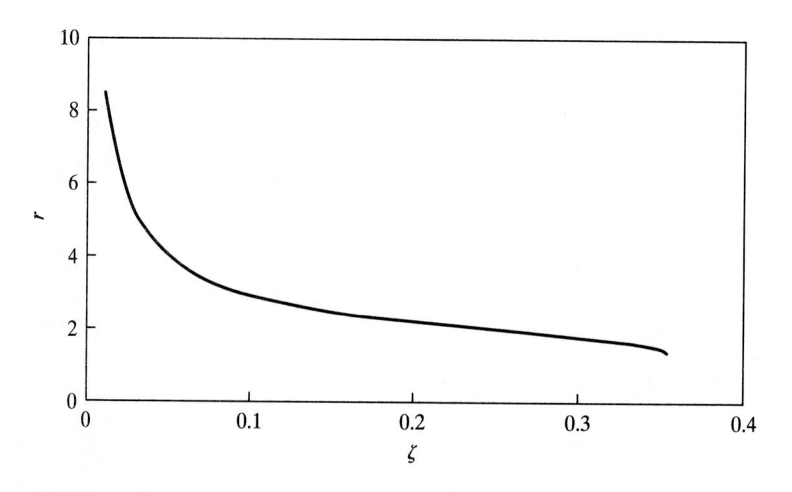

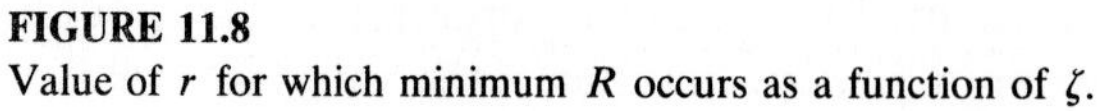

FIGURE 11.8
Value of r for which minimum R occurs as a function of ζ.

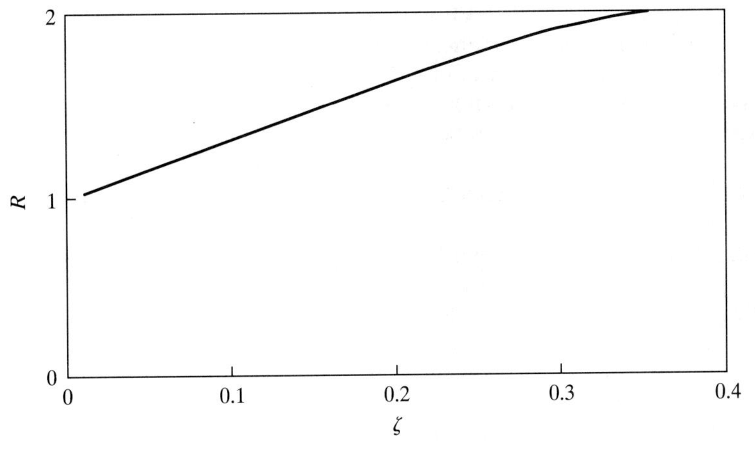

FIGURE 11.9

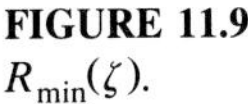
$R_{\min}(\zeta)$.

If the natural frequency of a system whose vibrations are due to a rotating unbalance is fixed, Fig. 11.7 shows that the transmitted force has a minimum for some value of r. If r exceeds this value, the force increases without bound as r increases. If ζ is small, the curve in the vicinity of the relative minimum is flat. The transmitted force varies little over a range of r. This suggests that for situations where vibrations must be isolated over a range of excitation frequencies, it is best to choose ω_n such that the value of r at the center of the operating range is near the value of r for which the relative minimum occurs.

The limit process used to develop Eq. (11.17) is performed for a fixed value of ω_n as ω is increased. Thus for a fixed ω_n the transmitted force approaches $m_0 e \omega \omega_n$. The limit of F_T as ω_n goes to zero for a fixed ω is zero. Thus decreasing the natural frequency decreases the magnitude of the transmitted force for a specific excitation frequency. Decreasing the natural frequency such that the minimum is to the left of the operating range reduces the magnitude of the repeating component of the transmitted force over a portion of the operating range. However, the transmitted force may vary greatly over the operating range.

Example 11.2. A 250-kg pump operates at speeds between 1000 rpm and 2400 rpm and has a rotating unbalance of 2.5 kg · m. The pump is placed at a location in an industrial plant where it has been determined that the maximum repeated force that should be applied to the floor is 15,000 N. Specify the stiffness of an isolator of damping ratio 0.1 that can be used to reduce the repeating component of the transmitted force to an acceptable level.

If the pump is placed directly on the floor, the repeating component of the transmitted force is 27,400 N at 1000 rpm and 157,800 N at 2400 rpm. Thus isolation is necessary.

From Fig. 11.8, for $\zeta = 0.1$ the minimum value of R occurs for $r = 2.94$. If ω_n is chosen such that $r = 2.94$ is at the center of the operating range, then

$$\omega_n = \frac{1700 \text{ rpm}}{2.94} = 578.2 \text{ rpm} = 60.55 \frac{\text{rad}}{\text{s}}$$

At the lower end of the operating range, the frequency ratio is 1.73 and the transmitted force is

$$F_T = m_0 e \omega_n^2 R(1.73, 0.1)$$

$$= 2.5 \text{ kg} \cdot \text{m}\left(60.55 \frac{\text{rad}}{\text{s}}\right)^2 (1.73)^2 \sqrt{\frac{1 + (0.346)^2}{\left[1 - (1.73)^2\right]^2 + (0.346)^2}}$$

$$= 14{,}350 \text{ N}$$

At the upper end of the operating range, the frequency ratio is 4.15 and the transmitted force is

$$F_T = m_0 e \omega_n^2 R(4.15, 0.1)$$

$$= (2.5 \text{ kg} \cdot \text{m})\left(60.55 \frac{\text{rad}}{\text{s}}\right)^2 (4.15)^2 \sqrt{\frac{1 + (0.830)^2}{\left[1 - (4.15)^2\right]^2 + (0.830)^2}}$$

$$= 12{,}630 \text{ N}$$

Vibration isolation of a system subject to a multifrequency excitation can be difficult, especially if the lowest frequency is very low. Consider a system subject to an excitation composed of n harmonics

$$F(t) = \sum_{i=1}^{n} F_i \sin(\omega_i t + \psi_i) \tag{11.19}$$

The principle of linear superposition is used to calculate the total response of the system due to this excitation. Equation (11.2) is used to calculate transmitted force leading to

$$F_T(t) = \sum_{i=1}^{n} T(r_i, \zeta) F_i \sin(\omega_i t + \psi_i - \lambda_i) \tag{11.20}$$

where

$$r_i = \frac{\omega_i}{\omega_n}$$

Since the harmonic terms of Eq. (11.20) are out of phase, their maxima occur at different times. A closed-form expression for the absolute maximum is difficult to attain. The following is used as an upper bound:

$$F_{T_{\max}} < \sum_{i=1}^{n} F_i T(r_i, \zeta) \tag{11.21}$$

Equation (11.21) can be viewed as an equation which provides an approximation to the upper bound of the natural frequency. If the natural frequency is less than the calculated upper bound, the repeating force transmitted to the floor is always less than the allowable force. An initial guess for the upper bound is obtained by determining the natural frequency such that the transmitted force due to the lowest-frequency harmonic only is reduced to F_T. Since additional forces at higher frequencies are present, greater isolation is required. The natural frequency can be systematically reduced from this initial guess, checking Eq. (11.21), until an upper bound is obtained.

Example 11.3. The 500-kg punch press of Example 4.16 is to be mounted on an isolator such that the maximum of the repeating force transmitted to the floor is 1000 N. Determine the required static deflection of an isolator assuming a damping ratio of 0.1. What is the resulting maximum deflection of the isolator during the punching operation?

From Example 4.16 the excitation force is periodic and is expressed using a Fourier series as

$$F(t) = 2000 + \frac{5000\sqrt{2}}{\pi} \sum_{i=1}^{\infty} \frac{1}{i} \sqrt{1 - \cos 0.8\pi i} \sin(4\pi i t + \psi_i) \text{ N}$$

The 2000-N term is the average force applied to the punch during one cycle. It contributes to the total static load applied to the floor and is not part of the repeating load. Application of Eq. (11.21) to the repeated components of loading gives

$$1000 > \frac{5000\sqrt{2}}{\pi} \sum_{i=1}^{\infty} \frac{1}{i} \sqrt{1 - \cos 0.8\pi i}\, T(r_i, \zeta)$$

where

$$r_i = \frac{4\pi i}{\omega_n} = i r_1$$

An initial guess for an upper bound for the natural frequency is obtained by calculating r_1 such that the transmitted force due to the lowest-frequency harmonic is less than 1000 N. This leads to

$$1000 = \frac{5000}{\pi} \sqrt{2(1 - \cos 0.8\pi)} \sqrt{\frac{1 + (0.2 r_1)^2}{(1 - r_1^2)^2 + (0.2 r_1)^2}}$$

which gives $r_1 = 1.54$. Defining

$$f(r_1) = \frac{5000\sqrt{2}}{\pi} \sum_{i=1}^{\infty} \frac{1}{i} \sqrt{1 - \cos 0.8\pi i}\, T(i r_1, \zeta)$$

it is desired to solve

$$f(r_1) = 1000$$

A lower bound on the value of r_1 that solves the preceding equation is 1.54. A trial-and-error solution is used to determine r_1:

r_1	$f(r_1)$
1.54	2459
1.56	2346
1.58	2242
1.60	2141
1.62	2043
1.64	1949
1.66	1858
1.68	1780
1.70	1709
1.72	1640
1.74	1572
1.76	1506
1.78	1443
1.80	1383
1.82	1325
1.84	1267
1.86	1210
1.88	1155
1.90	1106
1.92	1058
1.94	1010
1.96	964

Using $r_1 = 1.95$, an upper bound for the natural frequency is calculated as

$$\omega_n = \frac{\omega_1}{1.95} = \frac{4\pi}{1.95} = 6.44 \frac{\text{rad}}{\text{s}}$$

The required static deflection of the isolator is

$$\Delta_{\text{st}} = \frac{g}{\omega_n^2} = 236 \text{ mm}$$

This static deflection is excessive and requires a flexible foundation. In addition, the average force of 2000 N causes an additional deflection of the isolator during operation. If an isolator is used such that the system natural frequency is 6.44 rad/s, the maximum deflection of the isolator from its undeflected state is the static deflection plus the constant deflection due to the average load plus the sum of the amplitudes of the harmonic components

$$\Delta_{\max} < \Delta_{\text{st}} + \frac{1}{m\omega_n^2}\left(2000 + \frac{5000\sqrt{2}}{\pi}\sum_{i=1}^{\infty}\frac{1}{i}\sqrt{1-\cos 0.8\pi i}\,M(r_i,\zeta)\right)$$

or $\quad \Delta_{\max} < 247.1 \text{ mm}$

11.4 PRACTICAL ASPECTS OF VIBRATION ISOLATION

Vibration isolation is required in a variety of military and industrial applications. Isolation is required to reduce the force transmitted between a machine and its foundation during ordinary operation or to isolate a machine from vibrations of its surroundings. Motors are often isolated to protect mountings from forces arising from harmonic variation of torque and unbalanced rotors. Electrical components such as transformers and circuit breakers are isolated to protect surroundings from electromagnetic forces generated in solenoids or as a result of alternating current. Large harmonic inertia forces are developed by rotating components of single-cylinder reciprocating engines. Isolation is required to protect the engine mounting from these forces. Other machines with rotating components such as fans, pumps, and presses are often isolated to protect against inherent rotating unbalance.

The maximum stiffness of an isolator required for a particular application is calculated using the theory of Sec. 11.3. A one-degree-of-freedom system using an isolator is modeled as the simple mass-spring-dashpot system of Fig. 11.4.

Specifications provided in catalogs of commercially available isolators include allowable static deflections. If the isolated system of Fig. 11.4 has a minimum required natural frequency ω_n, the required minimum static deflection of the isolator is

$$\Delta_{st} = \frac{g}{\omega_n^2} \tag{11.22}$$

Isolation of low-frequency vibrations requires a small natural frequency, which leads to a large isolator static deflection.

The vibration amplitude of a machine during operation is calculated from Eq. (4.22),

$$\frac{m\omega_n^2 X}{F_0} = M(r, \zeta)$$

Multiplying both sides of the preceding equation by r^2 leads to

$$\frac{m\omega^2 X}{F_0} = r^2 M(r, \zeta) = \Lambda(r, \zeta) \tag{11.23}$$

where $\Lambda(r, \zeta)$ is defined in Eq. (4.30). Since vibration isolation requires $r > \sqrt{2}$ and $\Lambda(r, \zeta)$ decreases and approaches 1 as r increases, the steady-state amplitude decreases as isolation is improved. However, for fixed m, F_0, and ω the steady-state amplitude has a lower bound given by

$$X > \frac{F_0}{m\omega^2} \tag{11.24}$$

Equations (11.23) and (11.24) show that if an isolator is being designed to provide isolation over a range of frequencies, the steady-state amplitude is greatest at the lowest operating speed.

Since vibration isolation requires $r > \sqrt{2}$, the speed at which the maximum vibration amplitude occurs must be passed during start-up and stopping.

The maximum vibration amplitude for a fixed ω_n is obtained using Eq. (4.23) as

$$X_{\max} = \frac{F_0}{m\omega_n^2} \frac{1}{2\zeta\sqrt{1-\zeta^2}} \tag{11.25}$$

The smaller the natural frequency, the larger is the resonant amplitude. In addition, the larger the damping ratio, the smaller the resonant amplitude.

A large vibration amplitude can lead to ineffective operation of machinery. Large-amplitude vibrations of machines which must be properly aligned with devices that feed materials to the machine can lead to improper alignment and improper operation. Many machine tools are designed requiring a rigid foundation for effective operation. Equations (11.23) and (11.25) show that one way to reduce the amplitude of vibration during operation and the resonant amplitude is to increase the mass of the isolated system. Equation (11.24) shows that the only way to reduce the amplitude below a calculated value at a given operating speed is to increase the system mass. Increasing the mass allows a proportional increase in the stiffness required to achieve sufficient isolation.

The mass of a system can be increased by rigidly mounting the machine on a block of concrete. A small machine can be mounted above ground, while a large machine is usually mounted in a specially designed pit, The static load applied to the isolator and the mounting is increased when the mass of the system is increased.

Example 11.4. A milling machine of mass 450 kg operates at 1800 rpm and has an unbalance which causes a harmonic repeated force of magnitude 20,000 N. Design an isolation system to limit the transmitted force to 4000 N, the amplitude of vibration during operation to 1 mm, and the amplitude of vibration during start-up to 10 mm. Specify the required stiffness of the isolator and the minimum mass that should be added to the machine. Assume a damping ratio of 0.05.

The maximum allowable transmissibility is

$$T = \frac{4000 \text{ N}}{20{,}000 \text{ N}} = 0.2$$

The minimum frequency ratio is determined by solving

$$0.2 = \sqrt{\frac{1 + 0.01r^2}{1 - 1.99r^2 + r^4}}$$

which yields $r = 2.48$ and a maximum natural frequency of

$$\omega_n = \frac{\omega}{2.48} = 76.00 \frac{\text{rad}}{\text{s}}$$

The maximum amplitude during start-up for the 450-kg machine mounted on an isolator such that the system natural frequency is 76.00 rad/s is

$$X_{\max} = \frac{20{,}000 \text{ N}}{(450 \text{ kg})(76.0 \text{ rad/s})^2} \frac{1}{2(0.05)\sqrt{1-(0.05)^2}} = 76.9 \text{ mm}$$

The resonant amplitude can be decreased to 10 mm only by increasing the mass to

$$m = \frac{20{,}000\ \text{N}}{(0.01\ \text{m})(76.00\ \text{rad/s})^2} \frac{1}{2(0.05)\sqrt{1 - (0.05)^2}} = 3460\ \text{kg}$$

When the mass is increased to 3460 kg, the amplitude of vibration of the milling machine, when operating at 1800 rpm, is

$$X = \frac{20{,}000\ \text{N}}{(3460\ \text{kg})(76.00\ \text{rad/s})^2} \frac{1}{\sqrt{\left[1 - (2.48)^2\right]^2 + [2(0.05)(2.48)]^2}}$$

$$= 0.19\ \text{mm}$$

The isolator stiffness is calculated by

$$k = m\omega_n^2 = (3460\ \text{kg})(76.00\ \text{rad/s})^2 = 2.0 \times 10^7\ \text{N/m}$$

The milling machine should be mounted on a concrete block of mass 3010 kg and the system isolated by springs with an equivalent stiffness of 2×10^7 N/m.

There are three classes of isolators in general use. The choice of an isolator for a particular application depends upon the constraints noted previously (static deflection, vibration amplitude, and resonant amplitude), as well as other factors such as cost, weight, and space limitations, the amount of damping required, and environmental conditions.

Helical coil steel springs are used as isolators when large static deflections (>1 in. or 3 cm) are required and a flexible foundation is acceptable. This occurs when good isolation is required at low operating speeds. Hysteresis in steel springs is low, so discrete viscous dampers are used in parallel with the springs to provide adequate damping. Steel springs may be used in combination with other isolation methods when a machine must be mounted on a concrete block. These isolators can be designed for specific use or can be obtained commercially.

Isolators made of elastomers are used in applications where small static deflections are required. If used for larger static loads, the elastomers are subject to creep, reducing their effectiveness after a period of time. Caution should be taken in using these isolators in extreme temperatures. Hysteretic damping inherent in the isolators is usually sufficient. However, discrete dampers can be employed in conjunction with these isolators. The damping ratio of an isolator depends upon the elastomeric material from which it is made, the steady-state frequency and the amplitude. The damping ratio for isolators made of natural rubber varies little with amplitude but is highly dependent upon frequency. The damping ratio of a natural rubber isolator at 200 Hz is $\zeta = 0.03$ while $\zeta = 0.09$ at 1200 Hz.

Figure 11.10 is a copy of a page from the catalog of Stock Drive Products, a commercial manufacturer of vibration isolators. These isolators are usually used for small machines where the required static deflection does not exceed half an inch. The load rating given refers to the static load that the isolator can

FOR COMPRESSION LOADS TO 120 POUNDS
SHEAR LOADS TO 63 POUNDS

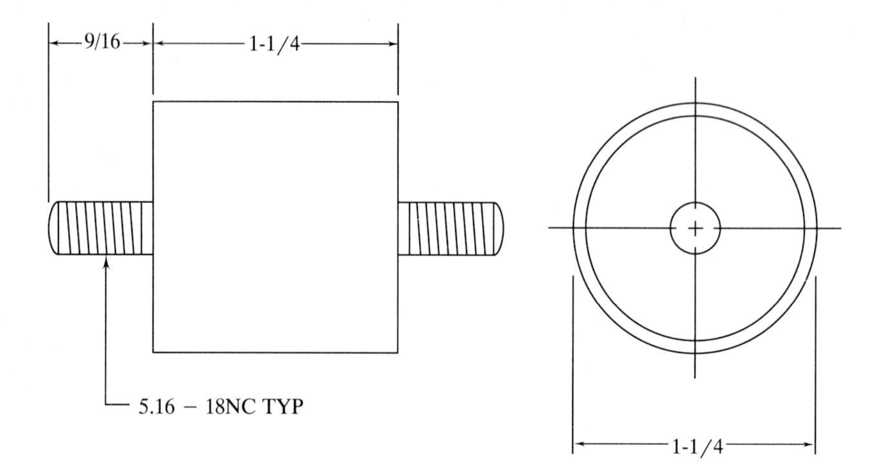

Compression		Minimum Load for 81% Isolation (lbs)									
		Forcing Frequency in Cycles per Minute									
Load Rating	Maximum Load (lbs)	600	850	950	1100	1250	1500	1750	2000	2500	3000
A	41				34.5	27.5	19	14	10	7	
B	64					48	32	24	17.5	12	8.5
C	90					80	55	41.5	30	20	14
D	120						89	70.5	53	38.5	26.5

Shear		Minimum Load for 81% Isolation (lbs)									
		Forcing Frequency in Cycles per Minute									
Load Rating	Maximum Load (lbs)	600	850	950	1100	1250	1500	1750	2000	2500	3000
A	21	20	11.0	8.5	6.7	5.5	*	*	*	*	*
B	31		18	14	10.5	8	5.5	*	*	*	*
C	48		31.5	25	19.5	15.5	11	8.5	*	*	*
D	63		50	41	32.6	27.5	20.5	16	14	8	*

*At these forcing frequencies lesser loads will yield 81% isolation.

FIGURE 11.10
Example page from *Stock Drive Products Catalog* showing vibration isolator and its properties. (*Courtesy of Stock Drive Products.*)

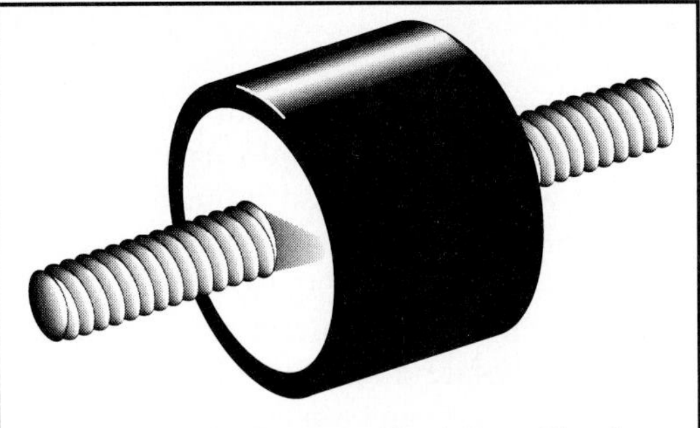

MATERIAL: Fastener—Steel, Brass Plated
Isolator—Natural Rubber

LOAD DEFLECTION GRAPHS

Deflections below the line X---X are considered safe practice for static loads; data above that line are useful for calculating deflections under dynamic loads.

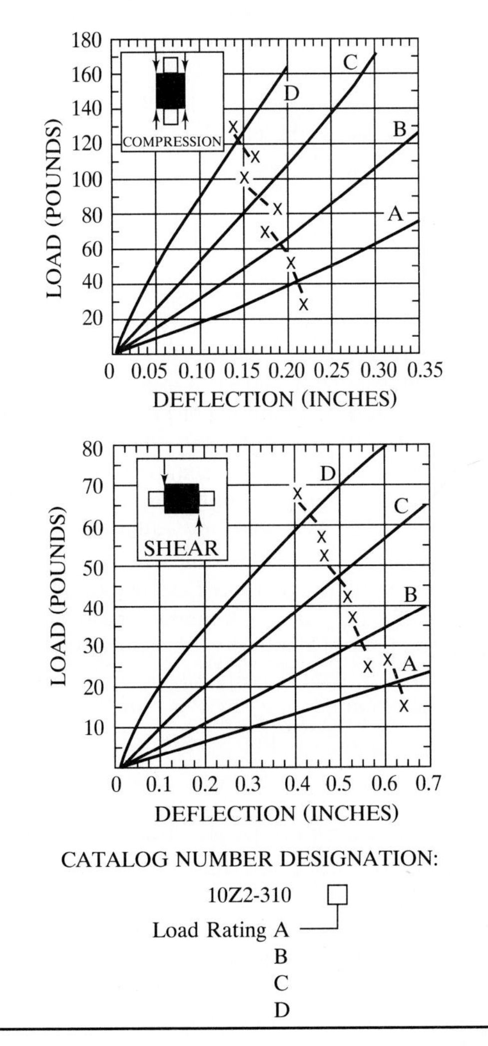

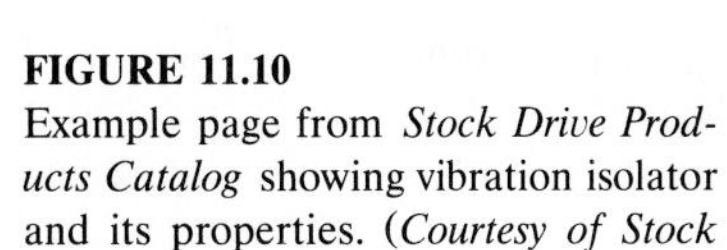

FIGURE 11.10
Example page from *Stock Drive Products Catalog* showing vibration isolator and its properties. (*Courtesy of Stock Drive Products.*)

tolerate. An isolator that can tolerate a larger static load can supply a smaller static deflection. Isolators with different load ratings are made from different structural rubber compounds. The isolators with a higher load rating are harder and have a higher damping ratio. The damping ratios generally range from 0.02 for those with load rating A to 0.1 for those with load rating D.

Pads made of materials such as cork, felt, or elastomeric resin are often used to isolate large machines. Pads used to isolate a specific machine can be cut from larger pads. Pads of prescribed thicknesses can be placed on top of one another, acting as springs in series, to provide increased flexibility.

Example 11.5. A small rotor balancing machine of weight 80 lb is to be placed on a table in a laboratory. The table vibrates due to operation of other equipment in the laboratory. An accelerometer was used to monitor the vibrations of the table. Analysis of its output showed that the vibrations of the table are made up of a number of components at different frequencies. However, the largest contribution is from the lowest frequency, an acceleration amplitude of 10 in./s^2 at a frequency of 150 rad/s. Accurate use of the rotor balancing machine requires that its base acceleration be limited to 1.5 in./s^2. If balanced four-point mounting is used, can the isolator of Fig. 11.10, mounted in compression, be used to sufficiently isolate the rotor balancing machine from the vibrations of the table?

The maximum transmissibility ratio is

$$T = \frac{1.5 \text{ in./s}^2}{10.0 \text{ in./s}^2} = 0.15$$

Assuming a damping ratio of 0.1, a frequency ratio greater than 2.95 is required. The maximum natural frequency is 50.85 rad/s and the required isolator static deflection is 0.150 in.

If four-point mounting is used, each isolator is subject to a static load of 20 lb. From the force-deflection curve for compression mounting of Fig. 11.10, a 20-lb load only produces an isolator deflection of 0.1 in. However, this isolator can tolerate a static deflection of 0.15 in., which will occur if the static load is 30 lb. This suggests that the rotor balancing machine can be mounted on a 40-lb block. Then the isolator of Fig. 11.10 can be used to sufficiently isolate the rotor balancing machine from the table vibrations.

Since an isolator of load rating A is used, the preceding calculations can be modified by assuming a damping ratio of 0.02. Then a frequency ratio of 2.78 sufficiently isolates vibrations. This leads to an isolator static deflection of 0.133 in. A static load of 26.6 lb on each isolator is required to provide the required deflection. Thus the machine needs to be mounted on a block of only 26.4 lb.

11.5 SHOCK ISOLATION

If the forge hammer of Fig. 11.2 is rigidly mounted to the foundation, the foundation is subject to a large impulsive force when the hammer impacts the anvil. An isolation system modeled as a spring and viscous damper in parallel can be designed to reduce the magnitude of the force to which the foundation is subject. The principles used in the design of a shock isolation system are similar to the principles used to design an isolation system to protect against harmonic excitation, but the equations are different.

If the duration, t_0, of a transient excitation, $F(t)$, is small, say $t_0 < T/3$ where T is the natural period of the system, then the system response can be adequately approximated by the response due to an impulse of magnitude

$$I = \int_0^{t_0} F(t)\, dt \tag{11.26}$$

If the system is at rest in equilibrium when a pulse of short duration is applied, the principle of impulse-momentum is used to calculate the velocity imparted to the mass as

$$v = \frac{I}{m} \tag{11.27}$$

The impulse provides external energy to initiate vibrations. Time is measured beginning immediately after the excitation is removed. The ensuing response is the free-vibration response due to an impulse providing the mass with an initial velocity, v.

$$x(t) = \frac{v}{\omega_d} e^{-\zeta\omega_n t} \sin \omega_d t \tag{11.28}$$

The maximum displacement occurs at a time

$$t_m = \tan^{-1}\left(\frac{\sqrt{1-\zeta^2}}{\zeta}\right) \tag{11.29}$$

and is equal to

$$x_{\max} = \frac{v}{\omega_n} \exp\left[-\frac{\zeta}{\sqrt{1-\zeta^2}} \tan^{-1}\left(\frac{\sqrt{1-\zeta^2}}{\zeta}\right)\right] \tag{11.30}$$

Equations (11.2) and (11.28) and trigonometric identities are used to calculate the force transmitted to the foundation through the isolator as

$$F_T(t) = \tilde{F} e^{-\zeta\omega_n t} \sin(\omega_d t - \beta) \tag{11.31}$$

where

$$\tilde{F} = \frac{m\omega_n v}{\sqrt{1-\zeta^2}} \tag{11.32}$$

and

$$\beta = -\tan^{-1}\left(\frac{2\zeta\sqrt{1-\zeta^2}}{1-2\zeta^2}\right) \tag{11.33}$$

The maximum value of the transmitted force is obtained by differentiating Eq. (11.31) with respect to time, solving for the smallest time for which the derivative is zero, and finding the transmitted force at this time. The time for which the maximum transmitted force occurs is

$$t_{m_F} = \frac{1}{\omega_d} \tan^{-1}\left[\frac{\sqrt{1-\zeta^2}(1-4\zeta^2)}{\zeta(3-4\zeta^2)}\right] \tag{11.34}$$

The corresponding maximum transmitted force is

$$F_{T_{\max}} = mv\omega_n \exp\left(-\frac{\zeta}{\sqrt{1-\zeta^2}} \tan^{-1}\left[\frac{\sqrt{1-\zeta^2}(1-4\zeta^2)}{\zeta(3-4\zeta^2)}\right]\right) \tag{11.35}$$

Equation (11.34) shows that the maximum transmitted force occurs at $t = 0$ for $\zeta = 0.5$. For $\zeta > 0.5$, the first time where $dF/dt = 0$ corresponds to a minimum. Thus, for $\zeta \geq 0.5$, the maximum transmitted force occurs at $t = 0$ and is given by

$$F_T(0) = cv = 2\zeta m\omega_n v \tag{11.36}$$

Equations (11.35) and (11.36) are combined to develop a nondimensional function $Q(\zeta)$ that is a measure of the maximum transmitted force, defined by

$$Q(\zeta) = \frac{F_T}{mv\omega_n} = \begin{cases} \exp\left(-\dfrac{\zeta}{\sqrt{1-\zeta^2}} \tan^{-1}\left[\dfrac{\sqrt{1-\zeta^2}(1-4\zeta^2)}{\zeta(3-4\zeta^2)}\right]\right) & \zeta < 0.5 \\ 2\zeta & 0.5 \leq \zeta < 1 \end{cases} \tag{11.37}$$

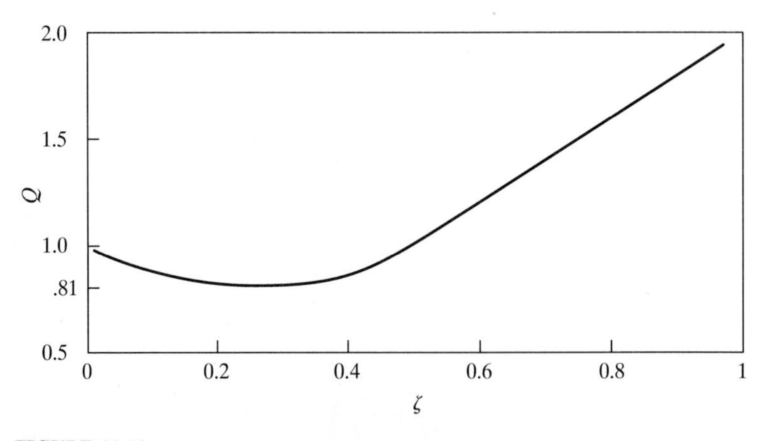

FIGURE 11.11
Shock isolation function for short duration pulse.

Figure 11.11 shows that $Q(\zeta)$ is flat and approximately equal to 0.81 for $0.23 < \zeta < 0.30$. If minimization of the transmitted force is the sole criterion for the isolator design, the isolator should have a damping ratio near 0.25.

Equation (11.37) shows that, for a given ζ, the transmitted force is proportional to the natural frequency. Thus a low natural frequency and large natural period is necessary and the short duration assumption is often valid.

Equation (11.30) shows that the maximum displacement varies inversely with the natural frequency. Thus requiring a small transmitted force leads to a large displacement. The natural frequency is eliminated between Eqs. (11.30) and (11.37), yielding

$$\frac{F_{T_{\max}} x_{\max}}{\frac{1}{2}mv^2} = S(\zeta) \tag{11.38}$$

where

$$S(\zeta) = \begin{cases} 2\exp\left(-\dfrac{\zeta}{\sqrt{1-\zeta^2}}\tan^{-1}\left[\dfrac{\zeta\sqrt{1-\zeta^2}(4-8\zeta^2)}{8\zeta^2-8\zeta^4-1}\right]\right) & \zeta \le 0.5 \\ 4\zeta\exp\left[-\dfrac{\zeta}{\sqrt{1-\zeta^2}}\tan^{-1}\left(\dfrac{\sqrt{1-\zeta^2}}{\zeta}\right)\right] & 0.5 < \zeta < 1 \end{cases} \tag{11.39}$$

The denominator of the nondimensional ratio of Eq. (11.38) is the initial kinetic energy of the system. The numerator is a measure of the work done by the transmitted force. The inverse of this ratio is the fraction of energy absorbed by the isolator, the isolator efficiency. Figure 11.12 shows that the maximum isolator efficiency occurs for $\zeta = 0.40$ where $S = 1.04$.

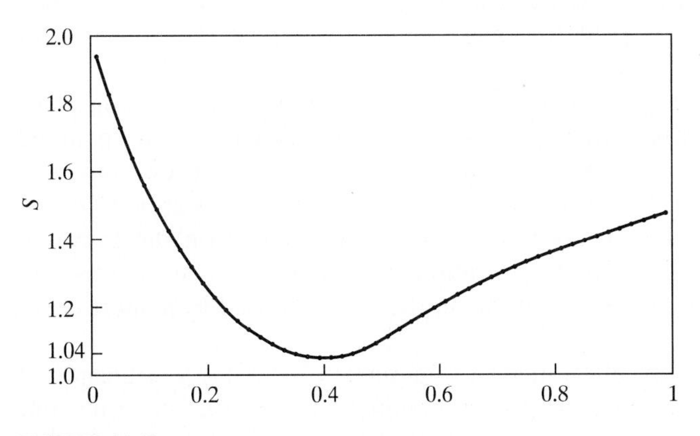

FIGURE 11.12
Energy absorption for short duration pulse.

Example 11.6. The 200-kg hammer of a 1000-kg forge hammer is dropped from a height of 1 m. Design an isolator to minimize the maximum displacement when the maximum force transmitted to the foundation is 20,000 N. What is the maximum displacement of the hammer when placed on this isolator?

The excitation is a result of the impact of the hammer with the anvil, and thus of short duration. The velocity of the anvil at the time of impact is

$$v = \sqrt{2(9.81 \text{ m/s}^2)(1 \text{ m})} = 4.43 \text{ m/s}$$

The velocity of the machine after impact is determined using the principle of impulse and momentum

$$v = \frac{(200 \text{ kg})(4.43 \text{ m/s})}{1000 \text{ kg}} = 0.886 \frac{\text{m}}{\text{s}}$$

The product of the maximum transmitted force and the maximum displacement is minimized by selecting $\zeta = 0.4$. Then if the transmitted force is limited to 20,000 N, the maximum displacement is obtained using Eq. (11.38)

$$x_{\max} = \frac{\frac{1}{2}mv^2}{F_{T_{\max}}}S(0.4) = \frac{\frac{1}{2}(1000 \text{ kg})(0.886 \text{ m/s})^2}{20{,}000 \text{ N}}1.04 = 0.02 \text{ m}$$

The natural frequency of the isolator is calculated using Eq. (11.37)

$$\omega_n = \frac{F_{T_{\max}}}{mvQ(0.4)} = \frac{20{,}000 \text{ N}}{(1000 \text{ kg})(0.886 \text{ m/s})(0.88)} = 25.65 \frac{\text{rad}}{\text{s}}$$

and the maximum isolator stiffness is calculated as

$$k = m\omega_n^2 = (1000 \text{ kg})(25.65 \text{ rad/s})^2 = 6.58 \times 10^5 \text{ N/m}$$

For pulses of longer duration, it is possible that the maximum displacement or transmitted force could occur while the excitation is being applied. In addition, the shape of the pulse has an effect on these extreme values. Shock isolation calculations, for pulses of duration longer than $T/3$, are made using the shock spectrum for the pulse. The shock spectrum, first introduced in Sec. 5.6, is a nondimensional plot of the maximum displacement as a function of t_0/T.

Shock spectra are often calculated only for undamped systems. Algebraic complexity usually prevents analytical determination of shock spectra for damped systems. The maximum response is obtained by either numerical evaluation of the exact expression for the displacement, or by numerical solution of the differential equation. Damping does not have as much effect on the transient response due to a pulse of longer duration as it does on the steady-state response due to a harmonic excitation or on the response due to a short-duration pulse.

Since shock isolation often involves minimizing the force transmitted between a system and its support, a plot similar to the shock spectrum, but involving the maximum value of the transmitted force, is useful. The vertical coordinate of the force spectrum is the ratio of the maximum value of the transmitted force to the maximum value of the excitation force. The transmitted force is calculated using Eq. (11.31). When the system is undamped, the transmitted force spectrum is the same as the shock spectrum.

Figures 11.13 through 11.18 present displacement spectra and force (acceleration) spectra for common pulse shapes. These spectra were obtained using a Runge-Kutta solution of the governing differential equation. A system with

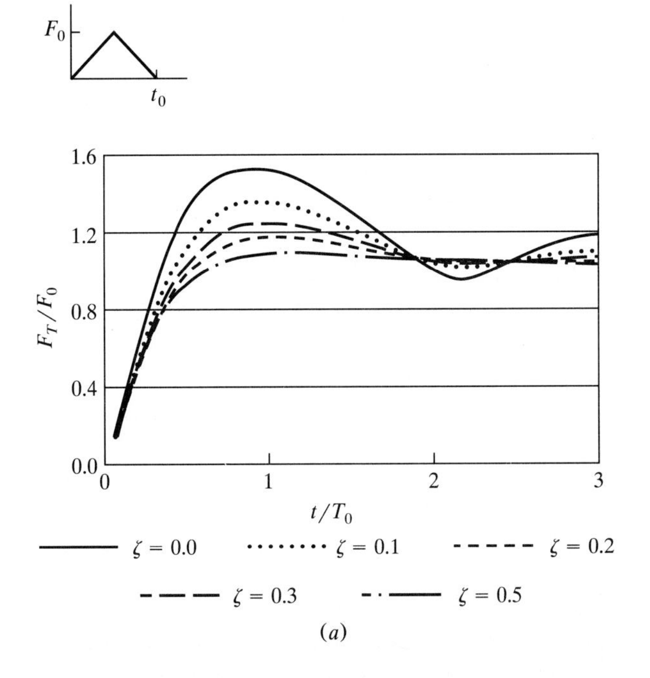

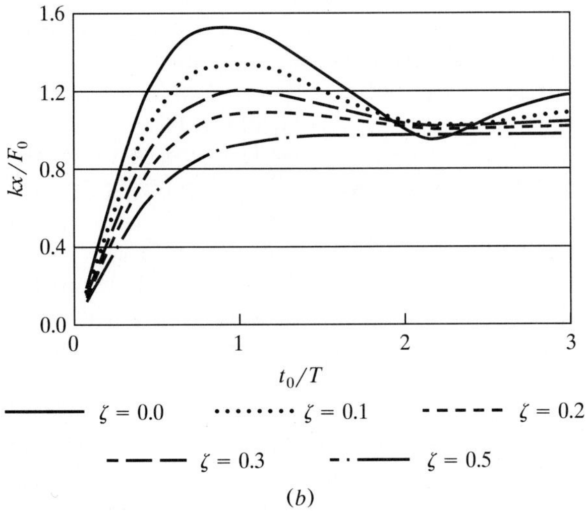

FIGURE 11.13
(*a*) Force spectrum triangular pulse; (*b*) displacement spectrum triangular pulse.

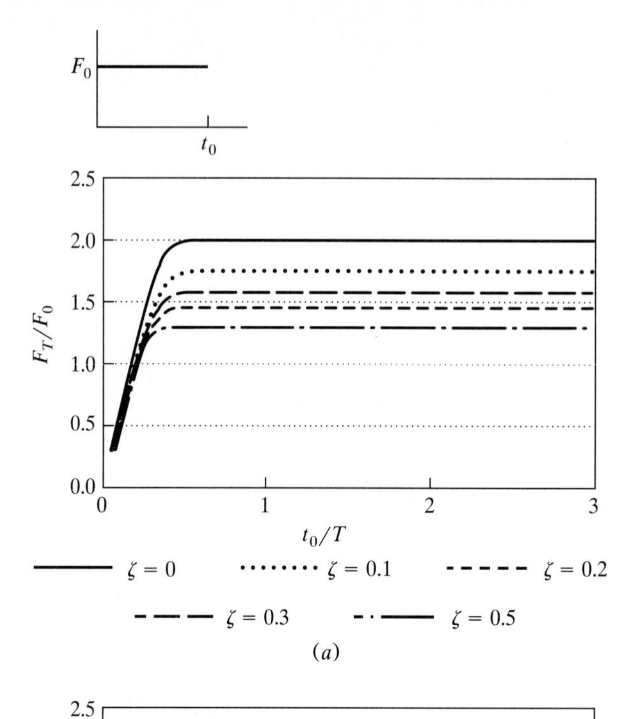

(*a*)

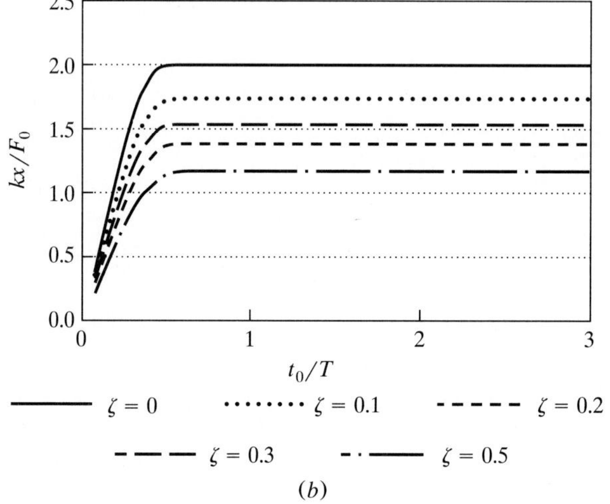

(*b*)

FIGURE 11.14
(*a*) Force spectrum rectangular pulse; (*b*) displacement spectrum rectangular pulse.

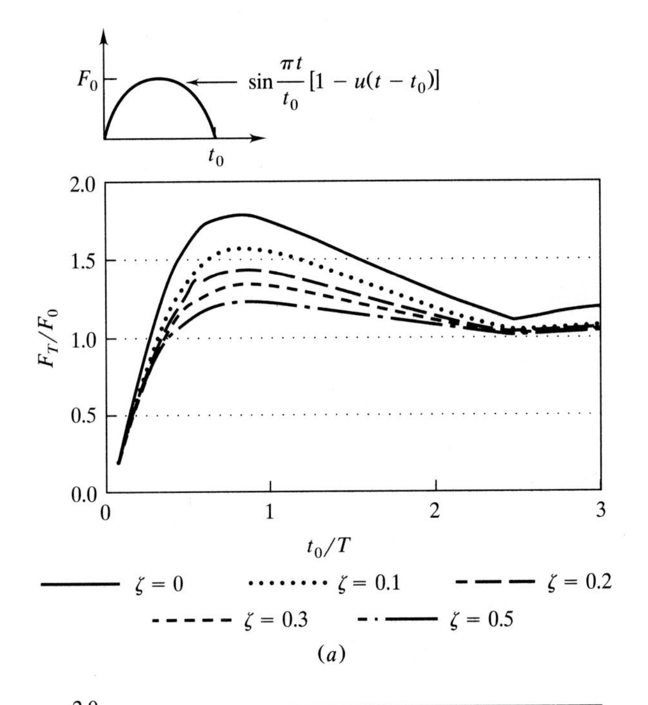

(*a*)

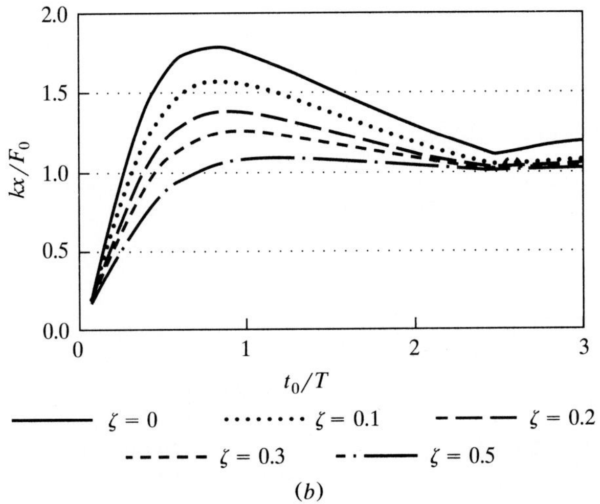

(*b*)

FIGURE 11.15
(*a*) Force spectrum sinusoidal pulse; (*b*) displacement spectrum sinusoidal pulse.

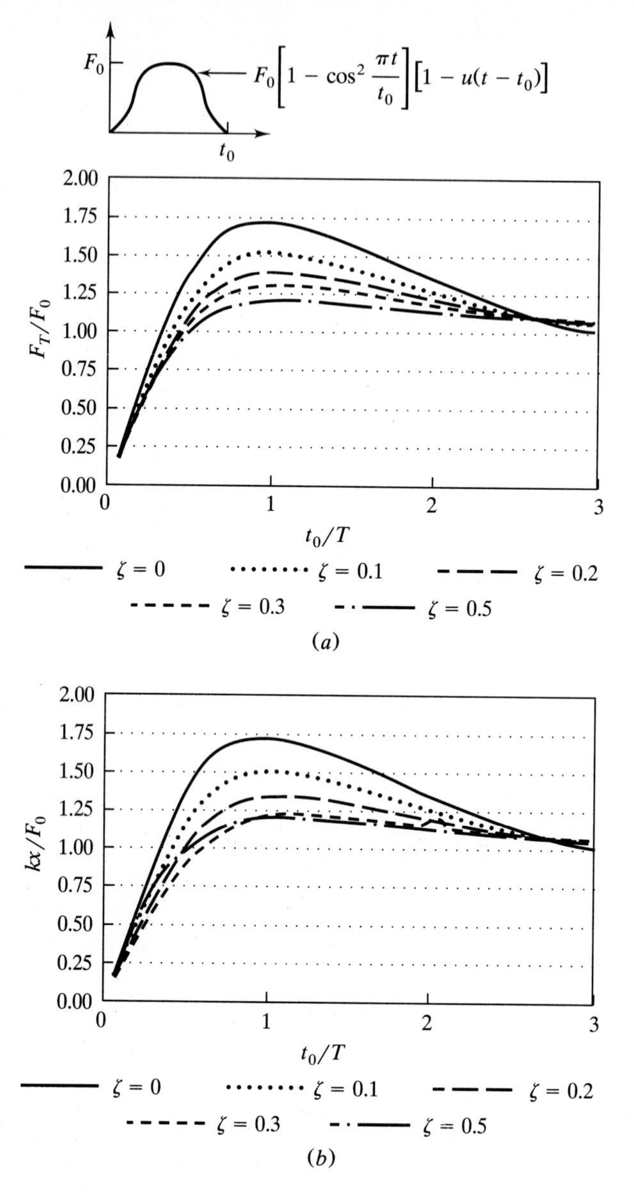

FIGURE 11.16
(*a*) Force spectrum versed sine pulse; (*b*) displacement spectrum versed sine pulse.

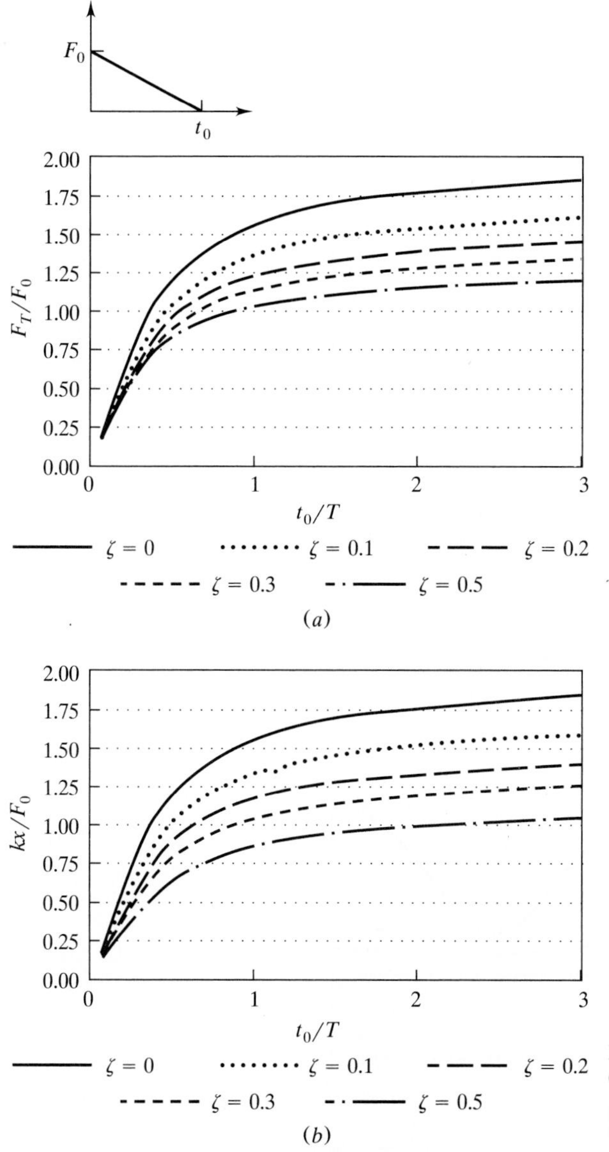

FIGURE 11.17
(*a*) Force spectrum negative slope ramp; (*b*) displacement spectrum negative slope ramp.

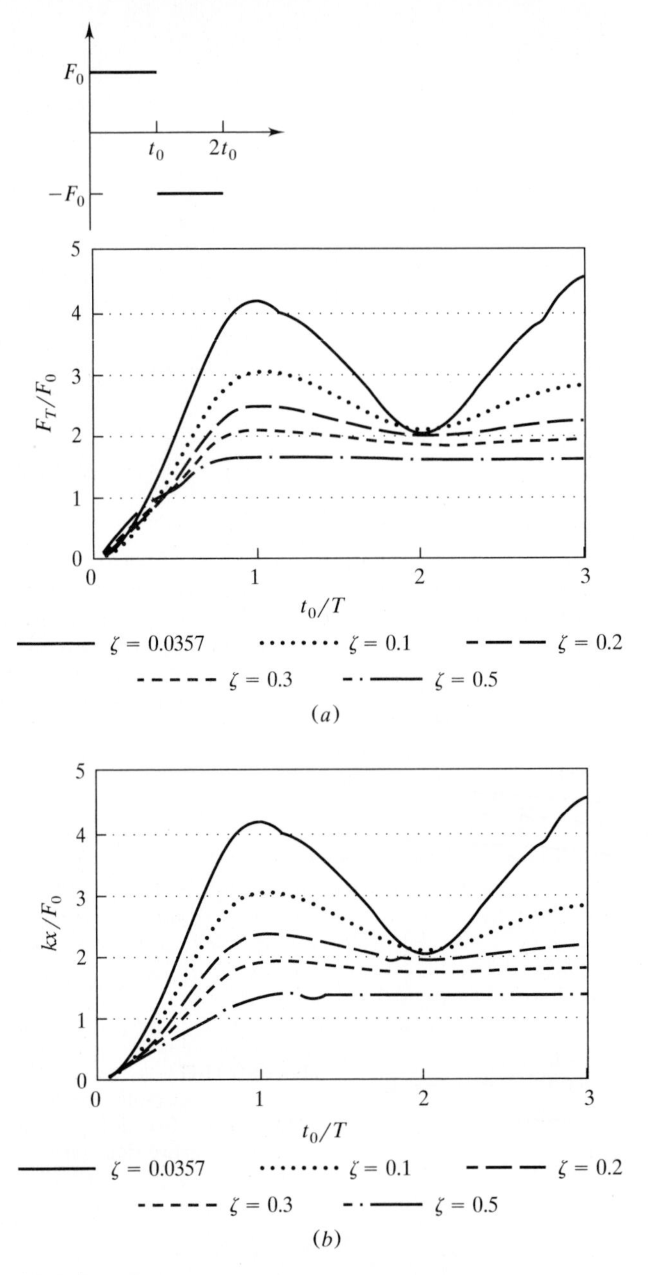

FIGURE 11.18
(*a*) Reversed step pulse force spectrum; (*b*) reversed step pulse displacement spectrum.

$\omega_n = 1$ and $m = 1$ was arbitrarily used. A time increment of the smaller of $t_0/50$ and $T/50$ was used. The Runge-Kutta solution was carried out until the larger of $4t_0$ or $4T$. The displacement and transmitted force were calculated at each time step and compared to maxima from the previous times. The spectra were developed for several values of ζ.

The force spectra for the rectangular pulse, the triangular pulse, the sinusoidal pulse, the versed sine pulse, the negative-slope ramp pulse, and the reversed loading pulse show that shock isolation is achieved only for small natural frequencies. The shock spectra for these excitations show that the nondimensional displacement is small for small natural frequencies. However, the dimensional displacement is calculated using the nondimensional displacement from

$$x_{\max} = \frac{F_0}{m\omega_n^2}\left(\frac{m\omega_n^2 x_{\max}}{F_0}\right) \tag{11.40}$$

Thus a small natural frequency leads to a large displacement.

Example 11.7. A 1000-kg machine is subject to a triangular pulse of duration 0.05 s and peak of 20,000 N. What is the range of isolator stiffness for an undamped isolator such that the maximum transmitted force is less than 8000 N and the maximum displacement is less than 3 cm?

The force spectrum for the triangular pulse shows that for $F_T/F_0 < 0.4$, $\omega_n t_0/(2\pi) < 0.16$ which gives

$$\omega_n < \frac{2\pi(0.16)}{0.05 \text{ s}} = 20.1 \frac{\text{rad}}{\text{s}}$$

The lower bound on the natural frequency is obtained by trial and error using the displacement spectrum for the triangular pulse. For a guessed value of ω_n, $\omega_n t_0/(2\pi)$ is calculated and the corresponding value of the maximum nondimensional displacement is found from the displacement spectrum. The maximum dimensional displacement is calculated from Eq. (11.40). If the displacement is greater than the allowable displacement, the guess for the lower bound must be increased. The calculations for this example are given in Table 11.1. The lower bound is calculated as 17 rad/s. Thus the allowable stiffness range is

$$2.89 \times 10^5 \text{ N/m} < k < 4.04 \times 10^5 \text{ N/m}$$

TABLE 11.1

ω_n (rad / s)	$\frac{\omega_n t_0}{2\pi}$	$m\omega_n^2 x_{\max} / F_0$	$x_{\max}$ (cm)
10	0.08	0.25	5.0
15	0.12	0.38	3.4
18	0.14	0.42	2.6
17	0.135	0.40	2.8

11.6 DYNAMIC VIBRATION ABSORBERS

Large-amplitude vibrations of a one-degree-of-freedom system subject to a harmonic excitation occur when an excitation frequency is near a natural frequency. The amplitude of response is reduced when the system properties are changed such that the natural frequency is away from the excitation frequency. Alternately, an additional degree of freedom can be added such that the natural frequencies of the resulting two-degrees-of-freedom system are away from the excitation frequency.

A vibration absorber is an auxiliary mass-spring system that when correctly tuned and attached to a vibrating body subject to a harmonic excitation causes steady-state motion of the point to which it is attached to cease. Figure 11.19 shows a schematic of a one-degree-of-freedom system with an absorber added. A mass m_1, called the primary mass, attached to a rigid foundation through a spring of stiffness k_1 is excited by a harmonic excitation, whose frequency is near $\sqrt{k_1/m_1}$. An absorber of mass m_2 is connnected to the original mass through an elastic element of stiffness k_2. The resulting system has two degrees of freedom. If correctly designed, the frequency response curve for the primary mass is altered such that the steady-state amplitude of the primary mass is small at the excitation frequency.

The solution for the steady-state vibrations of the resulting two-degree-of-freedom system is presented in Example 8.1. The steady-state amplitude of the primary mass is given by

$$\frac{k_1 X_1}{F_0} = \frac{1 - r_2^2}{r_1^2 r_2^2 - r_2^2 - (1 + \mu) r_1^2 + 1} \tag{11.41}$$

and the steady-state amplitude of the absorber mass is

$$\frac{k_1 X_2}{F_0} = \frac{1}{r_1^2 r_2^2 - r_2^2 - (1 + \mu) r_1^2 + 1} \tag{11.42}$$

where

$$r_1 = \omega \sqrt{\frac{m_1}{k_1}} = \frac{\omega}{\omega_{11}} \tag{11.43}$$

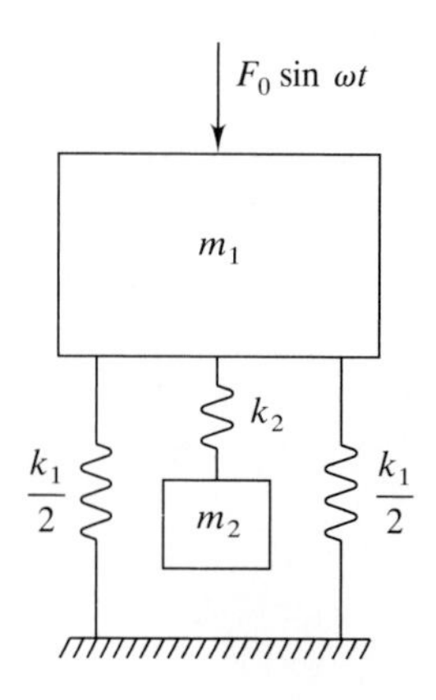

FIGURE 11.19

When $\omega \approx \sqrt{k_1/m_1}$, large amplitude vibrations of the primary system occur. The addition of an auxiliary mass-spring system (the vibration absorber) changes the system to two-degrees-of-freedom with natural frequencies away from $\sqrt{k_1/m_1}$.

is the ratio of the excitation frequency to the natural frequency of the original system,

$$r_2 = \omega\sqrt{\frac{m_2}{k_2}} = \frac{\omega}{\omega_{22}} \tag{11.44}$$

is the ratio of the excitation frequency to the natural frequency of the absorber system, if taken by itself, and

$$\mu = \frac{m_2}{m_1} \tag{11.45}$$

The natural frequencies of the two-degree-of-freedom system are the values of ω such that the denominator of Eqs. (11.41) and (11.42) is zero. They are obtained as

$$\omega_{1,2} = \frac{\omega_{11}}{\sqrt{2}}\sqrt{1 + q^2(1+\mu) \pm \sqrt{q^4(1+\mu)^2 + 2(\mu - 1)q^2 + 1}} \tag{11.46}$$

where

$$q = \frac{\omega_{22}}{\omega_{11}} \tag{11.47}$$

From Eq. (11.41) the steady-state amplitude of the original mass is zero when

$$\sqrt{\frac{k_2}{m_2}} = \omega \tag{11.48}$$

Under this condition the steady-state amplitude of the absorber mass is

$$X_2 = \frac{F_0}{k_2} \tag{11.49}$$

A vibration absorber can be used to eliminate unwanted steady-state vibrations from a one-degree-of-freedom freedom system if the natural frequency of the absorber is tuned to the excitation frequency. The vibration absorber has many practical applications. However, the following must be kept in mind when designing an undamped vibration absorber.

1. When the absorber is tuned to the excitation frequency, Eq. (11.46) shows that one of the two degree of freedom's natural frequencies is less than the absorber's natural frequency while the other is greater. Hence the lower natural frequency must be passed during start-up and stopping, leading to large-amplitude vibrations during these transient periods.
2. The steady-state vibrations of the original mass are eliminated only at a single operating speed. If the system operates over a range of frequencies, the steady-state amplitudes at frequencies away from the absorber frequency may be large. Figure 11.20 shows k_1X_1/F_0 as a function of r_1 for $q = 1$ and

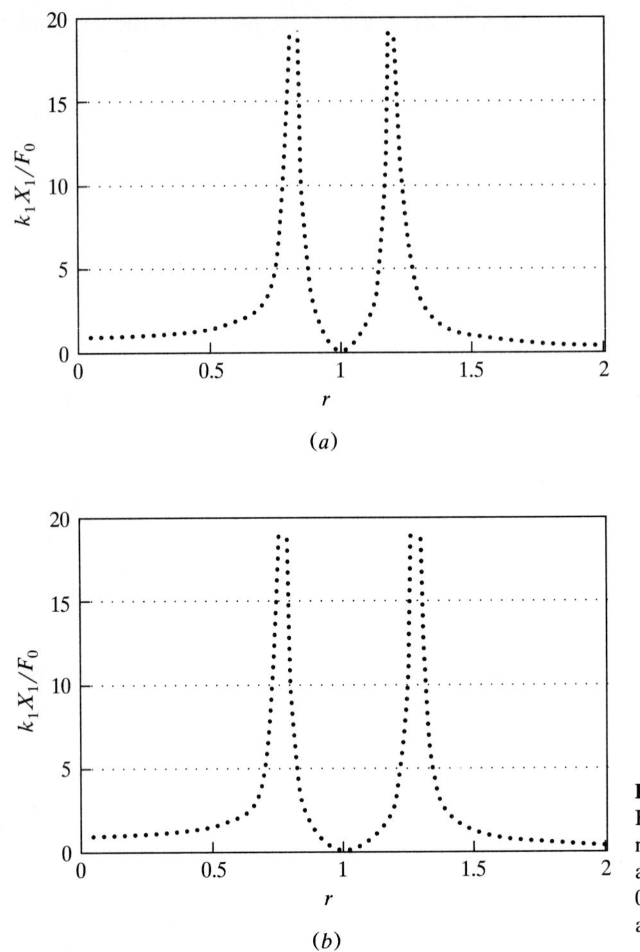

FIGURE 11.20
Frequency response of the primary mass with the addition of a vibration absorber: (a) $\mu = 0.15$ and $q = 1.0$; (b) $\mu = 0.25$, and $q = 1.0$.

$\mu = 0.15$ and $\mu = 0.25$. If r_1 is much less than or greater than q, the steady-state amplitude of the primary mass is large. An effective operating range should be defined for each application by limiting the amplitude of the vibrations or the transmitted force to an acceptable maximum.

3. Figure 11.21 shows the natural frequencies as a function of the mass ratio, μ, for fixed q. The separation of the two natural frequencies is small for small μ, resulting in a narrow operating range. The separation of natural frequencies and the effective operating range increases for larger μ. However, a small μ is usually desired for practical reasons.

4. If the absorber is tuned to the excitation frequency and a given mass ratio μ is not to be exceeded, then the maximum value of the absorber spring stiffness is

$$k_{2_{\max}} = \mu m_1 \omega \tag{11.50}$$

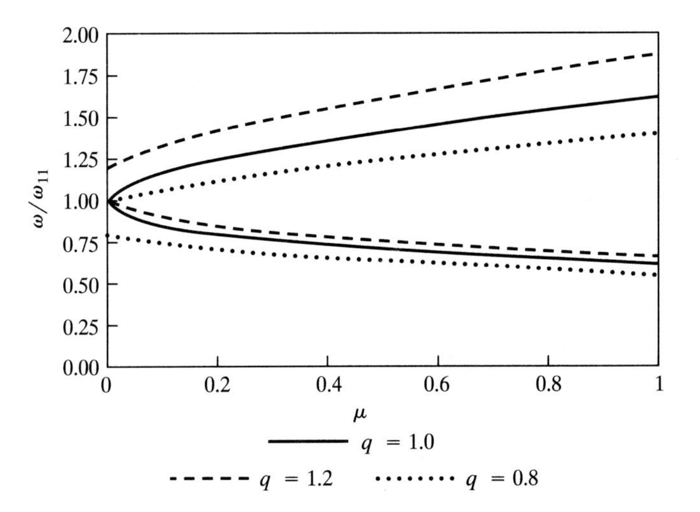

FIGURE 11.21
The natural frequencies of the system with absorber added are farther apart for larger mass ratios.

and the minimum steady-state amplitude of the absorber mass is

$$X_{2_{\min}} = \frac{F_0}{\mu m_1 \omega} \tag{11.51}$$

5. The preceding analysis is valid only for an undamped system. If damping is present in the absorber it is not possible to eliminate steady-state vibrations of the original mass. The amplitude of vibrations can only be reduced.

Example 11.8. A machine of mass 150 kg with a rotating unbalance of 0.5 kg · m is placed at the midspan of a 2-m-long simply supported beam. The machine operates at a speed of 1200 rpm. The beam has an elastic modulus of 210×10^9 N/m^2 and a cross-sectional moment of inertia of 2.1×10^{-6} m^4. Design a dynamic vibration absorber such that when attached to the midspan of the beam, the vibrations of the beam will cease and the steady-state absorber amplitude will be less than 20 mm. What are the system's natural frequencies with the absorber in place? What is the effective operating range such that the midspan deflection does not exceed 5 mm when the absorber is in place?

Modeling the beam vibrations using one degree of freedom and ignoring the mass of the beam, the stiffness and natural frequency of the original system are calculated as

$$k_1 = \frac{48EI}{L^3} = \frac{48(210 \times 10^9 N/m^2)(2.1 \times 10^{-6}\ \text{m}^4)}{(2\ \text{m})^3} = 2.65 \times 10^6\ \frac{\text{N}}{\text{m}}$$

$$\omega_{11} = \sqrt{\frac{k_1}{m_1}} = \sqrt{\frac{2.65 \times 10^6\ \text{N/m}}{150\ \text{kg}}} = 132.8\ \frac{\text{rad}}{\text{s}}$$

The operating speed is

$$\omega = (1200 \text{ rpm})\left(2\pi \frac{\text{rad}}{\text{rev}}\right)\left(1\frac{1 \text{ min}}{60 \text{ s}}\right) = 125.6 \frac{\text{rad}}{\text{s}}$$

Thus large-amplitude vibrations are expected to occur in the absence of an absorber. The natural frequency of the absorber is chosen to coincide with the operating speed

$$\omega_{22}\sqrt{\frac{k_2}{m_2}} = 125.6 \frac{\text{rad}}{\text{s}}$$

Under this condition the steady-state amplitude of the absorber mass is given by Eq. (11.49). Requiring this amplitude to be less than 20 mm leads to

$$k_2 > \frac{F_0}{X_{2_{\max}}} = \frac{(0.5 \text{ kg} \cdot \text{m})(125.6 \text{ rad/s})^2}{0.02 \text{ m}} = 3.94 \times 10^5 \frac{\text{N}}{\text{m}}$$

The required absorber mass is

$$m_2 = \frac{k_2}{\omega_{22}^2} = 25 \text{ kg}$$

The natural frequencies of the two-degree-of-freedom system are calculated from Eq. (11.46) with $\mu = 1/6$ and $q = 0.946$. They are

$$\omega_1 = 105.8 \text{ rad/s} \qquad \omega_2 = 157.6 \text{ rad/s}$$

The effective operating range is obtained from Eq. (11.41), setting $F_0 = 0.5\omega^2$, and using Eqs. (11.43) and (11.44). The denominator of Eq. (11.41) is negative for all ω between the two natural frequencies. When $r_2 < 1$, the numerator is positive and X_1 is set equal to -0.005 m. When $r_2 > 1$ the numerator is negative and X_1 is set equal to 0.005 m. The following quadratic equations in ω_2 are obtained to determine the bounds of the operating range:

$$\omega^4 - 2.79 \times 10^4\omega^2 + 1.67 \times 10^8 = 0 \qquad r_2 > 1$$

$$\omega^4 - 7.63 \times 10^4\omega^2 + 8.28 \times 10^8 = 0 \qquad r_2 < 1$$

The resulting operating range is

$$114.8 \text{ rad/s} < \omega < 138.5 \text{ rad/s}$$

The lower end of the operating range is closer to the lower natural frequency than the upper end is to the higher natural frequency because the excitation force is smaller for smaller excitation frequencies.

11.7 DAMPED VIBRATION ABSORBERS

Viscous damping may be added in parallel to the elastic element of a vibration absorber to limit the amplitude as the lower natural frequency is passed during system start-up and stopping or to increase the effective operating range of the resulting two-degree-of-freedom system. A steady-state analysis of the system of Fig. 11.22 when excited by a harmonic force is presented in Example 8.9. The

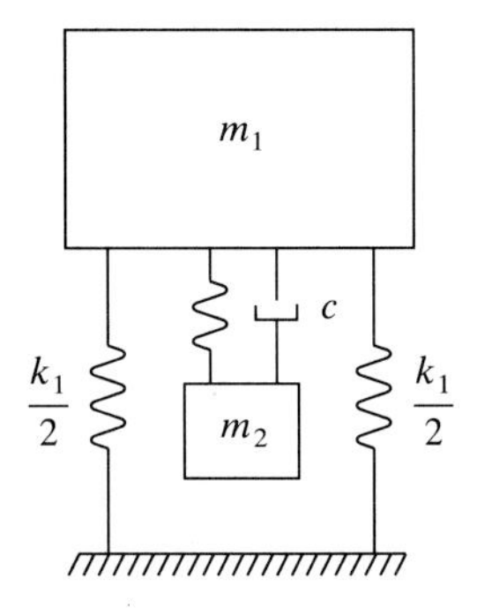

FIGURE 11.22
The damped vibration absorber consists of an auxiliary mass connected to the primary mass by an elastic element in parallel with a viscous damper.

equations for calculating the steady-state amplitudes are nondimensionalized, yielding

$$\frac{k_1 X_1}{F_0} = G(r_1, q, \zeta, \mu) \tag{11.52}$$

where

$$G = \sqrt{\frac{(2\zeta r_1 q)^2 + (r_1^2 - q^2)^2}{\{r_1^4 - [1 + (1+\mu)q^2]r_1^2 + q^2\}^2 + (2\zeta r_1 q)^2[1 - r_1^2(1+\mu)]^2}} \tag{11.53}$$

and

$$\frac{k_1 X_2}{F_0} = \sqrt{\frac{q^4 + (2\zeta q)^2}{\{r_1^4 - [1 + (1+\mu)q^2]r_1^2 + q^2\}^2 + (2\zeta r_1 q)^2[1 - r_1^2(1+\mu)]^2}} \tag{11.54}$$

where r_1 and q are defined in Eqs. (11.43) and (11.47), respectively, and

$$\zeta = \frac{c}{2\sqrt{k_2 m_2}} \tag{11.55}$$

The nondimensional function G defined by Eq. (11.53) is shown in Fig. 11.23 for $\mu = 0.25$ and $q = 1$ for several values of ζ. The steady-state motion of the original mass is not zero for any r_1. A minimum, near $r_1 = 1$, is reached between two peaks. An absorber using this choice of parameters is not effective because the peak at the lower resonant frequency is still large. It is noted that each curve, for different ζ, passes through the same two points.

G is plotted in Fig. 11.24 for $\mu = 0.25$ and $q = 0.8$. The peak at the lower resonant frequency is smaller than the peak at the higher resonant frequency. However, the higher peak occurs near $r_1 = 1$, which is the region where an absorber is usually needed. Also, the effective operating range is still small. It is

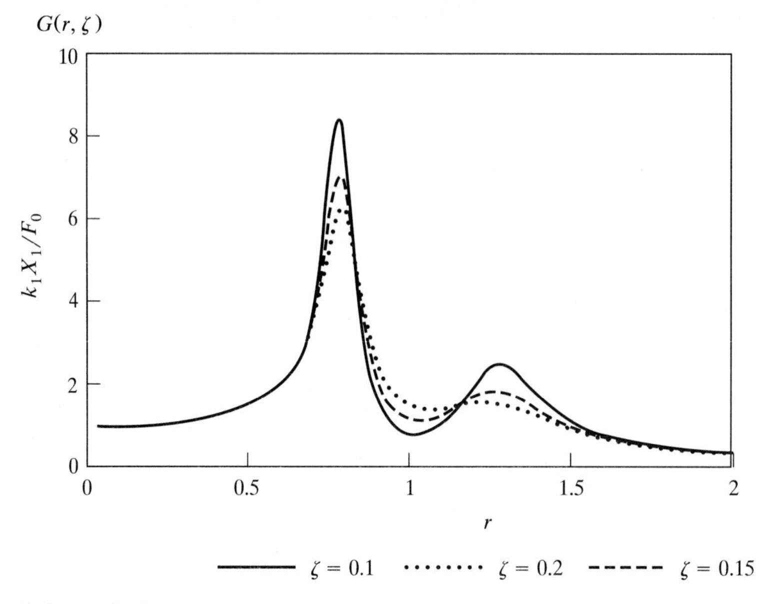

FIGURE 11.23
Frequency response of primary mass with additions of damped vibration absorber with $\mu = 0.25$, $q = 1.0$.

noted again, that there are two fixed points through which each curve passes. These fixed points are different than those in Figure 11.23.

Since it is not possible to eliminate steady-state motion of the original system when damping is present, a damped vibration absorber must be designed to reduce the peak at the lower resonant frequency and to widen the effective operating range. Absorbers using the parameters used to generate Fig. 11.23 and 11.24 are not suitable for these purposes.

Widening the operating range requires that the two peaks have approximately the same magnitude. Since the locations of the fixed points are dependent on q, it should be possible to tune the absorber such that the values of G at the fixed points are the same. Since curves for all values of ζ pass through the fixed points, it should be possible to find a value of ζ such that the fixed points are near the peaks.

For fixed values of μ and q, there are two values of r_1 which yield a value of G, independent of ζ. The value of G at these points is written as

$$G = \sqrt{\frac{A(\mu, q)\zeta^2 + B(\mu, q)}{C(\mu, q)\zeta^2 + D(\mu, q)}} \tag{11.56}$$

Since Eq. (11.56) holds for all ζ and powers of ζ are linearly independent

$$\frac{A}{C} = \frac{B}{D} \tag{11.57}$$

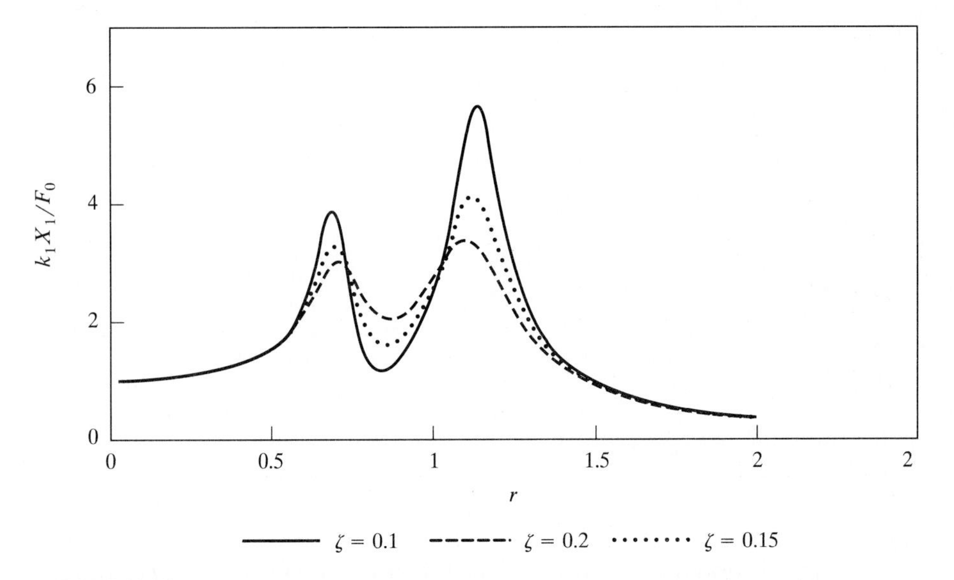

FIGURE 11.24
Frequency response of primary mass with addition of damped vibration absorber with $\mu = 0.25$, $q = 0.80$.

Using Eq. (11.53) to determine the forms of A, B, C, and D, substituting into Eq. (11.57), and rearranging leads to

$$r_1^4\left(1 + \frac{\mu}{2}\right) - \left[1 + \left(q^2(1 + \mu)\right]r_1^2 + q^2 = 0 \tag{11.58}$$

The solution of Eq. (11.58) places the fixed points at

$$r_1 = \sqrt{\frac{1 + (1 + \mu)q^2 \pm \sqrt{1 - 2q^2 + (1 + \mu)^2 q^4}}{2 + \mu}} \tag{11.59}$$

Since Eq. (11.56) yields the same value of G, independent of ζ for r_1 given by Eq. (11.59), letting $\zeta \to \infty$ gives

$$G = \sqrt{\frac{1}{\left[1 - r_1^2(1 + \mu)\right]^2}} \tag{11.60}$$

Requiring G to be the same at both fixed points leads to

$$q = \frac{1}{1 + \mu} \tag{11.61}$$

An optimum absorber could be designed with an appropriate value of ζ such that the smaller r_1 given by Eq. (11.59) corresponds to both a fixed point and a peak on the frequency response curve. The appropriate value of ζ is

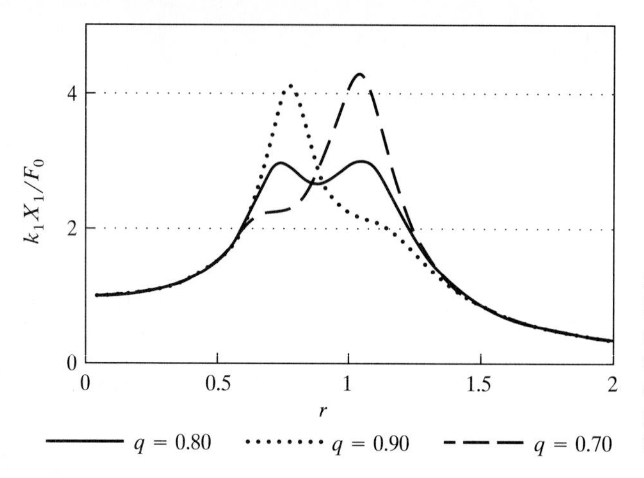

FIGURE 11.25
For $\mu = 0.25$, $\zeta_{opt} = 0.2739$ and $q_{opt} = 0.80$. This comparison shows that values of q other than q_{opt} lead to larger peaks and a narrower operating range.

obtained by setting $dG/d\zeta = 0$, using q from Eq. (11.61). The same procedure can be followed to yield the value of ζ such that the larger value of r_1 given by Eq. (11.59) corresponds to both a fixed point and a peak. Since the values of ζ are not equal, their average is usually used to define the optimum damping ratio,

$$\zeta_{opt} = \sqrt{\frac{3\mu}{8(1+\mu)}} \tag{11.62}$$

In summary, the optimum design of a damped vibration absorber requires that the absorber be tuned to the frequency calculated from Eq. (11.61) with the damping ratio of Eq. (11.62). For $\mu = 0.25$, Eq. (11.61) gives an optimum damping ratio of $\zeta = 0.2379$ and an optimum $q = 0.80$. Figure 11.25 shows G for these vales as a function of r_1. This figure also shows G for the same μ and ζ but with values of q, one on each side of the optimum. The curve corresponding to the optimum value of q has smaller resonant peaks and the value of G does not vary much between the peaks.

Example 11.9. A milling machine has a mass of 250 kg and a natural frequency of 120 rad/s and is subject to a harmonic excitation of magnitude 10,000 N at speeds between 95 rad/s and 120 rad/s. Design a damped vibration absorber of mass 50 kg such that the steady-state amplitude is no greater than 15 mm at all operating speeds.

The mass ratio is

$$\mu = \frac{50 \text{ kg}}{250 \text{ kg}} = 0.2$$

Since a wide operating range is required, the optimum absorber design is tried.

From Eqs. (11.61) and (11.62)

$$q = \frac{1}{1.2} = 0.833$$

$$\zeta = \sqrt{\frac{3(0.2)}{8(1.2)}} = 0.25$$

The steady-state amplitude at any operating speed for this absorber design is calculated using Eqs. (11.52) and (11.53). The results are used to generate the frequency response curve of Fig. 11.26. Since the extremes of the operating range lie between the peaks and the steady-state amplitudes at the extremes are

$$X(\omega = 95\ \text{rad/s}) = 9.13\ \text{mm} \qquad X(\omega = 120\ \text{rad/s}) = 8.86\ \text{mm}$$

and both are less than 15 mm, the optimum design is acceptable. The absorber stiffness and damping ratio are calculated from Eqs. (11.47) and (11.55), respectively, as

$$k_2 = m_2\omega_{22}^2 = \mu q^2 k_1 = (0.2)(0.833)^2(3.6 \times 10^6\ \text{N/m}) = 5.08 \times 10^5\ \text{N/m}$$

$$c = 2\zeta\sqrt{k_2 m_2} = 2500\ \text{N}\cdot\text{s/m}$$

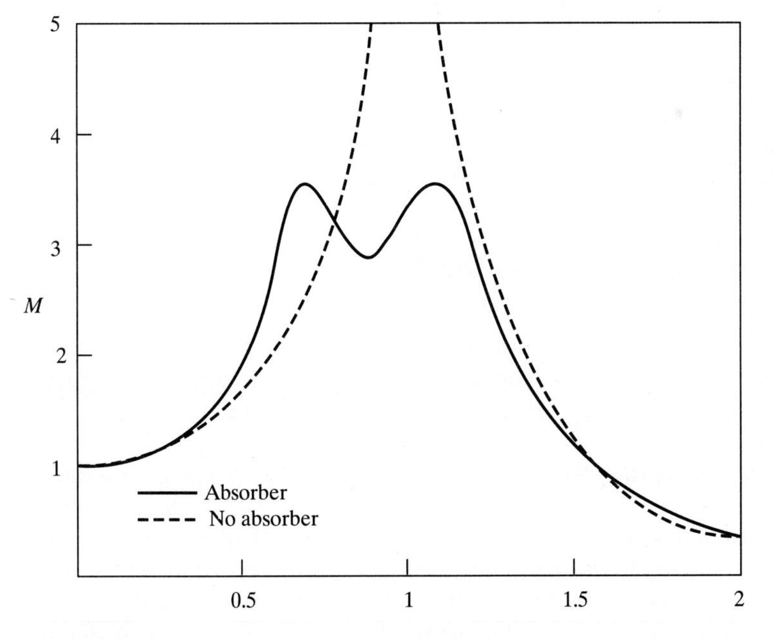

FIGURE 11.26
The addition of the optimum damped vibration absorber to the system of Example 11.9 reduces the steady-state amplitude to an acceptable level over the entire operating range.

The problem can also be solved using ABSORB.BAS, included on the accompanying diskette. The given parameters are used as input for the design of a damped absorber using ABSORB.BAS. The optimum absorber design is selected and the output follows, as well as a plot in Fig. 11.26, showing a comparison of the primary system response with and wiithout the absorber.

```
DAMPED VIBRATION ABSORBER ANALYSIS

PRIMARY SYSTEM PARAMETERS
     Mass= 2.500E+02 kg
     Stiffness= 3.600E+06 N/m
     Excitation frequency= 1.200E+02 rad/sec
     Excitation amplitude= 1.000E+04 N

ABSORBER PARAMETERS (DIMENSIONAL)
     Mass= 5.000E+01 kg
     Stiffness= 5.000E+05 N/m
     Damping coefficient= 2.500E+03 N-sec/m

ABSORBER PARAMETERS (NONDIMENSIONAL)
     Mass ratio= 2.000E-01
     Frequency ratio= 8.333E-01
     Damping ratio= 2.500E-01

STEADY-STATE AMPLITUDES AT SPECIFIED FREQUENCIES
     For omega= 9.500E+01 rad/sec, X= 9.127E-03 m
     For omega= 1.200E+02 rad/sec, X= 8.861E-03 m

MAXIMUM START-UP AMPLITUDE= 9.233E-03 m at OMEGA= 9.252E+01
  rad/sec
```

11.8 MULTI-DEGREE-OF-FREEDOM SYSTEMS

Large-amplitude responses occur for an n-degree-of-freedom system when a harmonic excitation frequency is near one of the system's n natural frequencies. Vibration absorbers can be designed to reduce the steady-state amplitudes near these resonances.

Consider an n-degree-of-freedom system subject to a harmonic excitation near one of the system's natural frequencies. Let $x_1, x_2, \ldots, x_n$ be the chosen generalized coordinates. Let k_{ij} and m_{ij} be the elements of the stiffness and mass matrices, respectively, for this choice of generalized coordinates. A vibration absorber of mass $\tilde{m}$ and stiffness $\tilde{k}$ is attached to the system at the particle whose displacement is given by x_1. The generalized coordinate x_{n+1} is defined as the displacement of the absorber mass. The mass and stiffness matrices for the resulting $n + 1$-degree-of-freedom system are

$$\mathbf{M} = \begin{bmatrix} m_{11} & m_{12} & \cdots & m_{1n} & 0 \\ m_{21} & m_{22} & \cdots & m_{2n} & 0 \\ \vdots & \vdots & \ddots & \vdots & \vdots \\ m_{n1} & m_{n2} & \cdots & m_{nn} & 0 \\ 0 & 0 & \cdots & 0 & \tilde{m} \end{bmatrix} \tag{11.63}$$

$$\mathbf{K} = \begin{bmatrix} k_{11} + \tilde{k} & k_{12} & \cdots & k_{1n} & -\tilde{k} \\ k_{21} & k_{22} & \cdots & k_{2n} & 0 \\ \vdots & \vdots & \ddots & \vdots & \vdots \\ k_{n1} & k_{n2} & \cdots & k_{nn} & 0 \\ -\tilde{k} & 0 & \cdots & 0 & \tilde{k} \end{bmatrix} \tag{11.64}$$

For a single-frequency harmonic excitation, the force vector takes the form

$$\mathbf{F} = \begin{bmatrix} A_1 \\ A_2 \\ \vdots \\ A_n \\ 0 \end{bmatrix} \sin \omega t \tag{11.65}$$

Using the methods of Sec. 8.2, the steady-state amplitude of the particle on the original system to which the absorber is attached is

$$X_1 = \frac{1}{D(\omega)} \det \begin{bmatrix} A_1 & k_{12} - m_{12}\omega^2 & k_{13} - m_{13}\omega^2 & \cdots & k_{1n} - m_{1n}\omega^2 & -\tilde{k} \\ A_2 & k_{22} - m_{22}\omega^2 & k_{23} - m_{23}\omega^2 & \cdots & k_{2n} - m_{2n}\omega^2 & 0 \\ A_3 & k_{32} - m_{32}\omega^2 & k_{33} - m_{33}\omega^2 & \cdots & k_{3n} - m_{3n}\omega^2 & 0 \\ \vdots & \vdots & \vdots & \ddots & \vdots & \vdots \\ A_1 & k_{n2} - m_{n2}\omega^2 & k_{n3} - m_{n3}\omega^2 & \cdots & k_{nn} - m_{nn}\omega^2 & 0 \\ 0 & 0 & 0 & \cdots & 0 & \tilde{k} - \tilde{m}\omega^2 \end{bmatrix} \tag{11.66}$$

where $D(\omega) = \det\{-\omega^2\mathbf{M} - \mathbf{K}\}$. The determinant of the numerator of Eq. (11.66) is evaluated by expanding by the last row. The result, using the notation of App. C is

$$X_1 = \frac{(\tilde{k} - \tilde{m}\omega^2)C_{n+1,n+1}}{D(\omega)} \tag{11.67}$$

If the absorber is tuned such that its frequency

$$\tilde{\omega} = \sqrt{\frac{\tilde{k}}{\tilde{m}}} \tag{11.68}$$

is equal to the excitation frequency, then $X_1 = 0$, from Eq. (11.67).

The preceding result is summarized as follows: Addition of a vibration absorber tuned to the excitation frequency of a multi-degree-of-freedom system leads to no steady-state motion of the system prticle to which it is attached. That particle becomes a node. The steady-state amplitude is nonzero for all other generalized coordinates.

The vibration absorber attached to a multi-degree-of-freedom system works by shifting the natural frequencies of the system away from the excitation frequency. The natural frequency near the excitation frequency is decreased and a natural frequency is added between the excitation frequency and the next higher natural frequency. The higher and lower natural frequencies are only slightly altered by the addition of a vibration absorber.

Steady-state motion of the particle to which the absorber is attached ceases when the excitation frequency coincides with the absorber frequency, but steady-state motion exists for all particles which are not rigidly connected to thic particle. However, since the excitation frequency is away from all natural frequencies when the absorber is added, the steady-state amplitudes corresponding to all other generalized coordinates are significantly reduced from when the absorber is not added.

The addition of damping to the absorber leads to a much more complicated analysis. The steady-state amplitude cannot be eliminated for any particle when damping is present. An optimum tuning frequency and optimum damping ratio can be determined for a specific system.

Example 11.10. The slender rod of Fig. 11.27 is subject to a harmonic excitation force at 75 rad/s. Design a dynamic vibration absorber of mass 3 kg such that steady-state motion of the mass center of the bar ceases when the absorber is added. What is the corresponding steady-state amplitude of the point where the force is applied? Draw frequency response curves for the steady-state amplitude of the mass center and the steady-state amplitude of angular oscillations with and without the absorber.

Let x_1, the displacement of the mass center, and θ, the clockwise angular rotation of the rod, be chosen as generalized coordinates to describe the motion of the original system. The governing differential equations are

$$\begin{bmatrix} m & 0 \\ 0 & m\frac{L^2}{12} \end{bmatrix} \begin{bmatrix} \dot{x}_1 \\ \ddot{\theta} \end{bmatrix} + \begin{bmatrix} 2k & -k\frac{L}{4} \\ -k\frac{L}{4} & \frac{5kL^2}{16} \end{bmatrix} \begin{bmatrix} x_1 \\ \theta \end{bmatrix} = \begin{bmatrix} F_0 \\ F_0\frac{L}{2} \end{bmatrix} \sin \omega t$$

Using the values given in Fig. 11.27, the natural frequencies are calculated as

$$\omega_1 = 74.02 \text{ rad/s} \qquad \omega_2 = 117.0 \text{ rad/s}$$

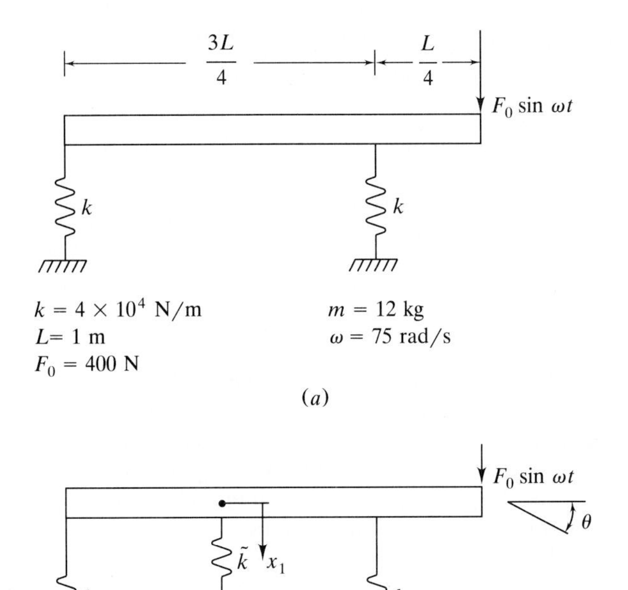

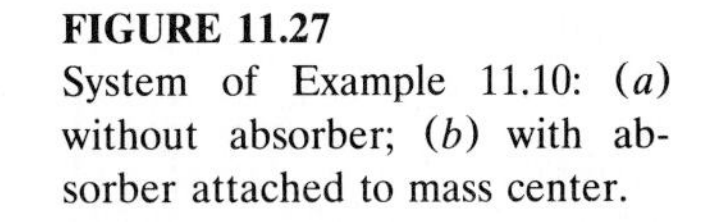

FIGURE 11.27
System of Example 11.10: (*a*) without absorber; (*b*) with absorber attached to mass center.

An absorber of mass 3 kg is attached to the center of mass of the bar. If the absorber is tuned to 75 rad/s, steady-state vibrations of the center of mass are eliminated, and the bar rotates about its mass center. This requires

$$\tilde{k} = (3\text{ kg})(75\text{ rad/s})^2 = 1.6 \times 10^4\text{ N/m}$$

Let x_2 be the absorber displacement. When the absorber is added to the system, the differential equations become

$$\begin{bmatrix} m & 0 & 0 \\ 0 & \frac{1}{12}mL^2 & 0 \\ 0 & 0 & \tilde{m} \end{bmatrix} \begin{bmatrix} \ddot{x}_1 \\ \ddot{\theta} \\ \ddot{x}_2 \end{bmatrix} + \begin{bmatrix} 2k + \tilde{k} & -k\frac{L}{4} & -\tilde{k} \\ -k\frac{L}{4} & 5k\frac{L^2}{16} & 0 \\ -\tilde{k} & 0 & \tilde{k} \end{bmatrix} \begin{bmatrix} x_1 \\ \theta \\ x_2 \end{bmatrix} = \begin{bmatrix} F_0 \\ F_0\frac{L}{2} \\ 0 \end{bmatrix} \sin \omega t$$

The natural frequencies are obtained by solving $D(\omega) = 0$ or

$$(2k + \tilde{k} - m\omega^2)\left(5k\frac{L^2}{16} - m\frac{L^2}{12}\omega^2\right)(\tilde{k} - \tilde{m}\omega^2)$$

$$- \tilde{k}^2\left(5k\frac{L^2}{16} - m\frac{L^2}{12}\omega^2\right) - \left(\frac{kL}{4}\right)^2(\tilde{k} - \tilde{m}\omega^2) = 0$$

Substitution of the given and previously calculated values leads to

$$\omega_1 = 58.98 \text{ rad/s} \qquad \omega_2 = 92.78 \text{ rad/s} \qquad \omega_3 = 118.86 \text{ rad/s}$$

and $$D(\omega) = -m\left(m\frac{L^2}{12}\right)\tilde{m}(\omega^2 - 58.98^2)(\omega^2 - 92.78^2)(\omega^2 - 118.86^2)$$

The steady-state amplitudes for an arbitrary value of ω are obtained using the method of Sec. 8.2 and Cramer's rule as

$$X_1 = \frac{F_0}{D(\omega)}(\tilde{k} - \tilde{m}\omega^2)\left(7k\frac{L^2}{16} - m\frac{L^2}{12}\omega^2\right)$$

and $$\Theta = \frac{F_0 L}{2D(\omega)}\left[\tilde{k}^2 + (\tilde{k} - \tilde{m}\omega^2)\left(\tfrac{5}{2}k + \tilde{k} - m\omega^2\right)\right]$$

The steady-state amplitude of the particle where the force is applied is $\Theta L/2$. Substituting the given values for the absorber frequency and the excitation frequency equal to 75 rad/s leads to the steady-state amplitude of A as 0.01 m.

Dimensional plots of the steady-state amplitudes as functions of the excitation frequency are shown in Figs. 11.28 and 11.29.

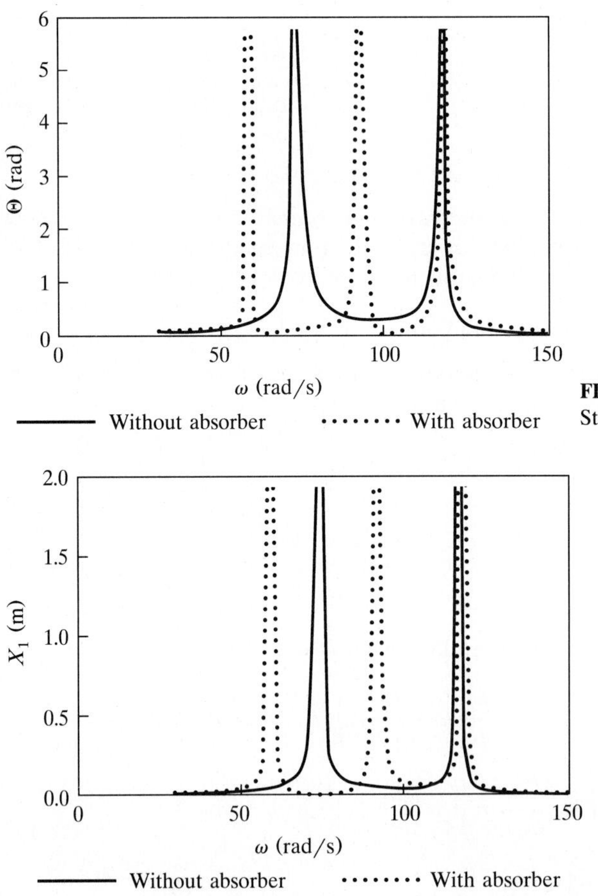

FIGURE 11.28
Steady-state Θ for Example 11.10.

FIGURE 11.29
Steady-state X_1 for Example 11.10.

11.9 USE OF NONLINEAR ELASTIC ELEMENTS

The use of a nonlinear spring in the design of a vibration isolator or vibration absorber has certain advantages over the use of a linear spring. However, the use of nonlinear springs has been limited because the resulting nonlinear system is much more difficult to analyze and only approximate solutions are available. Nonlinear phenomena such as secondary resonances may be present. Secondary resonances can lead to responses at frequencies other than the natural frequency which may be present even when damping is present.

This section considers the use of springs with cubic nonlinearities as isolators and in absorber systems. A cubic spring has a nondimensional spring force-displacement relationship of

$$F = x + \epsilon x^3 \qquad \epsilon > 0 \tag{11.69a}$$

for a hardening spring and

$$F = x - \alpha x^3 \qquad \alpha > 0 \tag{11.69b}$$

for a softening spring.

11.9.1 Isolation from Harmonic Excitation

Many commercially available vibration isolators such as rubber isolators, conical springs, and steel mesh isolators naturally have nonlinear characteristics.

The isolation effectiveness of a linear isolator is reduced as the mass to which it is attached is decreased because the natural frequency increases. For an isolator with a hardening spring, as the applied mass is decreased, the deflection is decreased, causing the stiffness to decrease, and hence the natural frequency decreases. Thus an isolator with a hardening spring can be used for a larger range of applied masses with the same efficiency.

11.9.2 Shock Isolation

Consider a system isolated with an undamped softening spring. The system is subject to an impulse producing an initial nondimensional velocity v, causing the spring to deflect to $x_{\max}$. Application of the principle of work-energy gives

$$\begin{aligned} \frac{1}{2}v^2 &= \int_0^{x_{\max}} F(x)\,dx \\ &= \int_0^{x_{\max}} (x - \alpha x^3)\,dx \\ &= \frac{x_{\max}^2}{2}\left(1 - \alpha\frac{x_{\max}^2}{2}\right) \end{aligned} \tag{11.70}$$

The maximum acceleration is the force in the spring at its maximum deflection,

$$\ddot{x}_{\max} = x_{\max}\left(1 - \alpha x_{\max}^2\right) \tag{11.71}$$

The energy absorption ratio defined in Sec. 11.5 is calculated by

$$S = \frac{\ddot{x}_{\max} x_{\max}}{\frac{1}{2}v^2} \tag{11.72}$$

Substituting Eqs. (11.70) and (11.71) into Eq. (11.72) leads to

$$S = 2\left(\frac{1 - \alpha x_{\max}^2}{1 - \alpha \dfrac{x_{\max}^2}{2}}\right) \tag{11.73}$$

Equation (11.73) and Fig. 11.30 show that a softening spring has greater energy absorption than a linear spring. The amount of energy absorption is dependent on the initial velocity.

Addition of viscous damping further enhances the energy absorption characteristics of a softening spring. The first approximation to the response of a system with a cubic nonlinearity and small viscous damping is given by Eq. (10.37)

$$x(t) = Ae^{-\zeta t}\sin\left[\left(1 - \tfrac{3}{8}\alpha A^2\right)t + \phi\right]$$

The maximum displacement and acceleration are determined in a manner

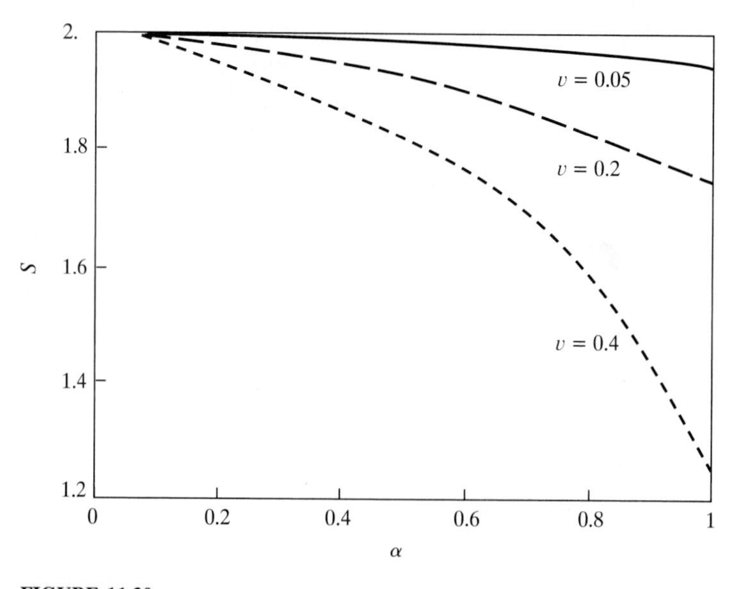

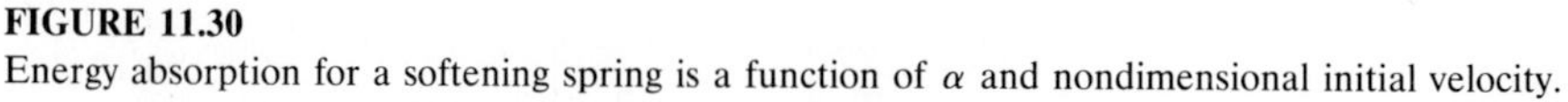
FIGURE 11.30
Energy absorption for a softening spring is a function of α and nondimensional initial velocity.

similar to that for a linear system as explained in Sec. 11.5. The calculated maximum acceleration is used as an approximation for the maximum transmitted force. The energy absorption ratio is determined as

$$S(\zeta) = \begin{cases} \dfrac{2(\lambda^2 + \zeta^2)^{3/2} e^{-\zeta(t_1+t_2)}}{\sqrt{\lambda^4(\lambda^2 + 3\zeta^2) + \zeta^4(\zeta^2 + 3\lambda^2)}}, & \zeta < \dfrac{1}{\sqrt{3}}\lambda \\[2ex] \dfrac{4\zeta e^{-\zeta t_1}}{\sqrt{\zeta^2 + \lambda^2}}, & \zeta > \dfrac{1}{\sqrt{3}}\lambda \end{cases} \tag{11.74}$$

where

$$\lambda = 1 - \tfrac{3}{8}\alpha A^2$$

$$t_1 = \frac{1}{\lambda}\tan^{-1}\left(\frac{\lambda}{\zeta}\right)$$

$$t_2 = \frac{1}{\lambda}\tan^{-1}\left(\frac{\zeta}{\lambda}\frac{\lambda^2 - 3\zeta^2}{3\lambda^2 - \zeta^2}\right)$$

Equation (11.74) is plotted in Fig. 11.31 for $\alpha = 0.2$ when the system is subject to a non-dimensional initial velocity of 0.2. The minimum value of S and the value of λ for which it occurs depends upon α and v. For a nondimensional velocity of 0.2 with $\alpha = 0.4$ a minimum S of 1.036 occurs for $\zeta = 0.043$. Thus the use of a nonlinear softening element with viscous damping does give some improvement in energy absorption over a linear element.

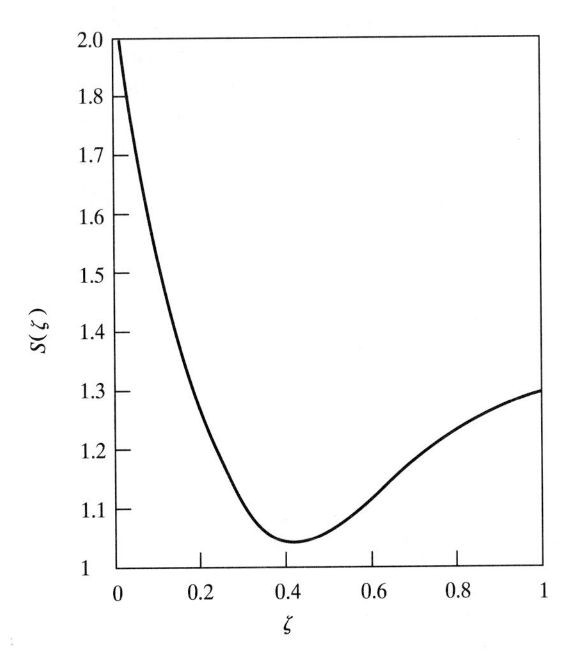

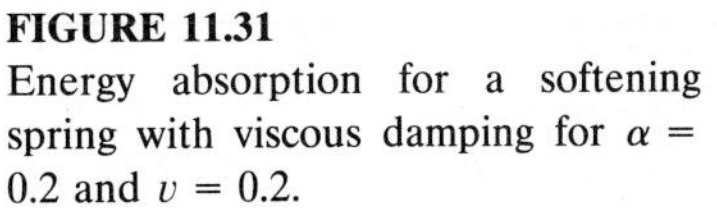
FIGURE 11.31
Energy absorption for a softening spring with viscous damping for $\alpha = 0.2$ and $v = 0.2$.

11.9.3 Vibration Absorbers

The vibration absorber of Fig. 11.32 consists of an auxiliary mass, m_2, connected to the primary system through an elastic element which has a cubic nonlinearity. The differential equations governing the motion of the system are

$$\begin{aligned} m_1\ddot{x}_1 + k_1 x - k_2(x_2 - x_1) - \lambda(x_2 - x_1)^3 &= F_0 \sin \omega t \\ m_2\ddot{x}_2 + k_2(x_2 - x_1) + \lambda(x_2 - x_1)^3 &= 0 \end{aligned} \tag{11.75}$$

Using the notation of Sec. 11.6, Eqs. (11.75) are nondimensionalized by introducing

$$x_i^* = \frac{x_i m_1 \omega_{11}^2}{F_0} \qquad t^* = \omega_{11} t \tag{11.76}$$

Substitution of Eqs. (11.76) into Eqs. (11.75) leads to

$$\begin{aligned} \ddot{x}_1 + x_1 - \mu q^2(x_2 - x_1) - \Lambda(x_2 - x_1)^3 &= \sin rt \\ \mu\ddot{x}_2 + \mu q^2(x_2 - x_1) + \Lambda(x_2 - x_1)^3 &= 0 \end{aligned} \tag{11.77}$$

where the * has been dropped from the nondimensional variables, and

$$\Lambda = \frac{\lambda F_0^2}{m_1^3 \omega_{11}^6} \tag{11.78}$$

An exact solution of Eqs. (11.77) is not available. The natural frequencies of the linearized system are given by Eq. (11.46). Using the theory of Chaps. 8 and 9, the corresponding mode shapes are calculated, the principal coordinates for the linearized system (p_1 and p_2) defined, and the linearized system uncoupled. When written using p_1 and p_2 as dependent variables, the governing differential equations are of the form

$$\begin{aligned} \ddot{p}_1 + \omega_1^2 p_1 + a_1(p_2 - b_1 p_1)^3 &= g_1 \sin rt \\ \ddot{p}_2 + \omega_2^2 p_2 + a_2(p_2 - b_2 p_1)^3 &= g_2 \sin rt \end{aligned} \tag{11.79}$$

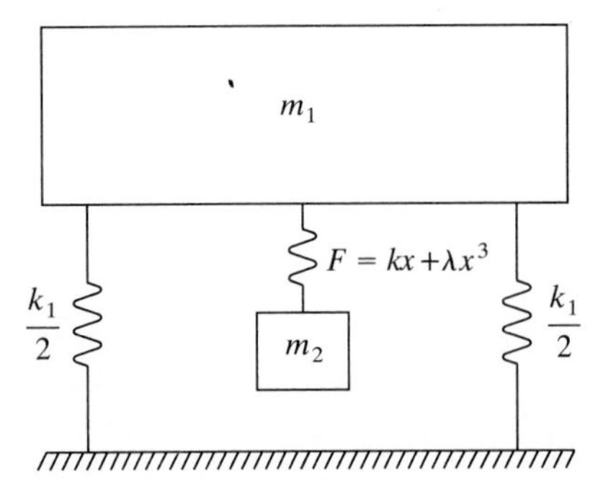

FIGURE 11.32
Absorber consists of auxiliary mass m_2 connected to primary mass through elastic element with a cubic nonlinearity.

Noting that a vibration absorber is only necessary when the excitation frequency is near the natural frequency of the original system, $r \approx 1$, and using Eqs. (11.79) and the information of Chap. 10, the following is noted about the response of the system with the nonlinear absorber:

1. The system has a primary resonance when the excitation frequency is near one of the system's natural frequencies. The absorber tunes the natural frequencies of the resulting two-degree-of-freedom system away from the natural frequency of the primary system. The separation of natural frequencies is greater for larger μ. Thus a large-amplitude response due to a primary resonance is not expected during the absorber operation. However, the frequency response near $r = 1$ is influenced by the nonlinear phenomena occurring for $r \approx \omega_1/\omega_{11}$ and $r \approx \omega_2/\omega_{11}$.
2. The system is free of subharmonic and superharmonic resonances at the desired operating speeds, but is subject to a combination resonance specific for multi-degree-of-freedom system with cubic nonlinearities. Consider, for example, a vibration absorber designed such that $q = 1$ and $\mu = 0.25$. Then from Eq. (11.46) the natural frequencies of the two-degree-of-freedom linear system are $\omega_1 = 0.781\omega_{11}$ and $\omega_2 = 1.256\omega_{11}$. Then $\omega_1 + \omega_2 \approx 2\omega_{11}$, and according to Eq. (10.47), the system has a combination resonance.
3. It is difficult to obtain the frequency response of a nonlinear system over a wide range of frequencies because of different resonances appearing at different frequencies. The true nonlinear response to a harmonic excitation should involve periodic response terms corresponding to all resonance conditions. An approximate frequency response is obtained by considering only terms in the response at the excitation frequency. To this end, assume

$$
\begin{aligned}
x_1(t) &= A \sin rt \\
x_2(t) &= B \sin rt
\end{aligned}
\tag{11.80}
$$

Substituting Eqs. (11.80) into Eqs. (11.79), using trigonmetric identities, ignoring terms with frequencies other than r, and collecting coefficients of $\sin(rt)$ terms leads to the following equations:

$$
\begin{aligned}
(1 - r^2)A - \mu q^2(B - A) - \tfrac{3}{4}\Lambda(B - A)^3 &= 1 \\
\mu(q^2 - r^2)B - \mu q^2 A + \tfrac{3}{4}\Lambda(B - A)^3 &= 0
\end{aligned}
\tag{11.81}
$$

Equations (11.81) are rearranged, yielding

$$
W^3 + \frac{4\mu^3 r^6}{3\Lambda}\left[\frac{q^2 - r^2}{r^2} - \frac{\mu r^2}{1 - (\mu + 1)r^2}\right]W - \frac{4\mu^4 r^8}{3\Lambda[1 - (\mu + 1)r^2]} = 0
\tag{11.82}
$$

where

$$
W = [1 - (\mu + 1)r^2]A - 1 \tag{11.83}
$$

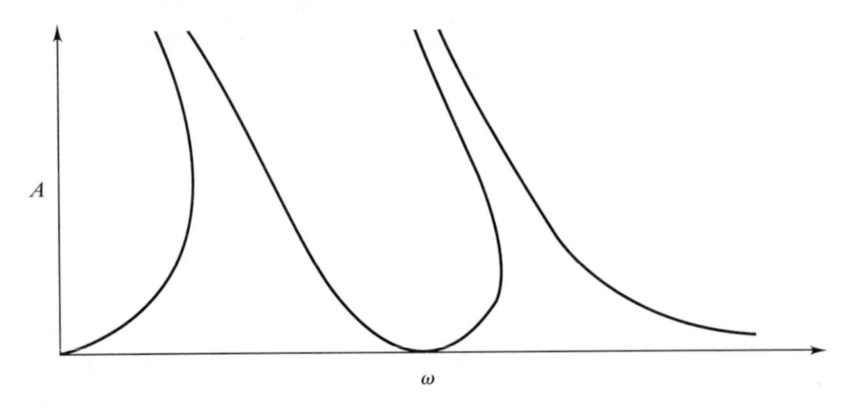

FIGURE 11.33
Possible frequency response curve for primary mass when softening spring is used in auxiliary system. Jump phenomenon limits amplitude during start up.

For fixed values of μ, q, and Λ, Eq. (11.82) is solved to determine W as a function of r. There are two ranges of values of r where three real solutions of Eq. (11.82) exist, and hence A is triple valued. When this occurs for a given value of r, only two solutions correspond to stable steady-state solutions, and hence the "jump" phenomenon occurs.

In contrast to a linear absorber, the amplitude of response is a function of the amplitude of excitation. The steady-state amplitude of the primary mass is zero for certain values of r, but these values are a dependent upon Λ and hence the amplitude of excitation.

Figure 11.33 shows the representation of the response of a primary mass when a softening spring is used in the absorber. The jump phenomenon can

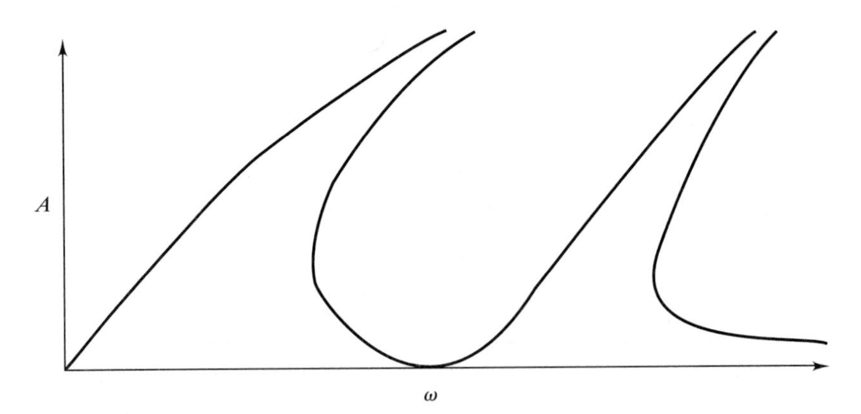

FIGURE 11.34
Possible frequency response curve for primary mass when hardening spring is used in auxiliary system.

occur for two ranges of frequencies. The start up amplitude is limited by the jump phenomena. Figure 11.34 shows the representation of the response of a primary mass when a hardening spring is used in the absorber. The amplitude during start up is larger than when a softening spring is used and the effective operating range is narrower.

PROBLEMS

The program ISOL.EXE on the accompanying diskette provides numerical solutions for vibration isolation problems. The program ABSORB.EXE aids in the design of damped and undamped vibration isolators. The program SPECT.BAS calculates the force and displacement spectra for a use defined excitation. These programs may be used in the solution of many of the problems in this chapter.

11.1. What is the minimum static deflection of an undamped isolator that provides 75% isolation to a 200-kg washing machine operating at 500 rpm?

11.2. What is the maximum allowable stiffness of an isolator of damping ratio 0.05 that provides 81% isolation to a 40-kg printing press operating at 850 rpm?

11.3. When set on a rigid foundation and operating at 800 rpm, a 100-kg machine tool provides a repeating force of magnitude 18,000 N to its foundation. An engineer has calculated that the maximum repeated force to which the foundation should be subject is 2600 N. What is the maximum stiffness of an undamped isolator that provides sufficient isolation between the tool and its foundation?

11.4. Repeat Prob. 11.3 if the damping ratio of the isolator is 0.11.

11.5. If an isolator with the maximum allowable stiffness is used in Prob. 11.4, what is the stead-state amplitude of the machine tool when placed on the isolator and what is the maximum amplitude during start-up?

11.6. A 65-kg mixer operates at speeds between 1400 and 2400 rpm. At each speed, a harmonic force of magnitude 15,000 N is developed. What is the maximum stiffness of an isolator of damping ratio 0.1 such that the force transmitted between the machine and its foundation is no larger than 6500 N at all operating speeds?

11.7. Cork pads of stiffness 6×10^5 N/m and damping ratio 0.2 are to be used to isolate a 40-kg machine tool from its foundation. The machine tool operates at 1400 rpm and produces a repeating force of magnitude 80,000 N. If the pads are placed in series, how many are required such that the force transmitted to the foundation is less than 20,000 N?

11.8. A 100-kg tumbler operates at 1200 rpm and produces a repeating force of magnitude 20,000 N. It is desired to reduce the magnitude of the repeating force transmitted to the foundation to 6000 N. However, the only isolator readily available has a stiffness of 5×10^5 N/m and a damping ratio of 0.15. Concrete blocks of 25 kg are available to mount the tumbler. How many blocks must be used to achieve the desired isolation?

11.9. A 100-kg machine operates at 1400 rpm and produces a repeating force of 80,000 N. The force transmitted to the foundation is to be reduced to 20,000 N by mounting the machine on four undamped isolators in parallel. What is the maximum stiffness of each of the isolators?

11.10. A 100-kg sewing machine operates as 550 rpm and produces a repeating force of 45,000 N. Design an isolation system such that all of the following are met:
(*a*) The transmitted force is reduced to 10,000 N.
(*b*) The system has the minimum possible static deflection.
(*c*) The steady-state amplitude is less than 1.5 mm.
(*d*) The maximum amplitude during start-up is 5.5 mm.
Specify the isolator stiffness, damping ratio, and mass that must be added to the sewing machine.

11.11. A 7-kg machine with a rotating unbalance of 0.1 kg · m is placed at the end of a 70-cm-long cantilever steel beam ($E = 210 \times 10^9$ N/m^2). The machine operates at 400 rpm. The beam has a square cross section of sides 15 mm. Using the Goodman fatigue failure criteria with the static loading due to the weight of the machine, an engineer calculates that the maximum stress due to the repeated loading should be 150×10^6 N/m^2. Design an undamped isolator such that when the isolator is placed between the machine and the beam the maximum repeating stress is less than 150×10^6 N/m^2.

11.12. A machine of mass 1000 kg has a rotating unbalance of 0.1 kg · m. The machine operates at speeds between 500 and 750 rpm. What is the maximum isolator stiffness of an undamped isolator that can be used to reduce the transmitted force to 300 N at all operating speeds?

11.13. A 10-kg laser flow-measuring device is used on a table in a laboratory. Due to operation of other equipment, the table is subject to external vibrations. Accelerometer measurements show that the dominant component of the table vibration is at 300 Hz and has an acceleration amplitude of 4.3 m/s^2. For effective operation the laser can be subject to an acceleration amplitude of only 0.7 m/s^2. Design an undamped isolator to reduce the transmitted acceleration.

11.14. A motorcycle travels over a road whose contour is approximately sinusoidal, $y(z) = 0.2\sin(0.4z)$ m, where z is measured in meters. Using a simple one-degree-of-freedom model, design a suspension system with a damping ratio of 0.1 such that the acceleration felt by the rider is less than 15 m/s^2 at all horizontal speeds between 30 and 80 m/s. The mass of the motorcycle and its rider is 225 kg.

11.15. Rough seas cause a ship to heave with an amplitude of 0.4 m at a frequency of 20 rad/s. Design an isolation system with a damping ratio of 0.13 such that the navigational computer is subject to an acceleration of only 200 m/s^2.

11.16. A sensitive computer is being transported by rail in a boxcar. Accelerometer measurements indicate that when the train is traveling at its normal speed of 85 m/s, the dominant component of the boxcar's vertical acceleration is 8.5 m/s^2 at a frequency of 36 rad/s. The crate in which the computer is being carried is tied to the floor of the boxcar. What is the required stiffness of an isolator with a damping ratio of 0.05 such that the acceleration amplitude of the 60-kg computer is less than 0.5 m/s^2. Using this isolator, what is the displacement of the computer relative to the crate?

11.17. A 54-kg air compressor operates at speeds between 800 and 2000 rpm and has a rotating unbalance of 0.23 kg · m. Design an isolator with a damping ratio of 0.15

such that the repeating part of the transmitted force is less than 1000 N at all operating speeds.

11.18. A 150-kg motor operates at speeds between 1200 and 2500 rpm and has a rotating unbalance of 0.35 kg · m. Design an isolator such that the transmitted force is less than 7500 N at all operating speeds and the maximum amplitude during start-up is 6 mm.

11.19. A 150-kg machine tool operates at speeds between 850 and 1700 rpm. The tool has a rotating unbalance of 0.5 kg at a distance 0.05ω m from the center of rotation where ω is in rad/s. Design an undamped isolator such that when the tool is placed on the isolator the transmitted force is no larger than 50,000 N at all operating speeds.

11.20. Repeat Prob. 11.19 if the isolator has a damping ratio of 0.07.

11.21. A 150-kg washing machine has a rotating unbalance of 0.45 kg · m. The machine is placed on isolators of equivalent stiffness 4×10^5 N/m and damping ratio 0.08. Over what range of operating speeds will the transmitted force between the washer and the floor be less than 3000 N?

11.22. When a machine tool is placed directly on a rigid floor, it provides an excitation of the form

$$F(t) = (4000 \sin 100t + 5100 \sin 150t) \text{ N}$$

to the floor. Determine the natural frequency of the system with an undamped isolator with the minimum possible static deflection such that when the machine is mounted on the isolator the maximum repeated force transmitted to the floor is 3500 N.

11.23. Rework Example 11.3 when the total period of the punching operation is 1.2 s and a force of 4000 N is applied over 0.3 s.

11.24. Use the force shown in Fig. P11.24 as an approximation to the force provided by the punch press during its operation and rework Example 11.3.

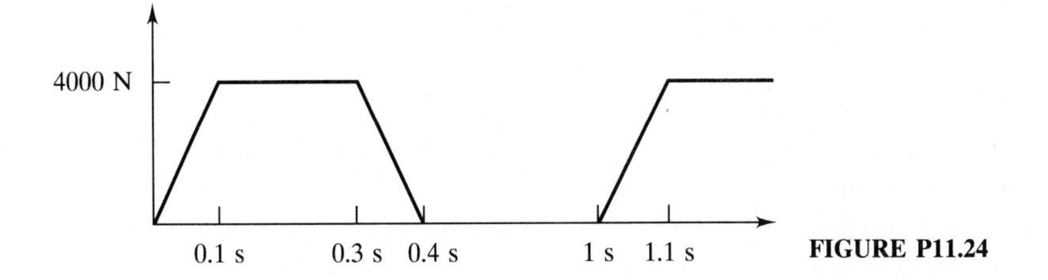

FIGURE P11.24

11.25. Consider the schematic of a single-cylinder reciprocating engine shown in Fig. P4.57 and discussed in Prob. 4.57. Write a FORTRAN program that will aid in the design of a vibration isolator for the engine. The parameters m_r, m_p, r, and l, the operating range $\omega_1 < \omega < \omega_2$, and the maximum allowable transmitted force, $F_{\max}$ are input variables. The program should output the maximum allowable stiffness as a function of the damping ratio for $\zeta = 0, 0.01, 0.02, \ldots, 0.20$.

11.26. Repeat Prob. 11.14 if the road contour is described by

$$y(z) = (0.15 \sin 0.4z + 0.04 \sin 0.8z) \text{ m}$$

11.27. A 35-kg air compressor normally operates at 1200 rpm and produces a harmonic force of magnitude 7500 N.

(*a*) Specify the maximum stiffness of an isolator of damping ratio 0.14 such that the force transmitted through the isolator is 1000 N during normal operation.

(*b*) If the compressor is mounted on the isolator designed in part (*a*) and subject to an impulse of magnitude 1000 N · s, what is the maximum displacement of the resulting motion and what is the maximum force transmitted through the isolator?

11.28. During its normal operation, a 144-kg machine tool is subject to a 15,000-N · s impulse. Design an efficient isolator such that the maximum force transmitted through the isolator is 2500 N and the maximum displacement is minimized.

11.29. A 110-kg pump is mounted on an isolator of stiffness 4×10^5 N/m and a damping ratio of 0.15. The pump is given a sudden velocity of 30 m/s. What is the maximum force transmitted through the isolator and what is the maximum displacement of the pump?

11.30. A ship is moored at a dock in rough seas and frequently impacts the dock. The maximum velocity change caused by the impact is 15 m/s. Design an isolator to protect a sensitive 80-kg navigational control system such that its maximum acceleration is 60 m/s^2.

11.31. During lift-off a 500-kg rocket reaches a velocity of 400 m/s in a very short time. Design an isolator to protect a 10-kg experiment being conducted in the rocket. The maximum acceleration felt by the experiment should be 600 m/s^2.

11.32. The isolation between a sensitive piece of equipment and its packaging is to be designed to minimize the acceleration felt by the equipment when the package is dropped. Suppose a package is dropped from a height h. Its displacement, measured from its initial position is

$$z(t) = \frac{1}{2} g t^2 [1 - u(t - t_0)], \qquad t_0 = \sqrt{\frac{2h}{g}}$$

The packaging material is modeled by a spring of stiffness k in parallel with a viscous damper of damping coefficient c. Answer the following without performing any calculations.

(*a*) Discuss a method to obtain the displacement of the equipment relative to the package and its absolute acceleration.

(*b*) Which time is of more concern in designing the isolation system; $t < t_0$ or $t > t_0$?

(*c*) Discuss a method to design the packaging for a specific situation.

11.33. A 110-kg machine is placed on the floor of an industrial plant. It has been determined that beyond the static load the maximum transmitted force should be 5000 N. During its operation the machine is to be subject to a variety of excitations. For each form of excitation shown in Fig. P11.33, determine the maximum allowable stiffness of an isolator of damping ratio 0.1. Calculate the resulting maximum displacement.

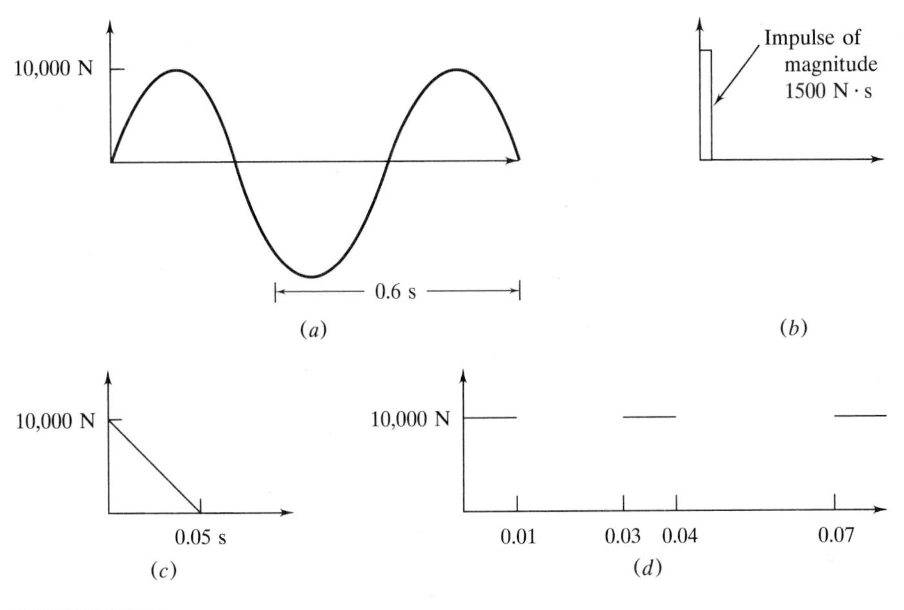

FIGURE P11.33

11.34. Repeat Example 11.7 if the isolator has a damping ratio of 0.1.

11.35. A 20-kg machine is on a foundation that is often subject to an acceleration that is modeled as a versed sine pulse of magnitude 20 m/s^2 and duration 0.4 s. Design an undamped isolator such that the maximum acceleration felt by the machine is 15 m/s^2. What is the maximum displacement of the machine relative to its foundation using this isolator?

11.36. During operation, a 100-kg machine tool is exposed to a force that is modeled as a sinusoidal pulse of magnitude 3100 N and duration 0.05 s. Design an isolator with a damping ratio 0.1 such that the maximum force transmitted through the isolator is 2000 N and the maximum displacement of the tool is 3 cm.

11.37. A 50-kg lathe machine mounted on an elastic foundation of stiffness 4×10^5 N/m has a vibration amplitude of 35 cm when the motor speed is 95 rad/s. Design an undamped dynamic vibration absorber such that steady-state vibrations are completely eliminated at 95 rad/s and the maximum displacement of the absorber mass at this speed is 5 cm.

11.38. What is the lowest natural frequency of the lathe of Prob. 11.37 with the absorber in place? What is the operating range around 95 rad/s such that the steady-state amplitude of the lathe is less than 1 cm?

11.39. What is the required stiffness of an undamped dynamic vibration absorber whose mass is 5 kg to eliminate vibrations of a 25-kg machine of natural frequency 125 rad/s when the machine operates at 110 rad/s?

11.40. A 35-kg machine is attached to the end of a cantilever beam of length 2 m, elastic modulus 210×10^9 N/m^2, and moment of inertia 1.3×10^{-7} m^4. The machine operates at 180 rpm and has a rotating unbalance of 0.3 kg · m.

(*a*) What is the required stiffness of an undamped absorber of mass 5 kg such that steady-state vibrations are eliminated at 180 rpm?

(*b*) With the absorber in place what are the natural frequencies of the system?
(*c*) For what range of operating speeds will the steady-state amplitude of the machine be less than 8 mm?

11.41. A 150-kg pump experiences large-amplitude vibrations when operating at 1500 rpm. Assuming this is the natural frequency of a one-degree-of-freedom system, design a dynamic vibration absorber such that the lower natural frequency of the two-degree-of-freedom system is less than 1300 rpm and the higher natural frequency is greater than 1700 rpm.

11.42. A solid disk of diameter 30 cm and mass 10 kg is attached to the end of a solid 3-cm-diameter, 1-m-long steel shaft ($G = 80 \times 10^9$ N/m^2). A torsional vibration absorber consists of a disk attached to a shaft that is then attached to the primary system. If the absorber disk has a mass of 3 kg and a diameter of 10 cm, what is the required diameter of a 50-cm-long absorber shaft to eliminate steady-state vibrations of the original system when excited at 500 rad/s?

11.43. A 200-kg machine is placed on a massless simply supported beam as shown in Fig. P11.43. The machine has a rotating unbalance of 1.41 kg · m and operates at 3000 rpm. The steady-state vibrations of the machine are to be absorbed by hanging a mass attached to a 40 cm steel cable from the location on the beam where the machine is attached. What is the required diameter of the cable such that machine vibrations are eliminated at 3000 rpm and the amplitude of the absorber mass is less than 50 mm?

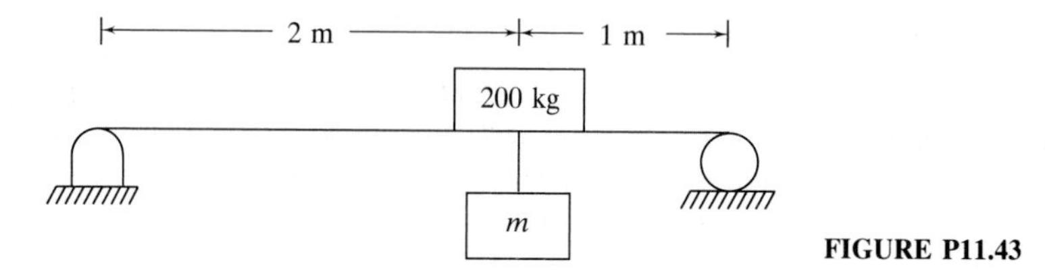

FIGURE P11.43

11.44. The disk in Fig. P11.44 rolls without slip and the pulley is massless. What is the mass of the block that should be hung from the cable such that steady-state vibrations of the cylinder are eliminated when $\omega = 120$ rad/s?

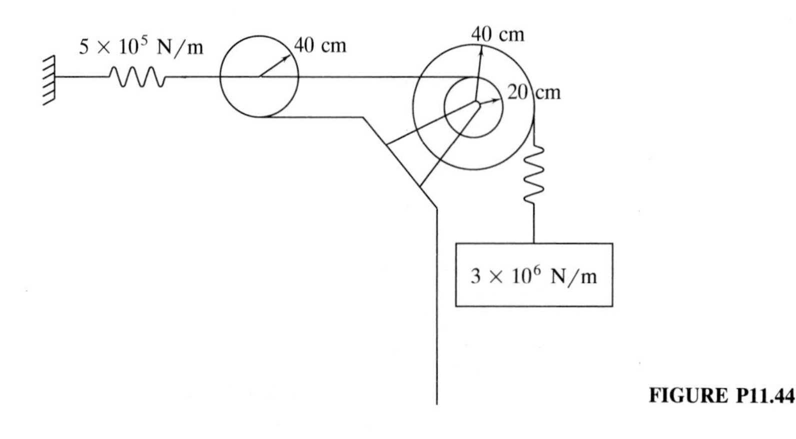

FIGURE P11.44

11.45. Vibration absorbers are used in boxcars to protect sensitive cargo from large accelerations due to periodic excitations provided by rail joints. For a particular railway joints are spaced 5 m apart. The boxcar, when empty has a mass of 25,000 kg. Two absorbers, each of mass 12,000 kg, are used. Absorbers for a particular boxcar are designed to eliminate vibrations of the main mass when the boxcar is loaded with a 12,000-kg cargo and travels at 100 m/s. The natural frequency of the unloaded boxcar is 165 rad/s.

(*a*) At what speeds will resonance occur for the boxcar with a 12,000-kg cargo?

(*b*) What is the best speed for the boxcar when it is loaded with a 25,000-kg cargo?

11.46. A 500-kg reciprocating machine is mounted on a foundation of equivalent stiffness 5×10^6 N/m. When operating at 800 rpm, the machine produces an unbalanced harmonic force of magnitude 50,000 N. Two cantilever beams with end masses are added to the machine to act as absorbers. The beams are made of steel ($E = 210 \times 10^9$ N/m^2) and have a moment of inertia of 4×10^{-6} m^4. A 10-kg mass is attached to each beam. The absorbers are adjustable in that the location of the mass on the absorber can be varied.

(*a*) How far away from the support should the masses be located when the machine is operating at 800 rpm? What is the amplitude of the absorber mass?

(*b*) If the compressor operates at 1000 rpm and produces a harmonic force of amplitude 100,000 N, where should the absorber masses be placed and what is their vibration amplitude?

11.47. A 100-kg machine is placed at the midspan of a 2-m-long cantilever beam ($E = 210 \times 10^9$ N/m^2, $I = 2.3 \times 10^{-6}$ m^4). The machine produces a harmonic force of amplitude 60,000 N. Design a damped vibration absorber of mass 30 kg such that when hung from the beam at midspan, the steady-state amplitude of the machine is less than 8 mm at all speeds between 1300 and 2000 rpm.

11.48. Repeat Prob. 11.47 if the excitation is due to a rotating unbalance of magnitude 0.33 kg · m.

11.49. For the absorber designed in Prob. 11.47, what is the minimum steady-state amplitude of the machine and at what speed does it occur?

11.50. Determine values of k and c such that the steady-state amplitude of the center of the cylinder is less than 4 mm for 60 rad/s $< \omega <$ 110 rad/s?

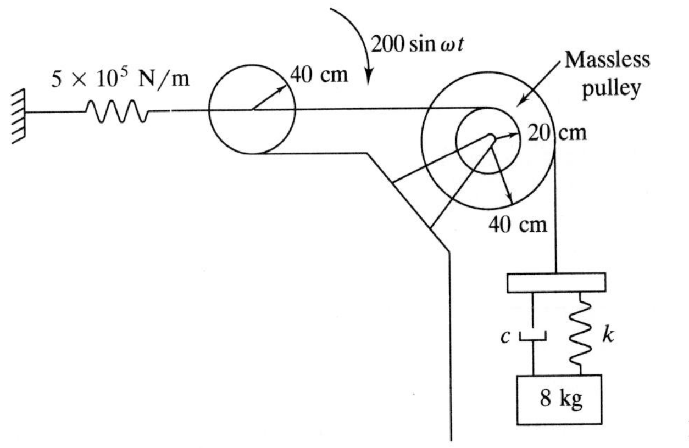

FIGURE P11.50

11.51. Use the Laplace transform method to analyze the situation of an undamped absorber attached to a viscously damped system, as shown in Fig. P11.51.

(*a*) Determine the steady-state amplitude of the mass m_1.

(*b*) Use the results of part (*a*) to design an absorber for a 123-kg machine of natural frequency 87 rad/s and damping ratio 0.13. Use an absorber mass of 35 kg.

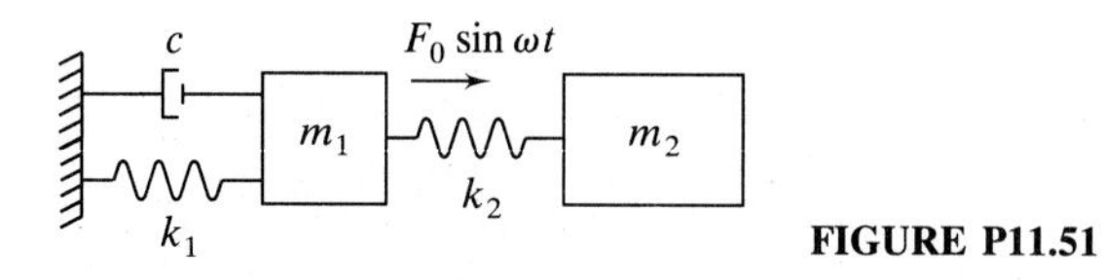

FIGURE P11.51

11.52. What is the steady-state amplitude of the mass center of the bar of Example 11.10 if an adsorber is added to eliminate steady-state vibrations of the right end of the bar?

11.53. Design an undamped absorber such that steady-state the motion of the 25-kg machine component in Fig. P11.53 ceases when the absorber is added. What is the steady-state amplitude of the 31 kg component?

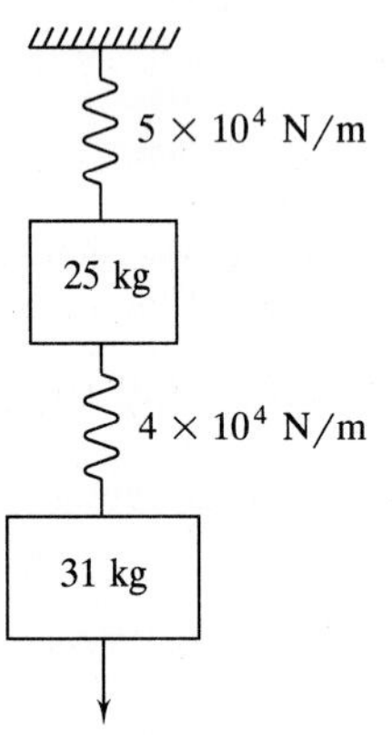

FIGURE P11.53

11.54. A 100-kg machine is subject to a harmonic excitation of magnitude 10,000 N in the frequency range from 1000 to 2700 rpm. The maximum transmitted force between the machine and its foundation should be 2000 N.

(*a*) Design an isolator with $\zeta = 0.12$ to reduce the transmitted force to exactly 2000 N.

(*b*) The maximum steady-state amplitude of the machine should be 10 cm. Design an isolator of minimum static deflection (and minimum added mass, if necessary) to satisfy force and displacement constraints.

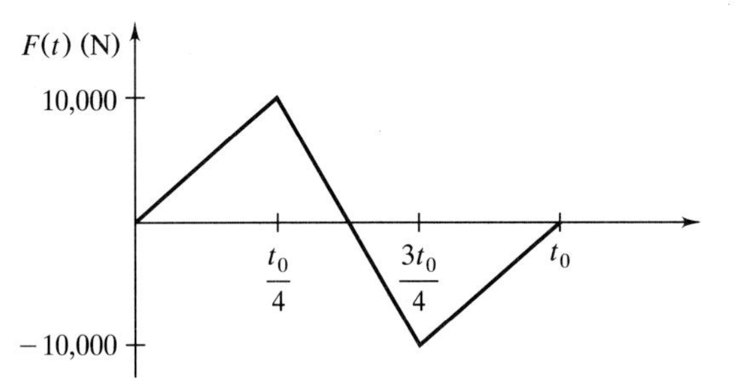

FIGURE P11.54

(*c*) Now suppose the machine is subject to the loading of Fig. P11.54. Using the isolator designed in part (*a*), what is the maximum steady-state displacement if (1) $t_0 = 0.05$ s; (2) $t_0 = 0.1$ s; (3) $t_0 = 0.2$ s; (4) $t_0 = 0.3$ s?

(*d*) Repeat part (*c*), determining the maximum force transmitted to the foundation.

REFERENCES

1. Arnold, F. R.: "Steady-State Behavior of Systems Provided With Nonlinear Dynamic Vibration Absorbers," *Journal of Applied Mechanics*, vol. 22, pp. 487–492, 1955.
2. Crede, C. M.: *Vibration and Shock Isolation*, Wiley, New York, 1951.
3. Ebrahimi, N. D.: "On Optimum Design of Vibration Absorbers," *Journal of Vibrations, Acoustics, Stress, and Reliability in Design*, vol. 109, pp. 214–216, 1987.
4. Frolov, K. V., and F. A. Furman: *Applied Theory of Vibration Isolation Systems*, Hemisphere, New York, 1990.
5. Harris, C. M., and C. E. Crede, eds.: *Shock and Vibration Handbook*, 2nd ed., McGraw-Hill, New York, 1976.
6. Hunt, J. B.: *Dynamic Vibration Absorbers*, Mechanical Engineering Publishers, London, 1979.
7. Ormondroyd, J., and J. P. DenHartog: "Theory of Dynamic Vibration Absorbers," *Transactions of the ASME*, vol. 50, PAPM-241, 1928.
8. Prakash, S. and Puri, V. K.: *Foundations for Machines: Analysis and Design*, Wiley, New York, 1988.
9. Rao, S. S.: *Mechanical Vibrations*, 2nd ed., Addison-Wesley, Reading, Mass., 1990.
10. Rivin, E. I.: "Vibration Isolation of Industrial Machines—Basic Considerations," *Sound and Vibration*, vol. 12, pp. 14–19, 1978.
11. Ruzicka, J. E.: "Fundamental Concepts of Vibration Control," *Journal of Sound and Vibration*, vol. 5, pp. 16–23, 1971.
12. Salerno, F. M., and R. M. Hochheiser: "How to Select Vibration Isolators for Use as Machinery Mounting," *Sound and Vibration*, vol. 7, pp. 22–28, 1973.
13. Soom, A., and M. Lee: "Optimal Design of Linear and Nonlinear Vibration Absorbers for Daped Systems," *Journal of Vibrations, Acoustics, Stress, and Reliability in Design*, vol. 105, pp. 112–119, 1983.
14. Stock Drive Products, *Vibration and Shock Mount Handbook*, *Product Catalog 814*, Stock Drive Products, New York, 1984.
15. Wang, B. P., L. Kitis, W. D. Pilkey, and A. Palazzolo: "Synthesis of Dynamic Vibration Absorbers," *Journal of Vibrations, Acoustics, Stress, and Reliability in Design*, vol. 107, pp. 161–164, 1985.

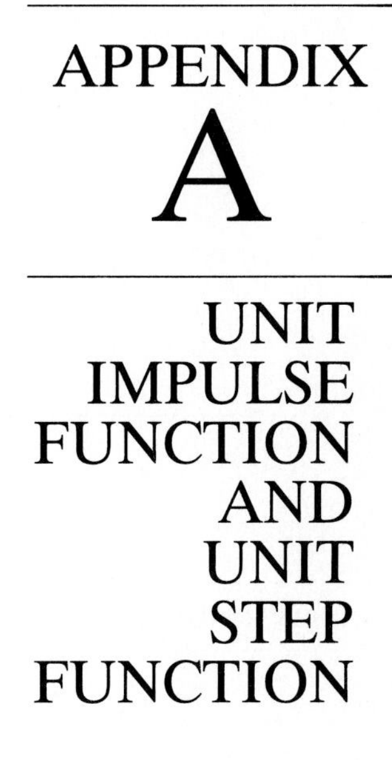

APPENDIX A

UNIT IMPULSE FUNCTION AND UNIT STEP FUNCTION

Consider the function, $f_\Delta(x; a)$, where $f_\Delta(x; a)$ as shown in Fig. A.1 is defined by

$$f_\Delta(x;a) = \begin{cases} 0 & -\infty < x < a - \dfrac{\Delta}{2} \\ \dfrac{1}{\Delta} & a - \dfrac{\Delta}{2} \le x \le a + \dfrac{\Delta}{2} \\ 0 & a + \dfrac{\Delta}{2} < x < \infty \end{cases} \tag{A.1}$$

The function has the unique property that

$$\int_{-\infty}^{\infty} f_\Delta(x;a)\,dx = 1 \tag{A.2}$$

Taking the limit of $f_\Delta(x; a)$ as $\Delta \to 0$ yields

$$\lim_{\Delta \to 0} f_\Delta(x;a) = \delta(x-a) = \begin{cases} 0 & x \ne a \\ \infty & x = a \end{cases} \tag{A.3}$$

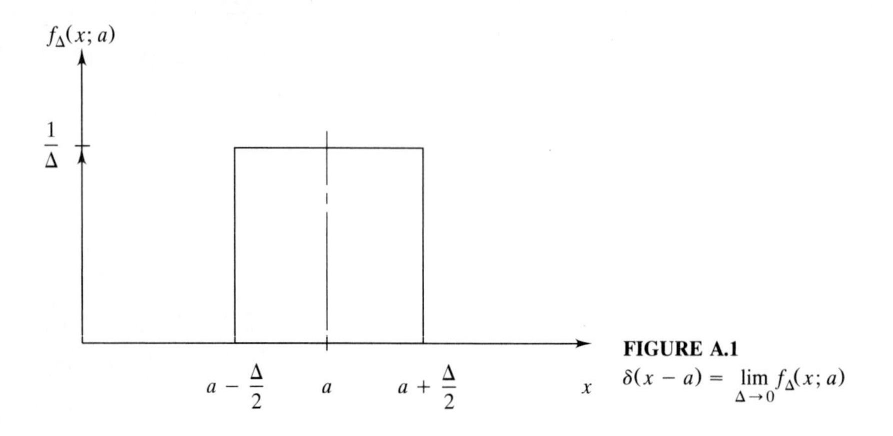

FIGURE A.1
$\delta(x-a) = \lim_{\Delta \to 0} f_\Delta(x; a)$

From Eq. (A.2)

$$\int_{-\infty}^{\infty} \delta(x-a)\, dx = 1 \tag{A.4}$$

The function defined in Eq. (A.3) and whose valuable property is given in Eq. (A.4) is called the unit impulse function. It has many applications in physics and engineering. It is used to mathematically represent the force that is applied to cause a unit impulse applied at a time $t = a$ in a mechanical system. It is used to represent a unit concentrated load applied at a location $x = a$ to a structure. The unit impulse function is also used to represent a unit heat source in a heat transfer problem.

Now define

$$\begin{aligned} u(x-a) &= \int_0^x \delta(x-a)\, dx \\ &= \int_0^x \lim_{\Delta \to 0} f_\Delta(x; a)\, dx \\ &= \begin{cases} 0 & x \le a \\ 1 & x > a \end{cases} \end{aligned} \tag{A.5}$$

The function defined in Eq. (A.5) is called the unit step function and is plotted in Fig. A.2. Differentiating Eq. (A.5) gives

$$\frac{du(x-a)}{dx} = \delta(x-a) \tag{A.6}$$

The definitions of the unit impulse function and unit step function can also be

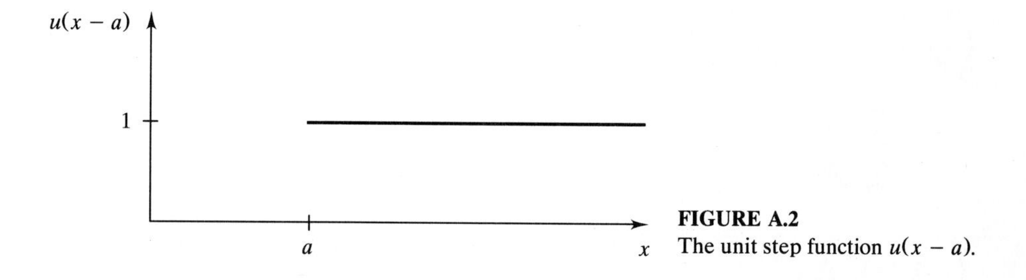

FIGURE A.2
The unit step function $u(x - a)$.

used to derive the following integral formulas. For any function $g(t)$,

$$\int_0^t \delta(\tau - a) g(\tau)\, d\tau = u(t - a) g(a) \tag{A.7}$$

and

$$\int_0^t u(\tau - a) g(\tau)\, d\tau = u(t - a) \int_a^t g(\tau)\, d\tau \tag{A.8}$$

APPENDIX B

LAPLACE TRANSFORMS

B.1 DEFINITION

The Laplace transform of a function $f(t)$ is defined as

$$\bar{f}(s) = \mathscr{L}\{f(t)\} = \int_0^\infty f(t)e^{-st}\,dt \tag{B.1}$$

If there exist values of α, M, and T such that

$$e^{-\alpha t}|f(t)| \langle M \quad \text{for all } t \rangle T \tag{B.2}$$

then $\bar{f}(s)$ exists for $s > \alpha$. Equation (B.2) is satisfied for all excitations and responses in this text.

The Laplace transform transforms a real-valued function into a function of a complex variable, s. For many functions the Laplace transform can be obtained by direct integration.

Example B.1. Determine the Laplace transform of $f(t) = e^{\alpha t}$.

$$\mathscr{L}\{e^{\alpha t}\} = \int_0^\infty e^{\alpha t}e^{-st}\,dt$$

$$= \frac{1}{\alpha - s}e^{(\alpha - s)t}\Big|_0^\infty$$

$$= \frac{1}{s - \alpha} \qquad s > \alpha$$

TABLE B.1

Number	$f(t)$	$\bar{f}(s)$
1	1	$\frac{1}{s}$
2	t^n	$\frac{n!}{s^{n+1}}$
3	$e^{\alpha t}$	$\frac{1}{s-\alpha}$
4	$\sin \omega t$	$\frac{\omega}{s^2+\omega^2}$
5	$\cos \omega t$	$\frac{s}{s^2+\omega^2}$
6	$\delta(t-a)$	e^{-as}
7	$u(t-a)$	$\frac{e^{-as}}{s}$

B.2 TABLE OF TRANSFORMS

Equation (B.1) is used to develop a table of Laplace transforms of common functions. Laplace transforms of other functions can be developed using Table B.1 in conjunction with properties of the transform.

B.3 LINEARITY

The Laplace transform operator is a linear operator. Let $\bar{f}(s = \mathscr{L}\{f(t)\}$ and $\bar{g}(s) = \mathscr{L}\{g(t)\}$ and let α and β be any real numbers. Then

$$\mathscr{L}\{\alpha f(t) + \beta g(t)\} = \alpha \bar{f}(s) + \beta \bar{g}(s) \tag{B.3}$$

B.4 TRANSFORM OF DERIVATIVES

The property of the Laplace transforms of derivatives allows easy application of the Laplace transform to the solution of differential equations

$$\mathscr{L}\left\{\frac{d^n f}{ds^n}\right\} = s^n \bar{f}(s) - s^{n-1} f(0) - s^{n-2} \dot{f}(0) - \cdots - s f^{(n-2)}(0) - f^{(n-1)}(0) \tag{B.4}$$

Example B.2. Use transform pair 5 from Table B.1 and Eq. (B.4) to determine $\mathscr{L}\{\sin 2t\}$.

Noting that

$$\sin 2t = -\frac{1}{2}\frac{d(\cos 2t)}{dt}$$

and applying properties (B.3) and (B.4) with $n = 1$ gives

$$\mathscr{L}\{\sin 2t\} = -\tfrac{1}{2}(s\mathscr{L}\{\cos 2t\} - 1)$$

Using transform pair 5 from Table B.1,

$$\mathscr{L}\{\sin 2t\} = -\frac{1}{2}\left(\frac{s^2}{s^2+4} - 1\right)$$
$$= \frac{2}{s^2+4}$$

B.5 FIRST SHIFTING THEOREM

If $\bar{f}(s) = \mathscr{L}\{f(t)\}$, then

$$\mathscr{L}\{e^{-at}f(t)\} = \bar{f}(s+a) \tag{B.5}$$

Example B.3. Use Table B.1 and the first shifting theorem to calculate $\mathscr{L}\{e^{-\zeta\omega_n t}\cos\omega_d t\}$ where $\omega_d = \omega_n\sqrt{(1-\zeta^2)}$.

Using the first shifting theorem and transform pair 5 from Table B.1,

$$\mathscr{L}\{e^{-\zeta\omega_n t}\cos\omega_d t\} = \left.\frac{s}{s^2+\omega_d^2}\right|_{s\to s+\zeta\omega_n}$$
$$= \frac{s+\zeta\omega_n}{(s+\zeta\omega_n)^2+\omega_d^2}$$
$$= \frac{s+\zeta\omega_n}{s^2+2\zeta\omega_n s+\omega_n^2}$$

B.6 SECOND SHIFTING THEOREM

If $\bar{f}\zeta(s) = \mathscr{L}f(t)\}$, then

$$\mathscr{L}\{f(t-a)u(t-a)\} = e^{-as}\bar{f}(s) \tag{B.6}$$

Example B.4. Use Table B.1 and the second shifting theorem to determine the Laplace transform of the function of Fig. B.1.

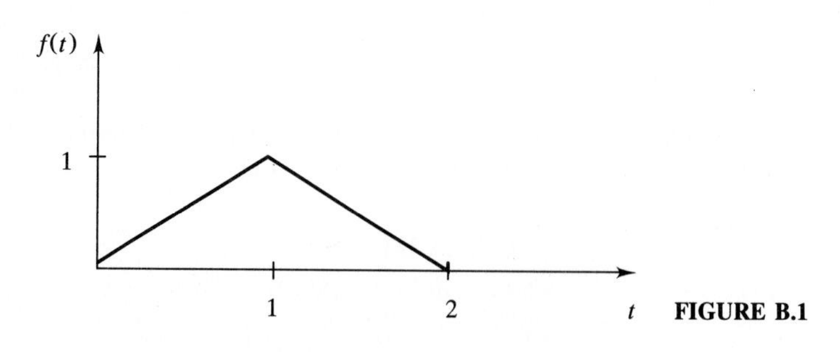

FIGURE B.1

The function of Fig. B.1 is written using unit step functions as

$$f(t) = t[u(t) - u(t-1)] + (2-t)[u(t-1) - u(t-2)]$$
$$= tu(t) - 2(t-1)u(t-1) + (t-2)u(t-2)$$

Use of transform pair 2 from Table B.1 with $n = 1$ and the second shifting theorem give

$$\mathscr{L}\{f(t)\} = \frac{1}{s} - e^{-s}\frac{2}{s} + e^{-2s}\frac{1}{s}$$
$$= \frac{1}{s}(1 - 2e^{-s} + e^{-2s})$$

B.7 INVERSION OF TRANSFORM

If $\bar{f}(s) = \mathscr{L}\{f(t)\}$, then $f(t) = \mathscr{L}^{-1}\{f(s)\}$ where

$$\mathscr{L}^{-1}\{\bar{f}(s)\} = \frac{1}{2\pi i}\int_{\gamma - i\infty}^{\gamma + i\infty} \bar{f}(s)e^{st}\,ds \tag{B.7}$$

is an integral carried out in the complex s plane. Inverse transforms are usually obtained by using Table B.1 in conjunction with transform properties.

Example B.5. If

$$\mathscr{L}^{-1}\left\{e^{-2s}\frac{s+5}{s^2+2s+5}\right\} = f(t)$$

find $f(t)$.

Completing the square of the denominator of $\bar{f}(s)$,

$$\bar{f}(s) = e^{-2s}\frac{s+5}{(s+1)^2+4}$$
$$= e^{-2s}\left[\frac{s+1}{(s+1)^2+4} + \frac{4}{(s+1)^2+4}\right]$$
$$= e^{-2s}\bar{g}(s)$$

Using linearity, the first shifting theorem, and transform pairs 4 and 5 from Table B.1,

$$g(t) = \mathscr{L}^{-1}\{\bar{g}(s)\} = e^{-t}(\cos 2t + 2\sin 2t)$$

Using the second shifting theorem,

$$f(t) = u(t-2)g(t-2) = e^{2-t}[\cos 2(t-2) + 2\sin 2(t-2)]u(t-2)$$

B.8 TABLE OF PROPERTIES

The useful properties of the Laplace transform are summarized in Table B.2.

TABLE B.2

Property name	Formula
Definition of transform	$\bar{f}(s) = \int_0^\infty e^{-st} f(t)\, dt$
Linearity	$\mathscr{L}\{\alpha f(t) + \beta g(t)\} = \alpha \bar{f}(s) + \beta \bar{g}(s)$
Transform of derivatives	$\mathscr{L}\left\{\frac{d^n f}{dt^n}\right\} = s^n \bar{f}(s) - s^{n-1} f(0) - \cdots - f^{(n-1)}(0)$
First shifting theorem	$\mathscr{L}\{e^{-at} f(t)\} = \bar{f}(s+a)$
Second shifting theorem	$\mathscr{L}\{f(t-a)u(t-a)\} = e^{-as}\bar{f}(s)$
Inverse transform	$f(t) = \frac{1}{2\pi i}\int_{\gamma - i\infty}^{\gamma + i\infty} \bar{f}(s) e^{st}\, ds$

REFERENCE

1. Churchill, R. V.: *Operational Mathematics*, McGraw-Hill, New York, 1972.

APPENDIX C

LINEAR ALGEBRA

C.1 DEFINITIONS

1. A matrix is a collection of numbers arranged in a specific order in rows and columns. If matrix **A** has n rows and n columns, then it is represented by

$$\mathbf{A} = \begin{bmatrix} a_{11} & a_{12} & a_{13} & \cdots & a_{1m} \\ a_{21} & a_{22} & a_{23} & \cdots & a_{2m} \\ a_{31} & a_{32} & a_{33} & \cdots & a_{3m} \\ \vdots & \vdots & \vdots & \ddots & \vdots \\ a_{n1} & a_{n2} & a_{n3} & \cdots & a_{nm} \end{bmatrix} \tag{C.1}$$

   Throughout this text a single capital letter in boldface is used to represent a matrix. The corresponding lowercase letter with two subscripts is used to refer to a specific element of the matrix. For example, the element a_{ij} resides in the ith row and jth column of **A**.

   A matrix with n rows and n columns is called an $m \times n$ matrix. A square matrix has the same number of rows and columns.

2. A column vector is a matrix with only one column. A row vector is a matrix with only one row. Usually, a single lowercase letter in boldface is used to represent a column vector or a row vector. The letter with a single subscript refers to a specific element of the vector. A column vector with n rows or a row vector with n columns is said to be an n-dimensional vector. If $\mathbf{x}$ is an n-dimensional column vector then x_i, $i \leq n$, is the element in the ith row of the vector.

3. A diagonal matrix is a square matrix with all off-diagonal elements equal to zero. That is $a_{ij} = 0$ if $i \neq j$.
4. An identity matrix is a diagonal matrix whose diagonal elements are all unity. That is $a_{ij} = \delta_{ij}$, where δ_{ij} is the Kronecker delta defined by

$$\delta_{ij} = \begin{cases} 1 & i = j \\ 0 & i \neq j \end{cases}$$

5. The transpose of the matrix **A**, denoted by $\mathbf{A}^T$, is the matrix obtained by interchanging the rows and columns of **A**. If $\mathbf{B} = \mathbf{A}^T$, then $b_{ij} = a_{ji}$. The transpose of a column vector is a row vector and vice versa.
6. A symmetric matrix is a square matrix whose transpose is equal to the matrix itself. If **A** is an $n \times n$ symmetric matrix, then $a_{ij} = a_{ji}$ for $i = 1, \ldots, n$ and $j = 1, \ldots, n$.

C.2 DETERMINANTS

The determinant of a square $n \times n$ matrix is a number associated with the matrix that is of great consequence in application. It is easiest to define the determinant of a 2×2 matrix and use this definition to define the determinant of larger matrices.

The determinant of the 2×2 matrix **A** is

$$\det\{\mathbf{A}\} = |\mathbf{A}| = \begin{vmatrix} a_{11} & a_{12} \\ a_{21} & a_{22} \end{vmatrix} = a_{11}a_{22} - a_{12}a_{21} \tag{C.2}$$

The minor corresponding to the element in the ith row and jth column of an $n \times n$ matrix **A**, denoted by M_{ij}, is the determinant of the $(n-1) \times (n-1)$ matrix obtained by deleting the ith row and jth column from **A**. The cofactor corresponding to the element in the ith row and jth column of **A**, denoted by C_{ij}, is

$$C_{ij} = (-1)^{i+j} M_{ij} \tag{C.3}$$

For an i, $i = 1, \ldots, n$, the determinant of **A** is obtained using the following row expansion:

$$|\mathbf{A}| = \sum_{j=1}^{n} a_{ij} C_{ij} \tag{C.4}$$

The value of the determinant is the same regardless of the value of i. The determinant can also be calculated using a column expansion according to the formula

$$|\mathbf{A}| = \sum_{j=1}^{n} a_{ji} C_{ji} \tag{C.5}$$

Since the minors themselves are determinants, row or columns expansions can be used to express each of the minors in terms of the minors of their corresponding matrix. These expansions continue until the remaining minors are 2×2 determinants.

Example C.1. Calculate the determinant of the 4×4 matrix $\mathbf{A}$ where

$$\mathbf{A} = \begin{bmatrix} 1 & 0 & 0 & 2 \\ 1 & 2 & -1 & 0 \\ 2 & -1 & 3 & 1 \\ 2 & 0 & -2 & 1 \end{bmatrix}$$

The determinant is evaluated using a first-row expansion, using Eq. (C.4),

$$|\mathbf{A}| = (1)\begin{vmatrix} 2 & -1 & 0 \\ -1 & 3 & 1 \\ 0 & -2 & 1 \end{vmatrix} - (2)\begin{vmatrix} 1 & 2 & -1 \\ 2 & -1 & 3 \\ 2 & 0 & -2 \end{vmatrix}$$

Expansion by the first row is used to evaluate each of the 3×3 determinants, resulting in

$$|\mathbf{A}| = (2)\begin{vmatrix} 3 & 1 \\ -2 & 1 \end{vmatrix} - (-1)\begin{vmatrix} -1 & 1 \\ 0 & 1 \end{vmatrix} - 2\left[(1)\begin{vmatrix} -1 & 3 \\ 0 & -2 \end{vmatrix} - (2)\begin{vmatrix} 2 & 3 \\ 2 & -2 \end{vmatrix} + (-1)\begin{vmatrix} 2 & -1 \\ 2 & 0 \end{vmatrix}\right]$$

The 2×2 determinants are evaluated using Eq. (C.2), yielding

$$\begin{aligned} |\mathbf{A}| &= (2)[(3)(1) - (1)(-2)] + [(-1)(1) - (1)(0)] \\ &\quad - (2)\{[(-1)(-2) - (3)(0)] - (2)[(2)(-2) - (3)(2)] \\ &\quad - [(2)(0) - (-1)(2)]\} \\ &= -31 \end{aligned}$$

The determinant of a matrix is zero if and only if the column vectors that form the matrix are linearly dependent. For example, the determinant of a matrix with a column of zeros is zero. A matrix whose determinant is zero is said to be singular. The row vectors of a singular matrix are also linearly dependent.

C.3 MATRIX OPERATIONS

If $\mathbf{C} = \mathbf{A} + \mathbf{B}$, then

$$c_{ij} = a_{ij} + b_{ij}. \tag{C.6}$$

If the number of columns of $\mathbf{A}$ equals the number of rows of $\mathbf{B}$, then the matrix $\mathbf{C} = \mathbf{AB}$ is defined as a matrix with the number of rows of $\mathbf{A}$ and the number of columns of $\mathbf{B}$ and c_{ij} is the sum of the products of the corresponding elements in the ith row of $\mathbf{A}$ and the jth column of $\mathbf{B}$. That is,

$$c_{ij} = \sum_{k=1}^{n} a_{ik} b_{kj} \tag{C.7}$$

Matrix multiplication is not commutative, but is associative and distributive. The transpose of the product has the following property. If $\mathbf{C} = \mathbf{AB}$, then

$$\mathbf{C}^T = (\mathbf{AB})^T = \mathbf{B}^T\mathbf{A}^T \tag{C.8}$$

Example C.2. Calculate $\mathbf{Ax}$ where

$$\mathbf{A} = \begin{bmatrix} 1 & 2 & 4 & -1 \\ 2 & 3 & 0 & 4 \\ 1 & 2 & 6 & 2 \\ 0 & 2 & 3 & 1 \end{bmatrix} \qquad \mathbf{x} = \begin{bmatrix} 1 \\ 4 \\ -1 \\ 2 \end{bmatrix}$$

The product of a 4×4 matrix and a four-dimensional column vector is a four-dimensional column vector,

$$\mathbf{Ax} = \begin{bmatrix} (1)(1) + (2)(4) + (4)(-1) + (-1)(2) \\ (2)(1) + (3)(4) + (0)(-1) + (4)(2) \\ (1)(1) + (2)(4) + (6)(-1) + (2)(2) \\ (0)(1) + (2)(4) + (3)(-1) + (1)(2) \end{bmatrix} = \begin{bmatrix} 3 \\ 22 \\ 7 \\ 7 \end{bmatrix}$$

C.4 SYSTEMS OF EQUATIONS

Consider the system of n simultaneous equations which are to be solved for the n unknowns $x_1, x_2, \ldots, x_n$,

$$\begin{aligned} a_{11}x_1 + a_{12}x_2 + \cdots + a_{1n}x_n &= y_1 \\ a_{21}x_1 + a_{22}x_2 + \cdots + a_{2n}x_n &= y_2 \\ \vdots \qquad\quad \vdots \qquad\qquad\quad &\ \ \vdots \\ a_{n1}x_1 + a_{n2}x_2 + \cdots + a_{nn}x_n &= y_n \end{aligned} \tag{C.9}$$

Using the definitions of matrix addition and matrix multiplication, the system of Eq. (C.9) is written in matrix form as

$$\mathbf{Ax} = \mathbf{y} \tag{C.10}$$

where $$\mathbf{A} = \begin{bmatrix} a_{11} & a_{12} & \cdots & a_{1n} \\ a_{21} & a_{22} & \cdots & a_{2n} \\ \vdots & \vdots & \ddots & \vdots \\ a_{n1} & a_{n2} & \cdots & a_{nn} \end{bmatrix} \qquad \mathbf{x} = \begin{bmatrix} x_1 \\ x_2 \\ \vdots \\ x_n \end{bmatrix} \qquad \mathbf{y} = \begin{bmatrix} y_1 \\ y_2 \\ \vdots \\ y_n \end{bmatrix}$$

Cramer's rule can be used to solve for the components of $\mathbf{x}$,

$$x_i = \frac{|\mathbf{B}_i|}{|\mathbf{A}|} \tag{C.11}$$

where $\mathbf{B}_i$ is the matrix obtained by replacing the ith column of $\mathbf{A}$ with $\mathbf{y}$. Thus if $\mathbf{A}$ is singular, a solution of Eq. (C.9) exists only for certain forms of $\mathbf{y}$. Since its rows are linearly dependent when the matrix is singular, the solution corresponding to special forms of $\mathbf{y}$ is not unique.

An equation in a system of equations can be replaced, without affecting the solution of the system, by an equation obtained by multiplying the equation by a scalar and adding or subtracting it from another equation. The equations can be so manipulated until one of the equations only has one unknown. This is the basis of the Gauss elimination method.

Matrix formulation of the equations expedites the application of Gauss elimination. The $n \times n$ coefficient matrix is augmented with the right-hand side vector to form an $n \times (n+1)$ matrix. Each row of the augmented matrix represents one equation. The Gauss elimination procedure is applied by performing manipulations on the rows of the augmented matrix such that coefficients below the diagonal become zero. The elimination procedure results in a coefficient matrix with all zeros below its diagonal. Back substitution is used to determine the solution.

Example C.3. Use Gauss elimination to determine the solution of the following set of simultaneous equations

$$3x_1 + 2x_2 - 2x_3 = 1$$

$$2x_1 - 3x_2 + 2x_3 = 4$$

$$x_1 + x_2 - 3x_3 = -3$$

The matrix formulation of these equations is

$$\begin{bmatrix} 3 & 2 & -2 \\ 2 & -3 & 2 \\ 1 & 1 & -3 \end{bmatrix} \begin{bmatrix} x_1 \\ x_2 \\ x_3 \end{bmatrix} = \begin{bmatrix} 1 \\ 4 \\ -3 \end{bmatrix}$$

The augmented matrix used in Gauss elimination is

$$\begin{bmatrix} 3 & 2 & -2 & 1 \\ 2 & -3 & 2 & 4 \\ 1 & 1 & -3 & -3 \end{bmatrix}$$

The second row is replaced with a row obtained by multiplying the second row by 3/2 and subtracting from the first row. The third row is replaced by a row obtained by multiplying the third row by 3 and subtracting from the first row. The resulting matrix is

$$\begin{bmatrix} 3 & 2 & -2 & 1 \\ 0 & \dfrac{13}{2} & -5 & -5 \\ 0 & -1 & 7 & 10 \end{bmatrix}$$

The new third row is replaced by a row obtained by multiplying by -2 and subtracting from the first row, resulting in

$$\begin{bmatrix} 3 & 2 & -2 & 1 \\ 0 & \dfrac{13}{2} & -5 & -5 \\ 0 & 0 & 12 & 21 \end{bmatrix}$$

Back substitution is used to determine the unknowns

$$12x_3 = 21 \to x_3 = \frac{7}{4}$$

$$\frac{13}{2}x_2 - 5x_3 = -5 \to x_2 = \frac{2}{13}\left[-5 + 5\left(\frac{7}{4}\right)\right] = \frac{15}{26}$$

$$3x_1 + 2x_2 - 2x_3 = 1 \to x_1 = \frac{1}{3}\left[1 - 2\left(\frac{15}{26}\right) + 2\left(\frac{7}{4}\right)\right] = \frac{29}{26}$$

C.5 INVERSE MATRIX

If **A** is a nonsingular $n \times n$ matrix, then a matrix $\mathbf{A}^{-1}$, called the inverse of **A**, exists such that

$$\mathbf{A}\mathbf{A}^{-1} = \mathbf{A}^{-1}\mathbf{A} = \mathbf{I} \tag{C.12}$$

If $\mathbf{A}^{-1}$ is known, Eq. (C.10) can be solved by premultiplying both sides by $\mathbf{A}^{-1}$,

$$\mathbf{A}^{-1}\mathbf{A}\mathbf{x} = \mathbf{x} = \mathbf{A}^{-1}\mathbf{y} \tag{C.13}$$

If **y** is a column vector with all zeros except $y_i = 1$, then $\mathbf{A}^{-1}\mathbf{y}$ is the ith column of $\mathbf{A}^{-1}$. This provides the basis of an extension of Gauss elimination which is used to determine $\mathbf{A}^{-1}$. The coefficient matrix is augmented by the $n \times n$ identity matrix. The procedure used in Gauss elimination is applied until the identity matrix appears in place of the original matrix. The matrix that augments the identity matrix is $\mathbf{A}^{-1}$.

Example C.4. Determine the inverse of

$$\mathbf{A} = \begin{bmatrix} 2 & -1 & 0 \\ -1 & 3 & -2 \\ 0 & -2 & 3 \end{bmatrix}$$

Gauss elimination is applied to the following matrix:

$$\begin{bmatrix} 2 & -1 & 0 & 1 & 0 & 0 \\ -1 & 3 & -2 & 0 & 1 & 0 \\ 0 & -2 & 3 & 0 & 0 & 1 \end{bmatrix}$$

Gauss elimination is used to develop zeros below the diagonal of the coefficient matrix

$$\begin{bmatrix} 2 & -1 & 0 & 1 & 0 & 0 \\ 0 & 5 & -4 & 1 & 2 & 0 \\ 0 & 0 & \frac{7}{2} & 1 & 2 & \frac{5}{2} \end{bmatrix}$$

The procedure of Gauss elimination is used to eliminate the zeros above the diagonal of the coefficient matrix. Each row is divided by the value of the element along the diagonal of the matrix that has taken the place of the original coefficient matrix. The result is

$$\begin{bmatrix} 1 & 0 & 0 & \frac{5}{7} & \frac{3}{7} & \frac{2}{7} \\ 0 & 1 & 0 & \frac{3}{7} & \frac{6}{7} & \frac{4}{7} \\ 0 & 0 & 1 & \frac{2}{7} & \frac{4}{7} & \frac{5}{7} \end{bmatrix}$$

Thus

$$\mathbf{A}^{-1} = \begin{bmatrix} \frac{5}{7} & \frac{3}{7} & \frac{2}{7} \\ \frac{3}{7} & \frac{6}{7} & \frac{4}{7} \\ \frac{2}{7} & \frac{4}{7} & \frac{5}{7} \end{bmatrix}$$

C.6 EIGENVALUE PROBLEMS

The eigenvalues of an $n \times n$ matrix, $\mathbf{A}$, are the values of λ such that the system of equations

$$\mathbf{Ax} = \lambda \mathbf{x} \tag{C.14}$$

has a nontrivial solution. The nontrivial solution corresponding to an eigenvalue is called an eigenvector. Equation (C.14) can be rewritten as

$$(\mathbf{A} - \lambda \mathbf{I})\mathbf{x} = \mathbf{0} \tag{C.15}$$

From Cramer's rule, Eq. (C.11), the solution for x_i is

$$x_i = \frac{0}{|\mathbf{A} - \lambda \mathbf{I}|} \qquad i = 1, \ldots, n$$

Thus, for each $i = 1, \ldots, n$, $x_i = 0$, unless

$$|\mathbf{A} - \lambda \mathbf{I}| = 0 \tag{C.16}$$

The determinant of Eq. (C.16) can be expanded using a row or column expansion. This yields an nth-order polynomial equation of the form

$$\lambda^n + C_1 \lambda^{n-1} + C_2 \lambda^{n-2} + \cdots + C_{n-1} \lambda + C_n = 0 \tag{C.17}$$

called the characteristic equation. Equation (C.17) has n roots, and $\mathbf{A}$ has n eigenvalues. Since the coefficients in Eq. (C.17) are all real, if complex eigenvalues occur, they occur as complex conjugate pairs.

If λ is an eigenvalue of $\mathbf{A}$, then Eq. (C.14) has a nontrivial solution, an eigenvector. From Eq. (C.16), the matrix $\mathbf{A} - \lambda I$ is singular. Thus the equations defining the components of the corresponding eigenvector are not all independent and the eigenvector is not unique. The eigenvector is unique only to an arbitrary multiplicative constant.

Example C.5. Determine the eigenvalues and eigenvectors of the matrix

$$\mathbf{A} = \begin{bmatrix} 2 & -1 & 0 \\ -1 & 3 & -2 \\ 0 & -2 & 3 \end{bmatrix}$$

The eigenvalues of **A** are determined by finding the values of λ satisfying Eq. (C.16), which for this example become

$$\begin{vmatrix} 2-\lambda & -1 & 0 \\ -1 & 3-\lambda & -2 \\ 0 & -2 & 3-\lambda \end{vmatrix} = 0$$

Expansion of the determinant by its first row gives

$$(2-\lambda)\begin{vmatrix} 3-\lambda & -2 \\ -2 & 3-\lambda \end{vmatrix} - (-1)\begin{vmatrix} -1 & -2 \\ 0 & 3-\lambda \end{vmatrix} = 0$$

When the 2×2 determinants are expanded using Eq. (C.2), the following cubic equation is obtained:

$$-\lambda^3 + 8\lambda^2 - 16\lambda + 7 = 0$$

The eigenvalues are the roots of the cubic equation which are 0.609, 2.227, and 5.164. The eigenvector corresponding to the smallest eigenvalue is obtained by solving

$$\begin{bmatrix} 1.391 & -1 & 0 \\ -1 & 2.391 & -2 \\ 0 & -2 & 2.391 \end{bmatrix}\begin{bmatrix} x_1 \\ x_2 \\ x_3 \end{bmatrix} = \begin{bmatrix} 0 \\ 0 \\ 0 \end{bmatrix}$$

The first equation gives $x_1 = 0.719x_2$. The third equation gives $x_3 = 0.836x_2$. When these relationships are substituted into the second equation, it is identically satisfied. Thus x_2 remains arbitrary and the eigenvector of **A** corresponding to $\lambda = 0.609$ is

$$C_1\begin{bmatrix} 0.719 \\ 1 \\ 0.836 \end{bmatrix}$$

where C_1 is an arbitrary constant. The same procedure is followed yielding the eigenvectors corresponding to the second and third eigenvalues. These are

$$C_2\begin{bmatrix} -4.41 \\ 1 \\ 2.59 \end{bmatrix} \qquad C_3\begin{bmatrix} -0.316 \\ 1 \\ -0.924 \end{bmatrix}$$

respectively.

If **A** is an $n \times n$ singular matrix, then one of its eigenvalues is zero. If **A** is nonsingular, then the eigenvalues of $\mathbf{A}^{-1}$ are the reciprocals of the eigenvalues of **A**. The eigenvectors of $\mathbf{A}^{-1}$ are the same as the eigenvectors of **A**.

C.7 SCALAR PRODUCTS

Let **u**, **v**, and **w** be arbitrary real n-dimensional column vectors. A scalar product is an operation among two of these vectors yielding a real value. The scalar product of **u** and **v** is denoted by $(\mathbf{u}, \mathbf{v})$. The scalar product must satisfy four requirements.

1. The scalar product is commutative. That is,

$$(\mathbf{u}, \mathbf{v}) = (\mathbf{v}, \mathbf{u}) \tag{C.18}$$

2. For any real α,

$$(\alpha \mathbf{u}, \mathbf{v}) = \alpha(\mathbf{u}, \mathbf{v}) \tag{C.19}$$

3. The scalar product is distributive

$$(\mathbf{u} + \mathbf{v}, \mathbf{w}) = (\mathbf{u}, \mathbf{w}) + (\mathbf{v}, \mathbf{w}) \tag{C.20}$$

4.

$$(\mathbf{u}, \mathbf{u}) \geq 0 \tag{C.21}$$

and

$$(\mathbf{u}, \mathbf{u}) = 0 \quad \text{if and only if } \mathbf{u} = \mathbf{0} \tag{C.22}$$

The definition of a scalar product is not unique. The standard scalar product is defined as

$$(\mathbf{u}, \mathbf{v}) = \mathbf{u}^T \mathbf{v} \tag{C.23}$$

Two vectors, **u** and **v**, are said to be orthogonal with respect to a scalar product if

$$(\mathbf{u}, \mathbf{v}) = 0 \tag{C.24}$$

A matrix **A** is said to be positive definite with respect to a scalar product if

$$(\mathbf{Au}, \mathbf{u}) \geq 0 \tag{C.25}$$

and

$$(\mathbf{Au}, \mathbf{u}) = 0 \quad \text{if and only if } \mathbf{u} = \mathbf{0} \tag{C.26}$$

Example C.6. Show that if **A** is a positive-definite symmetric matrix, then

$$(\mathbf{u}, \mathbf{v})_{\mathbf{A}} = (\mathbf{Au}, \mathbf{v}) \tag{C.27}$$

is a valid scalar product where $(\mathbf{u}, \mathbf{v})$ is the standard scalar product defined by Eq. (C.23).

In order for Eq. (C.27) to represent a valid scalar product, it is necessary to show that the four properties of Eqs. (C.18) through (C.22) are true knowing that they are true for the standard scalar product.

1.

$$\begin{aligned}
(\mathbf{u}, \mathbf{v})_A &= (\mathbf{Au})^T \mathbf{v} && \text{Eq. (C.27)} \\
&= \mathbf{u}^T \mathbf{A}^T \mathbf{v} && \text{Eq. (C.9)} \\
&= \mathbf{u}^T \mathbf{A} \mathbf{v} && \text{symmetry of } \mathbf{A} \\
&= (\mathbf{u}, \mathbf{Av}) && \text{Eq. (C.23)} \\
&= (\mathbf{Av}, \mathbf{u}) && \text{Eq. (C.18)} \\
&= (\mathbf{v}, \mathbf{u})_A && \text{Eq. (C.27)}
\end{aligned}$$

2. For any real α

$$\begin{aligned}
(\alpha \mathbf{u}, \mathbf{v})_A &= \alpha (\mathbf{Au})^T \mathbf{v} \\
&= \alpha (\mathbf{u}, \mathbf{v})_A
\end{aligned}$$

3.
$$
\begin{aligned}
(\mathbf{u}+\mathbf{v},\mathbf{w})_A &= [\mathbf{A}(\mathbf{u}+\mathbf{v})]^T\mathbf{w} \\
&= \left[(\mathbf{Au})^T + (\mathbf{Av})^T\right]\mathbf{w} \\
&= (\mathbf{Au})^T\mathbf{v} + (\mathbf{Av})^T\mathbf{w} \\
&= (\mathbf{u},\mathbf{w})_A + (\mathbf{v},\mathbf{w})_A
\end{aligned}
$$

4. The validity of property 4 for this definition of the scalar product follows directly from the positive definiteness of **A**, Eqs. (C.25) and (C.26).

The concept of scalar products can be extended to continuous functions. Any operation between two continuous functions that results in a scalar and the operation obeys Eqs. (C.18) through (C.22) is a valid scalar product. For example, for two functions $f(x)$ and $g(x)$ that are everywhere continuous between $x = 0$ and $x = 1$, a valid scalar product is

$$(f, g) = \int_0^1 f(x)g(x)\,dx \tag{C.28}$$

REFERENCES

1. Kreider, D. L., R. O. Kuller, D. R. Ostberg, and F. W. Perkins: *An Introduction to Linear Analysis*, 2nd ed., Addison-Wesley, Reading, Mass., 1966.
2. Rabenstein, A. L.: *Elementary Differential Equations with Linear Algebra*, 3rd ed., Academic Press, New York, 1982.
3. Strang, G.: *Linear Algebra and Its Applications*, Academic Press, New York, 1980.

APPENDIX D

DEFLECTION OF BEAMS SUBJECT TO CONCENTRATED LOADS

Consider a beam of total length L, subject to arbitrary end constraints. Let x be a coordinate along the neutral axis of the beam. The beam has n intermediate simple supports at $x = x_i$, $i = 1, 2, \ldots, n$. It is desired to calculate the deflection of the beam as a function of x due to a concentrated unit load applied at $x = a$. If $y(x)$ is the deflection of the beam, measured positive downward from the horizontal, then use of the usual assumptions of linear elastic beam theory leads to

$$EI\frac{d^4y}{dx^4} = w(x) \tag{D.1}$$

where $w(x)$ represents the loading, E is the elastic modulus of the beam, and I is the moment of inertia of the cross-sectional area about the neutral axis.

The intermediate supports are replaced by concentrated loads. The analysis is performed requiring the deflection to be zero at the intermediate supports.

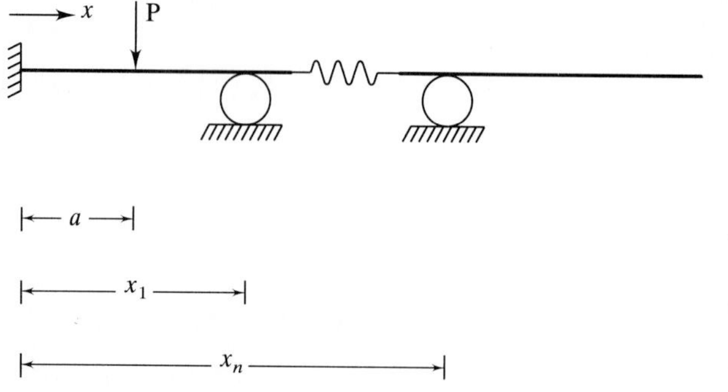

FIGURE D.1
Deflection equation for beam with intermediate supports due to a concentrated load is developed by representing the load and the support reactions using the unit impulse function.

The mathematical representation for a concentrated load of magnitude P applied at $x = b$ is $P\delta(x - b)$ where $\delta(x)$ is the unit impulse function. Thus the loading function $w(x)$ for the beam of Fig. D.1 is written as

$$w(x) = \delta(x - a) + \sum_{i=1}^{n} R_i \delta(x - x_i) \tag{D.2}$$

where R_i, $i = 1, \ldots, n$, are the reactions at the intermediate supports. Equation (D.2) is substituted into Eq. (D.1) and the resulting equation is integrated three times, using Eq. (A.5), giving

$$EI\frac{d^3y}{dx_3} = u(x - a) + \sum_{i=1}^{n} R_i u(x - x_i) + C_1 \tag{D.3}$$

$$EI\frac{d^2y}{dx^2} = (x - a)u(x - a) + \sum_{i=1}^{n} R_i(x - x_i)u(x - x_i) + C_1 x + C_2 \tag{D.4}$$

$$EI\frac{dy}{dx} = \frac{1}{2}(x - a)^2 u(x - a) + \frac{1}{2}\sum_{i=1}^{n} R_i(x - x_i)^2 u(x - x_i)$$

$$+ C_1\frac{x^2}{2} + C_2 x + C_3 \tag{D.5}$$

$$EIy = \frac{1}{6}(x - a)^3 u(x - a) + \frac{1}{6}\sum_{i=1}^{n} R_i(x - x_i)^3 u(x - x_i)$$

$$+ C_1\frac{x^3}{6} + C_2\frac{x^2}{2} + C_3 x + C_4 \tag{D.6}$$

TABLE D.1

End condition	Boundary condition	Boundary condition
Free	$EI\dfrac{d^2y}{dx^2} = 0$	$EI\dfrac{d^3y}{dx^3} = 0$
Fixed	$y = 0$	$\dfrac{dy}{dx} = 0$
Pinned	$y = 0$	$EI\dfrac{d^2y}{dx^2} = 0$

where C_1, C_2, C_3, and C_4 are constants of integration which are determined upon application of the appropriate boundary conditions.

The appropriate boundary conditions depend on the type of support at the boundaries. Table D.1 provides the boundary conditions for different types of support. Two boundary conditions are applied at each end of the beam. Thus $n + 4$ equations are applied to determine the $n + 4$ unknowns, n intermediate support reactions, and four constants of integration.

Example D.1. Determine the deflection of a beam fixed at $x = 0$ and pinned at $x = L$ due to a concentrated load applied at $x = a$, $0 < a < L$.

From Table D.1, the appropriate boundary conditions are

$$y(0) = 0 \quad (a) \qquad\qquad y(L) = 0 \quad (c)$$

$$\left.\frac{dy}{dx}\right|_{x=0} = 0 \quad (b) \qquad\qquad \left.\frac{d^2y}{dx^2}\right|_{x=L} = 0 \quad (d)$$

Application of (a) yields $C_4 = 0$. Application of (b) yields $C_3 = 0$. Application of (c) and (d) yields the following equations:

$$\frac{L^3}{6}C_1 + \frac{L^2}{2}C_2 = -\frac{1}{6}(L - a)^3$$

$$LC_1 + C_2 = -(L - a)$$

respectively. The preceding equations are solved simultaneously, yielding

$$C_1 = \frac{1}{2}\left(1 - \frac{a}{L}\right)\left[\left(\frac{a}{L}\right)^2 - 2\frac{a}{L} - 2\right]$$

$$C_2 = a\left(1 - \frac{a}{L}\right)\left(1 - \frac{a}{2L}\right)$$

TABLE D.2

The deflection, $y(x)$, of a uniform beam of elastic modulus E and cross-sectional moment of inertia I due to a unit concentrated load applied at $x = a$ is

$$y(x) = \frac{1}{EI}\left[\frac{1}{6}(x-a)^3 u(x-a) + \frac{1}{6}\sum_{i=1}^{n} R_i (x-x_i)^3 u(x-x_i) + C_1\frac{x^3}{6} + C_2\frac{x^2}{2} + C_3 x + C_4\right]$$

where R_i is the reaction at an intermediate support located at $x = x_i$. The forms of the constants and the intermediate reactions for common beams are given as follows.

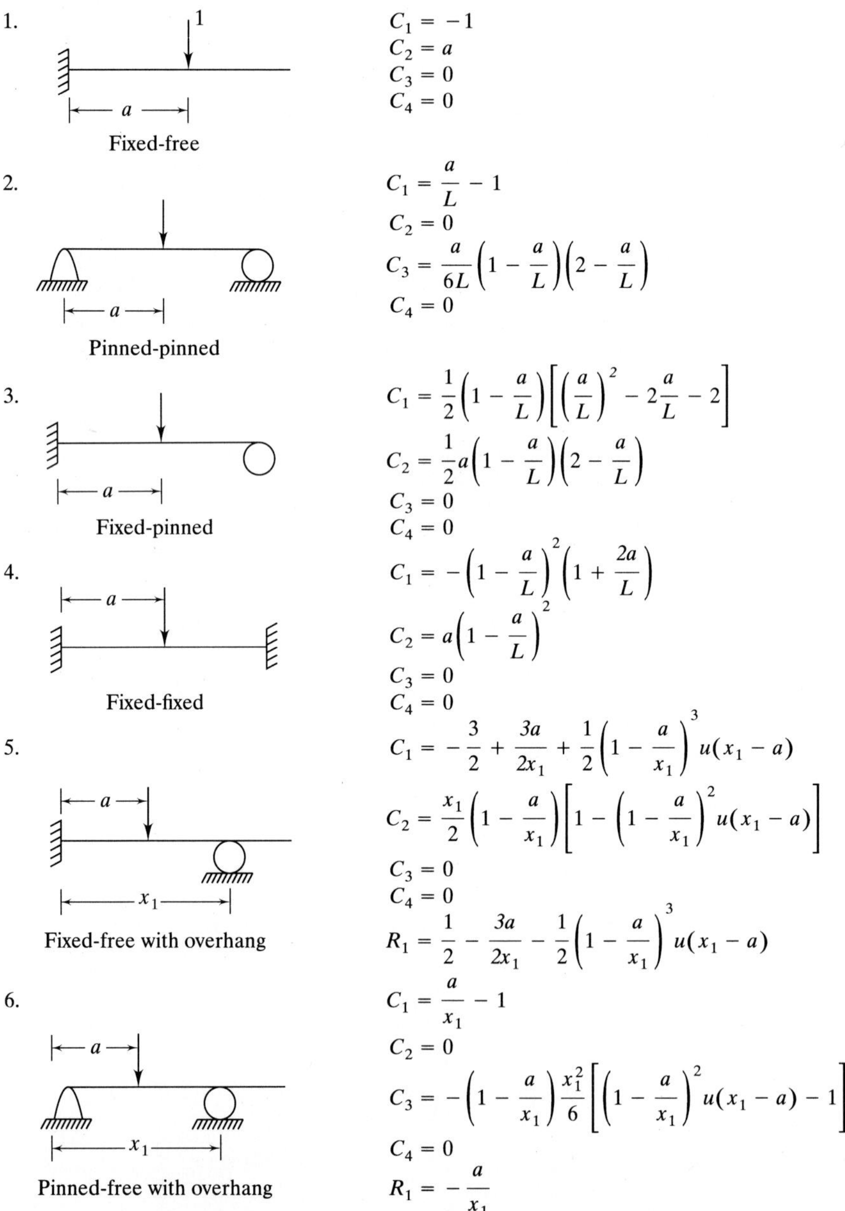

1. Fixed-free

$$C_1 = -1$$
$$C_2 = a$$
$$C_3 = 0$$
$$C_4 = 0$$

2. Pinned-pinned

$$C_1 = \frac{a}{L} - 1$$
$$C_2 = 0$$
$$C_3 = \frac{a}{6L}\left(1 - \frac{a}{L}\right)\left(2 - \frac{a}{L}\right)$$
$$C_4 = 0$$

3. Fixed-pinned

$$C_1 = \frac{1}{2}\left(1 - \frac{a}{L}\right)\left[\left(\frac{a}{L}\right)^2 - 2\frac{a}{L} - 2\right]$$
$$C_2 = \frac{1}{2}a\left(1 - \frac{a}{L}\right)\left(2 - \frac{a}{L}\right)$$
$$C_3 = 0$$
$$C_4 = 0$$

4. Fixed-fixed

$$C_1 = -\left(1 - \frac{a}{L}\right)^2\left(1 + \frac{2a}{L}\right)$$
$$C_2 = a\left(1 - \frac{a}{L}\right)^2$$
$$C_3 = 0$$
$$C_4 = 0$$

5. Fixed-free with overhang

$$C_1 = -\frac{3}{2} + \frac{3a}{2x_1} + \frac{1}{2}\left(1 - \frac{a}{x_1}\right)^3 u(x_1 - a)$$
$$C_2 = \frac{x_1}{2}\left(1 - \frac{a}{x_1}\right)\left[1 - \left(1 - \frac{a}{x_1}\right)^2 u(x_1 - a)\right]$$
$$C_3 = 0$$
$$C_4 = 0$$
$$R_1 = \frac{1}{2} - \frac{3a}{2x_1} - \frac{1}{2}\left(1 - \frac{a}{x_1}\right)^3 u(x_1 - a)$$

6. Pinned-free with overhang

$$C_1 = \frac{a}{x_1} - 1$$
$$C_2 = 0$$
$$C_3 = -\left(1 - \frac{a}{x_1}\right)\frac{x_1^2}{6}\left[\left(1 - \frac{a}{x_1}\right)^2 u(x_1 - a) - 1\right]$$
$$C_4 = 0$$
$$R_1 = -\frac{a}{x_1}$$

Boundary conditions are applied to the beams of Table D.2, resulting in the evaluation of constants and, if applicable, intermediate reactions for each beam. Equation (D.6) is used to calculate the deflection of the beam at any point.

REFERENCE

1. Higdon, A., E. Ohlsen, W. B. Stiles, J. A. Weese, and W. F. Riley: *Mechanics of Materials*, 4th ed., Wiley, New York, 1985.

APPENDIX E

THE VIBES SOFTWARE PACKAGE

The accompanying diskette contains software programs of three types:

1. Programs that illustrate concepts presented in the text.
2. Programs that allow the user to find solutions of vibration problems. Chapters 3 through 11 have problems that require the use of these programs to solve.
3. Programs that provide numerical methods.

The files on the diskette have two types of extension: The files with the .EXE extensions are compiled versions of programs written in the BASIC computer language and can be run directly from the DOS prompt. These are executable files. These files do require user input. The files with the .BAS extensions are programs written in the BASIC language and can be used with a QBASIC interpreter. These files must be accessed through the interpreter and require user-provided subprograms.

The names of the programs provided on the diskette, a brief description of their contents and use, and their correspondence with text material are listed. More information on each of these files is contained in the file README. EXE.

README.EXE is a file that should be executed before using any other program. It contains more detailed descriptions of each program and other information regarding use of the software.

FREE.EXE provides a simulation of the free vibration of a linear one-degree-of-freedom system. The system may be undamped or with viscous damping. The program illustrates material presented in Sections 3.4 and 3.5.

FORCED.EXE provides a simulation of the response of a linear one-degree-of-freedom system with viscous damping such that the system is underdamped to a single-frequency harmonic excitation. The program illustrates material presented in Sections 4.3 and 4.4.

COULOMB.EXE provides a simulation of the free or forced harmonic response of a one-degree-of-freedom system with Coulomb damping. The program illustrates material presented in Sections 3.6 and 4.10.

CONVOL.EXE uses numerical integration of the convolution integral to determine the forced response of a linear one-degree-of-freedom system with viscous damping to several forms of excitation provided in a library. The program illustrates material presented in Sections 5.2, 5.3, and 5.7.

CONVOL2.BAS uses numerical integration of the convolution integral to determine the forced response of a linear one-degree-of-freedom system with viscous damping to any form of excitation. The form of the excitation is specified in a user-provided subprogram. The program illustrates material presented in Sections 5.2, 5.3, and 5.7.

BEAM.EXE develops the flexibility matrix for a beam listed in App. D when its vibrations are modeled using a finite number of degrees of freedom. This program illustrates the material presented in Sections 6.6 and 6.9 and App. D.

MITER.EXE uses matrix iteration to determine the natural frequencies and normalized mode shaped for a multi-degree-of-freedom system. It illustrates the material presented in Sections 7.1 through 7.10.

MODAL.EXE uses modal analysis to determine the forced vibration response of the four-degree-of-freedom model of a suspension system as it traverses a pothole. This is a generalization of Example 8.8. The program illustrates material presented in Sections 8.3 through 8.7.

MODAL2.BAS uses modal analysis to determine the forced response of a multi-degree-of-freedom system with proportional damping due to any type of excitation. The excitation is specified in a user-provided subprogram. The program illustrates material presented in Sections 8.3 through 8.7.

CFREQ.EXE determines a specified number of natural frequencies and normalized mode shapes for any of the beams given in Table 9.4. The program illustrates material presented in Sections 9.1, 9.2, and 9.4.

CMODA.BAS uses modal analysis to determine the forced response of a beam subject to a harmonic excitation of the form $F(x, t) = f(x)\sin \omega t$, where $f(x)$ is specified in a user-provided subprogram. The program illustrates material presented in Sections 9.1, 9.2, and 9.4.

DUFF.EXE uses a fourth-order Runge-Kutta method to integrate the damped forced Duffing's equation, Eq. (10.9). The program provides the time history of motion, the state plane, and a Poincaré section. The program illustrates material presented in Sections 10.2, 10.5, 10.6, and 10.9.

ISOL.EXE designs a vibration isolator to meet specified conditions. The program illustrates material presented in Sections 11.2 through 11.4.

SPECT.BAS develops the displacement spectrum and force spectrum for a one-degree-of-freedom system with viscous damping subject to any form of excitation. The excitation is specified in a user-provided subprogram. The program illustrates material presented in Sections 5.6 and 11.5.

ABSORB.EXE designs an undamped or damped vibration absorber to use on a one-degree-of-freedom system. The program illustrates material presented in Sections 11.6 and 11.7.

INDEX